AF708631

Atomic Energy Levels of the Iron-Period Elements: Potassium through Nickel

Journal of

Physical and Chemical Reference Data

David R. Lide, Jr., Editor

The Journal of Physical and Chemical Reference Data (ISSN 0047-2689) is published quarterly by the American Chemical Society (1155 16th St., N. W., Washington, DC 20036) and the American Institute of Physics (335 E. 45th St., New York, NY 10017) for the National Bureau of Standards. Second-class postage paid at Washington, DC and additional mailing offices. POSTMASTER: Send address changes to Membership and Subscription Services, P. O. Box 3337, Columbus, Ohio 43210.

The objective of the Journal is to provide critically evaluated physical and chemical property data, fully documented as to the original sources and the criteria used for evaluation. Critical reviews of measurement techniques, whose aim is to assess the accuracy of available data in a given technical area, are also included. The Journal is not intended as a publication outlet for original experimental measurements such as are normally reported in the primary research literature, nor for review articles of a descriptive or primarily theoretical nature.

Supplements to the Journal are published at irregular intervals and are not included in subscriptions to the Journal. They contain compilations which are too lengthy for a journal format.

The Editor welcomes appropriate manuscripts for consideration by the Editorial Board. Potential contributors who are interested in preparing a compilation are invited to submit an outline of the nature and scope of the proposed compilation, with criteria for evaluation of the data and other pertinent factors, to:

David R. Lide, Jr., Editor
J. Phys. Chem. Ref. Data
National Bureau of Standards
Gaithersburg, MD 20899

One source of contributions to the Journal is The National Standard Reference Data System (NSRDS), which was established in 1963 as a means of coordinating on a national scale the production and dissemination of critically evaluated reference data in the physical sciences. Under the Standard Reference Data Act (Public Law 90-396) the National Bureau of Standards of the U.S. Department of Commerce has the primary responsibility in the Federal Government for providing reliable scientific and technical reference data. The Office of Standard Reference Data of NBS coordinates a complex of data evaluation centers, located in university, industrial, and other Government laboratories as well as within the National Bureau of Standards, which are engaged in the compilation and critical properties retrieved from the world scientific literature. The participants in this NBS-sponsored program, together with similar groups under private or other Government support which are pursuing the same ends, comprise the National Standard Reference Data System.

The primary focus of the NSRDS is on well-defined physical and chemical properties of well-characterized materials or systems. An effort is made to assess the accuracy of data reported in the primary research literature and to prepare compilations of critically evaluated data which will serve as reliable and convenient reference sources for the scientific and technical community.

Information for Contributors

Manuscripts submitted for publication must be prepared in accordance with *Instructions for Preparation of Manuscripts for the Journal of Physical and Chemical Reference Data,* available on request from the Editor.

Editorial Board

Term ending 31 December 1985
Pierre A. Ausloos, Lester Guttman, Thomas L. Isenhour
Term ending 31 December 1986
Gesina C. Carter, George H. Morrison, Louis R. Testardi
Term ending 31 December 1987
Josef Michl, Neil Olien, Thomas G. Trippe

Management Board

David R. Lide, Jr., Charles R. Bertsch, John T. Scott
Editorial Staff at NBS: Julian M. Ives, Joan Sauerwein
Editorial Staff at AIP: Kathleen Strum, Editorial Supervisor;
Doreene A. Berger, Chief Copy Editor;
Noëlle Falcier Benstock, Senior Copy Editor

New and renewal subscriptions should be sent with payment to the Office of the Controller at the American Chemical Society, 1155 Sixteenth Street, N.W., Washington, DC 20036. **Address changes**, with at least six weeks advance notice, should be sent to Manager, Membership and Subscription Services, American Chemical Society, P.O. Box 3337, Columbus, OH 43210. Changes of address must include both old and new addresses and ZIP codes and, if possible, the address label from the mailing wrapper of a recent issue. Claims for missing numbers will not be allowed: if loss was due to failure of the change-of-address notice to be received in the time specified; if claim is dated (a) North America: more than 90 days beyond issue date, (b) all other foreign: more than one year beyond issue date.

Members of AIP member and affiliate societies requesting member subscription rates should direct subscriptions, renewals, and address changes to American Institute of Physics, Dept. S/F, 335 E. 45th St., NY 10017.

Subscription Prices (1985) *(not including supplements)*	U.S.A.	Foreign (surface mail)	Optional air freight Europe Mideast N. Africa	Asia and Oceania
Members (of ACS, AIP, or affiliated society)	$ 50.00	$ 60.00	$ 75.00	$ 85.00
Regular rate	$200.00	$210.00	$225.00	$235.00

Rates above do not apply to nonmember subscribers in Japan, who must enter subscription orders with Maruzen Company Ltd., 3-10 Nihonbashi 2-chome, Chuo-ku, Tokyo 103, Japan. Tel: (03) 272-7211.

Back numbers are available at a cost of $60 per single copy and $230 per volume.

Orders for reprints, supplements, and back numbers should be addressed to the American Chemical Society, 1155 Sixteenth Street, N. W., Washington, DC 20036. Prices for reprints and supplements are listed at the end of this issue.

Copying Fees: The code that appears on the first page of articles in this journal gives the fee for each copy of the article made beyond the free copying permitted by AIP. (See statement under "Copyright" elsewhere in this journal.) If no code appears, no fee applies. The fee for pre-1978 articles is $0.25 per copy. With the exception of copying for advertising and promotional purposes, the express permission of AIP is not required provided the fee is paid through the *Copyright Clearance Center, Inc. (CCC), 21 Congress Street, Salem, MA 01970.* Contact the CCC for information on how to report copying and remit payment.

Microfilm subscriptions of the *Journal of Physical and Chemical Reference Data* are available on 16 mm and 35 mm. This journal also appears in Sec. I of *Current Physics Microform* (CPM) along with 26 other journals published by the American Institute of Physics and its member societies. A *Microfilm Catalog* is available on request.

Copyright 1985 by the U.S. Secretary of Commerce; copyright assigned to the American Institute of Physics (AIP) and the American Chemical Society (ACS). Individual teachers, students, researchers, and libraries acting for them are permitted to make copies of articles in this journal for their own use in research or teaching, including multiple copies for classroom or library reserve use, provided such copies are not sold. Copying for sale is subject to payment of copying fees. (See "Copying Fees" paragraph elsewhere in this journal.) Permission is granted to quote from this journal with the customary acknowledgment of the source. To reprint a figure, table, or other excerpt requires in addition the consent of one of the original authors and notification to AIP. Reproduction for advertising or promotional purposes, or republication in any form, is permitted only under license from AIP, which will normally require that the permission of one of the authors also be obtained. Direct inquiries to Office of Rights and Permissions, American Institute of Physics, 335 East 45th Street, New York, NY 10017. The right of the U.S. Government to unrestricted copying for its own use of copyrighted material originating in its laboratories or under its contracts is specifically recognized.

Journal of
Physical and Chemical Reference Data

Volume 14, 1985
Supplement No. 2

Atomic Energy Levels of the Iron-Period Elements: Potassium through Nickel

Jack Sugar and Charles Corliss

*National Measurement Laboratory,
National Bureau of Standards,
Gaithersburg, Maryland 20899*

Published by the **American Chemical Society**
and the **American Institute of Physics** for
the **National Bureau of Standards**

Copyright © 1985 by the U.S. Secretary of Commerce on behalf of the United States. This copyright will be assigned to the American Institute of Physics and the American Chemical Society, to whom all requests regarding reproduction should be addressed.

Library of Congress Catalog Card Number 85-72287

International Standard Book Number
0-88318-480-X

American Institute of Physics, Inc.

335 East 45th Street

New York, New York 10017-3483

Printed in the United States of America

Foreword

The *Journal of Physical and Chemical Reference Data* is published jointly by the American Institute of Physics and the American Chemical Society for the National Bureau of Standards. Its objective is to provide critically evaluated physical and chemical property data, fully documented as to the original sources and the criteria used for evaluation. One of the principal sources of material for the journal is the National Standard Reference Data System (NSRDS), a program coordinated by NBS for the purpose of promoting the compilation and critical evaluation of property data.

The regular issues of the *Journal of Physical and Chemical Reference Data* are published quarterly and contain compilations and critical data reviews of moderate length. Longer monographs, volumes of collected tables, and other material unsuited to a periodical format are published separately as *Supplements to the Journal.* This monograph, "Atomic Energy Levels of the Iron-Period Elements: Potassium through Nickel" by Jack Sugar and Charles Corliss is presented as Supplement No. 2 to Volume 14 of the *Journal of Physical and Chemical Reference Data.*

David R. Lide, Jr., Editor
Journal of Physical and Chemical Reference Data

Atomic Energy Levels of the Iron-Period Elements: Potassium through Nickel

Jack Sugar and Charles Corliss[a]

National Measurement Laboratory, National Bureau of Standards, Gaithersburg, MD 20899

Experimentally derived energy levels of the elements from potassium to nickel in all stages of ionization are critically compiled. The data for each level include its position in cm^{-1} (relative to the ground state), configuration, term designation, J-value, and, where available, the g-value and two leading percentages of the eigenvector composition in the most appropriate coupling scheme. For the He I and H I isoelectronic sequences, calculated level positions are given because they are considered more accurate than the measurements presently available. Ionization energies for each ion are derived either from Rydberg series, extrapolation, or calculation. Complete references are given for the compiled data.

Key words: calcium; chromium; cobalt; compilation; energy levels; manganese; nickel; potassium; scandium; titanium; vanadium.

[a]Present address: Forest Hills Laboratory, 2955 Albemarle St., NW, Washington, DC 20008.

Contents

Nickel (Continued)

Introduction

The NBS Atomic Energy Levels Data Center, in the Division of Atomic and Plasma Radiation, has undertaken to provide new compilations of atomic energy levels for all stages of ionization of each atom. A compilation of energy levels of the rare earth elements by Martin, Zalubas, and Hagan was issued in 1978. The recent program has concentrated on the elements H through Ni. The data for each element have been published separately as completed, with the intention of later revising and bringing together into one volume related groups of elements. The present volume is such a revision, containing the collection of all data for the elements K through Ni. It is based on the compilations of Ca, Sc, V, Cr, and Co by Sugar and Corliss (1979, 1980, 1978, 1977, 1981) and K, Ti, Mn, Fe, and Ni by Corliss and Sugar (1979, 1979, 1977, 1982, 1981), and Fe by Reader and Sugar (1975) with numerous revisions and additions taken from the published literature or received privately. Other works that have been issued in separate publications cover the elements Al, Mg, Na, and Si by Martin and Zalubas (1979, 1980, 1981, 1983). Having completed the elements of the iron period, the Data Center now intends to carry on the compilations for elements of the $n=4$ shell (Cu through Pd), and to complete the elements between H and Ar.

Generally, we have used only published papers as sources of data. Unpublished data are included when they constitute a substantial improvement over material in the literature. For many of the higher ions the original papers do not give energy level values, but only classifications of observed lines. In these cases we have derived the level values.

All energy levels are given in units of cm^{-1}, beginning with a value of zero for the ground level. Ionization energies found in the literature are usually given in eV or in cm^{-1}. The conversion factor, 8065.479(21) cm^{-1}/eV, given by Cohen and Taylor (1973), is used here. In a few cases where adequate data were available but the ionization energy had not been derived, we carried out the calculation. For a number of the ions, no suitable series are known. In these cases we have quoted values obtained by Lotz (1967) by a method of successive differences along isoelectronic sequences. Although uncertainties are not provided with these extrapolated values, we estimate that they are accurate to 0.2% by comparing them with recently determined values.

Nearly all of the data are based on observations of various types of laboratory light sources. However, the laboratory data are sometimes supplemented by data obtained from solar observations. This is particularly true where spin-forbidden lines are needed to establish the absolute energy of a system of excited levels and where parity-forbidden transitions between levels of a ground configuration are used to obtain accurate relative energies for the low levels. Whenever both solar data and equivalent laboratory data are available preference is generally given to the laboratory measurements.

When no observations are available to connect independent systems of levels, an estimate of the connecting energy is adopted. Those level values affected by the estimate are denoted by $+x$ following the value. The value of x is the systematic error of the estimate.

We have included under the heading "Leading Percentages" the results of calculations that express the eigenvector percentage composition of levels (rounded to the nearest %) in terms of the basis states of a single configuration, or more than one configuration where configuration interaction has been included. We give first the percentage of the basis state corresponding to the level's name; next the second largest percentage together with the related basis state. Sometimes the leading percentage in an alternative coupling scheme is given. Generally, when the leading percentage is less than 40%, no name is given. When the first and second resultant terms are the same and sum to $\geqq 40\%$, the first name is given. When the first and second resultant terms are the same but have different parentages, and their share of the eigenvector composition sums to 40% or more, the level will be named as the higher precentage term. In cases where these percentages differ by one or two units (an insignificant difference), either term may be selected for the level name, and the lower percentage may appear first. For the unnamed level, the term symbol follows the percentage. The user should of course bear in mind that the percentages are model dependent, so that the results of different calculations can yield notably different percentages. Percentages for the odd parity configurations of the neutral atoms of Mn, Fe, and Ni were obtained from Roth (1980). This publication gives the results of revised calculations intended to supersede those of Roth (1969, 1970) for Ca through Ni. We used other sources for the Ca and Sc calculations. In the case of Ti, V, Cr, and Co the 1969 and 1970 results by Roth are adopted. We intended to use his new calculations for these elements as well, but we found that the sum of percentages for a number of states exceeded 100 by significant amounts. We therefore have used the new results only for those cases where this error was not present.

For configurations of equivalent d-electrons, several terms of the same LS type may occur. These are theoretically distinguished by their seniority number. In the present compilations they are designated in the notation of Nielson and Koster (1963). For example, in the $3d^5$ configuration there are three ^{2}D terms with seniorities of 1, 3, and 5. These terms are denoted as ^{2}D1, ^{2}D2, and ^{2}D3, respectively, by Nielson and Koster. Martin, Zalubas, and Hagan (1978) give a complete summary of the coupling notations used here, tables of the allowed terms for equivalent electrons, etc.

The text for each ion does not include a complete review of the literature but is intended to credit the major contributions. In assembling the data for each spectrum, we referred to the following bibliographies:

i. Papers cited by Moore (1949, 1952)
ii. C. E. Moore (1968, 1969)
iii. L. Hagan and W. C. Martin (1972)
iv. L. Hagan (1977)
v. R. Zalubas and A. Albright (1980)
vi. Card file of publications since June 1979 maintained by the NBS Atomic Energy Levels Data Center

He I Isoelectronic Sequence

Spectra of K, Ca, Ti, and V were obtained by Aglitskii et al. (1974) with a laser-heated plasma in third and fifth orders of a crystal spectrograph. Reference lines of Mg XI and Al XII published by Flemberg (1942) were used, and an uncertainty of ± 0.0005 Å was reported for the lines of the He I isoelectronic sequence, which fall in the range of 2.3–3.6 Å. Flemberg's reference wavelengths were in x-units. The equivalence to Å that he used must be increased by 8 parts in 10^5, according to the more recent conversion determined by Deslattes and Henins (1973). With this correction, the data of Aglitskii et al. deviate randomly from the calculated wavelengths of Safronova (1981) by ± 0.0008 Å.

In a beam-foil experiment the He-like argon spectrum was observed by Briand et al. (1983a). Their wavelengths for the $1s^2\,{}^1S_0 - 1s2p\,{}^3P_1^\circ$ and ${}^1P_1^\circ$ transitions were 3.9693(3) Å and 3.9491(3) Å, in agreement with the calculated values by Safronova.

The $1s2s\,{}^3S_1 - 1s2p\,{}^3P_2^\circ$ transition has been measured in Ca XIX by Livingston (1983) and in Fe XXV by Buchet et al. (1982). The measured wavelengths are 466.78(8) Å for Ca and 271.04(10) Å for Fe. The corresponding energy differences are greater than those predicted by Safronova by 162(37) and 123(136) cm^{-1}, respectively, or 0.07% and 0.03% of the energy difference. A new calculation of these energies by Hata and Grant (1983) predicted values that were 60 cm^{-1} lower in Ca and 154 cm^{-1} lower in Fe than the observed values.

Because of the excellent agreement of Safronova's calculations with the best experimental data available and the paucity of these data, we have based our compilation of this sequence on her results. We quote her calculated energies for the $1s2s$ and $1s2p$ levels of the He I isoelectronic sequence and for the principal ionization energies (with correction to the Rydberg for finite atomic mass). The observed $1s2s\,{}^3S_1 - 1s2p\,{}^3P_2^\circ$ intervals in Ca XIX and Fe XXV mentioned above are incorporated in the respective level lists. For $n = 3$–5 we subtract the calculated binding energies reported by Ermolaev and Jones (1974) from the binding energy of the ground state by Safronova to arrive at energy level values. The uncertainty in the calculated energy levels and the ionization energies is assumed conservatively to be 2 parts in 10^4, corresponding to the deviations from the Aglitskii et al. (corrected) observations. (The deviation from the measurements in Ar is 1 part in 10^4.) The uncertainties in energy differences for levels of the same n-value are estimated to be 2 parts in 10^3. The deviation of the $1s2p\,{}^3P_1^\circ - {}^1P_1^\circ$ intervals measured by Aglitskii et al. with resonance lines differ randomly from the calculated values of Safronova by 3%.

The singlet-triplet mixing coefficients for the $1snp\,{}^{1,3}P^\circ$ states are quoted from Ermolaev and Jones.

H I Isoelectronic Sequence

No observations of $1s - np$ transitions have been sufficiently accurate to test the theoretical values. The best measurement available is for the $1s - 2p$ energies for Fe XXVI with an uncertainty of ± 5000 cm^{-1}, or 1 part in 10^4, by Briand, Tavernier, and Indelicato (1983b). Erickson (1977) has calculated the absolute binding energies for each of the levels through $n = 5$ and for the ns and np states through $n = 13$. An improved calculation of the Lamb-shift effects was reported by Mohr (1983), who gave the energy separations among the $n = 1$ and 2 levels. Gould and Marrus (1983) have measured the Lamb-shift of the $2s\,{}^2S_{1/2}$ state of Ar XVIII, obtaining the value 1264(13) cm^{-1}. Their result agrees with the value 1275.8(0.8) cm^{-1} calculated by Mohr and is three standard deviations lower than Erickson's value of 1301(2) cm^{-1}.

We have compiled Mohr's results for the energy separations of $n = 1$ and 2 levels, and Erickson's for $n = 3$–5 relative to the $2p\,{}^2P_{3/2}^\circ$ level. This increases Erickson's values for the levels, or, equivalently, increases the binding energy of the ground state (the ionization energy). Assuming that the uncertainty in these compiled values is mainly due to the error in the Lamb shift, we take the fractional error as equal to the experimental fractional error in the Ar measurement. This contribution to the level values relative to the ground state is about 4 parts in 10^6 for the iron period. This is about 10 times the error estimated by Mohr for his calculated $1s - 2p$ intervals. The corresponding intervals calculated by Erickson are lower than those of Mohr by about the same fractional amount.

Tables of Wavelengths

For general sources of wavelengths for the elements considered here we refer the reader to the compilation by Kelly (1985) for the range 1–2000 Å and by Kelly (1979) for the range 2000–3200 Å, to the tables of spectral lines in the CRC Handbook of Chemistry and Physics (1984–1985) from 40–40 000 Å, and to Tables of Spectral Line Intensities by Meggers, Corliss, and Scribner (1975) from 2000–9000 Å.

Acknowledgments

Throughout this work we have made extensive use of the bibliographical files and reprint collection maintained in the Atomic Energy Levels Data Center by Romuald Zalubas and Arlene Musgrove. Our thanks are extended to them for generous cooperation. The compilation has also benefited greatly from the preprints that were provided by many of our colleagues.

We thank W. C. Martin for his critical reading of the manuscript.

This work was supported by the U.S. Department of Energy, Division of Magnetic Fusion Energy, the National Aeronautics and Space Administration, Astronomy/Relativity Branch, and the Office of Standard Reference Data of the National Bureau of Standards.

References for Introduction

Aglitskii, E. V., Boiko, V. A., Zakharov, S. M., Pikuz, S. A., and Faenov, A. Y. (1974), Kvant. Elektron. **1**, 908.

Briand, J. P., Mossé, J. P., Indelicato, P., Chevallier, P., Girard-Vernhet, D., Chetioui, A., Ramos, M. T., and Desclaux, J. P. (1983a), Phys. Rev. **A28**, 1413.

Briand, J. P., Tavernier, M., and Indelicato, P. (1983b), Phys. Rev. Lett. **50**, 832.

Buchet, J. P., Buchet-Poulizac, M. C., Denis, A., Desequelles, J., Druetta, M., Knystautas, E. J., and Lecler, D. (1982), Nucl. Instrum. Meth. **202**, 79.

Cohen, E. R., and Taylor, B. N. (1973), J. Phys. Chem. Ref. Data **2**, 663.

Corliss, C., and Sugar, J. (1977), Energy Levels of Manganese, Mn I through Mn XXV, J. Phys. Chem. Ref. Data **6**, 1253.

Corliss, C., and Sugar, J. (1979a), Energy Levels of Titanium, Ti I through Ti XXII, J. Phys. Chem. Ref. Data **8**, 1.

Corliss, C., and Sugar, J. (1979b), Energy Levels of Potassium, K I through K XIX, J. Phys. Chem. Ref. Data **8**, 1109.

Corliss, C., and Sugar, J. (1981), Energy Levels of Nickel, Ni I through Ni XXVIII, J. Phys. Chem. Ref. Data **10**, 197.

Corliss, C., and Sugar, J. (1982), Energy Levels of Iron, Fe I through Fe XXVI, J. Phys. Chem. Ref. Data **11**, 135.

CRC Handbook of Chemistry and Physics, 65th Ed. (1984-1985), pp. E-187 to E-313, Line Spectra of the Elements, Ed. by J. Reader and C. H. Corliss (CRC Press, Inc., Boca Raton, FL).

Deslattes, R. D. and Henins, A. (1973), Phys. Rev. Lett. **31**, 972.

Erickson, G. W. (1977), J. Phys. Chem. Ref. Data **6**, 831.

Ermolaev, A. M., and Jones, M. (1974), J. Phys. **B7**, 199.

Flemberg, H. (1942), Ark. Mat. Astron. Fys. **28**, 1.

Gould, H., and Marrus, R. (1983), Phys. Rev. **28**, 2001.

Hagan, L. (1977), Bibliography on Atomic Energy Levels and Spectra, July 1971 through June 1975, Natl. Bur. Stand. (U.S.) Spec. Publ. 363, Suppl. 1 (U.S. Gov't Printing Office, Washington, D.C.).

Hagan, L., and Martin, W. (1972), Bibliography on Atomic Energy Levels and Spectra, July 1968 through June 1971, Natl. Bur. Stand. (U.S.) Spec. Publ. 363 (U.S. Gov't Printing Office, Washington, DC).

Hata, J., and Grant, I. P. (1983), J. Phys. **B16**, 523.

Kelly, R. L. (1985), Atomic and Ionic Spectrum Lines Below 2000 Angstroms Vol. II—Potassium through Zinc, J. Phys. Chem. Ref. Data.

Kelly, R. (1979), Atomic Emission Lines in the Near Ultraviolet; Hydrogen through Krypton, Sections I and II, NASA Technical Memorandum 80268.

Livingston, A. E. (1983), private communication.

Lotz, W. (1967), J. Opt. Soc. Am. *57*, 873.

Martin, W. C., Zalubas, R., and Hagan, L. (1978), Atomic Energy Levels—The Rare Earth Elements, Natl. Stand. Ref. Data Ser., Natl. Bur. Stand. (U.S.), 60.

Martin, W. C., and Zalubas, R. (1979), Energy Levels of Aluminum, Al I through Al XIII, J. Phys. Chem. Ref. Data **8**, 817.

Martin, W. C., and Zalubas, R. (1980), Energy Levels of Magnesium, Mg I through Mg XII, J. Phys. Chem. Ref. Data **9**, 1.

Martin, W. C., and Zalubas, R. (1981), Energy Levels of Sodium, Na I through Na XI, J. Phys. Chem. Ref. Data **10**, 153.

Martin , W. C., and Zalubas, R. (1983), Energy Levels of Silicon, Si I through Si XIV, J. Phys. Chem Ref. Data **12**, 323.

Meggers, W . F., Corliss, C. H., and Scribner, B. F. (1975), Natl. Bur. Stand. Monogr. 145.

Mohr, P. J. (1983), At. Data Nucl. Data Tables, **29**, 453.

Moore, C. E. (1952), Atomic Energy Levels, Natl. Bur. Stand. (U.S.) Circ. 467, Vol. II (reissued in 1971 as Natl. Bur. Stand. (U.S.) Natl. Stand. Ref. Data Ser. 35, Vol. II).

Moore, C. E. (1969), Bibliography on the Analyses of Optical Atomic Spectra, Section 2, Natl. Bur. Stand. (U.S.), Spec. Publ. 306-2 (U.S. Gov't Printing Office, Washington, DC).

Moore, C. E. (1968), Bibliography on the Analyses of Atomic Spectra, Section 1, Natl. Bur. Stand. (U.S.) Spec. Publ. 306 (U.S. Gov't Printing Office, Washington, DC).

Moore, C. E. (1949), Atomic Energy Levels, Natl. Bur. Stand. (U.S.) Circ. 467, Vol. I (reissued in 1971 as Natl. Bur. Stand. (U.S.) Natl. Stand. Ref. Data Ser. 35, Vol. I).

Nielson, C. W., and Koster, G. F., Spectroscopic Coefficients for the p^n, d^n, and f^n Configurations, 275 pp. (M.I.T. Press, Cambridge, MA, 1963).

Reader, J., and Sugar, J. (1975), Energy Levels of Iron, Fe I through Fe XXVI, J. Phys. Chem. Ref. Data **4**, 353.

Roth, C. (1969), J. Res. Natl. Bur. Stand. (U.S.) **73A**, 497.

Roth, C. (1970), J. Res. Natl. Bur. Stand. (U.S.) **74A**, 141.

Roth, C. (1980), At. Nucl. Data Tables **25**, 95.

Safronova, U. I. (1981), Phys. Scr. **23**, 241.

Sugar, J., and Corliss, C. (1977), Energy Levels of Chromium, Cr I through Cr XXIV, J. Phys. Chem. Ref. Data **6**, 317.

Sugar, J., and Corliss, C. (1978), Energy Levels of Vanadium, V I through V XXIII, J. Phys. Chem. Ref. Data **7**, 1191.

Sugar, J., and Corliss, C. (1979), Energy Levels of Calcium, Ca I through Ca XX, J. Phys. Chem. Ref. Data 8, 865.

Sugar, J., and Corliss, C. (1980), Energy Levels of Scandium, Sc I through Sc XXI, J. Phys. Chem. Ref. Data **9**, 473.

Sugar, J., and Corliss, C. (1981), Energy Levels of Cobalt, Co I through Co XXVII, J. Phys. Chem. Ref. Data **10**, 1097.

Zalubas, R., and Albright, A. (1980), Bibliography on Atomic Energy Levels and Spectra, July 1975 through June 1979, Natl. Bur. Stand. (U.S.) Spec. Publ. 363, Suppl. 2 (U.S. Gov't Printing Office, Washington, DC).

K I

$Z=19$

Ground state: $1s^2 2s^2 2p^6 3s^2 3p^6 4s\ ^2S_{1/2}$

Ionization energy $=$ 35 009.8140±0.0007 cm^{-1} (4.34069±0.00001 eV)

On the basis of measurements of 21 lines in the range 5300–9600 Å, Edlén (1936) established the series of terms $4s-8s$, $4p-6p$, $3d-6d$, and $4f-9f$. To connect the $3d$ and nf terms to the ground state he relied on earlier measurements of the forbidden $4s\ ^2S-3d\ ^2D$ doublet.

The np series was extended through $n=79$ in absorption observations by Kratz (1949), with resolved fine structure through $n=17$. Using frequency-doubled light from a dye laser, Lorenzen and Niemax (1983) observed the np series from $n=9$–21 with an accuracy of ±0.003 cm^{-1}. They note that the older data may be systematically shifted by pressure effects. Where they overlap, these data and the observations of Kratz differ by about 0.13 cm^{-1}, the latter being systematically higher in energy. We give the results of Lorenzen and Niemax to $n=21$. They derived a series formula which reproduces their observed levels within ±0.004 cm^{-1}. It requires only the insertion of the principal quantum number n, and may be used to calculate higher series members.

Risberg (1956) reobserved the spectrum from 3100–12 000 Å, using a hollow-cathode discharge. From these measurements and unpublished observations of I. Johansson beyond 12 000 Å, he determined the term values through $13s$, $10p$, $11d$, and $11f$. Higher members of the s and d series were observed by Harper, Wheatley and Levenson (1977) and by Shen and Curry (1977) by two-photon absorption from the ground state. The fine structure splitting of the nd states from $n=8$ to $n=19$ was measured by Harper and Levenson (1976). Gallagher and Cooke (1978) have measured the intervals of the $15d$, $16d$, $18d$, and $20d$ terms more accurately.

The $5d$ and $6d$ intervals were measured by Nilsson and Svanberg (1979) with an uncertainty of ±0.5 MHz. They combined their results with those of Gallagher and Cooke, who report approximately the same uncertainty, to derive a formula for all the nd splittings from $n=3$–20. They are compared below with the splittings measured by Lorenzen et al. (1981).

n	$\Delta E/h$ (MHz)	$\Delta E/hc$ (cm^{-1})	$\Delta E/hc$ Lorenzen et al.
3	71 967(5)	2.40051(2)	
4	32 356	1.07926	
5	15 102.9	0.50377	
6	7 965.5	0.26570	
7	4 655.0	0.15527	0.1541(7)
8	2 944.6	0.09822	0.0985
9	1 979.4	0.06602	0.0657
10	1 391.8	0.04642	0.0462
11	1 017.4	0.03394	0.0334
12	766.7	0.02557	0.0254
13	592.5	0.01976	0.0196
14	467.5	0.01559	0.0154
15	375.49	0.01252	0.0123
16	306.25	0.01022	0.0102
17	253.1	0.00844	0.0084
18	211.63	0.00706	0.0070
19	178.8	0.00596	0.0060
20	152.42	0.00508	0.0051

New observations of the energy levels of the ns-series ($n=9$–46) and nd-series ($n=7$–46) were made with two-photon absorption by Lorenzen, Niemax, and Pendril (1981) with an accuracy of ±0.0007 cm^{-1}. Their results, relative to the center of gravity of the hyperfine splitting of the ground state, are given here. Where they have skipped levels above $n=29$, we give values in brackets obtained by application of a Rydberg series formula fitted to their lower levels. Their value for the ionization energy relative to the center of gravity of the hyperfine splitting of the ground state is quoted (the uncertainty in eV is due to the uncertainty in the conversion factor). With a similar experiment Thompson, O'Sullivan, Stoicheff, and Xu (1983) obtained equivalent results for these series. They extended the ns series down to $n=6$ and up to $n=55$, and the nd series down to $n=5$ and up to 50. We have added their low ns ($n=6$–8) and nd ($n=5$–6) terms, adjusting them to the center of gravity of the ground state hyperfine splitting. The higher members of the nf series were observed by Bensoussan (1975) by means of continuum absorption from the $3d$ state, which was populated by dye laser pumping.

The $5g$ term is from Litzén (1970), who observed the $4f-5g$ transition at 40 158.37 Å. The three decimal term values for the $3d$, $4p$, $5s$, $5p$, and $4d$ levels are from infrared measurements of Johansson and Svendenius (1972).

The g factor of the ground state is from Vanden Bout et al. (1968), that for $5p\ ^2P^\circ_{1/2}$ from Fox and Series (1961), and those for the higher levels are from Belin, Holmgren, Lindgren, and Svanberg (1975).

The $3p^5 4s^2\ ^2P^\circ$ term was observed by Beutler and Guggenheimer (1933) at 653.31 and 662.38 Å in absorption from the ground state. Mansfield (1975) observed the absorption spectrum in the autoionizing region from 350–700 Å. He identified the $3p^5 4snd$ series to $n=20$ and the $3p^5 4sns$ series to $n=21$. We have added the $^2P^\circ$ term designation for the higher members of the s and d series starting with $8s$ and $7d$.

The single configuration identifications of these broad absorption lines in the autoionizing continuum should be regarded as a simplified approximation. There is clearly strong series mixing, as noted by Martin, Tech, and

Wilson (1969) in their analysis of the $3p^5\,3s3d$ configuration of K I. The position of the $3p^54s^2\,^2P°$ term is from Mansfield. Many other absorption features are tentatively identified in his paper, but are not included in the present compilation.

Arrangement of Tables

The first table of energy levels presented here for neutral potassium is arranged in the usual way; the terms are listed in order of increasing energy without regard to configuration assignments. Because many long Rydberg series have been observed in K I, we present a second table for this spectrum in which the series are listed separately, followed by their series limits. This table corresponds more closely to the character of the observed spectrum. The series are listed in order of increasing energy for the first series member. The series member with the largest value of n for each term type is followed by the limit of that series in K II.

References

Belin, G., Holmgren, L., Lindgren, I., and Svanberg, S. (1975), Phys. Scr. **12**, 287.
Bensoussan, P. (1975), Phys. Rev. **A11**, 1787.
Beutler, H., and Guggenheimer, K. (1933), Z. Phys. **87**, 188.
Edlén, B. (1936), Z. Phys. **98**, 445.
Fox, W. N., and Series, G. W. (1961), Proc. Phys. Soc. **77**, 1141.
Gallagher, T. F., and Cooke, W. E. (1978), Phys. Rev. **A18**, 2510.
Harper, C. D., and Levenson, M. D. (1976), Phys. Lett. **56A**, 361.
Harper, C. D., Wheatley, S. E., and Levenson, M. D. (1977), J. Opt. Soc. Am. **67**, 579.
Johansson, I., and Svendenius, N. (1972), Phys. Scr. **5**, 129.
Kratz, H. R. (1949), Phys. Rev. **75**, 1844.
Litzén, U. (1970), Phys. Scr. **1**, 253.
Lorenzen, C.-J., Niemax, K., and Pendril, L. R. (1981), Opt. Comm. **39**, 370.
Lorenzen, C.-J., and Niemax, K. (1983), Phys. Scr. **27**, 300.
Mansfield, M. W. D. (1975), Proc. R. Soc. London, **Ser. A346**, 539.
Martin, W. C., Tech, J. L., and Wilson, M. (1969), Phys. Rev. **181**, 66.
Nilsson, L., and Svanberg, S. (1979), Z. Phys. **A291**, 303.
Risberg, P. (1956), Ark. Fys. **10**, 583.
Shen, N. M., and Curry, S. M. (1977), Opt. Comm. **20**, 392.
Thompson, D. C., O'Sullivan, M. S., Stoicheff, B. P., and Xu, G.-X. (1983), Can. J. Phys. **61**, 949.
Vanden Bout, P., Aygun, E., Ehlers, V. J., Incesu, T., Saplakoglu, A., and Shugart, H. A. (1968), Phys. Rev. **165**, 88.

K I: Ordered by term values

Configuration	Term	J	Level (cm^{-1})	g
$3p^6(^1S)4s$	^2S	1/2	0.000	2.002295
$3p^6(^1S)4p$	$^2P°$	1/2	*12 985.170*	
		3/2	*13 042.876*	
$3p^6(^1S)5s$	^2S	1/2	21 026.551	
$3p^6(^1S)3d$	^2D	5/2	21 534.680	
		3/2	21 536.988	
$3p^6(^1S)5p$	$^2P°$	1/2	*24 701.382*	0.665
		3/2	*24 720.139*	
$3p^6(^1S)4d$	^2D	5/2	27 397.077	
		3/2	27 398.147	
$3p^6(^1S)6s$	^2S	1/2	27 450.7104	
$3p^6(^1S)4f$	$^2F°$	5/2,7/2	*28 127.85*	
$3p^6(^1S)6p$	$^2P°$	1/2	*28 999.27*	0.6663
		3/2	*29 007.71*	1.3337
$3p^6(^1S)5d$	^2D	5/2	30 185.2439	1.2004
		3/2	30 185.7476	0.7997
$3p^6(^1S)7s$	^2S	1/2	30 274.2487	2.0020
$3p^6(^1S)5f$	$^2F°$	5/2,7/2	*30 606.73*	
$3p^6(^1S)5g$	^2G	7/2,9/2	30 617.31	

K I: Ordered by term values—Continued

Configuration	Term	J	Level (cm^{-1})	g
$3p^6(^1S)7p$	$^2P°$	1/2	*31 069.90*	0.6659
		3/2	*31 074.40*	1.3336
$3p^6(^1S)6d$	^2D	5/2	31 695.9005	1.2013
		3/2	31 696.1661	0.7999
$3p^6(^1S)8s$	^2S	1/2	31 765.3767	2.0028
$3p^6(^1S)6f$	$^2F°$	5/2,7/2	*31 953.17*	
$3p^6(^1S)8p$	$^2P°$	1/2	*32 227.44*	
		3/2	*32 230.11*	
$3p^6(^1S)7d$	^2D	5/2	32 598.2881	
		3/2	32 598.4437	
$3p^6(^1S)9s$	^2S	1/2	32 648.3511	
$3p^6(^1S)7f$	$^2F°$	5/2,7/2	*32 764.80*	
$3p^6(^1S)9p$	$^2P°$	1/2	*32 940.2030*	
		3/2	*32 941.9262*	
$3p^6(^1S)8d$	^2D	5/2	33 178.1339	
		3/2	33 178.2324	
$3p^6(^1S)10s$	^2S	1/2	33 214.2267	
$3p^6(^1S)8f$	$^2F°$	5/2,7/2	*33 291.40*	
$3p^6(^1S)10p$	$^2P°$	1/2	*33 410.2306*	
		3/2	*33 411.3986*	
$3p^6(^1S)9d$	^2D	5/2	33 572.0592	
		3/2	33 572.1249	
$3p^6(^1S)11s$	^2S	1/2	33 598.5597	
$3p^6(^1S)9f$	$^2F°$	5/2,7/2	*33 652.32*	
$3p^6(^1S)11p$	$^2P°$	1/2	*33 736.4979*	
		3/2	*33 737.3284*	
$3p^6(^1S)10d$	^2D	5/2	33 851.5956	
		3/2	33 851.6418	
$3p^6(^1S)12s$	^2S	1/2	33 871.4788	
$3p^6(^1S)10f$	$^2F°$	5/2,7/2	*33 910.42*	
$3p^6(^1S)12p$	$^2P°$	1/2	*33 972.2064*	
		3/2	*33 972.8148*	
$3p^6(^1S)11d$	^2D	5/2	34 057.0051	
		3/2	34 057.0385	
$3p^6(^1S)13s$	^2S	1/2	34 072.2393	
$3p^6(^1S)11f$	$^2F°$	5/2,7/2	*34 101.36*	

(Continued)

K I: Ordered by term values—Continued

Configuration	Term	J	Level (cm^{-1})	g
$3p^6(^1S)13p$	$^2P^\circ$	1/2	*34 148.0284*	
		3/2	*34 148.4861*	
$3p^6(^1S)12d$	^2D	5/2	34 212.3139	
		3/2	34 212.3393	
$3p^6(^1S)14s$	^2S	1/2	34 224.2113	
$3p^6(^1S)12f$	$^2F^\circ$	5/2,7/2	*34 246.37*	
$3p^6(^1S)14p$	$^2P^\circ$	1/2	*34 282.6573*	
		3/2	*34 283.0181*	
$3p^6(^1S)13d$	^2D	5/2	34 332.5627	
		3/2	34 332.5823	
$3p^6(^1S)15s$	^2S	1/2	34 342.0150	
$3p^6(^1S)13f$	$^2F^\circ$	5/2,7/2	*34 359.36*	
$3p^6(^1S)15p$	$^2P^\circ$	1/2	*34 388.0315*	
		3/2	*34 388.3148*	
$3p^6(^1S)14d$	^2D	5/2	34 427.5513	
		3/2	34 427.5667	
$3p^6(^1S)16s$	^2S	1/2	34 435.1762	
$3p^6(^1S)14f$	$^2F^\circ$	5/2,7/2	*34 448.98*	
$3p^6(^1S)16p$	$^2P^\circ$	1/2	*34 472.0505*	
		3/2	*34 472.2798*	
$3p^6(^1S)15d$	^2D	5/2	34 503.8844	
		3/2	34 503.8967	
$3p^6(^1S)17s$	^2S	1/2	34 510.1190	
$3p^6(^1S)17p$	$^2P^\circ$	1/2	*34 540.1250*	
		3/2	*34 540.3088*	
$3p^6(^1S)16d$	^2D	5/2	34 566.1420	
		3/2	34 566.1522	
$3p^6(^1S)18s$	^2S	1/2	34 571.3017	
$3p^6(^1S)18p$	$^2P^\circ$	1/2	*34 596.0448*	
		3/2	34 596.1996	
$3p^6(^1S)17d$	^2D	5/2	34 617.5815	
		3/2	34 617.5899	
$3p^6(^1S)19s$	^2S	1/2	34 621.8976	
$3p^6(^1S)19p$	$^2P^\circ$	3/2	*34 642.6698*	
$3p^6(^1S)18d$	^2D	5/2	34 660.5702	
		3/2	34 660.5772	

K I: Ordered by term values—Continued

Configuration	Term	J	Level (cm^{-1})	g
$3p^6(^1S)20s$	^2S	1/2	34 664.2161	
$3p^6(^1S)20p$	$^2P°$	3/2	*34 681.7220*	
$3p^6(^1S)19d$	^2D	5/2	34 696.8629	
		3/2	34 696.8689	
$3p^6(^1S)21s$	^2S	1/2	34 699.9692	
$3p^6(^1S)21p$	$^2P°$	3/2	*34 714.8646*	
$3p^6(^1S)20d$	^2D	5/2	34 727.7798	
		3/2	34 727.7849	
$3p^6(^1S)22s$	^2S	1/2	34 730.4476	
$3p^6(^1S)21d$	^2D	5/2	34 754.3327	
		3/2	34 754.3371	
$3p^6(^1S)23s$	^2S	1/2	34 756.6407	
$3p^6(^1S)22d$	^2D	5/2	34 777.3058	
		3/2	34 777.3096	
$3p^6(^1S)24s$	^2S	1/2	34 779.3147	
$3p^6(^1S)23d$	^2D	5/2	34 797.3141	
		3/2	34 797.3174	
$3p^6(^1S)25s$	^2S	1/2	34 799.0740	
$3p^6(^1S)24d$	^2D	5/2	34 814.8472	
		3/2	34 814.8502	
$3p^6(^1S)26s$	^2S	1/2	34 816.3971	
$3p^6(^1S)25d$	^2D	5/2	34 830.2969	
		3/2	34 830.2994	
$3p^6(^1S)27s$	^2S	1/2	34 831.6690	
$3p^6(^1S)26d$	^2D	5/2	34 843.9804	
		3/2	34 843.9827	
$3p^6(^1S)28s$	^2S	1/2	34 845.2004	
$3p^6(^1S)27d$	^2D	5/2	34 856.1570	
		3/2	34 856.1590	
$3p^6(^1S)29s$	^2S	1/2	34 857.2470	
$3p^6(^1S)28d$	^2D	5/2	34 867.0404	
		3/2	34 867.0423	
$3p^6(^1S)30s$	^2S	1/2	34 868.0180	
$3p^6(^1S)29d$	^2D	5/2	[34 876.8070]	
		3/2	[34 876.8087]	
$3p^6(^1S)31s$	^2S	1/2	[34 877.6871]	

(Continued)

K I: Ordered by term values—Continued

Configuration	Term	J	Level (cm^{-1})	g
$3p^6(^1S)30d$	^2D	$5/2$	34 885.6048	
		$3/2$	34 885.6062	
$3p^6(^1S)32s$	^2S	$1/2$	34 886.3999	
$3p^6(^1S)31d$	^2D	$5/2$	[34 893.5578]	
		$3/2$	[34 893.5591]	
$3p^6(^1S)33s$	^2S	$1/2$	[34 894.2785]	
$3p^6(^1S)32d$	^2D	$5/2$	34 900.7707	
		$3/2$	34 900.7719	
$3p^6(^1S)34s$	^2S	$1/2$	34 901.4264	
$3p^6(^1S)33d$	^2D	$5/2$	[34 907.3326]	
		$3/2$	[34 907.3337]	
$3p^6(^1S)35s$	^2S	$1/2$	[34 907.9301]	
$3p^6(^1S)34d$	^2D	$5/2$	34 913.3194	
		$3/2$	34 913.3204	
$3p^6(^1S)36s$	^2S	$1/2$	34 913.8657	
$3p^6(^1S)35d$	^2D	$5/2$	[34 918.7968]	
		$3/2$	[34 918.7976]	
$3p^6(^1S)37s$	^2S	$1/2$	[34 919.2976]	
$3p^6(^1S)36d$	^2D	$5/2$	34 923.8203	
		$3/2$	34 923.8212	
$3p^6(^1S)38s$	^2S	$1/2$	34 924.2808	
$3p^6(^1S)37d$	^2D	$5/2$	[34 928.4396]	
		$3/2$	[34 928.4403]	
$3p^6(^1S)39s$	^2S	$1/2$	[34 928.8634]	
$3p^6(^1S)38d$	^2D	$5/2$	34 932.6961	
		$3/2$	34 932.6969	
$3p^6(^1S)40s$	^2S	$1/2$	34 933.0874	
$3p^6(^1S)39d$	^2D	$5/2$	[34 936.6273]	
		$3/2$	[34 936.6279]	
$3p^6(^1S)41s$	^2S	$1/2$	[34 936.9892]	
$3p^6(^1S)40d$	^2D	$5/2$	34 940.2653	
		$3/2$	34 940.2660	
$3p^6(^1S)42s$	^2S	$1/2$	34 940.6009	
$3p^6(^1S)41d$	^2D	$5/2$	[34 943.6386]	
		$3/2$	[34 943.6393]	
$3p^6(^1S)43s$	^2S	$1/2$	[34 943.9500]	

K I: Ordered by term values—Continued

Configuration	Term	J	Level (cm^{-1})	g
$3p^6(^1S)42d$	^2D	5/2	34 946.7726	
		3/2	[34 946.7730]	
$3p^{69}{}^1S)44s$	^2S	1/2	34 947.0620	
$3p^6(^1S)43d$	^2D	5/2	[34 949.6886]	
		3/2	[34 949.6892]	
$3p^6(^1S)45s$	^2S	1/2	[34 949.9585]	
$3p^6(^1S)44d$	^2D	5/2	[34 952.4071]	
		3/2	[34 952.4077]	
$3p^6(^1S)46s$	^2S	1/2	34 952.6589	
$3p^6(^1S)45d$	^2D	5/2	[34 954.9453]	
		3/2	[34 954.9458]	
$3p^6(^1S)46d$	^2D	5/2	34 957.3187	
		3/2	[34 957.3193]	
K II ($^1S_0$)	***Limit***		**35 009.8140**	
$3p^5\,4s^2$	$^2P°$	3/2	*151 008*	
		1/2	*153 085*	
$3p^5\,3d(^3P°)4s$	$^4P°$	1/2	*159 367*	
		3/2	*159 678*	
$3p^5\,3d(^3P°)4s$	$^2P°$	1/2	*162 404*	
		3/2	*163 006*	
$3p^5\,3d(^3D°)4s$	$^4D°$	3/2	*172 623*	
		1/2	*172 800*	
$3p^5\,3d(^1D°)4s$	$^2D°$	3/2	*173 043*	
$3p^5\,3d(^3D°)4s$	$^2D°$	3/2	*179 886*	
$3p^5\,3d(^1P°)4s$	$^2P°$	1/2	*180 551*	
		3/2	*180 791*	
$3p^5\,4s(^3P°)5s$	$^4P°$	3/2	*180 850*	
		1/2	*181 517*	
$3p^5\,4s(^3P°)5s$	$^2P°$	3/2	*182 152*	
$3p^5\,3d(^3P°)5s$	$^2P°$	1/2	*183 322*	
		3/2	*183 532*	
$3p^5\,4s(^1P°)5s$	$^2P°$	3/2	*184 342*	
		1/2	*185 153*	
$3p^5\,4d(^3D°)4s$	$^4D°$	1/2,3/2	*185 153*	
$3p^5\,4d(^1D°)4s$	$^2D°$	3/2	*186 656*	
$3p^5\,4d(^3D°)4s$	$^2D°$	3/2	*187 806*	

(Continued)

K I: Ordered by term values—Continued

Configuration	Term	J	Level (cm^{-1})	g
$3p^5\,4d(^1P°)4s$	$^2P°$	1/2,3/2	*188 565*	
$3p^5\,4s(^3P°)5d$	$^4P°$	3/2	*189 900*	
$3p^5\,4s(^3P°)5d$	$^2D°$	3/2	*190 434*	
$3p^5\,4s(^3P°)5d$	$^2P°$	1/2	*190 942*	
		3/2	*192 251*	
$3p^5\,4s(^3P°)5d$	$^4D°$	3/2	*191 359*	
$3p^5\,4s(^3P°)5d$	$^4F°$	3/2	*191 641*	
$3p^5\,4s(^1P°)6s$	$^2P°$	1/2,3/2	*193 065*	
$3p^5\,4s(^3P°)6d$	$^2D°$	3/2	*193 122*	
$3p^5\,4s(^3P°)6d$	$^2P°$	1/2	*193 244*	
		3/2	*194 275*	
$3p^5\,4s(^1P°)5d$	$^2D°$	3/2	*193 749*	
$3p^5\,4s(^1P°)5d$	$^2P°$	1/2	*193 948*	
		3/2	*194 068*	
$3p^5\,4s(^1P°)7s$	$^2P°$	1/2,3/2	*196 319*	
$3p^5\,4s(^1P°)6d$	$^2P°$	3/2	*196 362*	
		1/2	*196 718*	
$3p^5\,4s(^1P°)6d$	$^2D°$	3/2	*196 826*	
$3p^5\,4s(^1P°)7d$ $3p^5\,4s(^1P°)8s$	$^2P°$	1/2,3/2	*197 959*	
$3p^5\,4s(^1P°)8d$ $3p^5\,4s(^1P°)9s$	$^2P°$	1/2,3/2	*198 911*	
$3p^5\,4s(^1P°)9d$ $3p^5\,4s(^1P°)10s$	$^2P°$	1/2,3/2	*199 549*	
$3p^5\,4s(^1P°)10d$ $3p^5\,4s(^1P°)11s$	$^2P°$	1/2,3/2	*199 980*	
$3p^5\,4s(^1P°)11d$ $3p^5\,4s(^1P°)12s$	$^2P°$	1/2,3/2	*200 268*	
$3p^5\,4s(^1P°)12d$ $3p^5\,4s(^1P°)13s$	$^2P°$	1/2,3/2	*200 493*	
$3p^5\,4s(^1P°)13d$ $3p^5\,4s(^1P°)14s$	$^2P°$	1/2,3/2	*200 658*	
$3p^5\,4s(^1P°)14d$ $3p^5\,4s(^1P°)15s$	$^2P°$	1/2,3/2	*200 780*	
$3p^5\,4s(^1P°)15d$ $3p^5\,4s(^1P°)16s$	$^2P°$	1/2,3/2	*200 876*	

K I: Ordered by term values—Continued

Configuration	Term	J	Level (cm^{-1})	g
$3p^5\,4s(^1P^\circ)16d$ $3p^5\,4s(^1P^\circ)17s$	$^2P^\circ$	$1/2,3/2$	*200 955*	
$3p^5\,4s(^1P^\circ)17d$ $3p^5\,4s(^1P^\circ)18s$	$^2P^\circ$	$1/2,3/2$	*201 017*	
$3p^5\,4s(^1P^\circ)18d$ $3p^5\,4s(^1P^\circ)19s$	$^2P^\circ$	$1/2,3/2$	*201 074*	
$3p^5\,4s(^1P^\circ)19d$ $3p^5\,4s(^1P^\circ)20s$	$^2P^\circ$	$1/2,3/2$	*201 124*	
$3p^5\,4s(^1P^\circ)20d$ $3p^5\,4s(^1P^\circ)21s$	$^2P^\circ$	$1/2,3/2$	*201 152*	
K II ($^1P^\circ_1$)	*Limit*		201 471.3	

K I: Ordered by series

Configuration	Term	J	Level (cm^{-1})	g
$3p^6(^1S)4s$	^2S	$1/2$	0.000	2.002295
$3p^6(^1S)5s$	^2S	$1/2$	21 026.551	
$3p^6(^1S)6s$	^2S	$1/2$	27 450.7104	
$3p^6(^1S)7s$	^2S	$1/2$	30 274.2487	2.0020
$3p^6(^1S)8s$	^2S	$1/2$	31 765.3767	2.0028
$3p^6(^1S)9s$	^2S	$1/2$	32 648.3511	
$3p^6(^1S)10s$	^2S	$1/2$	33 214.2267	
$3p^6(^1S)11s$	^2S	$1/2$	33 598.5597	
$3p^6(^1S)12s$	^2S	$1/2$	33 871.4788	
$3p^6(^1S)13s$	^2S	$1/2$	34 072.2393	
$3p^6(^1S)14s$	^2S	$1/2$	34 224.2113	
$3p^6(^1S)15s$	^2S	$1/2$	34 342.0150	
$3p^6(^1S)16s$	^2S	$1/2$	34 435.1762	
$3p^6(^1S)17s$	^2S	$1/2$	34 510.1190	
$3p^6(^1S)18s$	^2S	$1/2$	34 571.3017	
$3p^6(^1S)19s$	^2S	$1/2$	34 621.8976	
$3p^6(^1S)20s$	^2S	$1/2$	34 664.2161	
$3p^6(^1S)21s$	^2S	$1/2$	34 699.9692	
$3p^6(^1S)22s$	^2S	$1/2$	34 730.4476	
$3p^6(^1S)23s$	^2S	$1/2$	34 756.6407	
$3p^6(^1S)24s$	^2S	$1/2$	34 779.3147	
$3p^6(^1S)25s$	^2S	$1/2$	34 799.0740	
$3p^6(^1S)26s$	^2S	$1/2$	34 816.3971	
$3p^6(^1S)27s$	^2S	$1/2$	34 831.6690	
$3p^6(^1S)28s$	^2S	$1/2$	34 845.2004	
$3p^6(^1S)29s$	^2S	$1/2$	34 857.2470	
$3p^6(^1S)30s$	^2S	$1/2$	34 868.0180	
$3p^6(^1S)31s$	^2S	$1/2$	[34 877.6871]	
$3p^6(^1S)32s$	^2S	$1/2$	34 886.3999	
$3p^6(^1S)33s$	^2S	$1/2$	[34 894.2785]	
$3p^6(^1S)34s$	^2S	$1/2$	34 901.4264	
$3p^6(^1S)35s$	^2S	$1/2$	[34 907.9301]	
$3p^6(^1S)36s$	^2S	$1/2$	34 913.8657	
$3p^6(^1S)37s$	^2S	$1/2$	[34 919.2976]	
$3p^6(^1S)38s$	^2S	$1/2$	34 924.2808	
$3p^6(^1S)39s$	^2S	$1/2$	[34 928.8634]	
$3p^6(^1S)40s$	^2S	$1/2$	34 933.0874	
$3p^6(^1S)41s$	^2S	$1/2$	[34 836.9892]	
$3p^6(^1S)42s$	^2S	$1/2$	34 940.6009	
$3p^6(^1S)43s$	^2S	$1/2$	[34 943.9500]	
$3p^6(^1S)44s$	^2S	$1/2$	34 947.0620	
$3p^6(^1S)45s$	^2S	$1/2$	[34 949.9585]	
$3p^6(^1S)46s$	^2S	$1/2$	34 952.6589	
K II ($^1S_0$)	***Limit***		**35 009.8140**	
$3p^6(^1S)4p$	$^2P°$	$1/2$	*12 985.170*	
		$3/2$	*13 042.876*	
$3p^6(^1S)5p$	$^2P°$	$1/2$	*24 701.382*	0.665
		$3/2$	*24 720.139*	
$3p^6(^1S)6p$	$^2P°$	$1/2$	*28 999.27*	0.6663
		$3/2$	*29 007.71*	1.3337
$3p^6(^1S)7p$	$^2P°$	$1/2$	*31 069.90*	0.6659
		$3/2$	*31 074.40*	1.3336

K I: Ordered by series—Continued

Configuration	Term	J	Level (cm^{-1})	g
$3p^6(^1S)8p$	$^2P°$	1/2	*32 227.44*	
		3/2	*32 230.11*	
$3p^6(^1S)9p$	$^2P°$	1/2	*32 940.2030*	
		3/2	*32 941.9262*	
$3p^6(^1S)10p$	$^2P°$	1/2	*33 410.2306*	
		3/2	*33 411.3986*	
$3p^6(^1S)11p$	$^2P°$	1/2	*33 736.4979*	
		3/2	*33 737.3284*	
$3p^6(^1S)12p$	$^2P°$	1/2	*33 972.2064*	
		3/2	*33 972.8148*	
$3p^6(^1S)13p$	$^2P°$	1/2	*34 148.0284*	
		3/2	*34 148.4861*	
$3p^6(^1S)14p$	$^2P°$	1/2	*34 282.6573*	
		3/2	*34 283.0181*	
$3p^6(^1S)15p$	$^2P°$	1/2	*34 388.0315*	
		3/2	*34 388.3148*	
$3p^6(^1S)16p$	$^2P°$	1/2	*34 472.0505*	
		3/2	*34 472.2798*	
$3p^6(^1S)17p$	$^2P°$	1/2	*34 540.1250*	
		3/2	*34 540.3088*	
$3p^6(^1S)18p$	$^2P°$	1/2	*34 596.0448*	
		3/2	*34 596.1996*	
$3p^6(^1S)19p$	$^2P°$	3/2	*34 642.6698*	
$3p^6(^1S)20p$	$^2P°$	3/2	*34 681.7220*	
$3p^6(^1S)21p$	$^2P°$	3/2	*34 714.8646*	
K II ($^1S_0$)	***Limit***		**35 009.8140**	
$3p^6(^1S)3d$	^2D	5/2	21 534.680	
		3/2	21 536.988	
$3p^6(^1S)4d$	^2D	5/2	27 397.077	
		3/2	27 398.147	
$3p^6(^1S)5d$	^2D	5/2	30 185.2439	1.2004
		3/2	30 185.7476	0.7997
$3p^6(^1S)6d$	^2D	5/2	31 695.9005	1.2013
		3/2	31 696.1661	0.7999
$3p^6(^1S)7d$	^2D	5/2	32 598.2881	
		3/2	32 598.4437	
$3p^6(^1S)8d$	^2D	5/2	33 178.1339	
		3/2	33 178.2324	
$3p^6(^1S)9d$	^2D	5/2	33 572.0592	
		3/2	33 572.1249	

(Continued)

K I: Ordered by series—Continued

Configuration	Term	J	Level (cm^{-1})	g
$3p^6(^1S)10d$	^2D	$^5/_2$	33 851.5956	
		$^3/_2$	33 851.6418	
$3p^6(^1S)11d$	^2D	$^5/_2$	34 057.0051	
		$^3/_2$	34 057.0385	
$3p^6(^1S)12d$	^2D	$^5/_2$	34 212.3139	
		$^3/_2$	34 212.3393	
$3p^6(^1S)13d$	^2D	$^5/_2$	34 332.5627	
		$^3/_2$	34 332.5823	
$3p^6(^1S)14d$	^2D	$^5/_2$	34 427.5513	
		$^3/_2$	34 427.5667	
$3p^6(^1S)15d$	^2D	$^5/_2$	34 503.8844	
		$^3/_2$	34 503.8967	
$3p^6(^1S)16d$	^2D	$^5/_2$	34 566.1420	
		$^3/_2$	34 566.1522	
$3p^6(^1S)17d$	^2D	$^5/_2$	34 617.5815	
		$^3/_2$	34 617.5899	
$3p^6(^1S)18d$	^2D	$^5/_2$	34 660.5702	
		$^3/_2$	34 660.5772	
$3p^6(^1S)19d$	^2D	$^5/_2$	34 696.8629	
		$^3/_2$	34 696.8689	
$3p^6(^1S)20d$	^2D	$^5/_2$	34 727.7798	
		$^3/_2$	34 727.7849	
$3p^6(^1S)21d$	^2D	$^5/_2$	34 754.3327	
		$^3/_2$	34 754.3371	
$3p^6(^1S)22d$	^2D	$^5/_2$	34 777.3058	
		$^3/_2$	34 777.3096	
$3p^6(^1S)23d$	^2D	$^5/_2$	34 797.3141	
		$^3/_2$	34 797.3174	
$3p^6(^1S)24d$	^2D	$^5/_2$	34 814.8472	
		$^3/_2$	34 814.8502	
$3p^6(^1S)25d$	^2D	$^5/_2$	34 830.2969	
		$^3/_2$	34 830.2994	
$3p^6(^1S)26d$	^2D	$^5/_2$	34 843.9804	
		$^3/_2$	34 843.9827	
$3p^6(^1S)27d$	^2D	$^5/_2$	34 856.1570	
		$^3/_2$	34 856.1590	
$3p^6(^1S)28d$	^2D	$^5/_2$	34 867.0404	
		$^3/_2$	34 867.0423	
$3p^6(^1S)29d$	^2D	$^5/_2$	[34 876.8070]	
		$^3/_2$	[34 876.8087]	

K I: Ordered by series—Continued

Configuration	Term	J	Level (cm^{-1})	g
$3p^6(^1S)30d$	^2D	$^5/_2$	34 885.6048	
		$^3/_2$	34 885.6062	
$3p^6(^1S)31d$	^2D	$^5/_2$	[34 893.5578]	
		$^3/_2$	[34 893.5591]	
$3p^6(^1S)32d$	^2D	$^5/_2$	34 900.7707	
		$^3/_2$	34 900.7719	
$3p^6(^1S)33d$	^2D	$^5/_2$	[34 907.3326]	
		$^3/_2$	[34 907.3337]	
$3p^6(^1S)34d$	^2D	$^5/_2$	34 913.3194	
		$^3/_2$	34 913.3204	
$3p^6(^1S)35d$	^2D	$^5/_2$	[34 918.7968]	
		$^3/_2$	[34 918.7976]	
$3p^6(^1S)36d$	^2D	$^5/_2$	34 923.8203	
		$^3/_2$	34 923.8212	
$3p^6(^1S)37d$	^2D	$^5/_2$	[34 928.4396]	
		$^3/_2$	[34 928.4403]	
$3p^6(^1S)38d$	^2D	$^5/_2$	34 932.6961	
		$^3/_2$	34 932.6969	
$3p^6(^1S)39d$	^2D	$^5/_2$	[34 936.6273]	
		$^3/_2$	[34 936.6279]	
$3p^6(^1S)40d$	^2D	$^5/_2$	34 940.2653	
		$^3/_2$	34 940.2660	
$3p^6(^1S)41d$	^2D	$^5/_2$	[34 943.6386]	
		$^3/_2$	[34 943.6393]	
$3p^6(^1S)42d$	^2D	$^5/_2$	34 946.7726	
		$^3/_2$	[34 946.7730]	
$3p^6(^1S)43d$	^2D	$^5/_2$	[34 949.6886]	
		$^3/_2$	[34 949.6892]	
$3p^6(^1S)44d$	^2D	$^5/_2$	34 952.4071	
		$^3/_2$	[34 952.4077]	
$3p^6(^1S)45d$	^2D	$^5/_2$	[34 954.9453]	
		$^3/_2$	[34 954.9458]	
$3p^6(^1S)46d$	^2D	$^5/_2$	34 957.3187	
		$^3/_2$	[34 957.3193]	
K II ($^1S_0$)	***Limit***		**35 009.8140**	
$3p^6(^1S)4f$	$^2F°$	$^5/_2,^7/_2$	*28 127.85*	
$3p^6(^1S)5f$	$^2F°$	$^5/_2,^7/_2$	*30 606.73*	
$3p^6(^1S)6f$	$^2F°$	$^5/_2,^7/_2$	*31 953.17*	
$3p^6(^1S)7f$	$^2F°$	$^5/_2,^7/_2$	*32 764.80*	
$3p^6(^1S)8f$	$^2F°$	$^5/_2,^7/_2$	*33 291.40*	
$3p^6(^1S)9f$	$^2F°$	$^5/_2,^7/_2$	*33 652.32*	

(Continued)

K I: Ordered by series—Continued

Configuration	Term	J	Level (cm^{-1})	g
$3p^6(^1S)10f$	$^2F°$	$5/2,7/2$	*33 910.42*	
$3p^6(^1S)11f$	$^2F°$	$5/2,7/2$	*34 101.36*	
$3p^6(^1S)12f$	$^2F°$	$5/2,7/2$	*34 246.37*	
$3p^6(^1S)13f$	$^2F°$	$5/2,7/2$	*34 359.36*	
$3p^6(^1S)14f$	$^2F°$	$5/2,7/2$	*34 448.98*	
K II ($^1S_0$)	***Limit***		**35 009.8140**	
$3p^6(^1S)5g$	^2G	$7/2,9/2$	30 617.31	
$3p^5\,4s^2$	$^2P°$	$3/2$	*151 008*	
		$1/2$	*153 085*	
$3p^5\,3d(^3P°)4s$	$^4P°$	$1/2$	*159 367*	
		$3/2$	*159 678*	
$3p^5\,3d(^3P°)4s$	$^2P°$	$1/2$	*162 404*	
		$3/2$	*163 006*	
$3p^5\,3d(^3D°)4s$	$^4D°$	$3/2$	*172 623*	
		$1/2$	*172 800*	
$3p^5\,3d(^1D°)4s$	$^2D°$	$3/2$	*173 043*	
$3p^5\,3d(^3D°)4s$	$^2D°$	$3/2$	*179 886*	
$3p^5\,3d(^1P°)4s$	$^2P°$	$1/2$	*180 551*	
		$3/2$	*180 791*	
$3p^5\,4s(^3P°)5s$	$^4P°$	$3/2$	*180 850*	
		$1/2$	*181 517*	
$3p^5\,4s(^3P°)5s$	$^2P°$	$3/2$	*182 152*	
$3p^5\,3d(^3P°)5s$	$^2P°$	$1/2$	*183 322*	
		$3/2$	*183 532*	
$3p^5\,4s(^1P°)5s$	$^2P°$	$3/2$	*184 342*	
		$1/2$	*185 153*	
$3p^5\,4s(^1P°)6s$	$^2P°$	$1/2,3/2$	*193 065*	
$3p^5\,4s(^1P°)7s$	$^2P°$	$1/2,3/2$	*196 319*	
$3p^5\,4s(^1P°)8s$	$^2P°$	$1/2,3/2$	*197 959*	
$3p^5\,4s(^1P°)9s$	$^2P°$	$1/2,3/2$	*198 911*	
$3p^5\,4s(^1P°)10s$	$^2P°$	$1/2,3/2$	*199 549*	
$3p^5\,4s(^1P°)11s$	$^2P°$	$1/2,3/2$	*199 980*	
$3p^5\,4s(^1P°)12s$	$^2P°$	$1/2,3/2$	*200 268*	
$3p^5\,4s(^1P°)13s$	$^2P°$	$1/2,3/2$	*200 493*	
$3p^5\,4s(^1P°)14s$	$^2P°$	$1/2,3/2$	*200 658*	
$3p^5\,4s(^1P°)15s$	$^2P°$	$1/2,3/2$	*200 780*	
$3p^5\,4s(^1P°)16s$	$^2P°$	$1/2,3/2$	*200 876*	
$3p^5\,4s(^1P°)17s$	$^2P°$	$1/2,3/2$	*200 955*	
$3p^5\,4s(^1P°)18s$	$^2P°$	$1/2,3/2$	*201 017*	
$3p^5\,4s(^1P°)19s$	$^2P°$	$1/2,3/2$	*201 074*	
$3p^5\,4s(^1P°)20s$	$^2P°$	$1/2,3/2$	*201 124*	
$3p^5\,4s(^1P°)21s$	$^2P°$	$1/2,3/2$	*201 152*	
K II ($^1P^\circ_1$)	***Limit***		**201 471.3**	

K I: Ordered by series—Continued

Configuration	Term	J	Level (cm^{-1})	g
$3p^5\,4d(^3D^\circ)4s$	$^4D^\circ$	$1/2, 3/2$	185 153	
$3p^5\,4d(^1D^\circ)4s$	$^2D^\circ$	$3/2$	186 656	
$3p^5\,4d(^3D^\circ)4s$	$^2D^\circ$	$3/2$	187 806	
$3p^5\,4d(^1P^\circ)4s$	$^2P^\circ$	$1/2, 3/2$	188 565	
$3p^5\,4s(^3P^\circ)5d$	$^4P^\circ$	$3/2$	189 900	
$3p^5\,4s(^3P^\circ)5d$	$^2D^\circ$	$3/2$	190 434	
$3p^5\,4s(^3P^\circ)5d$	$^2P^\circ$	$1/2$	190 942	
		$3/2$	192 251	
$3p^5\,4s(^3P^\circ)5d$	$^4D^\circ$	$3/2$	191 359	
$3p^5\,4s(^3P^\circ)5d$	$^4F^\circ$	$3/2$	191 641	
$3p^5\,4s(^3P^\circ)6d$	$^2D^\circ$	$3/2$	193 122	
$3p^5\,4s(^3P^\circ)6d$	$^2P^\circ$	$1/2$	193 244	
		$3/2$	194 275	
$3p^5\,4s(^1P^\circ)5d$	$^2D^\circ$	$3/2$	193 749	
$3p^5\,4s(^1P^\circ)5d$	$^2P^\circ$	$1/2$	193 948	
		$3/2$	194 068	
$3p^5\,4s(^1P^\circ)6d$	$^2P^\circ$	$3/2$	196 362	
		$1/2$	196 718	
$3p^5\,4s(^1P^\circ)6d$	$^2D^\circ$	$3/2$	196 826	
$3p^5\,4s(^1P^\circ)7d$	$^2P^\circ$	$1/2, 3/2$	197 959	
$3p^5\,4s(^1P^\circ)8d$	$^2P^\circ$	$1/2, 3/2$	198 911	
$3p^5\,4s(^1P^\circ)9d$	$^2P^\circ$	$1/2, 3/2$	199 549	
$3p^5\,4s(^1P^\circ)10d$	$^2P^\circ$	$1/2, 3/2$	199 980	
$3p^5\,4s(^1P^\circ)11d$	$^2P^\circ$	$1/2, 3/2$	200 268	
$3p^5\,4s(^1P^\circ)12d$	$^2P^\circ$	$1/2, 3/2$	200 493	
$3p^5\,4s(^1P^\circ)13d$	$^2P^\circ$	$1/2, 3/2$	200 658	
$3p^5\,4s(^1P^\circ)14d$	$^2P^\circ$	$1/2, 3/2$	200 780	
$3p^5\,4s(^1P^\circ)15d$	$^2P^\circ$	$1/2, 3/2$	200 876	
$3p^5\,4s(^1P^\circ)16d$	$^2P^\circ$	$1/2, 3/2$	200 955	
$3p^5\,4s(^1P^\circ)17d$	$^2P^\circ$	$1/2, 3/2$	201 017	
$3p^5\,4s(^1P^\circ)18d$	$^2P^\circ$	$1/2, 3/2$	201 074	
$3p^5\,4s(^1P^\circ)19d$	$^2P^\circ$	$1/2, 3/2$	201 124	
$3p^5\,4s(^1P^\circ)20d$	$^2P^\circ$	$1/2, 3/2$	201 152	
K II ($^1P^\circ_1$)	*Limit*		201 471.3	

K II

$Z=19$

Ar I isoelectronic sequence

Ground state: $1s^2 2s^2 2p^6 3s^2 3p^6\ {}^1S_0$

Ionization energy: 255 100±300 cm^{-1} (31.63±0.04 eV)

The early work of de Bruin (1926) and Bowen (1928) remains the only source of energy levels for this ion. The designations are those of Mansfield (1974), whose calculations show that *LS*-coupling is appropriate for all the known configurations except $3p^5 5s$. The percentage compositions are quoted from his paper.

The level at 223 124.1 cm^{-1} has not been designated by Mansfield because he considered it to be a strong mixture of the $3p^5 3d$ and $3p^5 4d$ configurations for the ${}^1P_1^\circ$ term. The level $3p^5 4p\ {}^1S_0$ at 194 776.1 cm^{-1} is questioned by Mansfield because it deviated from the calculated position by 977 cm^{-1} and is determined by only one line.

Semenov (1965) has measured the *g*-values quoted here. All but one, that for the level $3p^5 3d\ {}^3F_2^\circ$, agree with Mansfield's calculation.

The ionization energy was derived from a semi-empirical treatment of the complete iron period by Catalán and Rico (1958).

References

Bowen, I. S. (1928), Phys. Rev. **31**, 497.

Catalán, M. A., and Rico, F. R. (1958), An. R. Soc. Esp. Fis. Quim. **54A**, 289.

de Bruin, T. L. (1926), Z. Phys. **38**, 94.

Mansfield, M. W. D. (1974), Proc. R. Soc. London, **Ser. A341**, 277.

Semenov, R. I. (1965), Opt. Spect. **19**, 552.

K II

Configuration	Term	*J*	Level (cm^{-1})	*g*	Leading percentages		
$3p^6$	${}^1S$	0	0				
$3p^5\ 4s$	${}^3P^\circ$	2	*162 507.0*	1.48	92	8	$3d\ {}^3P^\circ$
		1	*163 237.0*	1.47	58	26	
		0	*165 149.5*		80	20	
$3p^5\ 3d$	${}^3P^\circ$	0	*163 436.3*		80	20	$4s\ {}^3P^\circ$
		1	*164 496.1*	1.38	64	35	
		2	*164 932.3*	1.47	92	8	
$3p^5\ 4s$	${}^1P^\circ$	1	*166 461.5*	1.07	83	10	$3d\ {}^3P^\circ$
$3p^5\ 3d$	${}^3F^\circ$	3	*170 835.4*	1.09	99		
		2	*171 526.8*	1.36	99		
$3p^5\ 4p$	${}^3S$	1	183 208.4	1.96	98		
$3p^5\ 4p$	${}^3D$	3	186 388.5	1.35	100		
		2	186 685.6	1.11	74	16	${}^3P$
		1	187 531.1	0.72	77	23	${}^1P$
$3p^5\ 4p$	${}^1D$	2	188 154.4	1.19	52	24	${}^3P$
$3p^5\ 4p$	${}^1P$	1	189 243.7	0.95	53	32	${}^3P$
$3p^5\ 4p$	${}^3P$	2	189 661.7	1.31	60	38	${}^1D$
		0	189 772.0		97		
		1	190 134.8	1.38	66	24	${}^1P$
$3p^5\ 4p$	${}^1S$	0	194 776.1?		97		

K II—Continued

Configuration	Term	J	Level (cm^{-1})	g	Leading percentages		
$3p^5(^2P^\circ_{3/2})5s$	$^2[^3\!/_2]^\circ$	2	*212 575.5*	1.51	100		
		1	*212 992.9*	1.19	97		
$3p^5(^2P^\circ_{1/2})5s$	$^2[^1\!/_2]^\circ$	0	*214 727.0*		100		
		1	*215 018.8*	1.25	97		
$3p^5\,4d$	$^3P^\circ$	1	*215 404.9*	1.44	97		
		2	*215 855.8*	1.46	89	11	$^3D^\circ$
$3p^5\,4d$	$^3F^\circ$	2	*217 066.3*		64	36	$^1D^\circ$
$3p^5\,4d$	$^3D^\circ$	3	*217 726.4*		66	26	$^1F^\circ$
		2	*219 196.2*		54	25	$^1D^\circ$
$3p^5\,4d$?	$^1P^\circ$	1	*223 124.1*				
K III ($^2P^\circ_{3/2}$)	***Limit***		**255 100**				

K III

$Z=19$

Cl I isoelectronic sequence

Ground state: $1s^2 2s^2 2p^6 3s^2 3p^5\,{}^2P^\circ_{3/2}$

Ionization energy $=$ 369 450±100 cm^{-1} (45.806±0.010 eV)

The initial work on the analysis of this spectrum was by Bowen (1928), who found the ground term splitting as well as the $3s3p^6\,{}^2S$ and $3p^4(^3P)4s\ {}^2P$ terms. The ground term interval given here is from Smitt, Svensson, and Outred (1976), with an uncertainty of ±0.7 cm^{-1}.

The analysis was extended by de Bruin (1929), who observed the $3p^44s-3p^44p$ transition array between 2500 and 3500 Å, and by Ram (1933) who found levels of the $3p^43d$, $4s$, and $5s$ configurations.

Edlén (1937) extended the analysis of $3p^44s$ and established the position of the ^4P term. Tsien (1939) changed some of Ram's assignments and found two new levels in $3p^43d$. Finally $3p^4(^3P)3d\ {}^2P$ and ^2D were established by Svensson and Ekberg (1968).

With the exception of the measurements of Bowen and of de Bruin, the levels below are based on the observations of Ekefors (1931).

The ionization energy was derived by Catalán and Rico (1958) from a treatment of data in the complete iron period.

References

Bowen, I. S. (1928), Phys. Rev. **31**, 497.

Catalán, M. A., and Rico, F. R. (1958), An. R. Soc. Esp. Fis. Quim. **54A**, 289.

de Bruin, T. L. (1929), Z. Phys. **53**, 658.

Edlén, B. (1937), Z. Phys. **104**, 407.

Ekefors, E. (1931), Z. Phys. **71**, 53.

Ram, M. (1933), Indian J. Phys. **8**, 151.

Smitt, R., Svensson, L. A., and Outred, M. (1976), Phys. Scr. **13**, 293.

Svensson, L. A., and Ekberg, J. O. (1968), Ark. Fys. **37**, 65.

Tsien, W.-Z. (1939), Chinese J. Phys. **3**, 117.

K III

Configuration	Term	J	Level (cm^{-1})	Configuration	Term	J	Level (cm^{-1})
$3s^2\,3p^5$	$^2P^\circ$	3/2	*0.0*	$3s^2\,3p^4(^1D)3d$	^2P	3/2	241 042
		1/2	*2 166.1*			1/2	242 549
$3s3p^6$	^2S	1/2	130 610	$3s^2\,3p^4(^1S)4s$	^2S	1/2	241 667
$3s^2\,3p^4(^3P)3d$	^2P	1/2	183 878	$3s^2\,3p^4(^3P)4p$	$^2D^\circ$	5/2	*243 120.6*
		3/2	185 276			3/2	*243 448.2*
$3s^2\,3p^4(^3P)3d$	^2D	3/2	190 917	$3s^2\,3p^4(^3P)4p$	$^2P^\circ$	3/2	*243 947.4*
		5/2	192 082			1/2	*245 382.3*
$3s^2\,3p^4(^3P)3d$	^2F	5/2	201 165	$3s^2\,3p^4(^1D)3d$	^2D	5/2	244 523
$3s^2\,3p^4(^3P)4s$	^4P	5/2	207 421.9			3/2	246 010
		3/2	208 687.8	$3s^2\,3p^4(^3P)4p$	$^4S^\circ$	3/2	*246 625.6*
		1/2	209 461.3				
				$3s^2\,3p^4(^1D)3d$	^2S	1/2	250 858
$3s^2\,3p^4(^3P)4s$	^2P	3/2	212 725.4				
		1/2	214 232.3	$3s^2\,3p^4(^3P)5s$	^2P	3/2	262 828
						1/2	263 770
$3s^2\,3p^4(^1D)4s$	^2D	5/2	225 051				
		3/2	225 084	$3s^2\,3p^4(^1D)5s$	^2D	5/2	289 400
						3/2	289 519
$3s^2\,3p^4(^3P)4p$	$^4P^\circ$	5/2	*237 512.0*				
		3/2	*237 912.2*	$3s^2\,3p^4(^1S)3d$	^2D	5/2	302 404
		1/2	*238 455.1*			3/2	303 902
$3s^2\,3p^4(^3P)4p$	$^4D^\circ$	7/2	*240 829.9*	K IV ($^3P_2$)	***Limit***		**369 450**
		5/2	*241 443.5*				
		3/2	*242 165.3*				
		1/2	*242 526.7*				

K IV

$Z=19$

S I isoelectronic sequence

Ground state: $1s^2 2s^2 2p^6 3s^2 3p^4\ {}^3P_2$

Ionization energy = 491 300±400 cm^{-1} (60.91±0.05 eV)

The analysis was initiated by Hopfield and Dieke (1926), who discovered the resonance triplet ($3s^2 3p^4 - 3s3p^5$). Smitt, Svensson and Outred (1976) remeasured this array and extended the analysis to include the present levels. Ram (1933), using the line-list of Ekefors (1931), reported the levels of $3p^3 3d$ and $4s$. His configuration assignments to $3d$ and $4s$ were interchanged by Bowen (1934), who also added seven new levels, including the ${}^3S°$ of $3p^3 5s$. Svensson and Ekberg (1968) confirmed the work of Bowen except for the substitution of a new level for $3p^3({}^2D°)3d\ {}^1P_1°$ and added $3p^3$ parent state identifications to the designations. Tsien (1939), also working from the line-list of Ekefors, reported the $3p^3({}^4S°)3d\ {}^3D°$ and $3p^3({}^2D°)3d\ {}^3P°$ terms.

The ionization energy was determined by Edlén (1937).

References

Bowen, I. S. (1934), Phys. Rev. **46**, 791.
Edlén, B. (1937), Z. Phys. **104**, 188.
Ekefors, E. (1931), Z. Phys. **71**, 53.
Hopfield, J. J., and Dieke, G. H. (1926), Phys. Rev. **27**, 638.
Ram, M. (1933), Ind. J. Phys. **8**, 151.
Smitt, R., Svensson, L. A., and Outred, M. (1976), Phys. Scr. **13**, 293.
Svensson, L. A., and Ekberg, J. O. (1968), Ark. Fys. **37**, 65.
Tsien, W.-Z. (1939), Chinese J. Phys. **3**, 117.

K IV

Configuration	Term	J	Level (cm^{-1})
$3s^2 3p^4$	${}^3P$	2	0.0
		1	1 671.4
		0	2 321.2
$3s^2 3p^4$	${}^1D$	2	16 384.1
$3s^2 3p^4$	${}^1S$	0	38 546.3
$3s3p^5$	${}^3P°$	2	*134 181.8*
		1	*135 658.3*
		0	*136 453.0*
$3s3p^5$	${}^1P°$	1	*171 139.5*
$3s^2 3p^3({}^4S°)3d$	${}^3D°$	3	*189 952*
		2	*191 203*
		1	*191 400*
$3s^2 3p^3({}^2D°)3d$	${}^1D°$	2	*216 387*
$3s^2 3p^3({}^2D°)3d$	${}^3P°$	2	*225 445*
		1	*226 082*
		0	*227 650*
$3s^2 3p^3({}^2D°)3d$	${}^1F°$	3	*242 475*
$3s^2 3p^3({}^2D°)3d$	${}^3S°$	1	*249 867*
$3s^2 3p^3({}^2P°)3d$	${}^3P°$	2	*256 032*
		1	*257 122*
		0	*257 809*
$3s^2 3p^3({}^2D°)3d$	${}^1P°$	1	*260 910*
$3s^2 3p^3({}^2P°)3d$	${}^3D°$	3	*261 623*
		2	*262 829*
		1	*263 658*
$3s^2 3p^3({}^2P°)3d$	${}^1D°$	2	*273 398*
$3s^2 3p^3({}^2D°)4s$	${}^3D°$	1	*277 792*
		2	*277 850*
		3	*277 986*
$3s^2 3p^3({}^2P°)3d$	${}^1F°$	3	*279 627*
$3s^2 3p^3({}^2D°)4s$	${}^1D°$	2	*282 371*
$3s^2 3p^3({}^2P°)4s$	${}^3P°$	0	*293 382*
		1	*293 471*
		2	*293 720*
$3s^2 3p^3({}^2P°)4s$	${}^1P°$	1	*298 132*
$3s^2 3p^3({}^2P°)3d$	${}^1P°$	1	*298 898*
$3s^2 3p^3({}^4S°)5s$	${}^3S°$	1	*367 888*
K V (${}^4S_{3/2}°$)	*Limit*		**491 300**

K v

$Z=19$

P I isoelectronic sequence

Ground state: $1s^22s^22p^63s^23p^3\,{}^4S^\circ_{3/2}$

Ionization energy = 666 700±1300 cm^{-1} (82.66±0.16 eV)

The analysis was begun by Ram (1933) with the classification of the $3s^23p^3-3s3p^4$ array from the measurements of Ekefors. Bowen (1934) supplemented these measurements and found many more terms of this spectrum, including the $^2P^\circ$ and $^2D^\circ$ of $3s^23p^3$, the ^2P of $3s3p^4$, and levels of $3s^23p^23d$ and $3s^23p^24s$. Tsien (1939) found the ^2S and ^2D of $3s3p^4$.

Bowen's (1955) measurements of nebular spectra provided him with the forbidden transition $^4S^\circ-{}^2D^\circ$ of $3s^23p^3$. This is the only observed connection between the doublets and quartets.

Using new laboratory measurements, Smitt, Svensson, and Outred (1976) have redetermined the level values for the $3s^23p^3$ and $3s3p^4$ configurations with an uncertainty of about ±2 cm^{-1}. We have combined these values with identifications given by Tsien (1939) and by Ekberg and Svensson (1970) of lines measured by Ekefors (1931), to derive new level values for the $3p^23d$ and $4s$ configurations. The uncertainty of these upper levels is about ±10 cm^{-1}. The $3p^2(^3P)3d\,{}^2F$ and ^2D terms of Tsien are rejected, as suggested by Martin (1959).

The ionization energy is from an extrapolation by Lotz (1967).

References

Bowen, I. S. (1934), Phys. Rev. **46**, 791.
Bowen, I. S. (1955), Astrophys. J. **121**, 306.
Ekberg, J. O., and Svensson, L. A. (1970), Phys. Scr. **2**, 283.
Ekefors, E. (1931), Z. Phys. **71**, 53.
Lotz, W. (1967), J. Opt. Soc. Am. **57**, 873.
Martin, W. C. (1959), J. Opt. Soc. Am. **49**, 1071.
Ram, M. (1933), Indian J. Phys. **8**, 151.
Smitt, R., Svensson, L. A., and Outred, M. (1976), Phys. Scr. **13**, 293.
Tsien, W.-Z. (1939), Chinese J. Phys. **3**, 117.

K v

Configuration	Term	J	Level (cm^{-1})
$3s^2\,3p^3$	$^4S^\circ$	3/2	*0.0*
$3s^2\,3p^3$	$^2D^\circ$	3/2	*24 012.5*
		5/2	*24 249.6*
$3s^2\,3p^3$	$^2P^\circ$	1/2	*39 758.1*
		3/2	*40 080.2*
$3s3p^4$	^4P	5/2	136 636.5
		3/2	138 037.5
		1/2	138 804.1
$3s3p^4$	^2D	3/2	169 579.5
		5/2	169 705.8
$3s3p^4$	^2P	3/2	194 805.1
		1/2	196 331.2
$3s3p^4$	^2S	1/2	205 799.9
$3s^2\,3p^2(^3P)3d$	^4F	3/2	206 720
		5/2	207 165
$3s^2\,3p^2(^3P)3d$	^4D	5/2	222 366
		3/2	222 711
$3s^2\,3p^2(^3P)3d$	^4P	5/2	257 865
		3/2	259 276
		1/2	259 726

Configuration	Term	J	Level (cm^{-1})
$3s^2\,3p^2(^3P)3d$	^2P	3/2	259 218
		1/2	260 882
$3s^2\,3p^2(^1D)3d$	^2D	5/2	280 585
		3/2	281 035
$3s^2\,3p^2(^1D)3d$	^2P	3/2	290 236
		1/2	290 784
$3s^2\,3p^2(^1D)3d$	^2F	5/2	292 497
		7/2	292 960
$3s^2\,3p^2(^1D)3d$	^2S	1/2	292 987
$3s^2\,3p^2(^1S)3d$	^2D	5/2	303 850
		3/2	304 476
$3s^2\,3p^2(^3P)4s$	^4P	1/2	336 628
		3/2	337 645
		5/2	339 172
$3s^2\,3p^2(^3P)4s$	^2P	1/2	343 740
		3/2	345 540
$3s^2\,3p^2(^1D)4s$	^2D	5/2	357 012
		3/2	357 050
$3s^2\,3p^2(^1S)4s$	^2S	1/2	380 994
K VI ($^3P_0$)	*Limit*		**666 700**

K VI

$Z=19$

Si I isoelectronic sequence

Ground state: $1s^2 2s^2 2p^6 3s^2 3p^2\ {}^3P_0$

Ionization energy = 802 000±1600 cm^{-1} (99.4±0.2 eV)

The early analysis is by Ram (1933) and Whitford (1934), who found most of the triplets, and by Robinson (1937), who found two singlets. Each used the measurements by Ekefors (1931).

The present level values for $3s^2 3p^2$ and $3s3p^3$ are taken from Smitt, Svensson, and Outred (1976), who report a level uncertainty of ±2 cm^{-1}. The values for the $3p3d$ and $3p4s$ configurations are derived from the measurements of Ekberg and Svensson (1970), between 374 and 726 Å. They have obtained the value for the ionization energy quoted here by extrapolation.

References

Ekberg, J. O., and Svensson, L. A. (1970), Phys. Scr. **2**, 283.
Ekefors, E. (1931), Z. Phys. **71**, 53.
Ram, M. (1933), Indian J. Phys. **8**, 151.
Robinson, H. A. (1937), Phys. Rev. **52**, 724.
Smitt, R., Svensson, L. A., and Outred, M. (1976), Phys. Scr. **13**, 293.
Whitford, A. E. (1934), Phys. Rev. **46**, 793.

K VI

Configuration	Term	J	Level (cm^{-1})	Configuration	Term	J	Level (cm^{-1})
$3s^2\,3p^2$	^{3}P	0	0.0	$3s^2\,3p3d$	^{3}P°	2	*252 327*
		1	1 133.4			1	*253 503*
		2	2 927.2			0	*254 037*
$3s^2\,3p^2$	^{1}D	2	18 977.8	$3s^2\,3p3d$	^{3}D°	1	*260 069*
						2	*260 503*
$3s^2\,3p^2$	^{1}S	0	43 358.8			3	*260 786*
$3s3p^3$	^{3}D°	1	*140 741.3*	$3s^2\,3p3d$	^{1}F°	3	*285 687*
		2	*140 795.4*				
		3	*140 995.7*	$3s^2\,3p3d$	^{1}P°	1	*293 723*
$3s3p^3$	^{3}P°	0	*163 421.3*	$3s^2\,3p4s$	^{3}P°	0	*387 423*
		1	*163 435.0*			1	*388 116*
		2	*163 438*			2	*390 496*
$3s3p^3$	^{1}D°	2	*178 872.9*	$3s^2\,3p4s$	^{1}P°	1	*394 420*
$3s3p^3$	^{3}S°	1	*218 317.3*	K VII ($^2P^\circ_{1/2}$)	***Limit***		**802 000**
$3s3p^3$	^{1}P°	1	*223 840.1*				

K VII

$Z=19$

Al I isoelectronic sequence

Ground state: $1s^2 2s^2 2p^6 3s^2 3p\ ^2P^\circ_{1/2}$

Ionization energy = 948 200±900 cm^{-1} (117.56±0.10 eV)

Using the wavelength measurements of Ekefors (1931), Whitford (1934) established the first known levels of the $3s^2 3p$, $3s3p^2$, $3p^3$, $3s^2 4s$, and $3s3p4s$ configurations. He found both doublet and quartet terms but no connection between them. The work was carried forward by Phillips (1939), who added levels of the configurations $3s^2 3d$, $3s^2 4d$, and $3s3p3d$ in both systems.

With new measurements between 397 and 673 Å, Ekberg and Svensson (1970) redetermined the energy levels and added the $3s3p3d\ ^4P^\circ$ term as well as the $3s^2 nf (n=4,5)$, $3s^2 nd (n=5,6)$, and $3s^2 ns (n=5,6)$ series members. Somewhat improved wavelength values for the transition array $3s^2 3p - 3s3p^2$ were given by Smitt, Svensson, and Outred (1976). No connection has been observed between the doublets and quartets.

The doublet terms of $3s^2 3p$ and $3s3p^2$ in this compilation are from Smitt, Svensson, and Outred (1976). The uncertainty of their measurements is ±2 cm^{-1}. The remaining terms are derived from the measurements of Ekberg and Svensson, which give a level uncertainty of ±6 cm^{-1}. They have given an extrapolated value for the position of the quartets.

The ionization energy was derived by Ekberg and Svensson from the $nf\ ^2F^\circ$ series. They estimated the error to be less than 1000 cm^{-1}.

References

Ekberg, J. O., and Svensson, L. A. (1970), Phys. Scr. **2**, 283.
Ekefors, E. (1931), Z. Phys. **71**, 53.
Phillips, L. W. (1939), Phys. Rev. **55**, 708.
Smitt, R., Svensson, L. A., and Outred, M. (1976), Phys. Scr. **13**, 293.
Whitford, A. E. (1934), Phys. Rev. **46**, 793.

K VII

Configuration	Term	J	Level (cm^{-1})	Configuration	Term	J	Level (cm^{-1})
$3s^2 3p$	$^2P^\circ$	1/2	*0.0*	$3s^2 4s$	^2S	1/2	439 322
		3/2	*3 134.0*				
				$3s3p4s$	$^4P^\circ$	1/2	*565 985+x*
$3s3p^2$	^4P	1/2	114 650+x			3/2	*567 062+x*
		3/2	115 786+x			5/2	*569 034+x*
		5/2	117 523+x				
				$3s^2 4d$	^2D	3/2	570 738
$3s3p^2$	^2D	3/2	151 883.9			5/2	570 922
		5/2	152 051.7				
				$3s^2 4f$	$^2F^\circ$	5/2	*608 532*
$3s3p^2$	^2S	1/2	193 084.5			7/2	*608 536*
$3s3p^2$	^2P	1/2	206 502.9	$3s^2 5s$	^2S	1/2	654 074
		3/2	208 432.5				
				$3s^2 5d$	^2D	3/2	716 949
$3s^2 3d$	^2D	3/2	250 663			5/2	716 986
		5/2	250 781				
				$3s^2 5f$	$^2F^\circ$	5/2,7/2	*732 500*
$3p^3$	$^4S^\circ$	3/2	*307 777+x*				
				$3s^2 6s$	^2S	1/2	754 539
$3s3p3d$	$^4P^\circ$	5/2	*362 492+x*				
		3/2	*363 321+x*	$3s^2 6d$	^2D	3/2	789 578
						5/2	789 600
$3s3p3d$	$^4D^\circ$	1/2	*365 688+x*				
		3/2	*366 101+x*				
		5/2	*366 409+x*	K VIII ($^1S_0$)	***Limit***		**948 200**
		7/2	*366 556+x*				

K VIII

$Z=19$

Mg I isoelectronic sequence

Ground state: $1s^2 2s^2 2p^6 3s^2\ ^1S_0$

Ionization energy = 1 249 200±400 cm^{-1} (154.88±0.05 eV)

The principal analysis is by Ekberg (1971), who lists 71 classified lines in the range of 91–927 Å. The wavelengths are from unpublished measurements of Bodén and from spectrograms taken earlier by Edlén. The resulting level uncertainty appears to be about ±40 cm^{-1}. Several of the classifications were made by earlier investigators. Fawcett (1970) classified 10 additional lines that provide the levels of the $3p3d$ configuration and the ^{1}D of $3p^2$. No experimental intersystem connection has been found; for the level $3s3p\ ^3P_1^\circ$ we use the value obtained by Finkenthal, Hinnov, Cohen, and Suckewer (1982) by interpolation. We estimate an uncertainty of ±200 cm^{-1} for this value.

The ionization energy calculated from the first three members of the $3snf\ ^1F^\circ$ series (n =4,5, and 6) by Ekberg has been corrrected for the new estimate of the intersystem interval.

References

Ekberg, J. O. (1971), Phys. Scr. **4**, 101.

Fawcett, B. C. (1970), J. Phys. **B3**, 1732.

Finkenthal, M., Hinnov, E., Cohen, S., and Suckewer, S. (1982), Phys. Lett. **91**, 284.

K VIII

Configuration	Term	J	Level (cm^{-1})
$3s^2$	^{1}S	0	0
$3s3p$	^{3}P°	0	*128 187+x*
		1	*129 299+x*
		2	*131 672+x*
$3s3p$	^{1}P°	1	*192 537*
$3p^2$	^{1}D	2	300 387
$3p^2$	^{3}P	0	304 890+x
		1	306 249+x
		2	308 826+x
$3p^2$	^{1}S	0	357 660
$3s3d$	^{3}D	1	368 197+x
		2	368 276+x
		3	368 407+x
$3s3d$	^{1}D	2	419 100
$3p3d$	^{3}F°	2	*503 877+x*
		3	*505 277+x*
		4	*507 077+x*
$3p3d$	^{1}D°	2	*510 990*
$3p3d$	^{3}P°	2	*535 417+x*
$3p3d$	^{3}D°	2	*538 637+x*
		3	*539 257+x*
$3p3d$	^{1}P°	1	*553 620*
$3s4s$	^{3}S	1	631 861+x
$3s4s$	^{1}S	0	644 451
$3s4p$	^{1}P°	1	*695 376*
$3s4d$	^{3}D	1	770 440+x
		2	770 496+x
		3	770 646+x
$3s4d$	^{1}D	2	773 844
$3p4s$	^{3}P°	0	*794 907+x*
		1	*795 814+x*
		2	*798 537+x*
$3s4f$	^{3}F°	2–4	*801 750+x*
$3s4f$	^{1}F°	3	*809 388*
$3p4p$	^{3}D	1	849 542+x
		2	850 564+x
		3	853 001+x
$3p4p$	^{3}P	0	856 054+x
		1	856 824+x
		2	858 616+x
$3p4p$	^{3}S	1	859 360+x
$3s5s$	^{3}S	1	885 149+x
$3s5s$	^{1}S	0	893 057

(Continued)

K VIII—Continued

Configuration	Term	*J*	Level (cm^{-1})	Configuration	Term	*J*	Level (cm^{-1})
3*s*5*p*	$^1P°$	1	*912 975*	3*s*6*p*	$^1P°$	1	*1 022 558*
3*s*5*d*	^1D	2	947 117	3*s*6*d*	^3D	1	1 043 075+*x*
						2	1 043 108+*x*
3*s*5*d*	^3D	3	948 579+*x*			3	1 043 166+*x*
		2	948 608+*x*				
		1	948 639+*x*	3*s*6*f*	$^3F°$	2–4	*1 051 544*+*x*
3*s*5*f*	$^3F°$	2–4	*963 880*+*x*	3*s*7*p*	$^1P°$	1	*1 097 914*
3*s*5*f*	$^1F°$	3	*966 616*	3*s*7*d*	^3D	1–3	1 099 352+*x*
3*s*6*s*	^3S	1	1 007 561+*x*	3*s*7*f*	$^3F°$	2–4	*1 104 323*+*x*
				K IX ($^2S_{1/2}$)	***Limit***		**1 249 200**

K IX

$Z=19$

Na I isoelectronic sequence

Ground state: $1s^2 2s^2 2p^6 3s\ ^2S_{1/2}$

Ionization energy: 1 418 063±20 cm^{-1} (175.8188±0.0030 eV)

The early work on this spectrum has been revised and considerably extended on the basis of new measurements in the range of 75–640 Å by Edlén and Bodén (1976). They state that the experimental uncertainty of their wavelengths is ±0.005 Å. We have rounded off their values for the energy levels accordingly. They also give calculated level values and intervals which are probably more accurate than the observed values for series members above $n=5$.

The 8*f*, 9*d*, and 10*d* terms are from Cohen and Behring (1976).

The value for the ionization energy was derived by Edlén (1978) from core polarization theory applied to the *nf* series.

References

Cohen, L., and Behring, W. E. (1976), J. Opt. Soc. Am. **66**, 899.
Edlén, B., and Bodén, E. (1976), Phys. Scr. **14**, 31.
Edlén, B. (1978), Phys. Scr. **17**, 565.

K IX

Configuration	Term	J	Level (cm^{-1})	Configuration	Term	J	Level (cm^{-1})
$2p^6(^1S)3s$	^2S	1/2	0	$2p^6(^1S)6d$	^2D	3/2	1 163 250
						5/2	1 163 310
$2p^6(^1S)3p$	$^2P°$	1/2	*157 152*				
		3/2	*160 913*	$2p^6(^1S)6f$	$^2F°$	5/2	*1 170 430*
						7/2	*1 170 440*
$2p^6(^1S)3d$	^2D	3/2	374 867				
		5/2	375 122	$2p^6(^1S)7s$	^2S	1/2	1 209 140
$2p^6(^1S)4s$	^2S	1/2	698 893	$2p^6(^1S)7p$	$^2P°$	1/2,3/2	*1 218 980*
$2p^6(^1S)4p$	$^2P°$	1/2	*758 262*	$2p^6(^1S)7d$	^2D	3/2	1 231 620
		3/2	*759 685*			5/2	1 231 640
$2p^6(^1S)4d$	^2D	3/2	836 869	$2p^6(^1S)7f$	$^2F°$	5/2	*1 236 130*
		5/2	837 003			7/2	*1 236 160*
$2p^6(^1S)4f$	$^2F°$	5/2	*860 849*	$2p^6(^1S)8p$	$^2P°$	1/2,3/2	*1 267 390*
		7/2	*860 889*				
				$2p^6(^1S)8d$	^2D	3/2	1 275 770
$2p^6(^1S)5s$	^2S	1/2	982 977			5/2	1 275 780
$2p^6(^1S)5p$	$^2P°$	1/2	*1 011 520*	$2p^6(^1S)8f$	$^2F°$	7/2	*1 278 860*
		3/2	*1 012 210*				
				$2p^6(^1S)9p$	$^2P°$	1/2,3/2	*1 300 170*
$2p^6(^1S)5d$	^2D	3/2	1 049 050				
		5/2	1 049 130	$2p^6(^1S)9d$	^2D	3/2	1 305 730
						5/2	*1 305 800*
$2p^6(^1S)5f$	$^2F°$	5/2	*1 061 400*				
		7/2	*1 061 410*	$2p^6(^1S)9f$	$^2F°$	5/2,7/2	*1 308 000*
$2p^6(^1S)6s$	^2S	1/2	1 126 510	$2p^6(^1S)10d$	^2D	5/2	1 327 340
$2p^6(^1S)6p$	$^2P°$	1/2	*1 142 410*	K X ($^1S_0$)	*Limit*		**1 418 063**
		3/2	*1 142 750*				

K X

$Z=19$

Ne I isoelectronic sequence

Ground state: $1s^22s^22p^6\,{}^1S_0$

Ionization energy $=4\,063\,000\pm5000$ cm^{-1} (503.8 ± 0.6 eV)

Only resonance lines between 29 and 42 Å are classified in this rare-gas-type system of energy levels by Edlén and Tyrén (1936), who identified 11 transitions from upper $J=1$ levels to the ${}^1S_0$ ground state.

Fawcett, Bromage, and Hayes (1979) give transition arrays for $2p^53d-4f$, $2p^53p-4d$, and $2p^53s-4p$, but they are not connected with the known system of levels.

We derived the ionization energy from the $2s^22p^5nd$ ${}^3D^\circ_1$ series for $n=3$ and 4, with the change in quantum defect, $n^*(3d)-n^*(4d)$, taken from Ti XIII. The value obtained by Lotz (1967) by extrapolation is in agreement with this result.

References

Edlén, B., and Tyrén, F. (1936), Z. Phys. **101**, 206.

Fawcett, B. C., Bromage, G. E., and Hayes, R. W. (1979), Mon. Not. R. Astron. Soc. **186**, 113.

Lotz, W. J. (1967), J. Opt. Soc. Am. **57**, 873.

K X

Configuration	Term	J	Level (cm^{-1})	Configuration	Term	J	Level (cm^{-1})
$2s^2\,2p^6$	${}^1S$	0	0	$2s2p^6\,3p$	${}^3P^\circ$	1	*3 219 400*
$2s^2\,2p^5\,3s$	${}^3P^\circ$	1	*2 407 260*	$2s^2\,2p^5\,4s$	${}^1P^\circ$	1	*3 232 400*
$2s^2\,2p^5\,3s$	${}^1P^\circ$	1	*2 430 250*	$2s2p^6\,3p$	${}^1P^\circ$	1	*3 237 600*
$2s^2\,2p^5\,3d$	${}^3P^\circ$	1	*2 760 200*	$2s^2\,2p^5\,4d$	${}^3D^\circ$	1	*3 356 400*
$2s^2\,2p^5\,3d$	${}^3D^\circ$	1	*2 794 900*	$2s^2\,2p^5\,4d$	${}^1P^\circ$	1	*3 379 700*
$2s^2\,2p^5\,3d$	${}^1P^\circ$	1	*2 832 300*	K XI (${}^2P^\circ_{3/2}$)	*Limit*		**4 063 000**
$2s^2\,2p^5\,4s$	${}^3P^\circ$	1	*3 205 100*				

K XI

$Z = 19$

F I isoelectronic sequence

Ground state: $1s^2 2s^2 2p^5\ ^2P^\circ_{3/2}$

Ionization energy $= 4\ 555\ 000 \pm 9000\ cm^{-1}$ (564.7 ± 1.0 eV)

The first work on this spectrum was by Edlén and Tyrén (1936), who classified 8 lines of the $2s^2 2p^5 - 2s^2 2p^4\ 3s$ and $3d$ transition arrays between 27 and 33 Å. This work was extended by Feldman, Doschek, Cowan, and Cohen (1973), from whose wavelengths, between 31 and 39 Å, the $3s$ and $3d$ levels are determined. Fawcett, Burgess, and Peacock (1967) identified the $2s^2 2p^5 - 2s 2p^6$ resonance doublet at ~155 Å. The $^2P^\circ$ ground state splitting was redetermined from new measurements of this doublet by Kaufman, Sugar, and Cooper (1982) with an uncertainty of $\pm 20\ cm^{-1}$.

The $2s 2p^5 3s\ ^2P^\circ$ term is from Feldman et al. (1973).

The ionization energy was obtained by extrapolation by Lotz (1967).

References

Edlén, B., and Tyrén, F. (1936), Z. Phys. **101**, 206.

Fawcett, B. C., Burgess, D. D., and Peacock, N. J. (1967), Proc. Phys. Soc. **91**, 970.

Feldman, U., Doschek, G. A., Cowan, R. D., and Cohen, L. (1973), J. Opt. Soc. Am. **63**, 1445.

Kaufman, V., Sugar, J., and Cooper, D. (1982), Phys. Scr. **25**, 623.

Lotz, W. J. (1967), J. Opt. Soc. Am. **57**, 873.

K XI

Configuration	Term	J	Level (cm^{-1})
$2s^2 2p^5$	$^2P^\circ$	3/2	*0*
		1/2	*23 530*
$2s2p^6$	^2S	1/2	655 901
$2s^2 2p^4(^3P)3s$	^4P	5/2	2 640 500
		3/2	2 653 000
		1/2	2 662 600
$2s^2 2p^4(^3P)3s$	^2P	3/2	2 671 500
		1/2	2 685 600
$2s^2 2p^4(^1D)3s$	^2D	5/2	2 727 800
		3/2	2 728 500
$2s^2 2p^4(^1S)3s$	^2S	1/2	2 811 800
$2s^2 2p^4(^3P)3d$	^4P	1/2	3 018 000
		3/2	3 022 300
		5/2	3 029 200
$2s^2 2p^4(^3P)3d$	^4F	5/2	3 020 500
$2s^2 2p^4(^3P)3d$	^2P	1/2	3 027 800
$2s^2 2p^4(^3P)3d$	^2F	5/2	3 034 300
$2s^2 2p^4(^3P)3d$	^2D	3/2	3 034 400
		5/2	3 048 000
$2s^2 2p^4(^1D)3d$	^2S	1/2	3 094 300
$2s^2 2p^4(^1D)3d$	^2F	5/2	3 098 100
$2s^2 2p^4(^1D)3d$	^2D	5/2	3 107 700
		3/2	3 115 400
$2s^2 2p^4(^1D)3d$	^2P	3/2	3 107 900
$2s^2 2p^4(^1S)3d$	^2D	5/2	3 176 300
		3/2	3 179 100
$2s2p^5(^3P^\circ)3s$	$^2P^\circ$	3/2	*3 265 800*
		1/2	*3 279 800*
K XII ($^3P_2$)	*Limit*		**4 555 000**

K XII

$Z = 19$

O I isoelectronic sequence

Ground state: $1s^2 2s^2 2p^4\ {}^3P_2$

Ionization energy $= 5\,077\,000 \pm 10\,000\ cm^{-1}$ (629.4 ± 1.2 eV)

The allowed lines of the transition array $2s^2 2p^4 - 2s2p^5$ were identified by Deutschman and House (1967), except for ${}^1S_0 - {}^1P^\circ_1$. Fawcett, Galanti, and Peacock (1974) reported the observation of $2s2p^5\ {}^1P_1 - 2p^6\ {}^1S_0$. These transition arrays were remeasured by Kaufman, Sugar, and Cooper (1982) who found several intersystem lines and reduced the energy level uncertainty to $\pm 50\ cm^{-1}$ for term positions and $\pm 20\ cm^{-1}$ for fine structure. The spectral lines fall in the range of 126–202 Å. Kaufman and Sugar (1982) provided percentage compositions for these levels. Their calculation included configuration interaction between $2s^2 2p^4$ and $2p^6$.

The $2p^3 3s$ levels were determined from the observations at 34 Å by Doschek, Feldman, and Cohen (1973). The $2p^3 3d$ levels are due to Fawcett and Hayes (1975). Some revisions of the $2p^3 3d$ classifications were given by Bromage and Fawcett (1977) on the basis of new calculations.

The ionization energy was obtained by Lotz (1967) by extrapolation.

References

Bromage, G. E., and Fawcett, B. C. (1977), Mon. Not. R. Astron. Soc. **178**, 591.

Deutschman, W. A., and House, L. L. (1967), Astrophys. J. **149**, 451.

Doschek, G. A., Feldman, U., and Cohen, L. (1973), J. Opt. Soc. Am. **63**, 1463.

Fawcett, B. C., Galanti, M., and Peacock, N. J. (1974), J. Phys. **B7**, 1149.

Fawcett, B. C., and Hayes, R. W. (1975), Mon. Not. R. Astron. Soc. **170**, 185.

Kaufman, V., and Sugar, J. (1982), private communication.

Kaufman, V., Sugar, J., and Cooper, D. (1982), Phys. Scr. **25**, 623.

Lotz, W. J. (1967), J. Opt. Soc. Am. **57**, 873.

K XII

Configuration	Term	J	Level (cm⁻¹)	Leading percentages		
$2s^2 2p^4$	${}^3P$	2	0	98	2	${}^1D$
		1	18 954	100		
		0	23 207	97	3	${}^1S$
$2s^2 2p^4$	${}^1D$	2	79 654	98	2	${}^3P$
$2s^2 2p^4$	${}^1S$	0	163 019	94	3	${}^3P$
$2s2p^5$	${}^3P^\circ$	2	*573 363*	100		
		1	*589 176*	100		
		0	*598 382*	100		
$2s2p^5$	${}^1P^\circ$	1	*789 574*	100		
$2p^6$	${}^1S$	0	1 336 760	97	3	$2s^2 2p^4\ {}^1S$
$2s^2 2p^3(^4S^\circ)3s$	${}^3S^\circ$	1	*2 930 400*			
$2s^2 2p^3(^2D^\circ)3s$	${}^3D^\circ$	1	*3 004 100*			
		2	*3 005 200*			
		3	*3 008 700*			
$2s^2 2p^3(^2D^\circ)3s$	${}^1D^\circ$	2	*3 024 600*			
$2s^2 2p^3(^2P^\circ)3s$	${}^1P^\circ$	1	*3 087 000*			
$2s^2 2p^3(^4S^\circ)3d$	${}^3D^\circ$	2	*3 265 700*			
		3	*3 271 100*			

K XII—Continued

Configuration	Term	*J*	Level (cm^{-1})	Leading percentages
$2s^2\,2p^3(^2D^\circ)3d$	$^3D^\circ$	1	*3 358 100*	
		2	*3 361 100*	
$2s^2\,2p^3(^2D^\circ)3d$	$^3P^\circ$	2	*3 370 700*	
$2s^2\,2p^3(^2D^\circ)3d$	$^1D^\circ$	2	*3 374 400*	
$2s^2\,2p^3(^2D^\circ)3d$	$^3S^\circ$	1	*3 383 100*	
$2s^2\,2p^3(^2D^\circ)3d$	$^1F^\circ$	3	*3 396 300*	
$2s^2\,2p^3(^2P^\circ)3d$	$^3P^\circ$	2	*3 405 200*	
		1	*3 410 000*	
$2s^2\,2p^3(^2P^\circ)3d$	$^1D^\circ$	2	*3 416 600*	
$2s^2\,2p^3(^2P^\circ)3d$	$^3D^\circ$	1,2	*3 427 000*	
		3	*3 428 400*	
$2s^2\,2p^3(^2P^\circ)3d$	$^1F^\circ$	3	*3 442 700*	
$2s^2\,2p^3(^2P^\circ)3d$	$^1P^\circ$	1	*3 474 100*	
K XIII ($^4S^\circ_{3/2}$)	***Limit***		**5 077 000**	

K XIII

$Z=19$

N I isoelectronic sequence

Ground state: $1s^2 2s^2 2p^3\ ^4S^\circ_{3/2}$

Ionization energy $= 5\ 764\ 000 \pm 12\ 000$ cm^{-1} (714.6 ± 1.4 eV)

The transition array $2s^2 2p^3 - 2s 2p^4$ was first observed by Fawcett, Burgess and Peacock (1967) and more completely by Boiko et al. (1970). Fawcett and Hayes (1975) have reported the $2s 2p^4 - 2p^5$ array. These lines were remeasured by Kaufman, Sugar, and Cooper (1982) in the range of 138–219 Å with an improved accuracy of ±0.01 Å. They have also provided the percentage compositions for these levels, which include mixing of $2s^2 2p^3$ and $2p^5$. An interpolated value for the $2s^2 2p^3\ ^4S^\circ_{3/2} - 2s 2p^4\ ^2P_{3/2}$ intersystem line is given by them, establishing the doublet system position with an uncertainty of ±100 cm^{-1} (denoted by x in the table).

The $2p^3 3d$ terms are from Fawcett and Hayes (1975), who observed the spectrum at 28 Å. They also reported the $2s 2p^4\ ^2D - 2p^5\ ^2P^\circ$ transitions. Some revisions of the classifications involving $2p^2 3d$ were proposed by Bromage and Fawcett (1977) on the basis of new calculations.

The ionization energy is from Lotz's (1967) extrapolation.

References

Boiko, V. A., Voinov, Y. P., Gribkov, V. A., and Sklizkov, G. V. (1970), Opt. Spectrosc. (USSR) **29**, 545.

Bromage, G. E., and Fawcett, B. C. (1977), Mon. Not. R. Astron. Soc. **179**, 683.

Fawcett, B. C., Burgess, D. D., and Peacock, N. J. (1967), Proc. Phys. Soc. London **91**, 970.

Fawcett, B. C., and Hayes, R. W. (1975), Mon. Not. R. Astron. Soc. **170**, 185.

Kaufman, V., Sugar, J., and Cooper, D. (1982), Phys. Scr. **26**, 163.

Lotz, W. (1967), J. Opt. Soc. Am. **57**, 873.

K XIII

Configuration	Term	J	Level (cm^{-1})	Leading percentages		
$2s^2 2p^3$	$^4S^\circ$	3/2	*0*	99	1	$^2P^\circ$
$2s^2 2p^3$	$^2D^\circ$	3/2	*100 647+x*	91	8	$^2P^\circ$
		5/2	*105 805+x*	100		
$2s^2 2p^3$	$^2P^\circ$	1/2	*160 743+x*	98	2	$2p^5\ ^2P^\circ$
		3/2	*168 177+x*	89	9	$2s^2 2p^3\ ^2D^\circ$
$2s(^2S)2p^4(^3P)$	^4P	5/2	480 524	100		
		3/2	496 266	100		
		1/2	503 717	99	1	$2s(^2S)2p^4(^1S)\ ^2S$
$2s(^2S)2p^4(^1D)$	^2D	3/2	661 810+x	99	1	$2s(^2S)2p^4(^3P)\ ^2P$
		5/2	662 880+x	100		
$2s(^2S)2p^4(^1S)$	^2S	1/2	769 780+x	91	9	$2s(^2S)2p^4(^3P)\ ^2P$
$2s(^2S)2p^4(^3P)$	^2P	3/2	801 500+x	99	1	$2s(^2S)2p^4(^1D)\ ^2D$
		1/2	822 670+x	91	8	$2s(^2S)2p^4(^1S)\ ^2S$
$2p^5$	$^2P^\circ$	3/2	*1 256 460+x*	98	2	$2s^2 2p^3\ ^2P^\circ$
		1/2	*1 281 760+x*	98	2	
$2s^2 2p^2(^3P)3d$	^2P	3/2	3 621 400+x			
$2s^2 2p^2(^3P)3d$	^4P	5/2	3 651 800			
		3/2	3 658 000			
$2s^2 2p^2(^3P)3d$	^2F	7/2	3 656 900+x			

K XIII—Continued

Configuration	Term	J	Level (cm^{-1})	Leading percentages
$2s^2\,2p^2(^3P)3d$	^2D	$5/2$	3 694 000+x	
$2s^2\,2p^2(^1D)3d$	^2D	$5/2$	3 726 000+x	
$2s^2\,2p^2(^1D)3d$	^2F	$7/2$	3 730 800+x	
		$5/2$	3 738 000+x	
$2s^2\,2p^2(^1D)3d$	^2P	$3/2$	3 748 000+x	
K XIV ($^3P_0$)	*Limit*		**5 764 000**	

K XIV

$Z = 19$

C I isoelectronic sequence

Ground state: $1s^2 2s^2 2p^2\ {}^3P_0$

Ionization energy = 6 345 000±13 000 cm^{-1} (786.6±1.6 eV)

Lines of the transition array $2s^2 2p^2 - 2s2p^3$ were identified by Boiko, Voinov, Gribkov, and Sklizkov (1970) in a laser-produced plasma. A more extensive analysis of this array was obtained by Fawcett and Hayes (1975) using a similar light source. The latter group also reported transitions between $2s^2 2p^2$ and $2s^2 2p3d$. New measurements in the range of 147–231 Å with an uncertainty of ±0.01 Å were reported by Sugar, Kaufman, and Cooper (1982). They include the intersystem transition $2s^2 2p^2\ {}^3P_2 - 2s2p^3\ {}^1D^\circ_2$ and provide several new levels of the $2s2p^3$ and $2p^4$ configurations. The energy levels and percentage compositions of the $2s^2 2p^2$, $2s2p^3$, and $2p^4$ configurations are taken from their work. Their calculation includes interaction between $2s^2 2p^2$ and $2p^4$. Predicted positions for the $2s2p^3\ {}^5S^\circ_2$ and $2p^4\ {}^3P_0$ levels in brackets are quoted from Edlén (1984) and from Sugar et al., respectively. The levels of $2s^2 2p3d$ and $2s2p^2 3d$ are from Fawcett and Hayes with some additions by Bromage and Fawcett (1977). Their measurements are at ~25 Å, and have a wavelength uncertainty of ±0.01 Å.

The ionization energy was obtained by Lotz (1967) by extrapolation.

References

Boiko, V. A., Voinov, Y. P., Gribkov, V. A., and Sklizkov, G. V. (1970), Opt. Spectrosc. (USSR) **29**, 545.

Bromage, G. E., and Fawcett, B. C. (1977), Mon. Not. R. Astron. Soc. **178**, 605.

Edlén, B. (1984), private communication.

Fawcett, B. C., and Hayes, R. W. (1975), Mon. Not. R. Astron. Soc. **170**, 185.

Lotz, W. (1967), J. Opt. Soc. Am. **57**, 873.

Sugar, J., Kaufman. V., and Cooper, D. (1982), Phys. Scr. **26**, 189.

K XIV

Configuration	Term	J	Level (cm^{-1})	Leading percentages		
$2s^2\,2p^2$	^{3}P	0	0	97	2	$2s^22p^2\ {}^1$S
		1	13 235	99	1	$2p^4\ {}^3$P
		2	28 225	95	4	$2s^22p^2\ {}^1$D
$2s^2\,2p^2$	^{1}D	2	95 913	95	4	$2s^22p^2\ {}^3$P
$2s^2\,2p^2$	^{1}S	0	178 914	93	5	$2p^4\ {}^1$S
$2s(^2S)2p^3(^4S)$	^{5}S	2	[250 640]	100		
$2s(^2S)2p^3(^2D)$	^{3}D°	2	*458 754*	97	3	$2s(^2S)2p^3(^2P^\circ)\ {}^3P^\circ$
		1	*459 498*	97	3	$2s(^2S)2p^3(^2P^\circ)\ {}^3P^\circ$
		3	*461 002*	100		
$2s(^2S)2p^3(^2P)$	^{3}P°	0	*537 402*	100		
		1	*538 032*	98	2	$2s(^2S)2p^3(^2D^\circ)\ {}^3D^\circ$
		2	*539 938*	96	3	$2s(^2S)2p^3(^2D^\circ)\ {}^3D^\circ$
$2s(^2S)2p^3(^2D)$	^{1}D°	2	*676 460*	99	1	$2s(^2S)2p^3(^2P^\circ)\ {}^3P^\circ$
$2s(^2S)2p^3(^4S)$	^{3}S°	1	*677 710*	97	3	$2s(^2S)2p^3(^2P^\circ)\ {}^1P^\circ$
$2s(^2S)2p^3(^2P)$	^{1}P°	1	*755 050*	97	3	$2s(^2S)2p^3(^4S)\ {}^3S^\circ$
$2p^4$	^{3}P	2	1 030 090	96	3	$2p^4\ {}^1$D
		1	1 050 620	99	1	$2s^22p^2\ {}^3$P
		0	[1 056 200]	97	2	$2p^4\ {}^1$S
$2p^4$	^{1}D	2	1 108 800	96	3	^{3}P

K XIV—Continued

Configuration	Term	J	Level (cm^{-1})	Leading percentages		
$2p^4$	^1S	0	1 254 810	94	5	$2s^22p^2\ ^1S$
$2s^2\,2p3d$	$^3F^\circ$	3	*3 871 700*			
$2s^2\,2p3d$	$^3D^\circ$	2	*3 901 000*			
		3	*3 911 700*			
$2s^2\,2p3d$	$^3P^\circ$	2	*3 919 300*			
$2s^2\,2p3d$	$^1F^\circ$	3	*3 955 600*			
$2s^2\,2p3d$	$^1P^\circ$	1	*3 958 800*			
$2s2p^2(^4P)3d$	^3F	4	4 190 100			
K XV ($^2P^\circ_{1/2}$)	***Limit***		**6 345 000**			

K XV

$Z=19$

B I isoelectronic sequence

Ground state: $1s^22s^22p\ ^2P^\circ_{1/2}$

Ionization energy $= 6\,945\,000\pm14\,000$ cm^{-1} (861.1 ± 1.7 eV)

The arrays $2s^22p-2s2p^2$ and $2s2p^2-2p^3$ were classified by Fawcett and Hayes (1975). They were remeasured by Sugar, Kaufman, and Cooper (1982) in the range of 166–240 Å with an accuracy of ±0.01 Å. Some additions to the analysis were made and a predicted value for the $2s2p^2\ ^4P_{5/2}$ level was given with an estimated uncertainty of ±100 cm^{-1}. The $2s2p^2-2p^3$ array was also given by Fawcett, Ridgeley, and Hatter (1980) with less accurate measurements but contained the additional levels $2s2p^2\ ^2D_{3/2}$ and $2p^3\ ^2D^\circ_{3/2}$, $^2D^\circ_{5/2}$, $^2P^\circ_{1/2}$. Percentage compositions for the levels of $2s2p^2$ were provided by Kaufman and Sugar (1982).

The configurations $2s^23d$, $2s2p3p$, and $2s2p3d$ are from Fawcett and Hayes from measurements at 23–25 Å. They estimate the wavelength uncertainty to be ±0.01 Å.

The ionization energy is from the extrapolation of Lotz (1967).

References

Fawcett, B. C., and Hayes, R. W. (1975), Mon. Not. R. Astron. Soc. **170**, 185.

Fawcett, B. C., Ridgeley, A., and Hatter, A. F. (1980), J. Opt. Soc. Am. **70**, 1349.

Kaufman, V., and Sugar, J. (1982), private communication.

Lotz, W. (1967), J. Opt. Soc. Am. **57**, 873.

Sugar, J., Kaufman, V., and Cooper, D. (1982), Phys. Scr. **26**, 293.

K XV

Configuration	Term	J	Level (cm^{-1})	Leading percentages		
$2s^2 2p$	$^2P^\circ$	1/2	*0*			
		3/2	*28 990*			
$2s2p^2$	^4P	1/2	248 320+x	100		
		3/2	259 630+x	100		
		5/2	274 200+x	99	1	^2D
$2s2p^2$	^2D	3/2	443 960	99	1	^2P
		5/2	445 510	99	1	^4P
$2s2p^2$	^2S	1/2	552 860	75	25	^2P
$2s2p^2$	^2P	1/2	588 260	75	25	^2S
		3/2	599 080	99	1	^2D
$2p^3$	$^4S^\circ$	3/2	*775 280*			
$2p^3$	$^2D^\circ$	3/2	*874 320*			
		5/2	*877 400*			
$2p^3$	$^2P^\circ$	1/2	*979 270*			
		3/2	*985 690*			
$2s^2 3d$	^2D	3/2	4 132 000			
		5/2	4 140 000			
$2s2p(^3P^\circ)3p$	^2P	1/2	4 239 000			
		3/2	4 260 000			
$2s2p(^3P^\circ)3p$	^2D	3/2	4 320 000			
		5/2	4 343 000			

K XV—Continued

Configuration	Term	J	Level (cm^{-1})	Leading percentages
$2s2p(^3P°)3d$	$^4D°$	7/2	*4 406 000+x*	
$2s2p(^3P°)3d$	$^4P°$	5/2	*4 418 000+x*	
$2s2p(^1P°)3d$	$^2D°$	5/2	*4 636 000*	
K XVI ($^1S_0$)	***Limit***		**6 945 000**	

K XVI

$Z=19$

Be I isoelectronic sequence

Ground state: $1s^2 2s^2\,{}^1S_0$

Ionization energy = 7 810 000±16 000 cm^{-1} (968±2 eV)

Fawcett and Hayes (1975) observed the resonance line $2s^2\,{}^1S_0 - 2s2p\,{}^1P^\circ_1$ at 206.27 Å and a group of transitions between 21 and 24 Å. They did not report any inter-system combinations; we used the interpolated value for the transition $2s^2\,{}^1S_0 - 2s2p\,{}^3P_1$ by Edlén (1983) to locate the triplet system relative to the ground level. The $2s2p - 2p^2$ array is given by Fawcett, Ridgeley, and Hatter (1980). The higher-lying configurations were found by Fawcett and Hayes.

The ionization energy was derived by Lotz (1967) by extrapolation.

References

Edlén, B. (1983), Phys. Scr. **28**, 51.

Fawcett, B. C., and Hayes, R. W. (1975), Mon. Not. R. Astron. Soc. **170**, 185.

Fawcett, B. C., Ridgeley, A., and Hatter, A. T. (1980), J. Opt. Soc. Am. **70**, 1349.

Lotz, W. (1967), J. Opt. Soc. Am. **57**, 873.

K XVI

Configuration	Term	J	Level (cm^{-1})	Configuration	Term	J	Level (cm^{-1})
$2s^2$	^1S	0	0	$2s3p$	$^1P^\circ$	1	*4 563 900*
$2s2p$	$^3P^\circ$	0	*243 520+x*	$2s3d$	^3D	2	4 636 000+x
		1	*252 520+x*			3	4 636 900+x
		2	*274 090+x*	$2s3d$	^1D	2	4 679 000
$2s2p$	$^1P^\circ$	1	*484 800*	$2p3p$	^3D	3	4 874 500+x
$2p^2$	^3P	0	646 720+x	$2p3d$	$^3D^\circ$	2	*4 958 000+x*
		1	660 170+x			3	*4 965 600+x*
		2	676 480+x	$2p3d$	$^1F^\circ$	3	*5 016 300*
$2p^2$	^1D	2	742 650+x	K XVII ($^2S_{1/2}$)	***Limit***		**7 810 000**
$2p^2$	^1S	0	902 760				

K XVII

$Z=19$

Li I isoelectronic sequence

Ground state: $1s^2 2s\ ^2S_{1/2}$

Ionization energy = 8 344 200±1000 cm^{-1} (1033.4±0.1 eV)

The $2p-3d$ transition was reported by Goldsmith, Feldman, Oren, and Cohen (1972). The value of the $2p\ ^2P^\circ$ term is from the $2s-2p$ transitions observed at 326.78 Å and 365.63 Å in a solar flare by Widing and Purcell (1976). They note, however, that the intensity ratio of the two lines is much too great. The wavelengths agree within ±0.01 Å with proposed values by Edlén (1983).

Boiko, Faenov, and Pikuz (1978) identified the $4p$, $4d$, and $5d$ terms. The $3p\ ^2P^\circ$ and $4f\ ^2F^\circ$ terms are from the identifications of the $3d-4f$ and $3p-4d$ transitions by Fawcett and Ridgeley (1981). The $1s2s2p$ and $1s2p^2$ levels were obtained from lines classified by Aglitskii, Boiko, Zakharov, Pikuz, and Faenov (1974) at 3.5 Å. The $1s2s3p$ and $1s2s3d$ levels are from observations by Boiko, Pikuz, Safronova, and Faenov (1978).

The ionization energy is from Edlén (1979).

References

Aglitskii, E. V., Boiko, V. A., Zakharov, S. M., Pikuz, S. A., and Faenov, A. Y. (1974), Sov. J. Quant. Electron. **4**, 500.

Boiko, V. A., Faenov, A. Y., and Pikuz, S. A. (1978), J. Quant. Spectrosc. Radiat. Transfer **19**, 11.

Boiko, V. A., Pikuz, S. A., Safronova, U. I., and Faenov, A. Y. (1978), Mon. Not. R. Astron. Soc. **185**, 789.

Edlén, B. (1979), Phys. Scr. **19**, 255.

Edlén, B. (1983), Phys. Scr. **28**, 51.

Fawcett, B. C., and Ridgeley, A. (1981), J. Phys. **B14**, 203.

Goldsmith, S., Feldman, U., Oren, L., and Cohen, L. (1972), Astrophys. J. **174**, 209.

Widing, K. G., and Purcell, J. D. (1976), Astrophys. J. **204**, L151.

K XVII

Configuration	Term	J	Level (cm^{-1})
$1s^2(^1S)2s$	^2S	1/2	0
$1s^2(^1S)2p$	$^2P^\circ$	1/2	*273 500*
		3/2	*306 020*
$1s^2(^1S)3p$	$^2P^\circ$	1/2	*4 778 700*
		3/2	*4 793 600*
$1s^2(^1S)3d$	^2D	3/2	4 814 800
		5/2	4 818 000
$1s^2(^1S)4p$	$^2P^\circ$	1/2,3/2	*6 347 200*
$1s^2(^1S)4d$	^2D	3/2	6 361 000
		5/2	6 367 700
$1s^2(^1S)4f$	$^2F^\circ$	5/2	*6 361 400*
		7/2	*6 362 200*
$1s^2(^1S)5d$	^2D	3/2	7 069 300
		5/2	7 073 800
K XVIII ($^1S_0$)	***Limit***		**8 344 200**
$1s(^2S)2s2p(^3P^\circ)$	$^4P^\circ$		*27 871 000*
$1s(^2S)2s2p(^3P^\circ)$	$^2P^\circ$	1/2,3/2	*28 079 000*
$1s(^2S)2p^2(^3P)$	^4P		28 160 000
$1s(^2S)2s2p(^1P^\circ)$	$^2P^\circ$	1/2,3/2	*28 182 000*
$1s(^2S)2p^2(^1D)$	^2D	3/2	28 318 000
		5/2	28 321 000
$1s(^2S)2p^2(^3P)$	^2P	1/2,3/2	28 384 000
$1s(^2S)2p^2(^1S)$	^2S	1/2	28 511 000
$1s2s(^1S)3p$	$^2P^\circ$		*32 863 000*
$1s2s3d$	^2D	3/2	32 910 000

K XVIII

$Z=19$

He I isoelectronic sequence

Ground state: $1s^2\ ^1S_0$

Ionization energy = 37 188 200±7000 cm^{-1} (4610.8±0.9 eV)

Because of the excellent agreement of the calculated energies of the $1s2l$ levels by Safronova (1981) with the few well-measured spectra in the He I sequence, we have compiled her results for the $n=2$ levels and for the ionization energy. Detailed comparisons are given in the Introduction. Levels of the $n=3$–5 shells are from the calculated binding energies by Ermolaev and Jones (1974) subtracted from Safronova's value for the $1s$ binding energy of the ground state. We have assumed an uncertainty of 2 parts in 10^4 for the excited levels relative to the ground state, and for the ionization energy. This is equal to the random deviation of the measurements by Aglitskii et al. (1974) from the calculations by Safronova (see Introduction). For differences between excited levels where $\Delta n=0$, we assumed an uncertainty of 2 parts in 10^3.

Corrected measurements (see Introduction) by Aglitskii et al. place the $1s2p\ ^3P^\circ_1$ level at 28 179 000 cm^{-1} and the $1s2p\ ^1P^\circ_1$ at 28 319 000 cm^{-1} with an estimated uncertainty of ±4000 cm^{-1}. These values are 6000 cm^{-1} higher than the values given by Safronova.

Percentage compositions are from Ermolaev and Jones.

References

Aglitskii, E. V., Boiko, V. A., Zakharov, S. M., Pikuz, S. A., and Faenov, A. Y. (1974), Sov. J. Quant. Electron. **4**, 500.
Ermolaev, A. M., and Jones, J. (1974), J. Phys. **B7**, 199.
Safronova, U. I. (1981), Phys. Scr. **23**, 241.

K XVIII

Configuration	Term	J	Level (cm^{-1})	Leading percentages		
$1s^2$	^1S	0	0			
$1s2s$	^3S	1	[28 004 980]			
$1s2p$	$^3P^\circ$	0	*[28 165 880]*			
		1	*[28 172 670]*	98	2	$^1P^\circ$
		2	*[28 200 800]*			
$1s2s$	^1S	0	[28 180 480]			
$1s2p$	$^1P^\circ$	1	*[28 312 910]*	98	2	$^3P^\circ$
$1s3s$	^3S	1	[33 151 930]			
$1s3p$	$^3P^\circ$	0	*[33 196 320]*			
		1	*[33 198 160]*	97	3	$^1P^\circ$
		2	*[33 206 560]*			
$1s3s$	^1S	0	[33 198 090]			
$1s3p$	$^1P^\circ$	1	*[33 237 140]*	97	3	$^3P^\circ$
$1s4s$	^3S	1	[34 929 710]			
$1s4p$	$^3P^\circ$	0	*[34 948 120]*			
		1	*[34 948 890]*	97	3	$^1P^\circ$
		2	*[34 952 440]*			
$1s4s$	^1S	0	[34 948 420]			
$1s4p$	$^1P^\circ$	1	*[34 964 920]*	97	3	$^3P^\circ$

K XVIII—Continued

Configuration	Term	J	Level (cm^{-1})	Leading percentages		
1s5s	^{3}S	1	[35 747 360]			
1s5p	^{3}P°	0	[*35 756 670*]			
		1	[*35 757 070*]	97	3	^{1}P°
		2	[*35 758 880*]			
1s5s	^{1}S	0	[35 756 710]			
1s5p	^{1}P°	1	[*35 765 190*]	97	3	^{3}P°
K XIX (^{2}S$_{1/2}$)	***Limit***		**37 188 200**			

K XIX

$Z = 19$

H I isoelectronic sequence

Ground state: $1s\ ^2S_{1/2}$

Ionization energy = 39 795 750±20 cm^{-1} (4934.084±0.013 eV)

No observations of this spectrum are reported. We give calculated values by Mohr (1983) for the $n=2$ shell and by Erickson (1977) for $n=3$–5 relative to the $2p\ ^2P^\circ_{3/2}$ level. Further details are given in the Introduction.

References

Erickson, G. W. (1977), J. Phys. Chem. Ref. Data **6**, 831.
Mohr, P. J. (1983), At. Data Nucl. Data Tables **29**, 453.

K XIX

Configuration	Term	J	Level (cm^{-1})
$1s$	^2S	1/2	0
$2p$	$^2P^\circ$	1/2	*[29 831 985]*
		3/2	*[29 880 266]*
$2s$	^2S	1/2	[29 833 524]
$3p$	$^2P^\circ$	1/2	*[35 372 804]*
		3/2	*[35 387 113]*
$3s$	^2S	1/2	[35 373 275]
$3d$	^2D	3/2	[35 387 088]
		5/2	[35 391 816]
$4p$	$^2P^\circ$	1/2	*[37 310 120]*
		3/2	*[37 316 154]*
$4s$	^2S	1/2	[37 310 320]
$4d$	^2D	3/2	[37 316 144]
		5/2	[37 318 139]
$4f$	$^2F^\circ$	5/2	*[37 318 135]*
		7/2	*[37 319 131]*
$5p$	$^2P^\circ$	1/2	*[38 205 981]*
		3/2	*[38 209 069]*
$5s$	^2S	1/2	[38 206 084]
$5d$	^2D	3/2	[38 209 064]
		5/2	[38 210 086]
$5f$	$^2F^\circ$	5/2	*[38 210 084]*
		7/2	*[38 210 593]*
$5g$	^2G	7/2	[38 210 593]
		9/2	[38 210 898]
	Limit		**39 795 750**

Ca I

$Z=20$

Ground state: $1s^2 2s^2 2p^6 3s^2 3p^6 4s^2\ ^1S_0$

Ionization energy = 49 305.95±0.08 cm^{-1} (6.11321±0.00002 eV)

Early History

The observation and analysis of the first spectrum of calcium have played an important role in the development of experimental methods and theoretical ideas in the field of atomic spectra. Saunders (1920) summarized the considerable early work on this spectrum which included observations of the 4*snp*, 4*pns*, 4*pnd*, and 3*dnf* series of triplets and singlets. The longest series was 4*snp* $^1P_1^\circ$ reaching $n=11$. Shortly thereafter Russell and Saunders (1925), and independently Bohr (1923), recognized the excitation of two electrons to form terms above the ionization energy. In the same paper Russell and Saunders proposed the now universally accepted designations for levels in *LS*-coupling. They also introduced the concept for ordering level values from zero for the level of highest binding energy. The paper reports an extension of the analysis of Ca I which classified the remaining strong lines of this spectrum. The ^3P terms of $4p^2$, $3d^2$, and 3*dnd* ($n=4$–6) are given, the latter occurring entirely above the principal ionization limit. For the configuration 3*d*4*p* the terms $^3P^\circ$, $^3D^\circ$, and $^3F^\circ$ were found, and for 3*d*5*p* the $^3D^\circ$ was observed above the limit. Several undesignated levels are included whose interpretation was provided by later workers. In a later paper Russell (1927) identified the 4*s*5*s* $^3S_1$ and $^1S_0$, the $4p^2\ ^1D$, and the 3*d*4*d* $^3S_1$ and $^3D_1$.

Measurements of the Zeeman effect are due to Back (1925), who observed the 3*d*4*s* − 3*d*4*p* transition array in a field of 39 000 gauss (3.9 T) and derived the following *g*-values for Ca I:

Config.	Term	*J*	*g*
3*d*4*s*	^3D	1	0.502
		2	1.162
		3	1.328
	^1D	2	1.007
3*d*4*p*	$^3F^\circ$	2	0.754
		3	1.076
		4	1.245
	$^1D^\circ$	2	0.893

An additional term of the $3d^2$ configuration, the ^3F, was discovered by Humphreys (1951) from observations in the infrared of the $3d4p-3d^2\ ^3F^\circ-{}^3F$ and $^3D^\circ-{}^3F$ multiplets.

Absorption measurements in the range of 1590–2400 Å by Garton and Codling (1965) produced 82 newly observed lines of calcium. Their classifications extended the known series as follows: 4*snp* $^1P_1^\circ$ to $n=33$, 3*dnp* $^1P_1^\circ$ to $n=29$, 3*dnp* $^3P_1^\circ$ to $n=25$, 3*dnp* $^3D_1^\circ$ to $n=19$, 3*dnf* $^1P_1^\circ$ to $n=15$, 3*dnf* $^3P_1^\circ$ to $n=10$, and 3*dnf* $^3D_1^\circ$ to $n=15$. A single line at 1740.32 Å was classified as $4s^2\ ^1S_0-4s5p\ ^1P_1^\circ$. Their classification of the level at 43 933 cm^{-1} as mainly 3*d*4*p* $^1P_1^\circ$ was accepted in several subsequent papers by other authors but was abandoned in the light of recent theoretical work cited below. It is now apparent that no level is preponderantly of this nature.

Sources of Level Values

The emission spectrum was completely reobserved by Risberg (1968) from 195–30 000 Å by means of a hollow cathode discharge. In this range 275 lines were measured, about 100 of which were observed for the first time. Autoionization and the low pressure of this light source prevented the detection of all but four levels above the ionization limit, $^3P_1$ and $^3D_{1,2,3}$ of the 3*d*4*d* configuration. The other members of the 3*d*4*d* and 3*d*5*d* configurations are derived from the classified lines of Russell and Saunders (1925) or C. E. Moore (1959) combined with the lower level values given by Risberg.

Several revisions and extensions are proposed in Risberg's paper. The earlier assignments of 4*s*6*s* $^1S_0$ and $4p^2\ ^1S_0$ to 41 786 cm^{-1} and 40 690 cm^{-1} respectively are interchanged. The $J=0$ level of 4*s*7*p* $^3P^\circ$ and the $J=1,2$ levels of 4*s*8*p* $^3P^\circ$ were found. The latter have been replaced by values found by Armstrong et al. (1979). The $^1D_2$ terms of 3*d*5*s* and $3d^2$ were also discovered, and in the 4*snf* series the $^1F_3^\circ$ of 4*s*12*f* was added.

Risberg has incorporated the interferometric measurements of the strong lines by Wagman (1937) and the accurate measurements of Grafenberger (1937) with her own extensive measurements to redetermine all the energy level values. Risberg's results are quoted here. Her estimated level uncertainty is ±0.016 cm^{-1}.

The data available for Ca I are particularly rich in long series. Those series for which the maximum observed value of *n* is 10 or more are listed on the following page.

Brown, Tilford, and Ginter (1973) observed absorption spectra of calcium series in the range of 1500–1770 Å and 2020–2090 Å at high dispersion. They recorded the 4*snp* $^1P_1^\circ$ series for $n=11$–79 and deduced from these data an ionization energy of 49 305.99 ±0.12 cm^{-1}.

Series in Ca I

	Config.	Term	n maximum	References
1.	$4snp$	$^3P°$	60	Armstrong et al. (1979)
2.	$4snp$	$^1P°$	79	Brown et al. (1973)
3.	$4snd$	^3D	60	Risberg (1968), Beigang et al. (1982)
4.	$4snd$	^1D	62	Borgström and Rubbmark (1977), Armstrong et al. (1977)
5.	$4sns$	^3S	36	Risberg (1968), Beigang et al. (1982)
6.	$4sns$	^1S	32	Borgström and Rubbmark (1977), Armstrong et al. (1977)
7.	$3dnp$	$^3D°$	58	Brown et al. (1973)
8.	$3dnp$	$^3P°$	33	Brown et al. (1973)
9.	$3dnp$	$^1P°$	61	Brown et al. (1973)
10.	$4snf$	$^3F°$	13	Risberg (1968), Camus (1974)
11.	$4snf$	$^1F°$	28	Borgström and Rubbmark (1977)
12.	$3dnf$	$^3D°$	30	Brown et al. (1973)
13.	$3dnf$	$^3P°$	38	Brown et al. (1973)
14.	$3dnf$	$^1P°$	39	Brown et al. (1973)
15.	$4sng$	^1G	14	Beigang and Wynne (1981)
16.	$4snh$	^1H	14	Beigang and Wynne (1981), Chang (1983)
17.	$4pns$	$^3P°$	16	Connerade et al. (1980)
18.	$4pns$	$^1P°$	15	Connerade et al. (1980)
19.	$4pnd$	$^1P°$	12	Connerade et al. (1980)
20.	$3p^5(^2P_{3/2})\ 4s^2nd$	$^2[3/2]°$	16	Mansfield and Newsom (1977)
21.	$3p^5(^2P°_{1/2})\ 4s^2nd$	$^2[3/2]°$	17	Mansfield and Newsom (1977)
22.	$3p^5(^2P°_{3/2})\ 4s^2ns$	$^2[3/2]°$	10	Mansfield and Newsom (1977)
23.	$3p^5(^2P°_{1/2})\ 4s^2ns$	$^2[1/2]°$	14	Mansfield and Newsom (1977)

Series of doubly excited configurations converging to the $3d\ ^2D$ limit were extended. The $3dnp\ ^3D°_1$ series was measured through $n=58$, the $3dnp\ ^1P_1$ series was measured through $n=58$, the $3dnp\ ^1P°_1$ through $n=61$, the $3dnp\ ^3P°_1$ through $n=33$, the $3dnf\ ^1P°_1$ through $n=39$, the $3dnf\ ^3P°_1$ through $n=38$ and the $3dnf\ ^3D°_1$ through $n=30$. They note that "the $3dnp\ ^1P°_1$ series lines are the strongest and most diffuse of the six observed series." Two levels that perturb this series at $n=6$ and $n=7$ were measured. They have been identified as $5s4p\ ^3P°_1$ and $^1P°_1$ by Newsom (1966). The results of Brown, Tilford, and Ginter are quoted here for series members beyond the observations of Risberg. Their estimated level uncertainty is ± 0.18 cm^{-1}.

In a paper by Armstrong, Esherick, and Wynne (1979), observations by means of multiphoton absorption experiments of the $4snp\ ^3P°$ series to $n=60$ were reported. This was previously given to $n=8$ by Risberg (1968). The estimated level uncertainty varies from ± 0.02 cm^{-1} to ± 0.2 cm^{-1}, with an average of ± 0.05 cm^{-1}.

Armstrong, Esherick, and Wynne (1977) greatly extended the observations of the $4sns\ ^1S_0$ series from $n=12$–32 and the $4snd\ ^1D_2$ series from $n=8$–62. In the latter series they substituted the level 46 199.23 cm^{-1} for $4s7d$, rejecting the level 46 308.257 given by Risberg. The level 47 449.083 cm^{-1} assigned by Risberg to $3d^2\ ^1D_2$ is renamed $3d5s\ ^1D_2$. This designation was used by Risberg for the level 48 083.383 cm^{-1}, which is now named $4s10d\ ^1D_2$ by Armstrong et al. These changes are confirmed by additional observations by Palenius and Risberg (1977). No level was found to replace the $3d^2\ ^1D_2$.

The estimated level uncertainty varies from ± 0.04 to ± 0.6 cm^{-1}, with most better than ± 0.1 cm^{-1}.

Borgström and Rubbmark (1977) observed absorption series from the $4s4p\ ^1P°_1$ level. They report values for $4sns\ ^1S_0$ from $n=12$–28 and $4snd\ ^1D_2$ from $n=7$–59. These two series have been extended to $n=32$ and $n=62$, respectively, by Armstrong, Esherick, and Wynne (1977). Borgström and Rubbmark observed the $4snf\ ^1F°_3$ series from $n=13$–28. They note that their level values are ~ 0.15 cm^{-1} lower than those of Armstrong et al. (1977) and claim an improved accuracy. We have compiled their results for $4sns\ ^1S_0$ ($n=12$–28) and for $4snd\ ^1D_2$ ($n=7$–59).

The $4sns\ ^3S_1$ series was observed from $n=10$ to $n=36$ and the $4snd\ ^3D_{1,2,3}$ from $n=11$ to $n=60$ by laser excitation from the metastable $4s5p\ ^3P$ levels by Beigang, Lücke, Schmidt, Timmerman, and West (1982). From these data they derived a value for the ionization energy of 49 305.94$\pm$0.04 cm^{-1}. Included below are the results of Risberg through $4s12s$ and $4s10d$. The higher series members are from Beigang et al. with an uncertainty of ± 0.04 cm^{-1}.

Five members of the $4sng\ ^1G_4$ series ($n=10$–14) have been observed by Beigang and Wynne (1981) in absorption from the $4s5f\ ^1F°_3$ level and, as reported by them, from a new level at 44 872 cm^{-1} that they could not identify. A measurement accuracy of ± 0.3 cm was given. Forsberg and Litzén (1982) reported the observation of the $4s5g\ ^1G_4$ level at 44 875.95 cm^{-1}. Chang (1983), using polarization theory, confirmed that this is the unknown level found by Beigang and Wynne. He identified the upper levels observed in absorption from this level as the $4s\,nh$ series ($n=10$–14) and designated them as $^1H_5$ levels.

Connerade et al. (1980) have reported observations of series in the doubly excited configurations $4pns$ and $4pnd$. They did not observe the three lines reported by Brown et al. at 62553, 62557, and 62562 cm^{-1} tentatively classified as $3s^2$–$4p4d$.

Mansfield and Newsom (1977) have observed absorption from the $3p$ shell between 320 and 500 Å. We have given the $3p^54s^2ns$ and $3p^54s^2nd$ series from their paper. The uncertainty of these level values is estimated to be ± 10 cm^{-1}.

Ionization Potential

Borgström and Rubbmark have determined a value of the ionization energy from the $4snf\ ^1F^\circ_3$ series equal to 49 305.92 ± 0.10 cm^{-1} which is within the error limits of the value of Brown et al., 49 305.99 $\pm$ 0.12 cm^{-1}. We have adopted the average of these values and that of Beigang et al. (1982). The error in the conversion factor, 8065.479 ± 0.021 cm^{-1} eV, determines the error in the value given in eV.

Perturbed Series

The identification of $3d4p\ ^1P^\circ_1$, which perturbs the $4snp\ ^1P^\circ_1$ series, has been the subject of much discussion in the literature. Russell and Shenstone (1932) decided from intensity and energy considerations that it must be the level at 36 731 cm^{-1}. Roth (1969) attributes to Racah his identification of this term with the level at 41 679 cm^{-1}. Friedrich and Trefftz (1969), on the basis of multiconfiguration calculations, retained it at 36 731 cm^{-1}. Garton and Codling (1965), from intensity considerations, placed it at 43 933 cm^{-1}. This assignment has been adopted by Risberg in her analysis of Ca I and by Moores (1966) in his multichannel quantum defect theory (MQDT) calculation of the $4snp\ ^1P^\circ_1$ series. The investigation of the $4snp\ ^1P^\circ_1$ series by Armstrong et al. (unpublished) using MQDT has shown that all of the above levels contain less than 30% $3dnp\ ^1P^\circ_1$ composition (summed over all n). The $3d4p\ ^1P^\circ_1$ state is diluted beyond recognition by mixing into the $4snp$ series. Among these levels the largest percentage of $3dnp\ ^1P^\circ_1$ (27%) is present in the level at 43 933 cm^{-1}. Two more calculations of this mixing have appeared: Victor, Stewart, and Laughlin (1976) using model potential calculations and Fisher and Hansen (1984) using multiconfiguration Hartree-Fock (MCHF) calculations have obtained the percentage distribution of $3d4p\ ^1P^\circ_1$ in the $4snp$ series. These results are summarized in the following table.

Level	% $4snp$ Fisher and Hansen		% $3d4p\ ^1P^\circ_1$ Fisher and Hansen	Victor
23652	84 $4s4p$		15.63	16.3
36731	74 $4s5p$	14 $4s4p$	9.27	12.7
41679	63 $4s6p$	25 $4s5p$	7.53	18.6
43933	52 $4s7p$	40 $4s6p$	7.16	23.0
45425	45 $4s7p$	41 $4s8p$	6.80	15.3
46479	55 $4s8p$	32 $4s9p$	5.99	7.4
47184	64 $4s9p$	24 $4s10p$	4.99	

Both calculations confirm the dilution of $3d4p$ in the $4snp$ series, though the largest percentage occurs in different levels. According to Fisher and Hansen's results the levels 43 933 and 45 425 each have about 50% $4s7p\ ^1P_1$. Since one of them needs to be removed from the series in order not to alter the principal quantum numbers of the higher series members, we arbitrarily choose 43 933, and label it $4snl\ ^1P^\circ_1$ in the table below. In the light of this mixing, Roth's calculation of the $(3d+4s)\,4p$ interaction is inappropriate and his results are not included here.

Another well-known perturbed series in Ca I is the $4snd\ ^3D$ series, strongly perturbed by $3d5s\ ^3D$ between $n=8$ and 9. This has been studied by Seaton (1966) using his MQDT method.

Some parity-forbidden series have been observed in Ca I. In Saunders' (1920) paper the first four members of $4s^2\ ^1S_0-4snd\ ^1D_2$ series are reported. Transitions to the ground state from $4p^2\ ^1S_0$ and from $4s7s$ and $4s8s\ ^1S_0$ are also given. McIlrath (1974) has observed lines in absorption between 2048 and 2091 Å which Palenius and Risberg have identified as transitions to $4s9d-14d\ ^1D_2$ and $4s13s-17s\ ^1S_0$ from the $4s^2\ ^1S_0$ ground state.

Arrangement of Tables

The first table of energy levels presented here for neutral calcium is arranged in the usual way; the terms are listed in order of increasing energy without regard to configuration assignments. Because many long Rydberg series have been observed in Ca I, we present a second table for this spectrum in which the series are listed separately, followed by their series limits. This table reveals more clearly the character of the observed spectrum. The series are listed in order of increasing first series member. The series member with the largest value of n for each term type is followed by the limit or limits of that series in Ca II. The assignment of the limits is discussed by Brown et al. and graphically displayed in their figure 10.

References

Armstrong, J. A., Esherick, P., and Wynne, J. J. (1977), Phys. Rev. **A15**, 180.

Armstrong, J. A., Esherick, P., and Wynne, J. J. (1979), J. Opt. Soc. Am. **69**, 211.

Back, E. (1925), Z. Phys. **33**, 579.

Beigang, R., Lücke, K., Schmidt, D., Timmerman, A., and West, P. J. (1982), Phys. Scr. **26**, 183.

Beigang, R., and Wynne, J. J. (1981), Opt. Lett. **6**, 295, 195.

Bohr, N. This discovery was noted by Wentzel, G. (1923)Z. Phys. **24**, 106.

Borgström, S. A., and Rubbmark, J. R. (1977), J. Phys. **B10**, 3607.

Brown, C. M., Tilford, S. G., and Ginter, M. J. (1973), J. Opt. Soc. Am. **63**, 1454.

Camus, P. (1974), unpublished data for $4s\,12f$ and $4s\,13f\ ^3F$ levels.

Chang, E. S. (1983), J. Phys. **B16**, L539.

Connerade, J. P., Baig, M. A., Garton, W. R. S., and Newsom, G. H. (1980), Proc. R. Soc. London, **Ser. A371**, 295.

Fisher, C. F., and Hansen, J. E. (1984), to be published.

Forsberg, P., and Litzén, U. (1982), Atomic Spectroscopy Annual Report, University of Lund.

Friedrich, H., and Trefftz, E. (1969), J. Quant. Spectrosc. Radiat. Transfer **9**, 333.
Garton, W. R. S., and Codling, K. (1965), Proc. Phys. Soc. **86**, 1067.
Grafenberger, H. (1937), Ann. Phys. [5] **30**, 267.
Humphreys, C. J. (1951), J. Res. Natl. Bur. Stand. **47**, 262.
Mansfield, M. W. D., and Newsom, G. H. (1977), Proc. R. Soc. London, Ser. **A357**, 77.
McIlrath, T. J. (1974), J. Phys. **B7**, 393.
Moore, C. E. (1959), "A Multiplet Table of Astrophysical Interest," Natl. Bur. Stand. Tech. Note 36, Washington, D.C. 20234.
Moores, D. L. (1966), Proc. Phys. Soc. **88**, 843.
Newsom, G. H. (1966), Proc. Phys. Soc. **87**, 975.
Palenius, H. P., and Risberg, G. (1977), J. Phys. **B10**, L435.
Risberg, G. (1968), Ark. Fys. **37**, 231.
Roth, C. (1969), J. Res. Natl. Bur. Stand. **73A**, 497.
Russell, H. N. (1927), Astrophys. J. 66, 184.
Russell, H. N., and Saunders, F. A. (1925), Astrophys. J. **61**, 38.
Russell, H. N., and Shenstone, A. G. (1932), Phys. Rev. **39**, 415.
Saunders, F. A. (1920), Astrophys. J. **52**, 256.
Seaton, M. J. (1966), Proc. Phys. Soc. **88**, 815.
Victor, G. A., Stewart, R. F., and Laughlin, C. (1976), in Beam-Foil Spectroscopy, Vol. I (Sellin, I. A. and Pegy, D. J., editors) Plenum, New York, 43-50.
Wagman, N. E. (1937), Univ. Pitts. Bull. **34**, 325.

Ca I: Ordered by term values

Configuration	Term	J	Level (cm^{-1})
$4s^2$	^1S	0	0.000
$4s4p$	$^3P°$	0	*15 157.901*
		1	*15 210.063*
		2	*15 315.943*
$3d\,4s$	^3D	1	20 335.360
		2	20 349.260
		3	20 371.000
$3d\,4s$	^1D	2	21 849.634
$4s4p$	$^1P°$	1	*23 652.304*
$4s5s$	^3S	1	31 539.495
$4s5s$	^1S	0	33 317.264
$3d\,4p$	$^3F°$	2	*35 730.454*
		3	*35 818.713*
		4	*35 896.889*
$3d\,4p$	$^1D°$	2	*35 835.413*
$4s5p$	$^3P°$	0	*36 547.688*
		1	*36 554.749*
		2	*36 575.119*
$4s5p$	$^1P°$	1	*36 731.615*
$4s4d$	^1D	2	37 298.287
$4s4d$	^3D	1	37 748.197
		2	37 751.867
		3	37 757.449
$3d\,4p$	$^3D°$	1	*38 192.392*
		2	*38 219.118*
		3	*38 259.124*
$4p^2$	^3P	0	38 417.543
		1	38 464.808
		2	38 551.558
$3d\,4p$	$^3P°$	0	*39 333.382*
		1	*39 335.322*
		2	*39 340.080*
$4s6s$	^3S	1	40 474.241
$3d\,4p$	$^1F°$	3	*40 537.893*
$4s6s$	^1S	0	40 690.435
$4p^2$	^1D	2	40 719.847
$4s6p$	$^1P°$	1	*41 679.008*
$4p^2$	^1S	0	41 786.276
$4s4f$	$^3F°$	2	*42 170.214*
		3	*42 170.558*
		4	*42 171.026*
$4s4f$	$^1F°$	3	*42 343.587*
$4s6p$	$^3P°$	0	*42 514.845*
		1	*42 518.708*
		2	*42 526.591*
$4s5d$	^3D	1	42 743.002
		2	42 744.716
		3	42 747.387
$4s5d$	^1D	2	42 919.053
$3d^2$	^3F	2	43 474.827
		3	43 489.119
		4	43 508.088
$4snp$	$^1P°$	1	*43 933.477*
$4s7s$	^3S	1	43 980.767
$4s7s$	^1S	0	44 276.538
$4s5f$	$^3F°$	2	*44 762.620*
		3	*44 762.839*
		4	*44 763.118*
$4s5f$	$^1F°$	3	*44 804.878*
$4s5g$	^3G	5	44 874.86
$4s5g$	^1G	4	44 875.95
$4s7p$	$^3P°$	0	*44 955.67*
		1	*44 957.655*
		2	*44 961.757*

Ca I: Ordered by term values—Continued

Configuration	Term	J	Level (cm^{-1})
4s6d	^{1}D	2	44 989.830
4s6d	^{3}D	1	45 049.073
		2	45 050.419
		3	45 052.374
4s7p	^{1}P°	1	*45 425.358*
4s8s	^{3}S	1	45 738.684
4s8s	^{1}S	0	45 887.200
4s6f	^{3}F°	2	*46 164.644*
		3	*46 164.785*
		4	*46 164.971*
4s6f	^{1}F°	3	*46 182.399*
4s7d	^{1}D	2	46 200.13
4s8p	^{3}P°	0	*46 284.12*
		1	*46 285.23*
		2	*46 287.63*
4s7d	^{3}D	1	46 301.973
		2	46 303.649
		3	46 306.059
4s8p	^{1}P°	1	*46 479.813*
4s9s	^{3}S	1	46 748.283
4s9s	^{1}S	0	46 835.055
4s8d	^{1}D	2	46 948.98
4s7f	^{3}F°	2	*47 006.194*
		3	*47 006.280*
		4	*47 006.400*
4s7f	^{1}F°	3	*47 015.141*
4s8d	^{3}D	1	47 036.225
		2	47 040.007
		3	47 045.241
4s9p	^{3}P°	1	*47 085.38*
		2	*47 086.99*
4s9p	^{1}P°	1	*47 184.370*
4s10s	^{3}S	1	47 382.048
4s10s	^{1}S	0	47 437.471
3d 5s	^{1}D	2	47 449.083
3d 5s	^{3}D	1	47 456.452
		2	47 466.014
		3	47 475.915
4s8f	^{3}F°	2	*47 550.214*
		3	*47 550.271*
		4	*47 550.371*
4s8f	^{1}F°	3	*47 555.23*
4s10p	^{3}P°	1	*47 604.75*
		2	*47 605.77*
4s10p	^{1}P°	1	*47 662.10*
4s9d	^{3}D	1	47 752.655
		2	47 757.286
		3	47 765.697
4s11s	^{3}S	1	47 806.20
4s9d	^{1}D	2	47 812.39
4s11s	^{1}S	0	47 843.76
4s9f	^{3}F°	2	*47 921.87*
		3	*47 921.981*
		4	*47 922.033*
4s9f	^{1}F°	3	*47 924.947*
4s11p	^{3}P°	1	*47 960.87*
		2	*47 961.53*
4s11p	^{1}P°	1	*47 997.49*
4s10d	^{3}D	1	48 031.58
		2	48 033.23
		3	48 036.212
4s10d	^{1}D	2	48 083.41
4s12s	^{3}S	1	48 104.02
4s12s	^{1}S	0	48 130.75
4s10f	^{3}F°	2	*48 187.045*
		3	*48 187.075*
		4	*48 187.118*
4s10f	^{1}F°	3	*48 188.990*
4s10g	^{1}G	4	48 202.2
4s10h	^{1}H°	5	*48 208*
4s12p	^{3}P°	1	*48 215.81*
		2	*48 216.36*
4s12p	^{1}P°	1	*48 240.53*
4s11d	^{3}D	1	48 258.30
		2	48 258.84
		3	48 260.29
4s11d	^{1}D	2	48 290.85
4s13s	^{3}S	1	48 321.14
4s13s	^{1}S	0	48 340.75

(Continued)

Ca I: Ordered by term values—Continued

Configuration	Term	J	Level (cm^{-1})
4s11f	$^3F^\circ$	2	48 382.70
		3	48 382.781
		4	48 382.801
4s11f	$^1F^\circ$	3	48 384.039
4s11g	^1G	4	48 394.3
6s11h	$^1H^\circ$	5	48 398.6
4s13p	$^3P^\circ$	1	48 404.57
		2	48 404.95
4s13p	$^1P^\circ$	1	48 422.09
4s12d	^3D	1	48 433.01
		2	48 433.46
		3	48 434.47
4s12d	^1D	2	48 451.73
4s14s	^3S	1	48 484.30
4s14s	^1S	0	48 499.14
$3d^2$	^3P	0	48 524.093
		1	48 537.623
		2	48 563.522
4s12f	$^3F^\circ$	3,4	48 531.04
4s12f	$^1F^\circ$	3	48 532.139
4s12g	^1G	4	48 540.1
4s12h	$^1H^\circ$	5	48 544.2
4s14p	$^3P^\circ$	1	48 548.30
		2	48 548.51
4s14p	$^1P^\circ$	1	48 561.10
4s13d	^3D	1	48 568.66
		2	48 568.95
		3	48 569.51
4s13d	^1D	2	48 578.32
4s15s	^3S	1	48 609.75
4s15s	^1S	0	48 621.53
4s13f	$^3F^\circ$	4	48 646.38
4s13f	$^1F^\circ$	3	48 647.30
4s13g	^1G	4	48 653.7
4s13h	$^1H^\circ$	5	48 657.0
4s15p	$^3P^\circ$	2	48 660.23
4s15p	$^1P^\circ$	1	48 669.83
4s14d	^3D	1	48 675.42
		2	48 675.65
		3	48 676.09
4s14d	^1D	2	48 678.97
4s16s	^3S	1	48 708.82
4s16s	^1S	0	48 718.02
4s14f	$^1F^\circ$	3	48 738.54
4s14g	^1G	4	48 743.4
4s14h	$^1H^\circ$	5	48 747.4
4s16p	$^3P^\circ$	2	48 749.04
4s16p	$^1P^\circ$	1	48 756.45
4s15d	^1D	2	48 760.14
4s15d	^3D	1	48 760.88
		2	48 761.20
		3	48 761.31
4s17s	^3S	1	48 788.03
4s17s	^1S	0	48 795.46
4s15f	$^1F^\circ$	3	48 812.09
4s17p	$^3P^\circ$	2	48 820.60
4s17p	$^1P^\circ$	1	48 826.54
4s16d	^1D	2	48 827.05
4s16d	^3D	1	48 829.87
		2	48 829.95
		3	48 830.47
4s18s	^3S	1	48 852.58
4s18s	^1S	0	48 858.59
4s16f	$^1F^\circ$	3	48 872.20
4s18p	$^3P^\circ$	2	48 879.31
4s17d	^1D	2	48 882.37
4s18p	$^1P^\circ$	1	48 884.06
4s17d	^3D	2	48 886.75
		1	48 886.90
		3	48 887.04
4s19s	^3S	1	48 905.60
4s19s	^1S	0	48 910.65
4s17f	$^1F^\circ$	3	48 921.95

Ca I: Ordered by term values—Continued

Configuration	Term	J	Level (cm^{-1})
4s19p	$^{3}P^{\circ}$	2	*48 927.93*
4s18d	^{1}D	2	48 928.79
4s19p	$^{1}P^{\circ}$	1	*48 931.82*
4s18d	^{3}D	2	48 934.04
		1	48 934.10
		3	48 934.32
4s20s	^{3}S	1	48 949.88
4s20s	^{1}S	0	48 954.13
4s18f	$^{1}F^{\circ}$	3	*48 963.67*
4s19d	^{1}D	2	48 968.10
4s20p	$^{3}P^{\circ}$	2	*48 968.67*
4s20p	$^{1}P^{\circ}$	1	*48 971.93*
4s19d	^{3}D	1	48 973.80
		3	48 973.92
		2	48 973.95
4s21s	^{3}S	1	48 987.35
4s21s	^{1}S	0	48 990.83
4s19f	$^{1}F^{\circ}$	3	*48 998.89*
4s20d	^{1}D	2	49 001.66
4s21p	$^{3}P^{\circ}$	2	*49 003.21*
4s21p	$^{1}P^{\circ}$	1	*49 005.92*
4s20d	^{3}D	2	49 007.58
		1	49 007.59
		3	49 007.70
4s22s	^{3}S	1	49 019.13
4s22s	^{1}S	0	49 022.02
4s20f	$^{1}F^{\circ}$	3	*49 028.93*
4s21d	^{1}D	2	49 030.55
4s22p	$^{3}P^{\circ}$	2	*49 032.70*
4s22p	$^{1}P^{\circ}$	1	*49 034.98*
4s21d	^{3}D	2	49 036.40
4s23s	^{3}S	1	49 046.22
4s23s	^{1}S	0	49 048.85
4s21f	$^{1}F^{\circ}$	3	*49 054.80*
4s22d	^{1}D	2	49 055.62
4s23p	$^{3}P^{\circ}$	2	*49 058.02*
4s23p	$^{1}P^{\circ}$	1	*49 060.02*
4s22d	^{3}D	2	49 061.23
4s24s	^{3}S	1	49 069.82
4s24s	^{1}S	0	49 071.99
4s22f	$^{1}F^{\circ}$	3	*49 077.20*
4s23d	^{1}D	2	49 077.48
4s24p	$^{3}P^{\circ}$	2	*49 080.04*
4s24p	$^{1}P^{\circ}$	1	*49 081.75*
4s23d	^{3}D	2	49 082.83
4s25s	^{3}S	1	49 090.40
4s25s	^{1}S	0	49 092.19
4s24d	^{1}D	2	49 096.52
4s23f	$^{1}F^{\circ}$	3	*49 096.74*
4s25p	$^{3}P^{\circ}$	2	*49 099.25*
4s25p	$^{1}P^{\circ}$	1	*49 100.72*
4s24d	^{3}D	2	49 101.51
4s26s	^{3}S	1	49 108.15
4s26s	^{1}S	0	49 109.85
4s25d	^{1}D	2	49 113.41
4s24f	$^{1}F^{\circ}$	3	*49 113.85*
4s26p	$^{3}P^{\circ}$	2	*49 116.07*
4s26p	$^{1}P^{\circ}$	1	*49 117.38*
4s25d	^{3}D	2	49 118.24
4s27s	^{3}S	1	49 124.02
4s27s	^{1}S	0	49 125.44
4s26d	^{1}D	2	49 128.34
4s25f	$^{1}F^{\circ}$	3	*49 129.00*
4s27p	$^{3}P^{\circ}$	2	*49 130.94*
4s27p	$^{1}P^{\circ}$	1	*49 132.09*
4s26d	^{3}D	2	49 132.77
4s28s	^{3}S	1	49 138.06
4s28s	^{1}S	0	49 139.31

(Continued)

Ca I: Ordered by term values—Continued

Configuration	Term	*J*	Level (cm^{-1})
4s27d	^{1}D	2	49 141.63
4s26f	^{1}F°	3	*49 142.38*
4s28p	^{3}P°	2	*49 144.19*
4s28p	^{1}P°	1	*49 145.15*
4s27d	^{3}D	2	49 145.66
4s29s	^{3}S	1	49 150.30
4s29s	^{1}S	0	49 151.59
4s28d	^{1}D	2	49 153.49
4s27f	^{1}F°	3	*49 154.29*
4s29p	^{3}P°	2	*49 155.90*
4s29p	^{1}P°	1	*49 156.78*
4s28d	^{3}D	2	49 157.39
4s30s	^{3}S	1	49 161.54
4s30s	^{1}S	0	49 162.47
4s29d	^{1}D	2	49 164.12
4s28f	^{1}F°	3	*49 164.99*
4s30p	^{3}P°	2	*49 166.43*
4s30p	^{1}P°	1	*49 167.20*
4s29d	^{3}D	2	49 167.55
4s31s	^{3}S	1	49 171.42
4s31s	^{1}S	0	49 172.43
4s30d	^{1}D	2	49 173.70
4s31p	^{3}P°	2	*49 175.87*
4s31p	^{1}P°	1	*49 176.58*
4s30d	^{3}D	2	49 177.13
4s32s	^{3}S	1	49 180.48
4s32s	^{1}S	0	49 181.49
4s31d	^{1}D	2	49 182.36
4s32p	^{3}P°	2	*49 184.37*
4s32p	^{1}P°	1	*49 185.02*
4s31d	^{3}D	2	49 185.24
4s33s	^{3}S	1	49 188.37
4s32d	^{1}D	2	49 190.17
4s33p	^{3}P°	2	*49 192.08*
4s33p	^{1}P°	1	*49 192.68*
4s32d	^{3}D	2	49 193.06
4s34s	^{3}S	1	49 195.64
4s33d	^{1}D	2	49 197.30
4s34p	^{3}P°	2	*49 199.09*
4s34p	^{1}P°	1	*49 199.62*
4s33d	^{3}D	2	49 199.93
4s35s	^{3}S	1	49 202.47
4s34d	^{1}D	2	49 203.78
4s35p	^{3}P°	2	*49 205.47*
4s35p	^{1}P°	1	*49 205.95*
4s34d	^{3}D	2	49 206.17
4s36s	^{3}S	1	49 208.53
4s35d	^{1}D	2	49 209.68
4s36p	^{3}P°	2	*49 211.34*
4s36p	^{1}P°	1	*49 211.72*
4s35d	^{3}D	2	49 212.01
4s36d	^{1}D	2	49 215.11
4s37p	^{3}P°	2	*49 216.61*
4s37p	^{1}P°	1	*49 217.02*
4s36d	^{3}D	2	49 217.18
4s37d	^{1}D	2	49 220.10
4s38p	^{3}P°	2	*49 221.51*
4s38p	^{1}P°	1	*49 221.87*
4s37d	^{3}D	2	49 222.06
4s38d	^{1}D	2	49 224.70
4s39p	^{3}P°	2	*49 226.02*
4s39p	^{1}P°	1	*49 226.35*
4s38d	^{3}D	2	49 226.56
4s39d	^{1}D	2	49 228.92

Ca I: Ordered by term values—Continued

Configuration	Term	J	Level (cm^{-1})
4s40*p*	$^{3}P^{\circ}$	2	*49 230.13*
4s40*p*	$^{1}P^{\circ}$	1	*49 230.45*
4s39*d*	^{3}D	2	49 230.54
4s40*d*	^{1}D	2	49 232.81
4s41*p*	$^{3}P^{\circ}$	2	*49 234.00*
4s41*p*	$^{1}P^{\circ}$	1	*49 234.28*
4s40*d*	^{3}D	2	49 234.47
4s41*d*	^{1}D	2	49 236.45
4s42*p*	$^{3}P^{\circ}$	2	*49 237.55*
4s41*d*	^{3}D	2	49 237.82
4s42*p*	$^{1}P^{\circ}$	1	*49 237.82*
4s42*d*	^{1}D	2	49 239.80
4s43*p*	$^{3}P^{\circ}$	2	*49 240.81*
4s42*d*	^{3}D	2	49 241.09
4s43*p*	$^{1}P^{\circ}$	1	*49 241.09*
4s43*d*	^{1}D	2	49 242.93
4s44*p*	$^{3}P^{\circ}$	2	*49 243.87*
4s44*p*	$^{1}P^{\circ}$	1	*49 244.13*
4s43*d*	^{3}D	2	49 244.26
4s44*d*	^{1}D	2	49 245.80
4s45*p*	$^{3}P^{\circ}$	2	*49 246.74*
4s45*p*	$^{1}P^{\circ}$	1	*49 246.98*
4s44*d*	^{3}D	2	49 247.04
4s45*d*	^{1}D	2	49 248.52
4s46*p*	$^{3}P^{\circ}$	2	*49 249.37*
4s45*d*	^{3}D	2	49 249.56
4s46*p*	$^{1}P^{\circ}$	1	*49 249.61*
4s46*d*	^{1}D	2	49 251.05
4s47*p*	$^{3}P^{\circ}$	2	*49 251.87*
4s47*p*	$^{1}P^{\circ}$	1	*49 252.08*
4s46*d*	^{3}D	2	49 252.10
4s47*d*	^{1}D	2	49 253.40
4s48*p*	$^{3}P^{\circ}$	2	*49 254.22*
4s48*p*	$^{1}P^{\circ}$	1	*49 254.41*
4s47*d*	^{3}D	2	49 254.43
4s48*d*	^{1}D	2	49 255.64
4s49*p*	$^{3}P^{\circ}$	2	*49 256.42*
4s48*d*	^{3}D	2	49 256.55
4s49*p*	$^{1}P^{\circ}$	1	*49 256.56*
4s49*d*	^{1}D	2	49 257.74
4s50*p*	$^{3}P^{\circ}$	2	*49 258.45*
4s50*p*	$^{1}P^{\circ}$	1	*49 258.60*
4s49*d*	^{3}D	2	49 258.65
4s50*d*	^{1}D	2	49 259.68
4s51*p*	$^{3}P^{\circ}$	2	*49 260.36*
4s51*p*	$^{1}P^{\circ}$	1	*49 260.51*
4s50*d*	^{3}D	2	49 260.64
4s51*d*	^{1}D	2	49 261.51
4s52*p*	$^{3}P^{\circ}$	2	*49 262.15*
4s52*p*	$^{1}P^{\circ}$	1	*49 262.30*
4s52*d*	^{1}D	2	49 263.31
4s51*d*	^{3}D	2	49 262.42
4s53*p*	$^{3}P^{\circ}$	2	*49 263.85*
4s53*p*	$^{1}P^{\circ}$	1	*49 263.99*
4s52*d*	^{3}D	3	49 264.19
4s53*d*	^{1}D	2	49 264.77
4s54*p*	$^{3}P^{\circ}$	2	*49 265.44*
4s54*p*	$^{1}P^{\circ}$	1	*49 265.59*
4s53*d*	^{3}D	2	49 265.60
4s54*d*	^{1}D	2	49 266.36
4s55*p*	$^{3}P^{\circ}$	2	*49 266.99*
4s55*p*	$^{1}P^{\circ}$	1	*49 267.10*
4s54*d*	^{3}D	2	49 267.18

(Continued)

Ca I: Ordered by term values—Continued

Configuration	Term	*J*	Level (cm^{-1})
4*s*55*d*	^{1}D	2	49 267.84
4*s*56*p*	^{3}P°	2	*49 268.40*
4*s*56*p*	^{1}P°	1	*49 268.52*
4*s*55*d*	^{3}D	2	49 268.57
4*s*56*d*	^{1}D	2	49 269.33
4*s*57*p*	^{3}P°	2	*49 269.77*
4*s*57*p*	^{1}P°	1	*49 269.88*
4*s*56*d*	^{3}D	2	49 270.01
4*s*57*d*	^{1}D	2	49 270.40
4*s*58*p*	^{3}P°	2	*49 271.01*
4*s*58*p*	^{1}P°	1	*49 271.14*
4*s*57*d*	^{3}D	2	49 271.25
4*s*58*d*	^{1}D	2	49 271.79
4*s*59*p*	^{3}P°	2	*49 272.24*
4*s*59*p*	^{1}P°	1	*49 272.35*
4*s*58*d*	^{3}D	2	49 272.49
4*s*59*d*	^{1}D	2	49 273.09
4*s*60*p*	^{3}P°	2	*49 273.37*
4*s*59*d*	^{3}D	2	49 273.53
4*s*60*p*	^{1}P°	1	*49 273.50*
4*s*60*d*	^{1}D	2	49 274.29
4*s*60*d*	^{3}D	2	49 274.52
4*s*61*p*	^{1}P°	1	*49 274.60*
4*s*61*d*	^{1}D	2	49 275.58
4*s*62*p*	^{1}P°	1	*49 275.62*
4*s*62*d*	^{1}D	2	49 275.87
4*s*63*p*	^{1}P°	1	*49 276.61*
4*s*64*p*	^{1}P°	1	*49 277.55*
4*s*65*p*	^{1}P°	1	*49 278.44*
4*s*66*p*	^{1}P°	1	*49 279.30*
4*s*67*p*	^{1}P°	1	*49 280.10*
4*s*68*p*	^{1}P°	1	*49 280.88*
4*s*69*p*	^{1}P°	1	*49 281.62*
4*s*70*p*	^{1}P°	1	*49 282.34*
4*s*71*p*	^{1}P°	1	*49 283.01*
4*s*72*p*	^{1}P°	1	*49 283.67*
4*s*73*p*	^{1}P°	1	*49 284.30*
4*s*74*p*	^{1}P°	1	*49 284.88*
4*s*75*p*	^{1}P°	1	*49 285.46*
4*s*76*p*	^{1}P°	1	*49 286.00*
4*s*77*p*	^{1}P°	1	*49 286.54*
4*s*78*p*	^{1}P°	1	*49 287.05*
4*s*79*p*	^{1}P°	1	*49 287.57*
Ca II ($^{2}S_{1/2}$)	***Limit***		**49 305.95**
3*d* 4*d*	^{3}D	1	51 351.74
		2	51 369.38
		3	51 396.32
3*d* 4*d*	^{3}G	3	51 553.6?
		4	51 579.0?
		5	51 611.4?
3*d* 4*d*	^{3}S	1	51 571.7
3*d* 5*p*	^{3}D°	1	*51 709.5*
		2	*51 734.6*
		3	*51 767.0*
3*d* 5*p*	^{3}P°	1	*51 908.*
3*d* 5*p*	^{1}P°	1	*53 100.*
3*d* 4*d*	^{3}F	4	53 200.4?
		2	53 214.6?
		3	53 247.9?
3*d* 4*d*	^{3}P	0	54 282.3
		1	54 288.74
		2	54 304.6
3*d*4*f*	^{3}D°	1	*55 902.8*
3*d*4*f*	^{3}P°	1	*55 946.6*
3*d*4*f*	^{1}P°	1	*55 982.3*
3*d* 6*p*	^{3}D°	1	*56 254.*
3*d* 5*d*	^{3}D	2	56 469.0?
3*d* 6*p*	^{3}P°	1	*56 532.63*
3*d* 5*d*	^{3}S	1	56 558.9
3*d* 6*p*	^{1}P°	1	*56 651.*
4*p*5*s*	^{3}P°	1	*57 462.*
3*d* 5*d*	^{3}P	0	57 611.2
		1	57 617.9
		2	57 638.4
4*p*5*s*	^{1}P°	1	*57 960.*
3*d*5*f*	^{3}D°	1	*58 431.31*

Ca I: Ordered by term values—Continued

Configuration	Term	J	Level (cm^{-1})
3d5f	$^3P^\circ$	1	58 491.91
3d5f	$^1P^\circ$	1	58 505.89
3d 7p	$^3D^\circ$	1	58 798.92
3d 7p	$^3P^\circ$	1	59 010.87
3d 7p	$^1P^\circ$	1	59 197.
3d 6d	^3P	1	59 368.
		2	59 391.
3d6f	$^3D^\circ$	1	59 802.21
3d6f	$^3P^\circ$	1	59 862.26
3d6f	$^1P^\circ$	1	59 878.52
3d 8p	$^3D^\circ$	1	60 046.13
3d 8p	$^3P^\circ$	1	60 150.26
3d 8p	$^1P^\circ$	1	60 300.
3d7f	$^3D^\circ$	1	60 632.18
3d7f	$^3P^\circ$	1	60 690.27
3d7f	$^1P^\circ$	1	60 709.54
3d 9p	$^3D^\circ$	1	60 807.28
3d 9p	$^3P^\circ$	1	60 869.28
3d 9p	$^1P^\circ$	1	60 973.6
3d8f	$^3D^\circ$	1	61 172.54
3d8f	$^3P^\circ$	1	61 228.65
3d8f	$^1P^\circ$	1	61 251.39
3d 10p	$^3D^\circ$	1	61 306.65
3d 10p	$^3P^\circ$	1	61 345.74
3d 10p	$^1P^\circ$	1	61 428.1
3d9f	$^3D^\circ$	1	61 543.95
3d9f	$^3P^\circ$	1	61 596.69
3d9f	$^1P^\circ$	1	61 622.60
3d 11p	$^3D^\circ$	1	61 652.25
3d 11p	$^3P^\circ$	1	61 676.62
3d 11p	$^1P^\circ$	1	61 747.2
3d 10f	$^3D^\circ$	1	61 810.68
3d 10f	$^3P^\circ$	1	61 859.22
3d 10f	$^1P^\circ$	1	61 888.22
3d 12p	$^3D^\circ$	1	61 901.04
3d 12p	$^3P^\circ$	1	61 916.38
3d 12p	$^1P^\circ$	1	61 980.49
3d 11f	$^3D^\circ$	1	62 009.17
3d 11f	$^3P^\circ$	1	62 052.49
3d 13p	$^3D^\circ$	1	62 084.92
3d 11f	$^1P^\circ$	1	62 084.92
3d 13p	$^3P^\circ$	1	62 097.08
3d 13p	$^1P^\circ$	1	62 154.62
3d 12f	$^3D^\circ$	1	62 161.61
3d 12f	$^3P^\circ$	1	62 198.92
3d 14p	$^3D^\circ$	1	62 223.68
3d 12f	$^1P^\circ$	1	62 234.47
3d 14p	$^3P^\circ$	1	62 237.12
3d 13f	$^3D^\circ$	1	62 279.33
3d 14p	$^1P^\circ$	1	62 288.02
3d 13f	$^3P^\circ$	1	62 312.63
3d 15p	$^3D^\circ$	1	62 331.04
3d 13f	$^1P^\circ$	1	62 346.47
3d 15p	$^3P^\circ$	1	62 351.93
3d 14f	$^3D^\circ$	1	62 375.29
3d 15p	$^1P^\circ$	1	62 392.01
3d 14f	$^3P^\circ$	1	62 403.22
3d 16p	$^3D^\circ$	1	62 416.47
3d 14f	$^1P^\circ$	1	62 433.38
3d 16p	$^3P^\circ$	1	62 443.64
3d 15f	$^3D^\circ$	1	62 453.85
3d 15f	$^3P^\circ$	1	62 472.14
3d 16p	$^1P^\circ$	1	62 478.26
3d 17p	$^3D^\circ$	1	62 486.45
3d 15f	$^1P^\circ$	1	62 506.62
3d 17p	$^3P^\circ$	1	62 511.97
3d 16f	$^3D^\circ$	1	62 520.06

(Continued)

Ca I: Ordered by term values—Continued

Configuration	Term	J	Level (cm⁻¹)
3d 16f	$^3P°$	1	62 533.90
3d 18p	$^3D°$	1	62 539.39
3d 17p	$^1P°$	1	62 548.16
4p4d	$^3D°$?	1	62 552.97
4p4d	$^3P°$?	1	62 557.45
4p4d	$^1P°$?	1	62 561.72
3d 17f	$^3D°$	1	62 565.68
3d 16f	$^1P°$	1	62 572.66
3d 17f	$^3P°$	1	62 578.91
3d 18p	$^3P°$	1	62 584.50
3d 19p	$^3D°$	1	62 587.83
3d 18p	$^1P°$	1	62 602.72
3d 18f	$^3D°$	1	62 610.75
3d 18f	$^3P°$	1	62 617.02
3d 17f	$^1P°$	1	62 625.37
3d 20p	$^3D°$	1	62 626.03
3d 19p	$^3P°$	1	62 628.76
3d 19f	$^3D°$	1	62 641.52
3d 19p	$^1P°$	1	62 648.37
3d 19f	$^3P°$	1	62 653.99
3d 21p	$^3D°$	1	62 660.15
3d 20p	$^3P°$	1	62 671.51
3d 18f	$^1P°$	1	62 671.51
3d 20f	$^3D°$	1	62 675.56
3d 20f	$^3P°$	1	62 683.89
3d 22p	$^3D°$	1	62 686.88
3d 20p	$^1P°$	1	62 690.32
3d 19f	$^1P°$	1	62 704.43
3d 21f	$^3D°$	1	62 704.43
3d 21f	$^3P°$	1	62 707.25
3d 21p	$^3P°$	1	62 711.37
3d 23p	$^3D°$	1	62 713.24
3d 22f	$^3D°$	1	62 720.41
3d 21p	$^1P°$	1	62 721.80
3d 22f	$^3P°$	1	62 729.19
3d 24p	$^3D°$	1	62 734.25
3d 22p	$^3P°$	1	62 736.10
3d 20f	$^1P°$	1	62 736.10
3d 23f	$^3D°$	1	62 743.11
3d 23f	$^3P°$	1	62 748.53
3d 22p	$^1P°$	1	62 750.58
3d 25p	$^3D°$	1	62 753.60
3d 23p	$^3P°$	1	62 762.86
3d 21f	$^1P°$	1	62 762.86
3d 24f	$^3P°$	1	62 764.10
3d 24f	$^3D°$	1	62 768.10
3d 26p	$^3D°$	1	62 769.47
3d 23p	$^1P°$	1	62 774.75
3d 25f	$^3P°$	1	62 780.30
3d 27p	$^3D°$	1	62 783.86
3d 22f	$^1P°$	1	62 785.02
3d 24p	$^3P°$	1	62 787.31
3d 26f	$^3D°$	1	62 790.08
3d 26f	$^3P°$	1	62 794.19
3d 24p	$^1P°$	1	62 795.55
3d 28p	$^3D°$	1	62 797.46
3d 23f	$^1P°$	1	62 804.38
3d 27f	$^3P°$	1	62 806.02
3d 25p	$^3P°$	1	62 807.90
3d 29p	$^3D°$	1	62 808.51
3d 28f	$^3D°$	1	62 812.00

Ca I: Ordered by term values—Continued

Configuration	Term	*J*	Level (cm^{-1})
3*d* 25*p*	$^{1}P^{\circ}$	1	*62 814.13*
3*d* 28*f*	$^{3}P^{\circ}$	1	*62 816.51*
3*d* 30*p*	$^{3}D^{\circ}$	1	*62 818.76*
3*d* 24*f*	$^{1}P^{\circ}$	1	*62 820.31*
3*d* 26*p*	$^{3}P^{\circ}$	1	*62 824.17*
3*d* 29*f*	$^{3}P^{\circ}$	1	*62 824.17*
3*d* 31*p*	$^{3}D^{\circ}$	1	*62 827.75*
3*d* 26*p*	$^{1}P^{\circ}$	1	*62 830.67*
3*d* 30*f*	$^{3}D^{\circ}$	1	*62 832.31*
3*d* 30*f*	$^{3}P^{\circ}$	1	*62 834.11*
3*d* 32*p*	$^{3}D^{\circ}$	1	*62 836.22*
3*d* 25*f*	$^{1}P^{\circ}$	1	*62 836.85*
3*d* 27*p*	$^{3}P^{\circ}$	1	*62 838.45*
3*d* 31*f*	$^{3}P^{\circ}$	1	*62 840.23*
3*d* 33*p*	$^{3}D^{\circ}$	1	*62 843.72*
3*d* 27*p*	$^{1}P^{\circ}$	1	*62 845.27*
3*d* 32*f*	$^{3}P^{\circ}$	1	*62 848.77*
3*d* 34*p*	$^{3}D^{\circ}$	1	*62 850.62*
3*d* 26*f*	$^{1}P^{\circ}$	1	*62 852.33*
3*d* 28*p*	$^{3}P^{\circ}$	1	*62 853.79*
3*d* 33*f*	$^{3}P^{\circ}$	1	*62 853.79*
3*d* 35*p*	$^{3}D^{\circ}$	1	*62 856.56*
3*d* 28*p*	$^{1}P^{\circ}$	1	*62 858.11*
3*d* 34*f*	$^{3}P^{\circ}$	1	*62 861.06*
3*d* 36*p*	$^{3}D^{\circ}$	1	*62 862.60*
3*d* 29*p*	$^{3}P^{\circ}$	1	*62 863.99*
3*d* 35*f*	$^{3}P^{\circ}$	1	*62 865.39*
3*d* 37*p*	$^{3}D^{\circ}$	1	*62 867.87*
3*d* 29*p*	$^{1}P^{\circ}$	1	*62 869.38*
3*d* 36*f*	$^{3}P^{\circ}$	1	*62 871.46*
3*d* 38*p*	$^{3}D^{\circ}$	1	*62 872.69*
3*d* 30*p*	$^{3}P^{\circ}$	1	*62 875.40*
3*d* 37*f*	$^{3}P^{\circ}$	1	*62 875.40*
3*d* 39*p*	$^{3}D^{\circ}$	1	*62 876.98*
3*d* 30*p*	$^{1}P^{\circ}$	1	*62 879.48*
3*d* 38*f*	$^{3}P^{\circ}$	1	*62 880.04*
3*d* 40*p*	$^{3}D^{\circ}$	1	*62 881.24*
3*d* 31*p*	$^{3}P^{\circ}$	1	*62 884.77*
3*d* 41*p*	$^{3}D^{\circ}$	1	*62 884.77*
3*d* 31*p*	$^{1}P^{\circ}$	1	*62 889.00*
3*d* 42*p*	$^{3}D^{\circ}$	1	*62 889.00*
3*d* 43*p*	$^{3}D^{\circ}$	1	*62 891.65*
3*d* 32*p*	$^{3}P^{\circ}$	1	*62 893.66*
3*d* 44*p*	$^{3}D^{\circ}$	1	*62 894.65*
3*d* 32*p*	$^{1}P^{\circ}$	1	*62 897.57*
3*d* 45*p*	$^{3}D^{\circ}$	1	*62 897.57*
3*d* 46*p*	$^{3}D^{\circ}$	1	*62 900.09*
3*d* 33*p*	$^{3}P^{\circ}$	1	*62 901.85*
3*d* 47*p*	$^{3}D^{\circ}$	1	*62 902.75*
3*d* 33*p*	$^{1}P^{\circ}$	1	*62 904.95*
3*d* 48*p*	$^{3}D^{\circ}$	1	*62 904.95*
3*d* 49*p*	$^{3}D^{\circ}$	1	*62 906.98*
3*d* 50*p*	$^{3}D^{\circ}$	1	*62 908.88*
3*d* 51*p*	$^{3}D^{\circ}$	1	*62 910.88*
3*d* 34*p*	$^{1}P^{\circ}$	1	*62 911.51*
3*d* 52*p*	$^{3}D^{\circ}$	1	*62 912.67*
3*d* 53*p*	$^{3}D^{\circ}$	1	*62 914.47*
3*d* 54*p*	$^{3}D^{\circ}$	1	*62 915.91*
3*d* 35*p*	$^{1}P^{\circ}$	1	*62 917.83*
3*d* 55*p*	$^{3}D^{\circ}$	1	*62 917.83*
3*d* 56*p*	$^{3}D^{\circ}$	1	*62 918.80*
3*d* 57*p*	$^{3}D^{\circ}$	1	*62 920.32*
3*d* 58*p*	$^{3}D^{\circ}$	1	*62 921.43*
3*d* 36*p*	$^{1}P^{\circ}$	1	*62 923.62*

(Continued)

Ca I: Ordered by term values—Continued

Configuration	Term	J	Level (cm^{-1})
3d 35f	$^1P^\circ$	1	62 926.36
3d 37p	$^1P^\circ$	1	62 928.81
3d 36f	$^1P^\circ$	1	62 931.29
3d 38p	$^1P^\circ$	1	62 933.61
3d 37f	$^1P^\circ$	1	62 935.88
3d 39p	$^1P^\circ$	1	62 938.02
3d 38f	$^1P^\circ$	1	62 939.79
3d 40p	$^1P^\circ$	1	62 942.10
3d 39f	$^1P^\circ$	1	62 943.85
3d 41p	$^1P^\circ$	1	62 945.88
3d 42p	$^1P^\circ$	1	62 949.34
3d 43p	$^1P^\circ$	1	62 952.58
3d 44p	$^1P^\circ$	1	62 955.57
Ca II ($^2D_{3/2}$)	**Limit**		**62 956.15**
3d 45p	$^1P^\circ$	1	62 958.38
3d 46p	$^1P^\circ$	1	62 961.01
3d 47p	$^1P^\circ$	1	62 963.43
3d 48p	$^1P^\circ$	1	62 965.72
3d 49p	$^1P^\circ$	1	62 967.84
3d 50p	$^1P^\circ$	1	62 969.89
3d 51p	$^1P^\circ$	1	62 971.78
3d 52p	$^1P^\circ$	1	62 973.55
3d 53p	$^1P^\circ$	1	62 975.23
3d 54p	$^1P^\circ$	1	62 976.81
3d 55p	$^1P^\circ$	1	62 978.26
3d 56p	$^1P^\circ$	1	62 979.72
3d 57p	$^1P^\circ$	1	62 981.03
3d 58p	$^1P^\circ$	1	62 982.35
3d 59p	$^1P^\circ$	1	62 983.48
3d 60p	$^1P^\circ$	1	62 984.63
3d 61p	$^1P^\circ$	1	62 985.70
Ca II ($^2D_{5/2}$)	**Limit**		**63 016.84**
4p6s	$^3P^\circ$	1	64 977.
4p6s	$^1P^\circ$	1	65 445.
4p5d	$^1P^\circ$	1	66 578.
4p7s	$^3P^\circ$	1	68 913.
4p6d	$^1P^\circ$	1	69 643.
4p7s	$^1P^\circ$	1	69 247.
4p8s	$^3P^\circ$	1	70 787.
4p8s	$^1P^\circ$	1	71 001.
4p7d	$^1P^\circ$	1	71 237.
4p9s	$^3P^\circ$	1	71 868.
4p9s	$^1P^\circ$	1	72 007.
4p8d	$^1P^\circ$	1	72 289.
4p10s	$^3P^\circ$	1	72 411.
4p10s	$^1P^\circ$	1	72 655.
4p11s	$^3P^\circ$	1	72 871.
4p9d	$^1P^\circ$	1	
4p11s	$^1P^\circ$	1	73 111.
4p12s	$^3P^\circ$	1	73 248.
4p10d	$^1P^\circ$	1	
4p12s	$^1P^\circ$	1	73 444.
4p13s	$^3P^\circ$	1	
4p11d	$^1P^\circ$	1	73 517.
4p14s	$^3P^\circ$	1	73 622.
4p13s	$^1P^\circ$	1	73 693.
4p15s	$^3P^\circ$	1	73 742.
4p12d	$^3P^\circ$	1	
4p14s	$^1P^\circ$	1	73 852.
4p16s	$^3P^\circ$	1	
4p15s	$^1P^\circ$	1	73 969.
Ca II ($^2P_{1/2}$)	**Limit**		**74 497.46**
Ca II ($^2P_{3/2}$)	**Limit**		**74 720.35**
$3p^5\,4s^2\,3d$	$^3P^\circ$	1	200 096
$3p^5\,4s^2\,3d$	$^3D^\circ$	1	218 991
$3p^5\,4s^2\,3d$	$^1P^\circ$	1	253 310
$3p^5\,4s^2\,4d$	$^3P^\circ$	1	255 022
$3p^5(^2P^\circ_{3/2})4s^2\,5s$	$^2[3/2]^\circ$	1	257 737
$3p^5(^2P^\circ_{1/2})4s^2\,5s$	$^2[1/2]^\circ$	1	260 193

Ca I: Ordered by term values—Continued

Configuration	Term	J	Level (cm^{-1})	Configuration	Term	J	Level (cm^{-1})
$3p^5(^2P^\circ_{3/2})4s^2\,6s$	$^2[3/2]^\circ$	1	*267 417*	$3p^5(^2P^\circ_{3/2})4s^2\,15d$	$^2[3/2]^\circ$	1	*276 214*
$3p^5\,4s^2\,4d$	$^3D^\circ$	1	*267 956*	$3p^5(^2P^\circ_{3/2})4s^2\,16d$	$^2[3/2]^\circ$	1	*276 283*
$3p^5(^2P^\circ_{1/2})4s^2\,6s$	$^2[1/2]^\circ$	1	*269 706*	$3p^5(^2P^\circ_{1/2})4s^2\,7d$	$^2[3/2]^\circ$	1	*276 631*
$3p^5\,4s^2\,4d$	$^1P^\circ$	1	*270 640*	Ca II $3p^54s^2\,^2P^\circ_{3/2}$	***Limit***		**276 750**
$3p^5(^2P^\circ_{3/2})4s^2\,5d$	$^2[3/2]^\circ$	1	*270 889*	$3p^5(^2P^\circ_{1/2})4s^2\,8d$	$^2[3/2]^\circ$	1	*277 430*
$3p^5(^2P^\circ_{3/2})4s^2\,7s$	$^2[3/2]^\circ$	1	*271 245*	$3p^5(^2P^\circ_{1/2})4s^2\,9d$	$^2[3/2]^\circ$	1	*277 917*
$3p^5(^2P^\circ_{3/2})4s^2\,6d$	$^2[3/2]^\circ$	1	*273 134*	$3p^5(^2P^\circ_{1/2})4s^2\,11s$	$^2[1/2]^\circ$	1	*277 991*
$3p^5(^2P^\circ_{1/2})4s^2\,7s$	$^2[1/2]^\circ$	1	*273 598*	$3p^5(^2P^\circ_{1/2})4s^2\,10d$	$^2[3/2]^\circ$	1	*278 228*
$3p^5(^2P^\circ_{3/2})4s^2\,7d$	$^2[3/2]^\circ$	1	*273 959*	$3p^5(^2P^\circ_{1/2})4s^2\,12s$	$^2[1/2]^\circ$	1	*278 302*
$3p^5(^2P^\circ_{1/2})4s^2\,5d$	$^2[3/2]^\circ$	1	*274 040*	$3p^5(^2P^\circ_{1/2})4s^2\,11d$	$^2[3/2]^\circ$	1	*278 469*
$3p^5(^2P^\circ_{3/2})4s^2\,9s$	$^2[3/2]^\circ$	1	*274 276*	$3p^5(^2P^\circ_{1/2})4s^2\,13s$	$^2[1/2]^\circ$	1	*278 534*
$3p^5(^2P^\circ_{3/2})4s^2\,8d$	$^2[3/2]^\circ$	1	*274 652*	$3p^5(^2P^\circ_{1/2})4s^2\,12d$	$^2[3/2]^\circ$	1	*278 647*
$3p^5(^2P^\circ_{3/2})4s^2\,10s$	$^2[3/2]^\circ$	1	*274 777*	$3p^5(^2P^\circ_{1/2})4s^2\,14s$	$^2[1/2]^\circ$	1	*278 702*
$3p^5(^2P^\circ_{3/2})4s^2\,9d$	$^2[3/2]^\circ$	1	*275 272*	$3p^5(^2P^\circ_{1/2})4s^2\,13d$	$^2[3/2]^\circ$	1	*278 796*
$3p^5(^2P^\circ_{3/2})4s^2\,10d$	$^2[3/2]^\circ$	1	*275 503*	$3p^5(^2P^\circ_{1/2})4s^2\,14d$	$^2[3/2]^\circ$	1	*278 900*
$3p^5(^2P^\circ_{3/2})4s^2\,11d$	$^2[3/2]^\circ$	1	*275 726*	$3p^5(^2P^\circ_{1/2})4s^2\,15d$	$^2[3/2]^\circ$	1	*278 983*
$3p^5(^2P^\circ_{3/2})4s^2\,12d$	$^2[3/2]^\circ$	1	*275 898*	$3p^5(^2P^\circ_{1/2})4s^2\,16d$	$^2[3/2]^\circ$	1	*279 050*
$3p^5(^2P^\circ_{3/2})4s^2\,13d$	$^2[3/2]^\circ$	1	*276 038*	$3p^5(^2P^\circ_{1/2})4s^2\,17d$	$^2[3/2]^\circ$	1	*279 117*
$3p^5(^2P^\circ_{3/2})4s^2\,14d$	$^2[3/2]^\circ$	1	*276 138*	Ca II $3p^54s^2\,^2P^\circ_{1/2}$	***Limit***		**279 530**

Ca I: Ordered by series

Configuration	Term	*J*	Level (cm^{-1})
$4s^2$	^1S	0	0.000
4s4p	$^3P^\circ$	0	*15 157.901*
		1	*15 210.063*
		2	*15 315.943*
4s5p	$^3P^\circ$	0	*36 547.688*
		1	*36 554.749*
		2	*36 575.119*
4s6p	$^3P^\circ$	0	*42 514.845*
		1	*42 518.708*
		2	*42 526.591*
4s7p	$^3P^\circ$	0	*44 955.67*
		1	*44 957.655*
		2	*44 961.757*
4s8p	$^3P^\circ$	0	*46 284.12*
		1	*46 285.23*
		2	*46 287.63*
4s9p	$^3P^\circ$	1	*47 085.38*
		2	*47 086.99*
4s10p	$^3P^\circ$	1	*47 604.75*
		2	*47 605.77*
4s11p	$^3P^\circ$	1	*47 960.87*
		2	*47 961.53*
4s12p	$^3P^\circ$	1	*48 215.81*
		2	*48 216.36*
4s13p	$^3P^\circ$	1	*48 404.57*
		2	*48 404.95*
4s14p	$^3P^\circ$	1	*48 548.30*
		2	*48 548.51*
4s15p	$^3P^\circ$	2	*48 660.23*
4s16p	$^3P^\circ$	2	*48 749.04*
4s17p	$^3P^\circ$	2	*48 820.60*
4s18p	$^3P^\circ$	2	*48 879.31*
4s19p	$^3P^\circ$	2	*48 927.93*
4s20p	$^3P^\circ$	2	*48 968.67*
4s21p	$^3P^\circ$	2	*49 003.21*
4s22p	$^3P^\circ$	2	*49 032.70*
4s23p	$^3P^\circ$	2	*49 058.02*
4s24p	$^3P^\circ$	2	*49 080.04*
4s25p	$^3P^\circ$	2	*49 099.25*
4s26p	$^3P^\circ$	2	*49 116.07*
4s27p	$^3P^\circ$	2	*49 130.94*
4s28p	$^3P^\circ$	2	*49 144.19*
4s29p	$^3P^\circ$	2	*49 155.90*
4s30p	$^3P^\circ$	2	*49 166.43*
4s31p	$^3P^\circ$	2	*49 175.87*
4s32p	$^3P^\circ$	2	*49 184.37*
4s33p	$^3P^\circ$	2	*49 192.08*
4s34p	$^3P^\circ$	2	*49 199.09*
4s35p	$^3P^\circ$	2	*49 205.47*
4s36p	$^3P^\circ$	2	*49 211.34*
4s37p	$^3P^\circ$	2	*49 216.61*
4s38p	$^3P^\circ$	2	*49 221.51*
4s39p	$^3P^\circ$	2	*49 226.02*
4s40p	$^3P^\circ$	2	*49 230.13*
4s41p	$^3P^\circ$	2	*49 234.00*
4s42p	$^3P^\circ$	2	*49 237.55*
4s43p	$^3P^\circ$	2	*49 240.81*
4s44p	$^3P^\circ$	2	*49 243.87*
4s45p	$^3P^\circ$	2	*49 246.74*
4s46p	$^3P^\circ$	2	*49 249.37*
4s47p	$^3P^\circ$	2	*49 251.87*
4s48p	$^3P^\circ$	2	*49 254.22*
4s49p	$^3P^\circ$	2	*49 256.42*
4s50p	$^3P^\circ$	2	*49 258.45*
4s51p	$^3P^\circ$	2	*49 260.36*
4s52p	$^3P^\circ$	2	*49 262.15*
4s53p	$^3P^\circ$	2	*49 263.85*
4s54p	$^3P^\circ$	2	*49 265.44*
4s55p	$^3P^\circ$	2	*49 266.99*
4s56p	$^3P^\circ$	2	*49 268.40*
4s57p	$^3P^\circ$	2	*49 269.77*
4s58p	$^3P^\circ$	2	*49 271.01*
4s59p	$^3P^\circ$	2	*49 272.24*
4s60p	$^3P^\circ$	2	*49 273.37*
Ca II ($^2S_{1/2}$)	***Limit***		**49 305.95**
4s4p	$^1P^\circ$	1	*23 652.304*
4s5p	$^1P^\circ$	1	*36 731.615*
4s6p	$^1P^\circ$	1	*41 679.008*
4s7p	$^1P^\circ$	1	*45 425.358*
4s8p	$^1P^\circ$	1	*46 479.813*
4s9p	$^1P^\circ$	1	*47 184.370*
4s10p	$^1P^\circ$	1	*47 662.10*
4s11p	$^1P^\circ$	1	*47 997.49*
4s12p	$^1P^\circ$	1	*48 240.53*
4s13p	$^1P^\circ$	1	*48 422.09*
4s14p	$^1P^\circ$	1	*48 561.10*
4s15p	$^1P^\circ$	1	*48 669.83*
4s16p	$^1P^\circ$	1	*48 756.45*
4s17p	$^1P^\circ$	1	*48 826.54*
4s18p	$^1P^\circ$	1	*48 884.06*
4s19p	$^1P^\circ$	1	*48 931.82*
4s20p	$^1P^\circ$	1	*48 971.93*
4s21p	$^1P^\circ$	1	*49 005.92*
4s22p	$^1P^\circ$	1	*49 034.98*
4s23p	$^1P^\circ$	1	*49 060.02*
4s24p	$^1P^\circ$	1	*49 081.75*
4s25p	$^1P^\circ$	1	*49 100.72*
4s26p	$^1P^\circ$	1	*49 117.38*
4s27p	$^1P^\circ$	1	*49 132.09*
4s28p	$^1P^\circ$	1	*49 145.15*
4s29p	$^1P^\circ$	1	*49 156.78*
4s30p	$^1P^\circ$	1	*49 167.20*
4s31p	$^1P^\circ$	1	*49 176.58*
4s32p	$^1P^\circ$	1	*49 185.02*

Ca I: Ordered by series—Continued

Configuration	Term	J	Level (cm^{-1})
4s33p	$^1P^\circ$	1	*49 192.68*
4s34p	$^1P^\circ$	1	*49 199.62*
4s35p	$^1P^\circ$	1	*49 205.95*
4s36p	$^1P^\circ$	1	*49 211.72*
4s37p	$^1P^\circ$	1	*49 217.02*
4s38p	$^1P^\circ$	1	*49 221.87*
4s39p	$^1P^\circ$	1	*49 226.35*
4s40p	$^1P^\circ$	1	*49 230.45*
4s41p	$^1P^\circ$	1	*49 234.28*
4s42p	$^1P^\circ$	1	*49 237.82*
4s43p	$^1P^\circ$	1	*49 241.09*
4s44p	$^1P^\circ$	1	*49 244.13*
4s45p	$^1P^\circ$	1	*49 246.98*
4s46p	$^1P^\circ$	1	*49 249.61*
4s47p	$^1P^\circ$	1	*49 252.08*
4s48p	$^1P^\circ$	1	*49 254.41*
4s49p	$^1P^\circ$	1	*49 256.56*
4s50p	$^1P^\circ$	1	*49 258.60*
4s51p	$^1P^\circ$	1	*49 260.51*
4s52p	$^1P^\circ$	1	*49 262.30*
4s53p	$^1P^\circ$	1	*49 263.99*
4s54p	$^1P^\circ$	1	*49 265.59*
4s55p	$^1P^\circ$	1	*49 267.10*
4s56p	$^1P^\circ$	1	*49 268.52*
4s57p	$^1P^\circ$	1	*49 269.88*
4s58p	$^1P^\circ$	1	*49 271.14*
4s59p	$^1P^\circ$	1	*49 272.35*
4s60p	$^1P^\circ$	1	*49 273.50*
4s61p	$^1P^\circ$	1	*49 274.60*
4s62p	$^1P^\circ$	1	*49 275.62*
4s63p	$^1P^\circ$	1	*49 276.61*
4s64p	$^1P^\circ$	1	*49 277.55*
4s65p	$^1P^\circ$	1	*49 278.44*
4s66p	$^1P^\circ$	1	*49 279.30*
4s67p	$^1P^\circ$	1	*49 280.10*
4s68p	$^1P^\circ$	1	*49 280.88*
4s69p	$^1P^\circ$	1	*49 281.62*
4s70p	$^1P^\circ$	1	*49 282.34*
4s71p	$^1P^\circ$	1	*49 283.01*
4s72p	$^1P^\circ$	1	*49 283.67*
4s73p	$^1P^\circ$	1	*49 284.30*
4s74p	$^1P^\circ$	1	*49 284.88*
4s75p	$^1P^\circ$	1	*49 285.46*
4s76p	$^1P^\circ$	1	*49 286.00*
4s77p	$^1P^\circ$	1	*49 286.54*
4s78p	$^1P^\circ$	1	*49 287.05*
4s79p	$^1P^\circ$	1	*49 287.57*
Ca II ($^2S_{1/2}$)	***Limit***		**49 305.95**
4snp	$^1P^\circ$	1	*43 933.477*
....			
4s3d	^3D	1	20 335.360
		2	20 349.260
		3	20 371.000
4s4d	^3D	1	37 748.197
		2	37 751.867
		3	37 757.449
4s5d	^3D	1	42 743.002
		2	42 744.716
		3	42 747.387
4s6d	^3D	1	45 049.073
		2	45 050.419
		3	45 052.374
4s7d	^3D	1	46 301.973
		2	46 303.649
		3	46 306.059
4s8d	^3D	1	47 036.225
		2	47 040.007
		3	47 045.241
4s9d	^3D	1	47 752.655
		2	47 757.286
		3	47 765.697
4s10d	^3D	1	48 031.58
		2	48 033.23
		3	48 036.212
4s11d	^3D	1	48 258.30
		2	48 258.84
		3	48 260.29
4s12d	^3D	1	48 433.01
		2	48 433.46
		3	48 434.47
4s13d	^3D	1	48 568.66
		2	48 568.95
		3	48 569.51
4s14d	^3D	1	48 675.42
		2	48 675.65
		3	48 676.09
4s15d	^3D	1	48 760.88
		2	48 761.20
		3	48 761.31
4s16d	^3D	1	48 829.87
		2	48 829.95
		3	48 830.47
4s17d	^3D	1	48 886.90
		2	48 886.75
		3	48 887.04

(Continued)

Ca I: Ordered by series—Continued

Configuration	Term	J	Level (cm^{-1})
4s18*d*	^{3}D	1	48 934.10
		2	48 934.04
		3	48 934.32
4s19*d*	^{3}D	1	48 973.80
		2	48 973.95
		3	48 973.92
4s20*d*	^{3}D	1	49 007.59
		2	49 007.58
		3	49 007.70
4s21*d*	^{3}D	2	49 036.40
4s22*d*	^{3}D	2	49 061.23
4s23*d*	^{3}D	2	49 082.83
4s24*d*	^{3}D	2	49 101.51
4s25*d*	^{3}D	2	49 118.24
4s26*d*	^{3}D	2	49 132.77
4s27*d*	^{3}D	2	49 145.66
4s28*d*	^{3}D	2	49 157.39
4s29*d*	^{3}D	2	49 167.55
4s30*d*	^{3}D	2	49 177.13
4s31*d*	^{3}D	2	49 185.24
4s32*d*	^{3}D	2	49 193.06
4s33*d*	^{3}D	2	49 199.93
4s34*d*	^{3}D	2	49 206.17
4s35*d*	^{3}D	2	49 212.01
4s36*d*	^{3}D	2	49 217.18
4s37*d*	^{3}D	2	49 222.06
4s38*d*	^{3}D	2	49 226.56
4s39*d*	^{3}D	2	49 230.54
4s40*d*	^{3}D	2	49 234.47
4s41*d*	^{3}D	2	49 237.82
4s42*d*	^{3}D	2	49 241.09
4s43*d*	^{3}D	2	49 244.26
4s44*d*	^{3}D	2	49 247.04
4s45*d*	^{3}D	2	49 249.56
4s46*d*	^{3}D	2	49 252.10
4s47*d*	^{3}D	2	49 254.43
4s48*d*	^{3}D	2	49 256.55
4s49*d*	^{3}D	2	49 258.65
4s50*d*	^{3}D	2	49 260.64
4s51*d*	^{3}D	2	49 262.42
4s52*d*	^{3}D	3	49 264.19
4s53*d*	^{3}D	2	49 265.60
4s54*d*	^{3}D	2	49 267.18
4s55*d*	^{3}D	2	49 268.57
4s56*d*	^{3}D	2	49 270.01
4s57*d*	^{3}D	2	49 271.25
4s58*d*	^{3}D	2	49 272.49
4s59*d*	^{3}D	2	49 273.53
4s60*d*	^{3}D	2	49 274.52
Ca II ($^{2}S_{1/2}$)	***Limit***		**49 305.95**
4s3*d*	^{1}D	2	21 849.634
4s4*d*	^{1}D	2	37 298.287
4s5*d*	^{1}D	2	42 919.053
4s6*d*	^{1}D	2	44 989.830
4s7*d*	^{1}D	2	46 200.13
4s8*d*	^{1}D	2	46 948.98
4s9*d*	^{1}D	2	47 812.39
4s10*d*	^{1}D	2	48 083.41
4s11*d*	^{1}D	2	48 290.85
4s12*d*	^{1}D	2	48 451.73
4s13*d*	^{1}D	2	48 578.32
4s14*d*	^{1}D	2	48 678.97
4s15*d*	^{1}D	2	48 760.14
4s16*d*	^{1}D	2	48 827.05
4s17*d*	^{1}D	2	48 882.37
4s18*d*	^{1}D	2	48 928.79
4s19*d*	^{1}D	2	48 968.10
4s20*d*	^{1}D	2	49 001.66
4s21*d*	^{1}D	2	49 030.55
4s22*d*	^{1}D	2	49 055.62
4s23*d*	^{1}D	2	49 077.48
4s24*d*	^{1}D	2	49 096.52
4s25*d*	^{1}D	2	49 113.41
4s26*d*	^{1}D	2	49 128.34
4s27*d*	^{1}D	2	49 141.63
4s28*d*	^{1}D	2	49 153.49
4s29*d*	^{1}D	2	49 164.12
4s30*d*	^{1}D	2	49 173.70
4s31*d*	^{1}D	2	49 182.36
4s32*d*	^{1}D	2	49 190.17
4s33*d*	^{1}D	2	49 197.30
4s34*d*	^{1}D	2	49 203.78
4s35*d*	^{1}D	2	49 209.68
4s36*d*	^{1}D	2	49 215.11
4s37*d*	^{1}D	2	49 220.10
4s38*d*	^{1}D	2	49 224.70
4s39*d*	^{1}D	2	49 228.92
4s40*d*	^{1}D	2	49 232.81
4s41*d*	^{1}D	2	49 236.45
4s42*d*	^{1}D	2	49 239.80
4s43*d*	^{1}D	2	49 242.93
4s44*d*	^{1}D	2	49 245.80
4s45*d*	^{1}D	2	49 248.52
4s46*d*	^{1}D	2	49 251.05
4s47*d*	^{1}D	2	49 253.40
4s48*d*	^{1}D	2	49 255.64
4s49*d*	^{1}D	2	49 257.74
4s50*d*	^{1}D	2	49 259.68
4s51*d*	^{1}D	2	49 261.51
4s52*d*	^{1}D	2	49 263.31
4s53*d*	^{1}D	2	49 264.77
4s54*d*	^{1}D	2	49 266.36
4s55*d*	^{1}D	2	49 267.84
4s56*d*	^{1}D	2	49 269.33
4s57*d*	^{1}D	2	49 270.40
4s58*d*	^{1}D	2	49 271.79
4s59*d*	^{1}D	2	49 273.09
4s60*d*	^{1}D	2	49 274.29
4s61*d*	^{1}D	2	49 275.58
4s62*d*	^{1}D	2	49 275.87
Ca II ($^{2}S_{1/2}$)	***Limit***		**49 305.95**

Ca I: Ordered by series—Continued

Configuration	Term	J	Level (cm^{-1})
$4s5s$	^3S	1	31 539.495
$4s6s$	^3S	1	40 474.241
$4s7s$	^3S	1	43 980.767
$4s8s$	^3S	1	45 738.684
$4s9s$	^3S	1	46 748.283
$4s10s$	^3S	1	47 382.048
$4s11s$	^3S	1	47 806.20
$4s12s$	^3S	1	48 104.02
$4s13s$	^3S	1	48 321.14
$4s14s$	^3S	1	48 484.30
$4s15s$	^3S	1	48 609.75
$4s16s$	^3S	1	48 708.82
$4s17s$	^3S	1	48 788.03
$4s18s$	^3S	1	48 852.58
$4s19s$	^3S	1	48 905.60
$4s20s$	^3S	1	48 949.88
$4s21s$	^3S	1	48 987.35
$4s22s$	^3S	1	49 019.13
$4s23s$	^3S	1	49 046.22
$4s24s$	^3S	1	49 069.82
$4s25s$	^3S	1	49 090.40
$4s26s$	^3S	1	49 108.15
$4s27s$	^3S	1	49 124.02
$4s28s$	^3S	1	49 138.06
$4s29s$	^3S	1	49 150.30
$4s30s$	^3S	1	49 161.54
$4s31s$	^3S	1	49 171.42
$4s32s$	^3S	1	49 180.48
$4s33s$	^3S	1	49 188.37
$4s34s$	^3S	1	49 195.64
$4s35s$	^3S	1	49 202.47
$4s36s$	^3S	1	49 208.53
Ca II ($^2S_{1/2}$)	***Limit***		**49 305.95**
$4s5s$	^1S	0	33 317.264
$4s6s$	^1S	0	40 690.435
$4s7s$	^1S	0	44 276.538
$4s8s$	^1S	0	45 887.200
$4s9s$	^1S	0	46 835.055
$4s10s$	^1S	0	47 437.471
$4s11s$	^1S	0	47 843.76
$4s12s$	^1S	0	48 130.75
$4s13s$	^1S	0	48 340.75
$4s14s$	^1S	0	48 499.14
$4s15s$	^1S	0	48 621.53
$4s16s$	^1S	0	48 718.02
$4s17s$	^1S	0	48 795.46
$4s18s$	^1S	0	48 858.59
$4s19s$	^1S	0	48 910.65
$4s20s$	^1S	0	48 954.13
$4s21s$	^1S	0	48 990.83
$4s22s$	^1S	0	49 022.02
$4s23s$	^1S	0	49 048.85
$4s24s$	^1S	0	49 071.99
$4s25s$	^1S	0	49 092.19
$4s26s$	^1S	0	49 109.85
$4s27s$	^1S	0	49 125.44
$4s28s$	^1S	0	49 139.31
$4s29s$	^1S	0	49 151.59
$4s30s$	^1S	0	49 162.47
$4s31s$	^1S	0	49 172.43
$4s32s$	^1S	0	49 181.49
Ca II ($^2S_{1/2}$)	***Limit***		**49 305.95**
$3d\,4p$	$^3F^\circ$	2	*35 730.454*
		3	*35 818.713*
		4	*35 896.889*
$3d\,4p$	$^1D^\circ$	2	*35 835.413*
$3d\,4p$	$^3D^\circ$	1	*38 192.392*
		2	*38 219.118*
		3	*38 259.124*
$3d\,4p$	$^3P^\circ$	0	*39 333.382*
		1	*39 335.322*
		2	*39 340.080*
$3d\,4p$	$^1F^\circ$	3	*40 537.893*
$3d\,5p$	$^3D^\circ$	1	*51 709.5*
		2	*51 734.6*
		3	*51 767.0*
$3d\,5p$	$^3P^\circ$	1	*51 908.*
$3d\,5p$	$^1P^\circ$	1	*53 100.*
$3d\,6p$	$^3D^\circ$	1	*56 254.*
$3d\,7p$	$^3D^\circ$	1	*58 798.92*
$3d\,8p$	$^3D^\circ$	1	*60 046.13*
$3d\,9p$	$^3D^\circ$	1	*60 807.28*
$3d\,10p$	$^3D^\circ$	1	*61 306.65*
$3d\,11p$	$^3D^\circ$	1	*61 652.25*
$3d\,12p$	$^3D^\circ$	1	*61 901.04*
$3d\,13p$	$^3D^\circ$	1	*62 084.92*
$3d\,14p$	$^3D^\circ$	1	*62 223.68*
$3d\,15p$	$^3D^\circ$	1	*62 331.04*
$3d\,16p$	$^3D^\circ$	1	*62 416.47*
$3d\,17p$	$^3D^\circ$	1	*62 486.45*
$3d\,18p$	$^3D^\circ$	1	*62 539.39*
$3d\,19p$	$^3D^\circ$	1	*62 587.83*
$3d\,20p$	$^3D^\circ$	1	*62 626.03*
$3d\,21p$	$^3D^\circ$	1	*62 660.15*
$3d\,22p$	$^3D^\circ$	1	*62 686.88*
$3d\,23p$	$^3D^\circ$	1	*62 713.24*
$3d\,24p$	$^3D^\circ$	1	*62 734.25*
$3d\,25p$	$^3D^\circ$	1	*62 753.60*
$3d\,26p$	$^3D^\circ$	1	*62 769.47*
$3d\,27p$	$^3D^\circ$	1	*62 783.86*
$3d\,28p$	$^3D^\circ$	1	*62 797.46*
$3d\,29p$	$^3D^\circ$	1	*62 808.51*
$3d\,30p$	$^3D^\circ$	1	*62 818.76*
$3d\,31p$	$^3D^\circ$	1	*62 827.75*
$3d\,32p$	$^3D^\circ$	1	*62 836.22*
$3d\,33p$	$^3D^\circ$	1	*62 843.72*

(Continued)

Ca I: Ordered by series—Continued

Configuration	Term	J	Level (cm^{-1})
3d 34p	^{3}D°	1	62 850.62
3d 35p	^{3}D°	1	62 856.56
3d 36p	^{3}D°	1	62 862.60
3d 37p	^{3}D°	1	62 867.87
3d 38p	^{3}D°	1	62 872.69
3d 39p	^{3}D°	1	62 876.98
3d 40p	^{3}D°	1	62 881.24
3d 41p	^{3}D°	1	62 884.77
3d 42p	^{3}D°	1	62 889.00
3d 43p	^{3}D°	1	62 891.65
3d 44p	^{3}D°	1	62 894.65
3d 45p	^{3}D°	1	62 897.57
3d 46p	^{3}D°	1	62 900.09
3d 47p	^{3}D°	1	62 902.75
3d 48p	^{3}D°	1	62 904.95
3d 49p	^{3}D°	1	62 906.98
3d 50p	^{3}D°	1	62 908.88
3d 51p	^{3}D°	1	62 910.88
3d 52p	^{3}D°	1	62 912.67
3d 53p	^{3}D°	1	62 914.47
3d 54p	^{3}D°	1	62 915.91
3d 55p	^{3}D°	1	62 917.83
3d 56p	^{3}D°	1	62 918.80
3d 57p	^{3}D°	1	62 920.32
3d 58p	^{3}D°	1	62 921.43
Ca II (^{2}D$_{3/2}$)	*Limit*		**62 956.15**
3d 6p	^{3}P°	1	56 532.63
3d 7p	^{3}P°	1	59 010.87
3d 8p	^{3}P°	1	60 150.26
3d 9p	^{3}P°	1	60 869.28
3d 10p	^{3}P°	1	61 345.74
3d 11p	^{3}P°	1	61 676.62
3d 12p	^{3}P°	1	61 916.38
3d 13p	^{3}P°	1	62 097.08
3d 14p	^{3}P°	1	62 237.12
3d 15p	^{3}P°	1	62 351.93
3d 16p	^{3}P°	1	62 443.64
3d 17p	^{3}P°	1	62 511.97
3d 18p	^{3}P°	1	62 584.50
3d 19p	^{3}P°	1	62 628.76
3d 20p	^{3}P°	1	62 671.51
3d 21p	^{3}P°	1	62 711.37
3d 22p	^{3}P°	1	62 736.10
3d 23p	^{3}P°	1	62 762.86
3d 24p	^{3}P°	1	62 787.31
3d 25p	^{3}P°	1	62 807.90
3d 26p	^{3}P°	1	62 824.17
3d 27p	^{3}P°	1	62 838.45
3d 28p	^{3}P°	1	62 853.79
3d 29p	^{3}P°	1	62 863.99
3d 30p	^{3}P°	1	62 875.40
3d 31p	^{3}P°	1	62 884.77
3d 32p	^{3}P°	1	62 893.66
3d 33p	^{3}P°	1	62 901.85
Ca II (^{2}D$_{5/2}$)	*Limit*		**63 016.84**

Configuration	Term	J	Level (cm^{-1})
3d 6p	^{1}P°	1	56 651.
3d 7p	^{1}P°	1	59 197.
3d 8p	^{1}P°	1	60 300.
3d 9p	^{1}P°	1	60 973.6
3d 10p	^{1}P°	1	61 428.1
3d 11p	^{1}P°	1	61 747.2
3d 12p	^{1}P°	1	61 980.49
3d 13p	^{1}P°	1	62 154.62
3d 14p	^{1}P°	1	62 288.02
3d 15p	^{1}P°	1	62 392.01
3d 16p	^{1}P°	1	62 478.26
3d 17p	^{1}P°	1	62 548.16
3d 18p	^{1}P°	1	62 602.72
3d 19p	^{1}P°	1	62 648.37
3d 20p	^{1}P°	1	62 690.32
3d 21p	^{1}P°	1	62 721.80
3d 22p	^{1}P°	1	62 750.58
3d 23p	^{1}P°	1	62 774.75
3d 24p	^{1}P°	1	62 795.55
3d 25p	^{1}P°	1	62 814.13
3d 26p	^{1}P°	1	62 830.67
3d 27p	^{1}P°	1	62 845.27
3d 28p	^{1}P°	1	62 858.11
3d 29p	^{1}P°	1	62 869.38
3d 30p	^{1}P°	1	62 879.48
3d 31p	^{1}P°	1	62 889.00
3d 32p	^{1}P°	1	62 897.57
3d 33p	^{1}P°	1	62 904.95
3d 34p	^{1}P°	1	62 911.51
3d 35p	^{1}P°	1	62 917.83
3d 36p	^{1}P°	1	62 923.62
3d 37p	^{1}P°	1	62 928.81
3d 38p	^{1}P°	1	62 933.61
3d 39p	^{1}P°	1	62 938.02
3d 40p	^{1}P°	1	62 942.10
3d 41p	^{1}P°	1	62 945.88
3d 42p	^{1}P°	1	62 949.34
3d 43p	^{1}P°	1	62 952.58
3d 44p	^{1}P°	1	62 955.57
3d 45p	^{1}P°	1	62 958.38
3d 46p	^{1}P°	1	62 961.01
3d 47p	^{1}P°	1	62 963.43
3d 48p	^{1}P°	1	62 965.72
3d 49p	^{1}P°	1	62 967.84
3d 50p	^{1}P°	1	62 969.89
3d 51p	^{1}P°	1	62 971.78
3d 52p	^{1}P°	1	62 973.55
3d 53p	^{1}P°	1	62 975.23
3d 54p	^{1}P°	1	62 976.81
3d 55p	^{1}P°	1	62 978.26
3d 56p	^{1}P°	1	62 979.72
3d 57p	^{1}P°	1	62 981.03
3d 58p	^{1}P°	1	62 982.35
3d 59p	^{1}P°	1	62 983.48
3d 60p	^{1}P°	1	62 984.63
3d 61p	^{1}P°	1	62 985.70
Ca II (^{2}D$_{5/2}$)	*Limit*		**63 016.84**

Ca I: Ordered by series—Continued

Configuration	Term	J	Level (cm^{-1})
$4p^2$	^{3}P	0	38 417.543
		1	38 464.808
		2	38 551.558
$4p^2$	^{1}D	2	40 719.847
$4p^2$	^{1}S	0	41 786.276
4s4f	^{3}F°	2	*42 170.214*
		3	*42 170.558*
		4	*42 171.026*
4s5f	^{3}F°	2	*44 762.620*
		3	*44 762.839*
		4	*44 763.118*
4s6f	^{3}F°	2	*46 164.644*
		3	*46 164.785*
		4	*46 164.971*
4s7f	^{3}F°	2	*47 006.194*
		3	*47 006.280*
		4	*47 006.400*
4s8f	^{3}F°	2	*47 550.214*
		3	*47 550.271*
		4	*47 550.371*
4s9f	^{3}F°	2	*47 921.87*
		3	*47 921.981*
		4	*47 922.033*
4s10f	^{3}F°	2	*48 187.045*
		3	*48 187.075*
		4	*48 187.118*
4s11f	^{3}F°	2	*48 382.70*
		3	*48 382.781*
		4	*48 382.801*
4s12f	^{3}F°	3,4	*48 531.04*
4s13f	^{3}F°	4	*48 646.38*
Ca II (^{2}S$_{1/2}$)	**Limit**		**49 305.95**
4s4f	^{1}F°	3	*42 343.587*
4s5f	^{1}F°	3	*44 804.878*
4s6f	^{1}F°	3	*46 182.399*
4s7f	^{1}F°	3	*47 015.141*
4s8f	^{1}F°	3	*47 555.23*
4s9f	^{1}F°	3	*47 924.947*
4s10f	^{1}F°	3	*48 188.990*
4s11f	^{1}F°	3	*48 384.039*
4s12f	^{1}F°	3	*48 532.139*
4s13f	^{1}F°	3	*48 647.30*
4s14f	^{1}F°	3	*48 738.54*
4s15f	^{1}F°	3	*48 812.09*
4s16f	^{1}F°	3	*48 872.20*
4s17f	^{1}F°	3	*48 921.95*
4s18f	^{1}F°	3	*48 963.67*
4s19f	^{1}F°	3	*48 998.89*
4s20f	^{1}F°	3	*49 028.93*
4s21f	^{1}F°	3	*49 054.80*
4s22f	^{1}F°	3	*49 077.20*
4s23f	^{1}F°	3	*49 096.74*
4s24f	^{1}F°	3	*49 113.85*
4s25f	^{1}F°	3	*49 129.00*
4s26f	^{1}F°	3	*49 142.38*
4s27f	^{1}F°	3	*49 154.29*
4s28f	^{1}F°	3	*49 164.99*
Ca II (^{2}S$_{1/2}$)	**Limit**		**49 305.95**
4s5g	^{3}G	5	44 874.86
4s5g	^{1}G	4	44 875.95
4s10g	^{1}G	4	48 202.2
4s11g	^{1}G	4	48 394.3
4s12g	^{1}G	4	48 540.1
4s13g	^{1}G	4	48 653.7
4s14g	^{1}G	4	48 743.4
Ca II (^{2}S$_{1/2}$)	**Limit**		**49 305.95**
4s10h	^{1}H°	5	*48 208*
4s11h	^{1}H°	5	*48 398.6*
4s12h	^{1}H°	5	*48 544.2*
4s13h	^{1}H°	5	*48 657.0*
4s14h	^{1}H°	5	*48 747.4*
Ca II (^{2}S$_{1/2}$)	**Limit**		**49 305.95**
$3d^2$	^{3}F	2	43 474.827
		3	43 489.119
		4	43 508.088
$3d^2$	^{3}P	0	48 524.093
		1	48 537.623
		2	48 563.522
3d 5s	^{1}D	2	47 449.083
3d 5s	^{3}D	1	47 456.452
		2	47 466.014
		3	47 475.915
3d 4d	^{3}D	1	51 351.74
		2	51 369.38
		3	51 396.32
3d 4d	^{3}G	3	51 553.6?
		4	51 579.0?
		5	51 611.4?

(Continued)

Ca I: Ordered by series—Continued

Configuration	Term	J	Level (cm^{-1})
3d 4d	^{3}S	1	51 571.7
3d 4d	^{3}F	4	53 200.4?
		2	53 214.6?
		3	53 247.9?
3d 4d	^{3}P	0	54 282.3
		1	54 288.74
		2	54 304.6
3d 5d	^{3}D	2	56 469.0?
3d 5d	^{3}S	1	56 558.9
3d 5d	^{3}P	0	57 611.2
		1	57 617.9
		2	57 638.4
3d 6d	^{3}P	1	59 368.
		2	59 391.
....			
3d4f	^{3}D°	1	*55 902.8*
3d5f	^{3}D°	1	*58 431.31*
3d6f	^{3}D°	1	*59 802.21*
3d7f	^{3}D°	1	*60 632.18*
3d8f	^{3}D°	1	*61 172.54*
3d9f	^{3}D°	1	*61 543.95*
3d 10f	^{3}D°	1	*61 810.68*
3d 11f	^{3}D°	1	*62 009.17*
3d 12f	^{3}D°	1	*62 161.61*
3d 13f	^{3}D°	1	*62 279.33*
3d 14f	^{3}D°	1	*62 375.29*
3d 15f	^{3}D°	1	*62 453.85*
3d 16f	^{3}D°	1	*62 520.06*
3d 17f	^{3}D°	1	*62 565.68*
3d 18f	^{3}D°	1	*62 610.75*
3d 19f	^{3}D°	1	*62 641.52*
3d 20f	^{3}D°	1	*62 675.56*
3d 21f	^{3}D°	1	*62 704.43*
3d 22f	^{3}D°	1	*62 720.41*
3d 23f	^{3}D°	1	*62 743.11*
3d 24f	^{3}D°	1	*62 768.10*
3d 26f	^{3}D°	1	*62 790.08*
3d 28f	^{3}D°	1	*62 812.00*
3d 30f	^{3}D°	1	*62 832.31*
Ca II (^{2}D$_{3/2}$)	***Limit***		**62 956.15**
3d4f	^{3}P°	1	*55 946.6*
3d5f	^{3}P°	1	*58 491.91*
3d6f	^{3}P°	1	*59 862.26*
3d7f	^{3}P°	1	*60 690.27*
3d8f	^{3}P°	1	*61 228.65*
3d9f	^{3}P°	1	*61 596.69*
3d 10f	^{3}P°	1	*61 859.22*
3d 11f	^{3}P°	1	*62 052.49*
3d 12f	^{3}P°	1	*62 198.92*
3d 13f	^{3}P°	1	*62 312.63*
3d 14f	^{3}P°	1	*62 403.22*
3d 15f	^{3}P°	1	*62 472.14*
3d 16f	^{3}P°	1	*62 533.90*
3d 17f	^{3}P°	1	*62 578.91*
3d 18f	^{3}P°	1	*62 617.02*
3d 19f	^{3}P°	1	*62 653.99*
3d 20f	^{3}P°	1	*62 683.89*
3d 21f	^{3}P°	1	*62 707.25*
3d 22f	^{3}P°	1	*62 729.19*
3d 23f	^{3}P°	1	*62 748.53*
3d 24f	^{3}P°	1	*62 764.10*
3d 25f	^{3}P°	1	*62 780.30*
3d 26f	^{3}P°	1	*62 794.19*
3d 27f	^{3}P°	1	*62 806.02*
3d 28f	^{3}P°	1	*62 816.51*
3d 29f	^{3}P°	1	*62 824.17*
3d 30f	^{3}P°	1	*62 834.11*
3d 31f	^{3}P°	1	*62 840.23*
3d 32f	^{3}P°	1	*62 848.77*
3d 33f	^{3}P°	1	*62 853.79*
3d 34f	^{3}P°	1	*62 861.06*
3d 35f	^{3}P°	1	*62 865.39*
3d 36f	^{3}P°	1	*62 871.46*
3d 37f	^{3}P°	1	*62 875.40*
3d 38f	^{3}P°	1	*62 880.04*
Ca II (^{2}D$_{3/2}$)	***Limit***		**62 956.15**
3d4f	^{1}P°	1	*55 982.3*
3d5f	^{1}P°	1	*58 505.89*
3d6f	^{1}P°	1	*59 878.52*
3d7f	^{1}P°	1	*60 709.54*
3d8f	^{1}P°	1	*61 251.39*
3d9f	^{1}P°	1	*61 622.60*
3d 10f	^{1}P°	1	*61 888.22*
3d 11f	^{1}P°	1	*62 084.92*
3d 12f	^{1}P°	1	*62 234.47*
3d 13f	^{1}P°	1	*62 346.47*
3d 14f	^{1}P°	1	*62 433.38*
3d 15f	^{1}P°	1	*62 506.62*
3d 16f	^{1}P°	1	*62 572.66*
3d 17f	^{1}P°	1	*62 625.37*
3d 18f	^{1}P°	1	*62 671.51*
3d 19f	^{1}P°	1	*62 704.43*
3d 20f	^{1}P°	1	*62 736.10*
3d 21f	^{1}P°	1	*62 762.86*
3d 22f	^{1}P°	1	*62 785.02*
3d 23f	^{1}P°	1	*62 804.38*
3d 24f	^{1}P°	1	*62 820.31*
3d 25f	^{1}P°	1	*62 836.85*
3d 26f	^{1}P°	1	*62 852.33*
3d 35f	^{1}P°	1	*62 926.36*
3d 36f	^{1}P°	1	*62 931.29*
3d 37f	^{1}P°	1	*62 935.88*
3d 38f	^{1}P°	1	*62 939.79*
3d 39f	^{1}P°	1	*62 943.85*
Ca II (^{2}D$_{5/2}$)	***Limit***		**63 016.84**

Ca I: Ordered by series—Continued

Configuration	Term	J	Level (cm^{-1})
$4p5s$	$^3P^\circ$	1	*57 462.*
$4p6s$	$^3P^\circ$	1	*64 977.*
$4p7s$	$^3P^\circ$	1	*68 913.*
$4p8s$	$^3P^\circ$	1	*70 787.*
$4p9s$	$^3P^\circ$	1	*71 868.*
$4p10s$	$^3P^\circ$	1	*72 411.*
$4p11s$	$^3P^\circ$	1	*72 871.*
$4p12s$	$^3P^\circ$	1	*73 248.*
$4p13s$	$^3P^\circ$	1	*73 444.*
$4p14s$	$^3P^\circ$	1	*73 622.*
$4p15s$	$^3P^\circ$	1	*73 742.*
$4p16s$	$^3P^\circ$	1	*73 852.*
Ca II ($^2P_{1/2}$)	***Limit***		**74 497.46**
$4p5s$	$^1P^\circ$	1	*57 960.*
$4p6s$	$^1P^\circ$	1	*65 445.*
$4p7s$	$^1P^\circ$	1	*69 247.*
$4p8s$	$^1P^\circ$	1	*71 001.*
$4p9s$	$^1P^\circ$	1	*72 007.*
$4p10s$	$^1P^\circ$	1	*72 655.*
$4p11s$	$^1P^\circ$	1	*73 111.*
$4p12s$	$^1P^\circ$	1	*73 444.*
$4p13s$	$^1P^\circ$	1	*73 693.*
$4p14s$	$^1P^\circ$	1	*73 852.*
$4p15s$	$^1P^\circ$	1	*73 969.*
Ca II ($^2P_{3/2}$)	***Limit***		**74 720.35**
$4p4d$	$^3D^\circ$?	1	*62 552.97*
$4p4d$	$^3P^\circ$?	1	*62 557.45*
....			
$4p4d$	$^1P^\circ$?	1	*62 561.72*
$3p5d$	$^1P^\circ$	1	*66 578.*
$4p6d$	$^1P^\circ$	1	*69 643.*
$4p7d$	$^1P^\circ$	1	*71 237.*
$4p8d$	$^1P^\circ$	1	*72 289.*
$4p9d$	$^1P^\circ$	1	*72 871.*
$4p10d$	$^1P^\circ$	1	*73 248.*
$4p11d$	$^1P^\circ$	1	*73 517.*
$4p12d$	$^1P^\circ$	1	*73 742.*
Ca II ($^2P_{3/2}$)	***Limit***		**74 720.35**
$3p^5\,4s^2\,3d$	$^3P^\circ$	1	*200 096*
$3p^5\,4s^2\,3d$	$^3D^\circ$	1	*218 991*
$3p^5\,4s^2\,3d$	$^1P^\circ$	1	*253 310*
$3p^5\,4s^2\,4d$	$^3P^\circ$	1	*255 022*
$3p^5\,4s^2\,4d$	$^3D^\circ$	1	*267 956*
$3p^5\,4s^2\,4d$	$^1P^\circ$	1	*270 640*
....			

Configuration	Term	J	Level (cm^{-1})
$3p^5(^2P^\circ_{3/2})4s^2\,5s$	$^2[3/2]^\circ$	1	*257 737*
$3p^5(^2P^\circ_{3/2})4s^2\,6s$	$^2[3/2]^\circ$	1	*267 417*
$3p^5(^2P^\circ_{3/2})4s^2\,7s$	$^2[3/2]^\circ$	1	*271 245*
$3p^5(^2P^\circ_{3/2})4s^2\,9s$	$^2[3/2]^\circ$	1	*274 276*
$3p^5(^2P^\circ_{3/2})4s^2\,10s$	$^2[3/2]^\circ$	1	*274 777*
Ca II $3p^54s^2\,^2P^\circ_{3/2}$	***Limit***		**276 750**
$3p^5(^2P^\circ_{1/2})4s^2\,5s$	$^2[1/2]^\circ$	1	*260 193*
$3p^5(^2P^\circ_{1/2})4s^2\,6s$	$^2[1/2]^\circ$	1	*269 706*
$3p^5(^2P^\circ_{1/2})4s^2\,7s$	$^2[1/2]^\circ$	1	*273 598*
$3p^5(^2P^\circ_{1/2})4s^2\,11s$	$^2[1/2]^\circ$	1	*277 991*
$3p^5(^2P^\circ_{1/2})4s^2\,12s$	$^2[1/2]^\circ$	1	*278 302*
$3p^5(^2P^\circ_{1/2})4s^2\,13s$	$^2[1/2]^\circ$	1	*278 534*
$3p^5(^2P^\circ_{1/2})4s^2\,14s$	$^2[1/2]^\circ$	1	*278 702*
Ca II $3p^54s^2\,^2P^\circ_{1/2}$	***Limit***		**279 530**
$3p^5(^2P^\circ_{3/2})4s^2\,5d$	$^2[3/2]^\circ$	1	*270 889*
$3p^5(^2P^\circ_{3/2})4s^2\,6d$	$^2[3/2]^\circ$	1	*273 134*
$3p^5(^2P^\circ_{3/2})4s^2\,7d$	$^2[3/2]^\circ$	1	*273 959*
$3p^5(^2P^\circ_{3/2})4s^2\,8d$	$^2[3/2]^\circ$	1	*274 652*
$3p^5(^2P^\circ_{3/2})4s^2\,9d$	$^2[3/2]^\circ$	1	*275 272*
$3p^5(^2P^\circ_{3/2})4s^2\,10d$	$^2[3/2]^\circ$	1	*275 503*
$3p^5(^2P^\circ_{3/2})4s^2\,11d$	$^2[3/2]^\circ$	1	*275 726*
$3p^5(^2P^\circ_{3/2})4s^2\,12d$	$^2[3/2]^\circ$	1	*275 898*
$3p^5(^2P^\circ_{3/2})4s^2\,13d$	$^2[3/2]^\circ$	1	*276 038*
$3p^5(^2P^\circ_{3/2})4s^2\,14d$	$^2[3/2]^\circ$	1	*276 138*
$3p^5(^2P^\circ_{3/2})4s^2\,15d$	$^2[3/2]^\circ$	1	*276 214*
$3p^5(^2P^\circ_{3/2})4s^2\,16d$	$^2[3/2]^\circ$	1	*276 283*
Ca II $3p^54s^2\,^2P^\circ_{3/2}$	***Limit***		**276 750**
$3p^5(^2P^\circ_{1/2})4s^2\,5d$	$^2[3/2]^\circ$	1	*274 040*
$3p^5(^2P^\circ_{1/2})4s^2\,7d$	$^2[3/2]^\circ$	1	*276 631*
$3p^5(^2P^\circ_{1/2})4s^2\,8d$	$^2[3/2]^\circ$	1	*277 430*
$3p^5(^2P^\circ_{1/2})4s^2\,9d$	$^2[3/2]^\circ$	1	*277 917*
$3p^5(^2P^\circ_{1/2})4s^2\,10d$	$^2[3/2]^\circ$	1	*278 228*
$3p^5(^2P^\circ_{1/2})4s^2\,11d$	$^2[3/2]^\circ$	1	*278 469*
$3p^5(^2P^\circ_{1/2})4s^2\,12d$	$^2[3/2]^\circ$	1	*278 647*
$3p^5(^2P^\circ_{1/2})4s^2\,13d$	$^2[3/2]^\circ$	1	*278 796*
$3p^5(^2P^\circ_{1/2})4s^2\,14d$	$^2[3/2]^\circ$	1	*278 900*
$3p^5(^2P^\circ_{1/2})4s^2\,15d$	$^2[3/2]^\circ$	1	*278 983*
$3p^5(^2P^\circ_{1/2})4s^2\,16d$	$^2[3/2]^\circ$	1	*279 050*
$3p^5(^2P^\circ_{1/2})4s^2\,17d$	$^2[3/2]^\circ$	1	*279 117*
Ca II $3p^54s^2\,^2P^\circ_{1/2}$	***Limit***		**279 530**

Ca II

$Z=20$

K I isoelectronic sequence

Ground state: $1s^22s^22p^63s^23p^64s\ ^2S_{1/2}$

Ionization energy $=$ 95 751.87±0.03 cm^{-1} (11.87181±0.00002 eV)

The strong resonance doublet of this spectrum, the Fraunhofer H and K lines, is used to measure the velocity of recession of galaxies in the farthest reaches of space and thereby provides us with the dimensions and age of the universe.

The analysis in its present state was nearly completed by Saunders and Russell (1925). The spectrum was reobserved between 3000 and 12 000 Å by Edlén and Risberg (1956), who recalculated the level values from their new grating measurements and the interferometric measurements of Wagman (1937) and found the $5g$, $9s$, and $10s$ terms. G. Risberg (1968) added $10d$, $8f$, $8h$, and $10h$ and revised the $6p\ ^2P^\circ_{1/2}$ level. The uncertainty of these levels is estimated to be ±0.05 cm^{-1}.

The $9-10f$ and $11-16d$ terms are derived from observations of Shenstone (1949) in the region 2890–3220 Å. Their uncertainty is probably ±0.5 cm^{-1}.

Edlén and Risberg derived the quoted ionization energy from the ng series.

The $3p^54s^2\ ^2P^\circ$ term position was deduced from the $3p^54s^2ns$ and nd series in Ca I by Mansfield and Newsom (1977).

References

Edlén, B., and Risberg, P. (1956), Ark. Fys. **10**, 553.

Mansfield, M. W. D., and Newsom, G. H. (1977), Proc. R. Soc. London, Ser. A **357**, 77.

Risberg, G. (1968), Ark. Fys. **37**, 231.

Saunders, F. A., and Russell, H. N. (1925), Astrophys. J. **62**, 1.

Shenstone, A. G. (1949), published in Atomic Energy Levels, Vol. I Natl. Bur. Stand. Circ. 467, Washington, D.C., 1949).

Wagman, N. E. (1937), Univ. Pitts. Bull. **34**, 325.

Ca II

Configuration	Term	J	Level (cm^{-1})
$3p^6(^1S)4s$	^2S	1/2	0.00
$3p^6(^1S)3d$	^2D	3/2	13 650.19
		5/2	13 710.88
$3p^6(^1S)4p$	$^2P^\circ$	1/2	*25 191.51*
		3/2	*25 414.40*
$3p^6(^1S)5s$	^2S	1/2	52 166.93
$3p^6(^1S)4d$	^2D	3/2	56 839.25
		5/2	56 858.46
$3p^6(^1S)5p$	$^2P^\circ$	1/2	*60 533.02*
		3/2	*60 611.28*
$3p^6(^1S)4f$	$^2F^\circ$	5/2,7/2	*68 056.91*
$3p^6(^1S)6s$	^2S	1/2	70 677.62
$3p^6(^1S)5d$	^2D	3/2	72 722.23
		5/2	72 730.93
$3p^6(^1S)6p$	$^2P^\circ$	1/2	*74 484.92*
		3/2	*74 521.75*
$3p^6(^1S)5f$	$^2F^\circ$	5/2,7/2	*78 034.39*
$3p^6(^1S)5g$	^2G	7/2,9/2	78 164.72
$3p^6(^1S)7s$	^2S	1/2	79 448.28
$3p^6(^1S)6d$	^2D	3/2	80 521.53
		5/2	80 526.16
$3p^6(^1S)6f$	$^2F^\circ$	5/2,7/2	*83 458.08*
$3p^6(^1S)6g$	^2G	7/2,9/2	83 540.00
$3p^6(^1S)8s$	^2S	1/2	84 300.89
$3p^6(^1S)7d$	^2D	3/2	84 933.65
		5/2	84 936.41
$3p^6(^1S)7f$	$^2F^\circ$	5/2,7/2	*86 727.06*
$3p^6(^1S)7g$	^2G	7/2,9/2	86 781.14
$3p^6(^1S)9s$	^2S	1/2	87 267.86
$3p^6(^1S)8d$	^2D	3/2	87 671.93
		5/2	87 673.72
$3p^6(^1S)8f$	$^2F^\circ$	5/2,7/2	*88 847.31*
$3p^6(^1S)8g$	^2G	7/2,9/2	88 884.54
$3p^6(^1S)8h$	$^2H^\circ$	9/2,11/2	*88 890.64*

Ca II—Continued

Configuration	Term	J	Level (cm^{-1})	Configuration	Term	J	Level (cm^{-1})
$3p^6(^1S)10s$	^2S	1/2	89 214.13	$3p^6(^1S)12d$	^2D	3/2,5/2	92 359.0
$3p^6(^1S)9d$	^2D	3/2 5/2	89 487.93 89 489.13	$3p^6(^1S)13d$	^2D	3/2,5/2	92 883.0
				$3p^6(^1S)14d$	^2D	3/2,5/2	93 297.6
$3p^6(^1S)9f$	$^2F°$	5/2,7/2	*90 300.0*	$3p^6(^1S)15d$	^2D	3/2,5/2	93 626.9
$3p^6(^1S)9g$	^2G	7/2,9/2	90 326.45	$3p^6(^1S)16d$	^2D	3/2,5/2	93 894.5
$3p^6(^1S)10d$	^2D	3/2 5/2	90 753.92 90 754.80	Ca III ($^1S_0$)	***Limit***		**95 751.87**
$3p^6(^1S)10f$	$^2F°$	5/2,7/2	*91 338.0*	$3p^5\,4s^2$	$^2P°$	3/2 1/2	*227 444* *230 224*
$3p^6(^1S)10h$	$^2H°$	9/2,11/2	*91 361.00*				
$3p^6(^1S)11d$	^2D	3/2,5/2	91 672.0				

Ca III

$Z = 20$

Ar I isoelectronic sequence

Ground state: $1s^2 2s^2 2p^6 3s^2 3p^6\ {}^1S_0$

Ionization energy = 410 642±2 cm^{-1} (50.9135±0.0002 eV)

The first observations and analysis of Ca III were carried out by Bowen (1928). He measured the spectrum in the range 400–4000 Å and determined levels in the configurations $3p^5 3d$, $4d$, $4s$, $4p$, and $5s$. A much more extensive analysis has since been completed by Borgström (1968, 1971), who has remeasured the spectrum from 440–9640 Å. His designations and level values are given here. The uncertainty in the values given with two decimal places is ±0.1 Å, and those with one decimal place, ±0.5 Å. He has also determined the ionization energy from the ng series.

Hansen, Persson, and Borgström (1975) found the $3s3p^6 3d$ configuration, which strongly interacts with the $3p^5 nf$ series, and added two levels to $3p^5 6p$. They reported the percentage compositions given here for the $3s^2 3p^5 5p$, $6p$, $4f$, $5f$, $6f$, and $3s3p^6 3d$ configurations with configuration interaction. The percentages for $3p^5 4p$ in *LS* coupling are from Borgström (1971).

The two ${}^1P^\circ_1$ levels from $3s3p^6 4p$ and $5p$ were determined by Kastner, Crooker, Behring, and Cohen (1977).

Schmitz, Breuckmann, and Mehlhorn (1976) reported the discovery of $3p^4 4s^2$ but their identification has been shown by Pejcev, Ottley, Rassi, and Ross (1978) to be incorrect.

References

Borgström, A. (1968), Ark. Fys. **38**, 243.

Borgström, A. (1971), Phys. Scr. **3**, 157.

Bowen, I. S. (1928), Phys. Rev. **31**, 497.

Hansen, J. E., Persson, W., and Borgström, A. (1975), Phys. Scr. **11**, 31.

Kastner, S. O., Crooker, A. M., Behring, W. E., and Cohen, L. (1977), Phys. Rev. **16A**, 577.

Pejcev, Y., Ottley, T. W., Rassi, D., and Ross, K. J. (1978), J. Phys. **B11**, 531.

Schmitz, W., Breuckmann, B., and Mehlhorn, W. (1976), J. Phys. **B9**, L493.

Ca III

Configuration	Term	*J*	Level (cm^{-1})	Leading percentages
$3s^2 3p^6$	${}^1S$	0	0.00	
$3s^2 3p^5 3d$	${}^3P^\circ$	0	*203 373.22*	
		1	*203 851.95*	
		2	*204 842.64*	
$3s^2 3p^5 3d$	${}^3F^\circ$	4	*212 310.04*	
		3	*213 379.40*	
		2	*214 334.06*	
$3s^2 3p^5 3d$	${}^1D^\circ$	2	*225 826.22*	
$3s^2 3p^5 3d$	${}^3D^\circ$	3	*226 333.56*	
		2	*227 388.56*	
		1	*227 432.11*	
$3s^2 3p^5 3d$	${}^1F^\circ$	3	*228 413.95*	
$3s^2 3p^5 4s$	${}^3P^\circ$	2	*242 547.19*	
		1	*243 930.44*	
		0	*245 611.88*	
$3s^2 3p^5 4s$	${}^1P^\circ$	1	*247 696.39*	
$3s^2 3p^5 4p$	${}^3S$	1	272 188.70	98

Ca III—Continued

Configuration	Term	J	Level (cm^{-1})	Leading percentages		
$3s^2\,3p^5\,4p$	^{3}D	3	277 022.40	100		
		2	277 380.86	77		
		1	278 621.01	83		
$3s^2\,3p^5\,3d$	^{1}P°	1	*279 353.64*			
$3s^2\,3p^5\,4p$	^{1}D	2	279 741.80	57		
$3s^2\,3p^5\,4p$	^{1}P	1	281 139.55	57		
$3s^2\,3p^5\,4p$	^{3}P	2	281 882.24	63		
		0	282 075.15	98		
		1	282 571.43	64		
$3s^2\,3p^5\,4p$	^{1}S	0	290 934.30	98		
$3s^2\,3p^5\,4d$	^{3}P°	0	*322 663.28*			
		1	*323 003.56*			
		2	*323 655.06*			
$3s^2\,3p^5\,4d$	^{3}F°	4	*324 110.24*			
		3	*324 660.47*			
		2	*325 467.67*			
$3s^2\,3p^5\,4d$	^{1}F°	3	*326 186.32*			
$3s^2\,3p^5(^2P^\circ_{3/2})5s$	$^2[3/2]^\circ$	2	*327 922.87*			
		1	*328 582.45*			
$3s^2\,3p^5\,4d$	^{3}D°	1	*327 962.11*			
		3	*328 588.76*			
		2	*328 606.78*			
$3s^2\,3p^5\,4d$	^{1}D°	2	*328 090.99*			
$3s^2\,3p^5(^2P^\circ_{1/2})5s$	$^2[1/2]^\circ$	0	*331 048.86*			
		1	*331 403.20*			
$3s^2\,3p^5\,4d$	^{1}P°	1	*336 749.11*			
$3s^2\,3p^5(^2P^\circ_{3/2})5p$	$^2[1/2]$	1	339 198.09	91	9	$(^2P^\circ_{1/2})\ ^2[1/2]$
		0	343 110.24	75	25	
$3s^2\,3p^5(^2P^\circ_{3/2})5p$	$^2[5/2]$	3	340 580.15	100		
		2	340 748.72	90	8	$(^2P^\circ_{3/2})\ ^2[3/2]$
$3s^2\,3p^5(^2P^\circ_{3/2})5p$	$^2[3/2]$	1	341 349.19	96	4	$(^2P^\circ_{1/2})\ ^2[3/2]$
		2	341 601.46	90	9	$(^2P^\circ_{3/2})\ ^2[5/2]$
$3s^2\,3p^5(^2P^\circ_{1/2})5p$	$^2[3/2]$	1	343 784.96	95	4	$(^2P^\circ_{3/2})\ ^2[3/2]$
		2	344 149.81	97		
$3s^2\,3p^5(^2P^\circ_{1/2})5p$	$^2[1/2]$	1	344 257.92	90	9	$(^2P^\circ_{3/2})\ ^2[1/2]$
		0	346 692.34	75	25	
$3s^2\,3p^5(^2P^\circ_{3/2})4f$	$^2[3/2]$	1	346 732.19	99		
		2	346 896.26	90	8	$(^2P^\circ_{3/2})\ ^2[5/2]$
$3s^2\,3p^5(^2P^\circ_{3/2})4f$	$^2[9/2]$	5	347 344.37	100		
		4	347 383.51	100		

(Continued)

Ca III—Continued

Configuration	Term	J	Level (cm^{-1})	Leading percentages		
$3s^2\,3p^5(^2P^\circ_{3/2})4f$	$^2[5/2]$	3	347 417.05	93	6	$(^2P^\circ_{1/2})\ ^2[5/2]$
		2	347 758.22	88	9	$(^2P^\circ_{3/2})\ ^2[3/2]$
$3s^2\,3p^5(^2P^\circ_{3/2})4f$	$^2[7/2]$	3	348 028.13	97		
		4	348 051.86	97		
$3s^2\,3p^5(^2P^\circ_{1/2})4f$	$^2[5/2]$	3	350 741.19	93	7	$(^2P^\circ_{3/2})\ ^2[7/2]$
		2	350 900.04	96	4	$(^2P^\circ_{3/2})\ ^2[5/2]$
$3s^2\,3p^5(^2P^\circ_{1/2})4f$	$^2[7/2]$	3	350 779.47	97		
		4	350 805.32	98		
$3s^2\,3p^5(^2P^\circ_{3/2})5d$	$^2[1/2]^\circ$	0	*358 940.91*			
		1	*359 156.83*			
$3s^2\,3p^5(^2P^\circ_{3/2})5d$	$^2[3/2]^\circ$	2	*359 520.60*			
		1	*361 794.08*			
$3s^2\,3p^5(^2P^\circ_{3/2})5d$	$^2[7/2]^\circ$	4	*359 543.08*			
		3	*359 800.57*			
$3s^2\,3p^5(^2P^\circ_{3/2})5d$	$^2[5/2]^\circ$	2	*360 207.16*			
		3	*360 388.68*			
$3s^2\,3p^5(^2P^\circ_{3/2})6s$	$^2[3/2]^\circ$	2	*361 154.07*			
		1	*361 353.31*			
$3s^2\,3p^5(^2P^\circ_{1/2})5d$	$^2[5/2]^\circ$	2	*362 893.47*			
		3	*363 118.75*			
$3s^2\,3p^5(^2P^\circ_{1/2})5d$	$^2[3/2]^\circ$	2	*363 021.27*			
		1	*365 363.88*			
$3s^2\,3p^5(^2P^\circ_{1/2})6s$	$^2[1/2]^\circ$	0	*364 229.89*			
		1	*364 343.85*			
$3s3p^6\,3d$	^3D	1	366 778.85	65	31	$3p^5(^2P^\circ_{3/2})5f\,^2[3/2]$
		2	366 926.33	55	17	$3p^5(^2P^\circ_{3/2})6p\ ^2[5/2]$
		3	367 472.91	53	24	$3p^5(^2P^\circ_{3/2})6p\ ^2[5/2]$
$3s^2\,3p^5(^2P^\circ_{3/2})6p$	$^2[5/2]$	3	367 026.65	75	18	$3s3p^6 3d\ ^3D$
$3s^2\,3p^5(^2P^\circ_{3/2})5f$	$^2[9/2]$	5	370 141.77	100		
		4	370 172.43	100		
$3s^2\,3p^5(^2P^\circ_{3/2})5f$	$^2[5/2]$	2	370 304.54?	58	34	$3p^5(^2P^\circ_{3/2})5f\,^2[3/2]$
		3	371 059.95?	80	17	$3s3p^6 3d\ ^3D$
$3s^2\,3p^5(^2P^\circ_{1/2})6p$	$^2[3/2]$	2	370 454.04	93	6	$3p^5(^2P^\circ_{3/2})5f\,^2[5/2]$
$3s^2\,3p^5(^2P^\circ_{3/2})5f$	$^2[7/2]$	3	370 515.36	99		
		4	370 533.07	99		
$3s^2\,3p^5(^2P^\circ_{3/2})5g$	$^2[5/2]^\circ$	2	*370 901.88*			
		3	*370 903.12*			
$3s^2\,3p^5(^2P^\circ_{3/2})5g$	$^2[11/2]^\circ$	6	*370 957.52*			
		5	*370 957.65*			

Ca III—Continued

Configuration	Term	J	Level (cm^{-1})	Leading percentages		
$3s^2\,3p^5(^2P^\circ_{3/2})5g$	$^2[7/2]^\circ$	3	*371 061.30*			
		4	*371 061.37*			
$3s^2\,3p^5(^2P^\circ_{3/2})5g$	$^2[9/2]^\circ$	4	*371 120.69*			
		5	*371 121.06*			
$3s^2\,3p^5(^2P^\circ_{3/2})5f$	$^2[3/2]$	2	371 274.48?	42	29	$3p^5(^2P^\circ_{3/2})5f\ ^2[5/2]$
		1	371 447.58?	67	31	$3s3p^6 3d\ ^3D$
$3s^2\,3p^5(^2P^\circ_{1/2})5f$	$^2[7/2]$	3	373 401.87	99		
		4	373 425.51	99		
$3s^2\,3p^5(^2P^\circ_{1/2})5f$	$^2[5/2]$	2	373 626.85	92	4	$3s3p^6 3d\ ^3D$
		3	373 880.37	91	9	
$3s^2\,3p^5(^2P^\circ_{1/2})5g$	$^2[9/2]^\circ$	4	*374 138.90*			
		5	*374 139.31*			
$3s^2\,3p^5(^2P^\circ_{1/2})5g$	$^2[7/2]^\circ$	4	*374 143.44*			
		3	*374 143.84*			
$3s^2\,3p^5(^2P^\circ_{3/2})6d$	$^2[7/2]^\circ$	4	*376 808.60*			
		3	*376 883.36*			
$3s3p^6\,3d$	^1D	2	377 168.1	93		
$3s^2\,3p^5(^2P^\circ_{3/2})6d$	$^2[5/2]^\circ$	2	*380 152.21*			
		3	*380 230.50*			
$3s^2\,3p^5(^2P^\circ_{3/2})6g$	$^2[9/2]^\circ$	5	*382 190.20*			
		4	*383 189.82*			
$3s^2\,3p^5(^2P^\circ_{3/2})6f$	$^2[9/2]$	5	382 565.1	100		
		4	382 587.9	100		
$3s^2\,3p^5(^2P^\circ_{3/2})6f$	$^2[7/2]$	3	382 784.7	100		
		4	382 798.5	99		
$3s^2\,3p^5(^2P^\circ_{3/2})6f$	$^2[5/2]$	3	382 791.5	99		
		2	382 852.3	79	19	$(^2P^\circ_{3/2})\ ^2[3/2]$
$3s^2\,3p^5(^2P^\circ_{3/2})6g$	$^2[5/2]^\circ$	2	*383 061.33*			
		3	*383 063.78*			
$3s^2\,3p^5(^2P^\circ_{3/2})6g$	$^2[11/2]^\circ$	6	*383 094.79*			
		5	*383 095.10*			
$3s^2\,3p^5(^2P^\circ_{3/2})6g$	$^2[7/2]^\circ$	4	*383 156.51*			
		3	*383 157.11*			
$3s^2\,3p^5(^2P^\circ_{1/2})6f$	$^2[7/2]$	3	385 757.6	100		
		4	385 775.5	100		
$3s^2\,3p^5(^2P^\circ_{1/2})6f$	$^2[5/2]$	3	385 867.2	99		
		2	385 906.9	99		
$3s^2\,3p^5(^2P^\circ_{1/2})6g$	$^2[9/2]^\circ$	4	*386 248.39*			
		5	*386 248.68*			

(Continued)

Ca III—Continued

Configuration	Term	J	Level (cm^{-1})	Leading percentages
$3s^2 3p^5(^2P^\circ_{1/2})6g$	$^2[7/2]^\circ$	4	*386 249.78*	
		3	*386 250.31*	
$3s^2 3p^5(^2P^\circ_{3/2})7f$	$^2[9/2]$	5	390 054.0	
$3s^2 3p^5(^2P^\circ_{3/2})7f$	$^2[7/2]$	4	390 207.6	
$3s^2 3p^5(^2P^\circ_{3/2})7g$	$^2[5/2]^\circ$	3	*390 392.62*	
$3s^2 3p^5(^2P^\circ_{3/2})7g$	$^2[11/2]^\circ$	6	*390 411.58*	
		5	*390 411.99*	
$3s^2 3p^5(^2P^\circ_{3/2})7g$	$^2[7/2]^\circ$	3	*390 451.97*	
$3s^2 3p^5(^2P^\circ_{3/2})7g$	$^2[9/2]^\circ$	4	*390 471.79*	
		5	*390 472.32*	
$3s^2 3p^5(^2P^\circ_{1/2})7f$	$^2[7/2]$	4	393 224.9	
$3s^2 3p^5(^2P^\circ_{1/2})7g$	$^2[9/2]^\circ$	5	*393 551.86*	
		4	*393 561.59*	
$3s^2 3p^5(^2P^\circ_{1/2})7g$	$^2[7/2]^\circ$	4	*393 552.31*	
		3	*393 552.81*	
Ca IV ($^2P^\circ_{3/2}$)	***Limit***		**410 642**	
$3s3p^6 4p$	$^1P^\circ$	1	*431 100*	
$3s3p^6 5p$	$^1P^\circ$	1	*492 850*	

Ca IV

$Z=20$

Cl I isoelectronic sequence

Ground state: $1s^2 2s^2 2p^6 3s^2 3p^5\ ^2P^\circ_{3/2}$

Ionization energy = 542 600±1000 cm^{-1} (67.27±0.13 eV)

The initial work on the analysis was by Bowen (1928), who identified the ground term, the $3s3p^6\ ^2S$ level, and the $3p^4(^3P)4s\ ^2P_{3/2}$ level. Kruger and Phillips (1937) also identified the $3p^4 4s$ terms from their observations below 350 Å.

Levels of $3p^4 3d\ ^4D$, ^2D, and ^2F terms and the $3p^4 4p$ and $5s$ levels were reported by Tsien (1939), who worked with the line list of Ekefors (1931). Svensson and Ekberg (1968), also using Ekefors' list, established the $3p^4(^3P)3d\ ^2P$ term.

A new analysis based on a new set of observations was provided by Smitt (1978), whose work is still in progress. He estimates an uncertainty of ±2 cm^{-1} for the levels relative to the ground term and less than ±1 cm^{-1} for the ground term splitting. The new analysis provides three times the number of previously known levels as well as some changes of designation and J-values. All levels compiled here are from this work except for the $3p^4 5s\ ^2D$ term retained from Tsien.

The ionization energy was obtained by extrapolation by Lotz (1967).

References

Bowen, I. S. (1928), Phys. Rev. **31**, 497.
Ekefors, E. (1931), Z. Phys. **71**, 53.
Kruger, P. G., and Phillips, L. W. (1937), Phys. Rev. **51**, 1087.
Lotz, W. (1967), J. Opt. Soc. Am. **57**, 873.
Smitt, R. (1978), Private communication.
Svensson, L. A., and Ekberg, J. O. (1968), Ark. Fys. **37**, 65.
Tsien, W. Z. (1939), Chinese J. Phys. **3**, 117.

Ca IV

Configuration	Term	J	Level (cm^{-1})
$3s^2 3p^5$	$^2P^\circ$	3/2	*0.0*
		1/2	*3 118.2*
$3s3p^6$	^2S	1/2	152 439.6
$3s^2 3p^4(^3P)3d$	^4D	7/2	201 505
		5/2	201 747
		3/2	202 031
		1/2	202 254
$3s^2 3p^4(^3P)3d$	^4F	9/2	218 383
		7/2	219 467
		5/2	220 240
		3/2	220 741
$3s^2 3p^4(^1D)3d$	^2P	1/2	219 991
		3/2	221 945
$3s^2 3p^4(^3P)3d$	^4P	1/2	227 214
		3/2	227 825
		5/2	228 694
$3s^2 3p^4(^1D)3d$	^2D	3/2	228 436
		5/2	230 119
$3s^2 3p^4(^3P)3d$	^2F	7/2	231 288
		5/2	233 851
$3s^2 3p^4(^1D)3d$	^2G	9/2	234 498
		7/2	234 642
$3s^2 3p^4(^1D)3d$	^2F	5/2	252 286
		7/2	252 903
$3s^2 3p^4(^1S)3d$	^2D	3/2	269 974
		5/2	270 532
$3s^2 3p^4(^3P)4s$	^4P	5/2	291 456
		3/2	292 864
		1/2	294 292
$3s^2 3p^4(^1D)3d$	^2S	1/2	293 009
$3s^2 3p^4(^3P)3d$	^2P	3/2	293 872
		1/2	295 133
$3s^2 3p^4(^3P)3d$	^2D	5/2	301 218
		3/2	303 850
$3s^2 3p^4(^3P)4s$	^2P	3/2	301 718
		1/2	303 604
$3s^2 3p^4(^1D)4s$	^2D	5/2	312 517
		3/2	312 650
$3s^2 3p^4(^3P)4p$	$^4P^\circ$	5/2	*330 693*
		3/2	*331 173*
		1/2	*331 969*

(Continued)

Ca IV—Continued

Configuration	Term	J	Level (cm^{-1})
$3s^2\,3p^4(^3\mathrm{P})4p$	$^4\mathrm{D}^\circ$	7/2	*335 122*
		5/2	*335 901*
		3/2	*336 958*
		1/2	*337 452*
$3s^2\,3p^4(^1\mathrm{S})4s$	$^2\mathrm{S}$	1/2	337 214
$3s^2\,3p^4(^3\mathrm{P})4p$	$^2\mathrm{D}^\circ$	5/2	*338 250*
		3/2	*340 377*
$3s^2\,3p^4(^3\mathrm{P})4p$	$^2\mathrm{P}^\circ$	1/2	*338 300*
		3/2	*338 959*
$3s^2\,3p^4(^3\mathrm{P})4p$	$^2\mathrm{S}^\circ$	1/2	*342 567*
$3s^2\,3p^4(^3\mathrm{P})4p$	$^4\mathrm{S}^\circ$	3/2	*342 915*
$3s^2\,3p^4(^1\mathrm{D})4p$	$^2\mathrm{F}^\circ$	5/2	*352 154*
		7/2	*352 616*
$3s^2\,3p^4(^1\mathrm{D})4p$	$^2\mathrm{D}^\circ$	3/2	*357 941*
		5/2	*358 306*
$3s^2\,3p^4(^1\mathrm{D})4p$	$^2\mathrm{P}^\circ$	3/2	*359 575*
		1/2	*361 170*
$3s3p^5(^3\mathrm{P}^\circ)3d$	$^4\mathrm{F}^\circ$	9/2	*367 174*
		7/2	*367 971*
		5/2	*368 660*
		3/2	*369 197*
$3s^2\,3p^4(^1\mathrm{S})4p$	$^2\mathrm{P}^\circ$	1/2	*379 774*
		3/2	*380 043*
$3s3p^5(^3\mathrm{P}^\circ)3d$	$^4\mathrm{D}^\circ$	7/2	*382 641*
		5/2	*382 974*
		3/2	*383 125*
		1/2	*383 176*
$3s^2\,3p^4(^1\mathrm{D})5s$	$^2\mathrm{D}$	5/2	399 755
		3/2	400 956
Ca V ($^3\mathrm{P}_2$)	***Limit***		**542 600**

Ca V

$Z=20$

S I isoelectronic sequence

Ground state: $1s^2 2s^2 2p^6 3s^2 3p^4\ {}^3P_2$

Ionization energy = 681 600±1300 cm^{-1} (84.50±0.17 eV)

Spectra of calcium from 1035–135 Å were obtained by Ekefors (1931) and were supplemented by longer exposures below 600 Å by Bowen (1934). From the combined line-lists Bowen derived levels of the configurations $3s^2 3p^4$, $3s3p^5$, $3s^2 3p^3 3d$, $3s^2 3p^3 4s$, and $3s^2 3p^3 5s$. By means of isoelectronic comparisons Svensson and Ekberg (1968) revised and extended the $3p^3 3d$ configuration.

New measurements of the transition array $3s^2 3p^4 - 3s3p^5$ by Smitt, Svensson, and Outred (1976) led to improved values for the levels of these configurations, with an uncertainty reported as less than 1 cm^{-1}. The uncertainty of the rest of the level values is ±5 cm^{-1}.

The ionization energy is an extrapolated value by Lotz (1967).

References

Bowen, I. S. (1934), Phys. Rev. **46**, 791.
Ekefors, E. (1931), Z. Phys. **71**, 53.
Lotz, W. (1967), J. Opt. Soc. Am. **57**, 873.
Smitt, R., Svensson, L. A., and Outred, M. (1976), Phys. Scr. **13**, 293.
Svensson, L. A., and Ekberg, J. O. (1968), Ark. Fys. **37**, 65.

Ca V

Configuration	Term	J	Level (cm^{-1})
$3s^2 3p^4$	^{3}P	2	0.0
		1	2 404.7
		0	3 275.6
$3s^2 3p^4$	^{1}D	2	18 830.3
$3s^2 3p^4$	^{1}S	0	43 836.5
$3s3p^5$	^{3}P°	2	*154 670.8*
		1	*156 760.2*
		0	*157 900.5*
$3s3p^5$	^{1}P°	1	*197 844.5*
$3s^2 3p^3(^2$D°$)3d$	^{1}D°	2	*254 124*
$3s^2 3p^3(^2$D°$)3d$	^{1}F°	3	*283 955*
$3s^2 3p^3(^2$D°$)3d$	^{3}S°	1	*293 785*
$3s^2 3p^3(^2$P°$)3d$	^{3}P°	2	*298 214*
		1	*299 534*
		0	*300 594*
$3s^2 3p^3(^2$D°$)3d$	^{1}P°	1	*302 184*
$3s^2 3p^3(^2$P°$)3d$	^{3}D°	3	*308 188*
		2	*309 831*
		1	*310 943*
$3s^2 3p^3(^2$P°$)3d$	^{1}D°	2	*318 741*
$3s^2 3p^3(^2$P°$)3d$	^{1}F°	3	*329 229*
$3s^2 3p^3(^4$S°$)4s$	^{3}S°	1	*350 914*
$3s^2 3p^3(^2$P°$)3d$	^{1}P°	1	*353 220*
$3s^2 3p^3(^2$D°$)4s$	^{3}D°	1	*369 590*
		2	*369 696*
		3	*369 959*
$3s^2 3p^3(^2$D°$)4s$	^{1}D°	2	*374 728*
$3s^2 3p^3(^2$P°$)4s$	^{3}P°	0	*387 039*
		1	*387 226*
		2	*387 652*
$3s^2 3p^3(^2$P°$)4s$	^{1}P°	1	*392 283*
$3s^2 3p^3(^4$S°$)5s$	^{3}S°	1	*501 127*
$3s^2 3p^3(^2$D°$)5s$	^{3}D°	1	*524 651*
		2	*524 770*
		3	*525 053*
$3s^2 3p^3(^2$D°$)5s$	^{1}D°	2	*526 523*
$3s^2 3p^3(^2$P°$)5s$	^{3}P°	1	*542 249*
		2	*542 650*
$3s^2 3p^3(^2$P°$)5s$	^{1}P°	1	*544 143*
Ca VI (^{4}S$^\circ_{3/2}$)	*Limit*		**681 600**

Ca VI

$Z = 20$

P I isoelectronic sequence

Ground state: $1s^2 2s^2 2p^6 3s^2 3p^3\ ^4S^\circ_{3/2}$

Ionization energy $= 877\ 400 \pm 2000\ \text{cm}^{-1}$ (108.78 ± 0.20 eV)

The present compilation is obtained from the work of Ekberg and Svensson (1970) and Smitt, Svensson, and Outred (1976). The level values for the $3s^2 3p^3$ and $3s3p^4$ configurations are taken from the latter paper. They have an uncertainty of about $\pm 2\ \text{cm}^{-1}$. We have combined these values with the measurements and classifications given by Ekberg and Svensson to derive new level values for the $3p^2 3d$ and $4s$ configurations. The uncertainty of these upper levels is about $\pm 10\ \text{cm}^{-1}$. Since no intersystem transitions have been observed, all of the doublets have an added systematic error x, relative to the ground term $^4S^\circ$. The value of x depends on the accuracy of calculations by Smitt, Svensson and Outred and is expected to fall within $\pm 20\ \text{cm}^{-1}$.

Most of the wavelengths used by Ekberg and Svensson are taken from Bowen (1934) or Ekefors (1931).

The ionization energy is from an extrapolation by Lotz (1967).

References

Bowen, I. S. (1934), Phys. Rev. **46**, 791.
Ekberg, J. O., and Svensson, L. A. (1970), Phys. Scr. **2**, 283.
Ekefors, E. (1931), Z. Phys. **71**, 53.
Lotz, W. (1967), J. Opt. Soc. Am. **57**, 873.
Smitt, R., Svensson, L. A., and Outred, M. (1976), Phys. Scr. **13**, 293.

Ca VI

Configuration	Term	J	Level (cm^{-1})
$3s^2 3p^3$	$^4S^\circ$	3/2	0.0
$3s^2 3p^3$	$^2D^\circ$	3/2	26 835.1+x
		5/2	27 246.6+x
$3s^2 3p^3$	$^2P^\circ$	1/2	44 586.7+x
		3/2	45 142.7+x
$3s3p^4$	^4P	5/2	155 786.5
		3/2	157 767.5
		1/2	158 830.5
$3s3p^4$	^2D	3/2	193 247.0+x
		5/2	193 444.8+x
$3s3p^4$	^2P	3/2	222 749.7+x
		1/2	224 773.3+x
$3s3p^4$	^2S	1/2	233 712.8+x
$3s^2 3p^2(^3P)3d$	^2P	3/2	294 630+x
		1/2	297 083+x
$3s^2 3p^2(^1D)3d$	^2D	5/2	320 919+x
		3/2	321 411+x
$3s^2 3p^2(^1D)3d$	^2P	1/2	331 968+x
		3/2	333 324+x
$3s^2 3p^2(^1D)3d$	^2F	5/2	336 219+x
		7/2	336 631+x
$3s^2 3p^2(^1S)3d$	^2D	5/2	348 819+x
		3/2	349 645+x
$3s^2 3p^2(^3P)4s$	^4P	1/2	433 849
		3/2	435 286
		5/2	437 392
$3s^2 3p^2(^3P)4s$	^2P	1/2	442 256+x
		3/2	444 724+x
$3s^2 3p^2(^1D)4s$	^2D	5/2	457 294+x
		3/2	457 358+x
$3s^2 3p^2(^1S)4s$	^2S	1/2	483 882+x
Ca VII ($^3P_0$)	*Limit*		877 400

Ca VII

$Z=20$

Si I isoelectronic sequence

Ground state: $1s^2 2s^2 2p^6 3s^2 3p^2\ ^3P_0$

Ionization energy = 1 026 000±2000 cm^{-1} (127.2±0.2 eV)

The level values for $3s^2 3p^2$ and $3s3p^3$ are taken from Smitt, Svensson, and Outred (1976) who report a level uncertainty of ±2 cm^{-1}. Ekberg and Svensson (1970) revised and extended the interpretation of the $3p^2-3p3d$, $4s$ arrays published by several earlier workers. We have combined these new classifications with the $3p^2$ levels given by Smitt et al. The uncertainty for the levels of $3p3d$ and $3p4s$ is ±10 cm^{-1}. Ekberg and Svensson have obtained the value for the ionization energy by extrapolation along the isoelectronic sequence.

References

Ekberg, J. O., and Svensson, L. A. (1970), Phys. Scr. **2**, 283.

Smitt, R., Svensson, L. A., and Outred, M. (1976), Phys. Scr. **13**, 293.

Ca VII

Configuration	Term	J	Level (cm^{-1})	Configuration	Term	J	Level (cm^{-1})
$3s^2 3p^2$	^3P	0	0.0	$3s^2 3p3d$	$^3P^\circ$	2	*286 224*
		1	1 624.9			1	*288 160*
		2	4 071.4			0	*289 004*
$3s^2 3p^2$	^1D	2	21 864.0	$3s^2 3p3d$	$^3D^\circ$	1	*295 138*
						2	*295 772*
$3s^2 3p^2$	^1S	0	48 981.4			3	*296 132*
$3s3p^3$	$^3D^\circ$	1	*160 157.5*	$3s^2 3p3d$	$^1F^\circ$	3	*324 885*
		2	*160 220.3*				
		3	*160 529.2*	$3s^2 3p3d$	$^1P^\circ$	1	*333 501*
$3s3p^3$	$^3P^\circ$	0	*185 356.6*	$3s^2 3p4s$	$^3P^\circ$	0	*490 059*
		1	*185 392.9*			1	*490 919*
		2	*185 412.2*			2	*494 262*
$3s3p^3$	$^1D^\circ$	2	*203 616.1*	$3s^2 3p4s$	$^1P^\circ$	1	*498 683*
$3s3p^3$	$^3S^\circ$	1	*245 240.5*	Ca VIII ($^2P^\circ_{1/2}$)	***Limit***		**1 026 000**
$3s3p^3$	$^1P^\circ$	1	*252 489.9*				

Ca VIII

$Z=20$

Al I isoelectronic sequence

Ground state: $1s^2 2s^2 2p^6 3s^2 3p\ ^2P^\circ_{1/2}$

Ionization energy = 1 187 600±1000 cm^{-1} (147.24±0.12 eV)

The doublet terms of $3s^2 3p$ and $3s3p^2$ are from Smitt, Svensson, and Outred (1976) who report a level uncertainty of 2 cm^{-1}.

The remaining terms are derived from the measurements and classifications of Ekberg and Svensson (1970). They obtained the position of the $3s3p^2\ ^4P$ term by extrapolation of the known data from Al I to Ar VI. Apart from the uncertainty in this extrapolation, the level uncertainty is ±10 cm^{-1}.

The ionization energy was determined by Ekberg and Svensson from the $nf\ ^2F^\circ$ series.

References

Ekberg, J. O., and Svensson, L. A. (1970), Phys. Scr. **2**, 283.

Smitt, R., Svensson, L. A., and Outred, M. (1976), Phys. Scr. **13**, 293.

Ca VIII

Configuration	Term	J	Level (cm^{-1})	Configuration	Term	J	Level (cm^{-1})
$3s^2 3p$	$^2P^\circ$	1/2	*0.0*	$3s^2 4s$	^2S	1/2	547 322
		3/2	*4 308.3*				
				$3s3p4s$	$^4P^\circ$	1/2	*688 747+x*
$3s3p^2$	^4P	1/2	129 100+x			3/2	*690 128+x*
		3/2	130 678+x			5/2	*692 833+x*
		5/2	133 042+x				
				$3s^2 4d$	^2D	3/2	698 232
$3s3p^2$	^2D	3/2	171 572.2			5/2	698 420
		5/2	171 830.7				
				$3s^2 4f$	$^2F^\circ$	5/2	*743 288*
$3s3p^2$	^2S	1/2	216 584.9			7/2	*743 330*
$3s3p^2$	^2P	1/2	231 016.3	$3s^2 5d$	^2D	3/2	885 693
		3/2	233 592.8			5/2	885 750
$3s^2 3d$	^2D	3/2	282 356	$3s^2 5f$	$^2F^\circ$	5/2	*905 052*
		5/2	282 577			7/2	*905 087*
$3p^3$	$^4S^\circ$	3/2	*345 274+x*	$3s^2 6d$	^2D	3/2	979 749
						5/2	980 089
$3s3p3d$	$^4P^\circ$	5/2	*408 227+x*	$3s^2 6f$	$^2F^\circ$	5/2	*991 023*
		3/2	*409 291+x*			7/2	*991 028*
$3s3p3d$	$^4D^\circ$	1/2	*411 816+x*	$3s^2 7f$	$^2F^\circ$	7/2	*1 043 207*
		3/2	*412 388+x*			5/2	*1 043 275*
		5/2	*412 772+x*				
		7/2	*412 881+x*	Ca IX ($^1S_0$)	*Limit*		1 187 600

Ca IX

$Z=20$

Mg I isoelectronic sequence

Ground state: $1s^2s^22p^63s^2\ ^1S_0$

Ionization energy = 1 520 640±500 cm^{-1} (188.54±0.02 eV)

Most of the levels for this spectrum are taken from the publication by Ekberg (1971) who estimates a level uncertainty of ±10 cm^{-1}. The identification of the $3s5f\ ^1F_3^\circ$ has been revised by Edlén and Bodén (1976).

The $3p^2\ ^1S$ term and the $3p3d$ levels are from Fawcett (1970). The $3p4f$ levels are from Fawcett (1976).

No intersystem transition has been observed. An interpolated value for the $3s^2\ ^1S_0-3s3p\ ^3P_1^\circ$ resonance line of Ca IX of 691.2 Å was given by Finkenthal et al. (1982), who observed this transition in a tokamak plasma for Sc X, Ti XI, V XII, Cr XIII, and Fe XV. The uncertainty for Ca IX is probably ±0.5 Å, or ±100 cm^{-1} for the $^3P_1^\circ$ level.

The ionization energy was calculated from the first three members of the $3snf\ ^3F^\circ$ series ($n=4$, 5, and 6).

References

Edlén, B., and Bodén, E. (1976), Phys. Scr. **14**, 31.

Ekberg, J. O. (1971), Phys. Scr. **4**, 101.

Fawcett, B. C. (1970), J. Phys. **B3**, 1732.

Fawcett, B. C. (1976), J. Opt. Soc. Am. **66**, 632.

Finkenthal, M. Hinnov, E., Cohen, S., and Suckewer, S. (1982), Phys. Lett. **91**, 284.

Ca IX

Configuration	Term	J	Level (cm^{-1})	Configuration	Term	J	Level (cm^{-1})
$3s^2$	^1S	0	0	$3s4s$	^1S	0	774 480
$3s3p$	$^3P^\circ$	0	*143 176+x*	$3s4p$	$^1P^\circ$	1	*832 314*
		1	*144 675+x*				
		2	*147 912+x*	$3s4d$	^3D	1	917 200+x
						2	917 314+x
$3s3p$	$^1P^\circ$	1	*214 482*			3	917 528+x
$3p^2$	^1D	2	336 245	$3s4d$	^1D	2	921 921
$3p^2$	^3P	0	339 963+x	$3p4s$	$^3P^\circ$	0	*941 471+x*
		1	341 872+x			1	*942 658+x*
		2	345 472+x			2	*946 378+x*
$3p^2$	^1S	0	398 900	$3s4f$	$^3F^\circ$	2–4	*954 594+x*
$3s3d$	^3D	1	412 078+x	$3s4f$	$^1F^\circ$	3	*963 050*
		2	412 191+x				
		3	412 405+x	$3p4p$	^3D	2	1 005 234+x
						3	1 008 574+x
$3s3d$	^1D	2	467 631				
				$3p4p$	^3P	0	1 010 894+x
$3p3d$	$^3F^\circ$	2	*563 714+x*			1	1 012 034+x
		3	*565 724+x*			2	1 014 384+x
		4	*568 194+x*				
				$3p4p$	^3S	1	1 015 624+x
$3p3d$	$^1D^\circ$	2	*571 900*				
				$3s5s$	^3S	1	1 068 804+x
$3p3d$	$^3D^\circ$	2	*601 204+x*				
		3	*602 704+x*	$3s5s$	^1S	0	1 076 110
$3p3d$	$^1P^\circ$	1	*618 520*	$3p4d$	$^3D^\circ$	1	*1 090 824+x*
$3s4s$	^3S	1	760 538+x	$3s5p$	$^1P^\circ$	1	*1 097 570*

(Continued)

Ca IX—Continued

Configuration	Term	J	Level (cm^{-1})
$3p4f$	^3G	5	1 127 004+x
$3p4f$	^3F	3	1 127 184+x
		4	1 130 494+x
$3s5d$	^1D	2	1 139 810
$3s5d$	^3D	1	1 145 234+x
		3	1 145 854+x
$3s5f$	$^3F^\circ$	2–4	*1 159 674*+x
$3s5f$	$^1F^\circ$	3	*1 162 610*
$3s6s$	^3S	1	1 219 784+x
$3s6p$	$^1P^\circ$	1	*1 235 830*
$3s6d$	^3D	1	1 259 644+x
		2	1 259 824+x
		3	1 260 016+x
$3s6d$	^1D	2	1 260 390
$3s6f$	$^3F^\circ$	2–4	*1 270 614*+x
$3s7p$	$^1P^\circ$	1	*1 315 300*
$3s7d$	^3D	1–3	1 331 324+x
$3s8d$	^3D	1–3	1 376 394+x
Ca X ($^2S_{1/2}$)	***Limit***		**1 520 640**

Ca X

$Z=20$

Na I isoelectronic sequence

Ground state: $1s^2 2s^2 2p^6 3s\ ^2S_{1/2}$

Ionization energy = 1 704 047±50 cm^{-1} (211.277±0.006 eV)

The publications by Edlén and Bodén (1976), by Fawcett (1976), and by Cohen and Behring (1976) provide considerable extensions of the early work of Kruger and Phillips (1939) on this spectrum. We have quoted level values from Edlén and Bodén and added the higher *nf* series members ($n=10$–11) from Cohen and Behring. The uncertainty in these level values is ±50 cm^{-1}. The measurements of the high members of the *nf* series by Fawcett do not agree well with the Ritz formulae given by Edlén and Bodén.

The value for the ionization energy was derived by Edlén (1978) from core polarization theory applied to the *nf* series.

References

Cohen, L., and Behring, W. E. (1976), J. Opt. Soc. Am. **66**, 899.
Edlén, B. (1978), Phys. Scr. **17**, 565.
Edlén, B., and Bodén, E. (1976), Phys. Scr. **14**, 31.
Fawcett, B. C. (1976), J. Opt. Soc. Am. **66**, 632.
Kruger, P. G., and Phillips, L. W. (1939), Phys. Rev. **55**, 352.

Ca X

Configuration	Term	J	Level (cm^{-1})
$2p^6(^1S)3s$	^2S	1/2	0
$2p^6(^1S)3p$	$^2P°$	1/2 3/2	*174 213* *179 287*
$2p^6(^1S)3d$	^2D	3/2 5/2	417 112 417 522
$2p^6(^1S)4s$	^2S	1/2	832 790
$2p^6(^1S)4p$	$^2P°$	1/2 3/2	*899 290* *901 200*
$2p^6(^1S)4d$	^2D	3/2 5/2	987 300 987 490
$2p^6(^1S)4f$	$^2F°$	5/2 7/2	*1 016 100* *1 016 150*
$2p^6(^1S)5s$	^2S	1/2	1 174 710
$2p^6(^1S)5p$	$^2P°$	1/2 3/2	*1 206 850* *1 207 760*
$2p^6(^1S)5d$	^2D	3/2 5/2	1 248 920 1 249 030
$2p^6(^1S)5f$	$^2F°$	5/2 7/2	*1 263 690* *1 263 720*
$2p^6(^1S)6s$	^2S	1/2	1 348 380
$2p^6(^1S)6p$	$^2P°$	1/2 3/2	*1 366 360* *1 366 890*
$2p^6(^1S)6d$	^2D	3/2 5/2	1 389 840 1 389 870
$2p^6(^1S)6f$	$^2F°$	5/2 7/2	*1 398 330* *1 398 440*
$2p^6(^1S)7s$	^2S	1/2	1 448 710
$2p^6(^1S)7p$	$^2P°$	1/2,3/2	*1 459 920*
$2p^6(^1S)7d$	^2D	3/2 5/2	1 474 040 1 474 090
$2p^6(^1S)7f$	$^2F°$	5/2 7/2	*1 479 470* *1 479 540*
$2p^6(^1S)8s$	^2S	1/2	1 511 780
$2p^6(^1S)8p$	$^2P°$	1/2,3/2	*1 519 200*
$2p^6(^1S)8d$	^2D	3/2 5/2	1 528 490 1 528 510
$2p^6(^1S)8f$	$^2F°$	5/2 7/2	*1 532 290* *1 532 390*
$2p^6(^1S)9p$	$^2P°$	1/2,3/2	*1 559 260*
$2p^6(^1S)9d$	^2D	3/2,5/2	1 565 730
$2p^6(^1S)9f$	$^2F°$	5/2 7/2	*1 568 390* *1 568 420*
$2p^6(^1S)10d$	^2D	5/2	1 592 260

(Continued)

Ca X—Continued

Configuration	Term	J	Level (cm^{-1})	Configuration	Term	J	Level (cm^{-1})
$2p^6(^1S)10f$	$^2F°$	$7/2$	*1 594 230*				
		$5/2$	*1 594 330*				
$2p^6(^1S)11f$	$^2F°$	$7/2$	*1 613 480*				
		$5/2$	*1 613 600*				
Ca XI ($^1S_0$)	***Limit***		**1 704 047**				

Ca XI

$Z=20$

Ne I isoelectronic sequence

Ground state: $1s^2 2s^2 2p^6\ {}^1S_0$

Ionization energy = 4 774 000±3000cm^{-1} (591.9±0.4 eV)

Only resonance lines between 25 and 36 Å are classified by this rare-gas-type system of energy levels. Edlén and Tyrén (1936) identified 11 transitions and derived a value for the ionization potential by extrapolation, which agrees well with the present value. Their level uncertainty is probably ±1000 cm^{-1}.

Kastner, Behring, and Cohen (1975) identified transitions between $2p^5 3p$ and $2p^5 4d$, but there is no connection with the levels given here.

We derived the ionization energy from the $2s^2 2p^5 nd\ {}^3D^\circ_1$ series for $n=3$ and 4, with the change in quantum defect Δn^* taken from Ti XIII.

References

Edlén, B., and Tyrén, F. (1936), Z. Phys. **101**, 206.

Kastner, S. O., Behring, W. E., and Cohen, L. (1975), Astrophys. J. **199**, 777.

Ca XI

Configuration	Term	J	Level (cm^{-1})	Configuration	Term	J	Level (cm^{-1})
$2s^2 2p^6$	^1S	0	0	$2s2p^6 3p$	$^1P^\circ$	1	*3 708 900*
$2s^2 2p^5 3s$	$^3P^\circ$	1	*2 810 900*	$2s^2 2p^5 4s$	$^3P^\circ$	1	*3 753 900*
$2s^2 2p^5 3s$	$^1P^\circ$	1	*2 839 900*	$2s^2 2p^5 4s$	$^1P^\circ$	1	*3 781 900*
$2s^2 2p^5 3d$	$^3P^\circ$	1	*3 199 300*	$2s^2 2p^5 4d$	$^3D^\circ$	1	*3 919 000*
$2s^2 2p^5 3d$	$^3D^\circ$	1	*3 239 700*	$2s^2 2p^5 4d$	$^1P^\circ$	1	*3 948 400*
$2s^2 2p^5 3d$	$^1P^\circ$	1	*3 284 300*	Ca XII ($^2P^\circ_{3/2}$)	*Limit*		**4 774 000**
$2s2p^6 3p$	$^3P^\circ$	1	*3 692 900*				

Ca XII

$Z=20$

F I isoelectronic sequence

Ground state: $1s^22s^22p^5\ ^2P^\circ_{3/2}$

Ionization energy $= 5\,301\,000 \pm 10\,000$ cm^{-1} (657.2 ± 1.3 eV)

The first work on this spectrum was by Edlén and Tyrén (1936), who classified 10 lines of the $2s^22p^5-2s^22p^43s$ and $3d$ transition arrays between 27 and 33 Å. This work was extended by Feldman, Doschek, Cowan, and Cohen (1973), from whose wavelengths the $3s$ and $3d$ levels are determined. The uncertainty in their levels is ± 1000 cm^{-1}. Fawcett, Burgess, and Peacock (1967) identified the $2s^22p^5-2s2p^6$ doublet at ~ 140 Å. The improved measurements by Kaufman, Sugar, and Cooper (1982), which give a level uncertainty of 50 cm^{-1}, are used here. The $2s^22p^5\ ^2P^\circ$ term interval is obtained from the solar flare line at 3327.5 ± 0.5 Å (in air) identified by Edlén (1942, 1976), giving an uncertainty in this interval of ± 5 cm^{-1}.

The $2s2p^53s\ ^2P^\circ$ term is from Feldman et al. (1973). The ionization energy was obtained by extrapolation by Lotz (1967).

References

Edlén, B. (1942), Z. Astrophys. **22**, 30.
Edlén, B. (1976), Mem. Soc. Roy. Sci. Liege **9**, 235.
Edlén, B., and Tyrén, F. (1936), Z. Phys. **101**, 206.
Fawcett, B. C., Burgess, D. D., and Peacock, N. J. (1967), Proc. Phys. Soc. **91**, 970.
Feldman, U., Doschek, G. A., Cowan, R. D., and Cohen, L. (1973), J. Opt. Soc. Am. **63**, 1445.
Kaufman, V., Sugar, J., and Cooper, D. (1982), Phys. Scr. **25**, 623.
Lotz, W. J. (1967), J. Opt. Soc. Am. **57**, 873.

Ca XII

Configuration	Term	J	Level (cm^{-1})
$2s^2\,2p^5$	$^2P^\circ$	3/2	*0*
		1/2	*30 041*
$2s2p^6$	^2S	1/2	709 030
$2s^2\,2p^4(^3P)3s$	^4P	5/2	3 062 300
		3/2	3 077 100
		1/2	3 089 300
$2s^2\,2p^4(^3P)3s$	^2P	3/2	3 097 800
		1/2	3 114 800
$2s^2\,2p^4(^1D)3s$	^2D	5/2	3 158 500
		3/2	3 159 300
$2s^2\,2p^4(^1S)3s$	^2S	1/2	3 249 600
$2s^2\,2p^4(^3P)3d$	^4P	1/2	3 475 800
		3/2	3 479 600
		5/2	3 489 400
$2s^2\,2p^4(^3P)3d$	^4F	5/2	3 480 000
$2s^2\,2p^4(^3P)3d$	^2P	1/2	3 486 700
		3/2	3 508 200
$2s^2\,2p^4(^3P)3d$	^2D	3/2	3 494 600
		5/2	3 511 500
$2s^2\,2p^4(^3P)3d$	^2F	5/2	3 494 900
$2s^2\,2p^4(^1D)3d$	^2S	1/2	3 559 300
$2s^2\,2p^4(^1D)3d$	^2F	5/2	3 562 800
$2s^2\,2p^4(^1D)3d$	^2P	3/2	3 574 900
$2s^2\,2p^4(^1D)3d$	^2D	5/2	3 574 900
		3/2	3 584 900
$2s^2\,2p^4(^1S)3d$	^2D	5/2	3 647 900
		3/2	3 652 300
$2s2p^5(^3P^\circ)3s$	$^2P^\circ$	3/2	*3 738 000*
		1/2	*3 755 900*
Ca XIII ($^3P_2$)	***Limit***		**5 301 000**

Ca XIII

$Z=20$

O I isoelectronic sequence

Ground state: $1s^22s^22p^4\ {}^3P_2$

Ionization energy = 5 861 000±12 000 cm^{-1} (726.6±1.4 eV)

The solar coronal line measured by Jefferies (1969) at 4087.1 Å was first classified by Edlén (1942) as the magnetic dipole transition $2s^22p^4\ {}^3P_2-{}^3P_1$. Laboratory observations of the $2s^22p^4-2s2p^5$ array were made by Fawcett, Burgess, and Peacock (1967) who identified the ${}^3P-{}^3P°$ multiplet. The $2p^6\ {}^1S_0$ level was reported by Fawcett, Galanti, and Peacock (1974).

New measurements of these arrays and identification of intersystem lines were made by Kaufman, Sugar, and Cooper (1982) who reported all the levels of the $2s^22p^4$, $2s2p^5$, and $2p^6$ configurations. Their wavelengths in the range of 110–190 Å are given with an uncertainty of ±0.01 Å and a level uncertainty of ±50 cm^{-1}. These are used along with the solar line, which determines the position of the $2p^4\ {}^3P_1$ level to ±3 cm^{-1}. The percentage compositions were provided by Kaufman and Sugar (1982). Their calculation includes configuration interaction between $2s^22p^4$ and $2p^6$.

The $2p^33s$ levels are from the observations of Doschek, Feldman, and Cohen (1973). Lines of the $2p^4-2p^33d$ array were classified by Fawcett and Hayes (1975). Subsequently, revisions of several line classifications were made by Bromage and Fawcett (1977) on the basis of new calculations. The uncertainty in the $3p^33s$ and $3p^33d$ levels is ±1000 cm^{-1}.

The ionization energy was obtained by Lotz (1967) by extrapolation.

References

Bromage, G. E., and Fawcett, B. C. (1977), Mon. Not. R. Astron. Soc. **178**, 591.

Doschek, G. A., Feldman, U., and Cohen, L. (1973), J. Opt. Soc. Am. **63**, 1463.

Edlén, B. (1942), Z. Astrophys. **22**, 30.

Fawcett, B. C., Burgess, D. D., and Peacock, N. J. (1967), Proc. Phys. Soc. **91**, 970.

Fawcett, B. C., Galanti, M., and Peacock, N. J. (1974), J. Phys. **B7**, 1149.

Fawcett, B. C., and Hayes, R. W. (1975), Mon. Not. R. Astron. Soc. **170**, 185.

Jefferies, J. T. (1969), Mem. Soc. Roy. Sci. Liege (5) **16**, 213.

Kaufman, V., and Sugar, J. (1982), private communication.

Kaufman, V., Sugar, J., and Cooper, D. (1982), Phys. Scr. **25**, 623.

Lotz, W.J. (1967), J. Opt. Soc. Am. **57**, 873

Ca XIII

Configuration	Term	J	Level (cm^{-1})	Leading percentages		
$2s^2\,2p^4$	^3P	2	0	97	3	^1D
		1	24 460	100		
		0	28 888	96	4	^1S
$2s^2\,2p^4$	^1D	2	88 208	97	3	^3P
$2s^2\,2p^4$	^1S	0	178 613	93	4	^3P
$2s2p^5$	$^3P°$	2	*618 268*	100		
		1	*638 238*	100		
		0	*650 105*	100		
$2s2p^5$	$^1P°$	1	*850 300*	100		
$2p^6$	^1S	0	1 440 320	97	3	$2s^22p^4\ {}^1S$
$2s^2\,2p^3(^4S°)3s$	$^3S°$	1	*3 374 600*			
$2s^2\,2p^3(^2D°)3s$	$^3D°$	1	*3 452 700*			
		2	*3 453 200*			
		3	*3 458 300*			
$2s^2\,2p^3(^2D°)3s$	$^1D°$	2	*3 474 600*			
$2s^2\,2p^3(^2P°)3s$	$^1P°$	1	*3 544 500*			

(Continued)

Ca XIII—Continued

Configuration	Term	J	Level (cm^{-1})	Leading percentages
$2s^2\,2p^3(^4S°)3d$	$^3D°$	2	*3 739 000*	
		3	*3 743 000*	
$2s^2\,2p^3(^2D°)3d$	$^3D°$	1	*3 828 000*	
		2	*3 838 000*	
		3	*3 841 000*	
$2s^2\,2p^3(^2D°)3d$	$^3P°$	2	*3 851 000*	
		1	*3 852 000*	
$2s^2\,2p^3(^2D°)3d$	$^3S°$	1	*3 864 000*	
$2s^2\,2p^3(^2P°)3d$	$^3P°$	2	*3 890 000*	
		1	*3 893 000*	
$2s^2\,2p^3(^2P°)3d$	$^1D°$	2	*3 905 000*	
$2s^2\,2p^3(^2P°)3d$	$^3D°$	1	*3 914 000*	
		3	*3 917 000*	
		2	*3 920 000*	
$2s^2\,2p^3(^2D°)3d$	$^1F°$	3	*3 929 000*	
$2s^2\,2p^3(^2P°)3d$	$^1P°$	1	*3 969 000*	
Ca XIV ($^4S°_{3/2}$)	***Limit***		5 861 000	

Ca XIV

$Z=20$

N I isoelectronic sequence

Ground state: $1s^2 2s^2 2p^3\ ^4S_{3/2}$

Ionization energy = 6 595 000±13 000 cm^{-1} (817.6±1.6 eV)

The transition arrays $2s^2 2p^3 - 2s2p^4$ and $2s2p^4 - 2p^5$ were analyzed by Kononov et al. (1976), but no intersystem lines were observed. Kaufman, Sugar, and Cooper (1982) reobserved these arrays and identified the intersystem line $2s^2 2p^3\ ^4S_{3/2} - 2s2p^4\ ^2P^\circ_{3/2}$. Their measured lines in the range of 116–216 Å, given with an uncertainty of ±0.010 Å, are used here to derive the levels, with an uncertainty of ±50 cm^{-1}. Kaufman and Sugar (1982) provided the calculated percentage compositions of the levels. Their calculation includes configuration interaction between $2s^2 2p^3$ and $2p^5$.

The $2p^2 3d$ levels are from the classifications of Fawcett and Hayes (1975) with some revised classifications by Bromage and Fawcett (1977) based on new calculations. The uncertainty in these levels is ±1000 cm^{-1}. They also reported the percentage compositions of these levels.

The ionization energy was obtained by Lotz (1967) by extrapolation.

References

Bromage, G. E., and Fawcett, B. C. (1977), Mon. Not. R. Astron. Soc. **179**, 683.

Fawcett, B. C., and Hayes, R. W. (1975), Mon. Not. R. Astron. Soc. **170**, 185.

Kaufman, V., and Sugar, J. (1982), private communication.

Kaufman, V., Sugar, J., and Cooper, D. (1982), Phys. Scr. **26**, 163.

Kononov, E. Y., Koshelev, K. N., Podobedova, L. I., and Churilov, S. S. (1976), Opt. Spectrosc. **40**, 121.

Lotz, W. (1967), J. Opt. Soc. Am. **57**, 873.

Ca XIV

Configuration	Term	J	Level (cm^{-1})	Leading percentages		
$2s^2\,2p^3$	$^4S^\circ$	3/2	*0*	98	2	$^2P^\circ$
$2s^2\,2p^3$	$^2D^\circ$	3/2	*105 870*	89	10	$^2P^\circ$
		5/2	*113 520*	100		
$2s^2\,2p^3$	$^2P^\circ$	1/2	*172 400*	98	2	$2p^5\ ^2P^\circ$
		3/2	*183 360*	86	11	$2p^3\ ^2D^\circ$
$2s(^2S)2p^4(^3P)$	^4P	5/2	515 800	99	1	$2s(^2S)2p^4(^1D)\ ^2D$
		3/2	535 870	100		
		1/2	545 090	99	1	$2s(^2S)2p^4(^1S)\ ^2S$
$2s(^2S)2p^4(^1D)$	^2D	3/2	710 710	99	1	$2s(^2S)2p^4(^3P)\ ^2P$
		5/2	712 500	99	1	$2s(^2S)2p^4(^3P)\ ^4P$
$2s(^2S)2p^4(^1S)$	^2S	1/2	825 050	88	11	$2s(^2S)2p^4(^3P)\ ^2P$
$2s(^2S)2p^4(^3P)$	^2P	3/2	858 240	99	1	$2s(^2S)2p^4(^1D)\ ^2D$
		1/2	885 610	89	11	$2s(^2S)2p^4(^1S)\ ^2S$
$2p^5$	$^2P^\circ$	3/2	*1 347 870*	98	2	$2s^2 2p^3\ ^2P^\circ$
		1/2	*1 380 110*	98	2	
$2s^2\,2p^2(^3P)3d$	^2P	3/2	4 114 000	57	28	$(^3P)\ ^4D$
$2s^2\,2p^2(^3P)3d$	^4P	5/2	4 143 700	76	19	$(^3P)\ ^4D$
		3/2	4 151 800	91		
$2s^2\,2p^2(^3P)3d$	^2F	7/2	4 154 100	57	20	$(^1D)\ ^2F$

(Continued)

Ca XIV—Continued

Configuration	Term	J	Level (cm^{-1})	Leading percentages		
$2s^2\,2p^2(^3P)3d$	^2D	$^3/_2$	4 188 000	89		
		$^5/_2$	4 199 400	78	14	$(^1D)\ ^2D$
$2s^2\,2p^2(^1D)3d$	^2F	$^7/_2$	4 230 600	60	32	$(^3P)\ ^2F$
		$^5/_2$	4 243 600	48	21	
$2s^2\,2p^2(^1D)3d$	^2D	$^5/_2$	4 242 000	67	18	$(^1D)\ ^2F$
$2s^2\,2p^2(^1D)3d$	^2P	$^3/_2$	4 251 200	90		
Ca XV $(^3P_0)$	***Limit***		**6 595 000**			

Ca XV

$Z=20$

C I isoelectronic sequence

Ground state: $1s^2 2s^2 2p^2\ ^3P_0$

Ionization energy $= 7\,215\,000 \pm 14\,000$ cm^{-1} (894.5 ± 1.8 eV)

The levels of the $2s^2 2p^2$, $2s2p^3$ and $2p^4$ configurations were determined by Kononov, Koshelev, Podobedova, and Churilov (1976), except for the intersystem connection. This was provided by the identification of the solar coronal line $2s^2 2p^2\ ^3P_2 - ^1D_2$ at 1375.95 Å by Sandlin, Brueckner, and Tousey (1977). Two more solar coronal lines at 5444 Å (in air) and 5693.6 Å (in air) were identified as transitions between levels of the ground term $^3P_1 - ^3P_2$ and $^3P_0 - ^3P_1$, respectively, by Edleń (1972). These coronal lines are used here to determine the positions of the levels of ^{3}P and ^{1}D in the $2s^2 2p^2$ configuration with an uncertainty of 2 cm^{-1}.

The spectrum was reobserved in the range of 125–217 Å with an accuracy of ± 0.01 Å by Sugar, Kaufman, and Cooper (1982) who identified several intercombination lines. The level values for $2s^2 2p^2\ ^1S_0$ and for all terms of $2s2p^3$ (except for $^5S_2^\circ$) and $2p^4$ are from their measurements and have an uncertainty of ± 50 cm^{-1}. The predicted position of $2s2p^3\ ^5S_2^\circ$ in brackets was given by Edlén (1984). Kaufman and Sugar (1982) provided the percentage compositions for these levels. Their calculation includes configuration interaction between $2s^2 2p^2$ and $2p^4$.

The levels of $2s^2 2p3d$ and $2s2p^2 3d$ are from the classifications of Fawcett and Hayes (1975) with additions by Bromage and Fawcett (1977). These measurements, at ~24 Å are not accompanied by an uncertainty estimate. Bromage and Fawcett provided the percentage compositions for the $2s^2 2p3d$ configuration.

The ionization energy was obtained by Lotz (1967) by extrapolation.

References

Bromage, G. E., and Fawcett, B. C. (1977), Mon. Not. R. Astron. Soc. **178**, 605.

Edlén, B. (1984), private communication.

Edlén, B. (1972), Solar Physics **24**, 356.

Fawcett, B. C., and Hayes, R. W. (1975), Mon Not. R. Astron. Soc. **170**, 185.

Kaufman, V., and Sugar, J. (1982), private communication.

Kononov, E. Y., Koshelev, K. N., Podobedova, L. I., and Churilov, S. S. (1976), Opt. Spectrosc. **40**, 121.

Lotz, W. (1967), J. Opt. Soc. Am. **57**, 873.

Sandlin, G. D., Brueckner, G. E., and Tousey, R. (1977), Astrophys. J. **241**, 898.

Sugar, J., Kaufman, V., and Cooper, D. (1982), Phys. Scr. **26**, 189.

Ca XV

Configuration	Term	J	Level (cm^{-1})	Leading percentages		
$2s^2\,2p^2$	^{3}P	0	0	96	3	$2s^22p^2\ ^1$S
		1	17 559	99	1	$2p^4\ ^3$P
		2	35 923	94	5	$2s^22p^2\ ^1$D
$2s^2\,2p^2$	^{1}D	2	108 600	94	5	$2s^22p^2\ ^3$P
$2s^2\,2p^2$	^{1}S	0	197 670	93	4	$2p^4\ ^1$S
$2s(^2S)2p^3(^4S^\circ)$	^{5}S°	2	[*275 900*]	100		
$2s(^2S)2p^3(^2D^\circ)$	^{3}D°	2	*496 680*	96	4	$2s(^2S)2p^3(^2P^\circ)\ ^3$P°
		1	*497 570*	97	2	$2s(^2S)2p^3(^2P^\circ)\ ^3$P°
		3	*500 230*	100		
$2s(^2S)2p^3(^2P^\circ)$	^{3}P°	0	*581 730*	100		
		1	*582 780*	97	2	$2s(^2S)2p^3(^2D^\circ)\ ^3$D°
		2	*585 670*	95	4	$2s(^2S)2p^3(^2D^\circ)\ ^3$D°
$2s(^2S)2p^3(^4S^\circ)$	^{3}S°	1	*728 880*	95	4	$2s(^2S)2p^3(^2P^\circ)\ ^1$P°
$2s(^2S)2p^3(^2D^\circ)$	^{1}D°	2	*729 650*	99	1	$2s(^2S)2p^3(^2P^\circ)\ ^3$P°
$2s(^2S)2p^3(^2P^\circ)$	^{1}P°	1	*814 380*	96	4	$2s(^2S)2p^3(^4S^\circ)\ ^3$S°

(Continued)

Ca XV —Continued

Configuration	Term	J	Level (cm^{-1})	Leading percentages		
$2p^4$	^{3}P	2	1 107 550	95	4	$2p^4\ ^1$D
		1	1 133 850	99	1	$2s^22p^2\ ^3$P
		0	1 139 970	96	2	$2p^4\ ^1$S
$2p^4$	^{1}D	2	1 195 120	95	4	^{3}P
$2p^4$	^{1}S	0	1 350 890	93	4	$2s^22p^2\ ^1$S
$2s^2\,2p3d$	^{3}F°	2	*4 363 300*	74	25	^{1}D°
		3	*4 379 400*	88		
$2s^2\,2p3d$	^{3}D°	1	*4 399 500*	78	15	^{3}P°
		2	*4 411 500*	45	24	^{1}D°
		3	*4 426 400*	88	10	^{3}F°
$2s^2\,2p3d$	^{3}P°	1	*4 434 500*	82	17	^{3}D°
		2	4 435 400	61	36	
$2s^2\,2p3d$	^{1}P°	1	*4 473 400*	92		
$2s^2\,2p3d$	^{1}F°	3	*4 475 000*	95		
$2s2p^2(^4$P$)3d$	^{3}F	4	4 726 800			
Ca XVI (^{2}P$^\circ_{1/2}$)	***Limit***		**7 215 000**			

Ca XVI

$Z=20$

B I isoelectronic sequence

Ground state: $1s^2 2s^2 2p\ ^2P^\circ_{1/2}$

Ionization energy = 7 860 000±16 000 cm^{-1} (974±2 eV)

Classification of the transition arrays $2s^2 2p - 2s2p^2$ and $2s2p^2 - 2p^3$ was made by Kononov, Koshelev, Podobedova, and Churilov (1975). The spectrum was remeasured by Sugar, Kaufman, and Cooper (1982) in the range of 154–224 Å with an accuracy of ±0.01 Å. The present $2s^2 2p$, $2s2p^2$, and $2p^3$ levels are determined from their data with an uncertainty of ±50 cm^{-1}. Their adopted value for the position of the level $2s2p^2\ ^4P_{5/2}$ with an estimated uncertainty of ±100 cm^{-1} is used here. Leading percentages for the $2s2p^2$ configuration were provided by Kaufman and Sugar (1982).

The higher lying configurations are from the classifications of the lines at 19–20 Å measured with an uncertainty of ±0.01 Å by Fawcett and Hayes (1975), and by Fawcett and Ridgeley (1981) who identified the $3d-4f$ doublet.

The ionization energy was obtained by Lotz (1967) by extrapolation.

References

Fawcett, B. C., and Hayes, R. W. (1975), Mon Not. R. Astron. Soc. **170**, 185.

Fawcett, B. C., and Ridgeley, A. (1981), J. Phys. **B14**, 203.

Kaufman, V., and Sugar, J. (1982), private communication.

Kononov, E. Y., Koshelev, K. N., Podobedova, L. I., and Churilov, S. S. (1975), Opt. Spectrosc. **39**, 458.

Lotz, W. (1967), J. Opt. Soc. Am. **57**, 873.

Sugar, J., Kaufman, V., and Cooper, D. (1982) Phys. Scr. **26**, 293.

Ca XVI

Configuration	Term	J	Level (cm^{-1})	Leading percentages		
$2s^2 2p$	$^2P^\circ$	1/2	*0*			
		3/2	*36 520*			
$2s2p^2$	^4P	1/2	267 990+x	99	1	^2S
		3/2	282 500+x	100		
		5/2	300 800+x	99	1	^2D
$2s2p^2$	^2D	3/2	479 420	99	1	^2P
		5/2	481 860	99	1	^4P
$2s2p^2$	^2S	1/2	592 180	67	32	^2P
$2s2p^2$	^2P	1/2	633 760	68	32	^2S
		3/2	645 660	99	1	^2D
$2p^3$	$^4S^\circ$	3/2	*834 860+x*			
$2p^3$	$^2D^\circ$	3/2	*940 000*			
		5/2	*944 700*			
$2p^3$	$^2P^\circ$	1/2	*1 052 700*			
		3/2	*1 062 030*			
$2s^2 3d$	^2D	3/2	4 662 000			
		5/2	4 664 000			
$2s2p(^3P^\circ)3p$	^2P	1/2	4 773 000			
		3/2	4 794 000			

(Continued)

Ca XVI—Continued

Configuration	Term	J	Level (cm^{-1})	Leading percentages
$2s2p(^3P^\circ)3d$	$^4D^\circ$	$3/2$	*4 930 000+x*	
		$5/2$	*4 934 000+x*	
		$7/2$	*4 952 000+x*	
$2s2p(^3P^\circ)3d$	$^4P^\circ$	$5/2$	*4 963 000+x*	
$2s2p(^3P^\circ)3d$	$^2F^\circ$	$5/2$	*5 000 000*	
		$7/2$	*5 022 000*	
$2s2p(^1P^\circ)3d$	$^2F^\circ$	$7/2$	*5 170 000*	
$2s2p(^1P^\circ)3d$	$^2D^\circ$	$5/2$	*5 198 000*	
$2s^2\,4f$	$^2F^\circ$	$5/2,7/2$	*6 396 000*	
Ca XVII ($^1S_0$)	***Limit***		**7 860 000**	

Ca XVII

$Z=20$

Be I isoelectronic sequence

Ground state: $1s^2 2s^2\ ^1S_0$

Ionization energy = 8 770 000±18 000 cm^{-1} (1087±2 eV)

The strong singlet resonance line $2s^2\ ^1S_0-2s2p\ ^1P^\circ_1$ at 192.86 Å has been observed in the laboratory by Kononov, Koshelev, Podobedova, and Churilov (1975), by Fawcett and Hayes (1975), and in spectra of solar flares. We used the wavelength of Kononov et al. to establish the value of the $^1P^\circ$ term with an uncertainty of ±50 cm^{-1}.

The intersystem transition $2s^2\ ^1S_0-2s2p\ ^3P^\circ_1$ has been identified in a solar flare spectrum by Sandlin, Brueckner, Scherrer, and Tousey (1976) at 371.11±0.03 Å. We have adopted their value to locate the triplet system relative to the singlets with an uncertainty of 20 cm^{-1}. The other ^3P levels of $2s2p$ and $2p^2$ and the ^1S of $2p^2$ are from Kononov et al. with an uncertainty of ±50 cm^{-1}. The $2p^2\ ^1D$ and $^3P_0$ are from Fawcett, Ridgeley, and Hatter (1980).

The levels of the higher configurations are derived from the line identifications by Fawcett and Hayes. Their level uncertainty is ±5000 cm^{-1}.

The ionization energy was obtained by Lotz (1967) by extrapolation.

References

Fawcett, B. C., and Hayes, R. W. (1975), Mon. Not. R. Astron. Soc. **170**, 185.

Fawcett, B. C., Ridgeley, A., and Hatter, A. T. (1980), J. Opt. Soc. Am. **70**, 1349.

Kononov, E. Y., Koshelev, K. N., Podobedova, L. I., and Churilov, S. S. (1975), Opt. Spectrosc. **39**, 458.

Lotz, W. (1967), J. Opt. Soc. Am. **57**, 873.

Sandlin, G. D., Brueckner, G. E., Scherrer, V. E., and Tousey, R. (1976), Astrophys. J. **105**, L47.

Ca XVII

Configuration	Term	J	Level (cm^{-1})	Configuration	Term	J	Level (cm^{-1})
$2s^2$	^1S	0	0	$2s3p$	$^1P^\circ$	1	*5 113 000*
$2s2p$	$^3P^\circ$	0	*258 290*	$2s3d$	^3D	2	5 186 000
		1	*269 460*			3	5 190 000
		2	*296 950*				
				$2s3d$	^1D	2	5 236 000
$2s2p$	$^1P^\circ$	1	*518 620*				
				$2p3p$	^3D	3	5 448 000
$2p^2$	^3P	0	689 080				
		1	706 680	$2p3d$	$^3D^\circ$	2	*5 533 000*
		2	726 450			3	*5 546 000*
$2p^2$	^1D	2	798 130	$2p3d$	$^1F^\circ$	3	*5 598 000*
$2p^2$	^1S	0	967 170	Ca XVIII ($^2S_{1/2}$)	*Limit*		8 770 000

Ca XVIII

$Z = 20$

Li I isoelectronic sequence

Ground state: $1s^2 2s\ ^2S_{1/2}$

Ionization energy $= 9\ 338\ 000 \pm 1300\ cm^{-1}$ (1157.8 ± 0.2 eV)

The $2s-3p$, $2s-4p$, $2s-5p$, $2p-3s$, $2p-3d$, $2p-4d$, $2p-5d$, and $2p-6d$, transitions were reported by Goldsmith, Feldman, Oren, and Cohen (1972) in the range of 12–20 Å with an uncertainty of ±0.005 Å. The value of the $2p\ ^2P°$ term is from the $2s-2p$ transitions observed at 302.19 Å and 344.76 Å in a solar flare spectrum by Widing and Purcell (1976) with an uncertainty of ±0.02 Å and a level uncertainty of $\pm 20\ cm^{-1}$.

Boiko, Faenov, and Pikuz (1978) confirmed the lines identified by Goldsmith et al. and added the $6p$, $7p$, $6d$, and $7d$ terms. The $3d-4f$ and $3s-4p$ transitions were reported by Fawcett and Ridgeley (1981). The levels derived from these data have an uncertainty of $\pm 2000\ cm^{-1}$. The K-shell excitations were observed by Aglitskii, Boiko, Zakharov, Pikuz, and Faenov (1974) at 3.2 Å in a laser-produced plasma. Levels derived from these data have an uncertainty of $\pm 5000\ cm^{-1}$.

Edlén (1979) derived the ionization energy from the *nd* Rydberg series.

References

Aglitskii, E. V., Boiko, V. A., Zakharov, S. M., Pikuz, S. A., and Faenov, A. Y. (1974), Kvantovaya Elektron. (Moscow) **1**, 908.

Boiko, V. A., Faenov, A. Y., and Pikuz, S. A. (1978), J. Quant. Spectrosc. Radiat. Transfer **19**, 11.

Edlén, B. (1979), Phys. Scr. **19**, 255.

Fawcett, B. C., and Ridgeley, A. (1981), J. Phys. **B14**, 203.

Goldsmith, S., Feldman, U., Oren, L., and Cohen, L. (1972), Astrophys. J. **174**, 209.

Widing, K. G., and Purcell, J. D. (1976), Astrophys. J. **204**, L151.

Ca XVIII

Configuration	Term	J	Level (cm^{-1})	Configuration	Term	J	Level (cm^{-1})
$1s^2\ 2s$	^2S	1/2	0	$1s^2\ 5d$	^2D	5/2	7 911 800
						3/2	7 913 200
$1s^2\ 2p$	$^2P°$	1/2	*290 057*				
		3/2	*330 918*	$1s^2\ 5p$	$^2P°$	1/2,3/2	*7 913 900*
$1s^2\ 3s$	^2S	1/2	5 276 800	$1s^2\ 6p$	$^2P°$	1/2,3/2	*8 341 000*
$1s^2\ 3p$	$^2P°$	1/2	*5 338 500*	$1s^2\ 6d$	^2D	3/2,5/2	8 346 000
		3/2	*5 350 200*				
				$1s^2\ 7p$	$^2P°$	1/2,3/2	*8 605 000*
$1s^2\ 3d$	^2D	3/2	5 381 200				
		5/2	5 384 200	$1s^2\ 7d$	^2D	3/2,5/2	8 610 000
$1s^2\ 4d$	^2D	3/2	7 112 300	Ca XIX ($^1S_0$)	***Limit***		**9 338 000**
		5/2	7 116 100				
				$1s(^2S)2s2p(^1P°)$	$^2P°$	3/2	*31 352 000*
$1s^2\ 4p$	$^2P°$	1/2	*7 112 700*				
		3/2	*7 118 100*	$1s2p^2$	^2D	3/2	31 484 000
						5/2	31 486 000
$1s^2\ 4f$	$^2F°$	5/2,7/2	*7 115 200*				
				$1s2p^2$	^2P	3/2	31 551 000

Ca XIX

$Z=20$

He I isoelectronic sequence

Ground state: $1s^2\ ^1S_0$

Ionization energy = 41 366 000±8000 cm^{-1} (5128.8±1.0 eV)

Because of the excellent agreement of the calculated energies of the $1s2l$ levels by Safronova (1981) with the few well-measured spectra in the He I sequence, we have compiled her results for these levels, except as noted below, and for the ionization energy. Detailed comparisons are given in the Introduction. Levels of the $n=3$–5 shells are from the calculated binding energies by Ermolaev and Jones (1974) subtracted from Safronova's value for the 1s binding energy of the ground state. We have assumed an uncertainty of 2 parts in 10^4 for the excited levels relative to the ground state, and for the ionization energy. For differences between excited levels where $\Delta n=0$, we assumed an uncertainty of 2 parts in 10^3 (see the Introduction).

The $1s2s\ ^3S_1-1s2p\ ^3P^\circ_2$ transition has been measured by Livingston (1983) in a beam foil experiment at 466.78±0.08 Å. Safronova predicts 466.43 Å for this wavelength. We used the observed value to obtain the position of the $1s2p\ ^3P^\circ_2$ level relative to $1s2s\ ^3S_1$.

Corrected measurements (see Introduction) by Aglitskii et al. (1974) place the $1s2p\ ^3P^\circ_1$ level at 31 320 000 cm^{-1} and the $1s2p\ ^1P^\circ_1$ at 31 480 000 cm^{-1} with an estimated uncertainty of ±5000 cm^{-1}.

Percentage compositions are from Ermolaev and Jones.

References

Aglitskii, E. V., Boiko, V. A., Zakharov, S. M., Pikuz, S. A., and Faenov, A. Y. (1974), Kvantovaya Elektron. (Moscow) **1**, 908.

Ermolaev, A. M., and Jones, M. (1974), J. Phys. **B7**, 199.

Livingston, A. E. (1983), private communication.

Safronova, U. I. (1981), Phys. Scr. **23**, 241.

Ca XIX

Configuration	Term	J	Level (cm^{-1})	Leading percentages		
$1s^2$	^1S	0	0			
1s2s	^3S	1	[31 142 150]			
1s2p	$^3P^\circ$	0	*[31 312 840]*			
		1	*[31 321 060]*	97	3	$^1P^\circ$
		2	*[31 356 380]*			
1s2s	^1S	0	[31 328 450]			
1s2p	$^1P^\circ$	1	*[31 473 810]*	97	3	$^3P^\circ$
1s3s	^3S	1	[36 870 940]			
1s3p	$^3P^\circ$	0	*[36 918 070]*			
		1	*[36 920 260]*	97	3	$^1P^\circ$
		2	*[36 930 850]*			
1s3s	^1S	0	[36 919 930]			
1s3p	$^1P^\circ$	1	*[36 962 850]*	97	3	$^3P^\circ$
1s4s	^3S	1	[38 850 670]			
1s4p	$^3P^\circ$	0	*[38 870 230]*			
		1	*[38 871 150]*	96	4	$^1P^\circ$
		2	*[38 875 630]*			
1s4s	^1S	0	[38 870 530]			
1s4p	$^1P^\circ$	1	*[38 888 680]*	96	4	$^3P^\circ$

(Continued)

Ca XIX—Continued

Configuration	Term	J	Level (cm^{-1})	Leading percentages		
1*s*5*s*	^{3}S	1	[39 761 380]			
1*s*5*p*	^{3}P°	0	[*39 771 270*]			
		1	[*39 771 750*]	96	4	^{1}P°
		2	[*39 774 040*]			
1*s*5*s*	^{1}S	0	[39 771 310]			
1*s*5*p*	^{1}P°	1	[*39 780 630*]	96	4	^{3}P°
Ca XX (^{2}S$_{1/2}$)	***Limit***		**41 366 000**			

Ca XX

$Z=20$

H I isoelectronic sequence

Ground state: $1s\ ^2S_{1/2}$

Ionization energy = 44 117 410±20 cm^{-1} (5469.906±0.013 eV)

The $1s\ ^2S-2p\ ^2P°$ transitions were observed in a solar flare spectrum by Feldman, Doschek, and Kreplin (1980), but no absolute wavelength calibration was available. We give calculated values by Mohr (1983) for the $n=2$ levels and by Erickson (1977) for $n=3$–5 relative to the $2p\ ^2P°_{3/2}$ level. Further details are given in the Introduction.

References

Erickson, G. W. (1977), J. Phys. Chem. Ref. Data, **6**, 831.

Feldman, V., Doschek, G. A., and Kreplin, R. W. (1980), Astrophys. J. **238**, 365.

Mohr, P. J. (1983), At. Data Nucl. Data Tables **29**, 453.

Ca XX

Configuration	Term	J	Level (cm^{-1})
1s	^2S	1/2	0
2p	$^2P°$	1/2	[*33 069 905*]
		3/2	[*33 129 258*]
2s	^2S	1/2	[33 071 743]
3p	$^2P°$	1/2	[*39 214 015*]
		3/2	[*39 231 607*]
3s	^2S	1/2	[39 214 580]
3d	^2D	3/2	[39 231 576]
		5/2	[39 237 382]
4p	$^2P°$	1/2	[*41 362 031*]
		3/2	[*41 369 449*]
4s	^2S	1/2	[41 362 270]
4d	^2D	3/2	[41 369 436]
		5/2	[41 371 887]
4f	$^2F°$	5/2	[*41 371 882*]
		7/2	[*41 373 104*]
5p	$^2P°$	1/2	[*42 355 223*]
		3/2	[*42 359 020*]
5s	^2S	1/2	[42 355 346]
5d	^2D	3/2	[42 359 013]
		5/2	[42 360 268]
5f	$^2F°$	5/2	[*42 360 265*]
		7/2	[*42 360 891*]
5g	^2G	7/2	[42 360 890]
		9/2	[42 361 265]
	Limit		**44 117 410**

Sc I

$Z=21$

Ground state: $1s^2 2s^2 2p^6 3s^2 3p^6 3d 4s^2\ ^2D_{3/2}$

Ionization energy = 52 922.0±0.5 cm^{-1} (6.56154±0.00006 eV)

The first extensive analysis of Sc I was carried out by Russell and Meggers (1927), who classified about 350 lines between 2690 and 8250 Å as combinations among 128 energy levels of nine configurations. The analysis was extended by Neufeld (1970) and Neufeld and Schrenk (1975), who added 150 newly classified lines and 22 new levels and measured g-values for 98 levels.

Observations of the absorption spectrum of Sc in the range 1200–3200 Å were made by Garton, Reeves, Tomkins, and Ercoli (1973) with a measurement accuracy of ±0.01 Å. They identified eight series arising from the ^2D ground term in combination with $4s^2nf\ ^2F°$, $3d4s(^1D)np\ ^2P°$, $3d4s(^1D)np\ ^2D°$, and $3d^2(^3F)np\ ^2L°$ (where L may be D, F, or G). By far the strongest series arises from the $3d4s^2\ ^2D-4s^2nf\ ^2F°$ transitions. The three series $3d4s^2\ ^2D-3d4s(^1D)np\ ^2P°$ were observed to much higher n and were used to derive the ionization energy. Two more series found to converge to the $3d4s(^1D)$ limit were identified tentatively as $3d4s^2\ ^2D-3d4s(^1D)np\ ^2D°$ with unresolved $^2D°$ splitting. A final series labeled $3d4s^2-3d^2(^3F)np\ ^2L°$ observed from $n=7$–15 and much stronger than the $^2D°$ series was reported. Since L may represent $^2D°$, $^2F°$, or $^2G°$ terms, the most likely candidate for a strong series is $^2F°$ and we have used this designation for the series. Many additional absorption features not compiled here were reported with no identification of the upper levels.

A new analysis of Sc I based on new observations between 2000 and 3300 Å, infrared measurements between 6500 and 33 900 Å, and Zeeman spectrograms has been published by Ben Ahmed and Verges (1977) together with a theoretical interpretation by Ben Ahmed (1977). They established 108 additional levels and rederived the complete system of known levels up to 49 000 cm^{-1}, classifying 1230 lines. In the analysis by Russell and Meggers, two unconnected systems of levels based on $3d^2(^3P)4s\ ^4P$ and ^2P were given. Ben Ahmed and Verges found the connection for the first, and rejected the second set, replacing it with newly found levels.

Our compilation is from the papers of Ben Ahmed and Verges, Ben Ahmed, and Garton et al. The accuracy of the level values must be judged from the number of significant decimal places given, probably ± a few units in the last digit. The theoretical treatment by Ben Ahmed of the odd configurations includes $3d^24p$, $3d4s4p$, $4s^24p$, $3d^25p$, and $3d4s5p$ with configuration interaction among them. An examination of the percentages shows that the configuration mixing is appreciable. Roth (1980) has calculated the mixture only of $3d^24p$, $3d4s4p$ and $3s^24p$. Although this is not as accurate a treatment as that of Ben Ahmed, Roth has chosen a better coupling scheme for $3d4s4p$; that is $3d(^2D)4s4p(^{1,3}P)$. We give the more complete calculation of Ben Ahmed. In his calculation of the even configurations Ben Ahmed has included $3d^24d$, $3d4s4d$, and $4s^24d$ with configuration interaction.

The g-values are from Ben Ahmed and Verges, except for those of the ground term, which are taken from the report of the magnetic resonance experiments by Childs (1971).

The ionization energy is from Garton et al. who derived their value from the $3d4s(^1D)np\ ^2P°_{3/2}$ series.

References

Ben Ahmed, Z., and Verges, J. (1977), Physica (Utrecht) **92C,** 113.

Ben Ahmed, A. (1977), Physica (Utrecht) **92C,** 122.

Childs, W. J. (1971), Phys. Rev. **A4,** 1767.

Garton, W. R. S., Reeves, E. M., Tomkins, F. S., and Ercoli, B. (1973), Proc. R. Soc. London, **Ser. A333,** 1.

Neufeld, L. W. (1970), Dissertation, Kansas State University.

Neufeld, L. W., and Schrenk, W. G. (1975), Spectrochim. Acta, Part B **30,** 45.

Roth, C. (1980), At. Data Nucl. Data Tables, **25,** 91.

Russell, H. N., and Meggers, W. F. (1927), Sci. Papers Natl. Bur. Stand. (U.S.) **22,** 329.

Sc I

Configuration	Term	J	Level (cm^{-1})	g	Leading percentages
$3d\ 4s^2$	^2D	3/2	0.00	0.79933	98
		5/2	168.34	1.20029	98
$3d^2(^3F)4s$	^4F	3/2	11 519.99	0.400	100
		5/2	11 557.69	1.026	100
		7/2	11 610.28	1.244	100
		9/2	11 677.38	1.325	100

Sc I—Continued

Configuration	Term	J	Level (cm^{-1})	g	Leading percentages		
$3d^2(^3F)4s$	^2F	5/2	14 926.07	0.861	100		
		7/2	15 041.92	1.134	100		
$3d\ 4s(^3D)4p$	$^4F°$	3/2	*15 672.58*	0.426	96		
		5/2	*15 756.57*	1.036	96		
		7/2	*15 881.75*	1.232	98		
		9/2	*16 026.62*	1.33	99		
$3d\ 4s(^3D)4p$	$^4D°$	1/2	*16 009.77*	0.002	97		
		3/2	*16 021.82*	1.039	92		
		5/2	*16 141.06*	1.336	79	9	$(^1D)\ ^2D°$
		7/2	*16 210.85*	1.43	96		
$3d\ 4s(^1D)4p$	$^2D°$	5/2	*16 022.73*	1.19	43	34	$(^3D)\ ^2D°$
		3/2	*16 096.90*	0.956	50	39	
$3d^2(^1D)4s$	^2D	5/2	17 012.76	1.226	94	4	$(^3P)\ ^4P$
		3/2	17 025.14	0.815	98		
$3d^2(^3P)4s$	^4P	1/2	17 226.04	2.662	100		
		3/2	17 255.07	1.719	99		
		5/2	17 307.08	1.575	96	4	$(^1D)\ ^2D$
$3d\ 4s(^3D)4p$	$^4P°$	1/2	*18 504.06*	2.529	83	7	$4s^24p\ ^2P°$
		3/2	*18 515.69*	1.698	88	4	
		5/2	*18 571.41*	1.600	98		
$4s^2\ 4p$	$^2P°$	1/2	*18 711.02*	0.777	37	31	$3d4s(^3D)4p\ ^2P°$
		3/2	*18 855.74*	1.356	37	33	
$3d^2(^1G)4s$	^2G	9/2	20 236.86	1.10	100		
		7/2	20 239.66	0.89	100		
$3d^2(^3P)4s$	^2P	1/2	20 681.43	0.668	99		
		3/2	20 719.86	1.331	99		
$3d\ 4s(^1D)4p$	$^2F°$	5/2	*21 032.75*	0.855	79	14	$(^3D)\ ^2F°$
		7/2	*21 085.85*	1.14	77	16	
$3d\ 4s(^1D)4p$	$^2P°$	1/2	*24 656.72*		74	17	$4s^24p\ ^2P°$
		3/2	*24 656.88*		68	17	
$3d\ 4s(^3D)4p$	$^2D°$	3/2	*24 866.17*	0.804	38	30	$3d^2(^3F)4p\ ^2D°$
		5/2	*25 014.21*	1.201	40	32	
$3d\ 4s(^3D)4p$	$^2F°$	5/2	*25 584.64*	0.857	64	21	$3d^2(^3F)4p\ ^2F°$
		7/2	*25 724.68*	1.138	62	22	
$3d^2(^1S)4s$	^2S	1/2	26 936.98		100		
$3d^2(^3F)4p$	$^4G°$	5/2	*29 022.82*	0.584	100		
		7/2	*29 096.18*	0.981	100		
		9/2	*29 189.84*	1.16	100		
		11/2	*29 303.51*		100		
$3d\ 4s(^3D)4p$	$^2P°$	1/2	*30 573.17*	0.680	44	31	$4s^24p\ ^2P°$
		3/2	*30 706.66*	1.332	43	32	

(Continued)

Sc I—Continued

Configuration	Term	J	Level (cm^{-1})	g	Leading percentages		
$3d^2(^3F)4p$	$^4F^\circ$	3/2	*31 172.70*	0.400	99		
		5/2	*31 215.81*	1.027	99		
		7/2	*31 275.39*	1.240	98		
		9/2	*31 350.84*	1.33	98		
$3d^2(^3F)4p$	$^4D^\circ$	1/2	*32 637.40*	0.03	96		
		3/2	*32 659.30*		96		
		5/2	*32 696.84*	1.373	95		
		7/2	*32 751.50*	1.414	95		
$3d^2(^3F)4p$	$^2G^\circ$	7/2	*33 055.98*	0.91	60	20	$(^3F)\ ^2F^\circ$
		9/2	*33 151.20*	1.06	90	10	$(^1G)\ ^2G^\circ$
$3d^2(^3F)4p$	$^2F^\circ$	5/2	*33 153.79*	0.853	62	13	$3d4s(^1D)4p\ ^2F^\circ$
		7/2	*33 278.40*	1.146	42	30	$3d^2(^3F)4p\ ^2G^\circ$
$3d^2(^3F)4p$	$^2D^\circ$	3/2	*33 614.88*	0.824	40	30	$3d^2(^3P)4p\ ^2D^\circ$
		5/2	*33 707.06*	1.186	39	30	
$3d^3$	^4F	3/2	33 763.53	0.395	100		
		5/2	33 798.64	1.026	100		
		7/2	33 846.59	1.23	100		
		9/2	33 906.38	1.33	100		
$3d\ 4s(^3D)5s$	^4D	1/2	34 390.25	0.00	100		
		3/2	34 422.83	1.192	100		
		5/2	34 480.00	1.370	100		
		7/2	34 567.19	1.43	100		
$3d^2(^3P)4p$	$^2S^\circ$	1/2	*35 346.35*	2.00	100		
$3d\ 4s(^3D)5s$	^2D	3/2	35 671.04		83	8	$3d^3\ ^2D1$
		5/2	35 745.62	1.212	82	12	$3d4s(^1D)5s\ ^2D$
$3d^3$	$^2D2$	3/2	36 276.63		40	34	$^2D1$
		5/2	36 330.59	1.196	42	35	
$3d^3$	^4P	1/2	36 492.64	2.634	98		
		3/2	36 515.76	1.712	95		
		5/2	36 572.77	1.59	96		
$3d^2(^1D)4p$	$^2F^\circ$	5/2	*36 666.42*		84	11	$(^3F)\ ^2F^\circ$
		7/2	*36 730.12*		82	10	
$3d^2(^3P)4p$	$^4D^\circ$	1/2	*36 764.20*	0.016	98		
		3/2	*36 793.65*	1.184	94		
		5/2	*36 860.20*	1.348	94		
		7/2	*36 959.03*		94	4	$(^1D)\ ^2F^\circ$
$3d^2(^1D)4p$	$^2D^\circ$	3/2	*36 933.91*	0.879	58	14	$3d4s(^3D)5p\ ^2D^\circ$
		5/2	*37 039.57*	1.207	59	14	
$3d^3$	^2G	7/2	36 977.51	0.89	100		
		9/2	37 054.51	1.110	99		
$3d^3$	^2P	1/2	37 085.84	0.682	96		
		3/2	37 148.22	1.328	93		

Sc I—Continued

Configuration	Term	J	Level (cm^{-1})	g	Leading percentages			
$3d^2(^1D)4p$	$^2P^\circ$	3/2	*37 086.02*	1.278	68		17	$(^3P)\ ^2P^\circ$
		1/2	*37 125.40*		70		17	
$3d^2(^3P)4p$	$^4S^\circ$	3/2	*37 486.86*	1.986	91		9	$^4P^\circ$
$3d\,4s(^1D)5s$	^2D	3/2	37 780.87	0.80	66		14	$(^3D)\ ^2D$
		5/2	37 855.61	1.18	63		17	
$3d^2(^3P)4p$	$^4P^\circ$	1/2	*37 877.78*	2.662	93		5	$3d4s(^3D)5p\ ^4P^\circ$
		3/2	*37 908.50*	1.731	85		9	$3d^2(^3P)4p\ ^4S^\circ$
		5/2	*37 964.89*	1.58	93		5	$3d4s(^3D)5p\ ^4P^\circ$
$3d\,4s(^3D)4d$	^2F	5/2	38 871.65	0.855	49		36	^4G
		7/2	38 959.16	1.14	46		37	^4D
$3d^2(^1G)4p$	$^2H^\circ$	9/2	*39 153.14*		98			
		11/2	*39 248.82*		100			
$3d^3$	^2H	9/2	39 164.11		100			
		11/2	39 225.33		100			
$3d^2(^1G)4p$	$^2G^\circ$	7/2	*39 392.79*	0.89	90		10	$(^3F)\ ^2G^\circ$
		9/2	*39 423.39*		88		10	
$3d\,4s(^3D)4d$	^4D	1/2	39 701.44	0.008	96			
		3/2	39 721.79	1.203	97			
		5/2	39 755.02	1.364	71		26	^4G
		7/2	39 799.99	1.439	41		47	^2F
$3d\,4s(^3D)4d$		5/2	39 861.37	0.555	45	^2F	36	^4G
$3d\,4s(^3D)4d$	^4G	7/2	39 902.75	0.968	78		20	^4D
		9/2	39 957.79	1.17	98			
		11/2	40 028.38	1.26	99			
$3d\,4s(^3D)5p$	$^4F^\circ$	3/2	*39 949.75*		96			
		5/2	*39 989.58*		93		6	$^4D^\circ$
		7/2	*40 048.72*		91		7	
		9/2	*40 145.90*		100			
$3d\,4s(^3D)5p$	$^4D^\circ$	1/2	*40 044.63*		97			
		3/2	*40 073.49*		95			
		5/2	*40 128.13*		89		6	$^4F^\circ$
		7/2	*40 210.88*		88		8	$^4F^\circ$
$3d\,4s(^3D)4d$	^2P	3/2	40 063.38	1.295	76		10	^2D
		1/2	40 070.30	0.660	94			
$3d\,4s(^3D)5p$	$^2F^\circ$	5/2	*40 104.19*		76		15	$3d^2(^1G)4p\ ^2F^\circ$
		7/2	*40 151.08*		75		16	
$3d\,4s(^3D)4d$	^2D	3/2	40 257.52	1.305	47		22	$(^3D)\ ^4S$
		5/2	40 334.31	1.196	66		17	$(^1D)\ ^2D$
$3d\,4s(^3D)4d$	^4S	3/2	40 282.16	1.535	72		13	^2P
$3d\,4s(^3D)5p$	$^2D^\circ$	3/2	*40 347.34*		68		16	$3d^2(^1D)4p\ ^2D^\circ$
		5/2	*40 351.30*		67		15	

(Continued)

Sc I—Continued

Configuration	Term	J	Level (cm^{-1})	g	Leading percentages		
$3d\,4s(^3D)4d$	^2G	$7/2$	40 418.55		69	27	$(^1D)\ ^2G$
		$9/2$	40 562.06		69	27	
$3d\,4s(^3D)5p$	$^2P°$	$1/2$	*40 499.71*		77	11	$3d^2(^3P)4p\ ^2P°$
		$3/2$	*40 594.07*		72	10	
$3d\,4s(^3D)4d$	^4F	$3/2$	40 521.27	0.401	99		
		$5/2$	40 554.99	1.035	99		
		$7/2$	40 603.95	1.235	99		
		$9/2$	40 670.87	1.336	100		
$3d\,4s(^3D)5p$	$^4P°$	$1/2$	*40 595.28*	2.570	94	5	$3d^2(^3P)4p\ ^4P°$
		$3/2$	*40 644.64*	1.687	88	5	$3d4s(^3D)5p\ ^2P°$
		$5/2$	*40 715.42*	1.58	94	5	$3d^2(^3P)4p\ ^4P°$
$3d^3$	^2F	$5/2$	40 802.76	0.843	77	22	$3d^2(^3F)5s\ ^2F$
		$7/2$	40 825.78	1.140	80	20	
$3d^2(^3P)4p$	$^2D°$	$3/2$	*41 153.42*		46	30	$3d4s(^1D)5p\ ^2D°$
		$5/2$	*41 162.52*		45	30	
$3d\,4s(^3D)4d$	^4P	$1/2$	41 446.85	2.659	88	5	^2S
		$3/2$	41 474.87	1.725	97		
		$5/2$	41 505.60	1.60	97		
$3d^2(^3F)5s$	^4F	$3/2$	41 921.89	0.395	100		
		$5/2$	41 960.97	1.021	100		
		$7/2$	42 015.58	1.237	100		
		$9/2$	42 085.18	1.32	100		
$3d\,4s(^1D)4d$	^2F	$5/2$	42 149.66		87	8	$3d^2(^1D)4d\ ^2F$
		$7/2$	42 198.84		86	8	
$3d\,4s(^1D)4d$	^2D	$5/2$	42 445.55		68	22	$(^3D)\ ^2D$
		$3/2$	42 466.39	0.802	69	23	
$3d\,4s(^1D)5p$	$^2P°$	$3/2$	*42 780.41*		69	10	$(^3D)\ ^2P°$
		$1/2$	*42 819.49*		67	11	
$3d\,4s(^3D)4d$	^2S	$1/2$	42 877.65	1.991	45	41	$(^1D)\ ^2S$
$3d^3$	$^2D1$	$5/2$	42 917.83	1.19	44	35	$^2D2$
		$3/2$	42 937.50	0.78	54	45	
$3d\,4s(^1D)5p$	$^2F°$	$5/2$	*42 938.79*		82	12	$3d^2(^1D)5p\ ^2F°$
		$7/2$	*42 978.81*		83	12	
$3d\,4s(^1D)4d$	^2G	$9/2$	42 942.51	1.01	63	30	$(^3D)\ ^2G$
		$7/2$	42 969.78	0.93	63	30	
$3d\,4s(^1D)5p$	$^2D°$	$3/2$	*43 170.45*		52	18	$3d^2(^3P)4p\ ^2P°$
		$5/2$	*43 252.56*		51	18	$3d^2(^3P)4p\ ^2D°$
$3d\,4s(^1D)4d$	^2P	$1/2$	43 429.68	0.680	87	7	$3d^2(^1D)4d\ ^2P$
		$3/2$	43 435.40	1.336	86	7	
$4s^2\,4d$	^2D	$3/2$	43 597.16		85	11	$3d4s(^3D)4d\ ^2D$
		$5/2$	43 658.53		86	11	

Sc I—Continued

Configuration	Term	*J*	Level (cm^{-1})	*g*	Leading percentages		
$3d\,4s(^3D)6s$	^4D	1/2	43 809.76	0.009			
		3/2	43 814.47				
		5/2	43 898.31				
		7/2	43 988.20	1.42			
$3d^2(^1G)4p$	$^2F^\circ$	5/2	*43 830.12*	0.845	72	16	$3d4s(^3D)5p\ ^2F^\circ$
		7/2	*43 860.12*	1.14	71	17	
$3d(^2D)4p^2(^3P)$	^4P	1/2	44 030.34	2.665			
		3/2	44 107.25	1.726			
		5/2	44 238.23	1.60			
$3d^2(^3P)4p$	$^2P^\circ$	1/2	*44 105.45*	0.668	52	16	$3d4s(^1D)5p\ ^2P^\circ$
		3/2	*44 189.29*	1.331	52	15	
$3d(^2D)4p^2(^3P)$	^2P	3/2	44 594.97	1.35			
		1/2	44 690.65	0.670			
$3d(^2D)4p^2(^3P)$	^4F	3/2	44 823.21	0.399			
		5/2	44 909.55	0.992			
		7/2	45 016.43	1.22			
		9/2	45 125.73	1.33			
$3d(^2D)4p^2(^3P)$	^2F	5/2	44 838.56	0.90			
		7/2	44 941.81	1.16			
$3d\,4s(^1D)4d$	^2S	1/2	45 514.98	2.00	49	49	$(^3D)\ ^2S$
$3d(^2D)4p^2(^3P)$	^4D	1/2	45 574.64	0.00			
		3/2	45 605.80	1.188			
		5/2	45 659.09	1.38			
		7/2	45 737.17	1.41			
$3d^2(^3F)5p$	$^4G^\circ$	5/2	*45 610.52*		100		
		7/2	*45 645.10*		100		
		9/2	*45 691.26*		100		
		11/2	*45 761.23*		100		
$3d^2(^3F)4d$	^4G	5/2	45 715.79		88	10	^2F
		7/2	45 752.28		80	18	^4H
		9/2	45 804.10	1.16	79	20	^4H
		11/2	45 870.92		82	17	^4H
$3d\,4s(^3D)5d$	^4D	1/2	45 875.18	0.00			
		3/2	45 900.04	1.183			
		5/2	45 945.09				
		7/2	46 016.63	1.4			
$3d^2(^3F)4d$	^4H	7/2	45 878.06	0.65	82	17	^4G
		9/2	45 925.09	1.00	80	20	
		11/2	45 985.91	1.14	83	16	
		13/2	46 054.28	1.21	100		
$3d\,4s(^3D)5d$	^4G	5/2	45 886.66	0.62			
		7/2	45 931.16	0.97			
		9/2	45 988.93	1.08			
		11/2	46 053.54	1.24			
		3/2	*45 898.61*				

(Continued)

Sc I—Continued

Configuration	Term	J	Level (cm^{-1})	g	Leading percentages		
$3d^2(^3F)4d$	^4D	1/2	45 927.81	0.00	97		
		3/2	45 947.35		97		
		5/2	45 983.23	1.33	96		
		7/2	46 042.69	1.43	91	7	^2F
$3d^2(^3F)5p$	$^4F^\circ$	3/2	*46 206.80*		100		
		5/2	*46 255.40*		99		
		7/2	*46 266.21*		99		
		9/2	*46 369.23*		100		
$3d\,4s(^3D)5d$	^4F	3/2	46 329.23	0.44			
		5/2	46 354.41	1.02			
		7/2	46 403.30	1.23			
		9/2	46 458.64	1.31			
	^2F	5/2	46 378.86	0.86			
		7/2	46 459.66	1.14			
$3d^2(^3F)5p$	$^4D^\circ$	1/2	*46 485.47*		100		
		3/2	*46 517.55*		99		
		5/2	*46 570.25*		98		
		7/2	*46 641.64*		97		
	^4P	5/2	46 623.54	1.60			
	^2D	3/2	46 914.54	0.83			
		5/2	46 989.52	1.13			
$3d^2(^3F)5p$	$^2D^\circ$	3/2	*47 229.54*		84	7	$3d4s(^1D)5p\ ^2D^\circ$
		5/2	*47 314.53*		84	7	
	^2D	3/2	47 375.66				
		5/2	47 425.46				
$3d\,4s(^3D)7s$	^4D	1/2	47 475.90				
		3/2	47 507.39				
		5/2	47 563.31				
		7/2	47 652.61				
	^4P	1/2	47 488.72				
		3/2	47 535.78				
		5/2	47 604.59				
	^2G	7/2	47 514.22				
		9/2	47 589.73				
	^4P	3/2	48 324.66	1.73			
		5/2	48 373.17	1.6			
	^4P	1/2	48 830.11	2.64			
		3/2	48 869.56	1.73			
		5/2	48 920.60	1.59			
$3d\,4s(^1D)7p$	$^2P^\circ$	3/2	*51 110.61*				
Sc II $3d4s(^3D_1)$	*Limit*		**52 922.0**				
$3d\,4s(^1D)9p$	$^2D^\circ$	3/2,5/2	*53 162.8*				

Sc I—Continued

Configuration	Term	J	Level (cm^{-1})	g	Leading percentages
$3d\,4s(^1D)9p$	$^2P°$	3/2	53 222.9		
$3d^2(^3F_3)7p$	$^2F°$	5/2,7/2	53 692.6		
$3d\,4s(^1D)10p$	$^2D°$	3/2,5/2	53 706.4		
$3d\,4s(^1D)10p$	$^2P°$	3/2	53 784.		
$3d\,4s(^1D)11p$	$^2D°$	3/2,5/2	54 080.0		
$3d\,4s(^1D)11p$	$^2P°$	3/2	54 111.8		
$3d\,4s(^1D)12p$	$^2D°$	3/2,5/2	54 343.6		
$3d\,4s(^1D)12p$	$^2P°$	3/2	54 366.6		
$3d\,4s(^1D)13p$	$^2D°$	3/2,5/2	54 539.3		
$3d\,4s(^1D)13p$	$^2P°$	1/2	54 553.4		
		3/2	54 556.6		
$3d\,4s(^1D)14p$	$^2D°$	3/2,5/2	54 687.6		
$3d\,4s(^1D)14p$	$^2P°$	3/2	54 712.1		
$3d\,4s(^1D)15p$	$^2D°$	3/2,5/2	54 803.2		
$3d\,4s(^1D)15p$	$^2P°$	3/2	54 822.0		
$3d\,4s(^1D)16p$	$^2D°$	3/2,5/2	54 895.0		
$3d^2(^3F_3)8p$	$^2F°$	5/2,7/2	54 903.5		
$3d\,4s(^1D)16p$	$^2P°$	1/2	54 906.2		
		3/2	54 908.4		
$3d\,4s(^1D)17p$	$^2D°$	3/2,5/2	54 971.4		
$3d\,4s(^1D)18p$	$^2D°$	3/2,5/2	55 030.7		
$3d\,4s(^1D)18p$	$^2P°$	1/2	55 036.4		
		3/2	55 038.2		
$3d\,4s(^1D)19p$	$^2P°$	1/2	55 084.2		
		3/2	55 086.0		
$3d\,4s(^1D)20p$	$^2P°$	1/2	55 125.0		
		3/2	55 126.6		
$3d\,4s(^1D)21p$	$^2P°$	1/2	55 159.2		
		3/2	55 160.8		
$3d\,4s(^1D)22p$	$^2P°$	3/2	55 190.2		
$3d\,4s(^1D)23p$	$^2P°$	1/2	55 214.3		
		3/2	55 215.4		
$3d\,4s(^1D)24p$	$^2P°$	1/2	55 236.3		
		3/2	55 237.4		

(Continued)

Sc I—Continued

Configuration	Term	J	Level (cm^{-1})	g	Leading percentages
$3d\,4s(^1D)25p$	$^2P°$	$1/2$	*55 255.6*		
		$3/2$	*55 256.2*		
$3d\,4s(^1D)26p$	$^2P°$	$1/2$	*55 272.6*		
		$3/2$	*55 273.1*		
$3d\,4s(^1D)27p$	$^2P°$	$1/2$	*55 287.4*		
		$3/2$	*55 288.0*		
$3d\,4s(^1D)28p$	$^2P°$	$3/2$	*55 301.2*		
$3d\,4s(^1D)29p$	$^2P°$	$3/2$	*55 312.6*		
$3d\,4s(^1D)30p$	$^2P°$	$3/2$	*55 323.2*		
$3d\,4s(^1D)31p$	$^2P°$	$3/2$	*55 332.8*		
$3d\,4s(^1D)32p$	$^2P°$	$3/2$	*55 341.4*		
$3d\,4s(^1D)33p$	$^2P°$	$3/2$	*55 349.0*		
$3d\,4s(^1D)34p$	$^2P°$	$3/2$	*55 356.8*		
$3d\,4s(^1D)35p$	$^2P°$	$3/2$	*55 362.8*		
$3d\,4s(^1D)36p$	$^2P°$	$3/2$	*55 368.7*		
$3d\,4s(^1D)37p$	$^2P°$	$3/2$	*55 373.6*		
$3d\,4s(^1D)38p$	$^2P°$	$3/2$	*55 378.1*		
Sc II $3d4s(^1D)$	***Limit***		**55 463**		
$3d^2(^3F)9p$	$^2F°$?	$5/2,7/2$	*55 657.3*		
$3d^2(^3F)10p$	$^2F°$?	$5/2,7/2$	*56 147.0*		
$3d^2(^3F)11p$	$^2F°$?	$5/2,7/2$	*56 486.9*		
$3d^2(^3F)12p$	$^2F°$?	$5/2,7/2$	*56 733.2*		
$3d^2(^3F)13p$	$^2F°$?	$5/2,7/2$	*56 916.3*		
$3d^2(^3F)14p$	$^2F°$?	$5/2,7/2$	*57 057.2*		
$3d^2(^3F)15p$	$^2F°$?	$5/2,7/2$	*57 167.5*		
$4s^2\,4f$	$^2F°$	$7/2$	*57 485.7*		
		$5/2$	*57 503.9*		
Sc II $3d^2(^3F_3)$	***Limit***		**57 805**		
$4s^2\,5f$	$^2F°$	$5/2,7/2$	*60 090.4*		
$4s^2\,6f$	$^2F°$	$5/2,7/2$	*61 498.0*		
$4s^2\,7f$	$^2F°$	$5/2,7/2$	*62 343.8*		
$4s^2\,8f$	$^2F°$	$5/2,7/2$	*62 886.2*		
$4s^2\,9f$	$^2F°$	$5/2,7/2$	*63 267.0*		
$4s^2\,10f$	$^2F°$	$5/2,7/2$	*63 533.2*		
$4s^2\,11f$	$^2F°$	$5/2,7/2$	*63 729.6*		
$4s^2\,12f$	$^2F°$	$5/2,7/2$	*63 878.6*		
$4s^2\,13f$	$^2F°$	$5/2,7/2$	*63 993.8*		
$4s^2\,14f$	$^2F°$	$5/2,7/2$	*64 086.2*		
$4s^2\,15f$	$^2F°$	$5/2,7/2$	*64 160.4*		
$4s^2\,17f$	$^2F°$	$5/2,7/2$	*64 271.2*		
$4s^2\,18f$	$^2F°$	$5/2,7/2$	*64 312.4*		
$4s^2\,19f$	$^2F°$	$5/2,7/2$	*64 348.0*		
$4s^2\,20f$	$^2F°$	$5/2,7/2$	*64 378.3*		
$4s^2\,21f$	$^2F°$	$5/2,7/2$	*64 404.8*		
$4s^2\,22f$	$^2F°$	$5/2,7/2$	*64 426.6*		
$4s^2\,23f$	$^2F°$	$5/2,7/2$	*64 445.8*		
Sc II $4s^2(^1S)$	***Limit***		**64 658**		

Sc II

$Z=21$

Ca I isoelectronic sequence

Ground state: $1s^22s^22p^63s^23p^63d4s\ ^3D_1$

Ionization energy $=$ 103 237.1±2 cm^{-1} (12.79987±0.00025 eV)

The first extensive analysis of Sc II was reported by Russell and Meggers (1927). They classified 142 lines in the range 2540–6600 Å as combinations among 53 energy levels of eight configurations. Many of these levels were determined to better accuracy by Neufeld (1970). Neufeld also provided g-values for 18 levels.

Johansson and Litzén (1980) have reobserved the spectrum from 1100–10 000 Å with a pulsed hollow cathode. They considerably extended the analysis, adding 11 new high-lying configurations. Their newly determined level values are quoted here, and are presumably accurate to a few units in the last decimal place given.

Wyart has calculated $3d4p$, $3d5p$, $3d4f$, $3d5f$, and $4s4p$ with configuration interaction. His results, as reported by Johansson and Litzén, are given here. The percentage compositions for the $3d5g$ configuration are from Goldschmidt (1982).

The ionization energy was derived by Johansson and Litzén from a polarization formula applied to the centers of gravity of the $3d4f$, $3d5f$, and $3d5g$ configurations.

References

Goldschmidt, Z. (1982), J. Phys. **B15**, 191.
Johansson, S., and Litzén, V. (1980), Phys. Scr. **22**, 49.
Neufeld, L. W. (1970), Dissertation, Kansas State University.
Russell, H. N., and Meggers, W. F. (1927), Sci. Papers Natl. Bur. Stand. (U.S.) **22**, 329.

Sc II

Configuration	Term	J	Level (cm^{-1})	g	Leading percentages
$3d\ 4s$	^3D	1	0.00	0.501	
		2	67.72	1.163	
		3	177.76	1.330	
$3d\ 4s$	^1D	2	2 540.95	0.998	
$3d^2$	^3F	2	4 802.87	0.665	
		3	4 883.57	1.083	
		4	4 987.79		
$3d^2$	^1D	2	10 944.56		
$4s^2$	^1S	0	11 736.36		
$3d^2$	^3P	0	12 074.10		
		1	12 101.50	1.487	
		2	12 154.42	1.481	
$3d^2$	^1G	4	14 261.32		
$3d^2$	^1S	0	25 955.2		
$3d\ 4p$	$^1D°$	2	*26 081.34*	0.999	99
$3d\ 4p$	$^3F°$	2	*27 443.71*	0.671	99
		3	*27 602.45*	1.086	99
		4	*27 841.35*	1.233	100
$3d\ 4p$	$^3D°$	1	*27 917.78*	0.506	100
		2	*28 021.29*	1.162	99
		3	*28 161.17*	1.323	99

(Continued)

Sc II—Continued

Configuration	Term	J	Level (cm^{-1})	g	Leading percentages		
3d 4p	^{3}P°	0	*29 736.27*		94	6	4s4p ^{3}P°
		1	*29 742.16*	1.484	92	6	
		2	*29 823.93*	1.495	94	6	
3d 4p	^{1}P°	1	*30 815.70*	1.006	93	5	4s4p ^{1}P°
3d 4p	^{1}F°	3	*32 349.98*		100		
4s4p	^{3}P°	0	*39 002.20*		94	6	3d4p ^{3}P°
		1	*39 115.04*		94	6	
		2	*39 345.52*		94	6	
4s4p	^{1}P°	1	*55 715.36*		91	5	3d4p ^{1}P°
3d 5s	^{3}D	1	57 551.88				
		2	57 614.40				
		3	57 743.92				
3d 5s	^{1}D	2	58 252.09				
3d 4d	^{1}F	3	59 528.42				
3d 4d	^{3}D	1	59 875.08				
		2	59 929.46				
		3	60 001.91				
3d 4d	^{3}G	3	60 267.16				
		4	60 348.46				
		5	60 457.12				
3d 4d	^{1}P	1	60 400.41				
3d 4d	^{3}S	1	61 071.43				
3d 4d	^{3}F	2	63 374.63				
		3	63 445.16				
		4	63 528.54				
3d 4d	^{1}D	2	64 366.68				
3d 4d	^{3}P	0	64 615.77				
		1	64 646.70				
		2	64 705.89				
3d 4d	^{1}G	4	65 236.04				
3d 5p	^{1}D°	2	*66 048.39*		94	3	^{3}F°
3d 5p	^{3}D°	1	*66 389.74*		100		
		2	*66 492.66*		60	35	^{3}F°
		3	*66 583.86*		78	22	^{3}F°
3d 5p	^{3}F°	2	*66 459.64*		62	37	^{3}D°
		3	*66 563.73*		78	22	^{3}D°
		4	*66 718.99*		100		
3d 4d	^{1}S	0	67 216.56				

Sc II—Continued

Configuration	Term	J	Level (cm^{-1})	g	Leading percentages		
$3d\,5p$	$^3P^\circ$	0	67 236.7		100		
		1	67 297.68		99		
		2	67 396.19		100		
$3d\,5p$	$^1F^\circ$	3	67 743.72		100		
$3d\,5p$	$^1P^\circ$	1	68 498.06		95	4	$4s4p\ ^1P^\circ$
$4p^2$	^1D	2	74 433.3				
$3d4f$	$^1G^\circ$	4	75 221.47		71	15	$^3G^\circ$
$3d4f$	$^3F^\circ$	2	75 308.13		89	8	$^1D^\circ$
		3	75 373.10		43	28	$^3G^\circ$
		4	75 470.24		65	22	$^3G^\circ$
$3d4f$	$^3G^\circ$	3	75 308.70		58	40	$^3F^\circ$
		4	75 390.49		43	20	$^3F^\circ$
		5	75 471.25		79	18	$^3H^\circ$
$3d4f$	$^1F^\circ$	3	75 552.46		65	15	$^3F^\circ$
$3d4f$	$^3H^\circ$	4	75 561.00		71	22	$^3G^\circ$
		5	75 610.83		76	21	$^3G^\circ$
		6	75 699.88		100		
$3d4f$	$^1D^\circ$	2	75 590.84		76	14	$^3D^\circ$
$3d4f$	$^3D^\circ$	1	75 650.90		92	4	$^1P^\circ$
		2	75 680.69		76	14	$^1D^\circ$
		3	75 715.75		83	14	$^1D^\circ$
$3d4f$	$^3P^\circ$	2	75 912.58		90	7	$^3D^\circ$
		1	75 951.88		77	16	$^1P^\circ$
		0	75 994.43		100		
$3d4f$	$^1H^\circ$	5	75 919.75		94	6	$^3H^\circ$
$3d4f$	$^1P^\circ$	1	76 073.28		79	19	$^3P^\circ$
$4p^2$	^3P	0	76 243.2				
		1	76 360.8				
		2	76 589.3				
$3d\,6s$	^3D	1	77 195.19				
		2	77 256.99				
		3	77 387.17				
$3d\,6s$	^1D	2	77 833.88				
$4s5s$	^3S	1	78 265.7				
$3d\,5d$	^1F	3	78 394.05				
$3d\,5d$	^3D	1	78 479.25				
		2	78 539.96				
		3	78 612.48				

(Continued)

Sc II—Continued

Configuration	Term	J	Level (cm^{-1})	g	Leading percentages			
$3d\,5d$	^{3}G	3	78 648.06					
		4	78 713.21					
		5	78 820.24					
$3d\,5d$	^{1}P	1	78 757.46					
$3d\,5d$	^{3}S	1	79 174.74					
$3d\,5d$	^{3}F	2	79 852.02					
		3	79 925.59					
		4	80 001.24					
$3d\,5d$	^{1}D	2	80 395.38					
$3d\,5d$	^{1}G	4	80 711.71					
$3d\,5d$	^{3}P	0	80 974.94					
		1	81 040.25					
		2	81 103.75					
$4s5s$	^{1}S	0	81 121.77					
$4s4d$	^{3}D	1	82 791.8					
		2	82 806.3					
		3	82 828.3					
$3d5f$	^{1}G°	4	*85 353.96*		58		20	^{3}F°
$3d5f$	^{3}G°	3	*85 405.39*		50		46	^{3}F°
		4	*85 605.99*		59		20	^{3}F°
		5	*85 636.05*		70		29	^{3}H°
$3d5f$	^{3}H°	4	*85 445.35*		69		21	^{3}G°
		5	*85 504.42*		61		28	^{1}H°
		6	*85 666.26*		100			
$3d5f$		3	*85 452.99*		31	^{1}F°	25	^{3}F°
$3d5f$	^{1}D°	2	*85 558.24*		40		31	^{3}D°
$3d5f$	^{3}F°	4	*85 580.3*		56		30	^{1}G°
$3d(^{2}D_{3/2})5g$	$^{2}[9/2]$	5	85 587.87		100			
		4	85 588.17		100			
$3d5f$	^{3}D°	1	*85 591.59*		75		14	^{1}P°
		2	*85 661.0*		41		39	^{1}D°
		3	*85 700.7*		71		24	^{1}F°
$3d(^{2}D_{3/2})5g$	$^{2}[7/2]$	3	85 611.4		99		1	$(^{2}D_{5/2})\ ^{2}[7/2]$
		4	85 611.55		99		1	
$3d5f$	^{1}F°	3	*85 622.6*		44		25	^{3}F°
$3d(^{2}D_{3/2})5g$	$^{2}[11/2]$	5	85 659.66		95		5	$(^{2}D_{5/2})\ ^{2}[11/2]$
		6	85 660.42		95		5	
$3d(^{2}D_{3/2})5g$	$^{2}[5/2]$	2,3	85 688.9		96		4	$(^{2}D_{5/2})\ ^{2}[5/2]$

Sc II—Continued

Configuration	Term	J	Level (cm^{-1})	g	Leading percentages		
3d5f	^{3}P°	2	*85 778.78*		70	23	^{3}D°
		1	*85 800.62*		65	22	^{3}D°
		0	*85 832.20*		100		
3d($^2D_{5/2}$)5g	$^2[9/2]$	5	85 788.01		100		
		4	85 788.22		100		
3d($^2D_{5/2}$)5g	$^2[11/2]$	5	85 803.49		95	5	($^2D_{3/2}$) $^2[11/2]$
		6	85 804.11		95	5	
3d5f	^{1}H°	5	*85 817.40*		89	10	^{3}H°
3d($^2D_{5/2}$)5g	$^2[7/2]$	3	85 819.8		99	1	($^2D_{3/2}$) $^2[7/2]$
		4	85 820.00		99	1	
3d($^2D_{5/2}$)5g	$^2[5/2]$	2,3	85 867.9		96	4	($^2D_{3/2}$) $^2[5/2]$
3d($^2D_{5/2}$)5g	$^2[13/2]$	7	85 877.52		100		
		6	85 878.48		100		
3d($^2D_{5/2}$)5g	$^2[3/2]$	1,2	85 915.9		100		
3d 7s	^{3}D	1	86 557.7				
		2	86 607.5				
		3	86 751.0				
3d 6d	^{3}G	3	87 233.09				
		4	87 295.23				
		5	87 405.48				
3s4d	^{1}D	2	89 940.27				
3d6f	^{3}G°	3	*90 821.1*				
		4	*91 074.4*				
		5	*91 094.7*				
3d6f	^{3}H°	5	*90 937.9*				
3d6f	^{3}F°	4	*91 062.5*				
3d6f	^{1}F°	3	*91 083.4*				
3d6f	^{3}D°	3	*91 128.0*				
3d6f	^{3}P°	2	*91 165.0*				
3d6f	^{1}H°	5	*91 197.0*				
Sc III ($^2D_{3/2}$)	***Limit***		**103 237.1**				

Sc III

$Z=21$

K I isoelectronic sequence

Ground state: $1s^2 2s^2 2p^6 3s^2 3p^6 3d\ ^2D_{3/2}$

Ionization energy = 199 677.37±0.1 cm^{-1} (24.75704±0.00005 eV)

The early work on Sc III was reported by Gibbs and White (1926), Smith (1927), and Russell and Lang (1927).

Two modern analyses of Sc III were published by Holmstrom (1972) and Van Deurzen, Conway, and Davis (1973). Holmstrom lists 64 observed lines between 557 and 8882 Å. He states the uncertainty of his level values as ±0.4 cm^{-1}. Van Deurzen, Conway, and Davis observed 93 lines between 557 and 9371 Å. Their level values have a reported uncertainty of less than 0.1 cm^{-1}. This compilation is taken from Van Deurzen et al. except for the ^{2}H terms given only by Holmstrom. They are derived here from Holmstrom's lines and the levels of Van Deurzen et al.

Van Deurzen et al. calculated the ionization energy from four members ($n=5$–8) of the ng series. Their value agrees with the value calculated by Holmstrom with a polarization formula. The uncertainty in eV is determined by the uncertainty in the conversion factor.

References

Gibbs, R. C., and White, H. E. (1926), Proc. Natl. Acad. Sci. **12**, 598.
Holmstrom, J. E. (1972), Phys. Scr. **5**, 249.
Russell, H. N., and Lang, R. J. (1927), Astrophys. J. **66**, 19.
Smith, S. (1927), Proc. Natl. Acad. Sci. **13**, 65.
Van Deurzen, C. H. H., Conway, J. G., and Davis, S. P. (1973), J. Opt. Soc. Am. **63**, 158.

Sc III

Configuration	Term	J	Level (cm^{-1})
$3p^6(^1S)3d$	^{2}D	3/2	0.00
		5/2	197.64
$3p^6(^1S)4s$	^{2}S	1/2	25 539.32
$3p^6(^1S)4p$	^{2}P°	1/2	*62 104.30*
		3/2	*62 578.18*
$3p^6(^1S)4d$	^{2}D	3/2	112 257.62
		5/2	112 302.95
$3p^6(^1S)5s$	^{2}S	1/2	114 862.48
$3p^6(^1S)5p$	^{2}P°	1/2	*128 107.12*
		3/2	*128 283.15*
$3p^6(^1S)4f$	^{2}F°	5/2	*136 873.87*
		7/2	*136 874.12*
$3p^6(^1S)5d$	^{2}D	3/2	148 130.03
		5/2	148 150.14
$3p^6(^1S)6s$	^{2}S	1/2	149 194.03
$3p^6(^1S)6p$	^{2}P°	1/2	*155 489.78*
		3/2	*155 575.20*
$3p^6(^1S)5f$	^{2}F°	5/2,7/2	*159 472.24*
$3p^6(^1S)5g$	^{2}G	7/2,9/2	160 072.18
$3p^6(^1S)6d$	^{2}D	3/2	165 592.55
		5/2	165 603.29
$3p^6(^1S)7s$	^{2}S	1/2	166 157.17
$3p^6(^1S)7p$	^{2}P°	1/2	*169 637.96*
		3/2	*169 685.9*
$3p^6(^1S)6f$	^{2}F°	5/2,7/2	*171 787.64*
$3p^6(^1S)6g$	^{2}G	7/2,9/2	172 177.41
$3p^6(^1S)6h$	^{2}H°	9/2,11/2	*172 224.57*
$3p^6(^1S)7d$	^{2}D	3/2	175 457.03
		5/2	175 463.56
$3p^6(^1S)8s$	^{2}S	1/2	175 795.73
$3p^6(^1S)7f$	^{2}F°	5/2,7/2	*179 214.70*
$3p^6(^1S)7g$	^{2}G	7/2,9/2	179 477.24
$3p^6(^1S)7h$	^{2}H°	9/2,11/2	*179 508.37*
$3p^6(^1S)8g$	^{2}G	7/2,9/2	184 214.61
Sc IV ($^1S_0$)	***Limit***		**199 677.37**

Sc IV

$Z = 21$

Ar I isoelectronic sequence

Ground state: $1s^2 2s^2 2p^6 3s^2 3p^6\ {}^1S_0$

Ionization energy = 592 732±3 cm^{-1} (73.4900±0.0004 eV)

Four resonance lines were classified by Kruger, Weissberg, and Phillips (1937). Smitt (1973) has carried out the extensive analysis quoted here, which confirmed only two of the four resonance lines. The uncertainty of his level values is estimated to be ±2 cm^{-1}. He calculated the ionization energy with a polarization formula, using $3p^5ng$, nh, and ni terms.

The $3s3p^6 5p\ {}^1P^{\circ}$ term was observed by Kastner, Crooker, Behring, and Cohen (1977) in absorption in a high voltage spark.

References

Kastner, S. O., Crooker, A. M., Behring, W. E., and Cohen, L. (1977), Phys. Rev. **A16**, 577.

Kruger, P. G., Weissberg, S. G., and Phillips, L. W. (1937), Phys. Rev. **51**, 1090.

Smitt, R. (1973), Phys. Scr. **8**, 292.

Sc IV

Configuration	Term	J	Level (cm^{-1})
$3s^2 3p^6$	^1S	0	0.0
$3s^2 3p^5 3d$	$^3P^{\circ}$	0	*239 723.2*
		1	*240 403.0*
		2	*241 814.0*
$3s^2 3p^5 3d$	$^3F^{\circ}$	4	*250 708.2*
		3	*252 108.3*
		2	*253 405.7*
$3s^2 3p^5 3d$	$^1D^{\circ}$	2	*267 424.4*
$3s^2 3p^5 3d$	$^3D^{\circ}$	3	*268 034.2*
		1	*269 426.5*
		2	*269 459.9*
$3s^2 3p^5 3d$	$^1F^{\circ}$	3	*271 055.4*
$3s^2 3p^5 4s$	$^3P^{\circ}$	2	*333 090.8*
		1	*334 405.1*
		0	*337 340.1*
$3s^2 3p^5 4s$	$^1P^{\circ}$	1	*337 483.5*
$3s^2 3p^5 3d$	$^1P^{\circ}$	1	*345 005.4*
$3s^2 3p^5 4p$	^3S	1	371 735.2
$3s^2 3p^5 4p$	^3D	3	378 077.0
		2	378 418.6
		1	380 148.3
$3s^2 3p^5 4p$	^3P	2	381 712.0
		0	384 817.8
		1	385 570.0
$3s^2 3p^5 4p$	^1P	1	383 527.1
$3s^2 3p^5 4p$	^1D	2	384 661.3
$3s^2 3p^5 4p$	^1S	0	397 510.7
$3s3p^6 3d$	^3D	1	429 695.7
		2	429 842.4
		3	430 078.5
$3s^2 3p^5 4d$	$^3P^{\circ}$	0	*440 517.9*
		1	*440 983.0*
		2	*441 927.3*
$3s3p^6 3d$	^1D	2	442 046.0
$3s^2 3p^5 4d$	$^3F^{\circ}$	4	*442 558.6*
		3	*443 191.3*
		2	*444 427.4*
$3s^2 3p^5 4d$	$^3D^{\circ}$	3	*445 207.5*
		1	*447 622.4*
		2	*448 725.1*
$3s^2 3p^5 4d$	$^1D^{\circ}$	2	*448 062.0*
$3s^2 3p^5 4d$	$^1F^{\circ}$	3	*448 607.8*
$3s^2 3p^5 4d$	$^1P^{\circ}$	1	*453 972.7*
$3s^2 3p^5(^2P^{\circ}_{3/2})5s$	$^2[3/2]^{\circ}$	2	*459 496.9*
		1	*460 426.9*
$3s^2 3p^5(^2P^{\circ}_{1/2})5s$	$^2[1/2]^{\circ}$	0	*463 768.9*
		1	*464 457.2*
$3s^2 3p^5(^2P^{\circ}_{3/2})5p$	$^2[1/2]$	1	474 764.4
		0	480 286.6

(Continued)

Sc IV—Continued

Configuration	Term	J	Level (cm^{-1})
$3s^2\,3p^5(^2P^\circ_{3/2})5p$	$^2[5/2]$	3	476 818.5
		2	477 018.0
$3s^2\,3p^5(^2P^\circ_{3/2})5p$	$^2[3/2]$	1	477 746.6
		2	478 213.2
$3s^2\,3p^5(^2P^\circ_{3/2})4f$	$^2[3/2]$	1	478 495.9
		2	478 767.7
$3s^2\,3p^5(^2P^\circ_{3/2})4f$	$^2[9/2]$	5	478 609.5
		4	478 768.4
$3s^2\,3p^5(^2P^\circ_{3/2})4f$	$^2[5/2]$	3	479 452.9
		2	480 097.9
$3s^2\,3p^5(^2P^\circ_{3/2})4f$	$^2[7/2]$	3	479 908.9
		4	479 933.2
$3s^2\,3p^5(^2P^\circ_{1/2})5p$	$^2[3/2]$	1	481 377.4
		2	481 975.2
$3s^2\,3p^5(^2P^\circ_{1/2})5p$	$^2[1/2]$	1	481 939.9
		0	484 972.9
$3s^2\,3p^5(^2P^\circ_{1/2})4f$	$^2[7/2]$	3	483 609.4
		4	483 708.0
$3s^2\,3p^5(^2P^\circ_{1/2})4f$	$^2[5/2]$	3	483 962.8
		2	484 257.6
$3s^2\,3p^5(^2P^\circ_{3/2})6s$	$^2[3/2]^\circ$	2	*511 228.2*
		1	*511 630.3*
$3s^2\,3p^5(^2P^\circ_{1/2})6s$	$^2[1/2]^\circ$	1	*515 738*
$3s^2\,3p^5(^2P^\circ_{3/2})5g$	$^2[5/2]^\circ$	2	*521 999.2*
		3	*522 004.8*
$3s^2\,3p^5(^2P^\circ_{3/2})5g$	$^2[11/2]^\circ$	6	*522 110.2*
		5	*522 110.7*
$3s^2\,3p^5(^2P^\circ_{3/2})5g$	$^2[7/2]^\circ$	4	*522 304.5*
		3	*522 307.3*
$3s^2\,3p^5(^2P^\circ_{3/2})5g$	$^2[9/2]^\circ$	4	*522 424.8*
		5	*522 425.7*
$3s^2\,3p^5(^2P^\circ_{1/2})5g$	$^2[9/2]^\circ$	4	*526 561.6*
		5	*526 563.1*
$3s^2\,3p^5(^2P^\circ_{1/2})5g$	$^2[7/2]^\circ$	4	*526 571.2*
		3	*526 574.0*
$3s^2\,3p^5(^2P^\circ_{3/2})7s$	$^2[3/2]^\circ$	1	*537 845*

Configuration	Term	J	Level (cm^{-1})
$3s^2\,3p^5(^2P^\circ_{1/2})7s$	$^2[1/2]^\circ$	1	*542 052*
$3s^2\,3p^5(^2P^\circ_{3/2})6g$	$^2[5/2]^\circ$	2	*543 642.4*
$3s^2\,3p^5(^2P^\circ_{3/2})6g$	$^2[11/2]^\circ$	6	*543 708.3*
$3s^2\,3p^5(^2P^\circ_{3/2})6g$	$^2[7/2]^\circ$	4	*543 824.9*
$3s^2\,3p^5(^2P^\circ_{3/2})6h$	$^2[7/2]$	3	543 843.1
		4	543 843.6
$3s^2\,3p^5(^2P^\circ_{3/2})6h$	$^2[13/2]$	6,7	543 869.8
$3s^2\,3p^5(^2P^\circ_{3/2})6g$	$^2[9/2]^\circ$	4	*543 892.1*
		5	*543 893.3*
$3s^2\,3p^5(^2P^\circ_{3/2})6h$	$^2[9/2]$	4,5	543 942.1
$3s^2\,3p^5(^2P^\circ_{3/2})6h$	$^2[11/2]$	5,6	543 971.2
$3s^2\,3p^5(^2P^\circ_{1/2})6g$	$^2[9/2]^\circ$	5	*548 104.3*
$3s^2\,3p^5(^2P^\circ_{1/2})6g$	$^2[7/2]^\circ$	4	*548 105.9*
$3s^2\,3p^5(^2P^\circ_{1/2})6h$	$^2[11/2]$	5,6	548 236.5
$3s^2\,3p^5(^2P^\circ_{1/2})6h$	$^2[9/2]$	4,5	548 237.8
$3s^2\,3p^5(^2P^\circ_{3/2})7h$	$^2[13/2]$	6,7	556 840.9
$3s^2\,3p^5(^2P^\circ_{3/2})7i$	$^2[9/2]^\circ$	4,5	*556 863.5*
$3s^2\,3p^5(^2P^\circ_{3/2})7i$	$^2[15/2]^\circ$	7,8	*556 871.3*
$3s^2\,3p^5(^2P^\circ_{3/2})7h$	$^2[9/2]$	4,5	556 886.9
$3s^2\,3p^5(^2P^\circ_{3/2})7i$	$^2[11/2]^\circ$	5,6	*556 900.8*
$3s^2\,3p^5(^2P^\circ_{3/2})7h$	$^2[11/2]$	5,6	556 903.5
$3s^2\,3p^5(^2P^\circ_{3/2})7i$	$^2[13/2]^\circ$	6,7	*556 910.7*
$3s^2\,3p^5(^2P^\circ_{1/2})7h$	$^2[11/2]$	5,6	561 191.0
$3s^2\,3p^5(^2P^\circ_{1/2})7h$	$^2[9/2]$	4,5	561 191.0
$3s^2\,3p^5(^2P^\circ_{1/2})7i$	$^2[11/2]^\circ$	5,6	*561 213.9*
$3s^2\,3p^5(^2P^\circ_{1/2})7i$	$^2[13/2]^\circ$	6,7	*561 213.9*
Sc V ($^2P^\circ_{3/2}$)	***Limit***		**592 732**
$3s3p^6\,5p$	$^1P^\circ$	1	*652 700*

Sc V

$Z=21$

Cl I isoelectronic sequence

Ground state: $1s^2 2s^2 2p^6 3s^2 3p^5\ {}^2P^\circ_{3/2}$

Ionization = 741 000±2000 cm^{-1} (91.9±0.2 eV)

Measurements by Beckman (1937) (±10 cm^{-1}) and by Kruger and Phillips (1937) (±5 cm^{-1}) between 220 and 590 Å in Sc V established the $3s^2 3p^5$, $3s3p^6$, and $3s^2 3p^4 4s$ configurations. The level values given here for the $3s^2 3p^4 4s$ configuration are averages of their determinations. For the ${}^4P_{5/2}$ and ${}^4P_{1/2}$ levels, the values of Kruger and Phillips are not in accord with the isoelectronic sequence and are not used.

The values for $3s^2 3p^5\ {}^2P^\circ$ and $3s3p^6\ {}^2S$ are from Smitt (1973) (±0.8 cm^{-1}). The $3p^4({}^3P)3d$ terms are taken from Svensson and Ekberg (1968) (±10 cm^{-1}) and the $3p^4({}^1D)3d$ terms from Fawcett and Gabriel (1966) (±20 cm^{-1}). The $3p^4 5s$ terms were identified by Fawcett, Peacock, and Cowan (1968), whose measurements at 180 Å are stated to be accurate to 0.03 Å (±90 cm^{-1}).

The ionization energy was determined by extrapolation by Lotz (1967).

References

Beckman, A. (1937), Dissertation, Uppsala.

Fawcett, B. C., and Gabriel, A. H. (1966), Proc. Phys. Soc. **88**, 262.

Fawcett, B. C., Peacock, N. J., and Cowan, R. D. (1968), J. Phys. **B1**, 295.

Kruger, P. G., and Phillips, L. W. (1937), Phys. Rev. **51**, 1087.

Lotz, W. (1967), J. Opt. Soc. Am. **57**, 873.

Smitt, R. (1973), Phys. Scr. **8**, 292.

Svensson, L. A., and Ekberg, J. O. (1968), Ark. Fys. **37**, 65.

Sc V

Configuration	Term	J	Level (cm^{-1})
$3s^2 3p^5$	${}^2P^\circ$	3/2	*0.0*
		1/2	*4 325.6*
$3s3p^6$	${}^2S$	1/2	174 412.0
$3s^2 3p^4({}^3P)3d$	${}^2P$	1/2	254 638
		3/2	257 284
$3s^2 3p^4({}^3P)3d$	${}^2D$	3/2	264 077
		5/2	266 633
$3s^2 3p^4({}^1D)3d$	${}^2S$	1/2	337 650
$3s^2 3p^4({}^1D)3d$	${}^2P$	3/2	345 320
		1/2	346 890
$3s^2 3p^4({}^1D)3d$	${}^2D$	3/2	355 880
$3s^2 3p^4({}^3P)4s$	${}^4P$	5/2	386 387
		3/2	388 862
$3s^2 3p^4({}^3P)4s$	${}^2P$	3/2	395 498
		1/2	398 440
$3s^2 3p^4({}^1D)4s$	${}^2D$	5/2	410 046
		3/2	410 122
$3s^2 3p^4({}^1S)4s$	${}^2S$	1/2	437 508
$3s^2 3p^4({}^3P)5s$	${}^2P$	1/2	550 800
		3/2	552 600
$3s^2 3p^4({}^1D)5s$	${}^2D$	5/2	556 020
		3/2	557 400
Sc VI (${}^3P_2$)	*Limit*		**741 000**

Sc VI

$Z=21$

S I isoelectronic sequence

Ground state: $1s^2 2s^2 2p^6 3s^2 3p^4\ ^3P_2$

Ionization energy = 892 700±400 cm^{-1} (110.68±0.05 eV)

The analysis of Sc VI was initiated by Beckman (1937) and by Kruger and Pattin (1937), who reported terms of the $3s^2 3p^4$, $3s3p^5$, and $3s^2 3p^3 4s$ configurations. The $3s3p^5\ ^1P°$ term was found by Edlén (1942).

The $3s^2 3p^3 3d$ configuration was observed by Svensson and Ekberg (1968) and the level values given here (with an uncertainty of about ±5 cm^{-1}) are derived from their observations.

The values for the two lower configurations ($3s^2 3p^4$ and $3s3p^5$) are from the more accurate observations of Smitt, Svensson and Outred (1976).

The $3p^3 4s$ levels are derived from Beckman (±5 cm^{-1}) and the $3p^3 4d$ and $5s$ levels are from Fawcett, Peacock, and Cowan (1968) (±100 cm^{-1}). Fawcett, Cowan, and Hayes (1972) have observed transitions in the $3p^3 3d - 3p^3 4f$ array, but they are not connected with the present system.

We derived the ionization energy from the $3p^3(^4S°)4s$ and $5s\ ^3S°$ terms, adopting a value for the change in the effective quantum number between them of 1.0247 obtained from the $3p^6 ns$ terms of Cr VI from the analysis of Ekberg (1973).

References

Beckman, A. (1937), Dissertation, Uppsala.

Edlén, B. (1942), Phys. Rev. **62**, 434.

Ekberg, J. O. (1973), Phys. Scr. **8**, 35.

Fawcett, B. C., Cowan, R. D., and Hayes, R. W. (1972), J. Phys. **B5**, 2143.

Fawcett, B. C., Peacock, N. J., and Cowan, R. D. (1968), J. Phys. **B1**, 295.

Kruger, P. G., and Pattin, H. S. (1937), Phys. Rev. **52**, 621.

Smitt, R., Svensson, L. A., and Outred, M. (1976), Phys. Scr. **13**, 293.

Svensson, L. A., and Ekberg, J. O. (1968), Ark. Fys. **37**, 65.

Sc VI

Configuration	Term	J	Level (cm^{-1})
$3s^2 3p^4$	^3P	2	0.0
		1	3 346.1
		0	4 457.1
$3s^2 3p^4$	^1D	2	21 393.0
$3s^2 3p^4$	^1S	0	49 224.6
$3s3p^5$	$^3P°$	2	*175 346.6*
		1	*178 202.1*
		0	*179 784.5*
$3s3p^5$	$^1P°$	1	*224 470.4*
$3s^2 3p^3(^2D°)3d$	$^1D°$	2	*288 026*
$3s^2 3p^3(^2D°)3d$	$^1F°$	3	*323 226*
$3s^2 3p^3(^2D°)3d$	$^3S°$	1	*335 099*
$3s^2 3p^3(^2P°)3d$	$^3P°$	2	*338 435*
		1	*339 807*
$3s^2 3p^3(^2D°)3d$	$^1P°$	1	*341 955*
$3s^2 3p^3(^2P°)3d$	$^3D°$	3	*351 787*
		2	*353 983*
		1	*355 461*
$3s^2 3p^3(^2P°)3d$	$^1D°$	2	*363 456*
$3s^2 3p^3(^2P°)3d$	$^1F°$	3	*375 740*
$3s^2 3p^3(^2P°)3d$	$^1P°$	1	*403 098*
$3s^2 3p^3(^4S°)4s$	$^3S°$	1	*452 070*
$3s^2 3p^3(^2D°)4s$	$^3D°$	1	*472 402*
		2	*472 566*
		3	*473 001*
$3s^2 3p^3(^2D°)4s$	$^1D°$	2	*478 354*
$3s^2 3p^3(^2P°)4s$	$^3P°$	0	*491 820*
		1	*492 086*
		2	*492 800*
$3s^2 3p^3(^2P°)4s$	$^1P°$	1	*497 990*
$3s^2 3p^3(^4S°)4d$	$^3D°$	1-3	*590 800*
$3s^2 3p^3(^2D°)4d$	$^1D°$	2	*619 600*
$3s^2 3p^3(^2D°)4d$	$^1F°$	3	*622 500*
$3s^2 3p^3(^4S°)5s$	$^3S°$	1	*648 100*
$3s^2 3p^3(^2D°)5s$	$^3D°$	3	*674 900*
$3s^2 3p^3(^2D°)5s$	$^1D°$	2	*676 700*
$3s^2 3p^3(^2P°)5s$	$^1P°$	1	*696 400*
Sc VII ($^4S°_{3/2}$)	***Limit***		**892 700**

Sc VII

$Z=21$

P I isoelectronic sequence

Ground state: $1s^2 2s^2 2p^6 3s^2 3p^3\ {}^4S^{\circ}_{3/2}$

Ionization energy = 1 113 000±2000 cm^{-1} (138.0±0.2 eV)

The levels are from the work of Ekberg and Svensson (1970) and Smitt, Svensson, and Outred (1976). The levels for the $3s^2 3p^3$ and $3s 3p^4$ configurations are taken from the latter paper and have an uncertainty of about ±2 cm^{-1}. We have combined these values with the measurements and classifications given by Ekberg and Svensson in the wavelength range of 182–598 Å to derive new level values for the $3p^2 3d$ and $4s$ configurations. Most of the wavelengths used by Ekberg and Svensson are taken from Beckman (1937), Kruger and Pattin (1937), and Fawcett (1970). The uncertainty of these upper levels is about ±10 cm^{-1}. Since no intersystem transitions have been observed, all of the doublets have an added systematic error "x," relative to the ground term ^{4}S°. The value of x depends on the accuracy of calculations by Smitt, Svensson and Outred and is expected to be less than ±20 cm^{-1}.

The ionization energy was obtained by extrapolation by Lotz (1967).

References

Beckman, A. (1937), Dissertation, Uppsala.
Ekberg, J. O., and Svensson, L. A. (1970), Phys. Scr. **2**, 283.
Fawcett, B. C. (1970), J. Phys. **B3**, 1732.
Kruger, P. G., and Pattin, H. S. (1937), Phys. Rev. **52**, 621.
Lotz, W. (1967), J. Opt. Soc. Am. **57**, 873.
Smitt, R., Svensson, L. A., and Outred, M. (1976), Phys. Scr. **13**, 293.

Sc VII

Configuration	Term	J	Level (cm^{-1})
$3s^2 3p^3$	^{4}S°	3/2	*0.0*
$3s^2 3p^3$	^{2}D°	3/2	*29 562.5+x*
		5/2	*30 239.9+x*
$3s^2 3p^3$	^{2}P°	1/2	*49 387.1+x*
		3/2	*50 300.9+x*
$3s3p^4$	^{4}P	5/2	175 055.0
		3/2	177 776.5
		1/2	179 197.3
$3s3p^4$	^{2}D	3/2	217 023.9+x
		5/2	217 326.5+x
$3s3p^4$	^{2}P	3/2	250 443.6+x
		1/2	252 995.3+x
$3s3p^4$	^{2}S	1/2	261 800.6+x
$3s^2 3p^2(^3P)3d$	^{2}P	3/2	329 516+x
		1/2	332 923+x
$3s^2 3p^2(^3P)3d$	^{4}P	5/2	334 944
		3/2	336 396
		1/2	337 224
$3s^2 3p^2(^1D)3d$	^{2}D	3/2	360 886+x
$3s^2 3p^2(^1D)3d$	^{2}P	1/2	372 840+x
		3/2	374 787+x
$3s^2 3p^2(^1D)3d$	^{2}F	5/2	378 080+x
		7/2	378 760+x
$3s^2 3p^2(^1D)3d$	^{2}S	1/2	382 195+x
$3s^2 3p^2(^1S)3d$	^{2}D	5/2	392 364+x
		3/2	393 384+x
$3s^2 3p^2(^3P)4s$	^{4}P	1/2	541 691
		3/2	543 600
		5/2	546 469
$3s^2 3p^2(^3P)4s$	^{2}P	1/2	551 487+x
		3/2	554 782+x
$3s^2 3p^2(^1D)4s$	^{2}D	5/2	568 431+x
		3/2	568 574+x
$3s^2 3p^2(^1S)4s$	^{2}S	1/2	597 518+x
Sc VIII ($^3P_0$)	*Limit*		**1 113 000**

Sc VIII

$Z=21$

Si I isoelectronic sequence

Ground state: $1s^2 2s^2 2p^6 3s^2 3p^2\ {}^3P_0$

Ionization energy = 1 275 000±2000 cm^{-1} (158.1±0.2 eV)

The study for this spectrum was initiated by Kruger and Phillips (1937), who classified 15 lines as transitions between the ground term and three odd terms $3s3p^3\ {}^3S°$, $3s^2 3p3d\ {}^3P°$ and $3s^2 3p4s\ {}^3P°$. Phillips (1939) found $3s^2 3p^2\ {}^1D_2$ and $3s3p^3\ {}^1P_1°$. Fawcett, Gabriel, and Saunders (1967) extended the $3p^2-p3d$ array; Fawcett (1970) added to the $3s^2 3p^2-3s3p^3$ array. Fawcett, Cowan, and Hayes (1972) identified lines in $3p3d-3p4f$ which are not connected with the other levels.

Ekberg and Svensson (1970) reanalyzed the spectrum using a compilation of wavelengths between 164 and 572 Å. Smitt, Svensson, and Outred (1976) made new observations between 362 and 640 Å. The level values for the $3s^2 3p^2$ and $3s3p^3$ configurations are taken from the more accurate data of Smitt et al. and the values for $3s^2 3p3d$ and $3s^2 3p4s$ are derived by combining those values with the wavelengths in Ekberg and Svensson. The uncertainty of the level values from Smitt et al. is about ±5 cm^{-1}. Four intersystem transitions have been observed.

The ionization energy was obtained by Ekberg and Svensson from an extrapolation formula.

References

Ekberg, J. O., and Svensson, L. A. (1970), Phys. Scr. **2**, 283.

Fawcett, B. C. (1970), J. Phys. **B3**, 1732.

Fawcett, B. C., Cowan, R. D., and Hayes, R. W. (1972), J. Phys. **B5**, 2143.

Fawcett, B. C., Gabriel, A. H., and Saunders, P. A. H. (1967), Proc. Phys. Soc. **90**, 863.

Kruger, P. G., and Phillips, L. W. (1937), Phys. Rev. **52**, 97.

Phillips, L. W. (1939), Phys. Rev. **55**, 708.

Smitt, R., Svensson, L. A., and Outred, M. (1976), Phys. Scr. **13**, 293.

Sc VIII

Configuration	Term	J	Level (cm^{-1})	Configuration	Term	J	Level (cm^{-1})
$3s^2\,3p^2$	^{3}P	0	0.0	$3s^2\,3p3d$	^{3}P°	2	*319 569*
		1	2 271.9			1	*322 541*
		2	5 507.7			0	*323 673*
$3s^2\,3p^2$	^{1}D	2	25 026.9	$3s^2\,3p3d$	^{3}D°	1	*329 862*
						2	*330 716*
$3s^2\,3p^2$	^{1}S	0	54 864.4			3	*331 153*
$3s3p^3$	^{3}D°	1	*179 962.3*	$3s^2\,3p3d$	^{1}F°	3	*363 462*
		2	*180 032.7*				
		3	*180 503.6*	$3s^2\,3p3d$	^{1}P°	1	*372 790*
$3s3p^3$	^{3}P°	0	*207 690.9*	$3s^2\,3p4s$	^{3}P°	0	*603 533*
		1	*207 761.0*			1	*604 609*
		2	*207 814.4*			2	*609 174*
$3s3p^3$	^{1}D°	2	*228 618.3*	$3s^2\,3p4s$	^{1}P°	1	*614 090*
$3s3p^3$	^{3}S°	1	*272 417.4*	Sc IX ($^2P_{1/2}°$)	***Limit***		**1 275 000**
$3s3p^3$	^{1}P°	1	*281 522.1*				

Sc IX

$Z=21$

Al I isoelectronic sequence

Ground state: $1s^2 2s^2 2p^6 3s^2 3p\ ^2P_{1/2}$

Ionization energy = 1 452 000±1000 cm^{-1} (180.03±0.12 eV)

The initial work on the analysis of this spectrum was by Kruger and Phillips (1937) and by Beckman (1937). About a third of Beckman's identifications were corrected by Fawcett (1970).

Using the earlier measurements, Ekberg and Svensson (1970) reanalyzed the spectrum between 90 and 540 Å and identified all the terms given here. They extrapolated the position of $3s3p^2\ ^4P$ along the isoelectronic sequence. Since no intersystem transitions have been observed, we use their extrapolation to establish the energy of $^4P_{1/2}$ relative to the ground level. The error is indicated by x.

Smitt, Svensson, and Outred (1976) remeasured the $3s^2 3p - 3s3p^2$ array between 380 and 540 Å and determined the doublet terms of those configurations with an uncertainty of ±4 cm^{-1}. We used their term values in combination with the earlier measurements of Ekberg and Svensson to establish the higher doublet term values.

The ionization energy was obtained by Ekberg and Svensson from the *nf* series.

References

Beckman, A. (1937), Dissertation, Uppsala.
Ekberg, J. O., and Svensson, L. A. (1970), Phys. Scr. **2,** 283.
Fawcett, B. C. (1970), J. Phys. **B3,** 1732.
Kruger, P. G., and Phillips, L. W. (1937), Phys. Rev. **52,** 97.
Smitt, R., Svensson, L. A., and Outred, M. (1976), Phys. Scr. **13,** 293.

Sc IX

Configuration	Term	J	Level (cm^{-1})	Configuration	Term	J	Level (cm^{-1})
$3s^2 3p$	$^2P°$	1/2	*0.0*	$3s3p(^3P°)4s$	$^4P°$	1/2	*822 056+x*
		3/2	*5 761.1*			3/2	*823 985+x*
						5/2	*827 620+x*
$3s3p^2$	^4P	1/2	143 500+x				
		3/2	145 622+x	$3s^2 4d$	^2D	3/2	837 212
		5/2	148 779+x			5/2	837 459
$3s3p^2$	^2D	3/2	191 609.3	$3s^2 4f$	$^2F°$	5/2	*889 330*
		5/2	191 987.1			7/2	*889 384*
$3s3p^2$	^2S	1/2	240 361.4	$3s^2 5s$	^2S	1/2	979 930
$3s3p^2$	^2P	1/2	255 829.4	$3s^2 5d$	^2D	3/2	1 070 740
		3/2	259 153.7			5/2	1 070 850
$3s^2 3d$	^2D	3/2	313 860	$3s^2 5f$	$^2F°$	5/2	*1 095 200*
		5/2	314 214			7/2	*1 095 250*
$3p^3$	$^4S°$	3/2	*383 047+x*	Sc X ($^1S_0$)	***Limit***		**1 452 000**
$3s^2 4s$	^2S	1/2	666 259				

Sc X

$Z=21$

Mg I isoelectronic sequence

Ground state: $1s^2 2s^2 2p^6 3s^2\ ^1S_0$

Ionization energy = 1 816 200±400 cm^{-1} (225.18±0.05 eV)

The initial work on the analysis was done by Beckman (1937) and Parker and Phillips (1940).

Ekberg (1971), using wavelengths between 76 and 470 Å, taken from the papers above, has redone the analysis and determined all the levels given in this compilation except the $3p^2\ ^1S_0$, and the levels of $3p3d$, $3p4d$, and $3p4f$. These are taken from Fawcett (1970, 1976) and from Kastner et al. (1978). The position of the triplet terms relative to the ground state is based on the observation of the transition $3s^2\ ^1S_0 - 3s3p\ ^3P_1^\circ$ at 624.5±0.15 Å in a tokamak plasma by Finkenthal et al. (1982).

We derived the value for the ionization energy from the $3snf\ ^3F^\circ$ series.

References

Beckman, A. (1937), Dissertation, Uppsala.

Ekberg, J. O. (1971), Phys. Scr. **4,** 101.

Fawcett, B. C. (1970), J. Phys. **B3,** 1732.

Fawcett, B. C. (1976), J. Opt. Soc. Am. **66,** 632.

Finkenthal, M., Hinnov, E., Cohen, S., and Suckewer, S. (1982), Phys. Lett. **A9,** 284.

Kastner, S. O., Swartz, M., Bhatia, A. K., and Lapides, J. (1978), J. Opt. Soc. Am. **68,** 1558.

Parker, W. L., and Phillips, L. W. (1940), Phys. Rev. **57,** 140.

Sc X

Configuration	Term	J	Level (cm^{-1})
$3s^2$	^1S	0	0
$3s3p$	$^3P^\circ$	0	*158 122*
		1	*160 128*
		2	*164 435*
$3s3p$	$^1P^\circ$	1	*236 490*
$3p^2$	^1D	2	372 398
$3p^2$	^3P	0	375 177
		1	377 778
		2	382 713
$3p^2$	^1S	0	440 480
$3s3d$	^3D	1	455 897
		2	456 084
		3	456 398
$3s3d$	^1D	2	516 218
$3p3d$	$^3F^\circ$	2	*623 686*
		3	*626 456*
		4	*629 776*
$3p3d$	$^1D^\circ$	2	*633 100*
$3p3d$	$^3D^\circ$	2	*665 016*
		3	*666 356*
$3p3d$	$^1P^\circ$	1	*683 850*
$3s4s$	^3S	1	900 155
$3s4s$	^1S	0	915 165
$3s4p$	$^1P^\circ$	1	*980 604*
$3s4d$	^3D	1	1 074 960
		2	1 075 140
		3	1 075 480
$3s4d$	^1D	2	1 081 820
$3p4s$	$^3P^\circ$	0	*1 098 960*
		1	*1 100 500*
		2	*1 105 490*
$3s4f$	$^3F^\circ$	2	*1 118 590*
		3	*1 118 610*
		4	*1 118 680*
$3s4f$	$^1F^\circ$	3	*1 128 150*
$3p4p$	^3D	2	1 171 140
		3	1 175 670
$3p4p$	^3P	0	1 176 770
		1	1 178 390
		2	1 181 500
$3p4p$	^3S	1	1 182 850
$3p4d$	$^3D^\circ$	1	*1 266 330*
$3p4d$	$^1F^\circ$	3	*1 282 250*
$3p4d$	$^1P^\circ$	1	*1 283 540?*

Sc X—Continued

Configuration	Term	*J*	Level (cm^{-1})	Configuration	Term	*J*	Level (cm^{-1})
$3p4d$	$^3P^\circ$	1	*1 284 310*	$3s5d$	^1D	2	1 350 870
$3p4f$	^3G	4	1 306 910	$3s5f$	$^3F^\circ$	4	*1 371 440*
		5	1 310 900			3	*1 371 610*
						2	*1 371 640*
$3p4f$	^3F	3	1 309 320				
		4	1 311 950	$3s6f$	$^3F^\circ$	4	*1 508 040*
						3	*1 508 140*
$3s5p$	$^1P^\circ$	1	*1 309 880*			2	*1 508 290*
$3p4f$	^3D	2	1 319 890	Sc XI ($^2S_{1/2}$)	***Limit***		**1 816 200**
$3s5d$	^3D	1	1 352 010				
		2,3	1 352 020				

Sc XI

$Z=21$

Na I isoelectronic sequence

Ground state: $1s^22s^22p^63s\ ^2S_{1/2}$

Ionization energy = 2 015 060±100 cm^{-1} (249.837±0.012 eV)

We have used the measurements and classifications of Kruger and Phillips (1939) between 97 and 523 Å to determine the $3p$, $3d$, $4d$, and $5f$ levels with an uncertainty of ±10–100 cm^{-1} and those of Edlén (1936) between 94 and 169 Å for the $4s$, $4p$, and $4f$ levels with an uncertainty of ±100 cm^{-1}. The $5s$, $5p$, $5d$, $6s$, $6p$, $6d$, $6f$, and $7d$ levels with an uncertainty of ±100 are from Beckman's (1937) measurements. The p, d, and f terms for $n=8,9$ are from Fawcett (1976) and the $7p$, $7f$, and $10p$ are from Cohen and Behring (1976). These have an estimated uncertainty of ±200 cm^{-1}.

The value for the ionization energy was derived by Edlén (1978) from core polarization theory applied to the nf series ($4f-7f$).

References

Beckman, A. (1937), Dissertation, Uppsala.
Cohen, L., and Behring, W. E. (1976), J. Opt. Soc. Am. **66,** 899.
Edlén, B. (1936), Z. Phys. **100,** 621.
Edlén, B. (1978), Phys. Scr. **17,** 565.
Fawcett, B. C. (1976), J. Opt. Soc. Am. **66,** 632.
Kruger, P. G., and Phillips, L. W. (1939), Phys. Rev. **55,** 352.

Sc XI

Configuration	Term	J	Level (cm^{-1})
$2p^6(^1S)3s$	^2S	1/2	0
$2p^6(^1S)3p$	$^2P^\circ$	1/2	*191 274*
		3/2	*197 974*
$2p^6(^1S)3d$	^2D	3/2	459 725
		5/2	460 337
$2p^6(^1S)4s$	^2S	1/2	997 720
$2p^6(^1S)4p$	$^2P^\circ$	1/2	*1 051 640*
		3/2	*1 054 170*
$2p^6(^1S)4d$	^2D	3/2	1 148 750
		5/2	1 149 040
$2p^6(^1S)4f$	$^2F^\circ$	5/2	*1 182 880*
		7/2	*1 182 980*
$2p^6(^1S)5s$	^2S	1/2	1 382 340
$2p^6(^1S)5p$	$^2P^\circ$	1/2	*1 418 260*
		3/2	*1 419 550*
$2p^6(^1S)5d$	^2D	3/2	1 465 010
		5/2	1 465 120
$2p^6(^1S)5f$	$^2F^\circ$	5/2	*1 482 340*
		7/2	*1 482 410*
$2p^6(^1S)6s$	^2S	1/2	1 589 040
$2p^6(^1S)6p$	$^2P^\circ$	3/2	*1 609 500*
$2p^6(^1S)6d$	^2D	3/2,5/2	1 635 270
$2p^6(^1S)6f$	$^2F^\circ$	5/2	*1 645 250*
		7/2	*1 645 270*
$2p^6(^1S)7p$	$^2P^\circ$	1/2,3/2	*1 721 700*
$2p^6(^1S)7d$	^2D	3/2,5/2	1 736 900
$2p^6(^1S)7f$	$^2F^\circ$	5/2,7/2	*1 743 800*
$2p^6(^1S)8p$	$^2P^\circ$	3/2	*1 792 500*
$2p^6(^1S)8d$	^2D	5/2	1 802 700
$2p^6(^1S)8f$	$^2F^\circ$	7/2	*1 807 700*
$2p^6(^1S)9p$	$^2P^\circ$	3/2	*1 840 100*
$2p^6(^1S)9d$	^2D	5/2	1 847 100
$2p^6(^1S)9f$	$^2F^\circ$	7/2	*1 858 500*
$2p^6(^1S)10p$	$^2P^\circ$	1/2,3/2	*1 875 000*
Sc XII ($^1S_0$)	*Limit*		**2 015 060**

Sc XII

$Z=21$

Ne I isoelectronic sequence

Ground state: $1s^2 2s^2 2p^6\ {}^1S_0$

Ionization energy = 5 543 900±1000 cm^{-1} (687.36±0.12 eV)

Only resonance lines between 20 and 31 Å are classified by this system of energy levels. Edlén and Tyrén (1936) identified transitions from the levels of the $2p^5 3s$ and $3d$ configurations. The uncertainty in these level values is probably ±500 cm^{-1}. Fawcett (1965) observed three transitions arising from $2p^5 4d$ and from $2s2p^6 3p\ {}^1P^\circ_1$. Feldman and Cohen (1967) observed eight transitions arising from the $2p^5 3p$, $4s$, $4d$, and $5d$ configurations. We have adopted the more accurate values of Feldman and Cohen, which give level uncertainties of ±1000 cm^{-1}.

To determine the leading percentages for the $2p^5 3s$ and $2p^5 3d$ levels we calculated these configurations, using Hartree-Fock radial integrals scaled according to fitted values in Al IV by Artru and Kaufman (1975).

Kastner, Behring, and Cohen (1975) identified transitions between $2p^5 3p$ and $2p^5 4d$, but there is no connection with the levels given here.

We derived the ionization energy by application of a Ritz formula to the $2s^2 2p^5(^2P^\circ_{3/2})nd\ ^2[3/2]^\circ$ series for $n=3$, 4, and 5.

References

Artru, M. C., and Kaufman, V. (1975), J. Opt. Soc. Am. **65**, 594.

Edlén, B., and Tyrén, F. (1936), Zeit. Phys. **101**, 210.

Fawcett, B. C. (1965), Proc. Phys. Soc. **86**, 1087.

Feldman, U., and Cohen, L. (1967), Astrophys. J. **149**, 265.

Kastner, S. O., Behring, W. E., and Cohen, L. (1975), Astrophys. J. **199**, 777.

Sc XII

Configuration	Term	J	Level (cm^{-1})	Leading percentages		
$2s^2 2p^6$	^1S	0	0			
$2s^2 2p^5(^2P^\circ_{3/2})3s$	$(3/2,1/2)^\circ$	1	*3 245 100*	96	4	$(1/2,1/2)^\circ$
$2s^2 2p^5(^2P^\circ_{1/2})3s$	$(1/2,1/2)^\circ$	1	*3 280 800*	96	4	$(3/2,1/2)^\circ$
$2s^2 2p^5 3d$	$^3P^\circ$	1	*3 668 400*	90	9	$^3D^\circ$
$2s^2 2p^5 3d$	$^3D^\circ$	1	*3 714 700*	60	37	$^1P^\circ$
$2s^2 2p^5 3d$	$^1P^\circ$	1	*3 767 300*	63	31	$^3D^\circ$
$2s2p^6 3p$	$^3P^\circ$	1	*4 198 000*			
$2s2p^6 3p$	$^1P^\circ$	1	*4 215 000*			
$2s^2 2p^5(^2P^\circ_{3/2})4s$	$(3/2,1/2)^\circ$	1	*4 339 300*			
$2s^2 2p^5(^2P^\circ_{1/2})4s$	$(1/2,1/2)^\circ$	1	*4 378 800*			
$2s^2 2p^5(^2P^\circ_{3/2})4d$	$^2[3/2]^\circ$	1	*4 521 000*			
$2s^2 2p^5(^2P^\circ_{1/2})4d$	$^2[3/2]^\circ$	1	*4 557 900*			
$2s^2 2p^5(^2P^\circ_{3/2})5d$	$^2[3/2]^\circ$	1	*4 892 800*			
$2s^2 2p^5(^2P^\circ_{1/2})5d$	$^2[3/2]^\circ$	1	*4 926 600*			
Sc XIII $(^2P^\circ_{3/2})$	***Limit***		**5 543 900**			

Sc XIII

$Z=21$

F I isoelectronic sequence

Ground state: $1s^22s^22p^5\ {}^2P^\circ_{3/2}$

Ionization energy = 6 103 000±12000 cm^{-1} (756.7±1.5 eV)

The first work on this spectrum was by Fawcett (1965), who classified lines of the $2s^22p^5-2s^22p^43s$ and $3d$ transition arrays between 24 and 28 Å. This work was revised and extended by Feldman, Doschek, Cowan, and Cohen (1973), from whose improved wavelengths the $3s$ and $3d$ levels are determined. Their estimated uncertainty of ±0.01 Å gives a level uncertainty of ±2000 cm^{-1}. The magnetic dipole transition $2s^22p^5\ {}^2P^\circ_{3/2}-{}^2P^\circ_{1/2}$ was observed at 2637.2±0.2 Å by Suckewer et al. (1980) from a tokamak plasma. Their value has been adopted for the ground term splitting. The $2s2p^6\ {}^2S$ term is from the measurements of Kaufman, Sugar, and Cooper (1982), which give a level uncertainty of ±50 cm^{-1}.

The composition of the $2p^43s$ and $2p^43d$ levels is from Chapman and Shadmi (1973).

The $2s2p^53s\ {}^2P^\circ$ term was identified by Feldman et al. (1973).

The ionization energy was obtained by Lotz (1967) by extrapolation .

References

Chapman, R. D., and Shadmi, Y. (1973), J. Opt. Soc. Am. **63**, 1440.

Fawcett, B. C. (1965), Proc. Phys. Soc. **86**, 1087.

Feldman, U., Doschek, G. A., Cowan, R. D., and Cohen, L. (1973), J. Opt. Soc. Am. **63**, 1445.

Kaufman, V., Sugar, J., and Cooper, D. (1982), Phys. Scr. **25**, 623.

Lotz, W. J. (1967), J. Opt. Soc. Am. **57**, 873.

Suckewer, S., Cecci, J. L., Cohen, S., Fonck, R., and Hinnov, E. (1980), Phys. Lett. **80A**, 259.

Sc XIII

Configuration	Term	J	Level (cm^{-1})	Leading percentages		
$2s^2 2p^5$	$^2P^\circ$	3/2	*0*			
		1/2	*37 908*			
$2s2p^6$	^2S	1/2	763 621			
$2s^2 2p^4(^3P)3s$	^4P	5/2	3 513 300	96		
		3/2	3 530 600	72	24	$(^3P)\ ^2P$
		1/2	3 547 200	95		
$2s^2 2p^4(^3P)3s$	^2P	3/2	3 554 800	70	26	$(^3P)\ ^4P$
		1/2	3 574 100	96		
$2s^2 2p^4(^1D)3s$	^2D	5/2	3 619 500	96		
		3/2	3 620 800	94		
$2s^2 2p^4(^1S)3s$	^2S	1/2	3 718 400	90		
$2s^2 2p^4(^3P)3d$	^4P	1/2	3 961 700	91		
		3/2	3 968 300	84		
		5/2	3 978 800	77		
$2s^2 2p^4(^3P)3d$	^4F	5/2	3 968 300	85		
$2s^2 2p^4(^3P)3d$	^2P	1/2	3 974 100	58	32	$(^1D)\ ^2P$
		3/2	4 000 300	48	28	
$2s^2 2p^4(^3P)3d$	^2D	3/2	3 984 200	42	25	$(^1D)\ ^2D$
		5/2	4 004 800	45	24	
$2s^2 2p^4(^3P)3d$	^2F	5/2	3 987 400	65	12	$(^3P)\ ^4P$
$2s^2 2p^4(^1D)3d$	^2S	1/2	4 054 200	92		

Sc XIII—Continued

Configuration	Term	J	Level (cm^{-1})	Leading percentages		
$2s^2 2p^4(^1D)3d$	^2F	$5/2$	4 057 100	92		
$2s^2 2p^4(^1D)3d$	^2D	$5/2$	4 071 700	57	23	$(^3P)\ ^2D$
		$3/2$	4 084 300	63	22	
$2s^2 2p^4(^1S)3d$	^2D	$5/2$	4 149 900	79		
		$3/2$	4 156 100	62	7	$(^3P)\ ^2D$
$2s2p^5(^3P°)3s$	$^2P°$	$3/2$	*4 242 100*			
		$1/2$	*4 264 200*			
Sc XIV ($^3P_2$)	***Limit***		**6 103 000**			

Sc XIV

$Z=21$

O I isoelectronic sequence

Ground state: $1s^2 2s^2 2p^4\ ^3P_2$

Ionization energy = 6 701 000±13000 cm^{-1} (830.8±1.6 eV)

The observed spectrum of Sc XIV consists of the strong transition array $2s^2 2p^4 - 2s 2p^5$, which lies between 109 and 176 Å, and the arrays $2p^4 - 2p^3 3s$ at 25 Å and $2p^4 - 2p^3 3d$ at 23 Å. The $^1S_0$ of $2p^6$ combines with $2s2p^5\ ^1P_1^{\circ}$ at 157 Å. The arrays at 25 Å and 23 Å were first observed by Goldsmith, Feldman, and Cohen (1971). The $J=0$ and 1 levels of the ground term could not be resolved at these wavelengths. Fawcett (1971) then observed the $2s^2 2p^4 - 2s 2p^5$ array and resolved the ground term. The $2s2p^5\ ^1P_1^{\circ} - 2p^6\ ^1S_0$ transition was reported by Fawcett, Galanti, and Peacock (1974). We have determined the levels of the $2s^2 2p^4$, $2s2p^5$, and $2p^6$ configurations from the improved measurements of Kaufman, Sugar, and Cooper (1982) who report a level uncertainty of 50 cm^{-1}. Transitions between terms of different spin were identified by these authors. The percentage compositions for these levels were provided by Kaufman and Sugar (1982). Their calculation includes configuration interaction between $2s^2 2p^4$ and $2p^6$. Suckewer et al. (1980) obtained the value 31 179±2 cm^{-1} for the $2s^2 2p^4\ ^3P_2 - ^3P_1$ interval from a magnetic dipole line observed in a tokamak plasma.

Improved measurements of the $2p^4 - 2p^3 3s$ array were obtained from Doschek, Feldman, and Cohen (1973), giving a level uncertainty of ±1000 cm^{-1}.

A revised analysis of $2p^4 - 2p^3 3d$ by Fawcett and Hayes (1975) is adopted here. A level uncertainty of ±3000 cm^{-1} is indicated. The subsequent revisions of this array proposed by Bromage and Fawcett (1977), following a new calculation, are included.

The ionization energy was evaluated by Lotz (1967) by extrapolation.

References

Bromage, G. E., and Fawcett, B. C. (1977), Mon. Not. R. Astron. Soc. **178,** 591.

Doschek, G. A., Feldman, U., and Cohen, L. (1973), J. Opt. Soc. Am. **63,** 1463.

Fawcett, B. C. (1971), J. Phys. **B4,** 981.

Fawcett, B. C., Galanti, M., and Peacock, N.J. (1974), J. Phys. **B7,** 1149.

Fawcett, B. C., and Hayes, R. W. (1975), Mon. Not. R. Astron. Soc. **170,** 185.

Goldsmith, S., Feldman, U., and Cohen, L. (1971), J. Opt. Soc. Am. **61,** 615.

Kaufman, V., and Sugar, J. (1982), private communication.

Kaufman, V., Sugar, J., and Cooper, D. (1982), Phys. Scr. **25,** 623.

Lotz, W. J. (1967), J. Opt. Soc. Am. **57,** 873.

Suckewer, S., Cecchi, J., Cohen, S., Fonck, R., and Hinnov, E. (1980), Phys. Lett. **A80,** 259.

Sc XIV

Configuration	Term	J	Level (cm^{-1})	Leading percentages		
$2s^2\,2p^4$	^3P	2	0	96	4	^1D
		1	31 174	100		
		0	35 351	94	6	^1S
$2s^2\,2p^4$	^1D	2	97 793	96	4	^3P
$2s^2\,2p^4$	^1S	0	195 985	92	6	^3P
$2s2p^5$	$^3P^{\circ}$	2	*664 483*	100		
		1	*689 445*	99	1	$^1P^{\circ}$
		0	*704 584*	100		
$2s2p^5$	$^1P^{\circ}$	1	*912 990*	99	1	$^3P^{\circ}$
$2p^6$	^1S	0	1 546 620	97	3	$2s^2 2p^4\ ^1S$
$2s^2\,2p^3(^4S^{\circ})3s$	$^3S^{\circ}$	1	*3 848 500*			
$2s^2\,2p^3(^2D^{\circ})3s$	$^3D^{\circ}$	1	*3 929 400*			
		2	*3 930 800*			
		3	*3 938 200*			

Sc XIV—Continued

Configuration	Term	J	Level (cm^{-1})	Leading percentages
$2s^2\,2p^3(^2D^\circ)3s$	$^1D^\circ$	2	*3 955 700*	
$2s^2\,2p^3(^2P^\circ)3s$	$^3P^\circ$	2	*4 004 600*	
$2s^2\,2p^3(^2P^\circ)3s$	$^1P^\circ$	1	*4 033 900*	
$2s^2\,2p^3(^4S^\circ)3d$	$^3D^\circ$	2	*4 245 000*	
		3	*4 249 000*	
$2s^2\,2p^3(^2D^\circ)3d$	$^3D^\circ$	1	*4 333 000*	
		2	*4 350 000*	
		3	*4 354 000*	
$2s^2\,2p^3(^2D^\circ)3d$	$^3P^\circ$	2	*4 362 000*	
		1	*4 368 000*	
$2s^2\,2p^3(^2D^\circ)3d$	$^1D^\circ$	2	*4 367 000*	
$2s^2\,2p^3(^2D^\circ)3d$	$^3S^\circ$	1	*4 378 000*	
$2s^2\,2p^3(^2D^\circ)3d$	$^1F^\circ$	3	*4 395 000*	
$2s^2\,2p^3(^2P^\circ)3d$	$^3P^\circ$	2	*4 407 000*	
		1	*4 419 000*	
$2s^2\,2p^3(^2P^\circ)3d$	$^1D^\circ$	2	*4 426 000*	
$2s^2\,2p^3(^2P^\circ)3d$	$^3D^\circ$	1	*4 432 000*	
		3	*4 433 000*	
		2	*4 445 000*	
$2s^2\,2p^3(^2P^\circ)3d$	$^1F^\circ$	3	*4 452 000*	
$2s^2\,2p^3(^2P^\circ)3d$	$^1P^\circ$	1	*4 493 000*	
Sc XV ($^4S^\circ_{3/2}$)	***Limit***		**6 701 000**	

Sc XV

$Z=21$

N I isoelectronic sequence

Ground state: $1s^2 2s^2 2p^3\ ^4S^\circ_{3/2}$

Ionization energy = 7 481 000±15000 cm^{-1} (927.5±2.0 eV)

The transition arrays $2s^2 2p^3 - 2s2p^4$ and $2s2p^4 - 2p^5$ were identified by Fawcett (1971). Connecting the quartet and doublet levels is the intersystem line $2s^2 2p^3\ ^4S^\circ_{3/2} - 2s2p^4\ ^2P_{3/2}$ at 109.084 Å found by Kaufman, Sugar, and Cooper (1982). They gave improved measurements of these arrays in the range of 109–204 Å with an accuracy of ±0.01 Å and a level uncertainty of 50 cm^{-1}. The percentage compositions of these levels were supplied by Kaufman and Sugar (1982). Their calculation includes configuration interaction between $2s^2 2p^3$ and $2p^5$.

The $2p^2 3d$ terms are from the interpretation by Fawcett and Hayes (1975) of the array at 22 Å. Some revisions were given by Bromage and Fawcett (1977) on the basis of new calculations. The uncertainty of these levels is ±2000 cm^{-1}.

The ionization energy is from Lotz's (1967) extrapolation.

References

Bromage, G. E., and Fawcett, B. C. (1977), Mon. Not. R. Astron. Soc. **179**, 683.

Fawcett, B. C. (1971), J. Phys. **B4**, 981.

Fawcett, B. C., and Hayes, R. W. (1975), Mon. Not. R. Astron. Soc. **170**, 185.

Kaufman, V., and Sugar, J. (1982), private communication.

Kaufman, V., Sugar, J., and Cooper, D. (1982), Phys. Scr. **26**, 163.

Lotz, W. (1967), J. Opt. Soc. Am. **57**, 873.

Sc XV

Configuration	Term	J	Level (cm^{-1})	Leading percentages		
$2s^2 2p^3$	$^4S^\circ$	3/2	*0*	98	2	$^2P^\circ$
$2s^2 2p^3$	$^2D^\circ$	3/2	*111 200*	87	12	$^2P^\circ$
		5/2	*121 960*	100		
$2s^2 2p^3$	$^2P^\circ$	1/2	*184 840*	98	2	$2p^5\ ^2P^\circ$
		3/2	*200 450*	83	13	$2p^3\ ^2D^\circ$
$2s(^2S)2p^4(^3P)$	^4P	5/2	551 980	99	1	$2s(^2S)2p^4(^1D)\ ^2D$
		3/2	577 220	100		
		1/2	588 360	99	1	$2s(^2S)2p^4(^1S)\ ^2S$
$2s(^2S)2p^4(^1D)$	^2D	3/2	761 250	98	1	$2s(^2S)2p^4(^3P)\ ^2P$
		5/2	763 990	99	1	$2s(^2S)2p^4(^3P)\ ^4P$
$2s(^2S)2p^4(^1S)$	^2S	1/2	881 890	86	13	$2s(^2S)2p^4(^3P)\ ^2P$
$2s(^2S)2p^4(^3P)$	^2P	3/2	916 720	99	1	$2s(^2S)2p^4(^1D)\ ^2D$
		1/2	951 660	87	13	$2s(^2S)2p^4(^1S)\ ^2S$
$2p^5$	$^2P^\circ$	3/2	*1 441 710*	98	2	$2s^2 2p^3\ ^2P^\circ$
		1/2	*1 482 190*	98	2	
$2s^2 2p^2(^3P)3d$	^2P	3/2	4 636 200			
$2s^2 2p^2(^3P)3d$	^4P	5/2	4 671 800			
		3/2	4 679 000			
$2s^2 2p^2(^3P)3d$	^2F	7/2	4 683 200			
$2s^2 2p^2(^3P)3d$	^2D	3/2	4 720 000			
		5/2	4 733 300			

Sc XV—Continued

Configuration	Term	J	Level (cm^{-1})	Leading percentages
$2s^2\,2p^2(^1D)3d$	^2D	$5/2$	4 758 000	
$2s^2\,2p^2(^1D)3d$	^2F	$7/2$	4 768 600	
		$5/2$	4 778 400	
$2s^2\,2p^2(^1D)3d$	^2P	$3/2$	4 789 200	
$2s^2\,2p^2(^1S)3d$	^2D	$3/2$	4 888 000	
Sc XVI ($^3P_0$)	*Limit*		**7 481 000**	

Sc XVI

$Z=21$

C I isoelectronic sequence

Ground state: $1s^2 2s^2 2p^2\ ^3P_0$

Ionization energy $=$ 8 140 000$\pm$16000 cm^{-1} (1009$\pm$2 eV)

The ^{3}P ground term splitting was observed through magnetic dipole transitions in a tokamak plasma by Suckewer et al. (1980). They obtained the values 4354.3$\pm$0.4 Å (in air) for $^3P_0-^3P_1$ and 4530.3$\pm$0.4 Å (in air) for $^3P_1-^3P_2$.

The $2s^2 2p^2-2s2p^3$ array was classified by Fawcett (1971) and by Fawcett and Hayes (1975), who also identified $2s2p^3-2p^4$ lines in the same region. These arrays were reobserved by Sugar, Kaufman, and Cooper (1982) in the range of 117–203 Å with an accuracy of $\pm$0.01 Å. They reported several intersystem lines giving the position of the singlet terms with respect to the ground term. Their results are used in combination with the tokamak observations to determine the levels of $2s^2 2p^2$, $2s2p^3$ and $2p^4$ with an uncertainty of $\pm$50 cm^{-1}. They have also provided the percentage compositions of these levels, including configuration interaction between $2s^2 2p^2$ and $2p^4$. The predicted value of $2s2p^3\ ^5S°$ in brackets is from this paper.

Levels of the $2s^2 2p3d$ configuration are from the classifications given by Bromage and Fawcett (1977) of wavelengths observed at ~20 Å. The configurations $2s^2 2p3s$ and $2s2p^2 3s$ were identified by Goldsmith, Feldman, Crooker, and Cohen (1972) from measurements at ~22 Å with an uncertainty of $\pm$0.005 Å. Levels of these configurations were obtained with an uncertainty of $\pm$2000 cm^{-1}.

The ionization energy is from the extrapolation by Lotz (1967).

References

Bromage, G. E., and Fawcett, B. C. (1977) Mon. Not. R. Astron. Soc. **178,** 605.

Fawcett, B. C. (1971), J. Phys. **B4,** 981.

Fawcett, B. C., and Hayes, R. W. (1975), Mon. Not. R. Astron. Soc. **170,** 185.

Goldsmith, S., Feldman, U., Crooker, A., and Cohen, L. (1972), J. Opt. Soc. Am. **62,** 260.

Lotz, W. (1967), J. Opt. Soc. Am. **57,** 873.

Suckewer, S., Cecci, J. L., Cohen, S., Fonck, R., and Hinnov, E. (1980), Phys. Lett. **80A,** 259.

Sugar, J., Kaufman, V., and Cooper, D. (1982), Phys. Scr. **26,** 189.

Sc XVI

Configuration	Term	J	Level (cm^{-1})	Leading percentages		
$2s^2\,2p^2$	^{3}P	0	0	95	4	$2s^2 2p^2\ ^1S$
		1	22 959	99	1	$2p^4\ ^3P$
		2	45 026	92	7	$2s^2 2p^2\ ^1D$
$2s^2\,2p^2$	^{1}D	2	123 360	92	7	^{3}P
$2s^2\,2p^2$	^{1}S	0	218 720	92	4	$2p^4\ ^1S$
$2s(^2S)2p^3(^4S°)$	^{5}S°	2	[*301 400*]	99	1	$2s(^2S)2p^3(^2P°)\ ^3P°$
$2s(^2S)2p^3(^2D°)$	^{3}D°	1	*536 610*	94	6	$2s(^2S)2p^3(^2P°)\ ^3P°$
		2	*537 720*	96	3	$2s(^2S)2p^3(^2P°)\ ^3P°$
		3	*542 030*	100		
$2s(^2S)2p^3(^2P°)$	^{3}P°	0	*628 600*	100		
		1	*630 250*	96	3	$2s(^2S)2p^3(^2D°)\ ^3D°$
		2	*634 430*	92	6	$2s(^2S)2p^3(^2D°)\ ^3D°$
$2s(^2S)2p^3(^4S°)$	^{3}S°	1	*782 360*	94	5	$2s(^2S)2p^3(^2P°)\ ^1P°$
$2s(^2S)2p^3(^2D°)$	^{1}D°	2	*785 740*	99	1	$2s(^2S)2p^3(^2P°)\ ^3P°$
$2s(^2S)2p^3(^2P°)$	^{1}P°	1	*877 000*	94	5	$2s(^2S)2p^3(^4S°)\ ^3S°$

Sc XVI—Continued

Configuration	Term	J	Level (cm^{-1})	Leading percentages		
$2p^4$	^3P	2	1 187 830	94	4	$2p^4\ ^1D$
		0	1 221 300	99	1	$2s^22p^2\ ^3P$
		1	1 227 760	95	3	$2p^4\ ^1S$
$2p^4$	^1D	2	1 285 480	94	5	^3P
$2p^4$	^1S	0	1 451 350	92	4	$2s^22p^2\ ^1S$
$2s^2\ 2p3s$	$^3P^\circ$	1	*4 606 000*			
		2	*4 646 000*			
$2s^2\ 2p3s$	$^1P^\circ$	1	*4 663 000*			
$2s^2\ 2p3d$	$^3F^\circ$	2	*4 900 000*			
$2s^2\ 2p3d$	$^3D^\circ$	1	*4 946 000*			
		2	*4 960 000*			
		3	*4 975 000*			
$2s^2\ 2p3d$	$^3P^\circ$	2	*4 983 000*			
		1	*4 986 000*			
$2s^2\ 2p3d$	$^1P^\circ$	1	*5 029 000*			
$2s^2\ 2p3d$	$^1F^\circ$	3	*5 028 000*			
$2s2p^2(^2D)3s$	^3D	2	5 120 000			
$2s2p^2(^4P)3d$	^3F	3	5 273 000			
$2s2p^2(^2D)3d$	^3F	4	5 430 000			
Sc XVII ($^2P^\circ_{1/2}$)	***Limit***		**8 140 000**			

Sc XVII

$Z=21$

B I isoelectronic sequence

Ground state: $1s^2 2s^2 2p\ ^2P^\circ_{1/2}$

Ionization energy = 8 820 000±20 000 cm^{-1} (1094±2 eV)

The ^{2}P ground term splitting was determined from a magnetic dipole transition at 2190.5±0.2 Å (in air) in a tokamak plasma by Suckewer et al. (1980).

Fawcett and Hayes (1975) classified the transition arrays $2s^2 2p - 2s2p^2$ and $2s2p^2 - 2p^3$. They were remeasured by Sugar, Kaufman, and Cooper (1982) in the range of 142–230 Å with an uncertainty of ±0.01 Å, or a level uncertainty of ±50 cm^{-1}. These authors identified the two new terms $2p^3\ ^2P^\circ$ and $^2D^\circ$ and the two intersystem lines $2s2p^2\ ^4P_{3/2,5/2} - 2p^3\ ^2D^\circ_{3/2,5/2}$ at 142.62 Å and 145.84 Å. Edlén (1983) suggested that these identifications are inconsistent with the intersystem connection found in a solar flare spectrum by Sandlin et al. (1976) for isoelectronic Fe XXII. Edlén proposes the values 142.576 Å and 145.773 Å for these lines. A direct connection to the ^{2}P ground term from ^{4}P was found by Denne and Hinnov (1984) in a tokamak plasma. They identified the line at 352.5 Å as the transition $2s^2 2p\ ^2P^\circ_{3/2} - 2s2p^2\ ^4P_{5/2}$. It is within 0.02 Å of Edlén's intersystem prediction and has been adopted here. The percentage composition of the mixed configurations $2s^2 2p + 2p^3$ and of the configuration $2s2p^2$ are given by Sugar et al.

The higher-lying configurations were identified by Fawcett and Hayes from their measurements of the spectrum at 17–18 Å with an uncertainty of ±0.01 Å, or a level uncertainty of ±3000 cm^{-1}.

The ionization energy is from an extrapolation by Lotz (1967).

References

Denne, B., and Hinnov, E. (1984), Phys. Rev. A **29**, 3442.

Edlén, B. (1983), Phys. Scr. **28**, 483.

Fawcett, B. C., and Hayes, R. W. (1975), Mon. Not. R. Astron. Soc. **170**, 185.

Lotz, W. (1967), J. Opt. Soc. Am. **57**, 873.

Sandlin, G. D., Brueckner, G. E., Scherrer, V. E., and Tousey, R. (1976), Astrophys. J. **205**, 247.

Suckewer, S., Cecci, J. L., Cohen, S., Fonck, R. and Hinnov, E. (1980), Phys. Lett. **80A**, 259.

Sugar, J., Kaufman, V., and Cooper, D. (1982), Phys. Scr. **26**, 293.

Sc XVII

Configuration	Term	J	Level (cm^{-1})	Leading percentages		
$2s^2 2p$	$^2P^\circ$	1/2	*0*	97	3	$2p^3\ ^2P^\circ$
		3/2	*45 637*	97	3	
$2s2p^2$	^4P	1/2	288 400	99	1	^2S
		3/2	306 680	100		
		5/2	329 320	99	1	^2D
$2s2p^2$	^2D	3/2	516 640	99	1	^2P
		5/2	520 630			
$2s2p^2$	^2S	1/2	632 370	60	39	^2P
$2s2p^2$	^2P	1/2	682 220	61	39	^2S
		3/2	694 950	99	1	^2D
$2p^3$	$^4S^\circ$	3/2	*896 870*	98	2	$^2P^\circ$
$2p^3$	$^2D^\circ$	3/2	*1 008 090*	94	5	$^2P^\circ$
		5/2	*1 015 250*	100		
$2p^3$	$^2P^\circ$	1/2	*1 129 040*	97	3	$2s^2 2p\ ^2P^\circ$
		3/2	*1 142 540*	90	6	$2p^3\ ^2D^\circ$
$2s^2 3d$	^2D	3/2	5 219 000			
		5/2	5 224 000			

Sc XVII—Continued

Configuration	Term	J	Level (cm^{-1})	Leading percentages
$2s2p(^3P°)3p$	^2P	1/2	5 325 000	
		3/2	5 356 000	
$2s2p(^3P°)3p$	^2D	3/2	5 431 000	
		5/2	5 457 000	
$2s2p(^3P°)3d$	$^2D°$	5/2	*5 529 000*	
$2s2p(^3P°)3d$	$^4D°$	7/2	*5 532 000*	
$2s2p(^3P°)3d$	$^4P°$	5/2	*5 532 000*	
$2s2p(^3P°)3d$	$^2F°$	5/2	*5 585 000*	
		7/2	*5 609 000*	
$2s2p(^1P°)3d$	$^2F°$	7/2	*5 765 000*	
$2s2p(^1P°)3d$	$^2D°$	5/2	*5 797 000*	
Sc XVIII ($^1S_0$)	***Limit***		**8 820 000**	

Sc XVIII

$Z = 21$

Be I isoelectronic sequence

Ground state: $1s^2 2s^2\ ^1S_0$

Ionization energy = 9 780 000±20000 cm^{-1} (1213±2 eV)

Fawcett and Hayes (1975) identified the resonance transition $2s^2\ ^1S_0 - 2s2p\ ^1P_1$. The transition array $2s2p - 2p^2$ was interpreted by Fawcett, Ridgeley, and Hatter (1980), including the intersystem transition $2s2p\ ^3P_2^\circ - 2p^2\ ^1D_2$. However, no connection of this $^1D_2$ level to the singlet system was given. The transition $2s^2\ ^1S_0 - 2s2p\ ^3P_1^\circ$ was observed in a tokamak plasma at 348.6 Å by Denne and Hinnov (1984). This agrees closely with the interpolated value for $2s2p\ ^3P_1^\circ$ given by Edlén (1983). Unpublished measurements for these transitions by Kaufman and Sugar with an uncertainty of ±0.01 Å are used to derive these levels with an uncertainty of ±50 cm^{-1}. The interval $2s2p\ ^3P_1^\circ - ^3P_2^\circ$ was observed as a magnetic dipole transition in a tokamak plasma at 2907.9±0.3 Å (in air) by Suckewer and Hinnov (1982).

The higher-lying configurations are from the classifications of the lines at ~18 Å measured with an uncertainty of ±0.01 Å by Fawcett and Hayes (1975), giving a level uncertainty of ±3000 cm^{-1}.

The ionization energy was obtained by Lotz (1967) by extrapolation.

References

Denne, B., and Hinnov, E. (1984), Phys. Rev. A **29**, 3442.

Edlén, B. (1983), Phys. Scr. **28**, 51.

Fawcett, B. C., and Hayes, R. W. (1975), Mon. Not. R. Astron. Soc. **170**, 185.

Fawcett, B. C., Ridgeley, A., and Hatter, A. T. (1980), J. Opt. Soc. Am. **70**, 1349.

Lotz, W. (1967), J. Opt. Soc. Am. **57**, 873.

Suckewer, S., and Hinnov, E., in Physics of Electronic and Atomic Collisions, (North-Holland Publ. Co., 1982).

Sc XVIII

Configuration	Term	J	Level (cm^{-1})
$2s^2$	^1S	0	0
$2s2p$	$^3P^\circ$	0	*273 200*
		1	*286 860*
		2	*321 240*
$2s2p$	$^1P^\circ$	1	*553 440*
$2p^2$	^3P	0	732 070
		1	754 860
		2	778 400
$2p^2$	^1D	2	856 160
$2p^2$	^1S	0	1 034 250
$2s3p$	$^1P^\circ$	1	*5 692 000*
$2s3d$	^3D	3	5 775 000
		2	5 777 000
$2p3p$	^3D	3	6 054 000
$2p3d$	$^3D^\circ$	3	*6 160 000*
$2p3d$	$^1F^\circ$	3	*6 218 000*
Sc XIX ($^2S_{1/2}$)	***Limit***		**9 780 000**

Sc XIX

$Z=21$

Li I isoelectronic sequence

Ground state: $1s^2 2s\ ^2S_{1/2}$

Ionization energy = 10 388 200±3000 cm^{-1} (1287.98±0.37 eV)

The $2s-2p$ transitions were observed in a tokamak plasma by Suckewer et al. (1980) at 279.8±0.2 Å and 326.0±0.3 Å, giving the $2p$ levels an uncertainty of 300 cm^{-1}. Goldsmith, Feldman, Oren, and Cohen (1972) identified the transitions $2s-3p$, $4p$, and $5p$ and $2p-3s$, $3d$, $4d$, and $5d$ in the range of 11–19 Å with a wavelength uncertainty of ±0.005 Å. These results were extended by Boiko, Faenov, and Pikuz (1978) to include $2s-7p$ and $2p-8d$. They also classified transitions from doubly excited configurations. The $3d-4f$ doublet was reported by Fawcett and Ridgeley (1981). These levels have an uncertainty of ±2000 cm^{-1}.

Edlén (1979) derived the value for the ionization energy by means of a polarization formula applied to the $2p-nd$ series.

References

Boiko, V. A., Faenov, A. Y., and Pikuz, S. A. (1978), J. Quant. Spectrosc. Radiat. Transfer **19**, 11.
Edlén, B. (1979), Phys. Scr. **19**, 255.
Fawcett, B. C., and Ridgeley, A. (1981), J. Phys. **B14**, 203.
Goldsmith, S., Feldman, U., Oren, L., and Cohen, L. (1972), Astrophys. J. **174**, 209.
Suckewer, S., Cecchi, J., Cohen, S., Fonck, R., and Hinnov, E. (1980), Phys. Lett. **A80**, 259.

Sc XIX

Configuration	Term	J	Level (cm^{-1})
$1s^2\,2s$	^2S	1/2	0
$1s^2\,2p$	$^2P°$	1/2	*306 700*
		3/2	*357 400*
$1s^2\,3s$	^2S	1/2	5 846 000
$1s^2\,3p$	$^2P°$	1/2	*5 931 000*
		3/2	*5 946 000*
$1s^2\,3d$	^2D	3/2	5 978 000
		5/2	5 982 000
$1s^2\,4s$	^2S	1/2	7 853 800
$1s^2\,4p$	$^2P°$	1/2	*7 890 000*
$1s^2\,4d$	^2D	3/2	7 909 000
		5/2	7 910 000
$1s^2\,4f$	$^2F°$	5/2,7/2	*7 910 000*
$1s^2\,5p$	$^2P°$	1/2,3/2	*8 790 000*
$1s^2\,5d$	^2D	3/2,5/2	8 800 000
$1s^2\,6p$	$^2P°$	1/2,3/2	*9 272 000*
$1s^2\,6d$	^2D	3/2,5/2	9 283 000
$1s^2\,7p$	$^2P°$	1/2,3/2	*9 576 000*
$1s^2\,7d$	^2D	3/2,5/2	9 578 000
$1s^2\,8d$	^2D	3/2,5/2	9 764 000
Sc XX ($^1S_0$)	***Limit***		**10 388 200**
$1s(^2S)2s2p(^1P°)$	$^2P°$	1/2,3/2	*34 664 000*
$1s2p^2$	^2D	3/2,5/2	34 840 000
$1s2p^2$	^2P	1/2,3/2	34 860 000
$1s2p^2$	$^2S°$	1/2	*35 047 000*

Sc XX

$Z=21$

He I isoelectronic sequence

Ground state: $1s^2\,^1S_0$

Ionization energy = 45 770 000±9000 cm^{-1} (5674.8±1.0 eV)

Because of the excellent agreement of the calculated energies of the $n=2$ shell by Safronova (1981) with the few well-measured spectra in the He I sequence, we have compiled her results for the $n=2$ levels and for the ionization energy. Detailed comparisons are given in the Introduction. Levels of the $n=3$–5 shells are from the calculated binding energies by Ermolaev and Jones (1974) subtracted from Safronova's value for the binding energy of the ground state. We have assumed an uncertainty of 2 parts in 10^4 for the excited levels relative to the ground state, and for the ionization energy. This is equal to the random deviation of the measurements by Aglitskii et al. (1974) from the calculations by Safronova (see Introducton). For differences between excited levels where $\Delta=0$, we assumed an uncertainty of 2 parts in 10^3.

Observations by Boiko et al. (1978) place the $1s2p\ ^3P^\circ_1$ level at 34 660 000 cm^{-1} and the $1s2p\ ^1P^\circ_1$ at 34 820 000 cm^{-1} with an estimated uncertainty of ±6000 cm^{-1}.

Percentage compositions are from Ermolaev and Jones.

References

Aglitskii, E. V., Boiko, V. A., Zakharov, S. M., Pikuz, S. A., and Faenov, A. Y. (1974), Sov. J. Quant. Electron. **4**, 500.

Boiko, V. A., Faenov, A. Y., and Pikuz, S. A. (1978), J. Quant. Spectrosc. Radiat. Transfer **19**, 11.

Ermolaev, A. M., and Jones, M. (1974), J. Phys. **B7**, 199.

Safronova, U. I. (1981), Phys. Scr. **23**, 241.

Sc XX

Configuration	Term	J	Level (cm^{-1})	Leading percentages		
$1s^2$	^1S	0	0			
1s2s	^3S	1	[34 448 120]			
1s2p	$^3P^\circ$	0	*[34 628 770]*			
		1	*[34 638 550]*	96	4	$^1P^\circ$
		2	*[34 682 810]*			
1s2s	^1S	0	[34 645 360]			
1s2p	$^1P^\circ$	1	*[34 805 000]*	96	4	$^3P^\circ$
1s3s	^3S	1	[40 790 620]			
1s3p	$^3P^\circ$	0	*[40 840 530]*			
		1	*[40 843 100]*	96	4	$^1P^\circ$
		2	*[40 856 320]*			
1s3s	^1S	0	[40 842 480]			
1s3p	$^1P^\circ$	1	*[40 889 690]*	96	4	$^3P^\circ$
1s4s	^3S	1	[42 983 370]			
1s4p	$^3P^\circ$	0	*[43 004 100]*			
		1	*[43 005 180]*	96	4	$^1P^\circ$
		2	*[43 010 770]*			
1s4s	^1S	0	[43 004 390]			
1s4p	$^1P^\circ$	1	*[43 024 380]*	96	4	$^3P^\circ$

Sc XX—Continued

Configuration	Term	J	Level (cm^{-1})	Leading percentages		
1*s*5*s*	^{3}S	1	[43 992 240]			
1*s*5*p*	^{3}P°	0	[*44 002 720*]	95	5	^{1}P°
		1	[*44 003 270*]			
		2	[*44 006 140*]			
1*s*5*s*	^{1}S	0	[44 002 740]			
1*s*5*p*	^{1}P°	1	[*44 013 010*]	95	5	^{3}P°
Sc XXI ($^2S_{1/2}$)	***Limit***		**45 770 000**			

Sc XXI

$Z=21$

H I isoelectronic sequence

Ground state: 1*s* $^2S_{1/2}$

Ionization energy = 48 665 520±20 cm^{-1} (6033.804±0.016 eV)

No observations of this spectrum are reported. We give calculated values by Mohr (1983) for the $n=2$ shell and by Erickson (1977) for $n=3$–5 relative to the 2*p* $^2P^\circ_{3/2}$ level. Further details are given in the Introduction.

References

Erickson, G. W. (1977), J. Phys. Chem. Ref. Data **6,** 831.
Mohr, P. J. (1983), At. Data Nucl. Data Tables **29,** 453.

Sc XXI

Configuration	Term	J	Level (cm^{-1})	Configuration	Term	J	Level (cm^{-1})
1*s*	^{2}S	1/2	0	4*f*	^{2}F°	5/2	*45 638 321*
						7/2	*45 639 807*
2*p*	^{2}P°	1/2	*36 477 160*				
		3/2	*36 549 404*	5*p*	^{2}P°	1/2	*46 721 966*
						3/2	*46 726 587*
2*s*	^{2}S	1/2	36 479 337				
				5*s*	^{2}S	1/2	46 722 112
3*p*	^{2}P°	1/2	*43 256 511*				
		3/2	*43 277 925*	5*d*	^{2}D	3/2	46 726 578
						5/2	46 728 104
3*s*	^{2}S	1/2	43 257 181				
				5*f*	^{2}F°	5/2	*46 728 101*
3*d*	^{2}D	3/2	43 277 887			7/2	*46 728 862*
		5/2	43 284 947				
				5*g*	^{2}G	7/2	46 728 861
4*p*	^{2}P°	1/2	*45 626 334*			9/2	46 729 317
		3/2	*45 635 363*				
					Limit		**48 665 520**
4*s*	^{2}S	1/2	45 626 617				
4*d*	^{2}D	3/2	45 635 346				
		5/2	45 638 326				

Ti I

$Z=22$

Ground state: $1s^2 2s^2 2p^6 3s^2 3p^6 3d^2 4s^2\ ^3F_2$

Ionization energy = 55 010±50 cm^{-1} (6.820±0.006 eV)

The structure of the arc spectrum of titanium was first studied by Kiess and Kiess (1923, 1924) who classified 400 lines as members of the triplet and quintet systems. The results are presented in the form of arrays of triplets and quintets and tables of observed and calculated Zeeman effects.

In 1927 Russell published his classical analyses of Ti I and Ti II. He extended Kiess' analysis of the triplets and quintets in Ti I and established the singlet system. His list of 142 terms is presented in nearly modern notation, with the energy level values measured from the ground state taken as zero. He also gave a table of wavelengths of 1394 classified lines between 9800 and 2100 Å and a table of observed and theoretical Zeeman patterns. Every term was assigned to a definite electron configuration. A key for translating Russell's notation into that used at present is given by Moore (1932). Russell determined an ionization limit at 55 100 cm^{-1}, very close to the value of Catalán and Velasco (1952) adopted here.

In 1928 Kiess calculated 3 decimal place term values for 62 levels of Ti I from interferometer measurements of Ti I wavelengths. This work was revised by Kiess and Thekaekara (1959), who give values for 152 Ti I levels with an uncertainty of ±0.005 cm^{-1}. These values will be found in the present compilation.

The analysis of Ti I was continued by Meggers and Kiess (1932), who observed the spectrum between 8377 and 10 775 Å. With these data they found the lowest term of $3d^4$, $a\ ^5D$, within 500 cm^{-1} of its position as predicted by Russell. The infrared observations were later extended to 11 974 Å by Kiess (1938), who found the next term of $3d^4$, $d\ ^3P$.

The g-values for the $a\ ^3F$ term were measured by Channapa and Pendlebury (1965). The rest were calculated by Moore (1949) from the Zeeman patterns quoted by Russell (1927). Moore has omitted the a $^1S_0$ level at 15 166 cm^{-1} given by Russell, following his advice. Russell questioned the reality of his level at 40 883 cm^{-1}, and his doubts were supported by the calculations of Roth (1969). We have omitted this level. The ^3H term was privately communicated to Moore (1949).

New measurements in the region 2117–3072 Å by Wilson and Thekaekara (1961) resulted in a revision of level values for 16 high odd levels and the discovery of two new triplet terms, $s\ ^2G°$ and $p\ ^3F°$, which have been assigned to $3d^3(^2F)4p$ by Smith and Siddall (1969). Smith and Siddall also assigned 7 terms to $3d^2 4s 5p$ or $3d^3 5p$.

Levels given with two decimal places from the sources mentioned above are assumed to have an uncertainty of ±0.05 cm^{-1}. The percentage compositions for the even parity configurations $3d^2 4s^2$, $3d^3 4s$, and $3d^4$ were calculated by Dembczynski (1980) with configuration interaction (CI). Those of the odd parity configurations $3d4s^2 4p$, $3d^2 4s4p$, and $3d^3 4p$ were calculated by Roth (1969) with CI. Both calculations were fit to the observed energy levels. Roth's designation changes for experimental levels are adopted here. Roth distinguished repeating terms of the $3d^3$ core by the letters a, b, etc. rather than by seniority. The percentages include the sum of seniority states contributing to the term. Roth included no experimental levels above 44 000 cm^{-1} in his calculation. We have quoted his results for four terms above 44 000 for which agreement with experiment was clear. A new calculation by Roth (1980) for the odd parity configurations could not be used because of errors in the eigenstate percentages, which in many cases summed to more than 100.

References

Catalán, M. A., and Velasco, R. (1952), An. R. Soc. Esp. Fis. Quim. **48A**, 247.

Channapa, K. H., and Pendlebury, M. J. (1965), Proc. Phys. Soc. **86**, 1145.

Dembczynski, J. (1980), Physica **100C**, 105.

Kiess, C. C., and Kiess, H. K. (1923), J. Wash. Acad. Sci. **13**, 270.

Kiess, C. C. (1924), J. Opt. Soc. Am. **8**, 607.

Kiess, C. C. (1928), J. Res. Natl. Bur. Stand. **1**, 75.

Kiess, C. C. (1938), J. Res. Natl. Bur. Stand. **20**, 33.

Kiess, C. C., and Thekaekara, M. P. (1959), Astrophys. J. **130**, 1008.

Meggers, W. F., and Kiess, C. C. (1932), J. Res. Natl. Bur. Stand. **9**, 309.

Moore, C. E. (1932), Astrophys. J. **75**, 222.

Moore, C. E. (1949), Atomic Energy Levels, Natl. Bur. Stand. (U.S.) Circ. 467, Vol. I (reissued in 1971 as Natl. Bur. Stand. (U.S.) Natl. Stand. Ref. Data Ser. 35,Vol. I).

Roth, C. (1969), J. Res. Natl. Bur. Stand. **73A**, 497.

Roth, C. (1980), At. Data Nucl. Data Tables **25**, 91.

Russell, H. N. (1927), Astrophys. J. **66**, 347.

Smith, G., and Siddall, J. (1969), J. Opt. Soc. Am. **59**, 419.

Wilson, C. M., and Thekaekara, M. P. (1961), J. Opt. Soc. Am. **51**, 289.

Ti I

Configuration	Term	J	Level (cm^{-1})	g	Leading percentages		
$3d^2\,4s^2$	$a\ ^3$F	2	0.000	0.66	100		
		3	170.132	1.08	100		
		4	386.874	1.25	100		
$3d^3(^4$F$)4s$	$a\ ^5$F	1	6 556.828	0.00	100		
		2	6 598.749	0.99	100		
		3	6 661.003	1.25	100		
		4	6 742.757	1.35	100		
		5	6 842.964	1.41	100		
$3d^2\,4s^2$	$a\ ^1$D	2	7 255.369	1.02	96	2	$3d^3(^2$D2$)4s\ ^1$D
$3d^2\,4s^2$	$a\ ^3$P	0	8 436.618		92	7	$3d^3(^2$P$)4s\ ^3$P
		1	8 492.421	1.50	92	7	
		2	8 602.340	1.49	90	7	
$3d^3(^4$F$)4s$	$b\ ^3$F	2	11 531.760	0.67	100		
		3	11 639.804	1.08	100		
		4	11 776.806	1.26	98	1	$3d^23s^2\ ^1$G
$3d^2\,4s^2$	$a\ ^1$G	4	12 118.394	0.98	90	8	$3d^3(^2$G$)4s\ ^1$G
$3d^3(^4$P$)4s$	$a\ ^5$P	1	13 981.75	2.50	100		
		2	14 028.47	1.82	100		
		3	14 105.68	1.66	100		
$3d^3(^2$G$)4s$	$a\ ^3$G	3	15 108.121	0.74	100		
		4	15 156.787	1.06	100		
		5	15 220.390	1.21	100		
$3d^2(^3$F$)4s4p(^3$P°$)$	$z\ ^5$G°	2	*15 877.18*	0.39	100		
		3	*15 975.59*	0.93	100		
		4	*16 106.08*	1.15	100		
		5	*16 267.51*	1.25	100		
		6	*16 458.71*	1.33	100		
$3d^2(^3$F$)4s4p(^3$P°$)$	$z\ ^5$F°	1	*16 817.19*	0.00	94		
		2	*16 875.19*		98		
		3	*16 961.42*	1.26?	98		
		4	*17 075.31*	1.34	98		
		5	*17 215.44*	1.42	98		
$3d^3(^2$D2$)4s$	$a\ ^3$D	1	17 369.59	0.49	67	30	$(^2$D1$)\ ^3$D
		2	17 423.853	1.17	67	30	
		3	17 540.205	1.34	67	30	
$3d^3(^2$P$)4s$	$b\ ^3$P	0	17 995.75		86	7	$3d^23s^2\ ^3$P
		1	18 061.54		87	7	
		2	18 145.40		87	7	
$3d^3(^2$H$)4s$	$a\ ^3$H	4	18 037.225	0.80	52	44	$(^2$G$)\ ^1$G
		5	18 141.229	1.02	100		
		6	18 192.577	1.17	100		
$3d^3(^2$G$)4s$	$b\ ^1$G	4	18 287.560	1.02	48	48	$(^2$H$)\ ^3$H

Ti I—Continued

Configuration	Term	J	Level (cm^{-1})	g	Leading percentages			
$3d^2(^3F)4s4p(^3P^\circ)$	$z\ ^5D^\circ$	0	*18 462.83*		94			
		1	*18 482.86*	1.65?	94			
		2	*18 525.07*	1.50	94			
		3	*18 593.99*	1.49	93			
		4	*18 695.23*	1.51	94			
$3d^3(^2P)4s$	$c\ ^3P$	0	18 818.23		94		5	$3d^3(^2P)4s\ ^3P$
		1	18 825.89	1.54?	94		3	$3d^3(^2P)4s\ ^3P$
		2	18 911.399	1.54	98		1	$3d^4\ ^3P2$
$3d^2(^3F)4s4p(^3P^\circ)$	$z\ ^3F^\circ$	2	*19 322.988*	0.67	88		8	$(^1D)(^3P^\circ)\ ^3F^\circ$
		3	*19 421.576*	1.07	88		7	
		4	*19 573.968*	1.26	88		7	
$3d^2(^3F)4s4p(^3P^\circ)$	$z\ ^3D^\circ$	1	*19 937.859*		84		8	$(^3P)(^3P^\circ)\ ^3D^\circ$
		2	*20 006.032*	1.16	83		8	
		3	*20 126.055*	1.34	83		8	
$3d^3(^2P)4s$	$a\ ^1P$	1	20 062.98	1.03	98		1	^3P
$3d^3(^2D2)4s$	$b\ ^1D$	2	20 209.444	1.01?	67		30	$(^2D1)\ ^1D^\circ$
$3d^3(^2H)4s$	$a\ ^1H$	5	20 795.599	1.01	100			
$3d^2(^3F)4s4p(^3P^\circ)$	$z\ ^3G^\circ$	3	*21 469.494*	0.75	95			
		4	*21 588.496*	1.05	95			
		5	*21 739.713*	1.21	95			
$3d^2(^3F)4s4p(^3P^\circ)$	$z\ ^1D^\circ$	2	*22 081.198*	1.00	86		10	$(^3P)(^3P^\circ)\ ^1D^\circ$
$3d^2(^3F)4s4p(^3P^\circ)$	$z\ ^1F^\circ$	3	*22 404.69*	1.00	97			
$3d^2(^3F)4s4p(^3P^\circ)$	$z\ ^1G^\circ$	4	*24 694.895*	0.97	94			
$3d^2(^3P)4s4p(^3P^\circ)$	$z\ ^3S^\circ$	1	*24 921.110*	1.99	90		7	$3d^3(^2P)4p\ ^3S^\circ$
$3d^2(^3P)4s4p(^3P^\circ)$	$z\ ^5S^\circ$	2	*25 102.88*	1.93	93			
$3d^2(^3F)4s4p(^1P^\circ)$	$y\ ^3F^\circ$	2	*25 107.417*		44		25	$3d^3(^4F)4p\ ^3F^\circ$
		3	*25 227.217*	1.06	43		25	
		4	*25 388.334*	1.21	41		23	
$3d^3(^4F)4p$	$y\ ^3D^\circ$	1	*25 317.813*	0.50	49		34	$3d^2(^3F)4s4p(^1P^\circ)\ ^3D^\circ$
$3d^2(^1D)4s4p(^3P^\circ)$		2	*25 438.898*	1.17	37	$^3P^\circ$	28	$3d^3(^4F)4p\ ^3D^\circ$
$3d^2(^1D)4s4p(^3P^\circ)$	$z\ ^3P^\circ$	2	*25 493.722*	1.47	49		20	$3d^3(^4F)4p\ ^3D^\circ$
		1	*25 537.276*	1.50	64		22	$3d^2(^3P)4s4p(^3P^\circ)\ ^5D^\circ$
$3d^2(^3P)4s4p(^3P^\circ)$	$y\ ^5D^\circ$	0	*25 605.03*		51		32	$3d^2(^1D)4s4p(^3P^\circ)\ ^3P^\circ$
		1	*25 635.74*		65		19	$3d^2(^1D)4s4p(^3P^\circ)\ ^3P^\circ$
		2	*25 699.95*		82		6	$3d^3(^4F)4p\ ^5D^\circ$
		3	*25 797.60*		56		19	$3d^3(^4F)4p\ ^3D^\circ$
		4	*25 926.771*	1.52	87		7	$3d^3(^4F)4p\ ^5D^\circ$
$3d^2(^3P)4s4p(^3P^\circ)$		3	*25 643.695*	1.33	32	$^5D^\circ$	32	$3d^3(^4F)4p\ ^3D^\circ$

(Continued)

Ti I—Continued

Configuration	Term	J	Level (cm^{-1})	g	Leading percentages			
$3d^3(^4\mathrm{F})4p$	$y\ ^5\mathrm{G}^\circ$	2	*26 494.322*	0.34	94			
		3	*26 564.385*	0.91	96			
		4	*26 657.409*	1.15	98			
		5	*26 772.965*	1.25	100			
		6	*26 910.705*	1.34	100			
$3d^2(^1\mathrm{D})4s4p(^3\mathrm{P}^\circ)$	$x\ ^3\mathrm{F}^\circ$	2	*26 803.417*	0.66	58		19	$(^3\mathrm{F})(^1\mathrm{P}^\circ)\ ^3\mathrm{F}^\circ$
		3	*26 892.926*	1.06	57		20	
		4	*27 025.652*	1.23	57		20	
$3d^2(^1\mathrm{D})4s4p(^3\mathrm{P}^\circ)$	$x\ ^3\mathrm{D}^\circ$	1	*27 355.042*		78		12	$(^3\mathrm{P})(^3\mathrm{P}^\circ)\ ^3\mathrm{D}^\circ$
		2	*27 418.015*	1.17	73		9	
		3	*27 480.047*	1.36	64		9	
$3d^2(^3\mathrm{F})4s4p(^1\mathrm{P}^\circ)$	$y\ ^3\mathrm{G}^\circ$	3	*27 498.975*	0.75	56		23	$3d^3(^4\mathrm{F})4p\ ^3\mathrm{G}^\circ$
		4	*27 614.667*	1.05	55		23	
		5	*27 750.124*	1.21	53		24	
$3d^2(^3\mathrm{P})4s4p(^3\mathrm{P}^\circ)$	$z\ ^5\mathrm{P}^\circ$	1	*27 665.57*		97			
		2	*27 740.19*		91			
		3	*27 887.74*		79		16	$(^1\mathrm{D})(^3\mathrm{P}^\circ)\ ^3\mathrm{D}^\circ$
$3d^2(^1\mathrm{D})4s4p(^1\mathrm{P}^\circ)$	$y\ ^1\mathrm{D}^\circ$	2	*27 907.026*	0.98	32		26	$3d4s^24p\ ^1\mathrm{D}^\circ$
$3d^3(^4\mathrm{F})4p$	$y\ ^5\mathrm{F}^\circ$	1	*28 596.293*	0.00	98			
		2	*28 638.832*	1.01	98			
		3	*28 702.768*	1.24	98			
		4	*28 788.372*	1.34	98			
		5	*28 896.062*	1.40	97			
$3d^4$	$a\ ^5\mathrm{D}$	0	28 772.86		100			
		1	28 791.62		100			
		2	28 828.51		100			
		3	28 882.44		100			
		4	28 952.10		100			
$3d^2(^3\mathrm{F})4s4p(^1\mathrm{P}^\circ)$	$w\ ^3\mathrm{D}^\circ$	1	*29 661.232*	0.51	33		24	$3d^3(^4\mathrm{F})4p\ ^3\mathrm{D}^\circ$
		2	*29 768.655*	1.16	29		21	$3d^3(^4\mathrm{F})4p\ ^3\mathrm{D}^\circ$
$3d^3(^2\mathrm{F})4s$	$a\ ^1\mathrm{F}$	3	29 818.31		98		1	$3d^4\ ^1\mathrm{F}$
$3d^3(^4\mathrm{F})4p$	$x\ ^5\mathrm{D}^\circ$	0	*29 829.097*		91			
		1	*29 855.248*	1.46	87		7	$3d^2(^3\mathrm{P})4s4p(^3\mathrm{P}^\circ)\ ^5\mathrm{D}^\circ$
		2	*29 907.273*	1.50	77		6	$3d^2(^3\mathrm{P})4s4p(^3\mathrm{P}^\circ)\ ^5\mathrm{D}^\circ$
		3	*29 986.185*	1.49	55		14	$3d^2(^3\mathrm{F})4s4p(^1\mathrm{P}^\circ)\ ^3\mathrm{D}^\circ$
		4	*30 060.328*	1.49	91			
$3d^3(^4\mathrm{F})4p$		3	*29 912.262*	1.34	35	$^5\mathrm{D}^\circ$	20	$3d^2(^3\mathrm{F})4s4p(^1\mathrm{P}^\circ)\ ^3\mathrm{D}^\circ$
$3d^2(^1\mathrm{G})4s4p(^3\mathrm{P}^\circ)$	$x\ ^3\mathrm{G}^\circ$	3	*29 914.720*		70		19	$(^3\mathrm{F})(^1\mathrm{P}^\circ)\ ^3\mathrm{G}^\circ$
		4	*29 971.078*		72		19	
		5	*30 039.211*	1.19	71		19	
$3d^2(^3\mathrm{P})4s4p(^3\mathrm{P}^\circ)$	$v\ ^3\mathrm{D}^\circ$	1	*31 184.021*	0.51	77		16	$(^1\mathrm{D})(^3\mathrm{P}^\circ)\ ^3\mathrm{D}^\circ$
		2	*31 190.631*	1.17	68		15	
		3	*31 205.985*	1.34	69		14	

Ti I—Continued

Configuration	Term	J	Level (cm^{-1})	g	Leading percentages		
$3d^3(^4F)4p$	$w\ ^3G°$	3	*31 373.801*	0.75	70	21	$3d^2(^3F)4s4p(^1P°)\ ^3G°$
		4	*31 489.451*	1.05	69	22	
		5	*31 628.668*	1.19	69	22	
$3d^2(^3P)4s4p(^3P°)$	$y\ ^3P°$	0	*31 685.90*		85	7	$3d^3(^2P)4p\ ^3P°$
		1	*31 725.75*	1.47	85	6	
		2	*31 805.94*		85	6	
$3d^2(^1G)4s4p(^3P°)$	$z\ ^3H°$	4	*31 829.972*	0.80	85	11	$3d^3(^2G)4p\ ^3H°$
		5	*31 914.277*	1.04	86	10	
		6	*32 013.534*	1.17	86	10	
$3d^2(^1D)4s4p(^1P°)$	$y\ ^1F°$	3	*32 857.721*	0.99	36	44	$3d^3(^2G)4p\ ^1F°$
$3d^3(^4P)4p$	$x\ ^3P°$	0	*33 085.153*		33	34	$3d^2(^3P)4s4p(^1P°)\ ^3P°$
		1	*33 090.492*	1.46	33	34	
		2	*33 114.412*	1.46	34	34	
$3d^3(^4F)4p$	$w\ ^3F°$	2	*33 655.853*	0.66	54	30	$3d^2(^3F)4s4p(^1P°)\ ^3F°$
		3	*33 680.130*	1.09	53	30	
		4	*33 700.874*	1.26	53	30	
$3d^2(^1D)4s4p(^1P°)$	$z\ ^1P°$	1	*33 660.671*	0.94	37	29	$(^3P)(^3P°)\ ^1P°$
$3d^2(^1G)4s4p(^3P°)$	$v\ ^3F°$	2	*33 980.639*	0.63	81	9	$3d4s^24p\ ^3F°$
		3	*34 078.580*	1.10	83	9	
		4	*34 204.971*	1.23	84	8	
$3d^4$	$d\ ^3P2$	0	34 170.95		56	41	$^3P1$
		1	34 327.96		56	41	
		2	34 535.04		56	41	
$3d^3(^2G)4p$	$z\ ^1H°$	5	*34 700.212*	1.02	58	26	$(^2H)\ ^1H°$
$3d^2(^3P)4s4p(^3P°)$	$y\ ^1P°$	1	*34 947.120*		54	26	$(^1D)(^1P°)\ ^1P°$
$3d^2(^3P)4s4p(^3P°)$	$x\ ^1D°$	2	*35 035.147*		56	12	$(^3F)(^3P°)\ ^1D°$
	$y\ ^3S°$	1	*35 439.228*	2.18			
$3d^3(^2G)4p$	$y\ ^3H°$	4	*35 454.051*	0.79	84	13	$3d^2(^1G)4s4p(^3P°)\ ^3H°$
		5	*35 559.627*	1.04	85	12	
		6	*35 685.160*	1.17	85	12	
$3d^3(^4P)4p$	$w\ ^5D°$	0	*35 503.40*		99		
		1	*35 527.76*	1.51	99		
		2	*35 577.14*	1.53	99		
		3	*35 652.95*	1.46	99		
		4	*35 757.51*	1.46	99		
$3d^2\ 4s(^4F)5s$	$e\ ^5F$	1	35 959.07	0.00			
		2	36 013.57	1.03?			
		3	36 096.47	1.24			
		4	36 208.92	1.34			
		5	36 351.43	1.42			
$3d^2(^1G)4s4p(^1P°)$	$y\ ^1G°$	4	*36 000.144*	1.00	45	32	$3d^3(^2G)4p\ ^1G°$

(Continued)

Ti I—Continued

Configuration	Term	J	Level (cm^{-1})	g	Leading percentages		
$3d^4$	$b\ ^3$G	3	36 065.75		100		
		4	36 132.21		100		
		5	36 200.94		100		
$3d^3(^4$P$)4p$	$y\ ^5$P°	1	*36 298.43*	2.47	97		
		2	*36 340.67*	1.81	97		
		3	*36 414.58*	1.66	98		
$3d^3(a\ ^2$D$)4p$	$w\ ^3$P°	0	*37 090.65*		35	35	$(^4$P$)\ ^3$P°
		1	*37 172.947*	1.53	36	35	
		2	*37 325.407*	1.48	33	33	
$3d^3(^4$P$)4p$	$y\ ^5$S°	2	*37 359.13*	1.99	90		
$3d^2\ 4s(^4$F$)5s$	$e\ ^3$F	2	37 538.804	0.67			
		3	37 659.927	1.11			
		4	37 824.748	1.27			
$3d^3(^2$G$)4p$	$v\ ^3$G°	3	*37 555.021*	0.77	77	7	$3d^2(^1$G$)4s4p(^3$P°$)\ ^3$G°
		4	*37 617.868*	1.05	81	7	
		5	*37 690.320*	1.20	85	7	
$3d\ 4s^2\ 4p$	$x\ ^1$F°	3	*37 622.573*	0.94	25	16	$3d^2(^1$G$)4s4p(^1$P°$)\ ^1$F°
$3d^3(a\ ^2$D$)4p$	$u\ ^3$F°	2	*37 654.77*	0.65	49	20	$(^2$G$)\ ^3$F°
		3	*37 743.933*	1.08	26	14	
		4	*37 852.434*	1.24	41	26	
$3d^2(^3$P$)4s4p(^1$P°$)$	$u\ ^3$D°	1	*37 852.021*	0.53	42	17	$3d^3(^2$P$)4p\ ^3$D°
		2	*37 976.78*	1.14?	41	17	
		3	*38 159.71*	1.35	39	17	
$3d^3(^2$P$)4p$	$z\ ^1$S°	0	*38 200.94*		80	18	$3d^2(^3$P$)4s4p(^3$P°$)\ ^1$S°
$3d^3(^2$G$)4p$	$t\ ^3$F°	2	*38 451.298*	0.66	58	27	$(a\ ^2$D$)\ ^3$F°
		3	*38 544.38*	1.08	55	30	
		4	*38 670.710*	1.25	52	34	
$3d^3(^2$H$)4p$	$z\ ^3$I°	5	*38 572.692*	0.81	100		
		6	*38 668.832*	1.02	100		
		7	*38 779.856*	1.15	100		
$3d^3(^2$P$)4p$	$t\ ^3$D°	1	*38 654.23*	0.54?	37	22	$3d4s^24p\ ^3$D°
		2	*38 699.767*		32	20	
		3	*38 764.832*	1.32	32	20	
$3d^3(^2$G$)4p$	$x\ ^1$G°	4	*38 959.499*	1.02	50	29	$(^2$H$)\ ^1$G°
$3d^3(a\ ^2$D$)4p$	$x\ ^1$P°	1	*39 077.713*		73	16	$(^2$P$)\ ^1$P°
$3d^3(^4$F$)5s$	$f\ ^5$F	1	39 107.25				
		2	39 149.26				
		3	39 214.38				
		4	39 302.36				
		5	39 412.78				
$3d^3(^2$H$)4p$	$x\ ^3$H°	4	*39 115.958*	0.88	93		
		5	*39 152.057*	1.02	85	12	$3d^2(^1$G$)4s4p(^1$P°$)\ ^3$H°
		6	*39 198.320*	1.18	85	12	$3d^2(^1$G$)4s4p(^1$P°$)\ ^3$H°

Ti I—Continued

Configuration	Term	J	Level (cm^{-1})	g	Leading percentages			
$3d^3(^2P)4p$	$w\ ^1D°$	2	*39 265.80*	1.06?	30		28	$3d4s^24p\ ^1D°$
$3d^3(^4F)5s$	$f\ ^3F$	2	39 526.89					
		3	39 640.98					
		4	39 785.94					
$3d^3(a\ ^2D)4p$	$s\ ^3D°$	1	*39 662.15*	0.52	54		14	$(^2P)\ ^3D°$
		2	*39 686.10*		60		17	
		3	*39 715.437*	1.31	55		20	
$3d^3(a\ ^2D)4p$	$w\ ^1F°$	3	*40 302.950*	1.05	82		7	$3d^2(^1D)4s4p(^1P°)\ ^1F°$
$3d^3(^2H)4p$	$z\ ^1I°$	6	*40 319.80*	1.03	99			
$3d\ 4s^2\ 4p$	$v\ ^3P°$	0	*40 369.76*		37		23	$3d^2(^3P)4s4p(^1P°)\ ^3P°$
		2	*40 466.979*		36		21	$3d^2(^3P)4s4p(^1P°)\ ^3P°$
$3d^3(^2P)4p$		1	*40 384.58*		30	$^3S°$	24	$3d4s^24p\ ^3P°$
$3d^3(^4P)4p$	$r\ ^3D°$	1	*40 556.07*	0.49	44		22	$(a\ ^2D)\ ^3D°$
		2	*40 670.60*		43		24	
		3	*40 844.19*		40		25	
$3d^3(^2P)4p$	$x\ ^3S°$	1	*40 844.19*		57		13	$3d4s^24p\ ^3P°$
$3d^3(^2H)4p$	$y\ ^1H°$	5	*41 039.874*	1.03	47		41	$(^2G)\ ^1H°$
$3d^2\ 4s(^2F)5s$	$e\ ^1F$	3	41 087.31	1.01				
$3d^2\ 4s(^4F)5p?$	$u\ ^3G°$	3	*41 170.003*	0.73				
		4	*41 255.400*	1.03				
		5	*41 341.553*	1.19				
$3d^2\ 4s(^4F)4d$	$e\ ^3G$	3	41 194.42					
		4	41 368.86					
		5	41 481.13					
$3d\ 4s^2\ 4p$	$s\ ^3F°$	2	*41 337.43*	0.66	66		18	$3d^3(a\ ^2D)4p\ ^3F°$
		3	*41 457.653*	1.09	67		19	
		4	*41 624.209*	1.24	68		18	
$3d^2\ 4s(^4F)4d$	$e\ ^3H$	4	41 515.09					
		5	41 556.33					
		6	41 615.02					
$3d^3(^2G)4p$	$v\ ^1F°$	3	*41 585.24*		43		31	$3d^2(^1D)4s4p(^1P°)\ ^1F°$
$3d^2\ 4s(^4F)4d$	$e\ ^5G$	2	41 714.35					
		3	41 757.47					
		4	41 818.70	1.12				
		5	41 903.48	1.24				
		6	42 019.22	1.34				
$3d^2\ 4s(^4F)5p$	$w\ ^3H°$	4	*41 780.95*					
		5	*41 895.15*					
		6	*41 995.39*					

(Continued)

Ti I—Continued

Configuration	Term	J	Level (cm^{-1})	g	Leading percentages		
$3d^2\,4s(^4F)5p$	$v\,^5D^\circ$	0	*41 822.99*				
		1	*41 854.01*				
		2	*41 906.51*				
		3	*41 985.83*				
		4	*42 092.42*				
$3d^2\,4s(^4F)4d$	$e\,^5H$	3	41 823.19				
		4	41 917.05				
		5	42 018.01	1.15			
		6	42 123.77	1.22			
		7	42 205.59	1.28			
$3d^2\,4s(^4F)4d$	$e\,^5D$	0	41 871.56				
		1	41 901.36				
		2	41 958.51				
		3	42 052.72				
		4	42 184.66				
$3d^2\,4s(^4F)4d$	$g\,^3F$	2	41 871.87				
		3	41 988.39				
		4	42 107.06				
$3d^3(^2P)4p$	$u\,^3P^\circ$	2	*41 928.528*		38	37	$(a\,^2D)\,^3P^\circ$
		1	*41 943.95*		36	37	
		0	*41 959.46*		36	36	
$3d\,4s^2\,4p$	$q\,^3D^\circ$	1	*42 146.39*		39	20	$3d^2(^3P)4s4p(^1P^\circ)\,^3D^\circ$
		2	*42 206.88*		27	14	
		3	*42 311.269*	1.32	39	20	
$3d^2\,4s(^4F)5p$?	$p\,^3D^\circ$	1	*42 194.04*				
		2	*42 269.78*				
		3	*42 376.45*				
$3d^2\,4s(^4F)4d$	$e\,^5P$	1	42 611.58				
		2	42 724.11				
		3	42 858.90	1.64			
	$w\,^1P^\circ$	1	*42 927.55*	1.00?			
$3d^2\,4s(^4F)4d$	$g\,^5F$	1	43 034.08				
		2	43 080.92				
		3	43 148.15				
		4	43 231.99				
		5	43 330.07				
$3d^2\,4s(^4F)5p$	$r\,^3F^\circ$	2	*43 467.55*				
		3	*43 583.14*				
		4	*43 744.55*				
$3d^3(^2F)4p$	$v\,^1G^\circ$	4	*43 674.130*	0.95	36	30	$(^2H)\,^1G^\circ$
	$v\,^1D^\circ$	2	*43 710.28*				
$3d^3(a\,^2D)4p$	$u\,^1D^\circ$	2	*43 799.455*	0.98	38	29	$3d^2(^1D)4s4p(^1P^\circ)\,^1D^\circ$

Ti I—Continued

Configuration	Term	J	Level (cm^{-1})	g	Leading percentages		
$3d^3(^4F)4d$	$f\,^5H$	3	43 843.82				
		4	43 901.630	0.91			
		5	43 971.513	1.11			
		6	44 051.333	1.21			
		7	44 134.639	1.29			
$3d^2\,4s(^4F)5p$	$o\,^3D^\circ$	1	*43 975.71*				
		2	*44 079.84*	1.18?			
		3	*44 233.65*				
$3d^2\,4s(^4F)5p$	$t\,^3G^\circ$	4	*44 162.44*				
		5	*44 375.57*				
$3d^2(^1G)4s4p(^1P^\circ)$	$x\,^1H^\circ$	5	*44 163.24*	1.03	72	27	$3d^3(^2H)4p\,^1H^\circ$
$3d^3(^4F)4d$	$f\,^5D$	3	44 254.39				
		4	44 381.17				
$3d^2\,4s(^2D)5s$	$e\,^1D$	2	44 581.16				
$3d^3(^4F)5p$	$q\,^3F^\circ$	2	*44 824.13*				
		3	*44 922.73*				
		4	*45 040.81*				
$3d^3(^4P)4p$	$w\,^3S^\circ$	1	*44 858.03*		59	35	$3d^2(^3P)4s4p(^1P^\circ)\,^3S^\circ$
$3d^3(^4F)5p$	$n\,^3D^\circ$	1	*44 966.39*				
		2	*45 063.80*				
		3	*45 206.27*				
	$t\,^3P^\circ$	0	*45 040.70*				
		1	*45 090.73*				
		2	*45 178.06*				
$3d^2\,4s(^2F)4d$	$e\,^1H$	5	45 485.35				
$3d^3(^4F)4d?$	$f\,^5G$	3	45 689.89				
		4	45 711.28				
		5	45 756.45?				
		6	45 904.73				
$3d^2\,4s(^2F)4d$	$f\,^3H$	4	45 721.878	0.80			
		5	45 832.50	1.03			
		6	45 960.439	1.17			
$3d^2\,4s(^4F)6s$	$h\,^5F$	1	45 764.71				
		2	45 813.01				
		3	45 893.26				
		4	46 007.62				
		5	46 157.76				
$3d^2\,4s(^2F)4d$	$e\,^1G$	4	46 068.04				
$3d^2\,4s(^4P)5s$	$e\,^3P$	2	46 244.60				
$3d^2\,4s(^2F)5p$	$u\,^1G^\circ$	4	*46 257.67*	0.95			
$3d^2\,4s(^4F)6s$	$h\,^3F$	4	46 530.45				

(Continued)

Ti I—Continued

Configuration	Term	J	Level (cm^{-1})	g	Leading percentages		
$3d^2\,4s(^2F)4d$	$f\,^1F$	3	46 650.26				
$3d^3(^2F)4p$	$s\,^3G°$	3	*46 725.42*		87	11	$(^2F)\ ^3F°$
		4	*46 838.09*		87	10	
		5	*46 974.65*		97		
$3d^2\,4p^2$	$g\,^5G$	2	46 943.91				
		3	47 030.28				
		4	47 139.86				
		5	47 280.69				
		6	47 446.84				
$3d^2\,4s(^2F)4d$	$i\,^3F$	3	47 038.16				
		4	47 194.68				
$3d^3(^2F)4p$	$p\,^3F°$	2	*47 187.54*		97		
		3	*47 281.90*		86	11	$(^2F)\ ^3G°$
		4	*47 463.06*		87	10	
$3d^3(^4F)6s$	$i\,^5F$	5	47 777.32				
$3d^2\,4s(^4F)5d$	$g\,^5H$	3	47 840.62				
		4	47 913.61				
		5	47 994.32				
		6	48 106.83				
		7	48 262.83				
$3d^2\,4s(^4F)5d$	$h\,^5G$	2	47 870.61				
		3	47 936.79				
		4	48 018.08				
		5	48 119.47				
		6	48 233.47				
$3d^2\,4p^2$	$j\,^5F$	1	48 058.85				
		2	48 107.42				
		3	48 208.87				
		4	48 328.81				
		5	48 462.11				
$3d^2\,4s(^4F)5d$	$g\,^5D$	3	48 059.82				
		4	48 186.11				
$3d^3(^2F)4p$	$u\,^1F°$	3	*48 365.09*		48	24	$3d4s^24p\ ^1F°$
$3d^2\,4s(^4F)5d$	$k\,^5F$	2	48 519.21				
		3	48 588.28				
		4	48 672.66				
		5	48 771.73				
$3d^2\,4p^2$	$e\,^3D$	2	48 724.34				
		1	48 724.83				
		3	48 839.74				

Ti I—Continued

Configuration	Term	*J*	Level (cm^{-1})	*g*	Leading percentages
$3d^2\,4p^2$	$h\ ^5D$	0	48 802.32		
		1	48 859.51		
		2	48 915.07		
		3	49 024.43		
		4	49 036.46		
	$f\ ^3D$	2	49 571.69		
		3	49 619.72		
	$f\ ^1D$	2	50 128.08		
	$f\ ^1G$	4	52 125.98		
	$e\ ^1P$	1	53 663.32		
Ti II ($^4F_{3/2}$)	***Limit***		**55 010**		

Ti II

$Z=22$

Sc I isoelectronic sequence

Ground state: $1s^22s^22p^63s^23p^63d^24s\ ^4F_{3/2}$

Ionization energy = 109 494±20 cm^{-1} (13.5756±0.0025 eV)

An analysis of this spectrum was carried out by Russell (1927). He reported 116 energy levels derived from 464 lines in the range of 1906–6717 Å. New observations by Huldt, Johansson, Litzén, and Wyart (1982) in the region of 1100–11 000 Å with a wavelength uncertainty of ±0.02 Å or better resulted in nearly doubling the known number of lines and energy levels. They have also calculated the percentage compositions for the even configurations $3d^3$, $3d^24s$, and $3d4s^2$ with configuration interaction (CI) as well as $3d^25s$ and $3d^24d$; and for the odd configurations $3d^24p$, $3d^25p$, and $3d4s4p$ with CI. Their results are quoted here. The level uncertainty is ±0.1 cm^{-1}.

Huldt et al. determined the value for the ionization energy from the $3d^2(^3F)ns\ 4f_{9/2}$ series for $n=4$, 5, and 6.

References

Huldt, S., Johansson, S., Litzén, U., and Wyart, J.-F. (1982), Phys. Scr. **25**, 401.

Moore, C. E. (1949), Atomic Energy Levels, Natl. Bur. Stand. (U.S.) Circ. 467, Vol. I (reissued in 1971 as Natl. Bur. Stand. (U.S.) Natl. Stand. Ref. Data Ser. 35, Vol. I).

Russell, H. N. (1927), Astrophys. J. **66**, 283.

Ti II

Configuration	Term	J	Level (cm^{-1})	g	Leading percentages		
$3d^2(^3F)4s$	$a\ ^4F$	3/2	0.00		100		
		5/2	94.10		100		
		7/2	225.73		100		
		9/2	393.44		100		
$3d^3$	$b\ ^4F$	3/2	908.02		100		
		5/2	983.89		100		
		7/2	1 087.32		100		
		9/2	1 215.84		100		
$3d^2(^3F)4s$	$a\ ^2F$	5/2	4 628.58		99		
		7/2	4 897.65		99		
$3d^2(^1D)4s$	$a\ ^2D$	3/2	8 710.44	0.80	72	17	$3d^3\ ^2$D2
		5/2	8 744.25		74	16	
$3d^3$	$a\ ^2G$	7/2	8 997.71		96		
		9/2	9 118.26		96		
$3d^3$	$a\ ^4P$	1/2	9 363.62	2.63	96		
		3/2	9 395.71	1.74	93		
		5/2	9 518.06		100		
$3d^3$	$a\ ^2P$	1/2	9 850.90	0.66	62	24	$3d^2(^3P)4s\ ^2P$
		3/2	9 975.92	1.33	48	27	$3d^2(^3P)4s\ ^4P$
$3d^2(^3P)4s$	$b\ ^4P$	1/2	9 872.73	2.60	90		
		3/2	9 930.69		72	18	$3d^3\ ^2P$
		5/2	10 024.73		99		
$3d^3$	$b\ ^2D$	3/2	12 628.73		53	26	^{2}D1
		5/2	12 758.11		54	24	

Ti II—Continued

Configuration	Term	J	Level (cm^{-1})	g	Leading percentages		
$3d^3$	$a\ ^2$H	$9/2$	12 676.97		100		
		$11/2$	12 774.69		100		
$3d^2(^1$G$)4s$	$b\ ^2$G	$9/2$	15 257.43		96		
		$7/2$	15 265.62		96		
$3d^2(^3$P$)4s$	$b\ ^2$P	$1/2$	16 515.86	0.66	72	28	$3d^3\ ^2$P
		$3/2$	16 625.11	1.33	72	28	
$3d^3$	$b\ ^2$F	$7/2$	20 891.66		99		
		$5/2$	20 951.62		99		
$3d\ 4s^2$	$c\ ^2$D	$3/2$	24 961.03		73	14	$3d^3\ ^2$D1
		$5/2$	25 192.79		71	15	
$3d^2(^3$F$)4p$	$z\ ^4$G°	$5/2$	*29 544.37*		98		
		$7/2$	*29 734.54*		99		
		$9/2$	*29 968.30*		100		
		$11/2$	*30 240.88*		100		
$3d^2(^3$F$)4p$	$z\ ^4$F°	$3/2$	*30 836.32*		97		
		$5/2$	*30 958.50*		98		
		$7/2$	*31 113.65*		98		
		$9/2$	*31 301.01*		99		
$3d^2(^3$F$)4p$	$z\ ^2$F°	$5/2$	*31 207.42*		85		
		$7/2$	*31 490.82*		88		
$3d^2(^3$F$)4p$	$z\ ^2$D°	$3/2$	*31 756.51*	0.92	82		
		$5/2$	*32 025.47*	1.20	77		
$3d^2(^1$S$)4s$	$a\ ^2$S	$1/2$	31 787.75		100		
$3d^3$	$d\ ^2$D1	$3/2$	32 275.32		58	25	$3d4s^2\ ^2$D
		$5/2$	32 332.73		57	27	
$3d^2(^3$F$)4p$	$z\ ^4$D°	$1/2$	*32 532.21*	0.00	97		
		$3/2$	*32 602.55*	1.20	94		
		$5/2$	*32 697.99*	1.37	90		
		$7/2$	*32 767.07*		96		
$3d^2(^3$F$)4p$	$z\ ^2$G°	$7/2$	*34 543.26*		95		
		$9/2$	*34 748.40*		95		
$3d^2(^3$P$)4p$	$z\ ^2$S°	$1/2$	*37 430.58*	2.09	99		
$3d^2(^1$D$)4p$	$y\ ^2$D°	$5/2$	*39 476.80*		55	25	$(^1$D$)\ ^2$F°
		$3/2$	*39 602.75*	1.21	48	36	$(^1$D$)\ ^2$P°
$3d^2(^1$D$)4p$	$z\ ^2$P°	$3/2$	*39 223.28*		58	36	$(^1$D$)\ ^2$D°
		$1/2$	*39 674.66*		98		
$3d^2(^1$D$)4p$	$y\ ^2$F°	$5/2$	*39 926.66*		63	25	$(^1$D$)\ ^2$D°
		$7/2$	*40 074.52*		85		
$3d^2(^3$P$)4p$	$z\ ^4$S°	$3/2$	*40 027.11*		95		

(Continued)

Ti II—Continued

Configuration	Term	J	Level (cm^{-1})	g	Leading percentages		
$3d^2(^3P)4p$	$y\ ^4D°$	1/2	40 330.16		97		
		3/2	40 425.59		96		
		5/2	40 581.49		96		
		7/2	40 798.30		93		
$3d^2(^3P)4p$	$z\ ^4P°$	1/2	41 996.57		97		
		3/2	42 068.52		97		
		5/2	42 208.59		97		
$3d^2(^1G)4p$	$y\ ^2G°$	7/2	43 740.65		95		
		9/2	43 780.79		95		
$3d^2(^3P)4p$	$x\ ^2D°$	5/2	44 902.29		82		
		3/2	44 914.70		80		
$3d^2(^3P)4p$	$y\ ^2P°$	1/2	45 472.27		95		
		3/2	45 548.76		94		
$3d^2(^1G)4p$	$z\ ^2H°$	9/2	45 673.62		99		
		11/2	45 908.53		100		
$3d^2(^1G)4p$	$x\ ^2F°$	7/2	47 466.54		91		
		5/2	47 624.88		92		
$3d(^2D)4s4p(^3P°)$	$^4F°$	3/2	52 330.33		95		
		5/2	52 472.12		93		
		7/2	52 705.16		94		
		9/2	53 096.83		99		
$3d(^2D)4s4p(^3P°)$	$^4D°$	1/2	52 339.34		97		
		3/2	52 459.33		94		
		5/2	52 631.00		91		
		7/2	52 846.65		92		
$3d(^2D)4s4p(^3P°)$	$w\ ^2D°$	5/2	53 554.76		88		
		3/2	53 597.10		88		
$3d(^2D)4s4p(^3P°)$	$y\ ^4P°$	1/2	56 222.81		97		
		3/2	56 249.10		97		
		5/2	56 325.70		97		
$3d(^2D)4s4p(^3P°)$	$w\ ^2F°$	5/2	59 322.65		94		
		7/2	59 468.02		95		
$3d(^2D)4s4p(^3P°)$	$^2P°$	3/2	59 387.68		83	16	$3d^2(^1S)4p\ ^2P°$
		1/2	59 439.94		79	20	
$3d^2(^3F)5s$	$e\ ^4F$	3/2	62 180.16		100		
		5/2	62 272.16		99		
		7/2	62 410.78		99		
		9/2	62 595.03		100		
$3d^2(^3F)5s$	$e\ ^2F$	5/2	63 169.02		99		
		7/2	63 445.88		99		
$3d^2(^1S)4p$	$^2P°$	1/2	63 276.60		76	20	$3d(^2D)4s4p(^3P°)\ ^2P°$
		3/2	63 375.08		81	16	

Ti II—Continued

Configuration	Term	J	Level (cm^{-1})	g	Leading percentages		
$3d^2(^3F)4d$	$e\ ^4G$	5/2	64 886.48		96		
		7/2	64 979.15		95		
		9/2	65 095.80		93		
		11/2	65 243.46		93		
$3d^2(^3F)4d$	$e\ ^4H$	7/2	65 186.75		96		
		9/2	65 308.30		93		
		11/2	65 446.27		93		
		13/2	65 590.19		100		
$3d^2(^3F)4d$	^4D	1/2	65 213.80		99		
		3/2	65 274.60		99		
		5/2	65 397.57		80	18	$3d^2(^3F)4d\ ^2F$
		7/2	65 598.73		49	49	$3d^2(^3F)4d\ ^2F$
$3d^2(^3F)4d$	$f\ ^2F$	5/2	65 314.27		75	19	^4D
		7/2	65 460.01		47	51	
$3d^2(^3F)4d$	^2P	1/2	66 521.01		93		
		3/2	66 794.01		91		
$3d^2(^3F)4d$	$e\ ^2G$	7/2	67 606.04		91		
		9/2	67 822.49		91		
$3d^2(^3F)4d$	$e\ ^2H$	9/2	68 331.02		95		
		11/2	68 584.28		95		
$3d^2(^3F)4d$	^2D	3/2	68 364.39		86	11	$3d^2(^1D)4d\ ^2D$
		5/2	68 482.41		85	12	
$3d^2(^3F)4d$	$f\ ^4F$	3/2	68 769.19		96		
		5/2	68 846.52		96		
		7/2	68 951.98		96		
		9/2	69 084.44		96		
$3d(^2D)4s4p(^1P^\circ)$	$v\ ^2D^\circ$	3/2	*69 327.52*		65	31	$3d^2(^3F)5p\ ^2D^\circ$
		5/2	*69 622.63*		65	31	
$3d(^3F)5p$	$v\ ^2F^\circ$	5/2	*70 607.53*		54	41	$3d(^2D)4s4p(^1P^\circ)\ ^2F^\circ$
		7/2	*70 892.79*		55	41	
$3d^2(^3F)5p$	$^4G^\circ$	5/2	*71 461.59*		98		
		7/2	*71 586.06*		99		
		9/2	*71 747.46*		100		
		11/2	*71 945.90*		100		
$3d^2(^3F)5p$	$^4F^\circ$	3/2	*71 728.62*		100		
		5/2	*71 825.81*		99		
		7/2	*71 960.29*		99		
		9/2	*72 126.70*		99		
$3d^2(^3F)5p$	$^4D^\circ$	1/2	*72 270.31*		100		
		3/2	*72 337.97*		99		
		5/2	*72 451.52*		99		
		7/2	*72 608.85*		99		
$3d^2(^3F)5p$	$^2G^\circ$	7/2	*72 884.04*		99		
		9/2	*73 133.35*		99		

(Continued)

Ti II—Continued

Configuration	Term	J	Level (cm^{-1})	g	Leading percentages		
$3d(^2D)4s4p(^1P^\circ)$	$^2F^\circ$	5/2	*73 521.67*		51	45	$3d^2(^3F)5p\ ^2F^\circ$
		7/2	*73 800.67*		52	44	
$3d(^2D)4s4p(^1P^\circ)$	$^2P^\circ$	1/2	*73 627.82*		86		
		3/2	*73 950.01*		86		
$3d^2(^3F)5p$	$^2D^\circ$	3/2	*74 378.67*		67	27	$3d(^2D)4s4p(^1P^\circ)\ ^2D^\circ$
		5/2	*74 645.08*		67	27	
$3d^2(^3F)4f$	$^4H^\circ$	7/2	*81 627.64*				
		9/2	*81 676.30*				
		11/2	*81 753.98*				
		13/2	*82 072.84*				
$3d^2(^3F)4f$	$^2G^\circ$	7/2	*81 685.81*				
		9/2	*81 842.09*				
$3d^2(^3F)4f$	$^4G^\circ$	5/2	*81 724.17*				
		7/2	*81 860.84*				
		9/2	*82 078.43*				
		11/2	*82 103.06*				
$3d^2(^3F)4f$	$^4I^\circ$	9/2	*81 738.13*				
		11/2	*81 889.46*				
		13/2	*81 919.58*				
		15/2	*82 216.91*				
$3d^2(^3F)4f$	$^4F^\circ$	5/2	*81 773.98*				
		3/2	*81 806.88*				
		7/2	*81 912.92*				
		9/2	*82 147.54*				
$3d^2(^3F)4f$	$^4P^\circ$	3/2	*81 867.70*				
		5/2	*81 996.83*				
		1/2	*82 043.85*				
$3d^2(^3F)4f$	$^2H^\circ$	9/2	*81 923.99*				
		11/2	*82 183.31*				
$3d^2(^3F)4f$	$^2F^\circ$	5/2	*81 943.25*				
		7/2	*82 151.31*				
$3d^2(^3F)4f$	$^2I^\circ$	11/2	*82 001.01*				
		13/2	*82 284.22*				
$3d^2(^3F)4f$	$^4D^\circ$	1/2	*82 008.55*				
		3/2	*82 020.94*				
		7/2	*82 133.99*				
		5/2	*82 246.37*				
$3d^2(^3F)4f$	$^2D^\circ$	3/2	*82 065.47*				
		5/2	*82 318.91*				
$3d^2(^3F)4f$	$^4S^\circ$	3/2	*82 304.25*				
$3d^2(^3F)4f$	$^2P^\circ$	3/2	*82 369.97*				

Ti II—Continued

Configuration	Term	J	Level (cm^{-1})	g	Leading percentages
$3d^2(^3F)6s$	^4F	3/2	82 839.08		
		5/2	82 914.09		
		7/2	83 058.33		
		9/2	83 257.04		
$3d^2(^3F)6s$	^2F	5/2	83 244.91		
		7/2	83 519.16		
$3d^2(^3F)5d$	^4H	7/2	84 165.71		
		9/2	84 334.14		
		11/2	84 490.76		
		13/2	84 722.08		
$3d^2(^3F)5d$	^4G	5/2	84 268.77		
		7/2	84 369.70		
		9/2	84 512.57		
		11/2	84 632.03		
Ti III ($^3F_2$)	***Limit***		**109 494**		

Ti III

$Z=22$

Ca I isoelectronic sequence

Ground state: $1s^2 2s^2 2p^6 3s^2 3p^6 3d^2\ ^3F_2$

Ionization energy $= 221\,735.6 \pm 2.0$ cm^{-1} (27.4919 ± 0.0002 eV)

The first study of Ti III was by Russell and Lang (1927). They identified the $3d^2$, $3d4s$, $3d4p$, $3d4d$, and $4s4p$ configurations. The spectrum has been remeasured and greatly extended by Edlén and Svensson (1975), who have identified the $4s^2$, $3d5p$, $3d5s$, $3d6s$, $3d5d$, $3d4f$, $3d5f$, $3d5g$, $3d6g$, $3d6h$, and $3d7h$ configurations. The levels have an estimated uncertainty of ± 0.1 cm^{-1} except for those of $3d6h$ whose uncertainty is ± 0.5 cm^{-1}. Their observations of $3d7h$ were fragmentary and for that configuration they gave calculated term values, which we have included here.

The composition of the $3d^2$ ground configuration was calculated by Pasternak and Goldschmidt (1972). The two mixed configurations $3d(4d+5s)$ and $3d(5d+6s)$ were calculated by Wyart (1975). He also calculated interaction of $4s4p$ with each of $3d(4p, 5p, 4f, 5f)$ plus the mixture of $3d5p$ and $3d4f$. The levels of $3d4s$ were calculated by Shadmi, Caspi, and Oreg (1969) but no percentages were given. Goldschmidt (1982) has provided the percentage compositions in J_1l-coupling for the $3d5g$ and $3d6g$ configurations.

Edlén and Svensson determined the ionization energy from $3d5g$, $3d6g$, and $3d6h$ by means of a polarization formula.

References

Edlén, B., and Svensson, J. W. (1975), Phys. Scr. **12**, 21.

Goldschmidt, Z. B. (1981), J. Phys. **B14**, 615.

Pasternak, A., and Goldschmidt, Z. B. (1972), Phys. Rev. **6A**, 55.

Russell, H. N., and Lang, R. J. (1927), Astrophys. J. **66**, 13.

Shadmi, Y., Caspi, E., and Oreg, J. (1969), J. Res. Natl. Bur. Stand. **73A**, 173.

Wyart, J.-F. (1975), Phys. Scr. **12**, 33.

Ti III

Configuration	Term	J	Level (cm^{-1})	Leading percentages		
$3d^2$	^{3}F	2	0.0	100		
		3	184.9	100		
		4	420.4	100		
$3d^2$	^{1}D	2	8 473.5	99		
$3d^2$	^{3}P	0	10 538.4	100		
		1	10 603.6	100		
		2	10 721.2	99		
$3d^2$	^{1}G	4	14 397.6	100		
$3d^2$	^{1}S	0	32 475.5	100		
3d 4s	^{3}D	1	38 064.35			
		2	38 198.95			
		3	38 425.99			
3d 4s	^{1}D	2	41 704.27			
3d 4p	^{1}D°	2	*75 198.21*	98		
3d 4p	^{3}D°	1	*77 000.23*	100		
		2	*77 167.43*	89	11	^{3}F°
		3	*77 424.45*	86	13	^{3}F°
3d 4p	^{3}F°	2	*77 421.86*	88	11	^{3}D°
		3	*77 746.44*	87	13	^{3}D°
		4	*78 158.61*	100		

Ti III—Continued

Configuration	Term	*J*	Level (cm^{-1})	Leading percentages		
3*d* 4*p*	^{3}P°	1	*80 939.19*	99		
		0	*80 944.87*	100		
		2	*81 024.47*	100		
3*d* 4*p*	^{1}F°	3	*83 116.93*	100		
3*d* 4*p*	^{1}P°	1	*83 796.86*	98		
4*s*^{2}	^{1}S	0	102 665.15			
3*d* 4*d*	^{1}F	3	127 790.57	98		
3*d* 4*d*	^{3}D	1	128 433.40	96	4	^{1}P
		2	128 546.38	100		
		3	128 689.67	98		
3*d* 4*d*	^{3}G	3	129 093.28	99		
		4	129 252.74	100		
		5	129 469.37	100		
3*d* 4*d*	^{1}P	1	129 253.41	96	4	^{3}D
3*d* 4*d*	^{3}S	1	130 739.82	99		
3*d* 4*d*	^{3}F	2	133 065.24	100		
		3	133 207.10	100		
		4	133 371.07	100		
3*d* 5*s*	^{3}D	1	133 898.50	100		
		2	133 999.79	91	8	^{1}D
		3	134 275.12	100		
3*d* 5*s*	^{1}D	2	134 557.84	60	32	3*d*4*d* ^{1}D
3*d* 4*d*	^{1}D	2	135 405.27	57	31	3*d*5*s* ^{1}D
3*d* 4*d*	^{3}P	0	135 541.46	100		
		1	135 601.47	100		
		2	135 721.51	88	9	^{1}D
3*d* 4*d*	^{1}G	4	136 339.74	100		
4*s*4*p*	^{3}P°	0	*137 258.9*	100		
		1	*137 487.8*	100		
		2	*137 961.2*	100		
3*d* 4*d*	^{1}S	0	140 019.24	100		
3*d* 5*p*	^{1}D°	2	*147 212.77*	90	5	^{3}D°
3*d* 5*p*	^{3}D°	1	*147 562.14*	99		
		2	*147 749.89*	92	5	^{3}F°
		3	*147 939.47*	99		
3*d* 5*p*	^{3}F°	2	*147 931.47*	92	5	^{1}D°
		3	*148 111.10*	99		
		4	*148 410.24*	100		

(Continued)

Ti III—Continued

Configuration	Term	J	Level (cm^{-1})	Leading percentages		
3d 5p	^{3}P°	1	*148 978.72*	71	26	^{1}P°
		0	*149 019.75*	100		
		2	*149 267.99*	99		
3d 5p	^{1}P°	1	*149 403.52*	64	28	^{3}P°
3d 5p	^{1}F°	3	*149 655.77*	99		
4s4p	^{1}P°	1	*157 204.16*	60	32	3d4f ^{1}P°
3d4f	^{1}G°	4	*158 285.34*	81	10	^{3}F°
3d4f	^{3}F°	2	*158 536.63*	93	6	^{1}D°
		3	*158 557.76*	67	31	^{3}G°
		4	*158 690.85*	46	28	^{3}G°
3d4f	^{3}G°	3	*158 740.92*	60	25	^{3}F°
		4	*158 865.03*	29	40	^{3}F°
		5	*158 903.55*	70	28	^{3}H°
3d4f	^{3}H°	4	*159 022.93*	58	35	^{3}G°
		5	*159 128.94*	69	30	^{3}G°
		6	*159 269.53*	100		
3d4f	^{1}D°	2	*159 123.78*	89	4	^{3}F°
3d4f	^{1}F°	3	*159 180.24*	75	6	^{3}D°
3d4f	^{3}D°	1	*159 394.89*	97		
		2	*159 403.91*	89	4	^{3}P°
		3	*159 481.95*	86	12	^{1}F°
3d4f	^{3}P°	2	*159 991.54*	92	6	^{3}D°
		1	*160 104.61*	97		
		0	*160 167.06*	100		
3d4f	^{1}H°	5	*160 054.90*	97		
3d4f	^{1}P°	1	*161 854.24*	67	28	4s4p ^{1}P°
3d 5d	^{1}F	3	167 724.09	92	5	^{3}D
3d 5d	^{3}D	1	167 905.19	86	14	^{1}P
		2	168 030.15	99		
		3	168 206.79	56	37	^{3}G
3d 5d	^{3}G	3	168 152.20	60	39	^{3}D
		4	168 307.06	99		
		5	168 520.52	100		
3d 5d	^{1}P	1	168 343.62	82	14	^{3}D
3d 5d	^{3}S	1	168 932.83	96	4	^{1}P
3d 5d	^{3}F	2	169 615.12	97		
		3	169 769.13	99		
		4	169 912.11	99		

Ti III—Continued

Configuration	Term	J	Level (cm^{-1})	Leading percentages		
3d 6s	^{3}D	1	169 875.52	100		
		2	169 930.80	72	25	^{1}D
		3	170 254.75	100		
3d 6s	^{1}D	2	170 270.02	47	26	^{3}D
3d 5d	^{3}P	0	170 579.96	99		
		1	170 659.72	99		
		2	170 666.94	56	25	^{1}D
3d 5d	^{1}D	2	170 840.80	46	43	^{3}P
3d 5d	^{1}G	4	171 141.93	99		
3d 5d	^{1}S	0	172 373.52	99		
3d5f	^{1}G°	4	*181 219.06*	68	17	^{3}F°
3d5f	^{3}F°	2	*181 339.27*	84	12	^{1}D°
		3	*181 368.45*	65	30	^{3}G°
		4	*181 611.79*	58	25	^{1}G°
3d5f	^{3}H°	4	*181 439.64*	69	20	^{3}G°
		5	*181 558.44*	73	21	^{3}G°
		6	*181 837.98*	100		
3d5f	^{3}G°	3	*181 507.92*	53	22	^{1}F°
		4	*181 758.34*	68	16	^{3}F°
		5	*181 821.83*	78	22	^{3}H°
3d5f	^{1}D°	2	*181 700.72*	61	19	^{3}D°
3d5f	^{1}F°	3	*181 860.55*	46	22	^{3}D°
3d5f	^{3}D°	2	*181 908.15*	58	23	^{1}D°
		3	*182 025.86*	66	32	^{1}F°
3d(^{2}D$_{3/2}$)5g	$^2[9/2]$	5	182 013.32	100		
		4	182 014.92	100		
3d(^{2}D$_{3/2}$)5g	$^2[7/2]$	3	182 067.49	98	2	(^{2}D$_{5/2}$) $^2[7/2]$
		4	182 068.87	98	2	
3d(^{2}D$_{3/2}$)5g	$^2[11/2]$	5	182 166.91	93	7	(^{2}D$_{5/2}$) $^2[11/2]$
		6	182 170.72	92	8	
3d5f	^{3}P°	2	*182 207.39*	76	19	^{3}D°
		1	*182 276.75*	84	12	^{3}D°
		0	*182 333.95*	100		
3d(^{2}D$_{3/2}$)5g	$^2[5/2]$	3	182 224.72	95	5	(^{2}D$_{5/2}$) $[5/2]$
		2	182 225.09	95	5	
3d5f	^{1}H°	5	*182 353.45*	94	5	^{3}H°
3d(^{2}D$_{5/2}$)5g	$^2[9/2]$	5	182 403.64	100		
		4	182 405.25	100		

(Continued)

Ti III—Continued

Configuration	Term	J	Level (cm^{-1})	Leading percentages		
$3d(^2D_{5/2})5g$	$^2[11/2]$	5	182 436.46	93	7	$(^2D_{3/2})\ ^2[11/2]$
		6	182 439.62	92	8	
$3d(^2D_{5/2})5g$	$^2[7/2]$	3	182 473.21	98	2	$(^2D_{3/2})\ ^2[7/2]$
		4	182 474.38	98	2	
$3d5f$	$^1P°$	1	*182 561.61*	89	8	$^3P°$
$3d(^2D_{5/2})5g$	$^2[5/2]$	3	182 587.12	95	5	$(^2D_{3/2})\ ^2[5/2]$
		2	182 587.62	95	5	
$3d(^2D_{5/2})5g$	$^2[13/2]$	7	182 596.87	100		
		6	182 601.96	100		
$3d(^2D_{5/2})5g$	$^2[3/2]$	1	182 680.27	100		
		2	182 680.53	100		
$3d(^2D_{3/2})6g$	$^2[9/2]$	5	194 168.25	100		
		4	194 169.81	100		
$3d(^2D_{3/2})6g$	$^2[7/2]$	3	194 200.46	99	1	$(^2D_{5/2})\ ^2[7/2]$
		4	194 201.57	99	1	
$3d(^2D_{3/2})6g$	$^2[11/2]$	5	194 261.00	98	2	$(^2D_{5/2})\ ^2[11/2]$
		6	194 264.82	98	2	
$3d(^2D_{5/2})6g$	$^2[9/2]$	4	194 556.67	100		
$3d(^2D_{5/2})6g$	$^2[11/2]$	5	194 567.08	97	2	$(^2D_{3/2})\ [11/2]$
		6	194 569.80	98	2	
$3d(^2D_{5/2})6g$	$^2[7/2]$	3	194 592.07	99	1	$(^2D_{3/2})\ ^2[7/2]$
		4	194 592.98	99	1	
$3d(^2D_{5/2})6g$	$^2[5/2]$	3	194 656.18	98	2	$(^2D_{3/2})\ ^2[5/2]$
$3d(^2D_{5/2})6g$	$^2[13/2]$	7	194 664.08	100		
		6	194 669.10	100		
$3d(^2D_{3/2})6h$	$^2[11/2]°$	5,6	*194 246.5*			
$3d(^2D_{3/2})6h$	$^2[9/2]°$	4,5	*194 261.8*			
$3d(^2D_{3/2})6h$	$^2[13/2]°$	6,7	*194 301.6*			
$3d(^2D_{3/2})6h$	$^2[7/2]°$	3,4	*194 317.2*			
$3d(^2D_{5/2})6h$	$^2[11/2]°$	5,6	*194 628.2*			
$3d(^2D_{5/2})6h$	$^2[13/2]°$	6,7	*194 639.2*			
$3d(^2D_{5/2})6h$	$^2[9/2]°$	4,5	*194 646.3*			
$3d(^2D_{5/2})6h$	$^2[7/2]°$	3,4	*194 680.2*			
$3d(^2D_{5/2})6h$	$^2[15/2]°$	7,8	*194 694.6*			
$3d(^2D_{5/2})6h$	$^2[5/2]°$	2,3	*194 717.4*			

Ti III—Continued

Configuration	Term	J	Level (cm^{-1})	Leading percentages
$3d(^2D_{3/2})7h$	$^2[^{11}/_2]°$	5,6	[*201 543.5*]	
$3d(^2D_{3/2})7h$	$^2[^9/_2]°$	4,5	[*201 553.5*]	
$3d(^2D_{3/2})7h$	$^2[^{13}/_2]°$	6,7	[*201 578.7*]	
$3d(^2D_{3/2})7h$	$^2[^7/_2]°$	3,4	[*201 588.7*]	
$3d(^2D_{5/2})7h$	$^2[^{11}/_2]°$	5,6	[*201 926.0*]	
$3d(^2D_{5/2})7h$	$^2[^{13}/_2]°$	6,7	[*201 932.7*]	
$3d(^2D_{5/2})7h$	$^2[^9/_2]°$	4,5	[*201 937.4*]	
$3d(^2D_{5/2})7h$	$^2[^7/_2]°$	3,4	[*201 958.3*]	
$3d(^2D_{5/2})7h$	$^2[^{15}/_2]°$	7,8	[*201 967.5*]	
$3d(^2D_{5/2})7h$	$^2[^5/_2]°$	2,3	[*201 981.7*]	
Ti IV ($^2D_{3/2}$)	***Limit***		**221 735.6**	

Ti IV

$Z=22$

K I isoelectronic sequence

Ground state: $1s^2 2s^2 2p^6 3s^2 3p^6 3d\ ^2D_{3/2}$

Ionization energy = 348 973.3±1.5 cm^{-1} (43.2675±0.0002 eV)

The initial study of the structure of Ti IV was by Gibbs and White (1929) who identified the low one-electron configurations $3d$, $4s$, and $4p$. The work was extended by Russell and Lang (1927) who recognized the higher configura tions $4d$, $5d$, $5s$, $6s$, $5p$, $4f$, and $5g$. New observations by Svensson and Edlén (1974) improved the accuracy of the earlier work and added the configurations $7s$, $6p$, $6d$, $7d$, $5f$, $6g$, $7g$, $6h$, $7h$, $7i$, and $8i$. They also discovered a level of $3p^5 3d^2$. The level uncertainty is estimated to be ±0.2 cm^{-1}. By application of the polarization formula to the hydrogen-like terms ng, nh, and ni they determined the ionization energy.

References

Gibbs, R. C., and White, H. E. (1926), Proc. Natl. Acad. Sci. **12**, 598.
Russell, H. N., and Lang, R. J. (1927), Astrophys. J. **66**, 13.
Svensson, J. W., and Edlén, B. (1974). Phys. Scr. **9**, 335.

Ti IV

Configuration	Term	J	Level (cm^{-1})
$3p^6\ 3d$	^2D	3/2	0.0
		5/2	382.1
$3p^6\ 4s$	^2S	1/2	80 388.93
$3p^6\ 4p$	$^2P°$	1/2	*127 921.36*
		3/2	*128 739.59*
$3p^6\ 4d$	^2D	3/2	196 804.27
		5/2	196 889.96
$3p^6\ 5s$	^2S	1/2	212 407.34
$3p^6\ 5p$	$^2P°$	1/2	*230 608.89*
		3/2	*230 924.38*
$3p^6\ 4f$	$^2F°$	5/2	*236 135.29*
		7/2	*236 142.30*
$3p^6\ 5d$	^2D	3/2	258 838.48
		5/2	258 877.08
$3p^6\ 6s$	^2S	1/2	265 847.42
$3p^6\ 6p$	$^2P°$	1/2	*274 726.29*
		3/2	*274 881.21*
$3p^5\ 3d^2$	$^2F°$	5/2	*274 839.82*
$3p^6\ 5f$	$^2F°$	5/2	*275 847.01*
		7/2	*275 861.94*
$3p^6\ 5g$	^2G	7/2	278 510.63
		9/2	278 511.23
$3p^6\ 6d$	^2D	3/2	289 185.99
		5/2	289 206.93
$3p^6\ 7s$	^2S	1/2	292 999.54
$3p^6\ 6g$	^2G	7/2	300 045.9
		9/2	300 046.2
$3p^6\ 6h$	$^2H°$	9/2, 11/2	*300 158.76*
$3p^6\ 7d$	^2D	3/2	306 395.69
		5/2	306 408.30
$3p^6\ 7g$	^2G	7/2	313 033.9
		9/2	313 034.1
$3p^6\ 7h$	$^2H°$	9/2, 11/2	*313 110.72*
$3p^6\ 7i$	^2I	11/2, 13/2	313 130.66
$3p^6\ 8i$	^2I	11/2, 13/2	321 531.3
Ti V ($^1S_0$)	*Limit*		**348 973.3**

Ti V

$Z=22$

Ground state: $1s^2 2s^2 2p^6 3s^2 3p^6\ ^1S_0$

Ionization energy $= 800\,900 \pm 100$ cm^{-1} (99.30 ± 0.01 eV)

Kruger and Weissberg (1935) and Kruger, Weissberg, and Phillips (1937) identified four resonance lines arising from the $3p^5 4s$ and $5s$ configurations. The $3p^5 3d$ configuration was identified by Gabriel, Fawcett, and Jordan (1966) and extended by Svensson and Ekberg (1968), who also identified the $3p^5 6s$ and $3s3p^6 4p$ configurations. In all, 10 resonance lines were classified.

New observations of the spectrum by Svensson (1976) in the range of 300–2500 Å enabled him to identify 231 more lines. His analysis completed the known configurations and provided the energy levels of the $3p^5 4p$, $3p^5 4d$, $3p^5 6s$, and $3s3p^6 3d$ configurations. The uncertainty in these level values is ± 2 cm^{-1}. He calculated the percentage compositions of the levels.

The $3s3p^6 np$ series was observed in a high voltage spark as absorption lines from the ground state by Kastner, Crooker, Behring, and Cohen (1977). They reported the series through $n=11$ and calculated the series limit at 997 500 cm^{-1}. The uncertainty in these level values is ± 50 cm^{-1}. This gives an ionization energy of 800 900 cm^{-1}.

References

Gabriel, A. H., Fawcett, B. C., and Jordan, C. (1966), Proc. Phys. Soc. **87**, 825.

Kastner, S. O., Crooker, A. M., Behring, W. E., and Cohen, L. (1977), Phys. Rev. **A16**, 577.

Kruger, P. G., and Weissberg, S. G. (1935), Phys. Rev. **48**, 659.

Kruger, P. G., Weissberg, S. G., and Phillips, L. W. (1937), Phys. Rev. **51**, 1090.

Svensson, L. A., and Ekberg, J. O. (1968), Ark. Fys. **37**, 65.

Svensson, L. A. (1976), Phys. Scr. **13**, 235.

Ti V

Configuration	Term	J	Level (cm^{-1})	Leading percentages		
$3s^2 3p^6$	^1S	0	0			
$3s^2 3p^5 3d$	$^3P°$	0	*274 439.7*	100		
		1	*275 371.9*	100		
		2	*277 310.6*	99		
$3s^2 3p^5 3d$	$^3F°$	4	*287 276.5*	100		
		3	*289 050.2*	98		
		2	*290 778.7*	98		
$3s^2 3p^5 3d$	$^1D°$	2	*306 874.5*	84	14	$^3D°$
$3s^2 3p^5 3d$	$^3D°$	3	*307 429.2*	80	20	$^1F°$
		1	*309 252.1*	100		
		2	*309 433.1*	85	14	$^1D°$
$3s^2 3p^5 3d$	$^1F°$	3	*311 433.8*	79	20	$^3D°$
$3s^2 3p^5 3d$	$^1P°$	1	*395 320.9*	100		
$3s^2 3p^5 4s$	$^3P°$	2	*434 339.4*	100		
		1	*436 849.8*	81	19	$^1P°$
		0	*440 065.2*	100		
$3s^2 3p^5 4s$	$^1P°$	1	*443 752.7*	81	19	$^3P°$
$3s^2 3p^5 4p$	^3S	1	481 987.7	97		
$3s^2 3p^5 4p$	^3D	3	487 974.6	65	35	$3s3p^6 3d\ ^3D$
		2	488 225.7	47	41	
		1	493 300.5	47	22	

(Continued)

Ti V—Continued

Configuration	Term	J	Level (cm^{-1})	Leading percentages			
$3s3p^6\,3d$	^{3}D	1	488 928.7	75		23	$3s^23p^54p\ ^3$D
		2	491 558.7	51		31	$3s^23p^54p\ ^1$D
		3	492 567.1	65		35	$3s^23p^54p\ ^3$D
$3s^2\,3p^5\,4p$		2	494 035.7	42	^{3}P	33	^{1}D
$3s^2\,3p^5\,4p$	^{1}P	1	496 890.7	50		29	^{1}D
$3s^2\,3p^5\,4p$	^{3}P	2	498 057.2	55		22	^{3}D
		0	498 176.4	98			
		1	499 336.2	66		31	^{1}P
$3s3p^6\,3d$	^{1}D	2	506 224.7	89		10	$3s^23p^54p\ ^1$D
$3s^2\,3p^5\,4p$	^{1}S	0	514 608.7	98			
$3s^2\,3p^5(^2P^\circ_{3/2})4d$	$^2[1/2]^\circ$	0	*568 698.5*	100			
		1	*569 304.5*	72		22	$(^2P^\circ_{3/2})\ ^2[3/2]^\circ$
$3s^2\,3p^5(^2P^\circ_{3/2})4d$	$^2[3/2]^\circ$	2	*570 597.8*	82		18	$(^2P^\circ_{1/2})\ ^2[3/2]^\circ$
		1	*577 249.8*	74		14	$(^2P^\circ_{3/2})\ ^2[1/2]^\circ$
$3s^2\,3p^5(^2P^\circ_{3/2})4d$	$^2[7/2]^\circ$	4	*571 401.1*	100			
		3	*572 093.9*	90			
$3s^2\,3p^5(^2P^\circ_{3/2})4d$	$^2[5/2]^\circ$	2	*573 838.3*	84		16	$(^2P^\circ_{1/2})\ ^2[5/2]^\circ$
		3	*574 683.8*	84			
$3s^2\,3p^5(^2P^\circ_{1/2})4d$	$^2[5/2]^\circ$	2	*578 698.5*	84		16	$(^2P^\circ_{3/2})\ ^2[5/2]^\circ$
		3	*579 334.6*	90			
$3s^2\,3p^5(^2P^\circ_{1/2})4d$	$^2[3/2]^\circ$	2	*579 584.2*	82		18	$(^2P^\circ_{3/2})\ ^2[3/2]^\circ$
		1	*582 836.5*	82		14	$(^2P^\circ_{3/2})\ ^2[1/2]^\circ$
$3s^2\,3p^5(^2P^\circ_{3/2})5s$	$^2[3/2]^\circ$	2	*607 033.0*	100			
		1	*608 100.7*	97			
$3s^2\,3p^5(^2P^\circ_{1/2})5s$	$^2[1/2]^\circ$	0	*612 793.2*	100			
		1	*613 558.2*	97			
$3s^2\,3p^5(^2P^\circ_{3/2})6s$	$^2[3/2]^\circ$	1	*680 748*				
$3s^2\,3p^5(^2P^\circ_{1/2})6s$	$^2[1/2]^\circ$	1	*685 940*				
$3s3p^6\,4p$	^{3}P°	1	*687 980*				
$3s3p^6\,4p$	^{1}P°	1	*691 797*				
Ti VI ($^2P^\circ_{3/2}$)	***Limit***		**800 900**				
$3s3p^6\,5p$	^{3}P°	1	*825 500*				
$3s3p^6\,5p$	^{1}P°	1	*827 650*				
$3s3p^6\,6p$	^{3}P°	1	*885 770*				
$3s3p^6\,6p$	^{1}P°	1	*888 930*				

Ti V—Continued

Configuration	Term	*J*	Level (cm^{-1})	Leading percentages
$3s3p^6\,7p$	$^3P^\circ$	1	*920 720*	
$3s3p^6\,7p$	$^1P^\circ$	1	*922 140*	
$3s3p^6\,8p$	$^3P^\circ$	1	*940 660*	
$3s3p^6\,8p$	$^1P^\circ$	1	*942 030*	
$3s3p^6\,9p$	$^3P^\circ$	1	*954 820*	
$3s3p^6\,9p$	$^1P^\circ$	1	*955 010*	
$3s3p^6\,10p$	$^3P^\circ$	1	*963 820*	
$3s3p^6\,10p$	$^1P^\circ$	1	*964 010*	
$3s3p^6\,11p$	$^1P^\circ$	1	*970 320*	
Ti VI ($^2S^\circ_{1/2}$)	***Limit***		**997 500**	

Ti VI

$Z = 22$

Cl I isoelectronic sequence

Ground state: $1s^2 2s^2 2p^6 3s^2 3p^5\ ^2P^\circ_{3/2}$

Ionization energy = 964 100±200 cm^{-1} (119.53±0.02 eV)

The first observation of Ti VI was made by Weissberg and Kruger (1936), who identified the resonance lines $3s^2 3p^5\ ^2P^\circ - 3s 3p^6\ ^2S$ at 508 and 524 Å. These have been remeasured by Svensson (1971). In 1937 Edlén reported the $3p^4 4s$ configuration. Gabriel, Fawcett, and Jordan (1966) and Fawcett and Gabriel (1966) reported terms in $3p^4 3d$. Fawcett, Peacock, and Cowan (1968) identified $3p^4 4d$, $5d$, and $5s$. Fawcett, Cowan, and Hayes (1972) observed $3p^4 3d - 3p^4 4f$ transitions at 235 Å but they do not involve any of the terms of the known system.

New measurements in the range of 125–354 Å were made by Svensson and Ekberg (1968), who improved the accuracy of the known levels and added to the known configurations. The uncertainty in their level values is ±10 cm^{-1}. They determined the ionization energy from an extrapolation formula.

References

Edlén, B. (1937), Z. Phys. **104,** 407.

Fawcett, B. C., Cowan, R. D., and Hayes, R. W. (1972), J. Phys. **B5,** 2143.

Fawcett, B. C., and Gabriel, A. H. (1966), Proc. Phys. Soc. **88,** 262.

Fawcett, B. C., Peacock, N. J., and Cowan, R. D. (1968) J. Phys. **B1,** 295.

Gabriel, A. H., Fawcett, B. C., and Jordan, C. (1966), Proc. Phys. Soc. **87,** 825.

Svensson, L. A. (1971), Phys. Scr. **4,** 111.

Svensson, L. A., and Ekberg, J. O. (1968), Ark, Fys. **37,** 65.

Weissberg, S. G., and Kruger, P. G. (1936), Phys. Rev. **49,** 872 A.

Ti VI

Configuration	Term	J	Level (cm^{-1})
$3s^2 3p^5$	$^2P^\circ$	3/2	*0*
		1/2	*5 829*
$3s3p^6$	^2S	1/2	196 628
$3s^2 3p^4(^3P)3d$	^2P	1/2	288 412
		3/2	291 890
$3s^2 3p^4(^3P)3d$	^2D	3/2	298 991
		5/2	302 386
$3s^2 3p^4(^3P)3d$	^4P	1/2	301 417
$3s^2 3p^4(^1D)3d$	^2F	5/2	331 221
$3s^2 3p^4(^1S)3d$	^2D	3/2	352 625
		5/2	354 340
$3s^2 3p^4(^1D)3d$	^2S	1/2	379 874
$3s^2 3p^4(^1D)3d$	^2P	3/2	391 583
		1/2	393 644
$3s^2 3p^4(^1D)3d$	^2D	5/2	399 231
		3/2	404 123
$3s^2 3p^4(^3P)4s$	^4P	5/2	492 126
		3/2	495 380
		1/2	497 389

Configuration	Term	J	Level (cm^{-1})
$3s^2 3p^4(^3P)4s$	^2P	3/2	502 571
		1/2	506 432
$3s^2 3p^4(^1D)4s$	^2D	5/2	518 797
		3/2	518 914
$3s^2 3p^4(^1S)4s$	^2S	1/2	548 995
$3s^2 3p^4(^3P)4d$	^2D	5/2	651 255
		3/2	651 960
$3s^2 3p^4(^3P)4d$	^4F	5/2	653 766
		3/2	654 503
$3s^2 3p^4(^3P)4d$	^2F	5/2	656 437
$3s^2 3p^4(^3P)4d$	^2P	3/2	658 339
$3s^2 3p^4(^1D)4d$	^2S	1/2	668 630
$3s^2 3p^4(^1D)4d$	^2P	3/2	671 096
		1/2	671 549
$3s^2 3p^4(^1D)4d$	^2D	5/2	674 297
		3/2	675 207
$3s^2 3p^4(^1S)4d$	^2D	3/2	704 270
		5/2	704 283

Ti VI—Continued

Configuration	Term	J	Level (cm^{-1})	Configuration	Term	J	Level (cm^{-1})
$3s^2 3p^4(^3P)5s$	^4P	$3/2$	708 652	$3s^2 3p^4(^3P)5d$	^4F	$5/2$	778 513
$3s^2 3p^4(^3P)5s$	^2P	$3/2$	712 034	$3s^2 3p^4(^1D)5d$	^2P	$3/2$	795 615
		$1/2$	714 742				
				$3s^2 3p^4(^1D)5d$	^2D	$5/2$	797 092
$3s^2 3p^4(^1D)5s$	^2D	$3/2$	731 453			$3/2$	797 406
		$5/2$	731 455				
$3s^2 3p^4(^3P)5d$	^2D	$5/2$	773 702	Ti VII ($^3P_2$)	***Limit***		**964 100**
		$3/2$	774 306				

Ti VII

Z=22

S I isoelectronic sequence

Ground state: $1s^2 2s^2 2p^6 3s^2 3p^4\ ^3P_2$

Ionization energy = 1 136 000±2000 cm^{-1} (140.8±0.2 eV)

Edlén (1937) initiated the analysis of this spectrum with identification of the three terms of the ground configuration and all the singlets and triplets of the $3p^3 4s$ configuration. He also determined the limit quoted here from an isoelectronic extrapolation. Kruger and Pattin (1937) observed the $3s^2 3p^4\ ^3P - 3s3p^5\ ^2P°$ multiplet. This has been remeasured by Svensson (1971), who has also observed the singlet transitions of the $3s^2 3p^4 - 3s3p^5$ array.

The $3p^3 3d$ configuration was first identified by Fawcett and Gabriel (1966) and Gabriel, Fawcett, and Jordon (1966). The higher configuration, $3p^3 4d$, was reported by Svensson and Ekberg (1968), who remeasured the spectrum between 128 and 332 Å. Fawcett, Cowan, and Hayes (1972) identified some lines in the $3p^3 3d - 3p^3 4f$ array but these are not connected to the known system.

Level-values for the triplet system of $3s^2 3p^4$ and $3s3p^5$ are from Svensson (1971). The position of the $^1D_2$ of $3s^2 3p^4$ is established by the intersystem transitions of the $3s^2 3p^4 - 3s^2 3p^3 3d$ array measured by Svensson and Ekberg (1968). The singlet transitions given by Svensson (1971) are then used to establish the $^1S_0$ of $3s^2 3p^4$ and the $^1P_1°$ of $3s3p^5$. A spin-forbidden resonance line arising from the $3s3p^5\ ^1P°$ level observed by Smitt, Svensson, and Outred (1976) at 398.075 Å was averaged with the data of Svensson (1971) to determine the position of this level.

The rest of the levels are derived from the wavelengths and identifications of Svensson and Ekberg (1968).

The uncertainty in the level values is estimated to be ±10 cm^{-1}.

References

Edlén, B. (1937), Z. Phys. **104**, 188.

Fawcett, B. C., Cowan, R. D., and Hayes, R. W. (1972), J. Phys. **B5**, 2143.

Fawcett, B. C., and Gabriel, A. H. (1966), Proc. Phys. Soc. **88**, 262.

Gabriel, A. H., Fawcett, B. C., and Jordan, C. (1966), Proc. Phys. Soc. **87**, 825.

Kruger, P. G., and Pattin, H. S. (1937), Phys. Rev. **52**, 621.

Smitt, R., Svensson, L. A., and Outred, M. (1976), Phys. Scr. **13**, 293.

Svensson, L. A. (1971), Phys. Scr. **4**, 111.

Svensson, L. A., and Ekberg, J. O. (1968), Ark. Fys. **37**, 65.

Ti VII

Configuration	Term	J	Level (cm^{-1})	Configuration	Term	J	Level (cm^{-1})
$3s^2\,3p^4$	^3P	2	0	$3s^2\,3p^3(^2P°)3d$	$^3D°$	3	*393 667*
		1	4 534			2	*396 572*
		0	5 888			1	*398 527*
$3s^2\,3p^4$	^1D	2	24 130	$3s^2\,3p^3(^2P°)3d$	$^1D°$	2	*407 703*
$3s^2\,3p^4$	^1S	0	54 801	$3s^2\,3p^3(^2P°)3d$	$^1F°$	3	*420 522*
$3s3p^5$	$^3P°$	2	*196 266*	$3s^2\,3p^3(^2P°)3d$	$^1P°$	1	*450 729*
		1	*200 059*				
		0	*202 202*	$3s^2\,3p^3(^4S°)4s$	$^3S°$	1	*564 217*
$3s3p^5$	$^1P°$	1	*251 214*	$3s^2\,3p^3(^2D°)4s$	$^3D°$	1	*586 092*
						2	*586 308*
$3s^2\,3p^3(^2D°)3d$	$^1D°$	2	*325 261*			3	*586 998*
$3s^2\,3p^3(^2D°)3d$	$^1F°$	3	*361 904*	$3s^2\,3p^3(^2D°)4s$	$^1D°$	2	*592 918*
$3s^2\,3p^3(^2D°)3d$	$^3S°$	1	*375 235*	$3s^2\,3p^3(^2P°)4s$	$^3P°$	0	*607 538*
						1	*607 982*
$3s^2\,3p^3(^2P°)3d$	$^3P°$	2	*377 614*			2	*609 116*
		1	*378 872*				
		0	*381 808*	$3s^2\,3p^3(^2P°)4s$	$^1P°$	1	*614 794*
$3s^2\,3p^3(^2D°)3d$	$^1P°$	1	*381 894*				

Ti VII—Continued

Configuration	Term	J	Level (cm^{-1})	Configuration	Term	J	Level (cm^{-1})
$3s^2\,3p^3(^4S^\circ)4d$	$^3D^\circ$	1	*726 277*	$3s^2\,3p^3(^2P^\circ)4d$	$^3D^\circ$	2	*775 416*
		2	*726 303*			1	*776 122*
		3	*726 424*			3	*779 699*
$3s^2\,3p^3(^2D^\circ)4d$	$^3P^\circ$	2	*752 850*	$3s^2\,3p^3(^2P^\circ)4d$	$^1D^\circ$	2	*780 853*
		1	*755 732*	$3s^2\,3p^3(^2P^\circ)4d$	$^1F^\circ$	3	*781 170*
$3s^2\,3p^3(^2D^\circ)4d$	$^3D^\circ$	3	*753 393*				
		2	*754 591*	$3s^2\,3p^3(^2P^\circ)4d$	$^1P^\circ$	1	*785 716*
$3s^2\,3p^3(^2D^\circ)4d$	$^3S^\circ$	1	*756 518*				
				Ti VIII ($^4S^\circ_{3/2}$)	***Limit***		**1 136 000**
$3s^2\,3p^3(^2D^\circ)4d$	$^1D^\circ$	2	*757 984*				
$3s^2\,3p^3(^2D^\circ)4d$	$^1F^\circ$	3	*760 504*				

Ti VIII

$Z=22$

P I isoelectronic sequence

Ground state: $1s^2 2s^2 2p^6 3s^2 3p^3\ {}^4S^\circ_{3/2}$

Ionization energy = 1 374 000±3000 cm^{-1} (170.4±0.4 eV)

This spectrum was initially studied by Kruger and Pattin (1937), who observed five multiplets of the $3s^2 3p^3 - 3s^2 3p^2 4s$ array in the region 150–162 Å. Fawcett, Gabriel, and Saunders (1967) observed the resonance lines from $3p^2(^3P)3d\ ^4P$ at 268 Å. Later Fawcett (1970) identified the $3s^2 3p^3 - 3s\,3p^4$ array in the range 423–514 Å. Fawcett, Cowan, and Hayes (1972) reported four lines of the $3p^2 3d - 3p^2 4f$ array, but they are not connected to the known levels.

The present compilation is based on the more complete and accurate work of Ekberg and Svensson (1970) and Smitt, Svensson, and Outred (1976). The level values for the $3s^2 3p^3$ and $3s\,3p^4$ configurations are taken from the latter paper. They have an uncertainty of about ±2 cm^{-1}. We have combined these values with the measurements of Ekberg and Svensson to derive new level values for the $3p^2 3d$ and $4s$ configurations. The uncertainty of these upper levels is about ±10 cm^{-1}. Since no intersystem transitions have been observed, all of the doublets have an added systematic error, x, relative to the ground term $^4S^\circ$. The value of x depends on the accuracy of calculations by Smitt, Svensson, and Outred and is expected to be ±20 cm^{-1}.

The ionization energy is from an extrapolation of Lotz (1967).

References

Ekberg, J. O., and Svensson, L. A. (1970), Phys. Scr. **2,** 283.

Fawcett, B. C. (1970), J. Phys. **B3,** 1732.

Fawcett, B. C., Cowan, R. D., and Hayes, R. W. (1972), J. Phys. **B5,** 2143.

Fawcett, B. C., Gabriel, A. H., and Saunders, P. A. H. (1967), Proc. Phys. Soc. **90,** 863.

Kruger, P.G., and Pattin, H. S. (1937), Phys. Rev. **52,** 624.

Lotz, W. (1967), J. Opt. Soc. Am. **57,** 873.

Smitt, R., Svensson, L. A., and Outred, M. (1976), Phys. Scr. **13,** 293.

Ti VIII

Configuration	Term	J	Level (cm^{-1})	Configuration	Term	J	Level (cm^{-1})
$3s^2\,3p^3$	$^4S^\circ$	3/2	*0.0*	$3s^2\,3p^2(^1D)3d$	^2P	1/2	412 858+x
						3/2	415 589+x
$3s^2\,3p^3$	$^2D^\circ$	3/2	*32 190.5*+x				
		5/2	*33 256.4*+x	$3s^2\,3p^2(^1D)3d$	^2F	5/2	418 873+x
						7/2	419 939+x
$3s^2\,3p^3$	$^2P^\circ$	1/2	*54 189.2*+x				
		3/2	*55 633.6*+x	$3s^2\,3p^2(^1D)3d$	^2S	1/2	423 834+x
$3s3p^4$	^4P	5/2	194 474.6	$3s^2\,3p^2(^1S)3d$	^2D	5/2	435 049+x
		3/2	198 097.9			3/2	436 270+x
		1/2	199 953.6				
				$3s^2\,3p^2(^3P)4s$	^4P	1/2	660 135
$3s3p^4$	^2D	3/2	240 971.6+x			3/2	662 835
		5/2	241 426.0+x			5/2	666 493
$3s3p^4$	^2P	3/2	278 037.7+x	$3s^2\,3p^2(^3P)4s$	^2P	1/2	671 405+x
		1/2	281 108.1+x			3/2	675 631+x
$3s3p^4$	^2S	1/2	290 233.6+x	$3s^2\,3p^2(^1D)4s$	^2D	5/2	690 446+x
						3/2	690 672+x
$3s^2\,3p^2(^3P)3d$	^2P	3/2	364 082+x				
		1/2	368 663+x	$3s^2\,3p^2(^1S)4s$	^2S	1/2	722 394+x
$3s^2\,3p^2(^3P)3d$	^4P	5/2	371 012				
		3/2	372 887	Ti IX ($^3P_0$)	***Limit***		**1 374 000**
		1/2	373 971				
$3s^2\,3p^2(^1D)3d$	^2D	5/2	399 323+x				
		3/2	399 772+x				

Ti IX

$Z = 22$

Si I isoelectronic sequence

Ground state: $1s^2 2s^2 2p^6 3s^2 3p^2\ {}^3P_0$

Ionization energy = 1 549 000±3000 cm^{-1} (192.1±0.4 eV)

The study of this spectrum was initiated by Phillips (1939) who classified seven lines as transitions between the ground term and two odd terms $3s3p^3\ {}^3S°$ and $3s^2 3p3d\ {}^3P°$. Fawcett, Gabriel, and Saunders (1967) extended the $3p^2 - 3p3d$ array; Fawcett and Peacock (1967) and Fawcett (1970) added to the $3s^2 3p^2 - 3s3p^3$ array. Fawcett, Cowan, and Hayes (1972) established terms in $3p4d$ and identified a line in $3p3d - 3p4f$ which is not connected with the known levels.

Ekberg and Svensson (1970) reobserved the spectrum between 136 and 400 Å. Smitt, Svensson, and Outred (1976) extended the new observations to 580 Å. The level values for the $3s^2 3p^2$ and $3s3p^3$ configurations are taken from the more accurate data of Smitt et al. and the values for $3s^2 3p3d$ and $3s^2 3p4s$ are derived by combining those values with the measurements of Ekberg and Svensson. The uncertainty of the level values is estimated by the authors to be about ±10 cm^{-1}. Six intersystem transitions have been observed.

The ionization energy was obtained by extrapolation by Ekberg and Svensson.

References

Ekberg, J. O., and Svensson, L. A. (1970), Phys. Scr. **2**, 283.
Fawcett, B. C. (1970), J. Phys. **B3**, 1732.
Fawcett, B. C., Cowan, R. D., and Hayes, R. W. (1972), J. Phys. **B5**, 2143.
Fawcett, B. C., Gabriel, A. H., and Saunders, P. A. H. (1967), Proc. Phys. Soc. **90**, 863.
Fawcett, B. C., and Peacock, N. J. (1967), Proc. Phys. Soc. **91**, 973.
Phillips, L. W. (1939), Phys. Rev. **55**, 708.
Smitt, R., Svensson, L. A., and Outred, M. (1976), Phys. Scr. **13**, 293.

Ti IX

Configuration	Term	J	Level (cm^{-1})	Configuration	Term	J	Level (cm^{-1})
$3s^2\ 3p^2$	^3P	0	0	$3s^2\ 3p3d$	$^3D°$	1	*364 414*
		1	3 119			2	*365 611*
		2	7 282			3	*366 074*
$3s^2\ 3p^2$	^1D	2	28 555	$3s^2\ 3p3d$	$^1F°$	3	*401 771*
$3s^2\ 3p^2$	^1S	0	61 100	$3s^2\ 3p3d$	$^1P°$	1	*411 820*
$3s3p^3$	$^3D°$	1	*200 209*	$3s^2\ 3p4s$	$^3P°$	0	*727 806*
		2	*200 293*			1	*729 111*
		3	*201 000*			2	*735 208*
$3s3p^3$	$^3P°$	0	*230 524*	$3s^2\ 3p4s$	$^1P°$	1	*740 648*
		1	*230 645*				
		2	*230 754*	$3s^2\ 3p4d$	$^1F°$	3	*926 660*
$3s3p^3$	$^1D°$	2	*254 028*	$3s^2\ 3p4d$	$^3D°$	3	*914 040*
$3s3p^3$	$^3S°$	1	*299 944*	Ti X ($^2P°_{1/2}$)	***Limit***		**1 549 000**
$3s3p^3$	$^1P°$	1	*311 087*				
$3s^2\ 3p3d$	$^3P°$	2	*352 632*				
		1	*356 962*				
		0	*358 427*				

Ti X

$Z=22$

Al I isoelectronic sequence

Ground state: $1s^22s^22p^63s^23p\ ^2P^\circ_{1/2}$

Ionization energy = 1 741 500±1000 cm^{-1} (215.92±0.12 eV)

The $3p$–$4d$ doublet of Ti X was identified by Edlén (1936) and the $3p-3d$ doublet by Gabriel, Fawcett, and Jordan (1966). Fawcett and Peacock (1967) identified the doublets of the $3s3p^2$ configuration. Fawcett (1970) later reported the $3s3p^2\,^4P-3p^3\,^4S^\circ$ multiplet.

Ekberg and Svensson (1970) remeasured the spectrum between 70 and 366 Å and identified all the remaining terms given here. They extrapolated the position of $3s3p^2\,^4P$ along the isoelectronic sequence. Since no intersystem transitions have been observed, we use their extrapolation to establish the energy of the quartet terms to the ground level. The error is indicated by x.

Smitt, Svensson, and Outred (1976) remeasured the $3s^23p-3s3p^2$ array between 350 and 490 Å and determined the doublet terms of those configurations with an uncertainty of ±4 cm^{-1}. We used their term values in combination with the earlier measurements of Ekberg and Svensson to establish the higher term values with an uncertainty of ±10 cm^{-1}.

The ionization energy was obtained by Ekberg and Svensson from the $3s^2nf\ ^2F^\circ$ series.

References

Edlén, B. (1936), Z. Phys. **103,** 536.
Ekberg, J. O., and Svensson, L. A. (1970), Phys. Scr. **2,** 283.
Fawcett, B. C. (1970), J. Phys. **B3,** 1732.
Fawcett, B. C., and Peacock, N. J. (1967), Proc. Phys. Soc. **91,** 973.
Gabriel, A. H., Fawcett, B. C., and Jordan, C. (1966), Proc. Phys. Soc. **87,** 825.
Smitt, R., Svensson, L. A., and Outred, M. (1976), Phys. Scr. **13,** 293.

Ti X

Configuration	Term	J	Level (cm^{-1})
$3s^2 3p$	$^2P^\circ$	1/2	*0*
		3/2	*7 543*
$3s3p^2$	^4P	1/2	157 850+x
		3/2	160 655+x
		5/2	164 764+x
$3s3p^2$	^2D	3/2	212 055
		5/2	212 606
$3s3p^2$	^2S	1/2	264 456
$3s3p^2$	^2P	1/2	281 045
		3/2	285 218
$3s^2 3d$	^2D	3/2	345 329
		5/2	345 857
$3p^3$	$^4S^\circ$	3/2	*421 188*+x
$3s3p3d$	$^4D^\circ$	3/2	*504 516*+x
		5/2	*505 134*+x
		7/2	*505 266*+x
$3s^2 4s$	^2S	1/2	797 113
$3s3p4s$	$^4P^\circ$	1/2	*966 176*+x
		3/2	*968 680*+x
		5/2	*973 441*+x
$3s^2 4d$	^2D	3/2	986 655
		5/2	986 919
$3s^2 4f$	$^2F^\circ$	5/2	*1 046 622*
		7/2	*1 046 694*
$3s^2 5s$	^2S	1/2	1 180 390
$3s^2 5d$	^2D	3/2	1 271 460
		5/2	1 271 680
$3s^2 5f$	$^2F^\circ$	5/2	*1 302 120*
		7/2	*1 302 170*
$3s^2 6d$	^2D	3/2	1 423 180
		5/2	1 423 470
$3s^2 6f$	$^2F^\circ$	5/2,7/2	*1 434 560*
Ti XI ($^1S_0$)	*Limit*		**1 741 500**

Ti XI

$Z=22$

Mg I isoelectronic sequence

Ground state: $1s^2 2s^2 2p^6 3s^2\ {}^1S_0$

Ionization energy = 2 137 900±500 cm^{-1} (265.07±0.06 eV)

Edlén (1936) reported three unconnected systems of levels for this ion: the resonance line $3s^2\ {}^1S_0 - 3s4p\ {}^1P_1^\circ$; the triplet system of $3s3p - 3s4s$, $3s4d$, $3s5d$ and the triplets of $3s3d - 3s4f$, $3s5f$. Fawcett and Peacock (1967) identified the terms $3p^2\ {}^3P$, ${}^1D$, and $3s3p\ {}^1P^\circ$. Fawcett (1970) identified $3p^2\ {}^1S$ and several terms in $3p3d$.

Svensson and Ekberg (1969) made extensive observations of titanium spark spectra in the region 50–425 Å. With these data and longer wavelength lines from Fawcett and Peacock, Ekberg (1971) determined all the levels given in this compilation with an uncertainty of ±40 cm^{-1}, except the $3p^2\ {}^1S$ and the $3p3d$ levels taken from Fawcett (1970) and the $3p4f$ levels from Fawcett (1976) whose level uncertainty is ±50 cm^{-1}. Some tentative classifications of $3p^2 - 3p4d$ and $3s3d - 3p4d$ are given by Kastner et al. (1978). The singlets and triplets have been connected by the observation of the $3s^2\ {}^1S_0 - 3s3p\ {}^3P_1^\circ$ line at 569.3±0.2 Å in a tokamak plasma by Finkenthal, Bell, and Moos (1982). An improved value of 568.98±0.04 Å was measured by Peacock, Stamp, and Silver (1984).

We calculated the ionization energy from the $3snf\ {}^3F^\circ$ series.

References

Edlén, B. (1936), Z. Phys. **103,** 536.
Ekberg, J. O. (1971), Phys. Scr. **4,** 101.
Fawcett, B. C., and Peacock, N. J. (1967), Prcc. Phys. Soc. **91,** 973.
Fawcett, B. C. (1970), J. Phys. **B3,** 1732.
Fawcett, B. C. (1976), J. Opt. Soc. Am. **66,** 632.
Finkenthal, M., Bell, R. E., and Moos, H. W. (1982), Phys. Lett. **88A,** 165.
Kastner, S. O., Swartz, M., Bhatia, A. K., and Lapides, J. (1978), J. Opt. Soc. Am. **68,** 1558.
Peacock, N. J., Stamp, M. F., and Silver, J. D. (1984), Phys. Scr. **T8,** 10.
Svensson, L. A., and Ekberg, J. O. (1969), Ark. Fys. **40,** 145.

Ti XI

Configuration	Term	J	Level (cm^{-1})
$3s^2$	^1S	0	0
$3s3p$	$^3P^\circ$	0	*173 200*
		1	*175 753*
		2	*181 400*
$3s3p$	$^1P^\circ$	1	*258 973*
$3p^2$	^3P	0	410 640
		1	415 150
		2	420 700
$3p^2$	^1D	2	408 820
$3p^2$	^1S	0	482 840
$3s3d$	^3D	1	499 840
		2	500 160
		3	500 650
$3s3d$	^1D	2	564 604
$3p3d$	$^3F^\circ$	2	*683 920*
		3	*687 580*
		4	*691 980*
$3p3d$	$^1D^\circ$	2	*694 610*

Configuration	Term	J	Level (cm^{-1})
$3p3d$	$^3D^\circ$	3	*730 560*
$3p3d$	$^1P^\circ$	1	*750 220*
$3s4s$	^3S	1	1 050 850
$3s4s$	^1S	0	1 065 780
$3s4p$	$^1P^\circ$	1	*1 139 920*
$3s4d$	^3D	1	1 243 920
		2	1 244 260
		3	1 244 630
$3s4d$	^1D	2	1 253 100
$3p4s$	$^3P^\circ$	0	*1 267 600*
		1	*1 269 840*
		2	*1 276 040*
$3s4f$	$^3F^\circ$	2	*1 293 870*
		3	*1 293 940*
		4	*1 294 040*
$3s4f$	$^1F^\circ$	3	*1 304 360*
$3p4p$	^3D	3	1 353 860

(Continued)

Ti XI—Continued

Configuration	Term	*J*	Level (cm^{-1})	Configuration	Term	*J*	Level (cm^{-1})
3*p*4*p*	^{3}P	0	1 353 940	3*s*5*d*	^{1}D	2	1 584 970
		1	1 356 280				
		2	1 360 140	3*s*5*f*	^{3}F°	2	*1 599 850*
						3	*1 599 940*
3*p*4*p*	^{3}S	1	1 361 780			4	*1 599 960*
3*s*5*s*	^{3}S	1	1 484 620	3*s*5*f*	^{1}F°	3	*1 603 140*
3*s*5*s*	^{1}S	0	1 491 740	3*s*6*p*	^{1}P°	1	*1 727 380*
3*p*4*f*	^{3}G	4	1 500 130	3*s*6*f*	^{3}F°	4	*1 765 260*
		5	1 505 620				
				3*s*7*p*	^{1}P°	1	*1 840 880*
3*p*4*f*	^{3}F	3	1 502 090				
		4	1 505 420	Ti XII ($^2S_{1/2}$)	***Limit***		**2 137 900**
3*s*5*p*	^{1}P°	1	*1 528 980*				
3*s*5*d*	^{3}D	1	1 577 670				
		2	1 577 820				
		3	1 577 980				

Ti XII

$Z = 22$

Na I isoelectronic sequence

Ground state: $1s^2 2s^2 2p^6 3s\ ^2S_{1/2}$

Ionization energy $= 2\ 351\ 080 \pm 100\ cm^{-1}$ (291.502 ± 0.012 eV)

The first observations were made in the range of 60–117 Å by Edlén (1936). His analysis of the spectrum provided three independent systems of doublet terms based on the $3s$, $3p$, and $3d$ terms. These were united by the identifications of the $3s-3p$ and $3p-3d$ multiplets by Fawcett and Peacock (1967) between 300 and 500 Å.

The new measurements by Ekberg and Svensson (1975) in the region of 52–960 Å enabled them to redetermine to known level values and extend the known series through $6s$, $6p$, $8d$, and $8f$ with an uncertainty of $\pm 20\ cm^{-1}$. Further extensions were made by Cohen and Behring (1976) through $7s$, $11p$, and $10d$. The level uncertainty is estimated to be $\pm 500\ cm^{-1}$.

Transitions $2p^6 3s - 2p^5 3s^2$, $2p^6 3p - 2p^5 3s 3p$, and $2p^6 3d - 2p^5 3s 3d$ were observed in the range of 26–28 Å with an accuracy of ± 0.01 Å by Burkhalter, Cohen, Cowan, and Feldman (1979). Classifications are made on the basis of calculated wavelength and intensities. The uncertainty in the values of the core-excited levels is $\pm 2000\ cm^{-1}$. Percentage compositions are given in two alternate coupling schemes.

The value for the ionization energy was derived by Edlén (1978) from core polarization theory applied to the nf series.

References

Burkhalter, P. G., Cohen, L., Cowan, R. D., and Feldman, U. (1979), J. Opt. Soc. Am. **69,** 1133.

Cohen, L., and Behring, W. E. (1976), J. Opt. Soc. Am. **66,** 899.

Edlén, B. (1936), Z. Phys. **100,** 621.

Edlén, B. (1978), Phys. Scr. **17,** 565.

Ekberg, J. O., and Svensson, L. A. (1975), Phys. Scr. **12,** 116.

Fawcett, B. C., and Peacock, N. J. (1967), Proc. Phys. Soc. **91,** 973.

Ti XII

Configuration	Term	J	Level (cm^{-1})	Leading percentages
$2p^6(^1S)3s$	^2S	1/2	0	
$2p^6(^1S)3p$	$^2P^\circ$	1/2	*208 385*	
		3/2	*217 042*	
$2p^6(^1S)3d$	^2D	3/2	501 922	
		5/2	502 814	
$2p^6(^1S)4s$	^2S	1/2	1 133 573	
$2p^6(^1S)4p$	$^2P^\circ$	1/2	*1 214 390*	
		3/2	*1 217 700*	
$2p^6(^1S)4d$	^2D	3/2	1 321 430	
		5/2	1 321 870	
$2p^6(^1S)4f$	$^2F^\circ$	5/2	*1 360 310*	
		7/2	*1 360 470*	
$2p^6(^1S)5s$	^2S	1/2	1 606 160	
$2p^6(^1S)5p$	$^2P^\circ$	1/2	*1 645 760*	
		3/2	*1 647 440*	
$2p^6(^1S)5d$	^2D	3/2	1 697 110	
		5/2	1 697 320	
$2p^6(^1S)5f$	$^2F^\circ$	5/2	*1 716 840*	
		7/2	*1 716 920*	

(Continued)

Ti XII—Continued

Configuration	Term	J	Level (cm^{-1})	Leading percentages
$2p^6(^1S)6s$	^2S	1/2	1 848 640	
$2p^6(^1S)6p$	$^2P°$	1/2	*1 870 660*	
		3/2	*1 871 490*	
$2p^6(^1S)6d$	^2D	3/2	1 899 500	
		5/2	1 899 540	
$2p^6(^1S)6f$	$^2F°$	5/2	*1 910 630*	
		7/2	*1 910 680*	
$2p^6(^1S)7s$	^2S	1/2	1 989 000	
$2p^6(^1S)7p$	$^2P°$	1/2,3/2	*2 003 500*	
$2p^6(^1S)7d$	^2D	3/2	2 020 620	
		5/2	2 020 690	
$2p^6(^1S)7f$	$^2F°$	5/2	*2 027 700*	
		7/2	*2 027 730*	
$2p^6(^1S)8p$	$^2P°$	1/2,3/2	*2 087 400*	
$2p^6(^1S)8d$	^2D	3/2,5/2	2 098 880	
$2p^6(^1S)8f$	$^2F°$	5/2	*2 103 590*	
		7/2	*2 103 610*	
$2p^6(^1S)9p$	$^2P°$	1/2,3/2	*2 144 000*	
$2p^6(^1S)9d$	^2D	3/2,5/2	2 152 300	
$2p^6(^1S)10p$	$^2P°$	1/2,3/2	*2 184 200*	
$2p^6(^1S)10d$	^2D	3/2,5/2	2 190 500	
$2p^6(^1S)11p$	$^2P°$	1/2,3/2	*2 214 000*	
Ti XIII ($^1S_0$)	***Limit***		**2 351 080**	
$2p^5(^2P°_{1/2})3s^2(^1S_0)$	(1/2,0)°	1/2	*3 687 000*	100 or 100 $2p^5(^2P°)3s^2\ ^2P°$
$2p^5(^2P°_{3/2})3s3p(^3P°_1)$	(3/2,1)	5/2	3 792 000	97 or 74 $2p^53s(^3P°)3p\ ^4D$
		1/2	3 805 000	84 or 43 $2p^53s(^3P°)3p\ ^4D$
$2p^5(^2P°_{3/2})3s3p(^3P°_0)$	(3/2,0)	3/2	3 797 000	71 or 57 $2p^53s(^3P°)3p\ ^4D$
$2p^5(^2P°_{3/2})3s3p(^3P°_2)$	(3/2,2)	5/2	3 812 000	98 or 69 $2p^53s(^3P°)3p\ ^4P$
		3/2	3 813 000	56 or 37 $2p^53s(^1P°)3p\ ^2D$
		1/2	3 826 000	58 or 58 $2p^53s(^3P°)3p\ ^4P$
$2p^5(^2P°_{1/2})3s3p(^3P°_2)$	(1/2,2)	5/2	3 854 000	96 or 42 $2p^53s(^1P°)3p\ ^2D$
$2p^5(^2P°_{1/2})3s3p(^3P°_1)$	(1/2,1)	1/2	3 862 000	42 or 62 $2p^53s(^1P°)3p\ ^2S$
$2p^5(^2P°_{3/2})3s3p(^1P°_1)$	(3/2,1)	5/2	3 890 000	99 or 69 $2p^53s(^3P°)3p\ ^2D$
$2p^5(^2P°_{1/2})3s3p(^1P°_1)$	(1/2,1)	3/2	3 942 000	98 or 54 $2p^53s(^3P°)3p\ ^2D$

Ti XII—Continued

Configuration	Term	J	Level (cm^{-1})	Leading percentages			
$2p^5(^2P^\circ_{3/2})3s3d(^3D_2)$	$(^3/_2,2)^\circ$	$^7/_2$	*4 105 000*	96	or	82	$2p^53s(^3P^\circ)3d\ ^4F^\circ$
		$^5/_2$	*4 124 000*	69	or	40	$2p^53s(^1P^\circ)3d\ ^2F^\circ$
		$^1/_2$	*4 127 000*	48	or	53	$2p^53s(^1P^\circ)3d\ ^2P^\circ$
$2p^5(^2P^\circ_{1/2})3s3d(^3D_2)$	$(^1/_2,^5/_2)^\circ$	$^7/_2$	*4 120 000*	99	or	68	$2p^53s(^3P^\circ)3d\ ^4D^\circ$
		$^5/_2$	*4 165 000*	41	or	61	$2p^53s(^1P^\circ)3d\ ^2D^\circ$
$2p^5(^2P^\circ_{3/2})3s3d(^3D_3)$	$(^3/_2,3)^\circ$	$^3/_2$	*4 134 000*	51	or	51	$2p^53s(^1P^\circ)3d\ ^2P^\circ$
$2p^5(^2P^\circ_{3/2})3s3d(^1D_2)$	$(^3/_2,2)^\circ$	$^7/_2$	*4 145 000*	99	or	76	$2p^53s(^3P^\circ)3d\ ^2F^\circ$
$2p^5(^2P^\circ_{1/2})3s3d(^3D_3)$	$(^1/_2,3)^\circ$	$^5/_2$	*4 159 000*	55	or	45	$2p^53s(^3P^\circ)3d\ ^4D^\circ$
$2p^5(^2P^\circ_{1/2})3s3d(^1D_2)$	$(^1/_2,2)^\circ$	$^3/_2$	*4 186 000*	36	or	44	$2p^53s(^3P^\circ)3d\ ^2D^\circ$
$2p^5(^2P^\circ_{1/2})3s3d(^1D_2)$	$(^1/_2,2)^\circ$	$^5/_2$	*4 199 000*	94	or	43	$2p^53s(^3P^\circ)3d\ ^2F^\circ$
$2p^5(^2P^\circ_{1/2})3s3d(^1D_2)$	$(^1/_2,2)^\circ$	$^3/_2$	*4 224 000*	58	or	73	$2p^53s(^3P^\circ)3d\ ^2P^\circ$

Ti XIII

$Z=22$

Ne I isoelectronic sequence

Ground state: $1s^2 2s^2 2p^6\ ^1S_0$

Ionization energy = 6 354 300±500 cm^{-1} (787.84±0.06 eV)

Resonance lines between 17 and 27 Å are classified by this system of energy levels. Edlén and Tyrén (1936) identified transitions from $2p^5 3s$ and $3d$. Fawcett (1965) observed three transitions arising from $2p^5 4d$ and from $2s2p^6 3p\ ^1P^\circ_1$. Feldman and Cohen (1967) observed nine transitions, including those reported by Fawcett. We have adopted the more accurate values of Feldman and Cohen, whose level uncertainty is ±2000 cm^{-1}.

The percentage compositions were given by Bogdanovich et al. (1980).

Kastner, Behring, and Cohen (1975) identified transitions between $2p^5 3p$ and $2p^5 4d$, but there is no connection with the levels given here. Three levels of the $2p^5 4p$ configuration were obtained from the classified lines of the $2p^5 3s - 2p^5 4p$ array by Fawcett, Bromage, and Hayes (1979). They have also observed lines of $2p^5 3d - 2p^5 4f$, $5f$, and $2p^5 3p - 2p^5 4s$, $4d$ but they are not connected with known levels.

We derived the ionization energy from the $2p^5 nd\ ^3D^\circ_1$ series for $n=3$, 4, and 5.

References

Bogdanovich, P. O., Bogdanovichene, M. I., Grudzinskas, I. I., Rudzikas, Z. B., Tutlis, V. I., and Shadzhjuvene, S. D., "Spectroscopy of Multiply Charged Ions," USSR Acad. of Sci., Sci. Counc. on Spectra (1980).

Edlén, B., and Tyrén, F. (1936), Z. Phys. **101**, 206.

Fawcett, B. C. (1965), Proc. Phys. Soc. **86**, 1087.

Fawcett, B. C. Bromage, G. E., and Hayes, R. W. (1979), Mon. Not. R. Astron. Soc. **186**, 113.

Feldman, U., and Cohen, L. (1967), Astrophys. J. **149**, 265.

Kastner, S. O., Behring, W. E., and Cohen, L. (1975), Astrophys. J. **199**, 777.

Ti XIII

Configuration	Term	J	Level (cm^{-1})	Leading percentages		
$2s^2 2p^6$	^1S	0	0	100		
$2s^2 2p^5 3s$	$^3P^\circ$	1	*3 709 200*	55	45	$^1P^\circ$
$2s^2 2p^5 3s$	$^1P^\circ$	1	*3 753 600*	55	45	$^3P^\circ$
$2s^2 2p^5 3d$	$^3P^\circ$	1	*4 168 200*	94		
$2s^2 2p^5 3d$	$^3D^\circ$	1	*4 219 800*	86		
$2s^2 2p^5 3d$	$^1P^\circ$	1	*4 281 600*	90		
$2s2p^6 3p$	$^3P^\circ$	1	*4 733 300*	96	4	$^1P^\circ$
$2s2p^6 3p$	$^1P^\circ$	1	*4 754 000*	96	4	$^3P^\circ$
$2s^2 2p^5 4s$	$^3P^\circ$	1	*4 966 500*	59	41	$^1P^\circ$
$2s^2 2p^5 4s$	$^1P^\circ$	1	*5 014 300*	59	41	$^3P^\circ$
$2s^2 2p^5(^2P^\circ_{3/2})4p$	$^2[5/2]$	2	5 047 200			
$2s^2 2p^5(^2P^\circ_{3/2})4p$	$^2[3/2]$	1	5 049 900			
$2s^2 2p^5(^2P^\circ_{1/2})4p$	$^2[3/2]$	2	5 097 300			
$2s^2 2p^5 4d$	$^3D^\circ$	1	*5 163 700*	61	34	$^1P^\circ$
$2s^2 2p^5 4d$	$^1P^\circ$	1	*5 207 200*	66	28	$^3D^\circ$

Ti XIII—Continued

Configuration	Term	*J*	Level (cm^{-1})	Leading percentages
$2s^2\,2p^5\,5d$	$^3D^\circ$	1	*5 596 300*	
$2s^2\,2p^5\,5d$	$^1P^\circ$	1	*5 641 100*	
Ti XIV ($^2P^\circ_{3/2}$)	***Limit***		**6 354 300**	

Ti XIV

$Z=22$

F I isoelectronic sequence

Ground state: $1s^2 2s^2 2p^5\ {}^2P^\circ_{3/2}$

Ionization energy = 6 961 000 ± 14 000 cm^{-1} (863.1 ± 1.7 eV)

The first work on this spectrum was done by Fawcett (1965), who classified many lines of the $2s^2 2p^5 - 2s^2 2p^4 3s$ and $3d$ transition arrays between 21 and 25 Å. This work was revised and extended by Feldman, Doschek, Cowan, and Cohen (1973), from whose improved wavelengths all of the $3s$ and $3d$ levels are determined with an uncertainty of ±4000 cm^{-1}. The ground term $2s^2 2p^5\ {}^2P^\circ$ interval was confirmed by Fawcett (1971) from his identification of the $2s^2 2p^5 - 2s2p^6$ doublet at ~125 Å. A more accurate value for the ground term splitting was determined by the observation of the magnetic dipole transition $2s^2 2p^5\ {}^2P^\circ_{3/2} - {}^2P^\circ_{1/2}$ in tokamak plasmas by Suckewer, Fonck, and Hinnov (1980) and by Lawson, Peacock, and Stamp (1981). The value given here is from these observations, with an uncertainty of ±4 cm^{-1}. The $2s2p^6\ {}^2S$ term is derived from the improved measurements of Kaufman, Sugar, and Cooper (1982) with an uncertainty of ±50 cm^{-1}.

The $2s2p^5 3s\ {}^2P^\circ_{1/2}$ term is from Feldman et al. (1973).

Bogdanovich et al. (1980) calculated the percentage compositions of the levels.

The ionization energy was obtained by extrapolation by Lotz (1967).

References

Bogdanovich, P. O., Bogdanovichene, M. I., Grudzinskas, I. I., Rudzikas, Z. B., Tutlis, V. I., and Shadzhjuvene, S. D., "Spectroscopy of Multiply Charged Ions," USSR Acad. of Sci., Sci. Counc. on Spectr. (1980).

Fawcett, B. C. (1965), Proc. Phys. Soc. **86**, 1087.

Fawcett, B. C. (1971), J. Phys. **B4**, 981.

Feldman, U., Doschek, G. A., Cowan, R. D., and Cohen, L. (1973), J. Opt. Soc. Am. **63**, 1445.

Kaufman, V., Sugar, J., and Cooper, D. (1982), Phys. Scr. **25**, 623.

Lawson, K. D., Peacock, N. J., and Stamp, M. F. (1981), J. Phys. **B14**, 1929.

Lotz, W. J. (1967), J. Opt. Soc. Am. **57**, 873.

Suckewer, S., Fonck, R., and Hinnov, E. (1980), Phys. Rev. **22**, 2278.

Ti XIV

Configuration	Term	J	Level (cm^{-1})	Leading percentages		
$2s^2 2p^5$	$^2P^\circ$	3/2	*0*			
		1/2	*47 219*			
$2s2p^6$	^2S	1/2	819 772			
$2s^2 2p^4(^3P)3s$	^4P	5/2	3 996 000	96		
		3/2	4 014 700	59	37	$(^3P)\ ^2P$
		1/2	4 035 900	96		
$2s^2 2p^4(^3P)3s$	^2P	3/2	4 043 600	56	41	$(^3P)\ ^4P$
		1/2	4 065 500	96		
$2s^2 2p^4(^1D)3s$	^2D	5/2	4 112 700	96		
		3/2	4 113 600	94		
$2s^2 2p^4(^1S)3s$	^2S	1/2	4 221 000	94		
$2s^2 2p^4(^3P)3d$	^4P	1/2	4 478 700	83		
		3/2	4 488 300	74	10	$(^3P)\ ^2D$
		5/2	4 501 500	55	24	$(^3P)\ ^4F$
$2s^2 2p^4(^3P)3d$	^4F	5/2	4 488 500	66	16	$(^3P)\ ^2F$
$2s^2 2p^4(^3P)3d$	^2P	1/2	4 494 600	55	24	$(^1D)\ ^2P$
		3/2	4 525 500	50	23	
$2s^2 2p^4(^3P)3d$	^2D	3/2	4 506 400	38	19	$(^1D)\ ^2D$
		5/2	4 531 900	55	18	

Ti XIV—Continued

Configuration	Term	J	Level (cm^{-1})	Leading percentages		
$2s^2\,2p^4(^3P)3d$	^2F	5/2	4 512 200	56	27	$(^3P)\ ^4P$
$2s^2\,2p^4(^1D)3d$	^2S	1/2	4 583 000	92		
$2s^2\,2p^4(^1D)3d$	^2D	5/2	4 601 500	64	24	$(^3P)\ ^2D$
		3/2	4 617 200	69	28	
$2s^2\,2p^4(^1D)3d$	^2P	3/2	4 601 400	71	25	$(^3P)\ ^2P$
$2s^2\,2p^4(^1S)3d$	^2D	5/2	4 685 800	94		
		3/2	4 693 800	88		
$2s2p^5(^3P°)3s$	$^2P°$	3/2	*4 778 600*	88		
		1/2	*4 806 100*	94		
Ti XV ($^3P_2$)	***Limit***		**6 961 000**			

Ti XV

$Z=22$

O I isoelectronic sequence

Ground state: $1s^22s^22p^4\ ^3P_2$

Ionization energy = 7 597 000±15 000 cm^{-1} (941.9±1.9 eV)

The transition array $2s^22p^4-2s2p^5$ was analyzed by Fawcett (1971) and with improved resolution by Doschek, Feldman, Cowan, and Cohen (1974). The $2p^6\ ^1S_0$ level was identified by Kasyanov et al. (1974). More accurate measurements of this spectral region (102–165 Å) were given by Kaufman, Sugar, and Cooper (1982) who also observed intersystem lines. These measurements are used to determine the energy levels with an uncertainty of ±50 cm^{-1}. Kaufman and Sugar (1982) provided the percentage compositions for these levels. Their calculation includes configuration interaction between $2s^22p^4$ and $2p^6$. The $2s^22p^4\ ^3P_2-{}^3P_1$ magnetic dipole transition was observed in a Tokamak plasma by Lawson, Peacock, and Stamp (1981) and by Suckewer, Fonck, and Hinnov (1980). They obtained the wavenumber intervals 39291 ±1 cm^{-1} and 39 284±2 cm^{-1}, respectively. Peacock, Stamp, and Silver (1984) measured the transition $^3P_2-{}^1D_2$ at 919.73±0.08 Å in a tokamak plasma. These magnetic dipole lines are used to determine the positions of the $^3P_1$ and $^1D_2$ levels.

The $2s^22p^4-2s^22p^33s$ array in the range of 22.4–23.2 Å was measured by Doschek, Feldman, and Cohen (1973) with an accuracy of ±0.01 Å. Their analysis is used to obtain the levels of the $2s^22p^33s$ configuration with an uncertainty of ±2000 cm^{-1}. Fawcett and Hayes (1975) give measurements of the $2s^22p^4-2s^22p^33d$ array in the range of 20.0–21.1 Å. The uncertainty of wavelength appears to be ±0.02–0.03 Å. We obtained the $2s^22p^33d$ levels from their classifications with an uncertainty of ±7000 cm^{-1}. Subsequent revisions by Bromage and Fawcett (1977) based on new calculations are included.

The ionization energy is an extrapolated value by Lotz (1967).

References

Bromage, G. E., and Fawcett, B. C. (1977), Mon. Not. R. Astron. Soc. **178,** 591.

Doschek, G. A., Feldman, U., and Cohen, L. (1973), J. Opt. Soc. Am. **63,** 1463.

Doschek, G. A., Feldman, U., Cowan, R. D., and Cohen, L. (1974), Astrophys. J. **188,** 417.

Fawcett, B. C. (1971), J. Phys. **B4,** 981.

Fawcett, B. C., and Hayes, R. W. (1975), Mon. Not. R. Astron. Soc. **170,** 185.

Kasyanov, Y. S., Kononov, E. Y., Korabkin, V. V., Koshelev, K. N., Ryabtsev, A. N., Serov, R. V., and Skokan, E. V. (1974), Opt. Spectrosc. **36,** 4.

Kaufman, V., and Sugar, J. (1982), private communication.

Kaufman, V., Sugar, J., and Cooper, D. (1982) Phys. Scr. **25,** 623.

Lawson, K. D., Peacock, N. J., and Stamp, M. F. (1981), J. Phys. **B14,** 1929.

Lotz, W. J. (1967), J. Opt. Soc. Am. **57,** 873.

Peacock, N. J., Stamp, M. F., and Silver, J. D. (1984), Phys. Scr. **T8,** 10.

Suckewer, S., Fonck, R., and Hinnov, E. (1980), Phys. Rev. **22,** 2278.

Ti XV

Configuration	Term	J	Level (cm^{-1})	Leading percentages		
$2s^2\,2p^4$	^3P	2	0	95	5	^1D
		1	39 288	100		
		0	42 345	92	8	^1S
$2s^2\,2p^4$	^1D	2	108 730	95	5	^3P
$2s^2\,2p^4$	^1S	0	215 528	90	8	^3P
$2s2p^5$	$^3P^\circ$	2	*712 278*	100		
		1	*742 877*	99	1	$^1P^\circ$
		0	*762 056*	100		
$2s2p^5$	$^1P^\circ$	1	*978 037*	99	1	$^3P^\circ$
$2p^6$	^1S	0	1 656 300	98	2	$2s^22p^4\ ^1S$
$2s^2\,2p^3(^4S^\circ)3s$	$^3S^\circ$	1	*4 354 100*			

Ti XV—Continued

Configuration	Term	J	Level (cm^{-1})	Leading percentages
$2s^2\,2p^3(^2D^\circ)3s$	$^3D^\circ$	1	*4 440 200*	
		2	*4 440 400*	
		3	*4 451 600*	
$2s^2\,2p^3(^2D^\circ)3s$	$^1D^\circ$	2	*4 469 100*	
$2s^2\,2p^3(^2P^\circ)3s$	$^3P^\circ$	2	*4 523 000*	
$2s^2\,2p^3(^2P^\circ)3s$	$^1P^\circ$	1	*4 557 300*	
$2s^2\,2p^3(^4S^\circ)3d$	$^3D^\circ$	2	*4 780 000*	
		3	*4 785 000*	
$2s^2\,2p^3(^2D^\circ)3d$	$^3D^\circ$	1	*4 873 000*	
		2	*4 891 000*	
		3	*4 898 000*	
$2s^2\,2p^3(^2D^\circ)3d$	$^3P^\circ$	2	*4 905 000*	
		1	*4 965 000*	
$2s^2\,2p^3(^2D^\circ)3d$	$^1D^\circ$	2	*4 911 000*	
$2s^2\,2p^3(^2D^\circ)3d$	$^1F^\circ$	3	*4 940 000*	
$2s^2\,2p^3(^2P^\circ)3d$	$^3P^\circ$	2	*4 950 000*	
$2s^2\,2p^3(^2P^\circ)3d$	$^1D^\circ$	2	*4 962 000*	
$2s^2\,2p^3(^2P^\circ)3d$	$^3D^\circ$	1	*4 984 000*	
		3	*4 987 000*	
		2	*5 006 000*	
$2s^2\,2p^3(^2P^\circ)3d$	$^1F^\circ$	3	*5 006 000*	
$2s^2\,2p^3(^2P^\circ)3d$	$^1F^\circ$	3	*5 014 000*	
$2s^2\,2p^3(^2P^\circ)3d$	$^1P^\circ$	1	*5 046 000*	
Ti XVI ($^4S^\circ_{3/2}$)	***Limit***		**7 597 000**	

Ti XVI

$Z=22$

N I isoelectronic sequence

Ground state: $1s^22s^22p^3\ ^4S^\circ_{3/2}$

Ionization energy = 8 420 000±17 000 cm^{-1} (1044±2 eV)

The transition arrays $2s^22p^3-2s2p^4$ and $2s2p^4-2p^5$ were identified by Fawcett (1971). The intersystem line $2s^22p^3\ ^4S^\circ_{3/2}-2s2p^4\ ^2P_{3/2}$ at 102.393 Å was identified by Kaufman, Sugar, and Cooper (1982) who remeasured the spectrum in the range of 102–178 Å with an uncertainty of ±0.01 Å. This provided a level uncertainty of ±50 cm^{-1}. Kaufman and Sugar (1982) provided the percentage compositions for these levels. Their calculation includes configuration interaction between $2s^2\,2p^3$ and $2p^5$.

Fawcett and Hayes (1975) classified the lines of the $2s^22p^3-2s^22p^23d$ array at 19 Å. Two additional classifications were made by Bromage and Fawcett (1977). The level uncertainty for the $2s^22p^23d$ configuration is ±2000 cm^{-1}.

The ionization energy was obtained by extrapolation by Lotz (1967).

References

Bromage, G. E. and Fawcett, B. C. (1977), Mon. Not. R. Astron. Soc. **179,** 683.

Fawcett, B. C. (1971), J. Phys. **B4,** 981.

Fawcett, B. C., and Hayes, R. W. (1975), Mon. Not. R. Astron. Soc. **170,** 185.

Kaufman, V., and Sugar, J. (1982), private communication.

Kaufman, V., Sugar, J., and Cooper, D. (1982), Phys. Scr. **26,** 163.

Lotz, W. (1967), J. Opt. Soc. Am. **57,** 873.

Ti XVI

Configuration	Term	J	Level (cm^{-1})	Leading percentages		
$2s^2\,2p^3$	$^4S^\circ$	3/2	*0*	97	3	$^2P^\circ$
$2s^2\,2p^3$	$^2D^\circ$	3/2	*116 030*	84	14	$^2P^\circ$
		5/2	*130 720*	100		
$2s^2\,2p^3$	$^2P^\circ$	1/2	*197 700*	98	2	$2p^5\ ^2P^\circ$
		3/2	*219 250*	81	15	$2p^3\ ^2D^\circ$
$2s(^2S)2p^4(^3P)$	^4P	5/2	589 140	99	1	$2s(^2S)2p^4(^1D)\ ^2D$
		3/2	620 470	100		
		1/2	633 660	98	2	$2s(^2S)2p^4(^3P)\ ^2S$
$2s(^2S)2p^4(^1S)$	^2D	3/2	813 080	98	2	$2s(^2S)2p^4(^3P)\ ^2P$
		5/2	817 210	99	1	$2s(^2S)2p^4(^3P)\ ^4P$
$2s(^2S)2p^4(^1S)$	^2S	1/2	939 920	83	16	$2s(^2S)2p^4(^1S)\ ^2P$
$2s(^2S)2p^4(^3P)$	^2P	3/2	976 650	98	2	$2s(^2S)2p^4(^1D)\ ^2D$
		1/2	1 020 500	84	15	$2s(^2S)2p^4(^1S)\ ^2S$
$2p^5$	$^2P^\circ$	3/2	*1 537 660*	98	2	$2s^22p^3\ ^2P^\circ$
		1/2	*1 587 830*	92	2	
$2s^2\,2p^2(^3P)3d$	^2P	3/2	5 194 100			
$2s^2\,2p^2(^3P)3d$	^4P	5/2	5 232 300			
		3/2	5 238 600			
$2s^2\,2p^2(^3P)3d$	^2F	7/2	5 245 500			
$2s^2\,2p^2(^3P)3d$	^2D	3/2	5 287 000			
		5/2	5 293 300			

Ti XVI—Continued

Configuration	Term	J	Level (cm^{-1})	Leading percentages
$2s^2\,2p^2(^1D)3d$	^2F	$7/2$	5 336 300	
		$5/2$	5 348 100	
$2s^2\,2p^2(^1D)3d$	^2P	$3/2$	5 361 000	
Ti XVII ($^3P_0$)	*Limit*		8 420 000	

Ti XVII

$Z=22$

C I isoelectronic sequence

Ground state: $1s^2 2s^2 2p^2\ ^3P_0$

Ionization energy $= 9\,120\,000 \pm 18\,000$ cm^{-1} (1131 ± 2 eV)

The ground term splitting $2s^2 2p^2\ ^3P_0 - ^3P_1$ was obtained from a magnetic dipole transition observed in a tokamak plasma at 3370.8±0.2 Å (in air) by Suckewer, Fonck, and Hinnov (1980). They also reported the $2s^2 2p^2\ ^3P_1 - ^3P_2$ line at 3834.4±0.2 Å (in air). These results are used here to determine the ^3P levels with an uncertainty of ±2 cm^{-1}.

The $2s^2 2p^2 - 2s2p^3$ array was classified by Fawcett, Galanti, and Peacock (1974), and by Kasyanov et al. (1974). New measurements and some revised classifications were given by Fawcett and Hayes (1975), who also identified $2s2p^3 - 2p^4$ lines. These arrays were re-observed by Sugar, Kaufman, and Cooper (1982) in the range of 109–188 Å with an accuracy of ±0.01 Å, including several intersystem lines. The levels of $2s^2 2p^2$, $2s2p^3$, and $2p^4$ are derived from these data with a level value uncertainty of ±50 cm^{-1}. None of the above investigations revealed the position of the $2s2p^3\ ^5S_2^\circ$ term. Denne and Hinnov (1984) have observed a faint line in a tokamak plasma that they classified as $2s^2 2p^2\ ^3P_2 - 2s2p^3\ ^5S_2^\circ$. We adopted their result. Kaufman and Sugar (1982) provided the percentage compositions of these levels. Their calculation includes configuration interaction between $2s^2 2p^2$ and $2p^4$.

The levels of $2s^2 2p3d$ are from the classifications by Fawcett and Hayes (1972) with additions by Bromage and Fawcett (1977). The levels of $2s^2 2p3s$, $2s2p^2 3s$ and $2s2p^2 3d$ are due to Goldsmith et al. (1972). They identified two quintet transitions which are not connected to the known levels. The uncertainty in the level values is ±2000 cm^{-1}.

The ionization energy is an extrapolated value by Lotz (1967).

References

Bromage, G. E., and Fawcett, B. C. (1977), Mon. Not. R. Astron. Soc. **178,** 605.

Denne, B., and Hinnov, E. (1984), Phys. Rev. **A29,** 3442.

Fawcett, B. C., Galanti, M., and Peacock, N. J. (1974), J. Phys. **B7,** 1149.

Fawcett, B. C., and Hayes, R.W. (1975), Mon. Not. R. Astron. Soc. **170,** 185.

Goldsmith, S., Feldman, U., Crooker, A., and Cohen, L. (1972), J. Opt. Soc. Am. **62,** 260.

Kasyanov, Y. S., Kononov, E. Y., Korobkin, V. V., Koshelev, K. N., Ryabtsev, A. N., Serov, R. V., and Skokan, E. V. (1974), Opt. Spectrosc. **36,** 4.

Kaufman, V., and Sugar, J. (1982), private communication.

Lotz, W. (1967), J. Opt. Soc. Am. **57,** 873.

Suckewer, S., Fonck, R., and Hinnov, E. (1980), Phys. Rev. **A22,** 2278.

Sugar, J., Kaufman, V., and Cooper, D. (1982), Phys. Scr. **26,** 189.

Ti XVII

Configuration	Term	J	Level (cm^{-1})	Leading percentages		
$2s^2 2p^2$	^3P	0	0	94	4	$2s^2 2p^2\ ^1S$
		1	29 658	99	1	$2p^4\ ^3P$
		2	55 730	89	10	$2s^2 2p^2\ ^1D$
$2s^2 2p^2$	^1D	2	140 660	89	10	^3P
$2s^2 2p^2$	^1S	0	242 180	91	4	$2p^4\ ^1S$
$2s(^2S)2p^3(^4S^\circ)$	$^5S^\circ$	2	*333 660*	99	1	$2s(^2S)2p^3(^2P^\circ)\ ^3P^\circ$
$2s(^2S)2p^3(^2D^\circ)$	$^3D^\circ$	2	*578 890*	93	7	$2s(^2S)2p^3(^2P^\circ)\ ^3P^\circ$
		1	*580 110*	95	5	$2s(^2S)2p^3(^2P^\circ)\ ^3P^\circ$
		3	*586 760*	100		
$2s(^2S)2p^3(^2P^\circ)$	$^3P^\circ$	0	*678 450*	100		
		1	*680 910*	94	5	$2s(^2S)2p^3(^2D^\circ)\ ^3D^\circ$
		2	*686 780*	90	7	$2s(^2S)2p^3(^2D^\circ)\ ^3D^\circ$
$2s(^2S)2p^3(^4S^\circ)$	$^3S^\circ$	1	*832 340*	92	7	$2s(^2S)2p^3(^2P^\circ)\ ^1D^\circ$

Ti XVII—Continued

Configuration	Term	J	Level (cm^{-1})	Leading percentages		
$2s(^2S)2p^3(^2D°)$	$^1D°$	2	*845 140*	98	2	$2s(^2S)2p^3(^2P°)\ ^3P°$
$2s(^2S)2p^3(^2P°)$	$^1P°$	1	*943 500*	92	7	$2s(^2S)2p^3(^4S°)\ ^3S°$
$2p^4$	^3P	2	1 271 380	93	6	$2p^4\ ^1D$
		1	1 313 280	99	1	$2s^22p^2\ ^3P$
		0	1 319 740	94	5	$2p^4\ ^1S$
$2p^4$	^1D	2	1 380 290	93	6	^3P
$2p^4$	^1S	0	1 556 810	91	4	$2s^22p^2\ ^1S$
$2s^2\,2p3s$	$^3P°$	1	*5 144 000*			
		2	*5 193 000*			
$2s^2\,2p3s$	$^1P°$	1	*5 200 000*			
$2s^2\,2p3d$	$^3F°$	2	*5 472 000*			
$2s^2\,2p3d$	$^1D°$	2	*5 502 000*			
$2s^2\,2p3d$	$^3D°$	1	*5 519 000*			
		2	*5 542 000*			
		3	*5 557 000*			
$2s^2\,2p3d$	$^3P°$	2	*5 568 000*			
$2s^2\,2p3d$	$^1P°$	1	*5 612 000*			
$2s^2\,2p3d$	$^1F°$	3	*5 614 000*			
$2s2p^2(^2D)3s$	^3D	2	5 707 000			
$2s2p^2(^4P)3d$	^3F	3	5 859 000			
$2s2p^2(^2D)3d$	^3F	4	6 025 000			
Ti XVIII ($^2P°_{1/2}$)	***Limit***		**9 120 000**			

Ti XVIII

$Z=22$

B I isoelectronic sequence

Ground state: $1s^2 2s^2 2p\ ^2P^\circ_{1/2}$

Ionization energy $= 9\,850\,000 \pm 20\,000$ cm^{-1} (1221 ± 2 eV)

The $^2P^\circ$ ground term splitting was obtained from a magnetic dipole transition at 1778.1±0.1 Å in a tokamak discharge by Suckewer, Fonck, and Hinnov (1980).

Kasyanov et al. (1974) and Fawcett and Hayes (1975) analyzed the transition arrays $2s^2 2p - 2s2p^2$ and $2s2p^2\ ^4P - 2p^3\ ^4S^\circ$ found between 130 Å and 200 Å. They were remeasured by Sugar, Kaufman, and Cooper (1982) with a wavelength uncertainty of ±0.01 Å, providing a level value uncertainty of ±50 cm^{-1}. Denne and Hinnov (1984) have tentatively classified a line at 322.6 Å observed in a tokamak plasma as the $2s^2 2p\ ^2P_{1/2} - 2s2p^2\ ^4P_{1/2}$ transition. We use this wavelength to determine the position of the quartet system. They also tentatively classify a fainter line at 361.1 Å as $2s^2 2p\ ^2P_{3/2} - 2s2p^2\ ^4P_{3/2}$, which we use. The $^2P^\circ$ and $^2D^\circ$ terms of $2p^3$ were found by Fawcett, Ridgeley, and Hatter (1980). The percentage composition for these levels was provided by Kaufman and Sugar (1982). Their calculation includes configuration interaction between $2s^2 2p$ and $2p^3$.

The higher lying levels are from the classifications of the spectrum at 16–17 Å by Fawcett and Hayes, who obtained a level value uncertainty of ±3000 cm^{-1}.

The value for the ionization energy was obtained by Lotz (1967) by extrapolation.

References

Denne, B., and Hinnov, E. (1984), Phys. Rev. A**29**, 3442.

Fawcett, B. C., and Hayes, R. W. (1975), Mon. Not. R. Astron. Soc. **170**, 185.

Fawcett, B. C., Ridgeley, A., and Hatter, A. T. (1980), J. Opt. Soc. Am. **70**, 1349.

Kaufman, V., and Sugar J. (1982), private communication.

Kasyanov, Y. S., Kononov, E. Y., Korobkin, V. V., Koshelev, K. N., Ryabtsev, A. N., Serov, R. V., and Skokan, E. V. (1974), Opt. Spectrosc. **36**, 4.

Lotz, W. (1967), J. Opt. Soc. Am. **57**, 873.

Suckewer, S., Fonck, R., and Hinnov, E. (1980), Phys. Rev. A**22**, 2278.

Sugar, J., Kaufman, V., and Cooper, D. (1982), Phys. Scr. **26**, 293.

Ti XVIII

Configuration	Term	J	Level (cm^{-1})	Leading percentages		
$2s^2\,2p$	$^2P^\circ$	1/2	*0*	97	3	$2p^3\ ^2P^\circ$
		3/2	*56 240*	97	3	
$2s2p^2$	^4P	1/2	309 980	99	1	^2S
		3/2	333 170	100		
		5/2	360 960	98	2	^2D
$2s2p^2$	^2D	3/2	555 860	98	2	^2P
		5/2	561 700	98	2	^4P
$2s2p^2$	^2S	1/2	673 680	52	47	^2P
$2s2p^2$	^2P	1/2	733 750	53	47	^2S
		3/2	747 070	98	2	^2D
$2p^3$	$^4S^\circ$	3/2	*962 100*	98	2	$^2P^\circ$
$2p^3$	$^2D^\circ$	3/2	*1 078 800*	92	7	$^2P^\circ$
		5/2	*1 088 900*	100		
$2p^3$	$^2P^\circ$	1/2	*1 208 800*	97	3	$2s^2 2p\ ^2P^\circ$
		3/2	*1 227 700*	88	8	$2p^3\ ^2D^\circ$
$2s^2\,3d$	^2D	3/2	5 807 000			
		5/2	5 815 000			

Ti XVIII—Continued

Configuration	Term	J	Level (cm^{-1})	Leading percentages
$2s2p(^3P°)3p$	^2P	1/2	5 917 000	
		3/2	5 960 000	
$2s2p(^3P°)3p$	^2D	3/2	6 038 000	
		5/2	6 072 000	
$2s2p(^3P°)3d$	$^2D°$	5/2	*6 142 000*	
$2s2p(^3P°)3d$	$^4D°$	7/2	*6 143 000*	
$2s2p(^3P°)3d$	$^4P°$	5/2	*6 148 000*	
$2s2p(^3P°)3d$	$^2F°$	5/2	*6 201 000*	
		7/2	*6 234 000*	
$2s2p(^1P°)3d$	$^2F°$	7/2	*6 393 000*	
$2s2p(^1P°)3d$	$^2D°$	5/2	*6 433 000*	
Ti XIX ($^1S_0$)	***Limit***		9 850 000	

Ti XIX

$Z=22$

Be I isoelectronic sequence

Ground state: $1s^2 2s^2\ ^1S_0$

Ionization energy = 10 860 000±22 000 cm^{-1} (1346±3 eV)

The resonance line $2s^2\ ^1S_0-2s2p\ ^1P_1^\circ$ was identified at 169.59 Å by Kasyanov et al. (1974). They also give the $2s2p\ ^3P^\circ-2p^2\ ^3P$ multiplet. A more complete array was identified by Fawcett, Ridgeley, and Hatter (1980), including the intersystem line $2s2p\ ^3P_2^\circ-2p^2\ ^1D_2$. Their wavelength uncertainty is ±0.05 Å giving a level value uncertainty of ±200 cm^{-1}. A new measurement of the $2s^2\ ^1S_0-2s2p\ ^1P_1^\circ$ transition was given by Stamp and Peacock (1982) at 169.580±0.002 Å. They also give a value for the $^1S_0-{}^3P_1^\circ$ line which, however, is shown to be incompatible with related data by Edlén (1983). Denne and Hinnov (1984) observed this line in a tokamak plasma at 328.3 Å, in close agreement with Edlén's prediction. We have used this value for the intersystem interval.

The $2s2p\ ^3P_1-{}^3P_2$ interval is obtained from the magnetic dipole transition observed by Lawson, Peacock, and Stamp (1981) at 2344.6±0.2 Å (in air) in a tokamak plasma.

The higher-lying configurations are from the classifications of Fawcett and Hayes (1975) and Boiko et al. (1977).

The ionization energy was obtained by Lotz (1967) by extrapolation.

References

Boiko, V. A., Pikuz, S. A., Safronova, U. L., and Faenov, A. Y. (1977), J. Phys. **B10**, 1253.
Denne, B., and Hinnov, E. (1984), Phys. Rev. A**29**, 3442.
Edlén, B. (1983), Phys. Scr. **28**, 51.
Fawcett, B. C., and Hayes, R. W. (1975), Mon. Not. R. Astron. Soc. **170**, 185.
Fawcett, B. C., Ridgeley, A., and Hatter, A. T. (1980), J. Opt. Soc. Am. **70**, 1349.
Lawson, K. D., Peacock, N. J., and Stamp, M. F. (1981), J. Phys. **B14**, 1929.
Lotz, W. (1967), J. Opt. Soc. Am. **57**, 873.
Stamp, M. F., and Peacock, N. J. (1982), J. Phys. **B15**, L703.
Kasyanov, Y. S., Kononov, E. Y., Korobkin, V. V., Koshelev, K. N., Ryabtsev, A. N., Serov, R. V., and Skokan, E. V. (1974), Opt. Spectrosc. **36**, 4.

Ti XIX

Configuration	Term	J	Level (cm^{-1})
$2s^2$	^1S	0	0
$2s2p$	$^3P^\circ$	0	*288 190*
		1	*304 600*
		2	*347 240*
$2s2p$	$^1P^\circ$	1	*589 692*
$2p^2$	^3P	0	775 810
		1	804 890
		2	832 410
$2p^2$	^1D	2	917 580
$2p^2$	^1S	0	1 104 170
$2s3s$	^3S	1	6 160 800
$2s3p$	$^1P^\circ$	1	*6 303 200*
$2s3d$	^3D	1	6 380 600
		2	6 389 200
		3	6 402 700
$2s3d$	^1D	2	6 445 900
$2p3p$	^3P	0	6 685 800
$2p3p$	^3D	3	6 699 700
$2p3p$	^3S	1	6 701 300
$2p3d$	$^1D^\circ$	2	*6 738 000*
$2p3d$	$^3D^\circ$	2	*6 789 100*
		3	*6 807 600*
$2p3p$	^1D	2	6 770 900
$2p3d$	$^3P^\circ$	1,2	*6 813 700*
$2p3d$	$^1P^\circ$	1	*6 866 000*
$2p3d$	$^1F^\circ$	3	*6 871 700*
Ti XX ($^2S_{1/2}$)	***Limit***		**10 860 000**

Ti XX

$Z=22$

Li I isoelectronic sequence

Ground state: $1s^2 2s\ ^2S_{1/2}$

Ionization energy = 11 497 000±4000 cm^{-1} (1425.4±0.4 eV)

The $2s-2p$ doublet was observed in a tokamak plasma at 259.272±0.004 Å and 309.072±0.010 Å by Stamp and Peacock (1982).

The $2s-3p$, $2p-3s$, and $2p-3d$ transitions near 16 Å were reported by Goldsmith, Feldman, Oren and Cohen (1972) with an uncertainty of ±0.005 Å. Aglitskii, Boiko, Pikuz, and Faenov (1974) confirmed the lines identified by Goldsmith et al. and added the $4p$ to $9p$ and $4d$ to $8d$ terms. The $3d4f$ transition was reported by Fawcett and Ridgeley (1981). These levels have an uncertainty of ±2000 cm^{-1}.

The doubly excited levels were obtained by Aglitskii, Boiko, Zakharov, Pikuz, and Faenov (1974) from lines observed at 2.6 Å in a laser-produced plasma. The uncertainty in these level values is ±7000 cm^{-1}.

We derived the ionization energy from the first three members of the nd series.

References

Aglitskii, E. V., Boiko, V. A., Zakharov, S. M., Pikuz, S. A., and Faenov, A. Y. (1974), J.E.T.P. Letters **19**, 8.

Aglitskii, E. V., Boiko, V. A., Pikuz, S .A., and Faenov, A.Y. (1974) Soviet J. Quant. Elec. **4**, 956.

Fawcett, B. C., and Ridgeley, A. (1981), J. Phys. **B14**, 203.

Goldsmith, S., Feldman, U., Oren, L., and Cohen, L. (1972), Astrophys. J. **174**, 209.

Stamp, M. F., and Peacock, N. J. (1982), J. Phys. **B15**, L703.

Ti XX

Configuration	Term	J	Level (cm^{-1})	Configuration	Term	J	Level (cm^{-1})
$1s^2\,2s$	^2S	1/2	0	$1s^2\,8d$	^2D	3/2, 5/2	10 812 000
$1s^2\,2p$	$^2P°$	1/2	*323 549*	$1s^2\,8p$	$^2P°$	1/2, 3/2	*10 815 000*
		3/2	*385 695*				
				$1s^2\,9p$	$^2P°$	1/2, 3/2	*10 955 000*
$1s^2\,3s$	^2S	1/2	6 466 000				
$1s^2\,3p$	$^2P°$	1/2	*6 556 000*	Ti XXI ($^1S_0$)	*Limit*		**11 497 000**
		3/2	*6 574 000*				
				$1s(^2S)2s2p(^3P°)$	$^4P°$		*37 764 000*
$1s^2\,3d$	^2D	3/2	6 610 000				
		5/2	6 616 000	$1s(^2S)2s2p(^3P°)$	$^2P°$	1/2, 3/2	*38 030 000*
$1s^2\,4p$	$^2P°$	1/2, 3/2	*8 732 000*	$1s2p^2$	^4P		38 117 000
$1s^2\,4d$	^2D	3/2	8 747 000	$1s(^2S)2s2p(^1P°)$	$^2P°$	1/2, 3/2	*38 162 000*
		5/2	8 748 000				
				$1s2p^2$	^2D	3/2	38 319 000
$1s^2\,4f$	$^2F°$	5/2, 7/2	*8 753 000*			5/2	38 329 000
$1s^2\,5p$	$^2P°$	1/2, 3/2	*9 730 000*	$1s2p^2$	^2P	3/2	38 416 000
$1s^2\,5d$	^2D	3/2, 5/2	9 740 000	$1s2p^2$	^2S	1/2	38 550 000
$1s^2\,6p$	$^2P°$	1/2, 3/2	*10 274 000*	$1s2p3p$			44 820 000
$1s^2\,6d$	^2D	3/2, 5/2	10 278 000				
$1s^2\,7p$	$^2P°$	1/2, 3/2	*10 600 000*				
$1s^2\,7d$	^2D	3/2, 5/2	10 601 000				

Ti XXI

$Z=22$

He I isoelectronic sequence

Ground state: $1s^2\,{}^1S_0$

Ionization energy = 50 401 000±10 000 cm^{-1} (6249.0±1.0 eV)

Because of the excellent agreement of the calculated energies of the $n=2$ shell by Safronova (1981) with the few well-measured spectra in the He I sequence, we have compiled her results for the $n=2$ levels and for the ionization energy. Detailed comparisons are given in the Introduction. Levels of the $n=3$–5 shells are from the calculated binding energies by Ermolaev and Jones (1974) subtracted from Safronova's value for the binding energy of the ground state. We have assumed an uncertainty of 2 parts in 10^4 for the excited levels relative to the ground state, and for the ionization energy. This is equal to the random deviation of the measurements by Aglitskii et al. (1974) from the calculations by Safronova (see Introduction). For differences between excited levels where $\Delta n=0$, we assumed an uncertainty of 2 parts in 10^3.

Corrected measurements (see Introduction) by Aglitskii et al. (1974) place the $1s2p\ {}^3P_1^\circ$ level at 38 123 000 cm^{-1} and the $1s2p\ {}^1P_1^\circ$ at 38 310 000 cm^{-1} with an estimated uncertainty of ±7000 cm^{-1}.

Percentage compositions are from Ermolaev and Jones.

References

Aglitskii, E. V., Boiko, V. A., Zakharov, S. M., Pikuz, S. A., and Faenov, A. Y. (1974), Sov. J. Quantum Electron. **4**, 500.

Ermolaev, A. M., and Jones, M. (1974), J. Phys. **B7**, 199.

Safronova, U. I. (1981), Phys. Scr. **23**, 241.

Ti XXI

Configuration	Term	J	Level (cm^{-1})	Leading percentages		
$1s^2$	^1S	0	0			
1s2s	^3S	1	[37 923 330]			
1s2p	$^3P^\circ$	0	*[38 114 090]*			
		1	*[38 125 540]*	96	4	$^1P^\circ$
		2	*[38 180 210]*			
1s2s	^1S	0	[38 131 630]			
1s2p	$^1P^\circ$	1	*[38 307 120]*	96	4	$^3P^\circ$
1s3s	^3S	1	[44 911 530]			
1s3p	$^3P^\circ$	0	*[44 964 290]*			
		1	*[44 967 230]*	95	5	$^1P^\circ$
		2	*[44 983 570]*			
1s3s	^1S	0	[44 966 300]			
1s3p	$^1P^\circ$	1	*[45 018 280]*	95	5	$^3P^\circ$
1s4s	^3S	1	[47 328 420]			
1s4p	$^3P^\circ$	0	*[47 350 340]*			
		1	*[47 351 570]*	94	6	$^1P^\circ$
		2	*[47 358 480]*			
1s4s	^1S	0	[47 350 610]			
1s4p	$^1P^\circ$	1	*[47 372 630]*	94	6	$^3P^\circ$

Ti XXI—Continued

Configuration	Term	J	Level (cm^{-1})	Leading percentages		
1s5s	^{3}S	1	[48 440 550]			
1s5p	^{3}P°	0	[*48 451 630*]			
		1	[*48 452 270*]	94	6	^{1}P°
		2	[*48 455 810*]			
1s5s	^{1}S	0	[48 451 640]			
1s5p	^{1}P°	1	[*48 462 950*]	94	6	^{3}P°
Ti XXII ($^2S_{1/2}$)	***Limit***		**50 401 000**			

Ti XXII

$Z=22$

H I isoelectronic sequence

Ground state: $1s\ ^2S_{1/2}$

Ionization energy = 53 440 800±100 cm^{-1} (6625.87±0.01 eV)

The two $1s$–$2p$ transitions have been observed by Bitter et al. (1982) in a tokamak discharge, but without absolute wavelength calibration. We give calculated values by Mohr (1983) for the $n=2$ shell and by Erickson (1977) for $n=3$-5 relative to the $2p\ ^2P^\circ_{3/2}$ level. Further details are given in the Introduction. Relative to the ground state, the level uncertainty is estimated to be 5 parts in 10^7. The uncertainty in the excited states relative to $2p\ ^2P^\circ_{3/2}$ is 1 part in 10^6.

References

Bitter, M., Cohen, S., Hill, K. W., Sesnic, S., Tenney, F., Timberlake, J., von Goeler, S., and Steenman-Clark, L. (1982), presented at 24th Annual Meeting of the Division of Plasma Physics, New Orleans, Louisiana.

Erickson, G. W. (1977), J. Phys. Chem. Ref. Data **6**, 831.

Mohr, P. J. (1983), At. Data Nucl. Data Tables **29**, 453.

Ti XXII

Configuration	Term	J	Level (cm^{-1})	g	Leading percentages
$1s$	^2S	1/2	0		
$2p$	$^2P^\circ$	1/2	[*40 054 190*]		
		3/2	[*40 141 335*]		
$2s$	^2S	1/2	[40 056 747]		
$3p$	$^2P^\circ$	1/2	[*47 500 870*]		
		3/2	[*47 526 701*]		
$3s$	^2S	1/2	[47 501 659]		
$3d$	^2D	3/2	[47 526 656]		
		5/2	[47 535 163]		
$4p$	$^2P^\circ$	1/2	[*50 103 648*]		
		3/2	[*50 114 539*]		
$4s$	^2S	1/2	[50 103 982]		
$4d$	^2D	3/2	[50 114 519]		
		5/2	[50 118 110]		
$4f$	$^2F^\circ$	5/2	[*50 118 104*]		
		7/2	[*50 119 890*]		
$5p$	$^2P^\circ$	1/2	[*51 306 846*]		
		3/2	[*51 312 419*]		
$5s$	^2S	1/2	[51 307 017]		
$5d$	^2D	3/2	[51 312 409]		
		5/2	[51 314 247]		
$5f$	$^2F^\circ$	5/2	[*51 314 244*]		
		7/2	[*51 315 161*]		
$5g$	^2G	7/2	[51 315 159]		
		9/2	[51 315 709]		
Limit			**53 440 800**		

V I

$Z = 23$

Ground state: $1s^2 2s^2 2p^6 3s^2 3p^6 3d^3 4s^2\ ^4F_{3/2}$

Ionization energy = 54 360±16 cm^{-1} (6.740±0.002 eV)

The compilation of levels for this spectrum is based on the analysis by Meggers and Russell (1936). The wavelengths of the lines between 2080 and 2756 Å and between 4494 and 11 911 Å were measured by Meggers at the National Bureau of Standards; those in the intermediate range, 2756–4494 Å, were mostly from the earlier literature. The g-values for the $a\ ^4$F and $a\ ^6$D terms were measured by Childs and Goodman (1969). The rest were measured by H. D. Babcock and included in the publication of Meggers and Russell. The spectrum was reobserved from 2495–8541 Å with an accuracy of ±0.005 Å by Davis and Andrew (1978) and from 1.0–5.6 μm by Davis, Andrew, and Verges (1978). These data were used by Davis (1976) to redetermine the energy levels with an uncertainty of ±0.05 cm^{-1}.

The alphabetic prefixing of terms with lower case letters for distinguishing repeating terms of the same type has been retained from Meggers and Russell except where the levels were reinterpreted by Roth (1970) on the basis of his theoretical treatment.

The percentage compositions for the even parity configurations $3d^3 4s^2$, $3d^4 4s$, and $3d^5$ were calculated by Dembczynski (1980) with configuration interaction (CI). Those of the odd parity configurations $3d^2 4s^2 4p$, $3d^3 4s4p$, and $3d^4 4p$ were calculated by Roth (1970) with CI. Both calculations were fit to the observed energy levels. Roth concluded that most of the terms above 43 000 cm^{-1} are strongly perturbed by $(3d+4s)^4 5p$ levels and he therefore did not include them in his least squares fit. His percentage compositions and designation changes for the experimental levels are adopted here. Roth distinguished repeating terms of the $3d^n$ core by the letters a, b, ... rather than by seniority. The percentages include the sum of seniority states contributing to the term.

Subsequent to the work of Meggers and Russell, Moore (1939) published seven new high odd terms and 33 miscellaneous high odd levels which could not be assigned to terms. With the exception of the level at 46 707 cm^{-1}, which combines nine times, and the $q\ ^4$G,° $n\ ^4$D°, and $r\ ^2$F° terms, all the levels make few combinations and many are of uncertain J-value. We therefore conclude that they need further confirmation. The $3d^5\ c^4$D term is from the unpublished results of Humphreys and Kostkowski (1950). The term $3d^4(^3G)4s\ ^2$G was reported by Dembczynski and Rudnicka-Szuba (1982).

Meggers and Russell determined the ionization energy from the $3d^3ns$ and $3d^3nd$ series (n =4,5).

References

Childs, W. J., and Goodman, L. S. (1967), Phys. Rev. **A156,** 64.

Davis, D., and Andrew, K. L. (1978), J. Opt. Soc. Am. **68,** 206.

Davis, D., Andrew, K. L., and Verges, J. (1978), J. Opt. Soc. Am. **68,** 235.

Davis, D. (1976), Dissertation, Purdue University.

Dembczynski, J. (1980), Physica **100C,** 105.

Dembczynski, J., and Rudnicka-Szuba, H. (1982), Physica **115C,** 101.

Humphreys, C. J., and Kostkowski, H. J. (1950), J. Opt. Soc. Am. **40,** 801(A).

Meggers, W. F., and Russell, H. N. (1936), J. Res. Natl. Bur. Stand. **17,** 125.

Moore, C. E. (1939), Phys. Rev. **55,** 710.

Roth, C. (1970), J. Res. Natl. Bur. Stand. **74A,** 141.

V I

Configuration	Term	J	Level (cm^{-1})	g	Leading percentages		
$3d^3\ 4s^2$	$a\ ^4$F	3/2	0.00	0.39848	98	1	$3d^4(^3F2)4s\ ^4F$
		5/2	137.38	1.02839	98	1	
		7/2	323.46	1.23821	98	1	
		9/2	552.96	1.33360	98	1	
$3d^4(^5D)4s$	$a\ ^6$D	1/2	2 112.28	3.33847	100		
		3/2	2 153.21	1.86829	100		
		5/2	2 220.11	1.65831	100		
		7/2	2 311.36	1.58833	100		
		9/2	2 424.78	1.55647	100		
$3d^4(^5D)4s$	$a\ ^4$D	1/2	8 413.00	0.00	100		
		3/2	8 476.23	1.19	100		
		5/2	8 578.53	1.35	100		
		7/2	8 715.76	1.39	100		

(Continued)

V I—Continued

Configuration	Term	*J*	Level (cm^{-1})	*g*	Leading percentages		
$3d^3\,4s^2$	$a\,^4$P	1/2	9 544.63	2.59	81	13	$3d^4(^3$P2$)4s\,^4$P
		3/2	9 637.03	1.70	81	13	
		5/2	9 824.61	1.55	81	13	
$3d^3\,4s^2$	$a\,^2$G	7/2	10 892.51	0.88	96	2	$3d^4(^3$G$)4s\,^2$G
		9/2	11 100.59	1.13	96	2	
$3d^3\,4s^2$	$a\,^2$P	3/2	13 801.54	1.20	59	40	$3d^34s^2\,^2$D2
		1/2	13 810.94	0.64	85	8	$3d^4(^3$P2$)4s\,^2$P
$3d^3\,4s^2$	$a\,^2$D2	3/2	14 514.76	0.97	49	25	^{2}P
		5/2	14 548.81	1.17	71	22	^{2}D1
$3d^4(^3$H$)4s$	$a\,^4$H	7/2	14 909.97	0.65	100		
		9/2	14 949.37	0.94	100		
		11/2	15 000.94	1.10	100		
		13/2	15 062.96	1.18	100		
$3d^4(^3$P2$)4s$	$b\,^4$P	1/2	15 078.39	2.60	46	37	$(^3$P1$)\,^4$P
		3/2	15 270.58	1.68	46	37	
		5/2	15 572.03	1.54	46	37	
$3d^3\,4s^2$	$a\,^2$H	9/2	15 103.78	0.90	90	7	$3d^4(^3$H$)4s\,^2$H
		11/2	15 264.82	1.07	90	7	
$3d^4(^3$F2$)4s$	$b\,^4$F	3/2	15 664.81	0.39	76	24	$(^3$F1$)\,^4$F
		5/2	15 688.87	1.05	76	24	
		7/2	15 724.23	1.22	76	24	
		9/2	15 770.78	1.31	76	24	
$3d^3(^4$F$)4s4p(^3$P°$)$	$z\,^6$G°	3/2	*16 361.50*	0.00	100		
		5/2	*16 449.90*	0.78	100		
		7/2	*16 572.63*	1.10	100		
		9/2	*16 728.81*	1.22	100		
		11/2	*16 917.25*	1.26	100		
		13/2	*17 136.56*	1.43	100		
$3d^4(^3$G$)4s$	$a\,^4$G	5/2	17 054.96	0.59	98		
		7/2	17 116.97	0.96	98		
		9/2	17 182.08	1.14	98		
		11/2	17 242.08	1.27	98		
$3d^3(^4$F$)4s4p(^3$P°$)$	$z\,^6$D°	1/2	*18 085.95*	3.20	90	8	$3d^4($ 5d$)4p\,^6$D°
		3/2	*18 126.23*	1.76	89	8	
		5/2	*18 198.08*	1.58	89	8	
		7/2	*18 302.26*	1.56	89	8	
		9/2	*18 438.02*	1.55	89	8	
$3d^3(^4$F$)4s4p(^3$P°$)$	$z\,^6$F°	1/2	*18 120.10*	-0.44	95		
		3/2	*18 174.06*	1.14	95		
		5/2	*18 258.89*	1.28	95		
		7/2	*18 372.39*	1.28	95		
		9/2	*18 513.37*	1.38	95		
		11/2	*18 680.03*	1.42	96		
$3d^4(^3$P2$)4s$	$b\,^2$P	1/2	18 805.06	0.67	50	32	$(^3$P1$)\,^2$P
		3/2	19 189.33	1.37	50	32	
$3d^4(^3$H$)4s$	$b\,^2$H	9/2	19 023.52	0.91	92	7	$3d^34s^2\,^2$H
		11/2	19 145.13	1.08	92	7	

V I—Continued

Configuration	Term	J	Level (cm^{-1})	g	Leading percentages		
$3d^4(^3\mathrm{F}2)4s$	$a\ ^2\mathrm{F}$	5/2	19 026.33	0.86	72	22	$(^3\mathrm{F}1)\ ^2\mathrm{F}$
		7/2	19 078.11	1.14	72	22	
$3d^5$	$a\ ^6\mathrm{S}$	5/2	20 202.47		100		
$3d^3(^4\mathrm{F})4s4p(^3\mathrm{P}^\circ)$	$z\ ^4\mathrm{D}^\circ$	1/2	*20 606.50*	−0.04	90		
		3/2	*20 687.76*	1.21	90		
		5/2	*20 828.48*	1.35	90		
		7/2	*21 032.51*	1.45	90		
$3d^4(^3\mathrm{D})4s$	$b\ ^4\mathrm{D}$	7/2	20 767.62	1.45	99	1	$3d^5\ ^4\mathrm{D}$
		5/2	20 789.10	1.25	99	1	
		3/2	20 813.10	1.20	99	1	
		1/2	20 830.34	0.10	99	1	
$3d^4(^3\mathrm{G})4s$	$^2\mathrm{G}$	7/2	21 101.61		86	8	$(^1\mathrm{G})\ ^2\mathrm{G}$
		9/2	21 275.65				
$3d^4(^1\mathrm{G}2)4s$	$b\ ^2\mathrm{G}$	9/2	21 603.07	1.11	59	32	$3d^4(^1\mathrm{G}1)4s\ 2$
		7/2	21 646.42	0.86	59	32	
$3d^3(^4\mathrm{F})4s4p(^3\mathrm{P}^\circ)$	$z\ ^4\mathrm{G}^\circ$	5/2	*21 841.42*	0.55	93		
		7/2	*21 963.45*	0.96	93		
		9/2	*22 121.07*	1.16	93		
		11/2	*22 313.80*	1.24	93		
$3d^3(^4\mathrm{F})4s4p(^3\mathrm{P}^\circ)$	$z\ ^4\mathrm{F}^\circ$	3/2	*23 088.06*	0.39?	94		
		5/2	*23 210.54*	0.98?	94		
		7/2	*23 353.10*	1.23	95		
		9/2	*23 519.87*	1.31	95		
$3d^3(^4\mathrm{F})4s4p(^3\mathrm{P}^\circ)$	$z\ ^2\mathrm{D}^\circ$	3/2	*23 608.77*	0.76	89		
		5/2	*23 935.12*	1.32?	89		
$3d^4(^5\mathrm{D})4p$	$z\ ^6\mathrm{P}^\circ$	3/2	*24 648.11*	2.34	71	29	$3d^3(^4\mathrm{P})4s4p(^3\mathrm{P}^\circ)\ ^6\mathrm{P}^\circ$
		5/2	*24 727.86*	1.86	71	29	
		7/2	*24 838.55*	1.67	71	29	
$3d^4(^5\mathrm{D})4p$	$z\ ^4\mathrm{P}^\circ$	1/2	*24 770.68*	2.54	89		
		3/2	*24 915.15*	1.71	89		
		5/2	*25 131.00*	1.59	89		
$3d^4(^5\mathrm{D})4p$	$y\ ^6\mathrm{F}^\circ$	1/2	*24 789.38*	−0.58	95		
		3/2	*24 830.23*	1.02	95		
		5/2	*24 898.77*	1.23	95		
		7/2	*24 992.88*	1.37	96		
		9/2	*25 111.47*	1.41	96		
		11/2	*25 253.43*	1.41	96		
$3d^4(^5\mathrm{D})4p$	$y\ ^4\mathrm{F}^\circ$	3/2	*25 930.55*	0.42	76	17	$3d^3(^4\mathrm{F})4s4p(^1\mathrm{P}^\circ)\ ^4\mathrm{F}^\circ$
		5/2	*26 004.23*	0.98	73	16	
		7/2	*26 122.08*	1.15	67	15	
		9/2	*26 171.92*	1.23	64	14	
$3d^3(^4\mathrm{F})4s4p(^3\mathrm{P}^\circ)$	$z\ ^2\mathrm{G}^\circ$	7/2	*26 021.92*	0.92	93		
		9/2	*26 344.90*	1.13	90		

(Continued)

V I—Continued

Configuration	Term	J	Level (cm^{-1})	g	Leading percentages		
$3d^4(^5D)4p$	$y\ ^4D°$	1/2	26 182.63	-0.06	65	30	$3d^3(^4F)4s4p(^1P°)\ ^4D°$
		3/2	26 249.48	1.17	65	29	
		5/2	26 352.65	1.34	65	28	
		7/2	26 480.29	1.39	63	27	
$3d^4(^5D)4p$	$y\ ^6D°$	1/2	26 397.54	3.25	82	10	$3d^3(^4F)4s4p(^3P°)\ ^6D°$
		3/2	26 437.64	1.86	82	10	$3d^3(^4F)4s4p(^3P°)\ ^6D°$
		5/2	26 505.93	1.59	81	10	$3d^3(^4F)4s4p(^3P°)\ ^6D°$
		7/2	26 604.80	1.58	78	10	$3d^3(^4F)4s4p(^3P°)\ ^6D°$
		9/2	26 738.32	1.50	72	10	$3d^4(^5D)4p\ ^4F°$
$3d^3(^4F)4s4p(^3P°)$	$z\ ^2F°$	5/2	27 187.76	1.07?	97		
		7/2	27 470.79	1.01	97		
$3d^3(^4P)4s4p(^3P°)$	$x\ ^6D°$	1/2	28 313.65	3.23	91		
		3/2	28 368.75	1.82	91		
		5/2	28 462.17	1.58	91		
		7/2	28 595.62	1.52	91		
		9/2	28 768.09	1.47	91		
	$z\ ^4S°$	3/2	28 621.27				
$3d^3(^4P)4s4p(^3P°)$	$y\ ^6P°$	3/2	29 202.75	2.32	71	29	$3d^4(^5D)4p\ ^6P°$
		5/2	29 296.38	1.76	71	29	
		7/2	29 418.07	1.62	71	29	
$3d^3(^4P)4s4p(^3P°)$	$y\ ^4P°$	1/2	30 021.58	2.67	80	9	$(a\ ^2D)(^3P°)\ ^4P°$
		3/2	30 094.57	1.74	82	8	
		5/2	30 120.81	1.67	83	7	
$3d^3(^2G)4s4p(^3P°)$	$y\ ^4G°$	5/2	30 635.59	0.53	65	30	$(^4F)(^1P°)\ ^4G°$
		7/2	30 694.35	0.93	68	27	
		9/2	30 771.73	1.13	70	25	
		11/2	30 864.27	1.21	73	22	
$3d^3(^4P)4s4p(^3P°)$	$z\ ^6S°$	5/2	30 832.62		96		
$3d^3(^2G)4s4p(^3P°)$	$x\ ^4F°$	3/2	31 200.15	0.38	79	7	$(a\ ^2D)(^3P°)\ ^4F°$
		5/2	31 229.03	1.01	80	7	
		7/2	31 268.11	1.21	80	6	
		9/2	31 317.44	1.32	81	6	
$3d^3(^4F)4s4p(^1P°)$	$x\ ^4G°$	5/2	31 397.83	0.53	62	23	$(^2G)(^3P°)\ ^4G°$
		7/2	31 541.15	0.95	64	20	
		9/2	31 721.71	1.12	66	18	
		11/2	31 937.27	1.20	69	15	
$3d^3(^2P)4s4p(^3P°)$	$^4P°$	1/2	31 786.18	2.30	63	10	$(a\ ^2D)(^3P°)\ ^4P°$
$3d^4(a\ ^3P)4p$	$y\ ^2S°$	1/2	31 962.18	2.21	33	34	$3d^3(^2P)4s4p(^3P°)\ ^2S°$
$3d^3(^2P)4s4p(^3P°)$	$x\ ^4D°$	1/2	32 348.96	0.08	56	16	$(^4P)(^3P°)\ ^4D°$
		3/2	32 456.58	1.17	58	16	
		5/2	32 660.40	1.29	59	14	
		7/2	32 891.01	1.35	60	12	

V I—Continued

Configuration	Term	J	Level (cm^{-1})	g	Leading percentages		
$3d^3(^2\mathrm{H})4s4p(^3\mathrm{P}^\circ)$	z $^4\mathrm{H}^\circ$	7/2	*32 692.06*	0.68	62	22	$3d^4(^3\mathrm{H})4p\ ^4\mathrm{H}^\circ$
		9/2	*32 788.18*	0.98	64	23	
		11/2	*32 897.91*	1.11	68	25	
		13/2	*32 963.93*	1.21	69	25	
$3d^3(^4\mathrm{P})4s4p(^3\mathrm{P}^\circ)$	z $^2\mathrm{P}^\circ$	1/2	*32 724.87*	0.73?	53	15	$3d^4(a\ ^3\mathrm{P})4p\ ^2\mathrm{P}^\circ$
		3/2	*32 767.95*	1.22	64	18	
$3d^3(^4\mathrm{F})4s4p(^1\mathrm{P}^\circ)$	w $^4\mathrm{F}^\circ$	3/2	*32 738.13*	0.52	61	20	$3d^4(^5\mathrm{D})4p\ ^4\mathrm{F}^\circ$
		5/2	*32 846.82*	1.01	61	19	
		7/2	*32 988.84*	1.18	61	19	
		9/2	*33 155.30*	1.30	61	19	
$3d^3(^2\mathrm{G})4s4p(^3\mathrm{P}^\circ)$	y $^2\mathrm{G}^\circ$	9/2	*33 306.89*	1.03	84	5	$(^2\mathrm{H})(^3\mathrm{P}^\circ)\ ^4\mathrm{H}^\circ$
		7/2	*33 360.28*	0.91	81	6	
$3d^3(^2\mathrm{G})4s4p(^3\mathrm{P}^\circ)$	y $^2\mathrm{F}^\circ$	7/2	*33 481.44*	1.11	76	8	$(a\ ^2\mathrm{D})(^3\mathrm{P}^\circ)\ ^2\mathrm{F}^\circ$
		5/2	*33 527.68*	0.85	79	8	
$3d^3(^2\mathrm{G})4s4p(^3\mathrm{P}^\circ)$	z $^2\mathrm{H}^\circ$	9/2	*33 640.28*	0.92	85	8	$(^2\mathrm{H})(^3\mathrm{P}^\circ)\ ^2\mathrm{H}^\circ$
		11/2	*33 695.32*	1.09	87	8	
$3d^3(^4\mathrm{P})4s4p(^3\mathrm{P}^\circ)$	w $^4\mathrm{D}^\circ$	1/2	*33 966.83*	0.09	68	14	$(^2\mathrm{P})(^3\mathrm{P}^\circ)\ ^4\mathrm{D}^\circ$
		3/2	*33 976.07*	0.80	57	12	
		5/2	*34 065.72*	1.30	63	13	
		7/2	*34 127.92*	1.35	68	12	
$3d^3(^2\mathrm{P})4s4p(^3\mathrm{P}^\circ)$	$^4\mathrm{S}^\circ$	3/2	*34 019.16*		70	6	$(^4\mathrm{P})(^3\mathrm{P}^\circ)\ ^4\mathrm{S}^\circ$
$3d^3(^2\mathrm{P})4s4p(^3\mathrm{P}^\circ)$	$^2\mathrm{D}^\circ$	3/2	*34 030.06*	0.86	26	16	$(a\ ^2\mathrm{D})(^3\mathrm{P}^\circ)\ ^2\mathrm{D}^\circ$
		5/2	*34 167.89*	1.32?	38	24	
$3d^5$	$^4\mathrm{G}$	5/2	34 129.20				
$3d^5$	c $^4\mathrm{D}$	1/2	34 329.05		96	2	$3d^5\ ^4\mathrm{P}$
		3/2	34 343.22		88	6	$3d^5\ ^4\mathrm{P}$
		5/2	34 359.26		81	12	$3d^5\ ^4\mathrm{P}$
		7/2	34 366.84		99	1	$3d^4(^3\mathrm{D})4s\ ^4\mathrm{D}$
$3d^3(a\ ^2\mathrm{D})4s4p(^3\mathrm{P}^\circ)$	$^4\mathrm{F}^\circ$	7/2	*34 374.88*	1.21	81	9	$3d^4(^3\mathrm{G})4p\ ^4\mathrm{F}^\circ$
		3/2	*34 428.80*	0.73	62	8	$3d^3(^2\mathrm{P})4s4p(^3\mathrm{P}^\circ)\ ^2\mathrm{D}^\circ$
		5/2	*34 486.75*	1.18	66	7	$3d^3(^2\mathrm{P})4s4p(^3\mathrm{P}^\circ)\ ^2\mathrm{D}^\circ$
		9/2	*34 529.84*	1.41	82	9	$3d^4(^3\mathrm{G})4p\ ^4\mathrm{F}^\circ$
$3d^3(^4\mathrm{F})4s4p(^1\mathrm{P}^\circ)$	v $^4\mathrm{D}^\circ$	1/2	*34 477.40*	0.00	24	18	$3d^4(a\ ^3\mathrm{P})4p\ ^4\mathrm{D}^\circ$
		3/2	*34 537.29*	1.05	25	16	
		5/2	*34 619.60*	1.28	27	17	
		7/2	*34 747.12*	1.35	28	15	
$3d^3(a\ ^2\mathrm{D})4s4p(^3\mathrm{P}^\circ)$	u $^4\mathrm{D}^\circ$	1/2	*35 013.26*		58	16	$(^4\mathrm{F})(^1\mathrm{P}^\circ)\ ^4\mathrm{D}^\circ$
		3/2	*35 092.52*	1.12	58	15	
		5/2	*35 225.01*	1.32	63	16	
		7/2	*35 379.30*	1.33	67	16	
$3d^3(^4\mathrm{P})4s4p(^3\mathrm{P}^\circ)$	y $^4\mathrm{S}^\circ$	3/2	*36 408.40*	1.85	41	25	$(a\ ^2\mathrm{D})(^3\mathrm{P}^\circ)\ ^4\mathrm{P}^\circ$
$3d^3(^4\mathrm{P})4s4p(^3\mathrm{P}^\circ)$	x $^2\mathrm{D}^\circ$	3/2	*36 416.40*	0.89	38	17	$3d^4(a\ ^3\mathrm{P})4p\ ^2\mathrm{D}^\circ$
		5/2	*36 700.81*	1.13	36	14	

(Continued)

V I—Continued

Configuration	Term	J	Level (cm^{-1})	g	Leading percentages			
$3d^4(a\,^3F)4p$		7/2	*36 461.33*	0.85	21	$^2G^\circ$	13	$(^3G)\ ^2G^\circ$
$3d^4(a\,^3P)4p$		1/2	*36 477.74*	0.74	32	$^2P^\circ$	31	$3d^3(a\,^2D)4s4p(^3P^\circ)\ ^2P^\circ$
$3d^4(a\,^3F)4p$	$^2G^\circ$	9/2	*36 538.56*	1.05	31		19	$(^3G)\ ^2G^\circ$
$3d^4(a\,^3P)4p$	$^2P^\circ$	3/2	*36 580.45*	1.17	31		29	$3d^3(a\,^2D)4s4p(^3P^\circ)\ ^2P^\circ$
$3d^3(a\,^2D)4s4p(^3P^\circ)$	$x\ ^4P^\circ$	5/2	*36 611.81*	1.54	59		15	$(^2P)(^3P^\circ)\ ^4P^\circ$
		1/2	*36 695.66*	2.51	69		18	$(^2P)(^3P^\circ)\ ^4P^\circ$
		3/2	*36 814.78*	1.77	44		24	$(^4P)(^3P^\circ)\ ^4S^\circ$
	$w\ ^2G^\circ$	7/2	*36 628.91*	0.65?				
		9/2	*36 828.33*					
$3d^3(^2H)4s4p(^3P^\circ)$	$w\ ^4G^\circ$	5/2	*36 763.48*		42		19	$3d^4(a\,^3F)4p\ ^4G^\circ$
		7/2	*36 822.82*	1.06	27		11	
		9/2	*36 897.97*	1.17	46		14	
		11/2	*36 938.42*	1.26	54		15	
$3d^4(a\,^3F)4p$	$x\ ^2F^\circ$	5/2	*36 766.05*	0.89	26		25	$3d^3(^2G)4s4p(^3P^\circ)\ ^2F^\circ$
		7/2	*36 925.89*	1.05	24		22	
$3d^5$	$e\ ^4F$	3/2	36 983.63		98		1	$3d^34s^2\ ^4F$
		5/2	36 989.20		98		1	
		7/2	37 025.60		98		1	
		9/2	37 075.57		98		1	
$3d^4(^5D)5s$	$e\ ^6D$	1/2	37 116.68	3.08				
		3/2	37 158.60	1.87				
		5/2	37 227.46	1.61				
		7/2	37 322.15	1.64				
		9/2	37 440.74	1.48				
$3d^4(^3H)4p$	$v\ ^2G^\circ$	7/2	*37 174.69*	0.99	60		11	$3d^3(^2H)4s4p(^3P^\circ)\ ^2G^\circ$
		9/2	*37 361.95*	1.05	52		20	$3d^4(^3H)4p\ ^4I^\circ$
$3d^3(^2H)4s4p(^3P^\circ)$	$y\ ^2H^\circ$	9/2	*37 180.96*	0.73	67		9	$3d^4(^3H)4p\ ^2H^\circ$
		11/2	*37 210.82*	1.08	74		10	
$3d^4(^3H)4p$	$z\ ^4I^\circ$	9/2	*37 285.07*	0.87	76		14	$(^3H)\ ^2G^\circ$
		11/2	*37 315.93*	0.96	98			
		13/2	*37 404.20*	1.08	99			
		15/2	*37 518.44*	1.15	100			
$3d^3(a\,^2D)4s4p(^3P^\circ)$	$w\ ^2F^\circ$	5/2	*37 342.56*	0.84	34		16	$3d^4(a\,^3F)4p\ ^2F^\circ$
		7/2	*37 475.12*	1.12	37		20	
$3d^3\,4s(^5F)5s$	$e\ ^6F$	1/2	37 375.18	-0.72				
		3/2	37 423.25	1.05				
		5/2	37 503.28	1.30				
		7/2	37 615.03	1.33				
		9/2	37 758.16	1.43				
		11/2	37 931.49	1.52				

V I—Continued

Configuration	Term	J	Level (cm^{-1})	g	Leading percentages		
$3d^4(a\ ^3F)4p$	$w\ ^2D^\circ$	3/2	*37 457.58*	0.80	30	27	$3d^3(^4P)4s4p(^3P^\circ)\ ^2D^\circ$
		5/2	*37 752.55*	1.18	35	26	
	$y\ ^4H^\circ$	7/2	*37 481.51*	0.76			
		9/2	*37 516.97*	1.05			
		11/2	*37 565.88*	1.09			
		13/2	*37 626.10*	1.24			
$3d^4(a\ ^3F)4p$	$v\ ^4G^\circ$	5/2	*37 498.82*	0.60	62	30	$3d^3(\ 2h)4s4p(^3P^\circ)\ ^4G^\circ$
		7/2	*37 556.04*	1.02	57	26	
		9/2	*37 644.46*	1.15	57	24	
		11/2	*37 764.94*	1.22	58	23	
$3d^4(^3H)4p$	$z\ ^2I^\circ$	11/2	*37 530.29*	0.94	45	44	$3d^3(^2H)4s4p(^3P^\circ)\ ^2I^\circ$
		13/2	*37 606.37*	1.06	42	46	
$3d^4(a\ ^3P)4p$	$t\ ^4D^\circ$	1/2	*37 757.30*	0.01	46	14	$3d^3(^4F)4s4p(^1P^\circ)\ ^4D^\circ$
		3/2	*37 835.08*	1.18	46	14	
		5/2	*37 959.70*	1.33	45	13	
		7/2	*38 115.69*	1.35	44	15	
$3d^4(^5D)5s$	$e\ ^4D$	1/2	37 940.21				
		3/2	38 004.03				
		5/2	38 106.38				
		7/2	38 242.53				
$3d^4(^3H)4p$	$x\ ^2H^\circ$	9/2	*38 123.79*	0.88	55	17	$3d^3(^2H)4s4p(^3P^\circ)\ ^2H^\circ$
		11/2	*38 220.65*	1.10	54	18	
$3d^4(^3H)4p$	$x\ ^4H^\circ$	7/2	*38 245.83*	0.67	46	27	$(^3G)\ ^4H^\circ$
		9/2	*38 323.93*	0.93	43	24	
		11/2	*38 405.06*	1.11	45	23	
		13/2	*38 483.05*	1.22	52	27	
$3d^3(^2H)4s4p(^3P^\circ)$	$u\ ^2G^\circ$	9/2	*38 529.79*	0.99	78	9	$(^2F)(^3P^\circ)\ ^2G^\circ$
		7/2	*38 610.90*	0.88?	78	9	
$3d^3(^2H)4s4p(^3P^\circ)$	$y\ ^2I^\circ$	11/2	*39 008.66*	0.92	53	46	$3d^4(^3H)4p\ ^2I^\circ$
		13/2	*39 081.26*	1.06	50	49	
$3d^3\ 4s(^5F)5s$	$f\ ^4F$	3/2	39 127.19	0.46?			
		5/2	39 241.39	1.03			
		7/2	39 398.91	1.22?			
		9/2	39 596.99	1.33?			
$3d^4(a\ ^3P)4p$	$w\ ^4P^\circ$	1/2	*39 237.20*	2.57	72	16	$3d^3(^2P)4s4p(^3P^\circ)\ ^4P^\circ$
		3/2	*39 248.92*	1.60	43	16	$3d^4(a\ ^3P)4p\ ^4S^\circ$
		5/2	*39 422.76*	1.52	52	11	$3d^3(^2P)4s4p(^3P^\circ)\ ^4P^\circ$
$3d^4(a\ ^3F)4p$	$u\ ^4F^\circ$	3/2	*39 266.68*	0.54	47	13	$3d^4(a\ ^3P)4p\ ^4S^\circ$
		5/2	*39 300.56*	1.00	67	9	$3d^3(^2F)4s4p(^3P^\circ)\ ^4F^\circ$
		7/2	*39 341.79*	1.21	66	9	$3d^3(^2F)4s4p(^3P^\circ)\ ^4F^\circ$
		9/2	*39 391.08*	1.30	66	9	$3d^3(^2F)4s4p(^3P^\circ)\ ^4F^\circ$
$3d^4(a\ ^3P)4p$	$x\ ^4S^\circ$	3/2	*39 847.34*	2.00	40	14	$(a\ ^3P)\ ^4P^\circ$
$3d^4(a\ ^3F)4p$	$s\ ^4D^\circ$	1/2	*39 877.80*	0.01	63	18	$(a\ ^3P)\ ^4D^\circ$
		3/2	*39 935.18*	1.10	59	18	
		5/2	*39 999.95*	1.33	46	16	
		7/2	*40 125.88*	1.38	44	18	

(Continued)

V I—Continued

Configuration	Term	J	Level (cm^{-1})	g	Leading percentages			
$3d^4(a\,^3P)4p$	$v\,^2D°$	3/2	*39 884.54*	0.92	40		17	$(a\,^3F)\,^2D°$
		5/2	*40 119.22*	1.14	40		16	
$3d^4(^3H)4p$	$u\,^4G°$	5/2	*39 962.18*	0.53	32		21	$(^3G)\,^4G°$
		7/2	*40 001.21*	0.99	38		25	
		9/2	*40 038.99*	1.19	42		27	
		11/2	*40 063.89*	1.23	42		28	
$3d^3(a\,^2D)4s4p(^3P°)$	$^2D°$	5/2	*40 153.51*		43		27	$(^2P)(^3P°)\,^2D°$
		3/2	*40 225.44*	0.70	37		21	
	$x\,^2S°$	1/2	*40 299.87*					
$3d^4(^3G)4p$	$w\,^4H°$	7/2	*40 314.90*	0.65	59		22	$(^3H)\,^4H°$
		9/2	*40 378.82*	0.92	66		24	
		11/2	*40 452.47*	1.08	67		23	
		13/2	*40 535.70*	1.22	70		23	
$3d^4(^3G)4p$	$^2F°$	5/2	*40 325.73*	1.12	40		12	$(a\,^3F)\,^4D°$
		7/2	*40 587.35*	1.01	44		14	
$3d^3(^2P)4s4p(^3P°)$	$x\,^2P°$	1/2	*40 328.61*		51		21	$(a\,^2D)(^3P°)\,^2P°$
		3/2	*40 437.36*	1.52	35		14	
	$w\,^2P°$	3/2	*40 693.83*					
$3d^4(^3G)4p$	$w\,^2H°$	11/2	*40 919.75*	0.96	71		10	$3d^3(^2G)4s4p(^3P°)\,^2H°$
		9/2	*40 980.67*	0.99	66		11	
$3d^4(^3G)4p$	$t\,^4F°$	3/2	*41 389.68*	0.42	47		20	$(^3D)\,^4F°$
$3d^4(^3G)4p$		5/2	*41 428.98*	0.89	27	$^4G°$	25	$(^3G)\,^4F°$
$3d^4(a\,^3F)4p$	$t\,^2G°$	7/2	*41 436.62*	0.90	44		30	$(^3G)\,^2G°$
		9/2	*41 539.18*	1.04	42		30	
$3d^4(^3G)4p$		7/2	*41 492.40*	1.15	26	$^4G°$	26	$(^3G)\,^4F°$
$3d^3(^2G)4s4p(^1P°)$	$v\,^2H°$	9/2	*41 501.45*	0.87	43		28	$3d^4(^1I)4p\,^2H°$
		11/2	*41 659.69*	1.05	42		30	
$3d^4(^3G)4p$		9/2	*41 599.44*	1.23	27	$^4F°$	23	$(^3G)\,^4G°$
$3d^4(^3G)4p$		5/2	*41 654.79*	0.58	27	$^4G°$	22	$(^3G)\,^4F°$
$3d^3(^4P)4s4p(^1P°)$	$v\,^4P°$	1/2	*41 751.93*	2.56	43		39	$3d^4(^3D)4p\,^4P°$
		3/2	*41 848.63*	1.62	39		38	
		5/2	*42 009.92*	1.48	38		40	
$3d^4(^3G)4p$		7/2	*41 758.37*	1.03	26	$^4G°$	26	$(^3G)\,^4F°$
$3d^4(^3G)4p$		9/2	*41 860.66*	1.20	29	$^4G°$	18	$(^3G)\,^4F°$
$3d^4(^3G)4p$	$t\,^4G°$	11/2	*41 918.26*	1.20	52		30	$(^3H)\,^4G°$
$3d^3(^4P)4s4p(^1P°)$	$r\,^4D°$	1/2	*41 928.48*	0.04	53		17	$3d^2(^3F)4s^24p\,^4D°$
		3/2	*41 999.26*	1.20	49		16	
		5/2	*42 138.06*	1.33	51		16	
		7/2	*42 245.45*	1.36	58		18	

V I—Continued

Configuration	Term	J	Level (cm^{-1})	g	Leading percentages		
$3d^4(^3D)4p$	$u\ ^2F^\circ$	5/2	*41 950.41*	0.84	41	19	$(a\ ^3F)\ ^2F^\circ$
		7/2	*42 020.91*	1.11	43	13	
$3d^4(^5D)4d$	$e\ ^6G$	3/2	42 033.84				
		5/2	42 070.05				
		7/2	42 114.34	1.08			
		9/2	42 177.66	1.23			
		11/2	42 257.39	1.32			
		13/2	42 353.37	1.35			
	$u\ ^2H^\circ$	9/2	*42 079.16*	0.85			
		11/2	*42 220.69*	1.06			
$3d^4(^5D)4d$	$e\ ^6P$	7/2	42 164.67	1.44?			
	2°	7/2	*42 236.61*				
$3d^4(^3D)4p$	$v\ ^2P^\circ$	3/2	*42 318.52*	1.34	35	30	$3d^3(^2P)4s4p(^3P^\circ)\ ^2P^\circ$
		1/2	*42 480.63*	1.14	36	28	
$3d^3(^4P)4s4p(^3P^\circ)$	$w\ ^2S^\circ$	1/2	*42 362.10*	1.50?	83	10	$3d^4(a\ ^3P)4p\ ^2S^\circ$
$3d^4(^5D)4d$	$f\ ^6F$	7/2	42 363.83				
		9/2	42 506.41				
		11/2	42 578.08	1.39			
$3d^4(^5D)4d$	$f\ ^6D$	7/2	42 404.36				
		9/2	42 553.56	1.61			
	$w\ ^6D^\circ$	5/2	*42 480.31*				
		7/2	*42 587.41*				
		9/2	*42 725.33*				
$3d^3(^4D)4s4p(^1P^\circ)$	$w\ ^4S^\circ$	3/2	*42 969.24*	1.94	70	14	$3d^4(a\ ^3P)4p\ ^4S^\circ$
	$s\ ^4F^\circ$	3/2	*42 981.69*				
		5/2	*43 051.55*				
		7/2	*43 147.28*				
		9/2	*43 266.35*				
	$q\ ^4D^\circ$	1/2	*43 249.50*				
		3/2	*43 309.07*				
		5/2	*43 410.99*				
		7/2	*43 555.32*	1.46			
	$u\ ^4P^\circ$	1/2	*43 443.53*				
		3/2	*43 504.25*				
		5/2	*43 585.80*				
$3d^3\ 4s(^5F)4d$	$e\ ^6H$	5/2	43 649.31	0.38			
		7/2	43 706.88	0.88			
		9/2	43 787.56	1.11			
		11/2	43 894.16	1.18			
		13/2	44 028.33	1.30			
		15/2	44 189.96	1.38			

(Continued)

V I—Continued

Configuration	Term	J	Level (cm^{-1})	g	Leading percentages		
	$x\ ^6F°$	1/2	43 708.39				
		3/2	43 845.80				
		5/2	43 959.62				
		7/2	44 026.51				
		9/2, 11/2	44 202.72				
$3d^3\ 4s(^5F)4d$	$f\ ^6G$	3/2	43 817.90	0.38?			
		5/2	43 847.09	0.78			
		7/2	43 911.88	1.12			
		9/2	44 005.07	1.26			
		11/2	44 139.57	1.34			
		13/2	44 327.09	1.35			
$3d^4(a\ ^1G)4p$	$t\ ^2F°$	7/2	43 873.74	1.04?	37	25	$(a\ ^3F)\ ^2F°$
		5/2	43 875.34	0.86	40	28	
$3d^3\ 4s(^3F)5s$	$e\ ^2F$	5/2	43 918.58	0.89			
		7/2	44 065.98	1.18			
	$x\ ^6P°$	7/2	43 988.31				
	$s\ ^4G°$	5/2	43 999.76				
		7/2	44 043.39	0.98			
		9/2	44 104.56	1.26			
		11/2	44 178.56	1.34			
$3d^4(a\ ^1G)4p$	$t\ ^2H°$	9/2	44 145.72	0.90	58	10	$(^3H)\ ^2H°$
		11/2	44 184.13	1.06?	40	38	$(^1I)\ ^2I°$
$3d^3\ 4s(^5F)4d$	$f\ ^6P$	3/2	44 443.66				
		5/2	44 532.60				
		7/2	44 690.47				
$3d^4(^3F)4p$	$s\ ^2G°$	9/2	44 463.27	1.09	21	19	$3d^4(^3G)4p\ ^2G°$
		7/2	44 495.50	0.91	25	24	
$3d^4(^3D)4p$	$p\ ^4D°$	1/2	44 514.55		77	16	$3d^3(a\ ^2D)4s4p(^3P°)\ ^4D°$
		3/2	44 554.58	1.22	45	28	$3d^4(^3D)4p\ ^2D°$
		5/2	44 616.92	1.37?	70	12	$3d^3(a\ ^2D)4s4p(^3P°)\ ^4D°$
		7/2	44 701.14	1.32?	75	13	$3d^3(a\ ^2D)4s4p(^3P°)\ ^4D°$
$3d^3\ 4s(^5F)4d$	$g\ ^6D$	3/2	44 844.84				
		5/2	44 920.85	1.55?			
		7/2	45 056.65				
		9/2	45 157.60				
	$r\ ^4F°$	3/2	44 973.48	0.58?			
		5/2	45 049.17				
		7/2	45 058.54	0.97			
		9/2	45 145.11	1.26			
	$q\ ^4F°$	3/2	45 066.49	0.59			
		5/2	45 107.24	0.93			
		7/2	45 157.77	1.05			
		9/2	45 237.19	1.22			
	$u\ ^2P°$	3/2	45 159.22	1.66?			

V I—Continued

Configuration	Term	J	Level (cm^{-1})	g	Leading percentages			
$3d^4(a\,^1G)4p$	$r\,^2G°$	7/2	45 175.87	0.98	42		16	$3d^3(^2H)4s4p(^1P°)\,^2G°$
		9/2	45 361.41	1.14	36		17	
$3d^3\,4s(^5F)4d$	$g\,^6F$	3/2	45 638.71					
		5/2	45 700.23					
		7/2	45 743.62	1.26				
		9/2	45 813.25					
		11/2	46 034.49					
$3d^3(^2F)4s4p(^3P°)$	$p\,^4F°$	3/2	45 648.97	0.60	83		9	$3d^4(^3G)4p\,^4F°$
		5/2	45 688.32		82		9	
		7/2	45 760.07	1.02	82		9	
		9/2	45 891.59	1.32	81		9	
	$t\,^2P°$	3/2	45 654.36	1.04?				
		1/2	45 946.66					
	$o\,^4D°$	1/2	45 702.35					
		3/2	45 762.26	0.96?				
		5/2	45 838.10					
		7/2	45 937.06	1.45				
	$r\,^4G°$	5/2	46 052.72	0.56				
		7/2	46 139.06	0.96				
		9/2	46 243.74	1.15				
		11/2	46 363.46	1.19				
	6°	3/2	46 707.51					
	$t\,^4P°$	1/2	46 851.19					
		3/2	46 862.78					
		5/2	46 868.08					
$3d^3(^2F)4s4p(^3P°)$		5/2	46 996.87		28	$^4G°$	19	$3d^3(a\,^2D)4s4p(^1P°)\,^2F°$
$3d^4(a\,^1D)4p$		7/2	47 143.14	1.02	19	$^2F°$	16	$3d^3(c\,^2D)4s4p(^1P°)\,^2F°$
	3°	3/2	47 423.34					
$3d^3(^2G)4s4p(^1P°)$	$^2H°$	9/2	47 611.84	1.01?	25		19	$3d^4(^1I)4p\,^2H°$
		11/2	47 701.82	0.94	31		23	
	$q\,^4G°$	5/2	47 690.53					
		7/2	47 824.23					
		9/2	48 014.22					
		11/2	48 191.04					
$3d^3(^2F)4s4p(^3P°)$	$q\,^2G°$	7/2	47 959.87	0.89	63		13	$3d^4(^1F)4p\,^2G°$
		9/2	48 160.60	1.08	62		15	
	$v\,^2S°$	1/2	48 844.67	2.03				
	$n\,^4D°$	1/2	49 189.99					
		3/2	49 283.79					
		5/2	49 439.94					
		7/2	49 583.83					
$3d^3(^2P)4s4p(^1P°)$	$t\,^2D°$	5/2	49 688.86	1.25	39		15	$3d^4(^3D)4p\,^2D°$
		3/2	49 722.72		38		15	

(Continued)

V I—Continued

Configuration	Term	J	Level (cm^{-1})	g	Leading percentages		
$3d^3 4s(^5F)4d$	$f\,^6H$	5/2	49 717.57				
		7/2	49 796.88				
		9/2	49 875.12				
		11/2	49 983.38				
		13/2	50 164.55				
		15/2	50 301.55				
$3d^3 4s(^5F)4d$	$g\,^6G$	7/2	49 789.08				
		9/2	49 932.11				
		11/2	50 114.49				
		13/2	50 209.05				
$3d^3(^2H)4s4p(^1P^\circ)$	$x\,^2I^\circ$	11/2	*49 978.18*	0.91	86	11	$3d^4(^1I)4p\,^2I^\circ$
		13/2	*50 120.90*	1.06	87	11	
	$r\,^2F^\circ$	5/2	*50 403.85*				
		7/2	*50 539.14*				
$3d^3 4p^2$	$h\,^6G$	3/2	50 584.53				
		5/2	50 654.82				
		7/2	50 751.90				
		9/2	50 876.17				
		11/2	51 026.23				
		13/2	51 201.06				
	$p\,^2G^\circ$	7/2	*52 774.09*				
		9/2	*52 947.96*				
	$r\,^2H^\circ$	9/2	*54 081.51*				
		11/2	*54 251.45*				
V II ($^5D_0$)	***Limit***		**54 360**				
	$s\,^2P^\circ$	3/2	*57 561.36*				
		1/2	*57 744.12*				

V II

$Z = 23$

Ti I isoelectronic sequence

Ground state: $1s^2 2s^2 2p^6 3s^2 3p^6 3d^4\ ^5D_0$

Ionization energy = 118 200±200 cm^{-1} (14.66±0.02 eV)

The first regularities in this spectrum were discovered by Meggers, Kiess, and Walters (1924). Subsequent work ultimately led to an extensive analysis by Meggers and Moore (1940). That paper also contains H. D. Babcock's previously unpublished Zeeman-effect data from which the present g-values were obtained.

Velasco and Gullon (1968) found a number of new even terms and miscellaneous odd levels, including two new $3d^3 4d$ and $5s$ terms. The $3d^3 5s$ terms were used by Russell (1950) to calculate a value for the ionization energy. The analysis was continued by Iglesias (1977) who has provided 23 new terms, including most of the terms of $3d^2 4s^2$, $3d^3(^4F)4d$ and $3d^3 5s$, and gave revised values for nearly all of the levels which are adopted here. She has replaced the earlier levels $3d^2 4s^2\ e\ ^3F$ and $3d^2 4s 4p\ y\ ^5G°$. The uncertainty in the level values is estimated to be ±0.05 cm^{-1}. Iglesias, Rico, and Garcia-Riquelme (1984) reported 44 levels of the $3d^3 4f$ configuration, based on the $3d^3\ ^4F$ parent term. No level uncertainty is given; we estimate it to be ±0.05 to ±0.10 cm^{-1}. The authors give the percentage composition of the levels in $J_1 l$-coupling.

The alphabetic prefixing of terms with lower case letters for distinguishing repeating terms of the same type has been retained from Meggers and Moore except where the levels were reinterpreted by Roth (1969) on the basis of his theoretical treatment.

Roth has calculated the odd-parity configurations $3d^3 4p$ and $3d^2 4s 4p$ with configuration interaction. His percentage compositions and designation changes for the experimental levels are adopted here. Roth distinguished repeating terms of the $3d^n$ core by the letters a, b, ... rather than by seniority. The percentages include the sum of seniority states contributing to the term.

The compositions of levels of the even configurations $3d^4$, $3d^3 4s$, and $3d^2 4s^2$ were calculated by Shadmi, Oreg, and Stein (1968). Percentages are given only for cases where the mixing is large. We have applied Roth's designation of repeating core terms to this work as well.

References

Iglesias, L. (1977), Opt. Pur. Apl. **10,** 267.

Iglesias, L., Rico, F. R., and Garcia-Riquelme, 0. (1984), Opt. Pur. Apl. **17,** 93.

Meggers, W. F., Kiess, C. C., and Walters, F. M. (1924), J. Opt. Soc. Am. **9,** 355.

Meggers, W. F., and Moore, C. E. (1940), J. Res. Natl. Bur. Stand. **25,** 83.

Roth, C. (1969), J. Res. Natl. Bur. Stand. **73A,** 159.

Russell, H. N. (1950), J. Opt. Soc. Am. **40,** 618.

Shadmi, Y., Oreg, J., and Stein, J. (1968), J. Opt. Soc. Am. **58,** 909.

Velasco, R., and Gullon, N. (1968), Opt. Pur. Apl. **1,** 93.

V II

Configuration	Term	J	Level (cm^{-1})	g	Leading percentages
$3d^4$	$a\ ^5D$	0	0.00		
		1	36.05		
		2	106.63		
		3	208.89		
		4	339.21		
$3d^3(^4F)4s$	$a\ ^5F$	1	2 604.82		
		2	2 687.01	0.97	
		3	2 808.72	1.20	
		4	2 968.22	1.30?	
		5	3 162.80	1.28?	
$3d^3(^4F)4s$	$a\ ^3F$	2	8 640.21	0.65	
		3	8 841.97	1.04	
		4	9 097.81	1.22	
$3d^4$	$a\ ^3P$	0	11 295.57		
		1	11 514.76	1.48	
		2	11 908.27	1.49	

(Continued)

V II—Continued

Configuration	Term	J	Level (cm^{-1})	g	Leading percentages		
$3d^4$	$a\ ^3\mathrm{H}$	4	12 545.14	0.83?			
		5	12 621.55	1.02			
		6	12 706.25	1.27?			
$3d^4$	$b\ ^3\mathrm{F}$	2	13 490.89	0.59			
		3	13 542.67	1.06			
		4	13 608.96	1.19			
$3d^3(^4\mathrm{P})4s$	$a\ ^5\mathrm{P}$	1	13 511.75	2.39			
		2	13 594.73	1.78			
		3	13 741.61	1.62			
$3d^4$	$a\ ^3\mathrm{G}$	3	14 461.75	0.74	57	42	$3d^3(^2\mathrm{G})4s\ ^3\mathrm{G}$
		4	14 556.09	1.00	56	42	
		5	14 655.63	1.17	58	42	
$3d^3(^2\mathrm{G})4s$	$b\ ^3\mathrm{G}$	3	16 340.97	0.76	58	42	$3d^4\ ^3\mathrm{G}$
		4	16 421.51	1.03	57	42	
		5	16 533.00	1.16	58	42	
$3d^4$	$a\ ^1\mathrm{G}$	4	17 910.94	0.95			
$3d^4$	$a\ ^3\mathrm{D}$	1	18 269.49	0.49	52	47	$3d^3(^2\mathrm{D})4s\ ^3\mathrm{D}$
		2	18 293.88	1.13	54	44	
		3	18 353.88	1.30	60	40	
$3d^3(^2\mathrm{G})4s$	$b\ ^1\mathrm{G}$	4	19 112.93	0.98			
$3d^3(^2\mathrm{P})4s$	$b\ ^3\mathrm{P}$	2	19 132.75	1.38			
		0	19 161.37				
		1	19 166.27	1.40			
$3d^4$	$a\ ^1\mathrm{I}$	6	19 191.40	0.96?			
$3d^4$	$a\ ^1\mathrm{S}$	0	19 902.57		59	38	$3d^3(^4\mathrm{P})4s\ ^3\mathrm{P}$
$3d^3(^4\mathrm{P})4s$	$c\ ^3\mathrm{P}$	1	20 089.60	1.35	80	7	$3d^3(^2\mathrm{P})4s\ ^3\mathrm{P}$
		0	20 156.64		51	40	$3d^4\ ^1\mathrm{S}$
		2	20 343.01	1.36	64	10	$3d^4\ ^3\mathrm{D}$
$3d^3(^2\mathrm{H})4s$	$b\ ^3\mathrm{H}$	4	20 242.37	0.82			
		5	20 280.23	1.01			
		6	20 363.23	1.14			
$3d^3(a\ ^2\mathrm{D})4s$	$b\ ^3\mathrm{D}$	1	20 522.11	0.58	51	43	$3d^4\ ^3\mathrm{D}$
		2	20 617.03	1.25	32	20	
		3	20 622.95	1.26	60	40	
$3d^3(a\ ^2\mathrm{D})4s$	$a\ ^1\mathrm{D}$	2	20 980.92	1.02	33	23	$3d^4\ ^1\mathrm{D}$
$3d^3(^2\mathrm{P})4s$	$a\ ^1\mathrm{P}$	1	22 273.60	0.97			
$3d^3(^2\mathrm{H})4s$	$a\ ^1\mathrm{H}$	5	23 391.14	1.04			
$3d^4$	$b\ ^1\mathrm{D}$	2	25 191.04	0.99	56	30	$3d^3(a\ ^2\mathrm{D})4s\ ^1\mathrm{D}$
$3d^4$	$a\ ^1\mathrm{F}$	3	26 839.77	0.97			

V II—Continued

Configuration	Term	J	Level (cm^{-1})	g	Leading percentages		
$3d^3(^2F)4s$	$c\ ^3F$	2	30 267.47	0.67	56	33	$3d^4\ ^3F$
		3	30 306.38	1.06	59	31	
		4	30 318.55	1.25	62	29	
$3d^4$	$d\ ^3F$	4	30 613.92	1.23	60	36	$3d^3(^2F)4s\ ^3F$
		3	30 641.76	1.05	57	39	
		2	30 673.08	0.67	54	41	
$3d^4$	$d\ ^3P$	2	32 040.64	1.38			
		1	32 299.27	1.48			
		0	32 420.04				
$3d^3(^2F)4s$	$b\ ^1F$	3	34 228.82	1.00			
$3d^3(^4F)4p$	$z\ ^5G°$	2	*34 592.72*	0.31	100		
		3	*34 745.72*	0.93	100		
		4	*34 946.55*	1.14	100		
		5	*35 193.13*	1.16	100		
		6	*35 483.39*		100		
$3d^4$	$c\ ^1G$	4	36 424.92	0.96			
$3d^3(^4F)4p$	$z\ ^3D°$	1	*36 489.36*	0.35	51	42	$(^4F)\ ^5F°$
		2	*37 041.12*	1.08	44	44	$(^4F)\ ^5F°$
		3	*37 204.98*	1.32	55	21	$(^4F)\ ^5D°$
$3d^3(^4F)4p$	$z\ ^5F°$	2	*36 673.48*	1.08	54	38	$(^4F)\ ^3D°$
		3	*36 919.21*	1.24	81	13	
		1	*36 954.63*	0.24	56	37	
		4	*37 150.51*		98		
		5	*37 352.45*	1.40?	98		
$3d^3(^4F)4p$	$z\ ^5D°$	0	*37 201.35*		97		
		1	*37 259.39*	1.39	94		
		2	*37 369.03*	1.39	87	10	$(^4F)\ ^3D°$
		3	*37 520.57*	1.47	73	23	
		4	*37 531.08*	1.44	97		
$3d^2\ 4s^2$	$e\ ^3F$	2	37 937.65				
		3	38 192.94				
		4	38 517.06				
$3d^3(^4F)4p$	$z\ ^3G°$	3	*39 234.05*	0.84	91	7	$(^2G)\ ^3G°$
		4	*39 403.74*	1.03	91	7	
		5	*39 612.96*	1.19	91	7	
$3d^3(^4F)4p$	$z\ ^3F°$	2	*40 001.70*	0.65	94		
		3	*40 195.52*	1.02	94		
		4	*40 430.04*	1.22	94		
$3d^3(b\ ^2D)4s$	$c\ ^3D$	3	44 098.46	1.27?			
		2	44 159.53	1.14?			
		1	44 201.85	0.50?			
$3d^2\ 4s^2$	$c\ ^1D$	2	44 657.94		57	39	$3d^4\ ^1D$

(Continued)

V II—Continued

Configuration	Term	J	Level (cm^{-1})	g	Leading percentages		
$3d^3(^4P)4p$	$y\ ^5D^\circ$	0	*46 586.37*		48	40	$(^4P)\ ^3P^\circ$
		1	*46 690.42*	1.44	55	33	$(^4P)\ ^3P^\circ$
		2	*47 101.89*	1.47	63	26	$(^4P)\ ^3P^\circ$
		3	*47 181.21*	1.48?	95		
		4	*47 420.25*	2.28	97		
$3d^3(^4P)4p$	$z\ ^3P^\circ$	2	*46 739.99*	1.48	44	34	$(^4P)\ ^5P^\circ$
		0	*47 027.95*		39	48	$(^4P)\ ^5D^\circ$
		1	*47 107.99*	1.43	48	41	$(^4P)\ ^5D^\circ$
$3d^3(^4P)4p$	$z\ ^5P^\circ$	1	*46 754.59*	2.28	94		
		2	*46 879.98*	1.68	62	20	$(^4P)\ ^5D^\circ$
		3	*47 051.86*	1.55	98		
$3d^3(^2G)4p$	$z\ ^3H^\circ$	4	*47 056.32*	0.78	88	12	$(^2H)\ ^3H^\circ$
		5	*47 297.04*	1.01	87	12	
		6	*47 607.79*	1.13	87	12	
$3d^3(b\ ^2D)4s$	$d\ ^1D$	2	47 324.31				
$3d^3(^2P)4p$	$z\ ^1S^\circ$	0	*48 258.22*		90	7	$(^4P)\ ^3P^\circ$
$3d^3(^2G)4p$	$y\ ^3G^\circ$	3	*48 579.96*	0.67	80	7	$(^4F)\ ^3G^\circ$
		4	*48 730.72*	1.02	87	7	
		5	*48 853.04*	1.22	84	7	
$3d^2\ 4s^2$	$e\ ^3P$	0	48 898.0?				
		1	48 975.72				
		2	49 204.65				
$3d^3(^2G)4p$	$y\ ^3F^\circ$	2	*49 201.64*	0.63	78	15	$(a\ ^2D)\ ^3F^\circ$
		3	*49 210.80*	0.99	46	29	$(^2G)\ ^1F^\circ$
		4	*49 268.60*	1.18	60	28	$(^2G)\ ^1G^\circ$
$3d^3(^2G)4p$	$z\ ^1F^\circ$	3	*49 568.42*	0.97	49	34	$(^2G)\ ^3G^\circ$
$3d^3(^2G)4p$	$z\ ^1H^\circ$	5	*49 593.36*	0.95	72	21	$(^2H)\ ^1H^\circ$
$3d^3(^2G)4p$	$z\ ^1G^\circ$	4	*49 723.68*	0.96	69	23	$(^2G)\ ^3F^\circ$
$3d^3(^4P)4p$	$z\ ^5S^\circ$	2	*49 731.34*		96		
$3d^3(^2P)4p$	$z\ ^1D^\circ$	2	*49 898.19*	0.93	50	36	$(a\ ^2D)\ ^1D^\circ$
$3d^3(^2P)4p$	$y\ ^3D^\circ$	1	*50 473.73*	0.49	52	21	$(^4P)\ ^3D^\circ$
		2	*50 775.47*	1.11	58	27	
		3	*51 085.71*	1.27	55	32	
$3d^3(^2P)4p$	$y\ ^3P^\circ$	0	*50 662.27*		63	36	$(a\ ^2D)\ ^3P^\circ$
		1	*50 738.75*	1.39	50	29	
		2	*51 123.21*	1.51	58	38	
$3d^4$	$e\ ^1D$	2	50 951.7		59	32	$3d^23s^2\ ^1D^\circ$
$3d^3(^2H)4p$	$y\ ^3H^\circ$	4	*52 082.82*	0.70	86	12	$(^2G)\ ^3H^\circ$
		5	*52 153.36*	0.98	87	12	
		6	*52 252.60*	1.04?	87	12	

V II—Continued

Configuration	Term	J	Level (cm^{-1})	g	Leading percentages		
$3d^3(^2P)4p$	$z\ ^3S°$	1	*52 181.18*	1.85	82	11	$(^4P)\ ^3S°$
$3d^3(a\ ^2D)4p$	$x\ ^3F°$	2	*52 245.69*	0.68	74	12	$(^2G)\ ^3F°$
		3	*52 391.94*	1.07	58	22	$(^4P)\ ^3D°$
		4	*52 657.47*	1.18?	85	10	$(^2G)\ ^3F°$
$3d^3(^4P)4p$	$x\ ^3D°$	1	*52 604.20*	0.63	52	16	$(^2P)\ ^3D°$
		2	*52 700.05*	1.10	51	25	
		3	*52 767.30*	1.26	34	29	
$3d^3(a\ ^2D)4p$	$z\ ^1P°$	1	*52 803.75*	0.92	73	11	$(^2P)\ ^1P°$
$3d^3(^2H)4p$	$z\ ^3I°$	5	*52 877.89*	0.84?	99		
		6	*53 076.71*	0.98	100		
		7	*53 319.56*	1.11?	100		
$3d^2\ 4s^2$	$d\ ^1G$	4	53 607.2?				
$3d^3(a\ ^2D)4p$	$w\ ^3D°$	1	*53 751.48*	0.49?	76	12	$(^2P)\ ^3D°$
		2	*53 868.61*	1.10	80	11	
		3	*53 927.17*	1.37	84	7	
$3d^3(^2H)4p$	$y\ ^1G°$	4	*54 144.17*	1.00	82	11	$(^2F)\ ^1G°$
$3d^3(a\ ^2D)4p$	$x\ ^3P°$	2	*54 715.66*		48	32	$(^2P)\ ^3P°$
		1	*54 717.86*		48	28	
		0	*54 813.40*		52	30	
$3d^3(a\ ^2D)4p$	$y\ ^1F°$	3	*55 142.03*	0.94	53	34	$(^2H)\ ^3G°$
$3d^3(^2H)4p$	$x\ ^3G°$	5	*55 206.79*	1.15	79	10	$3d^3(^2H)4p\ ^1H°$
		4	*55 304.30*	1.02	88	6	$3d^2(^3F)4s4p(^3P°)\ ^3G°$
		3	*55 349.63*	0.82	57	33	$3d^3(a\ ^2D)4p\ ^1F°$
$3d^3(^2H)4p$	$z\ ^1I°$	6	*55 403.30*	1.01?	100		
$3d^3(^2H)4p$	$y\ ^1H°$	5	*55 499.34*	1.03?	66	20	$(^2G)\ ^1H°$
$3d^3(^4P)4p$	$y\ ^3S°$	1	*55 663.23*	1.92	55	22	$(^2P)\ ^1P°$
$3d^3(^2P)4p$	$y\ ^1P°$	1	*56 171.40*	1.05?	60	25	$(^4P)\ ^3S°$
$3d^3(a\ ^2D)4p$	$y\ ^1D°$	2	*57 342.56*	0.98	54	40	$(^2P)\ ^1D°$
$3d^3(^2F)4p$	$w\ ^3F°$	2	*62 084.94*	0.58?	85	7	$3d^2(^3F)4s4p(^1P°)\ ^3F°$
		3	*62 133.30*	1.00	80	9	$3d^2(^3F)4s4p(^3P°)\ ^5G°$
		4	*62 176.19*	1.36?	78	11	$3d^2(^3F)4s4p(^3P°)\ ^5G°$
$3d^2(^3F)4s4p(^3P°)$	$^5G°$	2	*62 285.8*		96		
		3	*62 444.1*		90	8	$3d^3(^2F)4p\ ^3F°$
		4	*62 682.0*		88	10	
		5	*62 987.6*		100		
		6	*63 357.3*		100		
$3d^2(^3F)4s4p(^3P°)$	$y\ ^5F°$	1	*63 549.2*		98		
		2	*63 656.8*		98		
		3	*63 816.7*		98		
		4	*64 026.2*		98		
		5	*64 286.6*		91	7	$3d^3(^2F)4p\ ^3G°$

(Continued)

V II—Continued

Configuration	Term	J	Level (cm^{-1})	g	Leading percentages		
$3d^3(^2F)4p$	$w\ ^3G°$	3	*64 057.54*	0.72?	95		
		4	*64 130.80*	1.02	94		
		5	*64 229.19*		89	7	$3d^2(^3F)4s4p(^3P°)\ ^5F°$
$3d^3(^2F)4p$	$x\ ^1D°$	2	*64 586.14*	1.03?	67	16	$3d^3(^2F)4p\ ^3D°$
$3d^3(^2F)4p$	$v\ ^3D°$	3	*64 603.47*	1.22?	88	6	$3d^2(^3P)4s4p(^3P°)\ ^3D°$
		2	*64 804.17*	1.02?	72	15	$3d^3(^2F)4p\ ^1D°$
		1	*64 930.69*	0.46?	88	6	$3d^2(^3P)4s4p(^3P°)\ ^3D°$
$3d^2(^3F)4s4p(^3P°)$	$x\ ^5D°$	0	*65 781.8*		94		
		1	*65 815.2*		94		
		2	*65 884.4*		93		
		3	*65 995.6*		93		
		4	*66 157.6*		94		
$3d^3(^2F)4p$	$x\ ^1G°$	4	*65 790.18*	0.94	87	10	$(^2H)\ ^1G°$
$3d^3(^2F)4p$	$x\ ^1F°$	3	*66 303.84*	0.95	74	22	$3d^2(^3F)4s4p(^3P°)\ ^1F°$
$3d^2(^3F)4s4p(^3P°)$	$v\ ^3F°$	2	*67 737.7*		77	11	$3d^2(^1D)4s4p(^3P°)\ ^3F°$
		3	*67 904.8*		77	11	
		4	*68 147.0*		78	11	
$3d^2(^3F)4s4p(^3P°)$	$u\ ^3D°$	1	*68 759.2*		81	4	$3d^2(^1D)4s4p(^3P°)\ ^3D°$
		2	*68 797.4*		80	4	
		3	*68 944.9*		80	4	
$3d^3(^4F)5s$	^5F	1	69 146.30				
		2	69 228.20				
		3	69 352.43				
		4	69 518.36				
		5	69 724.14	1.39			
$3d^2(^3F)4s4p(^3P°)$	$v\ ^3G°$	3	*69 643.9*		92		
		4	*69 911.8*		92		
		5	*70 227.6*		93		
$3d^3(^4F)5s$	^3F	2	70 415.40				
		3	70 629.71	1.06			
		4	70 898.43	1.23			
$3d^2(^3F)4s4p(^3P°)$	$^1D°$	2	*70 923.28*		82	9	$3d^2(^3P)4s4p(^3P°)\ ^1D°$
$3d^2(^3F)4s4p(^3P°)$	$^1F°$	3	*70 935.97*		73	19	$3d^3(^2F)4p\ ^1F°$
	$w\ ^1G°$	4	*72 292.4?*				
$3d^3(^4F)4d$	$e\ ^5H$	3	72 448.40				
		4	72 551.09				
		5	72 680.70				
		6	72 837.42				
		7	73 020.96				
$3d^3(^4F)4d$	$e\ ^5P$	1	72 518.35				
		2	72 674.58				
		3	72 908.79				

V II—Continued

Configuration	Term	J	Level (cm^{-1})	g	Leading percentages		
$3d^3(^4\mathrm{F})4d$	^{5}F	1	72 839.18				
		2	73 027.18				
		3	73 146.21				
		4	73 279.16				
		5	73 417.14				
$3d^3(^4\mathrm{F})4d$	^{5}G	2	72 877.85				
		3	72 951.38				
		4	73 063.53				
		5	73 223.17				
		6	73 499.60				
$3d^3(^4\mathrm{F})4d$	^{3}D	1	73 181.42				
		2	73 309.92				
		3	73 530.58				
$3d^3(^4\mathrm{F})4d$	^{3}P	0	74 949.48				
		1	75 080.58				
		2	75 335.70				
$3d^3(^4\mathrm{F})4d$	^{3}H	4	75 140.52				
		5	75 346.15				
		6	75 592.36				
$3d^3(^4\mathrm{F})4d$	^{3}G	3	75 422.77				
		4	75 615.29				
		5	75 854.10				
$3d^2(^1\mathrm{D})4s4p(^3\mathrm{P}^\circ)$	$t\ ^3\mathrm{D}^\circ$	1	*75 716.0*	0.50?	58	12	$3d^3(b\ ^2\mathrm{D})4p\ ^3\mathrm{D}^\circ$
		2	*75 758.2*	1.14?	56	13	$3d^3(b\ ^2\mathrm{D})4p\ ^3\mathrm{D}^\circ$
		3	*75 848.0*	1.27?	54	19	$3d^2(^3\mathrm{P})4s4p(^3\mathrm{P}^\circ)\ ^5\mathrm{P}^\circ$
$3d^3(^4\mathrm{F})4d$	^{3}F	2	75 813.38				
		3	75 966.00				
		4	76 142.95				
	$u\ ^3\mathrm{F}^\circ$	2	*76 220.4*				
		3	*76 385.8*				
		4	*76 643.5*				
$3d^3(^4\mathrm{F})4d$	^{5}D	0	76 281.20				
		1	76 322.47				
		2	76 403.48				
		3	76 521.20				
		4	76 672.95				
	2°	3	*76 405.4*				
$3d^3(b\ ^2\mathrm{D})4p$	$v\ ^1\mathrm{D}^\circ$	2	*77 603.4*		74	14	$3d^2(^3\mathrm{P})4s4p(^3\mathrm{P}^\circ)$
$3d^3(b\ ^2\mathrm{D})4p$	$t\ ^3\mathrm{F}^\circ$	3	*77 841.9*		85	8	$3d^2(^1\mathrm{G})4s4p(^3\mathrm{P}^\circ)$
		2	*77 857.0*		66	12	$3d^2(b\ ^2\mathrm{D})4s4p(^3\mathrm{P}^\circ)$
		4	*77 968.9*		86	8	$3d^2(^1\mathrm{G})4s4p(^3\mathrm{P}^\circ)$
$3d^3(b\ ^2\mathrm{D})4p$	$v\ ^3\mathrm{P}^\circ$	2	*78 416.8*		55	24	$3d^2(^3\mathrm{P})4s4p(^3\mathrm{P}^\circ)$
		1	*78 569.1*		66	23	
		0	*78 644.1*		66	33	
$3d^3(b\ ^2\mathrm{D})4p$	$u\ ^1\mathrm{F}^\circ$	3	*79 327.6*		93		

(Continued)

V II—Continued

Configuration	Term	J	Level (cm^{-1})	g	Leading percentages		
$3d^3(^4P)5s$	^5P	1	80 542.20				
		2	80 623.12				
		3	80 782.28				
$3d^3(^2G)5s$	^3G	3	81 263.49				
		4	81 342.88				
		5	81 483.15				
$3d^3(^4P)5s$	^3P	0	81 669.39				
		1	81 735.96				
		2	81 914.22				
$3d^3(^2G)5s$	^1G	4	82 025.60				
$3d^3(^2H)5s$	^3H	4	86 027.94				
		5	86 091.61				
		6	86 191.60				
$3d^3(^2H)5s$	^1H	5	86 766.62				
$3d^3(^4F_{3/2})4f$	${}^2[9/2]^\circ$	4	*90 182.36*		95		
		5	*90 228.05*		79	20	($^4F_{5/2}$) ${}^2[11/2]^\circ$
$3d^3(^4F_{3/2})4f$	${}^2[5/2]^\circ$	3	*90 238.75*		63	34	($^4F_{3/2}$) ${}^2[7/2]^\circ$
		2	*90 251.48*		96		
$3d^3(^4F_{3/2})4f$	${}^2[7/2]^\circ$	3	*90 280.22*		60	24	($^4F_{3/2}$) ${}^2[5/2]^\circ$
		4	*90 310.26*		87	12	($^4F_{5/2}$) ${}^2[9/2]^\circ$
$3d^3(^4F_{5/2})4f$	${}^2[11/2]^\circ$	6	*90 355.34*		91		
		5	*90 375.68*		64	20	($^4F_{3/2}$) ${}^2[9/2]^\circ$
$3d^3(^4F_{5/2})4f$	${}^2[1/2]^\circ$	1	*90 369.29*		96		
		0	*90 424.58*		98		
$3d^3(^4F_{5/2})4f$	${}^2[3/2]^\circ$	1	*90 379.25*		92		
		2	*90 386.74*		88		
$3d^3(^4F_{5/2})4f$	${}^2[7/2]$	3	*90 412.89*		70	27	($^4F_{5/2}$) ${}^2[5/2]^\circ$
		4	*90 433.15*		86		
$3d^3(^4F_{5/2})4f$	${}^2[9/2]$	5	*90 429.32*		83	14	($^4F_{5/2}$) $[11/2]^\circ$
		4	*90 462.26*		73	12	($^4F_{3/2}$) ${}^2[7/2]^\circ$
$3d^3(^4F_{5/2})4f$	${}^2[5/2]^\circ$	2	*90 497.90*		97		
		3	*90 535.05*		46	25	($^4F_{5/2}$) ${}^2[7/2]^\circ$
$3d^3(^4F_{5/2})4f$	${}^2[13/2]^\circ$	7	*90 529.65*		96		
		6	*90 610.53*		61	36	($^4F_{7/2}$) ${}^2[11/2]^\circ$
$3d^3(^4F_{7/2})4f$	${}^2[1/2]^\circ$	1	*90 552.45*		98		
		0	*90 584.6*		98		
$3d^3(^4F_{7/2})4f$	${}^2[11/2]^\circ$	6	*90 580.25*		62	30	($^4F_{7/2}$) ${}^2[13/2]^\circ$
		5	*90 666.22*		83		
$3d^3(^4F_{7/2})4f$	${}^2[3/2]^\circ$	2	*90 608.51*		85		
		1	*90 668.46*		94		

V II—Continued

Configuration	Term	J	Level (cm^{-1})	g	Leading percentages		
$3d^3(^4F_{7/2})4f$	$^2[5/2]°$	2	*90 611.82*		87		
		3	*90 613.31*		62	36	$(^4F_{7/2})\ ^2[7/2]°$
$3d^3(^4F_{7/2})4f$	$^2[7/2]°$	4	*90 615.78*		71	27	$(^4F_{7/2})\ ^2[9/2]°$
		3	*90 709.33*		53	25	$(^4F_{7/2})\ ^2[5/2]°$
$3d^3(^4F_{7/2})4f$	$^2[9/2]°$	5	*90 630.16*		90		
		4	*90 657.34*		67	22	$(^4F_{7/2})\ ^2[7/2]°$
$3d^3(^4F_{9/2})4f$	$^2[15/2]°$	8	*90 745.99*		100		
		7	*90 846.30*		82	17	$(^4F_{9/2})\ ^2[13/2]°$
$3d^3(^4F_{9/2})4f$	$^2[3/2]°$	2	*90 792.4*		94		
$3d^3(^4F_{9/2})4f$	$^2[5/2]°$	3	*90 795.91*		90		
$3d^3(^4F_{9/2})4f$	$^2[7/2]°$	3	*90 845.61*		86		
		4	*90 853.28*		87	12	$(^4F_{9/2})\ ^2[9/2]°$
$3d^3(^4F_{9/2})4f$	$^2[13/2]°$	7	*90 846.25*		83	15	$(^4F_{9/2})\ ^2[15/2]°$
		6	*90 917.12*		85	13	$(^4F_{9/2})\ ^2[11/2]°$
$3d^3(^4F_{9/2})4f$	$^2[11/2]°$	6	*90 860.90*		87	13	$(^4F_{9/2})\ ^2[13/2]°$
		5	*90 932.73*		85	10	$(^4F_{9/2})\ ^2[9/2]°$
$3d^3(^4F_{9/2})4f$	$^2[9/2]°$	5	*90 874.76*		89	11	$(^4F_{9/2})\ ^2[11/2]°$
		4	*90 881.62*		84	10	$(^4F_{9/2})\ ^2[7/2]°$
....	....	....	....				
V III $(^4F_{3/2})$	*Limit*		118 200				

V III

$Z=23$

Sc I isoelectronic sequence

Ground state: $1s^2 2s^2 2p^6 3s^2 3p^6 3d^3\ ^4F_{3/2}$

Ionization energy = 236 410±20 cm^{-1} (29.311±0.002 eV)

The initial analysis was by White (1929), who found terms of $3d^3$, $3d^2 4s$, $3d^2 4p$, and $3d^2 4d$.

Many more configurations were discovered by Iglesias (1962, 1969) utilizing a new set of wavelength measurements. Her results are presented here, including revised values for the levels found by White and her determination of the ionization energy from the $3d^2 ng$ series ($n=5$–7). The levels of $3d^3$ were determined from wavelengths in the range 1600–600 Å measured with an uncertainty of about ±0.03 Å. The level uncertainty is probably ±3 cm^{-1}. Transitions among the excited configurations are in the range of 7800–1600 Å. Above 2000 the measurement uncertainty is about ±0.005 Å, permitting an improved level uncertainty of about ±0.05 Å in most cases, relative to the $3d^2 4s$ configuration.

Theoretical confirmation of the $3d^3$ and $3d^2 4s$ levels has been provided by Shadmi, Caspi, and Oreg (1969). Percentage compositions for $3d^3$ are from Pasternak and Goldschmidt (1972). The compositions of the $3d^2 4d$ and $3d^2 5s$ levels are from an unpublished calculation by Wyart (1983).

The percentage compositions of the $3d^2 4p$ levels are from Roth (1968). He has interchanged Iglesias' designations for the (^{3}F) $^2D^\circ_{5/2}$ and (^{3}F) $^4D^\circ_{5/2}$ levels. The percentage compositions of $3d^2 4f$ levels were calculated by Spector (1970).

References

Iglesias, L. (1962), An. R. Soc. Esp. Fis. Quim. **58A,** 191.
Iglesias, L. (1969), Opt. Pur. Apl. **2,** 132.
Pasternak, A., and Goldschmidt, Z. B. (1972), Phys. Rev. **6A,** 55.
Roth. C. (1968), J. Res. Natl. Bur. Stand. **72A,** 505.
Shadmi, Y., Caspi, E., and Oreg. J. (1969), J. Res. Natl. Bur. Stand. **73A,** 173.
Spector, N. (1970), Opt. Pur. Apl. **3,** 33.
White, H. E. (1929), Phys. Rev. **33,** 672.
Wyart, J.-F. (1983), private communication.

V III

Configuration	Term	J	Level (cm^{-1})	Leading percentages		
$3d^3$	$a\ ^4$F	3/2	0.0	100		
		5/2	145.5	100		
		7/2	341.5	100		
		9/2	583.8	100		
$3d^3$	$a\ ^4$P	1/2	11 513.8	100		
		3/2	11 591.8	99		
		5/2	11 769.7	100		
$3d^3$	$a\ ^2$G	7/2	11 966.3	100		
		9/2	12 187.0	100		
$3d^3$	$a\ ^2$P	3/2	15 550.3	67	25	^{2}D2
		1/2	15 579.8	100		
$3d^3$	$a\ ^2$D2	3/2	16 330.5	52	32	^{2}P
		5/2	16 374.7	77	22	^{2}D1
$3d^3$	$a\ ^2$H	9/2	16 810.9	100		
		11/2	16 977.6	100		
$3d^3$	$a\ ^2$F	7/2	27 727.8	100		
		5/2	27 846.8	100		
$3d^3$	$b\ ^2$D1	5/2	42 267.4	77	22	^{2}D2
		3/2	42 371.2	77	23	

V III—Continued

Configuration	Term	J	Level (cm^{-1})	Leading percentages		
$3d^2(^3F)4s$	$b\ ^4F$	3/2	43 942.49			
		5/2	44 110.04			
		7/2	44 345.82			
		9/2	44 646.96			
$3d^2(^3F)4s$	$b\ ^2F$	5/2	49 327.74			
		7/2	49 805.29			
$3d^2(^1D)4s$	$c\ ^2D$	5/2	56 160.42			
		3/2	56 256.75			
$3d^2(^3P)4s$	$b\ ^4P$	1/2	56 529.30			
		3/2	56 669.05			
		5/2	56 922.50			
$3d^2(^3P)4s$	$b\ ^2P$	1/2	61 578.74			
		3/2	61 777.15			
$3d^2(^1G)4s$	$b\ ^2G$	9/2	63 303.12			
		7/2	63 315.05			
$3d^2(^3F)4p$	$z\ ^4G°$	5/2	*85 524.00*	98		
		7/2	*85 875.74*	99		
		9/2	*86 306.40*	100		
		11/2	*86 809.39*	100		
$3d^2(^1S)4s$	$a\ ^2S$	1/2	86 304.0			
$3d^2(^3F)4p$	$z\ ^4F°$	3/2	*86 716.84*	98		
		5/2	*86 938.01*	99		
		7/2	*87 218.92*	99		
		9/2	*87 544.46*	99		
$3d^2(^3F)4p$	$z\ ^2F°$	5/2	*87 880.85*	85	6	$(^1D)\ ^2F°$
		7/2	*88 327.96*	88	5	
$3d^2(^3F)4p$	$^2D°$	3/2	*88 559.08*	59	29	$(^3F)\ ^4D°$
		5/2	*89 457.67*	52	35	
$3d^2(^3F)4p$	$^4D°$	1/2	*89 005.64*	96	4	$(^3P)\ ^4D°$
		3/2	*89 193.47*	67	25	$(^3F)\ ^2D°$
		5/2	*88 944.38*	59	28	$(^3F)\ ^2D°$
		7/2	*89 417.50*	91	4	$(^3F)\ ^2F°$
$3d^2(^3F)4p$	$z\ ^2G°$	7/2	*91 710.43*	96		
		9/2	*92 052.55*	96		
$3d^2(^3P)4p$	$z\ ^2S°$	1/2	*94 714.25*	99		
$3d^2(^3P)4p$	$z\ ^4S°$	3/2	*97 512.00*	74	25	$(^1D)\ ^2P°$
$3d^2(^1D)4p$	$z\ ^2P°$	3/2	*98 062.22*	70	26	$(^3P)\ ^4S°$
		1/2	*98 399.42*	97		
$3d^2(^1D)4p$	$y\ ^2F°$	5/2	*98 383.53*	88	7	$(^3F)\ ^2F°$
		7/2	*98 825.19*	86	6	

(Continued)

V III—Continued

Configuration	Term	J	Level (cm^{-1})	Leading percentages		
$3d^2(^3P)4p$	$y\ ^4D°$	1/2	*99 073.23*	95	4	$(^3F)\ ^4D°$
		3/2	*99 181.50*	91	4	$(^3F)\ ^4D°$
		5/2	*99 440.10*	88	5	$(^1D)\ ^2D°$
		7/2	*99 941.20*	91	5	$(^1D)\ ^2F°$
$3d^2(^1D)4p$	$y\ ^2D°$	3/2	*99 508.50*	81	6	$(^3P)\ ^2D°$
		5/2	*99 804.57*	80	7	$(^3P)\ ^4D°$
$3d^2(^3P)4p$	$z\ ^4P°$	1/2	*101 645.61*	99		
		3/2	*101 785.81*	99		
		5/2	*102 075.00*	98		
$3d^2(^1G)4p$	$y\ ^2G°$	7/2	*102 961.28*	96	4	$(^3F)\ ^2G°$
		9/2	*103 034.65*	96	4	
$3d^2(^3P)4p$	$x\ ^2D°$	5/2	*105 282.65*	82	10	$(^3F)\ ^2D°$
		3/2	*105 320.18*	80	10	$(^1D)\ ^2D°$
$3d^2(^1G)4p$	$z\ ^2H°$	9/2	*106 441.31*	99		
		11/2	*106 903.35*	100		
$3d^2(^3P)4p$	$y\ ^2P°$	1/2	*107 060.41*	99		
		3/2	*107 165.90*	97		
$3d^2(^1G)4p$	$x\ ^2F°$	7/2	*109 854.87*	97		
		5/2	*110 181.47*	97		
$3d^2(^1S)4p$	$x\ ^2P°$	1/2	*129 397.4*	98		
		3/2	*129 998.3*	98		
$3d^2(^3F)4d$	$e\ ^4G$	5/2	140 751.47	87	13	$(^3F)\ ^2F$
		7/2	140 933.85	92	6	$(^3F)\ ^4H$
		9/2	141 144.86	89	11	$(^3F)\ ^4H$
		11/2	141 405.80	88	12	$(^3F)\ ^4H$
$3d^2(^3F)4d$	$e\ ^2F$	5/2	141 196.87	82	13	$(^3F)\ ^4G$
		7/2	141 508.26	87	7	$(^3F)\ ^4D$
$3d^2(^3F)4d$	$e\ ^4H$	7/2	141 271.10	94	6	$(^3F)\ ^4G$
		9/2	141 489.06	89	11	$(^3F)\ ^4G$
		11/2	141 735.42	88	12	$(^3F)\ ^4G$
		13/2	141 991.58	100		
$3d^2(^3F)4d$	$e\ ^4D$	1/2	141 421.87	98	1	$(^3F)\ ^2P$
		3/2	141 531.94	99		
		5/2	141 715.20	98	2	$(^3F)\ ^2F$
		7/2	141 987.95	93	6	$(^3F)\ ^2F$
$3d^2(^3F)4d$	$e\ ^2P$	1/2	143 169.0	91	6	$(^1D)\ ^2P$
		3/2	143 651.5	91	5	
$3d^2(^3F)4d$	$e\ ^4P$	1/2	144 544.13	90	9	$(^3P)\ ^4P$
		3/2	144 828.58	89	9	
		5/2	145 073.42	88	10	
$3d^2(^3F)4d$	$e\ ^2G$	7/2	144 772.60	88	9	$(^1D)\ ^2G$
		9/2	145 144.00	88	9	

V III—Continued

Configuration	Term	J	Level (cm^{-1})	Leading percentages			
$3d^2(^3\mathrm{F})5s$	$e\,^4\mathrm{F}$	3/2	145 822.76	99			
		5/2	145 978.99	98		2	$(^3\mathrm{F})\ ^2\mathrm{F}$
		7/2	146 223.74	98		2	$(^3\mathrm{F})\ ^2\mathrm{F}$
		9/2	146 550.32	100			
$3d^2(^3\mathrm{F})4d$	$e\,^2\mathrm{H}$	9/2	145 903.31	93		7	$(^1\mathrm{G})\ ^2\mathrm{H}$
		11/2	146 347.77	93		7	
$3d^2(^3\mathrm{F})4d$	$e\,^2\mathrm{D}$	3/2	145 953.51	84		13	$(^1\mathrm{D})\ ^2\mathrm{D}$
		5/2	146 140.08	82		14	
$3d^2(^3\mathrm{F})4d$	$f\,^4\mathrm{F}$	3/2	146 846.99	97		2	$(^3\mathrm{P})\ ^4\mathrm{F}$
		5/2	146 984.08	97		3	
		7/2	147 170.06	97		3	
		9/2	147 401.29	97		3	
$3d^2(^3\mathrm{F})5s$	$f\,^2\mathrm{F}$	5/2	147 127.46	98		2	$(^3\mathrm{F})\ ^4\mathrm{F}$
		7/2	147 607.08	98		2	$(^3\mathrm{F})\ ^4\mathrm{F}$
$3d^2(^1\mathrm{D})4d$	$e\,^2\mathrm{S}$	1/2	152 317.7	98		1	$(^3\mathrm{P})\ ^2\mathrm{P}$
$3d^2(^3\mathrm{P})4d$	$f\,^2\mathrm{P}$	3/2	152 861.2	56		44	$(^1\mathrm{D})\ ^2\mathrm{P}$
		1/2	153 266.4	55		43	
$3d^2(^1\mathrm{D})4d$	$g\,^2\mathrm{F}$	5/2	152 940.0	85		13	$(^3\mathrm{P})\ ^2\mathrm{F}$
		7/2	153 113.9	87		10	
$3d^2(^1\mathrm{D})4d$	$f\,^2\mathrm{G}$	9/2	154 548.9	84		6	$(^3\mathrm{F})\ ^2\mathrm{G}$
		7/2	154 572.7	86		7	
$3d^2(^3\mathrm{P})4d$	$g\,^4\mathrm{F}$	3/2	155 497.3	96		2	$(^3\mathrm{F})\ ^4\mathrm{F}$
		5/2	155 569.8	94		3	$(^3\mathrm{P})\ ^4\mathrm{D}$
		7/2	155 682.5	91		4	$(^3\mathrm{P})\ ^4\mathrm{D}$
		9/2	155 862.3	92		4	$(^1\mathrm{D})\ ^2\mathrm{G}$
$3d^2(^3\mathrm{P})4d$	$f\,^4\mathrm{D}$	3/2	155 802.0	92		5	$(^1\mathrm{D})\ ^2\mathrm{D}$
		1/2	155 829.9?	99			
		5/2	155 853.9	92		3	$(^1\mathrm{D})\ ^2\mathrm{D}$
		7/2	155 982.0	95		4	$(^3\mathrm{P})\ ^4\mathrm{F}$
$3d^2(^1\mathrm{D})4d$	$f\,^2\mathrm{D}$	3/2	156 344.6	63		13	$(^3\mathrm{P})\ ^2\mathrm{D}$
		5/2	156 671.7	64		13	$(^3\mathrm{F})\ ^2\mathrm{D}$
$3d^2(^3\mathrm{P})4d$	$h\,^2\mathrm{F}$	5/2	157 027.6	59		32	$(^1\mathrm{G})\ ^2\mathrm{F}$
		7/2	157 068.0	61		32	
$3d^2(^1\mathrm{D})5s$	$g\,^2\mathrm{D}$	5/2	157 066.8	99			
		3/2	157 080.1	99		1	$(^1\mathrm{D})\ ^2\mathrm{P}$
$3d^2(^3\mathrm{P})4d$	$f\,^4\mathrm{P}$	3/2	158 155.5	81		8	$(^3\mathrm{F})\ ^4\mathrm{P}$
		1/2	158 188.2	83		8	
		5/2	158 240.2	88		10	
$3d^2(^3\mathrm{P})5s$	$g\,^4\mathrm{P}$	1/2	158 844.4	100			
		3/2	158 927.4	99			
		5/2	159 147.6	100			
$3d^2(^1\mathrm{D})4d$		3/2	158 976.0?	26	$^2\mathrm{P}$	24	$(^3\mathrm{P})\ ^2\mathrm{D}$

(Continued)

V III—Continued

Configuration	Term	J	Level (cm^{-1})	Leading percentages			
$3d^2(^3\mathrm{P})4d$	$^2\mathrm{D}$	5/2	159 067.6	55		26	$(^1\mathrm{G})\ ^2\mathrm{D}$
$3d^2(^3\mathrm{P})4d$		3/2	159 250.5	30	$^2\mathrm{D}$	20	$(^1\mathrm{D})\ ^2\mathrm{P}$
$3d\,4s(^3\mathrm{D})4p$	$x\ ^4\mathrm{D}^\circ$	1/2	*159 684.4*				
		3/2	*159 787.8*				
		5/2	*159 996.0*				
		7/2	*160 351.5*				
$3d^2(^3\mathrm{P})5s$	$h\ ^2\mathrm{P}$	1/2	160 098.6?	97		1	$3d^2(^1\mathrm{D})4d\ ^2\mathrm{P}$
		3/2	160 308.4	96		1	
$3d^2(^3\mathrm{F})5p$	$y\ ^4\mathrm{F}^\circ$	3/2	*160 144.92*				
		5/2	*160 380.35*				
		7/2	*160 682.19*				
		9/2	*161 028.02*				
$3d^2(^3\mathrm{F})5p$	$y\ ^4\mathrm{G}^\circ$	5/2	*160 322.56*				
		7/2	*160 577.33*				
		9/2	*160 880.67*				
		11/2	*161 240.69*				
$3d^2(^1\mathrm{G})4d$	$g\ ^2\mathrm{G}$	7/2	161 017.78	93		4	$(^3\mathrm{F})\ ^2\mathrm{G}$
		9/2	161 085.59	92		5	
$3d^2(^1\mathrm{G})4d$	$e\ ^2\mathrm{I}$	11/2	161 051.82	100			
		13/2	161 101.91	100			
$3d^2(^3\mathrm{F})5p$	$w\ ^2\mathrm{F}^\circ$	5/2	*161 166.33*				
		7/2	*161 588.54*				
$3d^2(^3\mathrm{F})5p$	$w\ ^4\mathrm{D}^\circ$	1/2	*161 491.07*				
		3/2	*161 570.12*				
		5/2	*161 924.00*				
		7/2	*162 385.82*				
$3d^2(^3\mathrm{F})5p$	$w\ ^2\mathrm{D}^\circ$	3/2	*161 885.32*				
		5/2	*162 305.28*				
$3d^2(^3\mathrm{F})5p$	$x\ ^2\mathrm{G}^\circ$	7/2	*162 336.36*				
		9/2	*162 760.84*				
$3d^2(^1\mathrm{G})4d$	$f\ ^2\mathrm{H}$	9/2	162 513.19	93		7	$(^3\mathrm{F})\ ^2\mathrm{H}$
		11/2	162 543.12	93		7	
$3d^2(^1\mathrm{G})5s$	$h\ ^2\mathrm{G}$	9/2	164 398.44	100			
		7/2	164 399.32	100			
$3d^2(^1\mathrm{G})4d$	$i\ ^2\mathrm{F}$	7/2	165 208.47	67		24	$(^3\mathrm{P})\ ^2\mathrm{F}$
		5/2	165 267.23	66		24	
$3d^2(^1\mathrm{D})5p$	$v\ ^2\mathrm{F}^\circ$	5/2	*171 614.7*				
		7/2	*171 809.8*				
$3d^2(^1\mathrm{D})5p$	$w\ ^2\mathrm{P}^\circ$	3/2	*171 717.3*				
		1/2	*171 863.5*				
$3d^2(^1\mathrm{D})5p$	$v\ ^2\mathrm{D}^\circ$	3/2	*171 983.8*				
		5/2	*172 112.3*				

V III—Continued

Configuration	Term	J	Level (cm^{-1})	Leading percentages			
$3d^2(^3P)5p$	$y\ ^4S^\circ$	3/2	*172 804.9*				
$3d^2(^3F)4f$	$^4H^\circ$	7/2	*173 246.54*	77			
		9/2	*173 367.21*	76			
		11/2	*173 517.63*	64		27	$(^3F)\ ^4I^\circ$
		13/2	*173 788.14*	58		38	$(^3F)\ ^4I^\circ$
$3d^2(^3F)4f$	$^2G^\circ$	7/2	*173 388.5?*	61			
		9/2	*173 670.02*	56			
$3d^2(^3P)5p$	$v\ ^4D^\circ$	1/2	*173 458.5*				
		3/2	*173 556.6*				
		5/2	*173 741.4*				
		7/2	*173 966.3*				
$3d^2(^3F)4f$	$^4I^\circ$	9/2	*173 534.92*	64			
		11/2	*173 773.28*	65		23	$(^3F)\ ^4H^\circ$
		13/2	*174 045.8*	56		41	$(^3F)\ ^4H^\circ$
		15/2	*174 223.7*	100			
$3d^2(^3F)4f$	$^4G^\circ$	5/2	*173 568.27*	71			
		7/2	*173 786.11*	59			
		9/2	*173 990.35*	46		36	$(^3F)\ ^2H^\circ$
		11/2	*174 237.76*	76			
$3d^2(^3F)4f$	$^2F^\circ$	5/2	*173 659.5*	41		29	$(^3F)\ ^4F^\circ$
		7/2	*174 259.15*	53		20	
$3d^2(^3F)4f$		7/2	*173 905.49*	32	$^2F^\circ$	27	$(^3F)\ ^4F^\circ$
$3d^2(^3F)4f$		5/2	*173 954.6*	29	$^2F^\circ$	20	$(^3F)\ ^4D^\circ$
$3d^2(^3F)4f$		5/2	*174 059.6*	30	$^4F^\circ$	30	$(^3F)\ ^4P^\circ$
$3d^2(^3F)4f$		3/2	*174 069.3*	35	$^2P^\circ$	23	$(^3F)\ ^4D^\circ$
$3d^2(^3F)4f$	$^2H^\circ$	9/2	*174 170.6*	41		36	$(^3F)\ ^2G^\circ$
		11/2	*174 500.46*	67		23	$(^3F)\ ^2I^\circ$
$3d^2(^3F)4f$	$^2I^\circ$	11/2	*174 192.88*	55		26	$(^3F)\ ^2H^\circ$
		13/2	*174 720.37*	95			
$3d^2(^3F)4f$		3/2	*174 207.7*	37	$^2D^\circ$	27	$(^3F)\ ^4S^\circ$
$3d^2(^3F)4f$	$^4F^\circ$	9/2	*174 367.6*	79			
$3d^2(^3F)4f$	$^4D^\circ$	7/2	*174 458.96*	66		29	$(^3F)\ ^4F^\circ$
$3d^2(^3F)4f$		5/2	*174 467.3*	32	$^2D^\circ$	31	$(^3F)\ ^4D^\circ$
$3d^2(^3P)5p$	$y\ ^4P^\circ$	1/2	*174 507.3*				
		3/2	*174 696.4*				
		5/2	*174 938.9*				
$3d^2(^1G)5p$	$w\ ^2G^\circ$	7/2	*178 053.05*				
		9/2	*178 085.62*				
$3d^2(^1G)5p$	$y\ ^2H^\circ$	9/2	*179 467.60*				
		11/2	*179 631.11*				

(Continued)

V III—Continued

Configuration	Term	J	Level (cm^{-1})	Leading percentages
$3d^2(^1G)5p$	$u\ ^2F°$	7/2	*179 855.04*	
		5/2	*179 889.40*	
$3d^2(^3F)5d$	$f\ ^4H$	7/2	181 650.16	
		9/2	181 826.54	
		11/2	182 077.54	
		13/2	182 444.06	
$3d^2(^3F)5d$	$f\ ^4G$	9/2	182 035.67	
		11/2	182 287.66	
$3d^2(^3F)5d$	$g\ ^4G$	5/2	182 168.8	
		7/2	182 517.8	
$3d^2(^3F)6s$	$h\ ^4F$	3/2	183 316.75	
		5/2	183 445.06	
		7/2	183 699.40	
		9/2	184 047.67	
$3d^2(^3F)5d$	$i\ ^2G$	7/2	183 760.91	
		9/2	184 087.90	
$3d^2(^3F)6s$	$j\ ^2F$	5/2	183 999.50	
		7/2	184 476.36	
$3d^2(^1D)4f$	$^2F°$	5/2	*184 104.9*	
		7/2	*184 304.4*	
$3d^2(^3F)5d$	$g\ ^2H$	9/2	184 237.44	
		11/2	184 648.09	
$3d^2(^1D)4f$	$^2H°$	11/2	*184 403.2?*	98
		9/2	*184 413.0*	98
$3d^2(^1D)4f$	$^2G°$	7/2	*184 558.3*	96
		9/2	*184 589.6*	98
$3d^2(^3F)5d$	$i\ ^4F$	9/2	184 613.01	
$3d^2(^3P)4f$	$^4D°$	7/2	*186 457.5*	90
$3d^2(^3P)4f$	$^4G°$	9/2	*186 604.4?*	90
		11/2	*186 732.9?*	98
$3d^2(^3P)4f$	$^2G°$	7/2	*186 861.6?*	90
		9/2	*186 989.2?*	90
$3d^2(^3P)4f$	$^4F°$	7/2	*187 369.6?*	96
		9/2	*187 382.9?*	97
$3d^2(^1G)4f$	$^2H°$	9/2	*191 345.6*	100
		11/2	*191 348.25*	100
$3d^2(^1G)4f$	$^2I°$	13/2	*191 540.37*	100
		11/2	*191 542.50*	100
$3d^2(^1G)4f$	$^2G°$	9/2	*191 644.8?*	100
		7/2	*191 742.7*	100

V III—Continued

Configuration	Term	J	Level (cm^{-1})	Leading percentages
$3d^2(^1G)4f$	$^2K°$	15/2	*192 359.4?*	100
		13/2	*192 362.3?*	100
$3d^2(^3F)5f$	$^4I°$	13/2	*196 473.86*	
		15/2	*196 823.96*	
$3d^2(^3P)6s$	$h\ ^4P$	3/2	196 517.1?	
		5/2	196 707.4	
$3d^2(^3F_2)5g$	$^2[5]$	9/2	196 753.60	
		11/2	196 755.63	
$3d^2(^3F_2)5g$	$^2[6]$	11/2	196 811.7	
		13/2	196 814.4	
$3d^2(^3F_3)5g$	$^2[5]$	9/2	197 075.29	
		11/2	197 076.68	
$3d^2(^3F_3)5g$	$^2[4]$	9/2	197 089.5?	
		7/2	197 089.6	
$3d^2(^3F_3)5g$	$^2[6]$	11/2	197 092.6	
		13/2	197 093.3	
$3d^2(^3F_3)5g$	$^2[7]$	15/2	197 141.14	
		13/2	197 142.21	
$3d^2(^3F_4)5g$	$^2[6]$	11/2	197 477.25	
		13/2	197 477.56	
$3d^2(^3F_4)5g$	$^2[5]$	9/2	197 482.10	
		11/2	197 482.67	
$3d^2(^3F_4)5g$	$^2[7]$	15/2	197 507.64	
		13/2	197 510.77	
$3d^2(^3F_4)5g$	$^2[8]$	17/2	197 559.40	
		15/2	197 562.98	
$3d^2(^1G)6s$	$j\ ^2G$	7/2	201 733.1?	
		9/2	201 735.22	
$3d^2(^3F)7s$	$j\ ^4F$	7/2	201 836.5?	
		9/2	202 209.3?	
$3d^2(^3F_4)6g$	$^2[8]$	17/2	209 668.2	
$3d^2(^3P_2)5g$	$^2[6]$	13/2	210 193.8?	
		11/2	210 196.07?	
$3d^2(^3F_4)7g$	$^2[8]$	17/2	216 964.5	
V IV ($^3F_2$)	*Limit*		**236 410**	

V IV

$Z=23$

Ca I isoelectronic sequence

Ground state: $1s^22s^22p^63s^23p^63d^2\ ^3F_2$

Ionization energy = 376 730±40 cm^{-1} (46.709±0.005 eV)

The initial analysis was due to White (1929), who reported levels of $3d^2$, $3d4d$, and all the levels of the $3d4s$ and $3d4p$ configurations.

The spectrum has been completely reobserved from 675–5940 Å and analyzed by Iglesias (1968), whose results are given here. The level uncertainty is about ±0.5 cm^{-1}. She has added the $^1S_0$ term of $3d^2$, all the terms but one of $3d4d$, $3d5d$, and $3d4f$, and all the terms of $3d5s$, $3d6s$, and $3d5p$. The compositions of levels of $3d4d+3d5s$, $3d5d+3d6s$ with configuration interaction (CI), and $3d4p$ are quoted from Wyart (1975). He has also provided a calculation of the percentage compositions of the configurations $3d5p+3d4f$ with CI. The percentages for $3d5g$ are from Goldschmidt (1982).

The ionization energy was derived by Iglesias from the three member $3dns$ series.

References

Goldschmidt, Z. B. (1982), J. Phys. **B15,** L581.
Iglesias, L. (1968), J. Res. Natl. Bur. Stand. **72A,** 295.
White, H. E. (1929), Phys. Rev. **33,** 538.
Wyart, J.-F. (1975), Phys. Scr. **12,** 33.

V IV

Configuration	Term	J	Level (cm^{-1})	Leading percentages
$3d^2$	^3F	2	0.0	
		3	325.4	
		4	734.7	
$3d^2$	^1D	2	10 959.3	
$3d^2$	^3P	0	13 122.8	
		1	13 239.2	
		2	13 458.3	
$3d^2$	^1G	4	18 391.2	
$3d^2$	^1S	0	42 462.1	
$3d\,4s$	^3D	1	96 196.1	
		2	96 412.1	
		3	96 798.0	
$3d\,4s$	^1D	2	100 200.7	
$3d\,4p$	$^1D°$	2	*144 273.1*	96
$3d\,4p$	$^3D°$	1	*146 117.7*	99
		2	*146 429.3*	95
		3	*146 855.1*	93
$3d\,4p$	$^3F°$	2	*147 135.2*	94
		3	*147 656.5*	94
		4	*148 369.2*	100
$3d\,4p$	$^3P°$	1	*151 427.0*	98
		0	*151 449.1*	100
		2	*151 567.3*	99
$3d\,4p$	$^1F°$	3	*153 918.7*	99

V IV—Continued

Configuration	Term	J	Level (cm^{-1})	Leading percentages		
3d 4p	^{1}P°	1	*155 565.5*	98		
3d 4d	^{1}F	3	215 957.7	97		
3d 4d	^{3}D	1	216 905.0	94	6	^{1}P
		2	217 108.0	100		
		3	217 350.0	98		
3d 4d	^{3}G	3	217 836.3	99		
		4	218 100.0	100		
		5	218 463.6	100		
3d 4d	^{1}P	1	217 990.7	93	6	^{3}D
3d 4d	^{3}S	1	220 343.5	99		
3d 4d	^{3}F	2	222 794.6	99		
		3	223 033.0	100		
		4	223 304.6	100		
3d 4d	^{1}D	2	225 804.1	95	4	^{3}P
3d 4d	^{3}P	0	226 521.6	100		
		1	226 617.1	100		
		2	226 796.3	96	4	^{1}D
3d 4d	^{1}G	4	227 712.5	100		
3d 4d	^{1}S	0	234 121.8	100		
3d 5s	^{3}D	1	236 148.6	100		
		2	236 322.4	94	6	^{1}D
		3	236 766.9	100		
3d 5s	^{1}D	2	237 638.8	94	6	^{3}D
3d 5p	^{1}D°	2	*254 468.8*	85	9	^{3}D°
3d 5p	^{3}D°	1	*254 824.1*	99	1	^{1}P°
		2	*255 146.8*	88	6	^{3}F°
		3	*255 445.5*	99	1	^{1}F°
3d 5p	^{3}F°	2	*255 463.3*	88	8	^{1}D°
		3	*255 747.6*	99	1	^{1}F°
		4	*256 251.7*	100		
3d 5p	^{3}P°	0	*256 739.9*	100		
		1	*256 781.8*	94	6	^{1}P°
		2	*257 143.2*	99	1	^{1}D°
3d 5p	^{1}F°	3	*257 690.8*	99	1	^{3}D°
3d 5p	^{1}P°	1	*258 288.8*	93	6	^{3}P°
3d4f	^{1}G°	4	*263 111.4*	82	9	^{3}F°
3d4f	^{3}F°	2	*263 593.0*	88	10	^{1}D°
		3	*263 608.3*	53	45	^{3}G°
		4	*264 113.1*	53	31	^{3}H°

(Continued)

V IV—Continued

Configuration	Term	J	Level (cm^{-1})	Leading percentages			
3d4f		4	*263 822.4*	39	^{3}G°	31	^{3}F°
3d4f	^{3}G°	3	*263 902.3*	49		39	^{3}F°
		5	*264 161.8*	67		32	^{3}H°
3d4f	^{3}H°	4	*264 401.9*	48		41	^{3}G°
		5	*264 591.9*	66		33	^{3}G°
		6	*264 845.7*	100			
3d4f	^{1}D°	2	*264 482.8*	87		8	^{3}F°
3d4f	^{1}F°	3	*264 902.2*	60		28	^{3}D°
3d4f	^{3}D°	1	*265 019.7*	96		2	^{3}P°
		2	*265 067.4*	90		4	^{3}P°
		3	*265 271.6*	67		32	^{1}F°
3d4f	^{3}P°	2	*265 879.2*	94		5	^{3}D°
3d4f	^{1}H°	5	*266 600.3*	98		2	^{3}H°
3d 5d	^{1}F	3	283 459.4	89		6	^{3}D
3d 5d	^{3}D	1	283 722.7	84		16	^{1}P
		2	283 940.4	99			
		3	284 226.7	78		13	^{3}G
3d 5d	^{3}G	3	284 101.1	83		16	^{3}D
		4	284 340.1	99			
		5	284 699.3	100			
3d 5d	^{1}P	1	284 365.7	80		16	^{3}D
3d 5d	^{3}S	1	285 298.6	95		4	^{1}P
3d 5d	^{3}F	2	285 798.9	97			
		3	286 056.9	98			
		4	286 286.5	99			
3d 5d	^{1}D	2	287 221.4	83		14	^{3}P
3d 5d	^{3}P	2	287 733.4	86		14	^{1}D
3d 5d	^{1}G	4	288 127.6	99			
3d 6s	^{3}D	1	291 796.0	100			
		2	291 918.1	85		15	^{1}D
		3	292 417.6	100			
3d 6s	^{1}D	2	292 766.7	84		15	^{3}D
$3d(^2D_{3/2})5g$	$^2[7/2]$	3	306 323.1	98		2	$(^2D_{5/2})\ ^2[7/2]$
		4	306 327.7	98		2	
$3d(^2D_{5/2})5g$	$^2[9/2]$	5	306 871.0	100			
		4	306 876.3	100			
V V ($^2D_{3/2}$)	***Limit***		**376 730**				

V v

$Z = 23$

K I isoelectronic sequence

Ground state: $1s^22s^22p^63s^23p^63d\ ^2D_{3/2}$

Ionization energy = 526 532±1 cm^{-1} (65.2822±0.0002 eV)

The early work of Gibbs and White (1929) revealed six low-lying doublet terms arising from the $3p^63d$, $4s$, $4p$, $4f$, $5s$, and $6s$ configurations. The first identifications of lines due to $3p^53d^2$ were given by Gabriel, Fawcett, and Jordan (1966).

The compilation is from the recent paper by Van Deurzen (1977). Earlier work by Van Deurzen, Conway, and Davis (1974) and by Ekberg (1974) is incorporated in the paper. Van Deurzen has observed the entire spectrum from 200–8500 Å with a vacuum sliding spark. The uncertainty of most of the level values is estimated to be ±0.03 relative to $3p^64p\ ^2P^\circ_{3/2}$=207 660.00±0.3 cm^{-1}, which reflects the larger uncertainty of the $3d-4p$ resonance transitions at 483 Å.

Van Deurzen (1977) calculated the ionization energy from ng, nh, and ni series members by means of a polarization formula.

References

Ekberg, J. O. (1974), Phys. Scr. **9**, 96.

Gabriel, A. H., Fawcett, B. C., and Jordan, C (1966), Proc. Phys. Soc. **87**, 825.

Gibbs, R. C., and White, H. E. (1929), Phys. Rev. **33**, 157.

Van Deurzen, C. H. H., Conway, J. G., and Davis, S. P. (1974), J. Opt. Soc. Am. **64**, 498.

Van Deurzen, C. H. H. (1977), J. Opt. Soc. Am. **67**, 476.

V v

Configuration	Term	J	Level (cm^{-1})	Configuration	Term	J	Level (cm^{-1})
$3p^6(^1S)3d$	^2D	3/2	0.00	$3p^6(^1S)6p$	$^2P^\circ$	1/2	*415 420.10*
		5/2	624.87			3/2	*415 675.69*
$3p^6(^1S)4s$	^2S	1/2	148 143.35	$3p^6(^1S)5g$	^2G	7/2	416 360.29
$3p^6(^1S)4p$	$^2P^\circ$	1/2	*206 393.72*			9/2	416 361.78
		3/2	*207 660.00*	$3p^6(^1S)5f$	$^2F^\circ$	5/2	*417 699.10*
$3p^6(^1S)4d$	^2D	3/2	293 902.86			7/2	*418 187.47*
		5/2	294 047.24	$3p^6(^1S)6d$	^2D	3/2	434 303.77
$3p^5(^2P^\circ)3d^2(^1G)$	$^2F^\circ$	5/2	*319 106.19*			5/2	434 340.92
		7/2	*320 731.60*	$3p^5(^2P^\circ)3d^2(^3P)$	$^2P^\circ$	1/2	*438 018.3*
$3p^6(^1S)5s$	^2S	1/2	328 217.30			3/2	*439 442.7*
$3p^5(^2P^\circ)3d^2(^1D)$	$^2F^\circ$	7/2	*332 198.1*	$3p^6(^1S)7s$	^2S	1/2	443 074.72
		5/2	*337 012.59*	$3p^5(^2P^\circ)3d^2(^3F)$	$^2D^\circ$	5/2	*444 153.7*
$3p^6(^1S)4f$	$^2F^\circ$	7/2	*349 252.40*			3/2	*444 620.8*
		5/2	*349 675.57*	$3p^6(^1S)6f$	$^2F^\circ$	5/2	*449 370.81*
$3p^6(^1S)5p$	$^2P^\circ$	1/2	*351 500.51*			7/2	*449 422.47*
		3/2	*352 018.34*	$3p^6(^1S)7p$	$^2P^\circ$	1/2	*449 586.71*
$3p^6(^1S)5d$	^2D	3/2	387 977.07			3/2	*449 772.79*
		5/2	388 043.69	$3p^6(^1S)6g$	^2G	7/2	450 024.54
$3p^5(^2P^\circ)3d^2(^3F)$	$^2F^\circ$	5/2	*396 135.24*			9/2	450 025.20
		7/2	*397 993.66*	$3p^6(^1S)6h$	$^2H^\circ$	9/2, 11/2	*450 247.99*
$3p^6(^1S)6s$	^2S	1/2	403 855.12	$3p^6(^1S)7d$	^2D	5/2	460 719.7

(Continued)

V V—Continued

Configuration	Term	J	Level (cm^{-1})	Configuration	Term	J	Level (cm^{-1})
$3p^6(^1S)8s$	^2S	$1/2$	466 065.79	$3p^6(^1S)8i$	^2I	$11/2, 13/2$	483 650.82
$3p^6(^1S)7f$	$^2F°$	$5/2$	*469 706.4*	$3p^6(^1S)9f$	$^2F°$	$5/2$	*492 144.3*
		$7/2$	*469 721*			$7/2$	*492 201.8*
$3p^6(^1S)7g$	^2G	$7/2$	470 333.35	$3p^5\,3d(^1F°)4s$	$^2F°$	$5/2$	*496 296*
		$9/2$	470 333.75			$7/2$	*497 556*
$3p^6(^1S)7h$	$^2H°$	$9/2, 11/2$	*470 488.77*	$3p^5\,3d(^3D°)4s$	$^2D°$	$5/2$	*500 117*
$3p^6(^1S)7i$	^2I	$11/2, 13/2$	470 524.11			$3/2$	*500 502*
$3p^5\,3d(^3F°)4s$	$^2F°$	$7/2$	*475 531*	V VI ($^1S_0$)	**Limit**		**526 532**
		$5/2$	*478 566*				
$3p^6(^1S)8f$	$^2F°$	$5/2$	*483 019*				
		$7/2$	*483 039*				

V VI

$Z=23$

Ar I isoelectronic sequence

Ground state: $1s^2 2s^2 2p^6 3s^2 3p^6\ ^1S_0$

Ionization energy = 1 033 400±200 cm^{-1} (128.13±0.02 eV)

Most of the analysis given here is by Ekberg (1976), based on measurements of the spectrum from 100–2500 Å. The earlier papers, which are concerned mainly with resonance lines, are referred to in his work. An uncertainty of ±2.5 cm^{-1} is assigned to the level values. He has calculated the percentage compositions for the levels and designated those of the $3p^5 3d$, $4p$, and $3s3p^6 3d$ configurations in LS coupling. The *jl*-scheme is used for $3p^5 4s$, $5s$, $4d$, and $4f$. His two $3p^5 5d$ levels have been redesignated as $3s3p^6 4p$ levels by Kastner, Crooker, Behring, and Cohen (1977). They have classified seven resonance lines observed in absorption near 100 Å as transitions to the $3s3p^6 4p$ to $8p$ series and determined the ionization energy. The level uncertainty obtained from these data is ±100 cm^{-1}.

References

Ekberg, J. O. (1976), Phys. Scr. **13**, 111.

Kastner, S. O., Crooker, A. M., Behring, W. E., and Cohen, L. (1977), Phys. Rev. **A16**, 577.

V VI

Configuration	Term	J	Level (cm^{-1})	Leading percentages		
$3s^2 3p^6$	^{1}S	0	0.0			
$3s^2 3p^5 3d$	^{3}P°	0	*308 149.8*	100		
		1	*309 394.8*	100		
		2	*311 977.9*	99		
$3s^2 3p^5 3d$	^{3}F°	4	*322 773.6*	100		
		3	*324 958.0*	98		
		2	*327 214.9*	97		
$3s^2 3p^5 3d$	^{1}D°	2	*345 139.4*	81	17	^{3}D°
$3s^2 3p^5 3d$	^{3}D°	3	*345 516.5*	77	23	^{1}F°
		1	*347 899.9*	100		
		2	*348 325.3*	81	17	^{1}D°
$3s^2 3p^5 3d$	^{1}F°	3	*350 644.5*	76	22	^{3}D°
$3s^2 3p^5 3d$	^{1}P°	1	*445 435.6*	100		
$3s3p^6 3d$	^{3}D	1	549 538.0	100		
		2	549 863.6	100		
		3	550 384.6	100		
$3s3p^6 3d$	^{1}D	2	566 433.0	100		
$3s^2 3p^5(^2P^\circ_{3/2})4s$	^{2}[3/2]°	2	*546 284.0*	100		
		1	*549 298.8*	82	18	($^2P^\circ_{1/2}$) ^{2}[1/2]°
$3s^2 3p^5(^2P^\circ_{1/2})4s$	^{2}[1/2]°	0	*553 820.1*	100		
		1	*557 636.1*	82	18	($^2P^\circ_{3/2}$) ^{2}[3/2]°
$3s^2 3p^5 4p$	^{3}S	1	602 974.3	96		
$3s^2 3p^5 4p$	^{3}D	3	612 289.7	100		
		2	612 392.8	71	24	^{1}D
		1	615 177.8	63	22	^{1}P

(Continued)

V VI—Continued

Configuration	Term	J	Level (cm^{-1})	Leading percentages		
$3s^2\,3p^5\,4p$	^{3}P	2	617 490.0	57	40	^{1}D
		0	621 757.1	98		
		1	623 594.5	66	30	^{1}P
$3s^2\,3p^5\,4p$	^{1}P	1	620 509.2	48	36	^{3}D
$3s^2\,3p^5\,4p$	^{1}D	2	622 724.5	38	36	^{1}D
$3s^2\,3p^5\,4p$	^{1}S	0	641 800.3	98		
$3s^2\,3p^5(^2P^\circ_{3/2})4d$	$^2[1/2]^\circ$	0	*707 280.3*	100		
		1	*708 044.6*	73	21	$(^2P^\circ_{3/2})\ ^2[3/2]^\circ$
$3s^2\,3p^5(^2P^\circ_{3/2})4d$	$^2[3/2]^\circ$	2	*709 747.4*	83	16	$(^2P^\circ_{1/2})\ ^2[3/2]^\circ$
		1	*716 760.4*	78	20	$(^2P^\circ_{3/2})\ ^2[1/2]^\circ$
$3s^2\,3p^5(^2P^\circ_{3/2})4d$	$^2[7/2]^\circ$	4	*710 695.2*	100		
		3	*711 426.2*	92	6	$(^2P^\circ_{3/2})\ ^2[5/2]^\circ$
$3s^2\,3p^5(^2P^\circ_{3/2})4d$	$^2[5/2]^\circ$	2	*713 742.3*	86	13	$(^2P^\circ_{1/2})\ ^2[5/2]^\circ$
		3	*714 667.9*	86	8	$(^2P^\circ_{3/2})\ ^2[7/2]^\circ$
$3s^2\,3p^5(^2P^\circ_{1/2})4d$	$^2[5/2]^\circ$	2	*720 074.9*	86	13	$(^2P^\circ_{3/2})\ ^2[5/2]^\circ$
		3	*720 836.0*	91	8	$(^2P^\circ_{3/2})\ ^2[7/2]^\circ$
$3s^2\,3p^5(^2P^\circ_{1/2})4d$	$^2[3/2]^\circ$	2	*721 187.6*	83	16	$(^2P^\circ_{3/2})\ ^2[3/2]^\circ$
		1	*723 421.6*	93	7	$(^2P^\circ_{3/2})\ ^2[1/2]^\circ$
$3s^2\,3p^5(^2P^\circ_{3/2})5s$	$^2[3/2]^\circ$	2	*770 494.5*	100		
		1	*771 723.1*	98		
$3s^2\,3p^5(^2P^\circ_{3/2})4f$	$^2[3/2]$	1	777 549.4			
		2	778 194.1			
$3s^2\,3p^5(^2P^\circ_{1/2})5s$	$^2[1/2]^\circ$	1	*778 944.0*	98		
$3s^2\,3p^5(^2P^\circ_{3/2})4f$	$^2[5/2]$	3	779 550.9			
$3s^2\,3p^5(^2P^\circ_{3/2})4f$	$^2[9/2]$	5	781 295.9			
		4	782 345.4			
$3s^2\,3p^5(^2P^\circ_{3/2})4f$	$^2[7/2]$	3	783 852.1			
		4	785 705.4			
$3s^2\,3p^5(^2P^\circ_{1/2})4f$	$^2[7/2]$	4	791 839.6			
$3s3p^6\,4p$	^{3}P°	1	*841 980*			
$3s3p^6\,4p$	^{1}P°	1	*849 170*			
$3s3p^6\,5p$	^{3}P°	1	*1 017 100*			
$3s3p^6\,5p$	^{1}P°	1	*1 021 120*			
V VII $(^2P^\circ_{3/2})$	***Limit***		**1 033 400**			
$3s3p^6\,6p$	^{1}P°	1	*1 102 540*			

V VI—Continued

Configuration	Term	J	Level (cm^{-1})	Leading percentages
$3s3p^6\,7p$	$^1P°$	1	*1 148 030*	
$3s3p^6\,8p$	$^1P°$	1	*1 175 490*	

V VII

$Z=23$

Cl I isoelectronic sequence

Ground state: $1s^2 2s^2 2p^6 3s^2 3p^5\ ^2P^\circ_{3/2}$

Ionization energy = 1 215 000±2000 cm^{-1} (150.6±0.2 eV)

All the known levels are derived from transitions to the $3s^2 3p^5\ ^2P^\circ$ ground term. Edlén (1937) identified lines originating from all levels of the $3s^2 3p^4 4s$ configuration. Earlier, Weissberg and Kruger (1936) had reported lines from the ^2P term of this configuration as well as from the ^2S term of $3s3p^6$. The transitions from $3s^2 3p^4 3d$ were identified by Gabriel, Fawcett, and Jordan (1966) and by Fawcett and Gabriel (1966). The parent states of $3p^4 3d$ levels are determined from the calculation by Bromage (1980). The transitions from $3s^2 3p^4 4d$ were identified by Fawcett, Peacock, and Cowan (1968) and by Fawcett, Cowan, and Hayes (1972). Line identifications in the $3p^4 4d - 3p^4 4f$ transition array were also given in the latter paper but cannot be used to derive energy levels because they do not embrace levels known with respect to the ground term.

The recent measurements of the $3p^5 - 3s3p^6$ doublet by Smitt, Svensson, and Outred (1976) were used to determine the ground term interval, with an uncertainty of ±1.4 cm^{-1}, and the $3s3p^6\ ^2S_{1/2}$ with an uncertainty of ±10 cm^{-1}. The uncertainty in the rest of the level values is ±100 cm^{-1}.

The ionization energy is an extrapolated value by Lotz (1967).

References

Bromage, G. E. (1980), Astron. and Astrophys. Suppl. Series **41**, 79.
Edlén, B. (1937), Z. Phys. **104**, 407.
Fawcett, B. C., and Gabriel, A. H. (1966), Proc. Phys. Soc. **88**, 262.
Fawcett, B. C., Peacock, N. J., and Cowan, R. D. (1968), J. Phys. **B1**, 295.
Fawcett, B. C., Cowan, R. D., and Hayes, R. W. (1972), J. Phys. **B5**, 2143.
Gabriel, A. H., Fawcett, B. C., and Jordan, C. (1965), Nature **206**, 390.
Lotz, W. J. (1967), J. Opt. Soc. Am. **57**, 873.
Smitt, R., Svensson, L. A., and Outred, M. (1976), Phys. Scr. **13**, 293.
Weissberg, S. G., and Kruger, P. G. (1936), Phys. Rev. **49**, 872.

V VII

Configuration	Term	J	Level (cm^{-1})	Configuration	Term	J	Level (cm^{-1})
$3s^2\,3p^5$	$^2P^\circ$	3/2	*0*	$3s^2\,3p^4(^1D)4s$	^2D	5/2	638 540
		1/2	*7 668*			3/2	638 710
$3s3p^6$	^2S	1/2	219 162	$3s^2\,3p^4(^1S)4s$	^2S	1/2	671 570
$3s^2\,3p^4(^1D)3d$	^2S	1/2	421 050	$3s^2\,3p^4(^3P)4d$	^2D	5/2	793 650
						3/2	794 570
$3s^2\,3p^4(^3P)3d$	^2P	3/2	435 970				
		1/2	438 770	$3s^2\,3p^4(^1P)4d$	^2S	1/2	812 550
$3s^2\,3p^4(^3P)3d$	^2D	5/2	444 130	$3s^2\,3p^4(^1D)4d$	^2P	3/2	815 660
		3/2	450 550				
				$3s^2\,3p^4(^1D)4d$	^2D	5/2	820 010
$3s^2\,3p^4(^3P)4s$	^4P	5/2	608 640			3/2	820 440
		3/2	612 810				
		1/2	615 490	$3s^2\,3p^4(^1S)4d$	^2D	5/2	853 200
$3s^2\,3p^4(^3P)4s$	^2P	3/2	620 650				
		1/2	625 570	V VIII ($^3P_2$)	***Limit***		**1 215 000**

V VIII

$Z = 23$

S I isoelectronic sequence

Ground state: $1s^2 2s^2 2p^6 3s^2 3p^4\ ^3P_2$

Ionization energy = 1 399 000±3000 cm^{-1} (173.4±.4 eV)

Edlén (1937) identified the $3p^4 - 3p^3 4s$ array which occurs between 135 and 148 Å. He identified singlet and triplet levels and two intercombination lines with an uncertainty of ±50 cm^{-1}. The present values for the levels of the ground configuration and $3s3p^5$ were measured by Smitt, Svensson, and Outred (1976), with an uncertainty of ±5 cm^{-1}.

The $3p^3 3d$ configuration was reported by Gabriel, Fawcett, and Jordan (1966) and by Fawcett and Gabriel (1966).

The $3p^3 4d$ terms were identified by Fawcett, Cowan, and Hayes (1972). They also observed nine transitions in the $3d - 4f$ array that are not connected with the present system of levels. The uncertainty in the $3s$, $3d$, and $4d$ level values is ±100 cm^{-1}.

The ionization energy is an extrapolated value by Lotz (1967).

References

Edlén, B. (1937), Z. Phys. **104**, 188.

Fawcett, B. C., and Gabriel, A. H. (1966), Proc. Phys. Soc. **88**, 262.

Fawcett, B. C., Cowan, R. D., and Hayes, R. W. (1972), J. Phys. **B5**, 2143.

Gabriel, A. H., Fawcett, B. C., and Jordan, C. (1966), Proc. Phys. Soc. **87**, 825.

Lotz, W. (1967), J. Opt. Soc. Am. **57**, 873.

Smitt, R., Svensson, L. A., and Outred, M. (1976), Phys. Scr. **13**, 293.

V VIII

Configuration	Term	J	Level (cm^{-1})
$3s^2 3p^4$	^3P	2	0.0
		1	6 007.8
		0	7 579.6
$3s^2 3p^4$	^1D	2	27 072
$3s^2 3p^4$	^1S	0	60 641
$3s3p^5$	$^3P°$	2	217 486.3
		1	222 405.6
		0	225 241.6
$3s3p^5$	$^1P°$	1	278 200
$3s^2 3p^3(^2D°)3d$	$^3P°$	2	416 330
$3s^2 3p^3(^4S°)3d$	$^3D°$	3	434 560
		2	438 300
		1	440 800
$3s^2 3p^3(^2D°)3d$	$^1D°$	2	450 780
$3s^2 3p^3(^2D°)3d$	$^1F°$	3	464 380
$3s^2 3p^3(^4S°)4s$	$^3S°$	1	687 260
$3s^2 3p^3(^2D°)4s$	$^3D°$	1	710 600
		2	710 910
		3	711 990
$3s^2 3p^3(^2D°)4s$	$^1D°$	2	718 430
$3s^2 3p^3(^2P°)4s$	$^3P°$	0	734 250
		1	734 890
		2	736 640
$3s^2 3p^3(^2P°)4s$	$^1P°$	1	742 720
$3s^2 3p^3(^4S°)4d$	$^3D°$	2	872 410
		3	872 680
		1	872 780
$3s^2 3p^3(^2D°)4d$	$^3P°$	2	903 350
$3s^2 3p^3(^2D°)4d$	$^3D°$	3	904 570
		2	905 990
$3s^2 3p^3(^2D°)4d$	$^1D°$	2	907 350
$3s^2 3p^3(^2D°)4d$	$^1F°$	3	909 920
$3s^2 3p^3(^4S°)4d$	$^1P°$	1	938 450
V IX ($^4S°_{3/2}$)	*Limit*		**1 399 000**

V IX

$Z=23$

P I isoelectronic sequence

Ground state: $1s^2 2s^2 2p^6 3s^2 3p^3\ ^4S^\circ_{3/2}$

Ionization energy = 1 660 000±3000 cm^{-1} (205.8±0.4 eV)

The levels of the $3s^2 3p^3 - 3s3p^4$ array are from Smitt, Svensson, and Outred (1976) who improved the measurements and extended the classifications of the earlier work of Fawcett and Peacock (1967) and Fawcett (1970, 1971). No intersystem transition have been observed; the quartet-doublet separation is based on calculations by Smitt et al. with an uncertainty of ±20 cm^{-1}. The error in the levels with the same multiplicity is ±5 cm^{-1}.

The $3p^2 3d$ configuration is from Fawcett (1971), who greatly extended the earlier identifications of Gabriel, Fawcett, and Jordan (1966) and Fawcett, Gabriel, and Saunders (1967). The $3p^2 4s$ configuration is from the paper by Fawcett, Cowan, and Hayes (1972), who also identified two unconnected multiplets in the $3d-4f$ array, from the early work of Kruger and Pattin (1937). The uncertainty in the $3d$ and $4s$ levels is ±100 cm^{-1}.

The ionization energy was obtained by Lotz (1967) by extrapolation.

References

Fawcett, B. C. and Peacock, N. J. (1967), Proc. Phys. Soc. **91**, 973.
Fawcett, B. C., Gabriel, A. H., and Saunders, P. A. H. (1967), Proc. Phys. Soc. **90**, 863.
Fawcett, B. C. (1970), J. Phys. **B3**, 1732.
Fawcett, B. C. (1971), J. Phys. **B4**, 1577.
Fawcett, B. C., Cowan, R. D., and Hayes, R. W. (1972), J. Phys. **B5**, 2143.
Gabriel, A. H., Fawcett, B. C., and Jordan, C. (1965), Nature **206**, 390.
Kruger, P. G., and Pattin, H. S. (1937), Phys. Rev. **52**, 621.
Lotz, W. (1967), J. Opt. Soc. Am. **57**, 873.
Smitt, R., Svensson, L. A., and Outred, M. (1976), Phys. Scr. **13**, 293.

V IX

Configuration	Term	J	Level (cm^{-1})	Configuration	Term	J	Level (cm^{-1})
$3s^2 3p^3$	$^4S^\circ$	3/2	*0*	$3s^2 3p^2(^1D)3d$	^2D	5/2	438 070+x
						3/2	438 420+x
$3s^2 3p^3$	$^2D^\circ$	3/2	*34 708+x*				
		5/2	*36 319+x*	$3s^2 3p^2(^1D)3d$	^2P	1/2	452 470+x
						3/2	456 150+x
$3s^2 3p^3$	$^2P^\circ$	1/2	*59 028+x*				
		3/2	*61 224+x*	$3s^2 3p^2(^3P)3d$	^2F	7/2	460 550+x
$3s3p^4$	^4P	5/2	214 067	$3s^2 3p^2(^3P)3d$	^2D	5/2	477 370+x
		3/2	218 814				
		1/2	221 174	$3s^2 3p^2(^3P)4s$	^4P	1/2	789 070
						3/2	792 690
$3s3p^4$	^2D	3/2	265 160+x			5/2	797 320
		5/2	265 835+x				
				$3s^2 3p^2(^3P)4s$	^2P	1/2	802 220+x
$3s3p^4$	^2P	3/2	305 664+x			3/2	807 570+x
		1/2	309 210+x				
				$3s^2 3p^2(^1D)4s$	^2D	5/2	823 290+x
$3s3p^4$	^2S	1/2	319 184+x			3/2	823 570+x
$3s^2 3p^2(^3P)3d$	^2P	3/2	398 530+x				
		1/2	404 560+x				
				V X ($^3P_0$)	*Limit*		**1 660 000**
$3s^2 3p^2(^3P)3d$	^4P	5/2	408 350				
		3/2	409 060				
		1/2	410 540				

V X

$Z=23$

Si I isoelectronic sequence

Ground state: $1s^22s^22p^63s^23p^2\ ^3P_0$

Ionization energy = 1 859 000±4000 cm^{-1} (230.5±0.2 eV)

The $3s^23p^2-3s3p^3$ array was measured by Smitt, Svensson, and Outred (1976), who improved the measurements and extended the classifications of Fawcett and Peacock (1967) and Fawcett (1970, 1971), including the discovery of the intersystem connection. Their level value uncertainty is ±5 cm^{-1}.

The levels of $3s^23p3d$ were determined by Fawcett, Gabriel, and Saunders (1967) and Fawcett (1971) with an uncertainty of ±100 cm^{-1}. Those of $3s^23p4s$ and $3s^23p4d$ are due to Fawcett, Cowan, and Hayes (1972), whose level uncertainty is ±300 cm^{-1}.

The ionization energy was obtained by extrapolation by Lotz (1967).

References

Fawcett, B. C., and Peacock, N. J. (1967), Proc. Phys. Soc. **91,** 973.

Fawcett, B. C., Gabriel, A. H., and Saunders, P. A. H. (1967), Proc. Phys. Soc. **90,** 863.

Fawcett, B. C. (1970), J. Phys. **B3,** 1732.

Fawcett, B. C. (1971), J. Phys. **B4,** 1577.

Fawcett, B. C., Cowan, R. D., and Hayes, R. W. (1972), J. Phys. **B5,** 2143.

Lotz, W. J. (1967), J. Opt. Soc. Am. **57,** 873.

Smitt, R., Svensson, L. A., and Outred, M. (1976), Phys. Scr. **13,** 293.

V X

Configuration	Term	J	Level (cm^{-1})	Configuration	Term	J	Level (cm^{-1})
$3s^2\,3p^2$	^3P	0	0	$3s^2\,3p3d$	$^3P°$	2	*385 790*
		1	4 180			1	*391 340*
		2	9 421				
				$3s^2\,3p3d$	$^3D°$	1	*399 130*
$3s^2\,3p^2$	^1D	2	32 509			2	*400 740*
						3	*401 210*
$3s^2\,3p^2$	^1S	0	67 751				
				$3s^2\,3p3d$	$^1F°$	3	*440 090*
$3s3p^3$	$^3D°$	1	*220 984*				
		2	*221 072*	$3s^2\,3p4s$	$^3P°$	1	*865 200*
		3	*222 104*			2	*873 100*
$3s3p^3$	$^3P°$	0	*253 936*	$3s^2\,3p4s$	$^1P°$	1	*878 700*
		1	*254 147*				
		2	*254 337*	$3s^2\,3p4d$	$^3D°$	3	*1 070 600*
$3s3p^3$	$^1D°$	2	*279 969*	$3s^2\,3p4d$	$^1F°$	3	*1 085 600*
$3s3p^3$	$^3S°$	1	*327 902*				
$3s3p^3$	$^1P°$	1	*341 335*	V XI ($^2P^\circ_{1/2}$)	*Limit*		**1 859 000**

V XI

$Z=23$

Al I isoelectronic sequence

Ground state: $1s^2 2s^2 2p^6 3s^2 3p\ ^2P^\circ_{1/2}$

Ionization energy $= 2\,062\,000 \pm 4000$ cm^{-1} (255.7 ± 0.2 eV)

The $3p-4d$ transitions were identified by Edlén (1936). Observations of this spectrum by Gabriel, Fawcett, and Jordan (1966) resulted in the discovery of the $3p-3d$ lines. The uncertainty of the $3d$ and $4d$ level values is ± 100 cm^{-1}.

The doublet terms of $3s3p^2$ were first observed by Fawcett and Peacock (1967) in a laser-produced plasma. They were remeasured by Fawcett (1970), who also observed the $^4P-^4S^\circ$ multiplet of the $3s3p^2-3p^3$ array. The present values of the ground term and the doublet terms of $3s3p^2$ are from Smitt, Svensson, and Outred (1976), with an uncertainty of ± 5 cm^{-1}.

Calculated spectra enabled Fawcett, Cowan, Kononov, and Hayes (1972) to identify the $3d-4f$ and $3p-4s$ lines from a theta-pinch spectrum and to determine level values with an uncertainty of ± 300 cm^{-1}. They also identified the $3s3p^2\ ^4P-3s3p4s\ ^4P^\circ$ and $3s3p3d\ ^4F^\circ-3s3p4f\ ^4G$ multiplets, the second of which has not been connected to the present system of levels.

No intersystem transitions have been observed. We have based the quartet system on the $3s3p^2\ ^4P_{5/2}$ level, the position of which is estimated from our extrapolation beyond the sequence Al I–P III. The uncertainty in our extrapolation is about ± 2000 cm^{-1}.

The value for the ionization energy was obtained by extrapolation by Lotz (1967).

References

Edlén, B. (1936), Z. Phys. **103**, 536.

Fawcett, B. C., and Peacock, N. J. (1967), Proc. Phys. Soc. **91**, 973.

Fawcett, B. C. (1970), J. Phys. **B3**, 1732.

Fawcett, B. C., Cowan, R. D., Kononov, E. Y., and Hayes, R. W. (1972), J. Phys. **B5**, 1255.

Gabriel, A. H., Fawcett, B. C., and Jordan, C. (1966), Proc. Phys. Soc. **87**, 825.

Lotz, W. (1967), J. Opt. Soc. Am. **57**, 873.

Smitt, R., Svensson, L. A., and Outred, M. (1976), Phys. Scr. **13**, 293.

V XI

Configuration	Term	J	Level (cm^{-1})	Configuration	Term	J	Level (cm^{-1})
$3s^2 3p$	$^2P^\circ$	1/2	*0*	$3p^3$	$^4S^\circ$	3/2	*453 640+x*
		3/2	*9 696*				
				$3s^2 4s$	^2S	1/2	939 500
$3s3p^2$	^4P	1/2	166 300+x				
		3/2	169 600+x	$3s3p4s$	$^4P^\circ$	3/2	*1 118 400+x*
		5/2	175 000+x			5/2	*1 124 300+x*
$3s3p^2$	^2D	3/2	232 973	$3s^2 4d$	^2D	3/2	1 147 240
		5/2	233 778			5/2	1 147 770
$3s3p^2$	^2S	1/2	288 914	$3s^2 4f$	$^2F^\circ$	5/2	*1 215 300*
						7/2	*1 215 500*
$3s3p^2$	^2P	1/2	306 801				
		3/2	311 890				
$3s^2 3d$	^2D	3/2	376 920	V XII ($^1S_0$)	*Limit*		**2 062 000**
		5/2	377 680				

V XII

$Z=23$

Mg I isoelectronic sequence

Ground state: $1s^2 2s^2 2p^6 3s^2\ ^1S_0$

Ionization energy = 2 485 000±3000 cm^{-1} (308.1±0.4 eV)

Edlén (1936) reported three unconnected systems of levels for this ion: the resonance line $3s^2\ ^1S_0 - 3s4p\ ^1P_1^\circ$; the triplets of $3s3p - 3s4s$, $3s4d$, $3s5d$, and the triplets of $3s3d - 3s4f$, $3s5f$. The triplets were unified by the work of Fawcett (1970), and Fawcett, Cowan, and Hayes (1972) who identified the $3s3p\ ^3P^\circ - 3s3d\ ^3D$ multiplet. Fawcett and Peacock (1967) reported the $3s^2\ ^1S_0 - 3s3p\ ^1P_1^\circ$ line at 355.11±0.05 Å. The triplet system is connected to the singlets by the $3s^2\ ^1S_0 - 3s3p\ ^3P_1^\circ$ line observed at 522.4 ±0.2 Å in a tokamak plasma by Finkenthal, Bell, and Moos (1982). The uncertainty in the levels of these configurations is ±100 cm^{-1}.

The $3p^2\ ^3P$ and $3s3p\ ^1P^\circ$ terms were reported by Fawcett and Peacock (1967). Fawcett (1970) provided the ^1S and ^1D of $3p^2$ and the known terms of $3p3d$. However, his identification of the lines arising from the $3p^2\ ^1D$ and ^1S terms are disputed by Kastner and Bhatia (1979) on the basis of their isoelectronic study.

Fawcett, Cowan, and Hayes (1972) found the singlets of $3s3d$, $3s4d$, and $3s4f$. Their publication was accompanied by a supplementary report which provides extensions of the analysis. Here they identified the $3p4s\ ^3P^\circ$, the $3p4f$ and $3p4d$ configurations, the $3s5s$ to $3s7s\ ^3S$ series, the $3s5p$ to $3s10p\ ^1P^\circ$ series, the $3s6d$ to $3s8d\ ^3D_3$ series, and the $3s6f$ to $3s8f\ ^3F_4^\circ$ series. The uncertainty in these level values is ±400 cm^{-1}.

The terms of $3p4f$ given here are from Fawcett, Cowan, Kononov, and Hayes (1972).

We determined the ionization energy from the $3snf$ series.

References

Edlén, B. (1936), Z. Phys. **103**, 536.

Ekberg, J. O. (1971), Phys. Scr. **4**, 101.

Fawcett, B. C., and Peacock, N. J. (1967), Proc. Phys. Soc. **91**, 973.

Fawcett, B. C. (1970), J. Phys. **B3**, 1732.

Fawcett, B. C., Cowan, R. D., Kononov, E Y., and Hayes, R. W. (1972), J. Phys. **B5**, 1255.

Fawcett, B. C., Cowan, R. D., and Hayes, R. W. (1972), J. Phys. **B5**, 2143 and Supplementary Publication No. SUP70005 (National Lending Library, Boston Spa, Yorkshire, LS237BQ, G.B.)

Finkenthal, M., Bell, R. E., and Moos, H. W. (1982), Phys. Lett. **88A**, 165.

Kastner, S. O., and Bhatia, A. K. (1979), J. Opt. Soc. Am. **69**, 1391.

V XII

Configuration	Term	J	Level (cm^{-1})	Configuration	Term	J	Level (cm^{-1})
$3s^2$	^1S	0	0	$3p3d$	$^3D^\circ$	3	*795 100*
$3s3p$	$^3P^\circ$	0	*188 330*	$3p3d$	$^1P^\circ$	1	*816 300*
		1	*191 420*				
		2	*198 580*	$3s4s$	^3S	1	1 212 500
$3s3p$	$^1P^\circ$	1	*281 600*	$3s4s$	^1S	0	1 227 300
$3p^2$	^3P	0	446 170	$3s4p$	$^1P^\circ$	1	*1 310 500*
		1	450 770				
		2	459 270	$3s4d$	^3D	1	1 424 500
						2	1 424 820
$3p^2$	^1D	2	445 740?			3	1 425 380
$3p^2$	^1S	0	524 840?	$3s4d$	^1D	2	1 426 200
$3s3d$	^3D	1	544 060	$3p4s$	$^3P^\circ$	0	*1 447 100*
		2	544 390			2	*1 458 100*
		3	545 000				
				$3s4f$	$^3F^\circ$	3	*1 480 540*
$3s3d$	^1D	2	613 280			2	*1 480 560*
						4	*1 480 590*
$3p3d$	$^3F^\circ$	2	*744 300*				
		3	*749 100*	$3s4f$	$^1F^\circ$	3	*1 492 100*
		4	*754 700*				

(Continued)

V XII—Continued

Configuration	Term	J	Level (cm^{-1})
$3p4p$	^3D	3	1 544 100
$3p4p$	^3P	1	1 545 300
		2	1 550 500
$3p4p$	^3S	1	1 552 500
$3p4d$	$^1F^\circ$	3	*1 640 800*
$3p4d$	$^3D^\circ$	2	*1 657 900*
		1	*1 658 100*
		3	*1 662 200*
$3p4d$	$^3F^\circ$	3	*1 660 100*
$3p4d$	$^3P^\circ$	2	*1 678 500*
$3p4f$	^3G	3	1 700 500
		4	1 704 600
		5	1 712 100
$3p4f$	^3F	4	1 713 100
$3s5s$	^3S	1	1 716 800
$3p4f$	^3D	3	1 722 500
$3s5p$	$^1P^\circ$	1	*1 765 100*
$3s5d$	^3D	1	1 818 300
		2	1 818 600
		3	1 818 900
$3s5d$	^1D	2	1 822 000
$3s5f$	$^3F^\circ$	4	*1 844 400*
$3s6s$	^3S	1	1 958 600
$3s6p$	$^1P^\circ$	1	*1 997 800*
$3s6d$	^3D	2	2 026 500
		3	2 026 700
$3s6f$	$^3F^\circ$	4	*2 041 900*
$3s7s$	^3S	1	2 110 100
$3s7p$	$^1P^\circ$	1	*2 131 600*
$3s7d$	^3D	3	2 151 400
$3s7f$	$^3F^\circ$	4	*2 160 000*
$3s8p$	$^1P^\circ$	1	*2 218 700*
$3s8d$	^3D	3	2 230 100
$3s8f$	$^3F^\circ$	4	*2 237 300*
$3s9p$	$^1P^\circ$	1	*2 271 000*
$3s10p$	$^1P^\circ$	1	*2 306 400*
V XIII ($^2S_{1/2}$)	***Limit***		2 485 000

V XIII

$Z=23$

Na I isoelectronic sequence

Ground state: $1s^22s^22p^63s\ ^2S_{1/2}$

Ionization energy = 2 712 250±100 cm^{-1} (336.279±0.010 eV)

The $3p$ and $3d$ levels are determined from the observations of Fawcett, Cowan, and Hayes (1972) with an uncertainty of ±50 cm^{-1}. The $4s$, $4p$, $4d$, $4f$, $5p$, $5d$, and $5f$ levels are from the observations of Edlén (1936) between 50 and 100 Å with an uncertainty of ±100 cm^{-1}. The levels of $6s$ to $11s$, $10d$, $10f$, and $11f$ are from Fawcett, Cowan, and Hayes. The $2p^53s^2\ ^2P°$ was observed by Feldman and Cohen (1967) at 24 Å. The remaining levels are from the measurements and assignments of Cohen and Behring (1976) from 39–62 Å. The uncertainty in these levels is estimated to be ±500 cm^{-1}.

The value for the ionization energy was derived by Edlén (1978) from core polarization theory applied to the nf series.

References

Cohen, L., and Behring, W. E. (1976), J. Opt. Soc. Am. **66**, 899.
Edlén, B. (1936), Z. Phys. **100**, 621.
Edlén, B. (1978), Phys. Scr. **17**, 565.
Fawcett, B. C., Cowan, R. D., and Hayes, R. W. (1972), J. Phys. **B5**, 2143.
Feldman, U., and Cohen, L. (1967), J. Opt. Soc. Am. **57**, 1128.

V XIII

Configuration	Term	J	Level (cm^{-1})
$2p^6(^1S)3s$	^2S	1/2	0
$2p^6(^1S)3p$	$^2P°$	1/2	*225 530*
		3/2	*236 570*
$2p^6(^1S)3d$	^2D	3/2	544 640
		5/2	545 980
$2p^6(^1S)4s$	^2S	1/2	1 300 490
$2p^6(^1S)4p$	$^2P°$	1/2	*1 388 410*
		3/2	*1 392 780*
$2p^6(^1S)4d$	^2D	3/2	1 505 900
		5/2	1 506 480
$2p^6(^1S)4f$	$^2F°$	5/2	*1 549 430*
		7/2	*1 549 740*
$2p^6(^1S)5s$	^2S	1/2	1 846 100
$2p^6(^1S)5p$	$^2P°$	1/2	*1 889 360*
		3/2	*1 891 430*
$2p^6(^1S)5d$	^2D	3/2	1 946 230
		5/2	1 946 500
$2p^6(^1S)5f$	$^2F°$	5/2	*1 967 880*
		7/2	*1 967 990*
$2p^6(^1S)6s$	^2S	1/2	2 127 000
$2p^6(^1S)6p$	$^2P°$	1/2	*2 151 400*
		3/2	*2 152 400*
$2p^6(^1S)6d$	^2D	3/2	2 182 800
		5/2	2 183 000
$2p^6(^1S)6f$	$^2F°$	5/2	*2 194 900*
		7/2	*2 195 100*
$2p^6(^1S)7s$	^2S	1/2	2 290 400
$2p^6(^1S)7p$	$^2P°$	1/2,3/2	*2 305 700*
$2p^6(^1S)7d$	^2D	3/2	2 324 700
		5/2	2 325 000
$2p^6(^1S)7f$	$^2F°$	5/2	*2 332 500*
		7/2	*2 332 800*
$2p^6(^1S)8s$	^2S	1/2	2 393 000
$2p^6(^1S)8p$	$^2P°$	1/2,3/2	*2 404 100*
$2p^6(^1S)8d$	^2D	3/2	2 416 300
		5/2	2 416 500
$2p^6(^1S)8f$	$^2F°$	5/2,7/2	*2 421 500*
$2p^6(^1S)9s$	^2S	1/2	2 462 800
$2p^6(^1S)9p$	$^2P°$	1/2,3/2	*2 470 500*
$2p^6(^1S)9d$	^2D	3/2,5/2	2 479 000
$2p^6(^1S)9f$	$^2F°$	7/2	*2 483 200*
$2p^6(^1S)10s$	^2S	1/2	2 508 600
$2p^6(^1S)10p$	$^2P°$	1/2,3/2	*2 517 600*
$2p^6(^1S)10d$	^2D	5/2	2 522 800

(Continued)

V XIII—Continued

Configuration	Term	J	Level (cm^{-1})	Configuration	Term	J	Level (cm^{-1})
$2p^6(^1S)10f$	$^2F°$	7/2	*2 526 400*	$2p^6(^1S)11f$	$^2F°$	7/2	*2 560 400*
$2p^6(^1S)11s$	^2S	1/2	2 547 800				
$2p^6(^1S)11p$	$^2P°$	1/2,3/2	*2 552 300*	V XIV ($^1S_0$)	***Limit***		**2 712 250**
$2p^6(^1S)11d$	^2D	3/2 5/2	2 556 000 2 556 600	$2p^5\,3s^2$	$^2P°$	3/2 1/2	*4 079 000* *4 132 000*

V XIV

$Z=23$

Ne I isoelectronic sequence

Ground state: $1s^2 2s^2 2p^6\ ^1S_0$

Ionization energy = 7 227 000±3000 cm^{-1} (896.0±0.4 eV)

Only resonance lines between 15 and 24 Å are classified by this system of energy levels. Edlén and Tyrén (1936) identified transitions from $2p^5 3s$ and $3d$. Fawcett (1965) observed five more transitions from $2p^5 3d$ and $4d$ and from $2s2p^6 3p\ ^1P^\circ_1$. Feldman and Cohen (1967) observed nine transitions, including four of those observed by Fawcett. We have adopted their more accurate values in preference to those of Fawcett. The uncertainty in the levels of the above configurations is ±3000 cm^{-1}.

Kastner, Behring, and Cohen (1975) identified transitions between $2p^5 3p$ and $2p^5 4d$, but there is no connection with the levels given here.

We derived the ionization energy from the $2s^2 2p^5 nd\ ^3D^\circ_1$ series for $n=3$, 4, and 5.

References

Edlén, B., and Tyrén, F. (1936), Z. Phys. **101**, 206.

Fawcett, B. C. (1965), Proc. Phys. Soc. **86**, 1087.

Feldman, U. and Cohen, L. (1967), Astrophys. J. **149**, 265.

Kastner, S. O., Behring, W. E., and Cohen, L. (1975), Astrophys. J. **199**, 777.

V XIV

Configuration	Term	J	Level (cm^{-1})	Configuration	Term	J	Level (cm^{-1})
$2s^2 2p^6$	^1S	0	0	$2s^2 2p^5 4s$	$^1P^\circ$	1	*5 690 000*
$2s^2 2p^5 3s$	$^3P^\circ$	1	*4 202 700*	$2s^2 2p^5 4d$	$^3P^\circ$	1	*5 794 000*
$2s^2 2p^5 3s$	$^1P^\circ$	1	*4 257 100*	$2s^2 2p^5 4d$	$^3D^\circ$	1	*5 850 000*
$2s^2 2p^5 3d$	$^3P^\circ$	1	*4 696 000*	$2s^2 2p^5 4d$	$^1P^\circ$	1	*5 904 000*
$2s^2 2p^5 3d$	$^3D^\circ$	1	*4 757 800*	$2s^2 2p^5 5d$	$^3D^\circ$	1	*6 350 000*
$2s^2 2p^5 3d$	$^1P^\circ$	1	*4 827 200*	$2s^2 2p^5 5d$	$^1P^\circ$	1	*6 407 000*
$2s2p^6 3p$	$^3P^\circ$	1	*5 299 000*				
$2s2p^6 3p$	$^1P^\circ$	1	*5 324 000*	V XV ($^2P^\circ_{3/2}$)	*Limit*		**7 227 000**
$2s^2 2p^5 4s$	$^3P^\circ$	1	*5 632 000*				

V xv

$Z = 23$

F I isoelectronic sequence

Ground state: $1s^2 2s^2 2p^5\ ^2P^\circ_{3/2}$

Ionization energy = 7 870 000±20 000 cm^{-1} (976±2 eV)

The early work on this spectrum by Fawcett (1965) and by Cohen, Feldman, and Kastner (1968) classified many lines of the $2s^2 2p^5 - 2s^2 2p^4 3s$ and $3d$ transition arrays between 19 and 23 Å. This work was revised and extended by Feldman, Doschek, Cowan, and Cohen (1973), from whose wavelengths all of the $3s$ and $3d$ levels are determined with an uncertainty of ±3000 cm^{-1}. Fawcett (1971) identified the $2s^2 2p^5 - 2s 2p^6$ doublet at ~120 Å. The present value utilizes the improved measurements of Kaufman, Sugar, and Cooper (1982) which give a level uncertainty of ±50 cm^{-1}.

The $2s 2p^5 3s\ ^2P^\circ$ term is from Feldman et al. (1973).

The ionization energy was obtained by extrapolation by Lotz (1967).

References

Cohen, L., Feldman, U., and Kastner, S. O. (1968), J. Opt. Soc. Am. **58**, 331.

Fawcett, B. C. (1965), Proc. Phys. Soc. **86**, 1087.

Fawcett, B. C. (1971), J. Phys. **B4**, 981.

Feldman, U., Doschek, G. A., Cowan, R. D., and Cohen, L. (1973), J. Opt. Soc. Am. **63**, 1445.

Kaufman, V., Sugar, J., and Cooper, D. (1982), Phys. Scr. **25**, 623.

Lotz, W.J. (1967), J. Opt. Soc. Am. **57**, 873.

V XV

Configuration	Term	J	Level (cm^{-1})	Configuration	Term	J	Level (cm^{-1})
$2s^2 2p^5$	$^2P^\circ$	3/2 1/2	*0* *58 093*	$2s^2 2p^4(^3P)3d$	^2D	3/2 5/2	5 061 300 5 090 400
$2s2p^6$	^2S	1/2	877 732	$2s^2 2p^4(^3P)3d$	^2F	5/2	5 069 700
$2s^2 2p^4(^3P)3s$	^4P	5/2 3/2 1/2	4 506 100 4 528 400 4 556 100	$2s^2 2p^4(^1D)3d$ $2s^2 2p^4(^1D)3d$	^2P ^2S	3/2 1/2	5 163 000 5 143 200
$2s^2 2p^4(^3P)3s$	^2P	3/2 1/2	4 564 300 4 586 800	$2s^2 2p^4(^1D)3d$	^2D	5/2 3/2	5 163 700 5 181 800
$2s^2 2p^4(^1D)3s$	^2D	5/2 3/2	4 636 500 4 638 500	$2s^2 2p^4(^1S)3d$	^2D	5/2 3/2	5 255 400 5 265 600
$2s^2 2p^4(^1S)3s$	^2S	1/2	4 756 200	$2s2p^5(^3P^\circ)3s$	$^2P^\circ$	3/2 1/2	*5 347 000* *5 379 400*
$2s^2 2p^4(^3P)3d$	^4P	1/2 3/2 5/2	5 028 200 5 039 000 5 055 100	V XVI ($^3P_2$)	*Limit*		**7 870 000**
$2s^2 2p^4(^3P)3d$	^4F	5/2	5 039 300				
$2s^2 2p^4(^3P)3d$	^2P	1/2 3/2	5 048 600 5 082 500				

V XVI

$Z=23$

O I isoelectronic sequence

Ground state: $1s^2 2s^2 2p^4\ {}^3P_2$

Ionization energy = 8 549 000±17 000 cm^{-1} (1060±2 eV)

Goldsmith, Feldman, and Cohen (1971) interpreted the $2p^4-2p^3 3d$ and $2p^4-2p^3 3s$ arrays at 18 Å and 20 Å. Their measurement uncertainty is ±0.005 Å. Additional identifications of transitions from $2p^3 3s$ were made by Doschek, Feldman, and Cohen (1973) who claim slightly improved wavelengths. Additions to the $2p^3 3d$ levels are from Fawcett, Galanti, and Peacock (1974a), Fawcett and Hayes (1975), and Bromage and Fawcett (1977). The $3s$ and $3d$ level uncertainty is ±3000 cm^{-1}.

The $2s^2 2p^4-2s2p^5$ array was analyzed by Fawcett (1971). The $^1S_0$ level of $2p^6$ was discovered by Fawcett, Galanti, and Peacock (1974 b). Improved measurements for these transitions in the range of 95–156 Å and the discovery of intersystem transitions was reported by Kaufman, Sugar, and Cooper (1982). They determined the levels with an uncertainty of ±50 cm^{-1}. Percentage compositions for these levels were provided by Kaufman and Sugar (1982). Their calculation includes configuration interaction between $2s^2 2p^4$ and $2p^6$.

The ionization energy was obtained by extrapolation by Lotz (1967).

References

Bromage, G. E. and Fawcett, B. C. (1977), Mon. Not. R. Astron. Soc. **178,** 591.

Doschek, G. A., Feldman, U., and Cohen, L. (1973), J. Opt. Soc. Am. **63,** 1463.

Fawcett, B. C. (1971), J. Phys. **B4,** 981.

Fawcett, B. C., Galanti, M., and Peacock, N. J. (1974a), J. Phys. **B7,** L106.

Fawcett, B. C., Galanti, M., and Peacock, N. J. (1974b), J. Phys. **B7,** 1149.

Fawcett, B. C., and Hayes, R. W. (1975), Mon. Not. R. Astron. Soc. **170,** 185.

Goldsmith, S., Feldman, U., and Cohen L. (1971), J. Opt. Soc. Am. **61,** 615.

Kaufman, V. and Sugar, J. (1982), private communication.

Kaufman, V., Sugar, J., and Cooper, D. (1982), Phys. Scr. **25,** 623.

Lotz, W. J. (1967), J. Opt. Soc. Am. **57,** 873.

V XVI

Configuration	Term	J	Level (cm^{-1})	Leading percentages		
$2s^2\,2p^4$	^{3}P	2	0	94	6	^{1}D
		1	48 937	100		
		0	49 970	89	10	^{1}S
$2s^2\,2p^4$	^{1}D	2	121 039	94	6	^{3}P
$2s^2\,2p^4$	^{1}S	0	237 705	87	11	^{3}P
$2s2p^5$	^{3}P°	2	*761 824*	100		
		1	*798 899*	99	1	^{1}P°
		0	*822 961*	100		
$2s2p^5$	^{1}P°	1	*1 045 590*	99	1	^{3}P°
$2p^6$	^{1}S	0	1 769 360	98	2	$2s^2 2p^4$ ^{1}S
$2s^2\,2p^3(^4S°)3s$	^{3}S°	1	*4 891 000*			
$2s^2\,2p^3(^2D°)3s$	^{3}D°	2	*4 980 000*			
		1	*4 981 000*			
		3	*4 996 000*			
$2s^2\,2p^3(^2D°)3s$	^{1}D°	2	*5 012 000*			
$2s^2\,2p^3(^2P°)3s$	^{3}P°	2	*5 068 000*			
$2s^2\,2p^3(^2P°)3s$	^{1}P°	1	*5 113 000*			

(Continued)

V XVI—Continued

Configuration	Term	J	Level (cm^{-1})	Leading percentages
$2s^2\,2p^3(^4S^\circ)3d$	$^3D^\circ$	2	*5 343 000*	
		3	*5 351 000*	
$2s^2\,2p^3(^2D^\circ)3d$	$^3P^\circ$	2	*5 451 000*	
$2s^2\,2p^3(^2D^\circ)3d$	$^3D^\circ$	2	*5 457 000*	
		3	*5 475 000*	
$2s^2\,2p^3(^2P^\circ)3d$	$^3P^\circ$	2	*5 488 000*	
$2s^2\,2p^3(^2D^\circ)3d$	$^1D^\circ$	2	*5 489 000*	
$2s^2\,2p^3(^2D^\circ)3d$	$^1F^\circ$	3	*5 518 000*	
$2s^2\,2p^3(^2P^\circ)3d$	$^3D^\circ$	3	*5 552 000*	
$2s^2\,2p^3(^2P^\circ)3d$	$^1F^\circ$	3	*5 596 000*	
V XVII ($^4S^\circ_{3/2}$)	***Limit***		**8 549 000**	

V XVII

$Z=23$

N I isoelectronic sequence

Ground state: $1s^2 2s^2 2p^3\ ^4S^\circ_{3/2}$

Ionization energy $=$ 9 420 000$\pm$20 000 cm^{-1} (1168$\pm$2 eV)

The transition array $2s^2 2p^3 - 2s2p^4$ was first analyzed by Fawcett (1971) and the $2s2p^4 - 2p^5$ by Fawcett, Galanti, and Peacock (1974). The intersystem connection $2s^2 2p^3\ ^4S^\circ_{3/2} - 2s2p^4\ ^2P_{3/2}$ at 96.270 Å was found by Kaufman, Sugar, and Cooper (1982). They remeasured these arrays from 96–167 Å with an uncertainty of $\pm$0.01 Å, giving a level uncertainty of $\pm$50 cm^{-1}. The percentage compositions for the levels were provided by Kaufman and Sugar (1982). Their calculation includes configuration interaction between $2s^2 2p^3$ and $2p^5$.

The $2p^2 3d$ levels are from the classifications by Fawcett and Hayes (1975) of the array $2p^3 - 2p^2 3d$ at 17 Å, measured with an uncertainty of $\pm$3000 cm^{-1}.

The ionization energy was obtained by extrapolation by Lotz (1967).

References

Fawcett, B. C. (1971), J. Phys. **B4,** 981.
Fawcett, B.C., Galanti, M , and Peacock, N. J. (1974), J. Phys. **B7,** 1149.
Fawcett, B. C., and Hayes, R. W. (1975), Mon. Not. R. Astron. Soc. **170,** 185.
Kaufman, V., and Sugar, J. (1982), private communication.
Kaufman, V., Sugar, J., and Cooper, D. (1982), Phys. Scr. **26,** 163.
Lotz, W. (1967), J. Opt. Soc. Am. **57,** 873.

V XVII

Configuration	Term	J	Level (cm^{-1})	Leading percentages		
$2s^2\,2p^3$	$^4S^\circ$	3/2	*0*	95	4	$^2P^\circ$
$2s^2\,2p^3$	$^2D^\circ$	3/2	*120 930*	82	16	$^2P^\circ$
		5/2	*140 260*	100		
$2s^2\,2p^3$	$^2P^\circ$	1/2	*211 420*	98	2	$2p^5\ ^2P^\circ$
		3/2	*240 500*	78	18	$2p^3\ ^2D^\circ$
$2s(^2S)2p^4(^3P)$	^4P	5/2	627 560	98	2	$2s(^2S)2p^4(^1D)\ ^2D$
		3/2	666 210	99		
		1/2	681 580	98	2	$2s(^2S)2p^4(^1S)\ ^2S$
$2s(^2S)2p^4(^1D)$	^2D	3/2	866 880	97	2	$2s(^2S)2p^4(^3P)\ ^2P$
		5/2	872 820	98	2	$2s(^2S)2p^4(^3P)\ ^4P$
$2s(^2S)2p^4(^1S)$	^2S	1/2	999 840	80	18	$2s(^2S)2p^4(^3P)\ ^2P$
$2s(^2S)2p^4(^3P)$	^2P	3/2	1 038 740	97	3	$2s(^2S)2p^4(^1D)\ ^2D$
		1/2	1 093 200	82	18	$2s(^2S)2p^4(^1S)\ ^2S$
$2p^5$	$^2P^\circ$	3/2	*1 636 530*	98	2	$2s^2 2p^3\ ^2P^\circ$
		1/2	*1 698 100*	98	2	
$2s^2\,2p^2(^3P)3d$	^4P	3/2,5/2	5 828 000			
$2s^2\,2p^2(^3P)3d$	^2F	7/2	5 843 000			
$2s^2\,2p^2(^3P)3d$	^2D	5/2	5 897 000			
$2s^2\,2p^2(^1D)3d$	^2F	7/2	5 934 000			
$2s^2\,2p^2(^1D)3d$	^2D	5/2	5 958 000			
V XVIII ($^3P_0$)	*Limit*		**9 420 000**			

V XVIII

$Z=23$

C I isoelectronic sequence

Ground state: $1s^22s^22p^2\ ^2P_0$

Ionization energy = 10 160 000±20 000 cm^{-1} (1260±2 eV)

The $2s^22p^2-2s2p^3$ array has been interpreted by Fawcett, Galanti, and Peacock (1974) and the $2s2p^3-2p^4$ by Fawcett (1975). New measurements in this region of 102–176 Å were made by Sugar, Kaufman, and Cooper (1982) with an accuracy of ±0.01 Å, giving a level uncertainty of ±50 cm^{-1}. They found several intersystem lines giving the connection of the singlets with the ground state. Their predicted value for the $2s^22p^2\ ^1S_0$ and Edlén's predicted value (1984) for the $2s2p^3\ ^5S_2^\circ$ level are given in brackets. Calculated percentage compositions for these levels were provided by Kaufman and Sugar (1982). They included configuration interaction between $2s^22p^2$ and $2p^4$.

The $2s^22p3d$ levels are derived from the measurements of Goldsmith, Feldman, Crooker, and Cohen (1972) at 16–17 Å with an uncertainty of ±0.005 Å, or a level uncertainty of ±2000 cm^{-1}. The classifications follow the predictions of Bromage and Fawcett (1977).

The ionization energy was obtained by extrapolation by Lotz (1967).

References

Bromage, G. E., and Fawcett, B. C. (1977), Mon. Not. R. Astron. Soc. **178,** 605.
Edlén, B. (1984), private communication.
Fawcett, B. C. (1975), At. Data Nucl. Data Tables **16,** 135.
Fawcett, B. C., Galanti, M., and Peacock, N. J. (1974), J. Phys. **B7,** 1149.
Goldsmith, S., Feldman, U., Crooker, A., and Cohen, L. (1972) J. Opt. Soc. Am. **62,** 260.
Kaufman, V., and Sugar, J. (1982), private communication.
Lotz, W. (1967), J. Opt. Soc. Am. **57,** 873.
Sugar, J., Kaufman, V., and Cooper, D. (1982), Phys. Scr. **26,** 189.

V XVIII

Configuration	Term	J	Level (cm^{-1})	Leading percentages		
$2s^2\ 2p^2$	^3P	0	0	93	5	$2s^22p^2\ ^1S$
		1	37 960	99	1	$2p^4\ ^3P$
		2	68 190	86	13	$2s^22p^2\ ^1D$
$2s^2\ 2p^2$	^1D	2	160 910	86	12	^3P
$2s^2\ 2p^2$	^1S	0	[269 000]	90	6	$2s^22p^2\ ^3P$
$2s(^2S)2p^3(^4S^\circ)$	$^5S^\circ$	2	[*366 960*]	99	1	$2s(^2S)2p^3(^2P^\circ)\ ^3P^\circ$
$2s(^2S)2p^3(^2D^\circ)$	$^3D^\circ$	2	*623 860*	91	9	$2s(^2S)2p^3(^2P^\circ)\ ^3P^\circ$
		1	*625 040*	93	6	$2s(^2S)2p^3(^2P^\circ)\ ^3P^\circ$
		3	*634 950*	100		
$2s(^2S)2p^3(^2P^\circ)$	$^3P^\circ$	0	*731 870*	100		
		1	*735 420*	92	6	$2s(^2S)2p^3(^2D^\circ)\ ^3D^\circ$
		2	*743 350*	87	9	$2s(^2S)2p^3(^2D^\circ)\ ^3D^\circ$
$2s(^2S)2p^3(^4S^\circ)$	$^3S^\circ$	1	*897 330*	89	9	$2s(^2S)2p^3(^2P^\circ)\ ^1P^\circ$
$2s(^2S)2p^3(^2D^\circ)$	$^1D^\circ$	2	*908 420*	97	3	$2s(^2S)2p^3(^2P^\circ)\ ^3P^\circ$
$2s(^2S)2p^3(^2P^\circ)$	$^1P^\circ$	1	*1 014 420*	90	9	$2s(^2S)2p^3(^4S^\circ)\ ^3S^\circ$
$2p^4$	^3P	2	1 358 710	92	7	$2p^4\ ^1D$
		1	1 410 770	99	1	$2s^22p^2\ ^3P$
		0	1 416 110	92	6	$2p^4\ ^1S$
$2p^4$	^1D	2	1 480 330	92	7	$2p^4\ ^3P$
$2p^4$	^1S	0	1 668 300	90	7	^3P

V XVIII—Continued

Configuration	Term	J	Level (cm^{-1})	Leading percentages
$2s^2\,2p3s$	$^3P°$	1	*5 726 000*	
		2	*5 786 000*	
$2s^2\,2p3s$	$^1P°$	1	*5 805 000*	
$2s^2\,2p3d$	$^3D°$	2	*6 157 000*	
		3	*6 174 000*	
$2s^2\,2p3d$	$^3P°$	2	*6 188 000*	
		1	*6 195 000*	
$2s^2\,2p3d$	$^1F°$	3	*6 234 000*	
$2s2p^2(^2D)3s$	^3D	2	6 323 300	
$2s2p^2(^4P)3d$	^1F	3	6 500 000	
$2s2p^2(^2D)3d$	^3F	4	6 674 000	
V XIX ($^2P°_{1/2}$)	***Limit***		**10 160 000**	

V XIX

$Z=23$

B I isoelectronic sequence

Ground state: $1s^22s^22p\ ^2P^\circ_{1/2}$

Ionization energy $=$ 10 930 000$\pm$20 000 cm^{-1} (1355$\pm$3 eV)

The $2s^22p-2s2p^2$ transition array was measured by Fawcett, Galanti, and Peacock (1974a) in the range of 124–181 Å with an uncertainty of $\pm$0.02 Å. Their identification of the $2s2p^2\ ^2D$ term has been revised by Fawcett and Hayes (1975), who also observed the three lines of $2s2p^2\ ^4P-2p^3\ ^4S^\circ$. We estimate the position of $2s2p^2\ ^4P_{1/2}$ to be 332 180$\pm$200 cm^{-1}, using the method of Sugar, Kaufman, and Cooper (1982). The doublet transitions of $2s2p^2-2p^3$ were identified by Fawcett, Ridgeley, and Hatter (1980). The observed levels of these three configurations have an uncertainty of $\pm$100 cm^{-1}. The percentage composition for these levels was provided by Kaufman and Sugar (1982). Their calculation includes configuration interaction between $2s^22p$ and $2p^3$.

Measurements in the range of 13–19 Å were reported by Fawcett, Galanti, and Peacock (1974b) with an uncertainty of $\pm$0.007 Å, and interpreted as transitions from $2s2p3p$ and $2s2p3d$ levels. Levels are determined with an uncertainty of $\pm$3000 cm^{-1}.

The ionization energy is an extrapolated value obtained by Lotz (1967).

References

Fawcett, B. C., Galanti, M., and Peacock, N. J. (1974a), J. Phys. **B7,** 1149.

Fawcett, B. C., Galanti, M., and Peacock, N. J. (1974b), J. Phys. L106.

Fawcett, B. C., and Hayes, R. W. (1975), Mon. Not. R. Astron. Soc. **170,** 185.

Fawcett, B. C., Ridgeley, A., and Hatter, A. T. (1980), J. Opt. Soc. Am. **70,** 1349.

Kaufman, V., and Sugar, J. (1982) private communication.

Lotz, W. (1967), J. Opt. Soc. Am. **57,** 873.

Sugar, J., Kaufman, V., and Cooper, D., (1982), Phys. Scr. **26,** 293.

V XIX

Configuration	Term	J	Level (cm^{-1})	Leading percentages		
$2s^2\,2p$	$^2P^\circ$	1/2	*0*	97	3	$2p^3\ ^2P^\circ$
		3/2	*68 610*	97	3	
$2s2p^2$	^4P	1/2	332 180+x	98	2	^2S
		3/2	361 600+x	100		
		5/2	394 560+x	98	2	^2D
$2s2p^2$	^2D	3/2	597 590	98	2	^2P
		5/2	605 320	98	2	^4P
$2s2p^2$	^2P	1/2	716 370	51	48	^2S
		3/2	802 560	98	2	^2D
$2s2p^2$	^2S	1/2	788 850	50	49	^2P
$2p^3$	$^4S^\circ$	3/2	*1 030 850*+x	97	3	$^2P^\circ$
$2p^3$	$^2D^\circ$	3/2	*1 152 900*	91	8	$^2P^\circ$
		5/2	*1 166 100*	100		
$2p^3$	$^2P^\circ$	1/2	*1 292 800*	97	3	$2s^23p\ ^2P^\circ$
		3/2	*1 318 200*	86	9	$2p^3\ ^2D^\circ$
$2s^2\,3d$	^2D	3/2	6 427 000			
		5/2	6 437 000			
$2s2p(^3P^\circ)3p$	^2P	3/2	6 590 000			
$2s2p(^3P^\circ)3p$	^2D	5/2	6 718 000			

V XIX—Continued

Configuration	Term	J	Level (cm^{-1})	Leading percentages
$2s2p(^3P°)3d$	$^4D°$	7/2	*6 787 000+x*	
$2s2p(^3P°)3d$	$^4P°$	5/2	*6 792 000+x*	
$2s2p(^3P°)3d$	$^2F°$	5/2	*6 845 000*	
		7/2	*6 885 000*	
$2s2p(^1P°)3p$	^2D	5/2	6 901 000	
$2s2p(^1P°)3d$	$^2F°$	7/2	*7 059 000*	
V XX ($^1S_0$)	***Limit***		**10 930 000**	

V XX

$Z=23$

Be I isoelectronic sequence

Ground state: $1s^2 2s^2\ {}^1S_0$

Ionization energy = 11 990 000±24 000 cm^{-1} (1486±3 eV)

The $2s^2\ {}^1S_0 - 2s2p\ {}^1P_1^\circ$ transition is given by Fawcett (1975) at 159.36 Å. The intersystem transition $2s^2\ {}^1S_0 - 2s2p\ {}^3P_1^\circ$ has not been observed. However, the intersystem connection $2s2p\ {}^3P_2 - 2p^2\ {}^1D_2$ at 164.59 Å was reported by Fawcett, Ridgeley, and Hatter (1980) along with their complete analysis of the $2s2p - 2p^2$ transition array. They give a wavelength uncertainty of ±0.05 Å for their measurements. The level uncertainty is ±200 cm^{-1}.

Observations of the higher-lying configurations by Fawcett and Hayes (1975) and by Boiko et al. (1977) were extended by Bromage, Cowan, Fawcett, and Ridgeley (1978). They observed the spectrum of a laser-generated plasma in the range 10–14 Å. No uncertainty estimate is given for their three decimal place measurements. We estimate a level uncertainty of ±5000 cm^{-1}. The ionization energy was obtained by extrapolation by Lotz (1967).

References

Boiko, V. A., Pikuz, S. A., Safronova, U. I., and Faenov, A. Y. (1977), J. Phys. **B10**, 1253.

Bromage, G. E., Cowan, R. D., Fawcett, B. C., and Ridgeley, A. (1978), J. Opt. Soc. Am. **68**, 48.

Fawcett, B. C., and Hayes, R. W. (1975), Mon. Not. R. Astron. Soc. **170**, 185.

Fawcett, B. C. (1975), At. Nucl. Data Tables **16**, 135.

Fawcett, B. C., Ridgeley, A., and Hatter, A. T. (1980), J. Opt. Soc. Am. **70**, 1349.

Lotz, W. (1967), J. Opt. Soc. Am. **57**, 873.

V XX

Configuration	Term	J	Level (cm^{-1})	Configuration	Term	J	Level (cm^{-1})
$2s^2$	^1S	0	0	$2p3d$	$^1D^\circ$	2	*7 423 300*
$2s2p$	$^3P^\circ$	0	*303 100*	$2p3d$	$^3D^\circ$	1	*7 436 500*
		1	*322 600*			2	*7 473 300*
		2	*375 000*			3	*7 493 200*
$2s2p$	$^1P^\circ$	1	*627 500*	$2p3p$	^1D	2	7 453 900
$2p^2$	^3P	0	820 100	$2p3d$	$^3P^\circ$	2	*7 505 000*
		1	856 900				
		2	888 600	$2p3d$	$^1F^\circ$	3	*7 554 600*
$2p^2$	^1D	2	982 600	$2p3d$	$^1P^\circ$	1	*7 571 900*
$2p^2$	^1S	0	1 177 600	$2s4p$	$^1P^\circ$	1	*9 140 000*
$2s3p$	$^3P^\circ$	1	*6 943 800*	$2s4d$	^3D	2	9 217 000
						3	9 218 300
$2s3p$	$^1P^\circ$	1	*6 964 000*			1	9 219 700
$2s3d$	^3D	1	7 046 600	$2s4d$	^1D	2	9 237 100
		2	7 047 500				
		3	7 052 400	$2p4d$	$^3F^\circ$	3	*9 567 000*
$2s3d$	^1D	2	7 109 600	$2p4d$	$^3D^\circ$	2	*9 569 200*
						1	*9 571 300*
$2p3p$	^3D	2	7 325 900			3	*9 639 800*
		3	7 378 300	$2p4p$	^3D	3	9 601 800
$2p3p$	^3P	2	7 402 900	$2p4d$	$^3P^\circ$	2	*9 639 800*
$2p3d$	$^3F^\circ$	3	*7 409 200*				

V XX—Continued

Configuration	Term	*J*	Level (cm^{-1})	Configuration	Term	*J*	Level (cm^{-1})
$2p4d$	$^1D°$	2	*9 645 800*				
$2p4d$	$^1F°$	3	*9 660 900*				
V XXI ($^2S_{1/2}$)	***Limit***		**11 990 000**				

V XXI

$Z=23$

Li I isoelectronic sequence

Ground state: $1s^22s\ ^2S_{1/2}$

Ionization energy = 12 660 000±2400 cm^{-1} (1569.6±0.3 eV)

The $2s\ ^2S_{1/2}-2p\ ^2P^\circ_{3/2}$ transition was reported at 240.37±0.05 Å by Fawcett, Ridgeley, and Hatter (1980). They give a predicted value of 293.71 Å for the $^2S_{1/2}-^2P^\circ_{1/2}$ line of the doublet.

Goldsmith, Feldman, Oren, and Cohen (1972) identified the $2s-3p$, $4p$ and $2p-3s$, $3p$, $3d$, $4d$ transitions in the range of 10–14 Å. Their work was extended by Aglitskii, Boiko, Pikuz, and Faenov (1974) to include $7p$ and $8d$, with a measurement uncertainty of ±0.003 Å. We have used their measurements to determine the level values with an uncertainty of ±2000 cm^{-1}.

The doubly excited levels were obtained by Aglitskii, Boiko, Zakharov, Pikuz, and Faenov (1974) from lines observed at 2.4 Å in a laser-produced plasma with an uncertainty of ±9000 cm^{-1}.

The ionization energy was determined by Edlén (1979).

References

Aglitskii, E. V., Boiko, V. A., Zakharov, S. M., Pikuz, S. A., and Faenov, A. Y. (1974), J. E. T. P. Letters **19**, 8.

Aglitskii, E. V., Boiko, V. A., Pikuz, S. A., and Faenov, A. Y. (1974), Kvantovaya Elektron. (Moscow) **1**, 1731.

Edlén, B. (1979), Phys. Scr. **19**, 255.

Fawcett, B. C., Ridgeley A., and Hatter, A. T. (1980), J. Opt. Soc. Am. **70**, 1349.

Goldsmith, S., Feldman, U., Oren, L., and Cohen, L. (1972), Astrophys. J. **174**, 209.

V XXI

Configuration	Term	J	Level (cm^{-1})	Configuration	Term	J	Level (cm^{-1})
$1s^2\ 2s$	^2S	1/2	0	$1s^2\ 6d$	^2D	3/2,5/2	11 315 000
$1s^2\ 2p$	$^2P^\circ$	1/2	*340 470*	$1s^2\ 7p$	$^2P^\circ$	1/2,3/2	*11 660 000*
		3/2	*416 020*				
				$1s^2\ 7d$	^2D	3/2,5/2	11 675 000
$1s^2\ 3s$	^2S	1/2	7 114 400				
				$1s^2\ 8d$	^2D	3/2,5/2	11 906 000
$1s^2\ 3p$	$^2P^\circ$	1/2	*7 210 000*				
		3/2	*7 232 000*				
$1s^2\ 3d$	^2D	3/2	7 269 100	V XXII ($^1S_0$)	***Limit***		**12 660 000**
		5/2	7 275 700				
				$1s(^2S)2s2p(^1P^\circ)$	$^2P^\circ$	1/2,3/2	*41 829 000*
$1s^2\ 4p$	$^2P^\circ$	1/2,3/2	*9 603 000*				
				$1s2p^2$	^2D	3/2	41 985 000
$1s^2\ 4d$	^2D	3/2	9 627 200			5/2	42 001 000
		5/2	9 631 000				
				$1s2p^2$	^2P	3/2	42 097 000
$1s^2\ 5p$	$^2P^\circ$	1/2,3/2	*10 693 000*				
				$1s2p^2$	^2S	1/2	42 279 000
$1s^2\ 5d$	^2D	3/2,5/2	10 721 000				
$1s^2\ 6p$	$^2P^\circ$	1/2,3/2	*11 308 000*				

V XXII

$Z=23$

He I isoelectronic sequence

Ground state: $1s^2\ ^1\mathrm{S}_0$

Ionization energy = 55 259 000±10 000 cm^{-1} (6851.3±1.0 eV)

Because of the excellent agreement of the calculated energies of the $n=2$ shell by Safronova (1981) with the few well-measured spectra in the He I sequence, we have compiled her results for the $n=2$ levels and for the ionization energy. Detailed comparisons are given in the Introduction. Levels of the $n=3$–5 shells are from the calculated binding energies by Ermolaev and Jones (1974) subtracted from Safronova's value for the binding energy of the ground state. We have assumed an uncertainty of 2 parts in 10^4 for the excited levels relative to the ground state, and for the ionization energy. This is equal to the random deviation of the measurements by Aglitskii et al. from the calculations by Safronova. For differences between excited levels where $\Delta n=0$, we assumed an uncertainty of 2 parts in 10^3.

Corrected measurements (see Introduction) by Aglitskii et al. (1974) place the $1s2p\ ^3\mathrm{P}^{\circ}_1$ level at 41 769 000 cm^{-1} and the $1s2p\ ^1\mathrm{P}^{\circ}_1$ at 41 973 000 cm^{-1} with an estimated uncertainty of ±9000 cm^{-1}.

Percentage compositions are from Ermolaev and Jones.

References

Aglitskii, E. V., Boiko, V. A., Zakharov, S. M., Pikuz, S. A., and Faenov, A.Y. (1974), Sov. Quantum Electronics **1,** 908.

Ermolaev, A. M., and Jones, M. (1974), J. Phys. **B7,** 199.

Safronova, U. I. (1981), Phys. Scr. **23,** 241.

V XXII

Configuration	Term	J	Level (cm^{-1})	Leading percentages		
$1s^2$	^{1}S	0	0			
$1s2s$	^{3}S	1	[41 568 280]			
$1s2p$	^{3}P°	0	[*41 769 360*]			
		1	[*41 782 570*]	95	5	^{1}P°
		2	[*41 849 500*]			
$1s2s$	^{1}S	0	[41 787 820]			
$1s2p$	^{1}P°	1	[*41 980 920*]	95	5	^{3}P°
$1s3s$	^{3}S	1	[49 234 370]			
$1s3p$	^{3}P°	0	[*49 290 020*]			
		1	[*49 293 350*]	94	6	^{1}P°
		2	[*49 313 350*]			
$1s3s$	^{1}S	0	[49 292 080]			
$1s3p$	^{1}P°	1	[*49 349 370*]	94	6	^{3}P°
$1s4s$	^{3}S	1	[51 886 530]			
$1s4p$	^{3}P°	0	[*51 909 670*]			
		1	[*51 911 070*]	94	6	^{1}P°
		2	[*51 919 520*]			
$1s4s$	^{1}S	0	[51 909 920]			
$1s4p$	^{1}P°	1	[*51 934 210*]	94	6	^{3}P°
$1s5s$	^{3}S	1	[53 107 060]			

(Continued)

V XXII—Continued

Configuration	Term	J	Level (cm^{-1})	Leading percentages		
1s5s	^{1}S	0	[53 118 740]			
1s5p	^{3}P°	0	[*53 118 760*]			
		1	[*53 119 470*]	94	6	^{1}P°
		2	[*53 123 810*]			
1s5p	^{1}P°	1	[*53 131 220*]	94	6	^{3}P°
V XXIII (^{2}S^0)	***Limit***		**55 259 000**			

V XXIII

$Z=23$

H I isoelectronic sequence

Ground state: $1s\ ^2S_{1/2}$

Ionization energy = 58 443 900±200 cm^{-1} (7246.18±0.02 eV)

No observations of this spectrum are reported. We give calculated values by Mohr (1983) for the $n=2$ shell and by Erickson (1977) for $n=3$–5 relative to the $2p\ ^2P^\circ_{3/2}$ level. Further details are given in the Introduction.

References

Erickson, G. W. (1977), J. Phys. Chem. Ref. Data **6**, 831.

Mohr, P. J. (1983), At. Data Nucl. Data Tables **29**, 453.

V XXIII

Configuration	Term	J	Level (cm^{-1})
$1s$	^2S	1/2	0
$2p$	$^2P^\circ$	1/2	*[43 801 550]*
		3/2	*[43 905 810]*
$2s$	^2S	1/2	[43 804 530]
$3p$	$^2P^\circ$	1/2	*[51 947 814]*
		3/2	*[51 978 720]*
$3s$	^2S	1/2	[51 948 737]
$3d$	^2D	3/2	[51 978 665]
		5/2	[51 988 832]
$4p$	$^2P^\circ$	1/2	*[54 794 745]*
		3/2	*[54 807 776]*
$4s$	^2S	1/2	[54 795 136]
$4d$	^2D	3/2	[54 807 752]
		5/2	[54 812 044]
$4f$	$^2F^\circ$	5/2	*[54 812 036]*
		7/2	*[54 814 175]*
$5p$	$^2P^\circ$	1/2	*[56 110 653]*
		3/2	*[56 117 320]*
$5s$	^2S	1/2	[56 110 853]
$5d$	^2D	3/2	[56 117 308]
		5/2	[56 119 506]
$5f$	$^2F^\circ$	5/2	*[56 119 502]*
		7/2	*[56 120 597]*
$5g$	^2G	7/2	[56 120 595]
		9/2	[56 121 252]
	Limit		**58 443 900**

Cr I

$Z=24$

Ground state: $1s^22s^22p^63s^23p^63d^54s\ ^7S_3$

Ionization energy $= 54575.6\pm0.3$ cm^{-1} (6.76669 ± 0.00004 eV)

The early contributions to the analysis of this spectrum are summarized by Catalán and Sancho (1931), who give a list of over 700 classified lines. Kiess (1953) remeasured the spectrum in the range of 11 610–1899 Å with a wavelength accuracy varying from ±0.005 Å to ±0.05 Å. About 4400 lines are given, 80% of which are classified, including Zeeman patterns for 10% of them. This constituted a significant extension of the analysis and provided improved values for all the known levels with an estimated uncertainty of ±0.05 cm^{-1}. Kiess obtained a value for the ionization energy of 54 570 cm^{-1}.

An absorption spectrum below 2000 Å was observed by Huber, Sandeman, and Tubbs (1975) with an accuracy of ±0.004 Å. They identified the $3d^5(^6S)np\ ^7P°$ Rydberg series for $n=8$ and for $n=12$–38 with an uncertainty of ±0.1 cm^{-1}, from which they derived a value for the ionization energy of 54 575.6±0.3 cm^{-1}. Improved wavelengths are given for the multiplets $a\ ^5D-q\ ^5D°$, $a\ ^5D-o\ ^5F°$, and $a\ ^5D-r\ ^5P°$ reported by Kiess.

Further observations of the absorption spectrum at lower wavelengths were reported by Connerade, Baig, and Newsom (1981) with an accuracy of ±0.05 Å. They identified five $3d^5(^6S)4s\ ^7S-3d^4(^5D)4s\ (^6D)np$ series for $n=6$ to $n=28$. For $n=6$ the term structure is identified by means of a diagonalization of the energy matrices of the $3d^44s6p$ configuration, as well as by quantum defects. Their results contradict the assignment by Kiess of terms to $3d^44s6p$, which we have therefore dropped. The levels identified by Connerade, Baig, and Newsom with this configuration and their percentage compositions are given. They have also observed the principal series $3d^5(^6S)np\ ^7P°$ reported by Huber et al. and have identified the missing $7p$, $9p$, $10p$, and $11p$ terms.

Mansfield (1977) reported the observation in absorption of the $3p^5(^2P)3d^5(^6S)4s^2\ ^7P°$ term.

Most of the g-factors for the levels are obtained from Catalán and Sancho, supplemented by Kiess' three-decimal-place values. The more accurate values for the $3d^5(^6S)4p\ ^7P_3^°$ and $^7P_4^°$ levels were measured in an atomic beam by Budick, Goshen, and Marcus (1964), and those for $3d^5(^6S)4s\ ^7S$ and $3d^44s^2\ ^5D$ were obtained by Childs and Goodman (1965).

The alphabetic prefixing of terms with lower case letters for distinguishing repeating terms of the same type has been retained from Kiess except where the levels were reinterpreted by Roth (1970) on the basis of his theoretical treatment.

Roth has calculated the odd-parity configurations $3d^54p$, $3d^44s4p$, and $3d^34s^24p$ with configuration interaction. His percentage compositions and designation changes for the experimental levels are adopted here. Roth distinguished repeating terms of the $3d^n$ core by the letters, a, b, ... rather than by seniority. The percentages include the sum of seniority states contributing to the term.

Fischer, Hansen, and Barwell (1976) pointed out that an error in Roth's calculation arising from insufficient precision in his diagonalization routine resulted in an incorrect mixture of the $z\ ^7D°$ and $y\ ^7P°$ terms. Revised percentages were provided for these terms and for $z\,^7P°$ by Hansen.

Percentages for the configuration $3d^54s$ were taken from an ab initio calculation by Vizbaraite, Kupliauskis, and Tutlys (1968).

References

Budick, B., Goshen, R. J., and Marcus, S. (1964), Bul. Am. Phys. Soc. **9**, 90.

Catalán, M. A., and Sancho, P. M. (1931), An. R. Soc. Esp. Fis. Quim. **29**, 327.

Childs, W. J., and Goodman, L.S. (1965), Phys. Rev. **A140**, 447.

Connerade, J. P., Baig, M. A., and Newsom, G. H. (1981), Proc. R. Soc. London, **Ser. A378**, 445.

Fischer, C. F., Hansen, J., and Barwell, M. (1976), J. Phys. **B9**, 1841.

Huber, M. C. E., Sandeman, R. J., and Tubbs, E. F. (1975), Proc. R. Soc. London, **Ser. A342**, 431.

Kiess, C. C. (1953), J. Res. Natl. Bur. Stand. (U.S.) **51**, 247.

Mansfield, M. W. D. (1977), Proc. R. Soc. London, **Ser. A358**, 253.

Roth, C. (1970), J. Res. Natl. Bur. Stand. (U.S.) **74A**, 157.

Vizbaraite, J., Kupliauskis, Z., and Tutlys,V. (1968), Lietuvos Fizikos Rinkinys **8**, 497.

Cr I

Configuration	Term	J	Level (cm^{-1})	g	Leading percentages		
$3d^5(^6S)4s$	$a\ ^7S$	3	0.00	2.00183	100		
$3d^5(^6S)4s$	$a\ ^5S$	2	7 593.16	2.006	100		
$3d^4\ 4s^2$	$a\ ^5D$	0	7 750.78				
		1	7 810.82	1.50060			
		2	7 927.47	1.50060			
		3	8 095.21	1.50060			
		4	8 307.57	1.50060			
$3d^5(^4G)4s$	$a\ ^5G$	2	20 517.40	0.37	100		
		6	20 519.60	1.33	100		
		3	20 520.92	0.93	100		
		4	20 523.69	1.13	100		
		5	20 523.94	1.25	100		
$3d^5(^4P)4s$	$a\ ^5P$	3	21 840.84	1.6	98		
		2	21 847.88	1.847	98		
		1	21 856.94	2.500	100		
$3d^4\ 4s^2$	$a\ ^3P$	0	23 163.27				
		1	23 512.00				
		2	24 093.16				
$3d^5(^6S)4p$	$z\ ^7P^\circ$	2	*23 305.01*	2.334	67	33	$3d^4(^5D)4s4p(^3P^\circ)\ ^7P^\circ$
		3	*23 386.35*	1.9176	67	33	
		4	*23 498.84*	1.7510	67	31	
$3d^4\ 4s^2$	$a\ ^3H$	4	23 933.90				
		5	24 056.11				
		6	24 200.23				
$3d^5(^4D)4s$	$b\ ^5D$	0	24 277.06		100		
		4	24 282.34	1.51	100		
		1	24 286.54	1.48	100		
		2	24 299.89	1.51	98		
		3	24 303.94	1.55	98		
$3d^5(^4G)4s$	$a\ ^3G$	3	24 833.86		100		
		4	24 897.55		100		
		5	25 038.61		100		
$3d^4\ 4s^2$	$a\ ^3F$	2	24 940.61				
		3	25 106.34				
		4	25 177.39				
$3d^4(^5D)4s4p(^3P^\circ)$	$z\ ^7F^\circ$	0	*24 971.21*		100		
		1	*25 010.64*	1.52	100		
		2	*25 089.20*	1.50	100		
		3	*25 206.02*	1.49	100		
		4	*25 359.62*	1.51	100		
		5	*25 548.64*	1.51	100		
		6	*25 771.40*	1.53	100		
$3d^5(^6S)4p$	$z\ ^5P^\circ$	3	*26 787.50*	1.670	92		
		2	*26 796.28*	1.830	91		
		1	*26 801.93*	2.512	92		

(Continued)

Cr I—Continued

Configuration	Term	J	Level (cm^{-1})	g	Leading percentages		
$3d^5(^4P)4s$	$b\ ^3P$	0	27 163.20		100		
		1	27 176.22		100		
		2	27 223.05		98		
$3d^4(^5D)4s4p(^3P^\circ)$	$z\ ^7D^\circ$	1	*27 300.19*	3.01	99		
		2	*27 382.18*	1.99	99		
		3	*27 500.37*	1.76	99		
		4	*27 649.71*	1.66	99		
		5	*27 825.45*	1.61	100		
$3d^4\ 4s^2$	$b\ ^3G$	3	27 597.22				
		4	27 703.84				
		5	27 816.88				
$3d^4(^5D)4s4p(^3P^\circ)$	$y\ ^7P^\circ$	2	*27 728.87*	2.341	66	33	$3d^5(^6S)4p\ ^7P^\circ$
		3	*27 820.23*	1.929	66	33	
		4	*27 935.26*	1.761	67	32	
$3d^5(^4D)4s$	$a\ ^3D$	3	28 637.00		100		
		1	28 679.43		100		
		2	28 682.18		98		
$3d^4(^5D)4s4p(^3P^\circ)$	$y\ ^5P^\circ$	1	*29 420.90*	2.513	95		
		2	*29 584.62*	1.836	95		
		3	*29 824.75*	1.669	96		
$3d^4(^5D)4s4p(^3P^\circ)$	$z\ ^5F^\circ$	1	*30 787.30*	0.002	96		
		2	*30 858.82*	0.997	96		
		3	*30 965.46*	1.245	96		
		4	*31 106.37*	1.345	95		
		5	*31 280.35*	1.396	96		
$3d^5(^2D3)4s$	$b\ ^3D$	3	31 009.00		56	18	$(^2D1)\ ^3D$
		2	31 028.33		50	24	$(^4F)\ ^5F$
		1	31 048.85		45	41	$(^4F)\ ^5F$
$3d^5(^2I)4s$	$a\ ^3I$	7	31 048.00		100		
		6	31 049.33		100		
		5	31 055.35		100		
$3d^5(^4F)4s$	$a\ ^5F$	1	31 352.42		59	31	$(^2D3)\ ^3D$
		2	31 355.21		74	12	
		3	31 364.33		88	6	
		4	31 377.96		100		
		5	31 393.40		100		
$3d^4\ 4s^2$	$a\ ^1G$	4	31 987.06				
$3d^4\ 4s^2$	$a\ ^1I$	6	32 097.36				
$3d^5(^2F1)4s$	$b\ ^3F$	2	33 040.10		79	14	$(^2D3)\ ^3D$
		3	33 060.74		83	13	
		4	33 113.27		100		
$3d^4(^5D)4s4p(^3P^\circ)$	$z\ ^5D^\circ$	0	*33 338.20*		89	8	$^3P^\circ$
		1	*33 423.79*	1.499	93		
		2	*33 542.11*	1.497	96		
		3	*33 671.55*	1.497	97		
		4	*33 816.06*	1.499	97		

Cr I—Continued

Configuration	Term	J	Level (cm^{-1})	g	Leading percentages		
$3d^4(^5D)4s4p(^3P°)$	$z\ ^3P°$	0	*33 762.56*		88	8	$^5D°$
		1	*33 897.26*	1.49	92		
		2	*34 190.49*	1.55	95		
$3d^5(^2I)4s$	$b\ ^1I$	6	33 762.74		100		
$3d^4\ 4s^2$	$c\ ^3D$	1	33 906.65				
		3	33 934.88				
		2	33 935.65				
$3d^6$	$c\ ^5D$	4	35 398.02				
		3	35 501.26				
		2	35 572.94				
		1	35 618.51				
		0	35 640.69				
$3d^5(^4F)4s$	$c\ ^3F$	2	35 807.90		100		
		3	35 813.73		100		
		4	35 862.82		98		
$3d^5(^2H)4s$	$b\ ^3H$	4	35 870.53		98		
		5	35 884.40		98		
		6	35 934.02		100		
$3d^4(^5D)4s4p(^3P°)$	$z\ ^3F°$	2	*35 897.87*		95		
		3	*36 034.22*		95		
		4	*36 212.15*		95		
$3d^5(^2F2)4s$	$d\ ^3F$	3	36 552.13		100		
		2	36 558.55		100		
		4	36 577.73		100		
$3d^5(^6S)5s$	$e\ ^7S$	3	36 895.73				
$3d^5(^2G2)4s$	$c\ ^3G$	3	37 205.88		98		
		5	37 233.50		96		
		4	37 244.17		98		
$3d^5(^6S)5s$	$e\ ^5S$	2	37 883.34				
$3d^5(^2H)4s$	$a\ ^1H$	5	38 537.68		98		
$3d^4(^5D)4s4p(^3P°)$	$z\ ^3D°$	1	*38 597.06*		96		
		2	*38 730.67*		96		
		3	*38 911.33*		96		
$3d^5(^2G2)4s$	$b\ ^1G$	4	39 158.63		98		
$3d^4(^5D)4s4p(^1P°)$	$y\ ^5F°$	1	*40 906.46*	0.004	84	12	$3d^5(^4G)4p\ ^5F°$
		2	*40 971.29*	1.28	83	12	
		3	*41 086.26*	1.246	83	12	
		4	*41 224.78*	1.360	82	13	
		5	*41 393.47*		82	14	
$3d^4(^5D)4s4p(^1P°)$	$x\ ^5P°$	1	*40 930.31*	2.455	68	7	$3d^5(^4P)4p\ ^5P°$
		2	*40 982.97*	1.76	56	7	
		3	*41 043.35*	1.640	55	8	

(Continued)

Cr I—Continued

Configuration	Term	J	Level (cm^{-1})	g	Leading percentages			
$3d^4(^5D)4s4p(^1P^\circ)$	$y\ ^5D^\circ$	0	41 224.80		52		14	$3d^5(^4P)4p\ ^5D^\circ$
		1	41 289.17	1.503	54		14	
		2	41 409.03	1.504	56		13	
		3	41 575.10	1.503	58		13	
		4	41 782.19	1.500	62		14	
$3d^4(^3H)4s4p(^3P^\circ)$	$z\ ^5H^\circ$	3	42 025.60		61		20	$3d^5(^4G)4p\ ^5H^\circ$
		4	42 079.81		55		19	$3d^5(^4G)4p\ ^5H^\circ$
		5	42 153.74		48		17	$3d^5(^4G)4p\ ^5H^\circ$
		7	42 387.32		65		24	$3d^5(^4G)4p\ ^5H^\circ$
$3d^4(a\ ^3P)4s4p(^3P^\circ)$	$x\ ^5D^\circ$	0	42 218.37		49		26	$3d^4(^5D)4s4p(^1P^\circ)\ ^5D^\circ$
		1	42 292.96	1.501	47		26	
		2	42 438.82	1.494	43		21	
		3	42 648.26	1.498	40		22	
		4	42 908.57	1.497	45		20	
$3d^5(^6S)5p$	$x\ ^7P^\circ$	2	42 238.04					
		3	42 254.11					
		4	42 275.20					
$3d^4(^3H)4s4p(^3P^\circ)$		6	42 252.17		38	$^5H^\circ$	25	$3d^5(^4G)4p\ ^5G^\circ$
$3d^5(^6S)4d$	$e\ ^7D$	1	42 253.42					
		2	42 254.52					
		3	42 256.26					
		4	42 258.37					
		5	42 261.06	1.55				
$3d^5(^4G)4p$	$z\ ^5G^\circ$	2	42 515.35	0.35	61		20	$3d^4(^3H)4s4p(^3P^\circ)\ ^5G^\circ$
		3	42 538.81		57		19	$3d^4(^3H)4s4p(^3P^\circ)\ ^5G^\circ$
		4	42 564.85		53		18	$3d^4(^3H)4s4p(^3P^\circ)\ ^5G^\circ$
		5	42 589.25	1.23	47		16	$3d^4(^3H)4s4p(^3P^\circ)\ ^5G^\circ$
		6	42 605.81	1.32	37		27	$3d^4(^3H)4s4p(^3P^\circ)\ ^5H^\circ$
$3d^5(^4P)4p$	$z\ ^5S^\circ$	2	43 124.88	1.93	46		46	$3d^4(a\ ^3P)4s4p(^3P^\circ)\ ^5S^\circ$
$3d^5(^6S)4d$	$e\ ^5D$	4	44 050.87					
		3	44 068.72					
		2	44 080.90					
		1	44 088.92					
		0	44 092.80					
$3d^5(^6S)5p$	$w\ ^5P^\circ$	1	44 125.90	2.74				
		2	44 186.92	1.79				
		3	44 259.36	1.68				
$3d^4(^3H)4s4p(^3P^\circ)$	$z\ ^5I^\circ$	4	44 246.70		99			
		5	44 307.96		99			
		6	44 393.10		99			
		7	44 514.44		99			
		8	44 666.55		100			
$3d^4(a\ ^3F)4s4p(^3P^\circ)$	$y\ ^5G^\circ$	2	44 299.98	0.35	85			
		3	44 373.34	0.93	80			
		4	44 534.46		68		18	$^5F^\circ$
		5	44 591.46	1.25	46		29	$^5F^\circ$
		6	44 746.26	1.34	85		5	$3d^4(^3G)4s4p(^3P^\circ)\ ^5G^\circ$

Cr I—Continued

Configuration	Term	J	Level (cm^{-1})	g	Leading percentages			
$3d^4(a\,^3P)4s4p(^3P^\circ)$	$v\,^5P^\circ$	1	*44 666.74*	2.47	78		8	$3d^5(^4D)4p\,^5P^\circ$
		2	*44 875.19*		76		9	
		3	*45 113.22*	1.65	70		13	
$3d^4(a\,^3F)4s4p(^3P^\circ)$	$x\,^5F^\circ$	1	*45 201.84*		66		28	$3d^5(^4G)4p\,^5F^\circ$
		2	*45 225.20*		63		27	$3d^5(^4G)4p\,^5F^\circ$
		3	*45 255.51*		58		26	$3d^5(^4G)4p\,^5F^\circ$
		4	*45 286.08*		50		23	$3d^5(^4G)4p\,^5F^\circ$
$3d^4(a\,^3F)4s4p(^3P^\circ)$		5	*45 305.45*	1.41	37	$^5G^\circ$	34	$^5F^\circ$
$3d^5(^4G)4p$	$z\,^3H^\circ$	6	*45 348.73*		68		16	$^5H^\circ$
		5	*45 354.18*		49		32	
		4	*45 358.63*		59		25	
$3d^5(^4G)4p$	$y\,^5H^\circ$	3	*45 566.02*	0.52	76		22	$3d^4(^3H)4s4p(^3P^\circ)\,^5H^\circ$
		4	*45 614.88*		51		28	$3d^5(^4G)4p\,^3H^\circ$
		5	*45 663.28*		43		38	$3d^5(^4G)4p\,^3H^\circ$
		6	*45 707.36*		58		20	$3d^5(^4G)4p\,^3H^\circ$
		7	*45 741.49*	1.29	74		25	$3d^4(^3H)4s4p(^3P^\circ)\,^5H^\circ$
$3d^5(^6S)6s$	$f\,^7S$	3	45 643.38	2.05				
$3d^5(^4P)4p$	$y\,^3P^\circ$	1	*45 719.20*		45		24	$3d^4(a\,^3P)4s4p(^3P^\circ)\,^3P^\circ$
		0	*45 722.59*		39		31	
		2	*45 734.32*		49		20	
$3d^5(^4G)4p$	$y\,^3F^\circ$	2	*45 966.45*		75		8	$3d^4(a\,^3F)4s4p(^3P^\circ)\,^3F^\circ$
		3	*46 000.36*		80		8	
		4	*46 058.20*		83		7	
$3d^5(^6S)6s$	$f\,^5S$	2	45 967.81					
$3d^5(^4P)4p$		1	*46 077.09*		22	$^3D^\circ$	19	$3d^5(^4D)4p\,^5F^\circ$
$3d^4(a\,^3F)4s4p(^3P^\circ)$	$w\,^5D^\circ$	0	*46 081.27*		40		25	$3d^4(a\,^3P)4s4p(^3P^\circ)\,^5D^\circ$
		1	*46 298.32*		42		20	
		2	*46 349.50*		40		22	
		3	*46 368.35*		36		27	
		4	*46 422.46*		39		34	
$3d^5(^4P)4p$		2	*46 109.26*	1.24	23	$^3D^\circ$	14	$3d^5(^4D)4p\,^5F^\circ$
$3d^5(^4P)4p$		3	*46 174.40*	1.33	37	$^3D^\circ$	15	$3d^4(a\,^3F)4s4p(^3P^\circ)\,^5D^\circ$
$3d^4\,4s5s$	$f\,^7D$	1	46 448.60	2.99				
		2	46 524.84	1.99				
		3	46 637.21	1.77				
		4	46 783.06	1.63				
		5	46 958.98	1.61				
$3d^4(^3G)4s4p(^3P^\circ)$	$w\,^5F^\circ$	2	*46 677.06*		20		18	$3d^5(^4D)4p\,^5F^\circ$
		1	*46 678.35*		21		18	
		3	*46 688.24*	1.25	23		22	
		5	*46 704.98*	1.37	33		30	
		4	*46 720.54*		23		24	

(Continued)

Cr I—Continued

Configuration	Term	J	Level (cm^{-1})	g	Leading percentages			
$3d^4(^3\mathrm{H})4s4p(^3\mathrm{P}^\circ)$	$z\,^3\mathrm{G}^\circ$	3	*46 846.77*		42		20	$3d^4(a\,^3\mathrm{F})4s4p(^3\mathrm{P}^\circ)\,^3\mathrm{G}^\circ$
		4	*46 905.03*		40		21	$3d^4(a\,^3\mathrm{F})4s4p(^3\mathrm{P}^\circ)\,^3\mathrm{G}^\circ$
		5	*46 985.87*		34		20	$3d^5(^4\mathrm{G})4p\,^3\mathrm{G}^\circ$
$3d^5(^4\mathrm{P})4p$	$u\,^5\mathrm{P}^\circ$	3	*46 878.61*	1.68	47		20	$3d^5(^4\mathrm{D})4p\,^5\mathrm{P}^\circ$
		2	*46 967.70*	1.84	51		20	
		1	*47 021.75*	2.42	55		20	
$3d^4(^3\mathrm{H})4s4p(^3\mathrm{P}^\circ)$	$x\,^5\mathrm{G}^\circ$	2	*47 047.47*	0.45	67		15	$3d^5(^4\mathrm{G})4p\,^5\mathrm{G}^\circ$
		3	*47 125.70*	0.96	62		14	
		4	*47 189.87*		60		15	
		6	*47 222.27*	1.44	70		19	
		5	*47 228.80*	1.27	63		16	
$3d^5(^4\mathrm{G})4p$	$y\,^3\mathrm{G}^\circ$	3	*47 048.48*		73		12	$3d^4(^3\mathrm{H})4s4p(^3\mathrm{P}^\circ)\,^3\mathrm{G}^\circ$
		4	*47 054.91*		69		16	
		5	*47 055.31*		62		24	
$3d^4(a\,^3\mathrm{P})4s4p(^3\mathrm{P}^\circ)$	$z\,^3\mathrm{S}^\circ$	1	*47 088.40*		72		8	$3d^5(^4\mathrm{P})4p\,^3\mathrm{S}^\circ$
$3d^4(^3\mathrm{H})4s4p(^3\mathrm{P}^\circ)$	$z\,^3\mathrm{I}^\circ$	5	*47 586.06*		61		28	$3d^4(^3\mathrm{G})4s4p(^3\mathrm{P}^\circ)\,^5\mathrm{H}^\circ$
		6	*47 630.43*		57		34	
		7	*47 692.63*		52		40	
$3d^4(^3\mathrm{G})4s4p(^3\mathrm{P}^\circ)$	$x\,^5\mathrm{H}^\circ$	3	*47 621.31*		79		12	$3d^4(^3\mathrm{H})4s4p(^3\mathrm{P}^\circ)\,^5\mathrm{H}^\circ$
		4	*47 688.51*		82		11	$3d^4(^3\mathrm{H})4s4p(^3\mathrm{P}^\circ)\,^5\mathrm{H}^\circ$
		5	*47 793.82*		56		30	$3d^4(^3\mathrm{H})4s4p(^3\mathrm{P}^\circ)\,^3\mathrm{I}^\circ$
		6	*47 942.29*		53		35	$3d^4(^3\mathrm{H})4s4p(^3\mathrm{P}^\circ)\,^3\mathrm{I}^\circ$
		7	*48 140.18*		49		42	$3d^4(^3\mathrm{H})4s4p(^3\mathrm{P}^\circ)\,^3\mathrm{I}^\circ$
$3d^4(^5\mathrm{D})4s5p(^3\mathrm{P}^\circ)$	$v\,^5\mathrm{F}^\circ$	1	*47 629.66*					
		2	*47 631.51*					
		3	*47 636.25*					
		4	*47 639.84*	1.34				
		5	*47 644.76*					
$3d^5(^6\mathrm{S})6p$	$w\,^7\mathrm{P}^\circ$	2	*47 697.44*					
		3	*47 708.59*					
		4	*47 719.08*					
$3d^5(^6\mathrm{S})5d$	$g\,^7\mathrm{D}$	1	47 700.18					
		2	47 700.95					
		3	47 702.30					
		4	47 704.66					
		5	47 709.80					
$3d^5(^4\mathrm{P})4p$	$v\,^5\mathrm{D}^\circ$	1	*47 772.30*	1.37	49		26	$3d^4(a\,^3\mathrm{F})4s4p(^3\mathrm{P}^\circ)\,^5\mathrm{D}^\circ$
		2	*47 786.10*	1.39	49		26	
		0	*47 788.08*		46		27	
		3	*47 814.40*	1.53	49		25	
		4	*47 866.48*	1.50	52		23	
$3d^5(^4\mathrm{G})4p$		1	*47 877.55*	0.00	15	$^5\mathrm{F}^\circ$	15	$3d^4(a\,^3\mathrm{P})4s4p(^3\mathrm{P}^\circ)\,^3\mathrm{P}^\circ$
$3d^5(^4\mathrm{G})4p$	$u\,^5\mathrm{F}^\circ$	2	*47 917.93*	1.04	35		19	$3d^4(a\,^3\mathrm{F})4s4p(^3\mathrm{P}^\circ)\,^5\mathrm{F}^\circ$
		3	*47 974.53*	1.36	35		20	
		5	*47 985.76*	1.38	39		25	
		4	*48 014.40*		39		23	
		4	*48 042.80*					

Cr I—Continued

Configuration	Term	J	Level (cm^{-1})	g	Leading percentages		
$3d^4(a\,^3\mathrm{F})4s4p(^3\mathrm{P}^\circ)$	$^3\mathrm{D}^\circ$	1	*48 210.04*		59	7	$3d^5(^4\mathrm{G})4p\ ^5\mathrm{F}^\circ$
		2	*48 217.83*		61	6	
		3	*48 251.91*		64	5	
$3d^4(a\,^3\mathrm{P})4s4p(^3\mathrm{P}^\circ)$	$x\ ^3\mathrm{P}^\circ$	0	*48 226.36*		29	24	$3d^5(^4\mathrm{P})4p\ ^3\mathrm{P}^\circ$
		1	*48 331.30*		28	15	
		2	*48 458.67*		42	19	
$3d^4(^3\mathrm{H})4s4p(^3\mathrm{P}^\circ)$	$y\ ^3\mathrm{H}^\circ$	4	*48 288.37*		76		
		5	*48 310.39*		89		
		6	*48 445.35*		83		
$3d^4\ 4s5s$	$f\ ^5\mathrm{D}$	0	48 488.23				
		1	48 507.56				
		2	48 558.57				
		3	48 661.59	1.46			
		4	48 824.50	1.46			
	$x\ ^3\mathrm{G}^\circ$	3	*48 515.08*				
		4	*48 562.16*				
		5	*48 786.39*				
$3d^4(a\,^3\mathrm{F})4s4p(^3\mathrm{P}^\circ)$	$^3\mathrm{F}^\circ$	3	*48 636.14*		60	15	$3d^4(^3\mathrm{G})4s4p(^3\mathrm{P}^\circ)\ ^3\mathrm{F}^\circ$
$3d^4(a\,^3\mathrm{P})4s4p(^3\mathrm{P}^\circ)$	$x\ ^3\mathrm{D}^\circ$	1	*48 839.90*		38	20	$3d^5(^4\mathrm{P})4p\ ^3\mathrm{D}^\circ$
		2	*49 027.58*		46	14	
		3	*49 310.86*		61	7	
$3d^5(^6\mathrm{S})7s$	$g\ ^7\mathrm{S}$	3	49 177.83				
$3d^5(^6\mathrm{S})7s$	$g\ ^5\mathrm{S}$	2	49 321.51				
$3d^4(a\,^3\mathrm{F})4s4p(^3\mathrm{P}^\circ)$	$w\ ^3\mathrm{G}^\circ$	3	*49 370.70*		40	9	$3d^4(^3\mathrm{H})4s4p(^3\mathrm{P}^\circ)\ ^3\mathrm{G}^\circ$
		4	*49 453.94*		40	13	
		5	*49 538.06*		51	17	
$3d^4(^3\mathrm{G})4s4p(^3\mathrm{P}^\circ)$	$w\ ^5\mathrm{G}^\circ$	2	*49 466.77*		65	18	$3d^5(^4\mathrm{G})4p\ ^5\mathrm{G}^\circ$
		3	*49 519.72*	1.04	53	14	
		4	*49 573.03*		63	19	
		5	*49 617.61*		70	17	
		6	*49 635.16*	1.35	73	17	
$3d^5(^4\mathrm{P})4p$	$y\ ^3\mathrm{S}^\circ$	1	*49 477.04*		73	17	$3d^4(a\,^3\mathrm{P})4s4p(^3\mathrm{P}^\circ)\ ^3\mathrm{S}^\circ$
	$e\ ^3\mathrm{F}$	2	49 586.38				
		3	49 717.88				
		4	49 863.50				
$3d^5(^4\mathrm{D})4p$	$t\ ^5\mathrm{P}^\circ$	1	*49 588.97*	2.48	46	30	$3d^5(^4\mathrm{P})4p\ ^5\mathrm{P}^\circ$
		3	*49 812.46*	1.77	31	26	
$3d^4(a\,^3\mathrm{F})4s4p(^3\mathrm{P}^\circ)$		2	*49 598.08*	1.88	20 $^1\mathrm{D}^\circ$	14	$3d^5(^4\mathrm{D})4p\ ^5\mathrm{P}^\circ$
$3d^5(^4\mathrm{D})4p$	$x\ ^3\mathrm{F}^\circ$	4	*49 620.69*		56	15	$3d^4(^3\mathrm{G})4s4p(^3\mathrm{P}^\circ)\ ^3\mathrm{F}^\circ$
		3	*49 650.22*		43	18	
		2	*49 652.76*		37	18	
$3d^4(a\,^3\mathrm{P})4s4p(^3\mathrm{P}^\circ)$	$y\ ^5\mathrm{S}^\circ$	2	*49 822.59*	2.00	31	30	$3d^5(^4\mathrm{P})4p\ ^5\mathrm{S}^\circ$

(Continued)

Cr I—Continued

Configuration	Term	J	Level (cm^{-1})	g	Leading percentages		
$3d^5(^4D)4p$	$s\ ^5F°$	1	*50 018.80*		44	29	$3d^4(^3G)4s4p(^3P°)\ ^5F°$
		2	*50 057.61*		43	29	
		3	*50 102.04*	1.27	44	29	
		4	*50 210.87*	1.25	30	21	
		5	*50 253.27*	1.39	46	33	
$3d^5(^4D)4p$	$w\ ^3D°$	1	*50 105.54*		66	11	$3d^5(^4P)4p\ ^3D°$
		2	*50 184.10*		57	14	
		3	*50 264.48*		51	18	
$3d^5(^6S)7p$	$^7P°$	3	*50 185*				
		4	*50 197*				
$3d^5(^4D)4p$	$u\ ^5D°$	4	*50 557.56*	1.54	44	15	$3d^5(^4D)4p\ ^5F°$
		3	*50 628.11*	1.54	60	7	$3d^5(^4P)4p\ ^5D°$
		2	*50 654.76*	1.51	68	8	$3d^5(^4P)4p\ ^5D°$
		0	*50 661.20*		69	10	$3d^5(^4P)4p\ ^5D°$
		1	*50 662.77*	1.46	71	9	$3d^5(^4P)4p\ ^5D°$
$3d^4(^3G)4s4p(^3P°)$	$w\ ^3F°$	2	*50 890.15*		41	27	$3d^5(^4D)4p\ ^3F°$
		3	*50 950.42*		42	26	
		4	*51 059.79*		46	23	
$3d^5(^6S)8s$	$h\ ^5S$	2	51 035.68				
$3d^5(^4D)4p$	$w\ ^3P°$	0	*51 176.88*		71	11	$3d^4(a\ ^3P)4s4p(^3P°)\ ^3P°$
		1	*51 246.87*		68	8	$3d^4(a\ ^3P)4s4p(^3P°)\ ^3P°$
		2	*51 286.52*		45	32	$3d^4(a\ ^3P)4s4p(^3P°)\ ^1D°$
$3d^4(^3H)4s4p(^3P°)$	$z\ ^1H°$	5	*51 401.24*		68	11	$3d^5(^2I)4p\ ^1H°$
$3d^5(^6S)8p$	$^7P°$	2	*51 529.4*				
		3	*51 531.5*				
		4	*51 534.4*				
$3d^4(^3D)4s4p(^3P°)$	$t\ ^5D°$	0	*51 999.62*		94		
		1	*52 003.06*		94		
		2	*52 012.44*		93		
		3	*52 031.72*		92		
		4	*52 064.27*		92		
$3d^5(^6S)9p$	$^7P°$		*52 341*				
$3d^4(^1I)4s4p(^3P°)$	$y\ ^3I°$	5	*52 591.94*		48	33	$3d^5(^2I)4p\ ^3I°$
		6	*52 660.61*		58	39	
		7	*52 677.88*		59	39	
$3d^4(^3G)4s4p(^3P°)$	$^3G°$	4	*52 720.07*		72	14	$3d^4(^3H)4s4p(^3P°)\ ^3G°$
$3d^5(^6S)10p$	$^7P°$		*52 857.3*				
$3d^4(a\ ^1G)4s4p(^3P°)$	$x\ ^3H°$	5	*52 885.39*		57	27	$3d^5(^2I)4p\ ^3H°$
		6	*52 914.94*		68	26	
		4	*52 963.44*		47	26	
$3d^4(^3D)4s4p(^3P°)$	$r\ ^5F°$	1	*53 011.65*		85	6	$3d^5(^4D)4p\ ^5F°$
		2	*53 037.52*		84	6	
		3	*53 073.90*		83	6	
		4	*53 117.54*		81	6	
		5	*53 172.33*	1.42	85	7	

Cr I—Continued

Configuration	Term	J	Level (cm^{-1})	g	Leading percentages
$3d^4\,4s4d$	$e\ ^7$G	1	53 148.35		
		2	53 177.87		
		3	53 228.49		
		4	53 298.90		
		5	53 393.50		
		6	53 517.85		
		7	53 662.64		
$3d^4\,4s4d$	$h\ ^7$D	2	53 195.03		
		3	53 284.34		
		4	53 375.46		
		5	53 627.75		
$3d^4\,4s4d$	$e\ ^7$F	1	53 215.40		
		2	53 279.80		
		3	53 384.72		
		4	53 526.22		
		5	53 706.06		
		6	53 927.47		
$3d^5(^6$S$)11p$	^{7}P°		*53 217.0*		
$3d^5(^6$S$)12p$	^{7}P°		*53 484.5*		
$3d^4\,4s5p$	$s\ ^5$D°	2	*53 541.25*		
		3	*53 640.74*		
		4	*53 782.77*		
$3d^5(^6$S$)13p$	^{7}P°		*53 671.4*		
$3d^3(^4$F$)4s^2\,4p$	$v\ ^3$G°	3	*53 804.84*		28 22 $3d^5(^4$F$)4p\ ^3$G°
		4	*53 927.59*		26 22
		5	*54 078.13*		25 22
$3d^5(^6$S$)14p$	^{7}P°		*53 815.8*		
$3d^5(^6$S$)15p$	^{7}P°		*53 928.0*		
$3d^4\,4s5p$	$s\ ^5$P°	1	*53 963.05*		
		2	*54 032.63*		
		3	*54 132.88*		
$3d^5(^6$S$)16p$	^{7}P°		*54 017.1*		
$3d^5(^6$S$)17p$	^{7}P°		*54 089.0*		
$3d^5(^6$S$)18p$	^{7}P°		*54 147.7*		
$3d^5(^6$S$)19p$	^{7}P°		*54 196.6*		
$3d^4\,4s5p$	$q\ ^5$F°	1	*54 198.23*		
		2	*54 252.19*		
		3	*54 328.95*		
		4	*54 425.29*		
		5	*54 536.53*		
$3d^5(^6$S$)20p$	^{7}P°		*54 237.7*		
$3d^5(^6$S$)21p$	^{7}P°		*54 272.4*		

(Continued)

Cr I—Continued

Configuration	Term	J	Level (cm^{-1})	g	Leading percentages		
	$e\ ^5$F	1	54 296.76				
		2	54 383.36				
		3	54 476.29				
		4	54 572.84				
		5	54 660.31				
$3d^5(^6$S$)22p$	^{7}P°		*54 302.4*				
$3d^5(^2$I$)4p$	$z\ ^3$K°	6	*54 316.83*		56	42	$3d^4(^1$I$)4s4p(^3$P°$)\ ^3$K°
		7	*54 404.94*		54	46	
		8	*54 498.27*		59	41	
$3d^5(^6$S$)23p$	^{7}P°		*54 327.2*				
$3d^5(^6$S$)24p$	^{7}P°		*54 349.5*				
$3d^5(^6$S$)25p$	^{7}P°		*54 368.6*				
$3d^5(^6$S$)26p$	^{7}P°		*54 385.5*				
$3d^5(^6$S$)27p$	^{7}P°		*54 400.5*				
$3d^5(^6$S$)28p$	^{7}P°		*54 413.5*				
$3d^5(^6$S$)29p$	^{7}P°		*54 425.4*				
$3d^5(^6$S$)30p$	^{7}P°		*54 436.1*				
$3d^5(^6$S$)31p$	^{7}P°		*54 445.0*				
$3d^5(^6$S$)32p$	^{7}P°		*54 453.4*				
$3d^5(^6$S$)33p$	^{7}P°		*54 461.7*				
$3d^5(^6$S$)34p$	^{7}P°		*54 468.5*				
$3d^5(^6$S$)35p$	^{7}P°		*54 474.8*				
$3d^5(^6$S$)36p$	^{7}P°		*54 480.6*				
$3d^5(^6$S$)37p$	^{7}P°		*54 485.4*				
$3d^5(^6$S$)38p$	^{7}P°		*54 491.2*				
Cr II (^{6}S$_{5/2}$)	***Limit***		**54 575.6**				
$3d^4\ 4s5s$	$g\ ^5$D	0	54 646.20				
		1	54 671.90				
		2	54 818.55				
		3	54 986.82				
		4	55 209.01				
$3d^4(^3$H$)4s4p(^1$P°$)$	$w\ ^3$H°	4	*54 736.55*		42	20	$3d^5(^2$I$)4p\ ^3$H°
		5	*54 799.18*		41	21	
		6	*54 886.82*		42	24	
$3d^4(^1$I$)4s4p(^3$P°$)$	^{3}K°	6	*54 800.26*		55	40	$3d^5(^2$I$)4p\ ^3$K°
$3d^4\ 4s5s$	$e\ ^3$D	1	54 804.69				
		2	54 974.64				
		3	55 204.79				
$3d^5(^4$G$)5p$	$v\ ^3$H°	4	*54 810.94*				
		6	*54 866.57*				
		5	*54 929.72*				
$3d^4(^3$D$)4s4p(^3$P°$)$	$v\ ^3$D°	1	*54 956.59*		33	28	$3d^5(a\ ^2$D$)4p\ ^3$D°
		2	*55 152.63*		30	26	
		3	*55 451.64*		30	19	
$3d^5(^2$I$)4p$	$z\ ^1$K°	7	*54 970.23*		79	10	^{3}K°

Cr I—Continued

Configuration	Term	J	Level (cm^{-1})	g	Leading percentages			
$3d^5(a\ ^2D)4p$	$v\ ^3F°$	2	*54 992.93*		30		20	$3d^5(a\ ^2F)4p\ ^3F°$
		3	*55 101.87*		22		15	
		4	*55 207.40*		30		18	
$3d^5(^4G)5p$	$u\ ^3F°$	4	*55 120.77*					
		3	*55 352.72*					
		2	*55 473.67*					
$3d^5(^2I)4p$		6	*55 516.69*		35	$^1I°$	28	$3d^4(^1I)4s4p(^3P°)\ ^3H°$
$3d^5(^2I)4p$	$x\ ^3I°$	5	*55 686.46*		29		30	$3d^4(^1I)4s4p(^3P°)\ ^3I°$
		6	*55 741.11*		34		30	
		7	*55 799.10*		39		32	
$3d^4(^1I)4s4p(^3P°)$	$u\ ^3H°$	5	*55 874.98*		61		12	$3d^5(^2H)4p\ ^3H°$
		4	*55 915.50*		67		12	$3d^5(^2H)4p\ ^3H°$
$3d^4(^1I)4s4p(^3P°)$		6	*55 908.12*		37	$^3H°$	37	$3d^5(^2I)4p\ ^1I°$
$3d^5(^2I)4p$	$y\ ^1H°$	5	*55 945.08*		43		12	$3d^4(^3G)4s4p(^3P°)\ ^1H°$
$3d^5(^4F)4p$	$v\ ^5G°$	2	*56 155.12*		46		44	$3d^3(^4F)4s^24p\ ^5G°$
		3	*56 209.81*		43		38	
		4	*56 279.56*		51		40	
		5	*56 361.86*		55		38	
		6	*56 449.10*		59		34	
$3d^4(^3D)4s4p(^3P°)$	$v\ ^3P°$	2	*56 591.88*		42		18	$3d^5(a\ ^2D)4p\ ^3P°$
		1	*56 722.60*		40		10	$3d^4(^3D)4s4p(^3P°)\ ^1P°$
		0	*56 802.50*		56		19	$3d^5(a\ ^2D)4p\ ^3P°$
$3d^4(a\ ^3F)4s4p(^1P°)$		3	*56 985.67*		21	$^3G°$	16	$3d^4(^3D)4s4p(^3P°)\ ^3F°$
$3d^4(a\ ^3F)4s4p(^1P°)$	$u\ ^3G°$	4	*57 033.60*		24		15	$3d^4(^3H)4s4p(^3P°)\ ^3G°$
		5	*57 088.25*		25		23	$3d^4(^3H)4s4p(^3P°)\ ^3G°$
$3d^4(a\ ^1S)4s4p(^3P°)$	$u\ ^3P°$	2	*57 087.70*		43		22	$3d^5(a\ ^2D)4p\ ^3P°$
		1	*57 132.59*		40		21	
		0	*57 154.59*		40		22	
	$p\ ^5F°$	1	*57 096.62*					
		2	*57 100.66*					
		3	*57 186.60*					
		4	*57 237.50*					
		5	*57 327.66*					
		3	*57 141.85*					
$3d^4(^3D)4s4p(^3P°)$		2	*57 220.67*		24	$^3F°$	14	$3d^3(^4F)4s^24p\ ^3F°$
$3d^5(^4F)4p$		3	*57 276.42*		10	$^3F°$	10	$3d^3(^4F)4s^24p\ ^3F°$
$3d^5(^4F)4p$		4	*57 335.47*		16	$^3F°$	13	$3d^3(^4F)4s^24p\ ^3F°$
$3d^5(^4G)5s$	$e\ ^5G$	2	57 350.65					
		3	57 361.24					
		4	57 372.78					
		5	57 382.93					
		6	57 389.32					

(Continued)

Cr I—Continued

Configuration	Term	J	Level (cm^{-1})	g	Leading percentages			
$3d^5(a\ ^2F)4p$	$t\ ^3G°$	3	*57 557.03*		40		14	$3d^5(^4F)4p\ ^5F°$
		5	*57 702.36*		58		20	$3d^4(a\ ^3F)4s4p(^3P°)\ ^3G°$
$3d^5(a\ ^2F)4p$		4	*57 587.36*		21	$^3G°$	18	$3d^5(^4F)4p\ ^5F°$
$3d^4\ 4s5p$	$r\ ^5D°$	0	*57 958.42*					
		1	*57 995.04*					
		2	*58 063.80*					
		3	*58 147.76*					
		4	*58 292.62*					
	$e\ ^3G$	3	57 984.94					
		5	57 990.23					
		4	57 992.15					
$3d^5(a\ ^2F)4p$	$s\ ^3F°$	2	*58 162.84*		26		15	$3d^5(a\ ^2D)4p\ ^3F°$
		4	*58 167.89*		25		20	
		3	*58 202.65*		25		23	
$3d^5(^4F)4p$		1	*58 725.28*		33	$^5D°$	18	$3d^5(^4F)4p\ ^3D°$
$3d^5(^2H)4p$	$t\ ^3H°$	4	*58 728.29*		23		16	$3d^5(^2I)4p\ ^3H°$
		5	*58 754.58*		24		17	
		6	*58 775.36*		26		17	
$3d^5(^4P)5p$	$t\ ^3D°$	2	*58 772.03*					
		1	*58 870.20*					
		3	*58 924.12*					
$3d^5(^4F)4p$	$u\ ^3D°$	2	*58 860.23*		39		28	$3d^4(^3D)4s4p(^3P°)\ ^3D°$
		3	*59 122.15*		25		25	$3d^4(^3D)4s4p(^3P°)\ ^3D°$
$3d^4\ 4s(^6D)6p$	$^7F°$	2	*59 220*					
		3	*59 310*					
		4	*59 443*					
$3d^4\ 4s(^6D)6p$	$^7P°$	2	*59 280*					
		3	*59 442*					
		4	*59 659*					
$3d^4(a\ ^3F)4s4p(^1P°)$	$r\ ^3F°$	2	*59 357.90*		51		11	$3d^5(^4F)4p\ ^3F°$
		3	*59 417.01*		46		9	
		4	*59 487.71*		50		9	
$3d^4\ 4s(^6D)6p$	$^7D°$	2	*59 487*					
		3	*59 662*					
		4	*59 877*					
$3d^4(^3H)4s4p(^1P°)$	$w\ ^3I°$	5	*59 806.27*		64		16	$3d^5(^2H)4p\ ^3I°$
		6	*59 884.27*		74		13	
		7	*59 957.46*		77		13	
$3d^4\ 4s(^6D)6p$	$^5P°$	2	*59 946*					
		3	*60 197*					
$3d^5(a\ ^2G)4p$		5	*60 005.60*		20	$^3G°$	13	$3d^5(^2H)4p\ ^1H°$
$3d^4(a\ ^3P)4s4p(^1P°)$	$x\ ^3S°$	1	*60 084.09*		48		19	$^3P°$

Cr I—Continued

Configuration	Term	J	Level (cm^{-1})	g	Leading percentages			
	$q\ ^5\mathrm{D}^\circ$	0	*60 238.04*					
		1	*60 286.59*					
		2	*60 372.86*					
		3	*60 491.22*					
		4	*60 625.65*					
	$q\ ^3\mathrm{F}^\circ$	2	*60 253.00*					
		3	*60 326.04*					
		4	*60 367.38*					
$3d^5(^2\mathrm{H})4p$	$v\ ^3\mathrm{I}^\circ$	5	*60 427.63*		50		16	$3d^5(a\ ^2\mathrm{G})4p\ ^1\mathrm{H}^\circ$
		6	*60 527.55*		74		11	$3d^5(a\ ^2\mathrm{G})4p\ ^3\mathrm{H}^\circ$
		7	*60 656.97*		85		6	$3d^5(^2\mathrm{I})4p\ ^3\mathrm{I}^\circ$
$3d^5(^2\mathrm{H})4p$	$x\ ^1\mathrm{I}^\circ$	6	*60 441.42*		49		17	$3d^5(a\ ^2\mathrm{G})4p\ ^3\mathrm{H}^\circ$
$3d^5(a\ ^2\mathrm{G})4p$		5	*60 467.85*		30	$^3\mathrm{G}^\circ$	21	$3d^5(^2\mathrm{H})4p\ ^3\mathrm{I}^\circ$
$3d^5(a\ ^2\mathrm{G})4p$	$s\ ^3\mathrm{G}^\circ$	4	*60 503.94*		40		9	$3d^5(^2\mathrm{H})4p\ ^3\mathrm{G}^\circ$
$3d^5(a\ ^2\mathrm{G})4p$		3	*60 518.16*		33	$^3\mathrm{G}^\circ$	16	$3d^5(a\ ^2\mathrm{F})4p\ ^1\mathrm{F}^\circ$
$3d^5(^4\mathrm{F})4p$		3	*60 615.84*		24	$^3\mathrm{D}^\circ$	17	$3d^5(a\ ^2\mathrm{D})4p\ ^3\mathrm{D}^\circ$
$3d^5(a\ ^2\mathrm{D})4p$	$s\ ^3\mathrm{D}^\circ$	2	*60 629.87*		25		24	$3d^4(a\ ^1\mathrm{D})4s4p(^3\mathrm{P}^\circ)\ ^3\mathrm{D}^\circ$
		1	*60 678.12*		29		22	$3d^4(a\ ^1\mathrm{D})4s4p(^3\mathrm{P}^\circ)\ ^3\mathrm{D}^\circ$
	$o\ ^5\mathrm{F}^\circ$	1	*60 678.53*					
		2	*60 777.85*					
		3	*60 902.33*					
		4	*61 052.53*					
		5	*61 193.98*					
$3d^5(b\ ^2\mathrm{F})4p$	$p\ ^3\mathrm{F}^\circ$	3	*60 819.50*		40		21	$3d^4(^1\mathrm{F})4s4p(^3\mathrm{P}^\circ)\ ^3\mathrm{F}^\circ$
		4	*60 960.58*		39		24	
$3d^5(a\ ^2\mathrm{G})4p$	$s\ ^3\mathrm{H}^\circ$	4	*60 870.63*		52		13	$3d^4(^3\mathrm{H})4s4p(^1\mathrm{P}^\circ)\ ^3\mathrm{H}^\circ$
		5	*61 008.07*		41		11	$3d^4(^3\mathrm{H})4s4p(^1\mathrm{P}^\circ)\ ^3\mathrm{H}^\circ$
	$r\ ^5\mathrm{P}^\circ$	1	*61 065.96*					
		2	*61 107.95*					
		3	*61 198.68*					
$3d^5(^2\mathrm{H})4p$	$r\ ^3\mathrm{G}^\circ$	3	*61 078.28*		24		19	$3d^5(b\ ^2\mathrm{F})4p\ ^3\mathrm{G}^\circ$
		4	*61 123.20*		19		15	$3d^5(b\ ^2\mathrm{F})4p\ ^3\mathrm{G}^\circ$
$3d^5(^2\mathrm{H})4p$		5	*61 161.35*		15	$^3\mathrm{G}^\circ$	15	$3d^5(a\ ^2\mathrm{G})4p\ ^3\mathrm{H}^\circ$
$3d^5(a\ ^2\mathrm{G})4p$		6	*61 191.64*		35	$^3\mathrm{H}^\circ$	27	$3d^5(^2\mathrm{H})4p\ ^1\mathrm{I}^\circ$
$3d^4(a\ ^1\mathrm{D})4s4p(^3\mathrm{P}^\circ)$	$t\ ^3\mathrm{P}^\circ$	0	*61 387.86*		77		7	$3d^5(a\ ^2\mathrm{D})4p\ ^3\mathrm{P}^\circ$
		1	*61 527.34*		65		8	$3d^5(a\ ^2\mathrm{D})4p\ ^3\mathrm{P}^\circ$
		2	*61 675.72*		58		8	$3d^3(^4\mathrm{F})4s^24p\ ^5\mathrm{F}^\circ$
$3d^4\ 4s5s$	$e\ ^5\mathrm{P}$	1	61 558.17					
		2	61 687.56					
		3	61 850.17					

(Continued)

Cr I—Continued

Configuration	Term	J	Level (cm^{-1})	g	Leading percentages			
$3d^5(^4F)4p$	$q\ ^3G°$	3	*61 930.05*		48		15	$3d^4(^3G)4s4p(^1P°)\ ^3G°$
$3d^5(^4F)4p$		4	*61 976.50*		32	$^3G°$	20	$3d^3(^4F)4s^24p\ ^5F°$
$3d^4\ 4p^2$	$f\ ^7F$	3	62 034.44					
		4	62 188.83					
		5	62 472.80					
		6	62 658.38					
$3d^5(^4F)4p$		5	*62 037.60*		31	$^3G°$	15	$3d^4(^3G)4s4p(^1P°)\ ^3H°$
$3d^5(^4G)4d$	$f\ ^5G$	2	62 646.60					
		3	62 661.96					
		4	62 671.00					
		6	62 673.92					
		5	62 690.96					
$3d^4\ 4s4p$	$r\ ^3H°$	4	*62 762.06*					
		5	*62 830.26*					
		6	*62 903.03*					
	$q\ ^3H°$	4	*63 116.80*					
		5	*63 144.36*					
		6	*63 182.94*					
$3d^5\ 4p$	$p\ ^3H°$	4	*63 841.81*					
		5	*63 927.27*					
		6	*63 997.86*					
$3d^4\ 4s5s$	$e\ ^5H$	3	64 712.04					
		4	64 751.42					
		5	64 802.08					
		6	64 836.30					
		7	64 940.28					
$3d^5\ 4p$	$p\ ^3G°$	3	*66 008.95*					
		4	*66 094.06*					
		5	*66 180.34*					
$3p^5\ 3d^5\ 4s^2$	$^7P°$	4	*311 280*					
		3	*314 230*					
		2	*316 820*					

Cr II

$Z=24$

V I isoelectronic sequence

Ground state: $1s^2 2s^2 2p^6 3s^2 3p^6 3d^5\ ^6S_{5/2}$

Ionization energy = 132 966±10 cm^{-1} (16.4858±0.0010 eV)

Kiess (1951) carried out the principal analysis of this spectrum using his extensive measurements of wavelengths and Zeeman-effect data. Changes in several of his term designations for the $3d^4 4p$ configuration suggested by Roth (1969) on the basis of theoretical calculations have been adopted here.

Johansson (1983) has reobserved the spectrum, extending the observations into the infrared. All the previously known level values were redetermined, and the new configurations $3d^3 4s4p$, $3d^4 4f$, $5p$, $5d$, $5f$, $5g$, $6s$, $6p$, $6d$, $6g$, and $7s$ as well as a few levels of $3d^4 8s$ and $9s$ were found. In addition, new levels of $3d^4 4s$, $4p$, and $5s$ were added. Altogether 450 new levels were reported. The level value determinations are still in a preliminary stage. The results are given here with an uncertainty of ±0.05 cm^{-1}.

From the 6-member $3d^4(^4D)ns\ ^6D_{9/2}$ series, Johansson (1983) determined a new value for the ionization energy.

The percentage compositions for $3d^4 4p$ are by Roth. Repeating terms of the $3d^4$ core are labeled in alphabetical order rather than by seniority. Shadmi, Oreg, and Stein (1968) have calculated the $3d^5$, $3d^4 4s$, and $3d^3 4s^2$ configurations but give percentages only for $a\,^2$H and $b\,^2$H, apparently the only highly mixed terms. No changes of designations were made except for the $e\,^2$G and $e\,^2$D terms. The percentages represent the sum of seniority states contributing the same core term.

References

Johansson, S. (1983), private communication.
Kiess, C. C. (1951), J. Res. Natl. Bur. Stand. (U.S.) **47**, 385.
Roth, C. (1969), J. Res. Natl. Bur. Stand. (U.S.) **73A**, 125.
Russell, H. N. (1950), J. Opt. Soc. Am. **40**, 618.
Shadmi, Y., Oreg., J., and Stein, J. (1968), J. Opt. Soc. Am. **58**, 909.

Cr II

Configuration	Term	J	Level (cm^{-1})	g	Leading percentages
$3d^5$	$a\ ^6$S	5/2	0.00		
$3d^4(^5D)4s$	$a\ ^6$D	1/2	11 961.81	3.323	
		3/2	12 032.58	1.867	
		5/2	12 147.82	1.669	
		7/2	12 303.86	1.578	
		9/2	12 496.44	1.554	
$3d^4(^5D)4s$	$a\ ^4$D	1/2	19 528.25	0.000	
		3/2	19 631.17	1.192	
		5/2	19 797.88	1.370	
		7/2	20 024.01	1.427	
$3d^5$	$a\ ^4$G	5/2	20 512.06	0.599	
		11/2	20 512.10	1.278	
		7/2	20 517.83	0.994	
		9/2	20 519.33	1.161	
$3d^5$	$a\ ^4$P	5/2	21 822.52	1.590	
		1/2	21 823.84	2.693	
		3/2	21 824.11	1.717	
$3d^5$	$b\ ^4$D	7/2	25 033.70	1.432	
		1/2	25 035.40	-0.045	
		3/2	25 042.81	1.207	
		5/2	25 046.76	1.381	
$3d^4(^3P)4s$	$b\ ^4$P	1/2	29 951.88	2.685	
		3/2	30 307.44	1.756	
		5/2	30 864.46	1.572	

(Continued)

Cr II—Continued

Configuration	Term	J	Level (cm^{-1})	g	Leading percentages		
$3d^5$	$a\ ^2I$	11/2	30 143.30				
		13/2	30 149.83				
$3d^4(^3H)4s$	$a\ ^4H$	7/2	30 156.79	0.667			
		9/2	30 218.81	0.978			
		11/2	30 298.51	1.162			
		13/2	30 391.83	1.234			
$3d^4(^3F)4s$	$a\ ^4F$	3/2	31 082.94	0.418			
		5/2	31 117.39	1.032			
		7/2	31 168.58	1.246			
		9/2	31 219.35	1.340			
$3d^5$	$a\ ^2D$	5/2	31 350.90				
		3/2	31 531.24				
$3d^5$	$a\ ^2F$	7/2	32 355.68				
		5/2	32 603.40				
$3d^5$	$b\ ^4F$	7/2	32 836.68				
		3/2	32 844.76				
		9/2	32 854.31				
		5/2	32 854.95				
$3d^4(^3G)4s$	$b\ ^4G$	5/2	33 417.99	0.588			
		7/2	33 521.11	1.024			
		9/2	33 618.94	1.185			
		11/2	33 694.15	1.276			
$3d^4(^3H)4s$	$a\ ^2H$	9/2	34 630.95		62	35	$3d^5\ ^2H$
		11/2	34 812.95		57	41	
$3d^4(^3P)4s$	$a\ ^2P$	1/2	34 659.32	0.670			
		3/2	35 355.89	1.331			
$3d^4(^3F)4s$	$b\ ^2F$	5/2	35 569.20	0.876			
		7/2	35 607.50	1.144			
$3d^5$	$b\ ^2H$	9/2	35 610.35		61	37	$3d^4(^3H)4s\ ^2H$
		11/2	35 707.49		58	42	
$3d^5$	$a\ ^2G$	7/2	36 101.58				
		9/2	36 272.54				
$3d^4(^3D)4s$	$c\ ^4D$	7/2	38 269.59				
		5/2	38 314.86				
		3/2	38 362.43				
		1/2	38 396.23				
$3d^4(^3G)4s$	$b\ ^2G$	7/2	38 508.93	0.910			
		9/2	38 563.01	1.100			
$3d^4(a^1G)4s$	$c\ ^2G$	7/2	39 683.75				
		9/2	39 824.38				
$3d^5$	$c\ ^2F$	5/2	39 742.09				
		7/2	39 877.07				

Cr II—Continued

Configuration	Term	J	Level (cm^{-1})	g	Leading percentages		
$3d^4(^1I)4s$	$b\ ^2I$	13/2	40 202.12				
		11/2	40 228.33				
$3d^4(^1S)4s$	$a\ ^2S$	1/2	40 415.09				
$3d^4(^3D)4s$	$b\ ^2D$	5/2	42 897.99				
		3/2	42 986.62				
$3d^5$	$b\ ^2S$	1/2	44 307.09				
$3d^4(^1D)4s$	$c\ ^2D$	3/2	45 669.37				
		5/2	45 730.58				
$3d^4(^5D)4p$	$z\ ^6F°$	1/2	*46 823.39*	−0.689	100		
		3/2	*46 905.17*	1.124	100		
		5/2	*47 040.35*	1.314	100		
		7/2	*47 227.24*	1.378	100		
		9/2	*47 464.55*	1.416	100		
		11/2	*47 751.62*		100		
$3d^5$	$d\ ^2D$	5/2	47 354.44				
		3/2	47 372.53				
$3d^4(^5D)4p$	$z\ ^6P°$	3/2	*48 398.95*	2.382	83		
		5/2	*48 491.10*	1.875	98		
		7/2	*48 632.12*	1.710	100		
$3d^4(^5D)4p$	$z\ ^4P°$	1/2	*48 749.36*	2.844	67	31	$(^5D)\ ^6D°$
		3/2	*49 005.93*	1.802	55	42	
		5/2	*49 706.33*	1.624	71	27	
$3d^4(^5D)4p$	$z\ ^6D°$	5/2	*49 351.80*	1.628	73	26	$(^5D)\ ^4P°$
		1/2	*49 492.77*	3.155	69	31	
		3/2	*49 564.60*	1.824	58	41	
		7/2	*49 645.77*	1.577	99		
		9/2	*49 838.38*	1.570	98		
$3d^4(^1F)4s$	$d\ ^2F$	7/2	50 667.24				
		5/2	50 687.62				
$3d^4(^5D)4p$	$z\ ^4F°$	3/2	*51 584.15*	0.406	97		
		5/2	*51 669.48*	1.025	97		
		7/2	*51 788.88*	1.248	96		
		9/2	*51 942.70*	1.338	96		
$3d^5$	$d\ ^2G$	7/2	52 297.81				
		9/2	52 321.01				
$3d^3\ 4s^2$	$c\ ^4F$	3/2	53 051.35				
		5/2	53 271.09				
		7/2	53 566.28				
		9/2	53 923.60				
$3d^4(^5D)4p$	$z\ ^4D°$	1/2	*54 418.02*	0.007	98		
		3/2	*54 499.52*	1.178	98		
		5/2	*54 625.62*	1.376	98		
		7/2	*54 784.48*	1.430	98		

(Continued)

Cr II—Continued

Configuration	Term	J	Level (cm^{-1})	g	Leading percentages		
$3d^4(^3F)4s$	$d\ ^4F$	9/2	54 867.61				
		3/2	54 868.62				
		5/2	54 883.54				
		7/2	54 887.97				
$3d^4(^3P)4s$	$c\ ^4P$	5/2	55 023.10				
		3/2	55 398.74				
		1/2	55 626.21				
$3d^4(^3P)4s$	$b\ ^2P$	3/2	59 130.36				
		1/2	59 526.73				
$3d^4(^3F)4s$	$e\ ^2F$	7/2	59 570.10				
		5/2	59 577.66				
$3d^4(^3G)4s$	$e\ ^2G$	9/2	62 688.95				
		7/2	62 701.67				
$3d^4(^3H)4p$	$z\ ^4H°$	7/2	*63 600.91*	0.680	82	16	$(^3G)\ ^4H°$
		9/2	*63 706.30*	1.030	80	16	
		11/2	*63 848.74*	1.138	80	15	
		13/2	*64 030.53*	1.234	83	13	
$3d^4(a\ ^3P)4p$	$y\ ^4D°$	1/2	*63 801.88*	0.000	87	7	$(a\ ^3F)\ ^4D°$
		3/2	*64 061.73*	1.199	86	8	
		5/2	*64 448.81*	1.380	85	10	
		7/2	*64 924.52*	1.411	80	14	
$3d^4(a\ ^3P)4p$	$z\ ^2S°$	1/2	*65 029.43*		73	7	$(a\ ^3P)\ ^4P°$
$3d^4(a\ ^3F)4p$	$z\ ^4G°$	5/2	*65 156.56*	0.593	79	14	$(^3G)\ ^4G°$
		7/2	*65 256.88*	0.920	70	12	
		9/2	*65 383.95*	1.120	59	10	
		11/2	*65 709.50*	1.265	73	13	
$3d^4(^3H)4p$	$z\ ^4I°$	9/2	*65 217.55*		96		
		11/2	*65 419.56*		95		
		13/2	*65 617.99*		96		
		15/2	*65 812.56*		100		
$3d^4(^3H)4p$	$z\ ^2G°$	7/2	*65 542.93*		49	33	$(a\ ^3F)\ ^2G°$
		9/2	*65 670.08*		41	31	
$3d^3\ 4s^2$	$d\ ^4P$	1/2	65 882.58				
		3/2	66 010.39				
		5/2	66 256.75				
$3d^4(a\ ^3P)4p$	$y\ ^4P°$	1/2	*66 256.47*	2.545	76	13	$(a\ ^3P)\ ^2S°$
		3/2	*66 354.83*	1.671	90		
		5/2	*66 726.81*	1.502	92		
$3d^4(a\ ^3P)4p$	$z\ ^2P°$	3/2	*66 649.38*		53	15	$(a\ ^3F)\ ^2D°$
		1/2	*66 871.93*		79	14	$(a\ ^3P)\ ^2S°$
$3d^4(a\ ^3F)4p$	$y\ ^4F°$	5/2	*67 012.10*		71	13	$(a\ ^3F)\ ^2D°$
		3/2	*67 070.45*		51	21	$(a\ ^3P)\ ^2P°$
		7/2	*67 393.51*		76	10	$(^3H)\ ^4G°$
		9/2	*67 448.57*		63	18	$(^3H)\ ^4G°$

Cr II—Continued

Configuration	Term	J	Level (cm^{-1})	g	Leading percentages		
$3d^4(^3H)4p$	$y\ ^4G°$	7/2	67 333.83		60	18	$(a\ ^3F)\ ^4F°$
		5/2	67 344.03		66	13	$(a\ ^3F)\ ^4G°$
		9/2	67 353.29		51	32	$(a\ ^3F)\ ^4F°$
		11/2	67 369.14		65	22	$(a\ ^3F)\ ^4G°$
$3d^4(a\ ^3F)4p$	$z\ ^2D°$	3/2	67 379.47		28	32	$(a\ ^3F)\ ^4F°$
		5/2	67 387.16		50	18	
$3d^4(^3H)4p$	$z\ ^2I°$	11/2	67 506.13		93		
		13/2	67 588.90		95		
$3d^4(a\ ^3F)4p$	$x\ ^4D°$	1/2	67 859.61		88	11	$(a\ ^3P)\ ^4D°$
		5/2	67 867.82		77	13	
		3/2	67 870.24		84	12	
		7/2	67 875.42		69	19	
$3d^4(a\ ^3P)4p$	$z\ ^4S°$	3/2	68 305.64	1.978	70	17	$(a\ ^3P)\ ^2P°$
$3d^4(^3H)4p$	$z\ ^2H°$	9/2	68 476.92		81	12	$(a\ ^1G)\ ^2H°$
		11/2	68 737.82		84	10	
$3d^4(a\ ^3F)4p$	$z\ ^2F°$	5/2	68 583.34		50	21	$(^3G)\ ^2F°$
		7/2	68 759.89		59	18	
$3d^4(^3G)4p$	$y\ ^4H°$	7/2	68 843.30		83	16	$(^3H)\ ^4H°$
		9/2	68 992.40		82	14	
		11/2	69 170.39		82	13	
		13/2	69 388.25		85	13	
$3d^4(a\ ^3P)4p$	$^2D°$	3/2	69 348.18		65	28	$(a\ ^3F)\ ^2D°$
		5/2	69 954.09		66	20	
$3d^4(^3G)4p$	$^4F°$	5/2	69 477.95		71	11	$(^3D)\ ^4F°$
		9/2	69 498.28		60	13	$(a\ ^3F)\ ^2G°$
		7/2	69 506.08		60	11	$(a\ ^3F)\ ^2G°$
		3/2	69 638.63		81	13	$(^3D)\ ^4F°$
$3d^4(a\ ^3F)4p$	$y\ ^2G°$	7/2	69 903.55		42	25	$(^3H)\ ^2G°$
		9/2	70 107.67		37	25	
$3d^4(^3G)4p$	$x\ ^4G°$	5/2	70 316.90		60	17	$(^3H)\ ^4G°$
		7/2	70 427.05		61	19	$(^3H)\ ^4G°$
		9/2	70 679.15		44	22	$(^3G)\ ^2H°$
		11/2	70 879.80		41	38	$(^3G)\ ^2H°$
$3d^4(^3G)4p$	$y\ ^2H°$	9/2	70 394.20		57	19	$(^3G)\ ^4G°$
		11/2	70 398.87		47	34	
$3d^4(^3G)4p$	$y\ ^2F°$	5/2	70 584.47		45	32	$(a\ ^3F)\ ^2F°$
		7/2	70 852.23		59	20	
$3d^4(^3G)4p$	$x\ ^2G°$	7/2	72 648.52		79	11	$(^3H)\ ^2G°$
		9/2	72 716.72		75	13	
$3d^4(^3D)4p$	$w\ ^4D°$	1/2	73 407.00		96		
		3/2	73 411.96		91		
		5/2	73 436.17		87		
		7/2	73 485.66		89		

(Continued)

Cr II—Continued

Configuration	Term	J	Level (cm^{-1})	g	Leading percentages		
$3d^4(a\ ^1G)4p$	$x\ ^2F^\circ$	7/2	*74 114.39*		64	19	$(^3D)\ ^4F^\circ$
		5/2	*74 436.15*		77	7	$(^3D)\ ^2F^\circ$
$3d^4(^3D)4p$	$w\ ^4F^\circ$	3/2	*74 273.31*		82	13	$(^3G)\ ^4F^\circ$
		5/2	*74 318.83*		48	40	$(^3D)\ ^4P^\circ$
		7/2	*74 423.67*		89	6	$(^3D)\ ^4F^\circ$
		9/2	*74 504.17*		86	13	$(^3G)\ ^4F^\circ$
$3d^4(^1I)4p$	$^2I^\circ$	11/2	*74 421.80*		85	12	$(a\ ^1G)\ ^2H^\circ$
		13/2	*74 743.28*		64	35	$(^1I)\ ^2K^\circ$
$3d^4(^1I)4p$	$^2K^\circ$	13/2	*74 424.19*		65	33	$(^1I)\ ^2I^\circ$
		15/2	*74 958.80*		100		
$3d^4(a\ ^1G)4p$	$x\ ^2H^\circ$	9/2	*74 455.82*		82	11	$(^3H)\ ^2H^\circ$
		11/2	*74 707.48*		72	11	$(^1I)\ ^2I^\circ$
$3d^4(^3D)4p$	$x\ ^4P^\circ$	5/2	*74 483.96*		53	30	$(^3D)\ ^4F^\circ$
		3/2	*74 717.59*		93		
		1/2	*74 920.44*		96		
$3d^4(^3D)4p$	$y\ ^2P^\circ$	1/2	*74 853.85*		55	40	$(a\ ^1S)\ ^2P^\circ$
		3/2	*74 984.78*		61	33	
$3d^4(a\ ^1G)4p$	$w\ ^2G^\circ$	7/2	*75 716.55*		82	8	$(^3G)\ ^2G^\circ$
		9/2	*75 810.04*		80	12	
$3d^4(^3D)4p$	$w\ ^2F^\circ$	7/2	*76 879.03*		73	12	$(a\ ^1G)\ ^2F^\circ$
		5/2	*76 987.70*		72	13	
$3d^4(^1I)4p$	$w\ ^2H^\circ$	11/2	*77 078.92*		88	8	$(^3G)\ ^2H^\circ$
		9/2	*77 270.25*		90	8	
$3d^4(a\ ^1S)4p$	$x\ ^2P^\circ$	3/2	*77 713.28*		31	26	$(^3D)\ ^2D^\circ$
		1/2	*77 777.33*		48	32	$(^3D)\ ^2P^\circ$
$3d^4(^3D)4p$	$x\ ^2D^\circ$	5/2	*77 935.25*		65	26	$(a\ ^1D)\ ^2D^\circ$
		3/2	*78 109.56*		48	22	$(a\ ^1S)\ ^2P^\circ$
$3d^4(a\ ^1D)4p$	$w\ ^2D^\circ$	3/2	*80 288.05*		74	13	$(^3D)\ ^2D^\circ$
		5/2	*80 420.20*		65	21	
$3d^4(a\ ^1D)4p$	$v\ ^2F^\circ$	5/2	*81 232.92*		86		
		7/2	*81 432.29*		89		
$3d^3(^4F)4s4p(^3P^\circ)$	$y\ ^6D^\circ$	1/2	*81 649.19*				
		3/2	*81 707.87*				
		5/2	*81 816.29*				
		7/2	*81 978.08*				
		9/2	*82 192.59*				
$3d^2(^4F)4s4p(^3P^\circ)$	y^6F°	1/2	*81 735.02*				
		3/2	*81 824.40*				
		5/2	*81 962.29*				
		7/2	*82 143.15*				
		9/2	*82 362.19*				
		11/2	*82 612.69*				

Cr II—Continued

Configuration	Term	J	Level (cm^{-1})	g	Leading percentages		
$3d^4(^5D)5s$	$e\ ^6D$	1/2	82 692.06				
		3/2	82 763.75				
		5/2	82 881.17				
		7/2	83 041.38				
		9/2	83 240.41				
$3d^4(a\ ^1D)4p$	$w\ ^2P°$	1/2	*82 853.85*		89	9	$(^3D)\ ^2P°$
		3/2	*82 919.90*		78	11	
$3d^4(^5D)5s$	$e\ ^4D$	1/2	84 209.88				
		3/2	84 320.21				
		5/2	84 495.70				
		7/2	84 726.71				
$3d^4(^1F)4p$	$u\ ^2F°$	5/2	*84 604.84*		89		
		7/2	*84 677.13*		87		
$3d^3(^4F)4s4p(^3P°)$	$u\ ^4D°$	1/2	*85 486.24*				
		3/2	*85 586.60*				
		5/2	*85 778.69*				
		7/2	*86 078.90*				
$3d^4(^1F)4p$	$v\ ^2G°$	7/2	*85 573.17*		95		
		9/2	*85 938.96*		97		
$3d^4(^5D)4d$	$e\ ^4S$	3/2	86 165.30				
$3d^4(^1F)4p$	$v\ ^2D°$	5/2	*86 507.31*		75	13	$(b\ ^3P)\ ^2D°$
		3/2	*86 919.01*		79	12	
$3d^3(^4F)4s4p(^3P°)$	$u\ ^4G°$	5/2	*86 566.55*				
		7/2	*86 797.35*				
		9/2	*87 092.65*				
		11/2	*87 450.47*				
$3d^4(^5D)4d$	$e\ ^6G$	3/2	86 594.33				
		5/2	86 654.18				
		7/2	86 738.27				
		9/2	86 847.03				
		11/2	86 980.10				
		13/2	87 137.08				
$3d^4(^5D)4d$	$e\ ^6P$	3/2	86 667.73				
		5/2	86 691.55				
		7/2	86 782.04				
$3d^4(^5D)4d$	$f\ ^6D$	1/2	87 453.50				
		3/2	87 470.58				
		5/2	87 514.82				
		7/2	87 588.00				
		9/2	87 687.52				
$3d^4(^5D)4d$	$e\ ^6F$	1/2	87 594.58				
		3/2	87 666.26				
		5/2	87 759.02				
		7/2	87 858.56				
		9/2	87 948.55				
		11/2	88 001.36				

(Continued)

Cr II—Continued

Configuration	Term	J	Level (cm⁻¹)	g	Leading percentages		
$3d^3(^4F)4s4p(^3P^\circ)$	$u\ ^4F^\circ$	3/2	*87 628.74*				
		5/2	*87 766.61*				
		7/2	*87 916.67*				
		9/2	*88 073.49*				
$3d^3(^4F)4s4p(^3P^\circ)$	$u\ ^2D^\circ$	3/2	*88 604.34*				
		5/2	*89 164.64*				
$3d^4(^5D)4d$	$e\ ^4P$	1/2	89 254.56				
		3/2	89 277.95				
		5/2	89 336.89				
$3d^4(^5D)4d$	$e\ ^4G$	5/2	89 056.02				
		7/2	89 174.08				
		9/2	89 325.32				
		11/2	89 508.55				
$3d^4(b\ ^3P)4p$	$y\ ^4P^\circ$	3/2	*89 422.36*		88	6	$(b\ ^3P)\ ^4D^\circ$
		5/2	*89 453.21*		82	7	
		1/2	*89 507.94*		95		
$3d^4(^5D)4d$	$f\ ^4D$	1/2	89 651.66				
		3/2	89 724.27				
		5/2	89 812.42				
		7/2	89 885.08				
$3d^4(b\ ^3P)4p$	$v\ ^4D^\circ$	7/2	*90 218.49*		69	26	$(b\ ^3F)\ ^4D^\circ$
		5/2	*90 258.20*		51	20	
		3/2	*90 450.62*		61	24	
		1/2	*90 475.48*		70	26	
$3d^4(b\ ^3F)4p$	$v\ ^4F^\circ$	3/2	*90 262.13*		97		
		5/2	*90 441.78*		93		
		7/2	*90 489.86*		95		
		9/2	*90 588.59*		98		
$3d^4(^5D)4d$	$e\ ^4F$	3/2	90 512.56				
		5/2	90 608.99				
		7/2	90 725.87				
		9/2	90 850.96				
$3d^4(b\ ^3F)4p$	$t\ ^2F^\circ$	5/2	*90 706.82*		85	5	$(b\ ^3F)\ ^4G^\circ$
		7/2	*90 830.79*		87	5	$(b\ ^1G)\ ^2F^\circ$
$3d^4(b\ ^3F)4p$	$w\ ^4G^\circ$	5/2	*91 078.72*				
		7/2	*91 122.82*		95		
		9/2	*91 189.51*		98		
		11/2	*91 292.16*				
$3d^4(b\ ^3P)4p$	$t\ ^2D^\circ$	5/2	*91 426.06*		50	18	$(^1F)\ ^2D^\circ$
		3/2	*91 556.40*		53	21	$(b\ ^3F)\ ^2D^\circ$
$3d^3(^4F)4s4p(^3P^\circ)$	$t\ ^2G^\circ$	7/2	*91 752.27*				
		9/2	*92 144.24*				
$3d^4(^5D)4d$	$e\ ^6S$	5/2	91 955.39				
$3d^3(^4P)4s4p(^3P^\circ)$	$x\ ^6P^\circ$	3/2	*92 235.35*				
		5/2	*92 417.93*				
		7/2	*92 653.28*				

Cr II—Continued

Configuration	Term	J	Level (cm^{-1})	Leading percentages		
$3d^4(b\ ^3P)4p$	$y\ ^4S°$	3/2	92 612.28	97		
$3d^4(^5D)5p$	$^6F°$	1/2	92 988.83			
		3/2	93 047.31			
		5/2	93 143.88			
		7/2	93 276.86			
		9/2	93 444.17			
		11/2	93 643.40			
$3d^4(b\ ^3F)4p$	$t\ ^4D°$	7/2	93 531.69	73	27	$(b\ ^3P)\ ^4D°$
		5/2	93 671.00	72	72	
		3/2	93 770.10	72	27	
		1/2	93 800.39	72	27	
$3d^4(^5D)5p$	$^6D°$	3/2	93 574.44			
		5/2	93 776.15			
		7/2	93 966.45			
		9/2	94 177.18			
$3d^4(b\ ^3F)4p$	$u\ ^2G°$	9/2	93 641.50	98		
		7/2	93 801.46	98		
$3d^4(^5D)5p$	$^4P°$	3/2	93 740.40			
		5/2	93 974.03			
$3d^3(^4F)4s4p(^3P°)$	$s\ ^2F°$	5/2	93 890.64			
		7/2	94 218.66			
$3d^3(^4P)4s4p(^3P°)$	$x\ ^6D°$	1/2	93 968.70			
		3/2	94 098.13			
		5/2	94 265.99			
		7/2	94 452.57			
		9/2	94 656.24			
$3d^4(^5D)5p$	$^6P°$	3/2	94 002.56			
		5/2	94 144.43			
		7/2	94 363.51			
$3d^4(^5D)5p$	$^4F°$	3/2	94 256.07			
		5/2	94 365.19			
		7/2	94 522.31			
		9/2	94 749.20			
$3d^4(b\ ^3P)4p$	$v\ ^2P°$	3/2	94 383.20	94		
		1/2	94 624.72	92		
$3d^4(^5D)5p$	$^4D°$	1/2	94 839.27			
		3/2	94 932.95			
		5/2	95 076.72			
		7/2	95 250.69			
$3d^4(b\ ^3P)4p$	$y\ ^2S°$	1/2	96 245.32	98		
$3d^3(^2G)4s4p(^3P°)$	$t\ ^4G°$	5/2	97 071.15			
		7/2	97 187.28			
		9/2	97 333.28			
		11/2	97 493.70			

(Continued)

Cr II—Continued

Configuration	Term	J	Level (cm^{-1})	g	Leading percentages		
$3d^3(^4P)4s4p(^3P^\circ)$	$w\ ^4P^\circ$	1/2	*97 168.48*				
		5/2	*97 182.54*				
		3/2	*97 294.06*				
$3d^4(b\ ^1G)4p$	$v\ ^2H^\circ$	9/2	*97 480.08*		86	12	$(b\ ^1G)\ ^2G^\circ$
		11/2	*97 899.41*		97		
$3d^4(b\ ^1G)4p$	$s\ ^2G^\circ$	7/2	*97 728.25*		97		
		9/2	*97 904.44*		86	12	$(b\ ^1G)\ ^2H^\circ$
$3d^3(^4P)4s4p(^3P^\circ)$	$z\ ^6S^\circ$	5/2	*97 875.00*				
$3d^4(b\ ^3F)4p$	$s\ ^2D^\circ$	5/2	*98 207.36*		71	29	$(b\ ^3P)\ ^2D^\circ$
		3/2	*98 314.99*		70	29	
$3d^3(^2G)4s4p(^3P^\circ)$	$t\ ^4F^\circ$	3/2	*98 578.50*				
		5/2	*98 641.88*				
		7/2	*98 719.40*				
		9/2	*98 812.67*				
$3d^4(b\ ^1G)4p$	$r\ ^2F^\circ$	7/2	*99 069.29*		87	8	$(b\ ^3F)\ ^2F^\circ$
		5/2	*99 243.99*		86	6	
$3d^4(a\ ^3P)5s$	$f\ ^4P$	1/2	99 677.93				
		3/2	100 040.22				
		5/2	100 650.52				
$3d^4(^3H)5s$	$e\ ^4H$	7/2	100 068.86				
		9/2	100 135.82				
		11/2	100 221.64				
		13/2	100 322.13				
$3d^3(^2P)4s4p(^3P^\circ)$	$r\ ^4D^\circ$	3/2	*100 691.80*				
		5/2	*101 074.56*				
		7/2	*101 514.29*				
$3d^4(a\ ^3P)5s$	$e\ ^2P$	1/2	100 782.80				
		3/2	101 492.84				
$3d^4(^3H)5s$	$e\ ^2H$	9/2	101 021.84				
		11/2	101 194.83				
$3d^3(^4P)4s4p(^3P^\circ)$	$u\ ^2P^\circ$	3/2	*101 157.69*				
$3d^3(^2H)4s4p(^3P^\circ)$	$w\ ^4H^\circ$	7/2	*101 170.38*				
		9/2	*101 296.60*				
		11/2	*101 783.20*				
		13/2	*101 900.82*				
$3d^4(a\ ^3F)5s$	$f\ ^4F$	3/2	101 245.00				
		5/2	101 276.60				
		7/2	101 321.83				
		9/2	101 382.97				
$3d^3(^2G)4s4p(^3P^\circ)$	$u\ ^2H^\circ$	9/2	*101 696.20*				
		11/2	*101 932.27*				
$3d^3(^2G)4s4p(^3P^\circ)$	$p\ ^2F^\circ$	7/2	*101 864.17*				
		5/2	*102 145.65*				

Cr II—Continued

Configuration	Term	J	Level (cm^{-1})	g	Leading percentages
$3d^3(^2G)4s4p(^3P^\circ)$	$r\ ^2G^\circ$	9/2	*101 938.04*		
		7/2	*102 121.92*		
$3d^3(^2D)4s4p(^3P^\circ)$	$s\ ^4F^\circ$	3/2	*101 987.13*		
		5/2	*102 297.15*		
		7/2	*102 492.60*		
		9/2	*102 725.66*		
$3d^4(a\ ^3F)5s$	$f\ ^2F$	5/2	102 148.81		
		7/2	102 243.22		
$3d^3(^4P)4s4p(^3P^\circ)$	$q\ ^4D^\circ$	3/2	*102 602.40*		
		1/2	*102 619.58*		
		5/2	*102 655.95*		
		7/2	*102 831.62*		
$3d^3(^4F)4s4p(^1P^\circ)$	$s\ ^4G^\circ$	5/2	*102 679.00*		
		7/2	*102 915.01*		
		9/2	*103 199.80*		
		11/2	*103 513.67*		
$3d^3(^2P)4s4p(^3P^\circ)$	$x\ ^4S^\circ$	3/2	*102 684.02*		
$3d^4(^3G)5s$	$f\ ^4G$	9/2	103 627.02		
		11/2	103 736.99		
$3d^4\ 4d$	$f\ ^4I$	9/2	103 755.48		
		11/2	103 843.33		
		13/2	103 948.26		
		15/2	104 069.62		
$3d^4\ 4d$	$f\ ^4H$	7/2	103 949.27		
		9/2	104 023.95		
		11/2	104 106.35		
		13/2	104 190.63		
$3d^3(^2D)4s4p(^3P^\circ)$	$o\ ^4D^\circ$	3/2	*104 274.62*		
		5/2	*104 467.83*		
		7/2	*104 680.78*		
$3d^3(^4F)4s4p(^1P^\circ)$	$p\ ^4D^\circ$	1/2	*104 439.77*		
		3/2	*104 616.27*		
		5/2	*104 869.13*		
		7/2	*105 206.69*		
$3d^3(^4F)4s4p(^1P^\circ)$	$r\ ^4F^\circ$	3/2	*104 446.50*		
		5/2	*104 630.01*		
		7/2	*104 875.34*		
		9/2	*105 203.46*		
$3d^4\ 4d$	$a\ ^4K$	11/2	104 460.28		
		13/2	104 539.92		
		15/2	104 633.04		
		17/2	104 734.37		
$3d^4\ 5s$	$g\ ^2G$	7/2	104 543.17		
		9/2	104 666.42		

(Continued)

Cr II—Continued

Configuration	Term	J	Level (cm^{-1})	g	Leading percentages
$3d^4(^5D)6s$	^6D	1/2	105 098.94		
		3/2	105 168.82		
		5/2	105 285.37		
		7/2	105 447.05		
		9/2	105 650.58		
$3d^4(^5D)4f$	$^4P^{\circ}$	5/2	*105 122.26*		
		3/2	*105 283.47*		
		1/2	*105 392.48*		
$3d^4\,4d$	$a\ ^2K$	13/2	105 124.90		
		15/2	105 285.33		
$3d^4\,4f$	$^6P^{\circ}$	7/2	*105 173.47*		
		3/2	*105 408.72*		
$3d^4(^5D)4f$	$^6H^{\circ}$	7/2	*105 197.38*		
		9/2	*105 263.52*		
		11/2	*105 367.93*		
		13/2	*105 508.05*		
		15/2	*105 681.82*		
$3d^4\,4d$	$g\ ^4H$	7/2	105 198.98		
		9/2	105 255.34		
		11/2	105 337.97		
		13/2	105 434.42		
$3d^4(^5D)4f$	$^4H^{\circ}$	7/2	*105 282.58*		
		9/2	*105 406.99*		
		11/2	*105 559.58*		
		13/2	*105 742.62*		
$3d^4\,4d$	$g\ ^4G$	9/2	105 365.50		
		11/2	105 423.32		
$3d^4(^5D)4f$	$^6D^{\circ}$	9/2	*105 398.27*		
		7/2	*105 420.09*		
		5/2	*105 595.35*		
$3d^4(^5D)4f$	$^6G^{\circ}$	5/2	*105 438.32*		
		9/2	*105 609.47*		
		9/2	*105 623.64*		
		7/2	*105 765.04*		
		13/2	*105 895.59*		
		11/2	*105 898.12*		
$3d^4(^5D)4f$	$^4D^{\circ}$	7/2	*105 507.52*		
		5/2	*105 532.18*		
$3d^4(^5D)4f$	$^6F^{\circ}$	3/2	*105 577.19*		
$3d^4(^5D)4f$	$^6F^{\circ}$	11/2	*105 638.84*		
		7/2	*105 823.86*		
		9/2	*105 870.43*		
$3d^4(^5D)4f$	$^4G^{\circ}$	5/2	*105 677.49*		
		7/2	*105 724.77*		
		11/2	*106 032.24*		
		9/2	*106 045.32*		

Cr II—Continued

Configuration	Term	J	Level (cm^{-1})	g	Leading percentages
$3d^4(^5D)4f$	$^4F°$	9/2	*105 790.06*		
		5/2	*105 903.05*		
		7/2	*105 985.63*		
$3d^4(^5D)6s$	^4D	1/2	105 923.4		
		3/2	106 030.93		
		5/2	106 095.64		
		7/2	106 275.24		
$3d^4\,4d$	$f\,^2I$	11/2	106 145.26		
		13/2	106 342.95		
$3d^3(^2H)4s4p(^3P°)$	$t\,^2H°$	11/2	*106 163.16*		
		9/2	*106 165.3*		
$3d^3(^2H)4s4p(^3P°)$	$r\,^4G°$	5/2	*106 719.38*		
		7/2	*106 779.14*		
		9/2	*106 791.84*		
		11/2	*106 827.42*		
$3d^3(^4P)4s4p(^3P°)$	$w\,^4S°$	3/2	*106 726.06*		
$3d^4(^5D)5d$	^6G	3/2	106 877.20		
		5/2	106 929.42		
		7/2	107 006.29		
		9/2	107 111.84		
		11/2	107 246.87		
		13/2	107 412.09		
$3d^4\,4d$	$i\,^2G$	9/2	106 923.98		
$3d^4(^5D)5d$	^4S	3/2	106 924.84		
$3d^3(^2D)4s4p(^3P°)$	$o\,^2F°$	5/2	*107 022.33*		
		7/2	*107 153.15*		
$3d^4(^5D)5d$	^6P	5/2	107 025.34		
		3/2	107 056.53		
		7/2	107 114.75		
$3d^3(^4P)4s4p(^3P°)$	$q\,^2D°$	3/2	*107 212.29*		
		5/2	*107 355.52*		
$3d^4(^5D)5d$	^6F	3/2	107 259.87		
		5/2	107 309.38		
		7/2	107 386.22		
		9/2	107 455.55		
		11/2	107 701.34		
$3d^4(^5D)5d$	^4G	5/2	107 400.84		
		7/2	107 500.37		
		9/2	107 632.26		
		11/2	107 794.15		
$3d^4(^5D)5d$	^6D	3/2	107 414.68		
		5/2	107 519.44		
		7/2	107 627.40		
		9/2	107 696.31		

(Continued)

Cr II—Continued

Configuration	Term	J	Level (cm^{-1})	g	Leading percentages
$3d^4(^5D)5d$	^4F	3/2	107 516.77		
		5/2	107 726.82		
		7/2	107 850.60		
		9/2	107 948.05		
$3d^4(^5D)5d$	^4D	5/2	107 597.65		
		7/2	107 716.30		
$3d^4\,4d$	$g\,^4I$	9/2	107 706.84		
		11/2	107 760.85		
		13/2	107 846.75		
		15/2	107 981.81		
$3d^3(^2H)4s4p(^3P^\circ)$	$q\,^2G^\circ$	9/2	*107 739.20*		
		7/2	*107 918.49*		
$3d^4\,4d$	$h\,^4H$	7/2	107 829.54		
		9/2	107 922.41		
		11/2	108 017.98		
		13/2	108 104.01		
$3d^3(^2H)4s4p(^3P^\circ)$	$x\,^2I^\circ$	11/2	*107 850.50*		
		13/2	*108 031.16*		
$3d^4(^5D)5d$	^6S	5/2	109 394.47		
$3d^4(^5D)6p$	$^6F^\circ$	1/2	*109 564.97*		
		3/2	*109 611.24*		
		5/2	*109 694.38*		
		7/2	*109 812.06*		
		9/2	*109 965.71*		
		11/2	*110 154.04*		
$3d^4(^5D)6p$	$^4P^\circ$	3/2	*109 661.41*		
		5/2	*109 974.05*		
$3d^4(^5D)6p$	$^6P^\circ$	3/2	*109 772.35*		
		5/2	*109 864.95*		
		7/2	*110 097.11*		
$3d^3(^2D)4s4p(^3P^\circ)$	$p\,^2D^\circ$	3/2	*109 914.68*		
		5/2	*109 943.57*		
$3d^4(^5D)6p$	$^6D^\circ$	1/2	*109 923.45*		
		3/2	*110 007.57*		
		5/2	*110 138.26*		
		7/2	*110 272.15*		
		9/2	*110 385.80*		
$3d^4(^5D)6p$	$^4F^\circ$	5/2	*110 315.08*		
		7/2	*110 471.38*		
		9/2	*110 665.54*		
$3d^4(^5D)6p$	$^4D^\circ$	3/2	*110 931.72*		
		5/2	*111 082.38*		
		7/2	*111 269.22*		

Cr II—Continued

Configuration	Term	J	Level (cm^{-1})	g	Leading percentages
$3d^4(^5D)5f$	$^6H°$	5/2	*115 208.37*		
		9/2	*115 288.81*		
		7/2	*115 309.16*		
		11/2	*115 396.86*		
		13/2	*115 546.61*		
		15/2	*115 734.85*		
$3d^4(^5D)7s$	^6D	1/2	115 234.53		
		3/2	115 301.92		
		5/2	115 417.88		
		7/2	115 581.62		
		9/2	115 788.38		
$3d^4(^5D)5f$	$^6P°$	7/2	*115 249.31*		
		3/2	*115 747.63*		
		5/2	*115 767.13*		
$3d^4(^5D)5f$	$^4P°$	3/2	*115 298.53*		
$3d^4(^5D)5f$	$^6D°$	5/2	*115 309.64*		
		9/2	*115 393.47*		
		3/2	*115 430.93*		
		7/2	*115 591.20*		
$3d^4(^5D_0)5g$	$^2[4]$	9/2	115 371.26		
		7/2	115 371.39		
$3d^4(^5D)5f$	$^4H°$	7/2	*115 398.46*		
		9/2	*115 430.45*		
		11/2	*115 598.97*		
		13/2	*115 782.87*		
$3d^4(^5D)5f$	$^4G°$	5/2	*115 408.13*		
		7/2	*115 627.76*		
		11/2	*115 916.22*		
		9/2	*115 927.40*		
$3d^4(^5D_1)5g$	$^2[4]$	9/2	115 411.98		
		7/2	115 412.17		
$3d^4(^5D_1)5g$	$^2[5]$	11/2	115 444.00		
		9/2	115 444.02		
$3d^4(^5D)5f$	$^6G°$	7/2	*115 447.64*		
		5/2	*115 461.80*		
		11/2	*115 585.21*		
		9/2	*115 592.43*		
		13/2	*115 836.61*		
$3d^4(^5D_2)5g$	$^2[5]$	11/2	115 554.95		
		9/2	115 555.05		
$3d^4(^5D_2)5g$	$^2[4]$	9/2	115 556.30		
		7/2	115 556.44		
$3d^4(^5D_2)5g$	$^2[6]$	13/2	115 560.97		
		11/2	115 561.07		

(Continued)

Cr II—Continued

Configuration	Term	J	Level (cm^{-1})	g	Leading percentages
$3d^4(^5D)5f$	$^6F°$	5/2	*115 570.91*		
		3/2	*115 605.95*		
		7/2	*115 797.28*		
		9/2	*115 824.49*		
		11/2	*115 840.45*		
$3d^4(^5D)7s$	^4D	3/2	115 640.42		
		5/2	115 818.49		
		7/2	116 047.90		
$3d^4(^5D)5f$	$^4F°$	9/2	*115 672.84*		
		7/2	*115 882.21*		
$3d^4(^5D_3)5g$	$^2[7]$	15/2	115 723.70		
		13/2	115 723.89		
$3d^4(^5D_3)5g$	$^2[3]$	7/2	115 732.06		
		5/2	115 732.23		
$3d^4(^5D_3)5g$	$^2[4]$	7/2	115 738.43		
$3d^4(^5D_3)5g$	$^2[6]$	13/2	115 741.12		
		11/2	115 741.30		
$3d^4(^5D_3)5g$	$^2[5]$	11/2	115 742.39		
		9/2	115 742.55		
$3d^4(^5D)5f$	$^4D°$	5/2	*115 810.34*		
$3d^4(^5D_4)5g$	$^2[8]$	17/2	115 927.17		
		15/2	115 927.30		
$3d^4(^5D_4)5g$	$^2[2]$	5/2	115 926.73		
$3d^3(^4P)4s4p(^1P°)$	$^4P°$	1/2	*115 943.7*		
		3/2	*115 966.7*		
		5/2	*116 041.7*		
$3d^4(^5D_4)5g$	$^2[3]$	7/2	115 945.47		
$3d^4(^5D_4)5g$	$^2[4]$	9/2	115 962.72		
		7/2	115 962.84		
$3d^4(^5D_4)5g$	$^2[7]$	15/2	115 968.86		
		13/2	115 969.19		
$3d^4(^5D_4)5g$	$^2[5]$	11/2	115 975.22		
		9/2	115 975.44		
$3d^4(^5D_4)5g$	$^2[6]$	13/2	115 979.31		
		11/2	115 979.59		
$3d^4(^5D)6d$	^6G	3/2	116 171.71		
		5/2	116 213.38		
		7/2	116 281.95		
		9/2	116 388.95		
		11/2	116 531.26		
		13/2	116 708.67		

Cr II—Continued

Configuration	Term	J	Level (cm^{-1})	g	Leading percentages
$3d^4(^5D)6d$	^6P	3/2	116 253.35		
		5/2	116 295.06		
		7/2	116 385.67		
$3d^4(^5D)6d$	^4S	3/2	116 355.55		
$3d^4(^5D)6d$	^6F	1/2	116 360.98		
		3/2	116 429.23		
		5/2	116 477.51		
		7/2	116 572.42		
		9/2	116 601.65		
		11/2	116 829.01		
$3d^4(^5D)6d$	^6D	3/2	116 581.74		
		5/2	116 687.20		
		7/2	116 790.31		
		9/2	116 831.84		
$3d^4(^5D)6d$	^4G	5/2	116 877.15		
		7/2	116 985.30		
		9/2	117 141.58		
		11/2	117 342.41		
$3d^4(^5D)6d$	^4D	5/2	117 072.83		
		7/2	117 263.48		
$3d^4(^5D)6d$	^4F	5/2	117 228.51		
		9/2	117 488.50		
		7/2	117 520.75		
$3d^4(^5D)6d$	^4P	3/2	117 381.64		
		5/2	117 481.24		
$3d^4(^5D)6d$	^6S	5/2	117 672.56		
$3d^3(^4P)4s4p(^1P^\circ)$	$^4D^\circ$	3/2	*118 622.6*		
		5/2	*118 640.08*		
		1/2	*118 661.4*		
		7/2	*118 753.64*		
$3d^4(^5D)8s$	^6D	1/2	120 702.6		
		3/2	120 757.1		
		5/2	120 870.78		
		7/2	121 036.43		
		9/2	121 246.83		
$3d^4(^5D_1)6g$	$^2[5]$	11/2	120 820.04		
		9/2	120 820.09		
$3d^4(^5D_2)6g$	$^2[6]$	13/2	120 938.67		
$3d^4(^5D_3)6g$	$^2[7]$	15/2	121 105.81		
		13/2	121 106.04		
$3d^4(^5D_3)6g$	$^2[6]$	13/2	121 114.05		
		11/2	121 114.25		
$3d^4(^5D_3)6g$	$^2[5]$	11/2	121 115.31		
		9/2	121 115.45		

(Continued)

Cr II—Continued

Configuration	Term	J	Level (cm^{-1})	g	Leading percentages
$3d^4(^5D_4)6g$	$^2[8]$	$17/2$	121 315.57		
		$15/2$	121 315.83		
$3d^4(^5D_4)6g$	$^2[4]$	$9/2$	121 335.13		
		$7/2$	121 335.25		
$3d^4(^5D_4)6g$	$^2[7]$	$15/2$	121 338.36		
		$13/2$	121 338.68		
$3d^4(^5D_4)6g$	$^2[6]$	$13/2$	121 344.81		
		$11/2$	121 345.09		
$3d^4(^5D)9s$	^6D	$7/2$	124 310.74		
		$9/2$	124 523.96		
Cr III ($^5D_0$)	*Limit*		**132 966**		

Cr III

$Z=24$

Ti I isoelectronic sequence

Ground state: $1s^2 2s^2 2p^6 3s^2 3p^6 3d^4\ ^5D_0$

Ionization energy = 249 700±200 cm^{-1} (30.96±0.02 eV)

The analysis was begun by White (1929) who found systems of triplet and quintet terms, which were later unified by Bowen (1937). F. L. Moore considerably augmented this work and provided his unpublished results to C. E. Moore (1952) for inclusion in her compilation. Ekberg (1976) has remeasured the spectrum in the region of 750–2700 Å with an uncertainty of ±0.005 Å and established 76 new levels while rejecting 26 of those found by Moore.

We give the results of Ekberg for the $3d^4$, $3d^3 4s$, and $3d^3 4p$ configurations, including his calculated percentage compositions for the levels. The level uncertainty is ±0.1 cm^{-1}. The $3d^3(^4F)4d\ ^5H$ term is from White. The other terms of $3d^3 4d$ and those of $3d^3 5s$ are from the analysis by Moore. The uncertainty of these levels is ±0.5 cm^{-1}.

Johannson and Ekberg (1982) discovered three terms of the $3d^2 4s 4p$ configuration and gave their results in a preliminary report.

The ionization energy was derived by Catalán and Velasco (1952) from the 2-member $3d^3 ns$ series.

References

Bowen, I. S. (1937), Phys. Rev. **52**, 1153.

Catalán, M. A., and Velasco, R. (1952), An. R. Soc. Esp. Fis. Quim. **48A**, 247.

Ekberg, J. O. (1976), Phys. Scr. **14**, 109.

Johannson, S., and Ekberg, J. O. (1982), Atomic Spectroscopy Annual Report, University of Lund.

Moore, C. E. (1952), Atomic Energy Levels. Natl. Bur. Stand. (U.S.) Circ. 467, Vol. II (reissued in 1971 as Natl Bur. Stand. (U.S.) Natl. Stand. Ref. Data Ser. 35, Vol. II).

White, H. E. (1929), Phys. Rev. **33**, 914.

Cr III

Configuration	Term	J	Level (cm^{-1})	Leading percentages		
$3d^4$	^{5}D	0	0.00	100		
		1	62.22	100		
		2	183.16	100		
		3	356.55	100		
		4	576.08	100		
$3d^4$	^{3}P2	0	16 771.36	58	41	^{3}P1
		1	17 168.56	59	41	
		2	17 851.18	59	41	
$3d^4$	^{3}H	4	17 273.70	99		
		5	17 396.92	99		
		6	17 530.65	100		
$3d^4$	^{3}F2	2	18 451.84	77	22	^{3}F1
		3	18 511.18	77	22	
		4	18 583.39	77	21	
$3d^4$	^{3}G	3	20 703.64	98		
		4	20 852.95	98		
		5	20 996.04	99		
$3d^4$	^{1}G2	4	25 138.87	64	35	^{1}G1
$3d^4$	^{3}D	3	25 726.44	100		
		2	25 780.94	100		
		1	25 848.31	100		
$3d^4$	^{1}I	6	26 014.89	100		
$3d^4$	^{1}S2	0	27 372.32	77	22	^{1}S1

(Continued)

Cr III—Continued

Configuration	Term	J	Level (cm^{-1})	Leading percentages		
$3d^4$	^{1}D2	2	32 151.99	77	22	^{1}D1
$3d^4$	^{1}F	3	37 005.16	100		
$3d^4$	^{3}F1	4	43 286.71	79	21	^{3}F2
		2	43 304.53	77	22	
		3	43 322.17	78	22	
$3d^4$	^{3}P1	2	43 441.99	59	41	^{3}P2
		1	43 916.59	59	41	
		0	44 141.36	59	41	
$3d^3(^4F)4s$	^{5}F	1	49 492.46	100		
		2	49 628.25	100		
		3	49 828.91	100		
		4	50 091.17	100		
		5	50 410.06	100		
$3d^4$	^{1}G1	4	49 768.65	65	35	^{1}G2
$3d^3(^4F)4s$	^{3}F	2	56 651.37	100		
		3	56 993.08	100		
		4	57 423.40	100		
$3d^3(^4P)4s$	^{5}P	1	63 045.74	100		
		2	63 174.30	99		
		3	63 421.92	100		
$3d^4$	^{1}D1	2	65 763.21	78	22	^{1}D2
$3d^3(^2G)4s$	^{3}G	3	65 892.38	100		
		4	66 030.01	100		
		5	66 225.09	99		
$3d^3(^4P)4s$	^{3}P	0	69 601.50	59	41	$(^2P)\ ^3P$
		1	69 781.89	56	34	
		2	70 292.86	71	15	
$3d^3(^2G)4s$	^{1}G	4	69 659.74	97		
$3d^3(^2P)4s$	^{3}P	2	70 191.01	63	26	$(^4P)\ ^3P$
		1	70 345.56	53	41	
		0	70 487.01	59	41	
$3d^3(^2D2)4s$	^{3}D	1	70 981.26	65	19	$(^2D1)\ ^3D$
		3	71 323.06	79	21	$(^2D1)\ ^3D$
		2	71 323.27	59	21	$(^2P)\ ^3P$
$3d^3(^2H)4s$	^{3}H	4	71 677.19	97		
		5	71 737.56	99		
		6	71 870.17	100		
$3d^3(^2P)4s$	^{1}P	1	73 881.54	96		
$3d^3(^2D2)4s$	^{1}D	2	74 788.88	77	22	$(^2D1)\ ^3D$
$3d^3(^2H)4s$	^{1}H	5	75 351.63	100		

Cr III—Continued

Configuration	Term	J	Level (cm^{-1})	Leading percentages		
$3d^3(^2F)4s$	^3F	4	84 374.12	100		
		3	84 484.76	100		
		2	84 572.53	100		
$3d^3(^2F)4s$	^1F	3	87 770.68	100		
$3d^3(^4F)4p$	$^5G^\circ$	2	*93 766.21*	100		
		3	*94 029.99*	100		
		4	*94 376.29*	100		
		5	*94 801.70*	100		
		6	*95 306.09*	100		
$3d^3(^4F)4p$	$^5F^\circ$	1	*96 149.25*	52	35	$(^4F)\ ^3D^\circ$
		2	*96 386.31*	44	29	$(^4F)\ ^5D^\circ$
		3	*97 121.42*	61	37	$(^4F)\ ^5D^\circ$
		4	*97 359.81*	73	25	$(^4F)\ ^5D^\circ$
		5	*97 619.48*	98		
$3d^3(^4F)4p$	$^5D^\circ$	0	*96 693.97*	97		
		3	*96 714.04*	51	35	$(^4F)\ ^5F^\circ$
		1	*96 774.38*	79	18	
		2	*96 922.02*	56	42	
		4	*97 098.28*	72	26	
$3d^3(^4F)4p$	$^3D^\circ$	1	*97 077.96*	57	30	$(^4F)\ ^5F^\circ$
		2	*97 306.59*	67	13	$(^4F)\ ^5F^\circ$
		3	*97 683.99*	79	9	$(^4F)\ ^5D^\circ$
$3d^3(^4F)4p$	$^3G^\circ$	3	*99 841.67*	93	5	$(^2G)\ ^3G^\circ$
		4	*100 100.66*	93		
		5	*100 423.01*	92		
$3d^3(^4F)4p$	$^3F^\circ$	2	*101 444.57*	96		
		3	*101 746.21*	95		
		4	*102 100.76*	96		
$3d^3(^2D1)4s$	^3D	3	102 236.46	79	21	$(^2D2)\ ^3D$
		2	102 333.49	78	22	
		1	102 401.80	77	23	
$3d^3(^2D1)4s$	^1D	2	105 626.89	78	22	$(^2D2)\ ^1D$
$3d^3(^4P)4p$	$^5P^\circ$	1	*108 250.89*	99		
		2	*108 461.40*	98		
		3	*108 795.84*	99		
$3d^3(^4P)4p$	$^5D^\circ$	0	*108 697.63*	55	37	$(^4P)\ ^3P^\circ$
		1	*108 864.98*	60	34	
		2	*109 570.89*	59	33	
		3	*109 722.79*	96		
		4	*110 154.89*	97		
$3d^3(^4P)4p$	$^3P^\circ$	2	*108 972.90*	52	38	$(^4P)\ ^5D^\circ$
		0	*109 458.11*	47	41	
		1	*109 570.62*	54	37	
$3d^3(^2G)4p$	$^3H^\circ$	4	*109 534.25*	85	13	$(^2H)\ ^3H^\circ$
		5	*109 944.98*	83	14	
		6	*110 507.18*	85	15	

(Continued)

Cr III—Continued

Configuration	Term	J	Level (cm^{-1})	Leading percentages			
$3d^3(^2G)4p$	$^3G^\circ$	3	111 376.09	86		5	$(^4F)\ ^3G^\circ$
		4	111 644.76	89		5	
		5	111 856.97	86		5	
$3d^3(^2G)4p$	$^3F^\circ$	4	112 372.88	52		37	$(^2G)\ ^1G^\circ$
		2	112 399.84	73		13	$(^2D2)\ ^3F^\circ$
		3	112 467.01	70		8	$(^2D2)\ ^3F^\circ$
$3d^3(^2G)4p$	$^1G^\circ$	4	113 115.21	59		30	$(^2G)\ ^3F^\circ$
$3d^3(^2G)4p$	$^1F^\circ$	3	113 328.80	72		12	$(^2D2)\ ^1F^\circ$
$3d^3(^4P)4p$	$^5S^\circ$	2	113 357.04	95			
$3d^3(^2G)4p$	$^1H^\circ$	5	113 419.93	77		18	$(^2H)\ ^1H^\circ$
$3d^3(^2P)4p$	$^1D^\circ$	2	113 767.13	43		22	$(^2D2)\ ^1D^\circ$
$3d^3(^2P)4p$	$^3P^\circ$	0	113 861.47	62		22	$(^2D2)\ ^3P^\circ$
		1	113 899.43	57		23	
		2	114 599.14	47		22	
$3d^3(^2P)4p$	$^3D^\circ$	1	114 716.79	70		22	$(^4P)\ ^3D^\circ$
		2	115 182.15	59		21	
		3	115 554.28	48		31	
$3d^3(^2H)4p$	$^3H^\circ$	4	115 571.96	83		13	$(^2G)\ ^3H^\circ$
		5	115 670.74	85		14	
		6	115 844.58	85		15	
$3d^3(^2D2)4p$		1	116 372.92	30	$^1P^\circ$	28	$(^4P)\ ^3D^\circ$
$3d^3(^2D2)4p$	$^3F^\circ$	2	116 391.66	40		25	$(^4P)\ ^3D^\circ$
		4	117 101.58	68		17	$(^2D1)\ ^3F^\circ$
$3d^3(^2P)4p$	$^3S^\circ$	1	116 451.19	60		13	$(^4P)\ ^3D^\circ$
$3d^3(^4P)4p$	$^3D^\circ$	3	116 532.95	44		21	$(^2D2)\ ^3F^\circ$
		2	116 782.05	34		31	$(^2P)\ ^3D^\circ$
$3d^3(^2P)4p$		1	116 774.00	32	$^3S^\circ$	21	$(^4P)\ ^3D^\circ$
$3d^3(^2D2)4p$		3	116 969.09	37	$^3F^\circ$	32	$(^2P)\ ^3D^\circ$
$3d^3(^2H)4p$	$^3I^\circ$	5	117 145.85	98			
		6	117 488.95	99			
		7	117 923.89	100			
$3d^3(^2D2)4p$	$^3D^\circ$	1	118 165.15	69		17	$(^2D2)\ ^3F^\circ$
		2	118 422.99	72		16	
		3	118 599.11	71		15	
$3d^3(^2H)4p$	$^1G^\circ$	4	118 900.53	81		15	$(^2F)\ ^1G^\circ$
$3d^3(^2D2)4p$	$^3P^\circ$	2	119 421.42	41		38	$(^2P)\ ^3P^\circ$
		1	119 489.02	42		33	
		0	119 625.62	45		35	
$3d^3(^2H)4p$	$^1H^\circ$	5	119 612.80	78		18	$(^2G)\ ^1H^\circ$

Cr III—Continued

Configuration	Term	J	Level (cm^{-1})	Leading percentages		
$3d^3(^2D2)4p$	$^1F^\circ$	3	119 846.47	61	18	$(^2D1)\ ^1F^\circ$
$3d^3(^2H)4p$	$^1I^\circ$	6	120 432.99	99		
$3d^3(^2H)4p$	$^3G^\circ$	5	120 700.27	89		
		4	120 749.55	89		
		3	120 767.26	90		
$3d^3(^4P)4p$	$^3S^\circ$	1	121 446.77	87		
$3d^3(^2P)4p$	$^1P^\circ$	1	122 396.49	71	17	$(^2D2)\ ^1F^\circ$
$3d^3(^2D2)4p$	$^1D^\circ$	2	123 105.84	46	41	$(^2P)\ ^1D^\circ$
$3d^3(^2F)4p$	$^3F^\circ$	2	128 754.62	95		
		3	128 784.13	94		
		4	128 850.49	94		
$3d^3(^2F)4p$	$^3G^\circ$	3	131 116.36	92	5	$(^2H)\ ^3G^\circ$
		4	131 265.66	92	5	$(^2H)\ ^3G^\circ$
		5	131 450.16	95		
$3d^3(^2F)4p$	$^1D^\circ$	2	132 070.71	79	18	$(^2D1)\ ^1D^\circ$
$3d^3(^2F)4p$	$^3D^\circ$	3	132 117.50	92	5	$(^2D1)\ ^3D^\circ$
		2	132 499.78	91	5	
		1	132 734.22	94	5	
$3d^3(^2F)4p$	$^1G^\circ$	4	133 969.57	83	15	$(^2H)\ ^1G^\circ$
$3d^3(^2F)4p$	$^1F^\circ$	3	134 887.54	94		
$3d^3(^2D1)4p$	$^3D^\circ$	1	146 936.18	77	19	$(^2D2)\ ^3D^\circ$
		2	146 973.33	77	18	
		3	147 090.73	77	17	
$3d^3(^2D1)4p$	$^1D^\circ$	2	148 573.81	53	20	$(^2D2)\ ^1D^\circ$
$3d^3(^2D1)4p$	$^3F^\circ$	2	149 344.03	62	20	$(^2D2)\ ^3F^\circ$
		3	149 383.64	73	23	
		4	149 626.30	75	23	
$3d^3(^2D1)4p$	$^3P^\circ$	2	151 351.27	75	23	$(^2D2)\ ^3P^\circ$
		1	151 687.44	75	24	
		0	151 852.12	75	24	
$3d^3(^2D1)4p$	$^1F^\circ$	3	152 037.21	77	20	$(^2D)\ ^1F^\circ$
$3d^3(^4F)4d$	^5H	3	152 927.3			
		4	153 099.1			
		5	153 314.6			
		6	153 571.7			
		7	153 871.7			
$3d^3(^2D1)4p$	$^1P^\circ$	1	156 929.93	75	24	$(^2D2)\ ^1P^\circ$

(Continued)

Cr III—Continued

Configuration	Term	J	Level (cm^{-1})	Leading percentages
$3d^3(^4F)5s$	^5F	1	157 303.2	
		2	157 435.1	
		3	157 637.3	
		4	157 908.0	
		5	158 241.8	
$3d^3(^4F)4d$	^3G	3	158 066.7	
		4	158 442.9	
$3d^3(^4F)5s$	^3F	2	159 031.8	
		3	159 375.6	
		4	159 803.3	
$3d^3(^4F)4d$	^3F	3	158 463.6	
		4	158 623.7	
$3d^3(^2H)4d$	^3I	5	173 200.5	
		6	173 662.8	
$3d^2(^3F)4s4p(^3P^\circ)$	$^5G^\circ$	2	175 965.9	
		3	176 274.8	
		4	176 684.7	
		5	177 196.0	
		6	177 803.9	
$3d^2(^3F)4s4p(^3P^\circ)$	$^5F^\circ$	1	177 292.3	
		2	177 466.1	
		3	177 724.9	
		4	178 063.9	
		5	178 473.5	
$3d^2(^3F)4s4p(^3P^\circ)$	$^5D^\circ$	0	179 815.5	
		1	179 863.2	
		2	179 951.6	
		3	180 100.8	
		4	180 334.0	
Cr IV ($^4F_{3/2}$)	*Limit*		**249 700**	

Cr IV

$Z=24$

Sc I isoelectronic sequence

Ground state: $1s^2 2s^2 2p^6 3s^2 3p^6 3d^3\ ^4F_{3/2}$

Ionization energy = 396 500±400 cm^{-1} (49.16±0.05 eV)

The initial analysis was by White (1929) and was later extended by Bowen (1937). Ekberg (1973) reobserved the spectrum and extended the analysis but reported no levels above 211 574 cm^{-1}. With new observations Ekberg and Engström (1982) greatly increased the range of known energy levels by establishing 59 of the 67 possible $3d^2 4d$ levels and 15 of the 16 $3d^2 5s$ levels, as well as the $3d^2(^1S)4s\ ^2S_{1/2}$ level. The uncertainty of the energy level values is ±0.4 cm^{-1}. Percentage compositions for all the known levels were obtained by these authors.

From the $3d^2(^3F)ns$ terms for $n=4,5$ and an extrapolated value for the change in effective quantum number between them, Ekberg and Engström determined the value for the ionization energy.

References

Bowen, I. S. (1937), Phys. Rev. **52**, 1153.
Ekberg, J. O. (1973), Phys. Scr. **7**, 55.
Ekberg, J. O., and Engström, L. (1982), Phys. Scr. **25**, 611.
White, H. E. (1929), Phys. Rev. **33**, 672.

Cr IV

Configuration	Term	J	Level (cm^{-1})	Leading percentages		
$3d^3$	^4F	3/2	0.0	100		
		5/2	237.4	100		
		7/2	556.4	100		
		9/2	945.5	100		
$3d^3$	^4P	1/2	14 059.9	100		
		3/2	14 177.5	99		
		5/2	14 472.2	100		
$3d^3$	^2G	7/2	15 053.6	100		
		9/2	15 402.4	100		
$3d^3$	^2P	3/2	19 439.4	93		
		1/2	19 520.8	100		
$3d^3$	$^2D2$	3/2	20 651.0	72	22	$^2D1$
		5/2	20 665.5	78	22	
$3d^3$	^2H	9/2	21 066.9	100		
		11/2	21 321.1	100		
$3d^3$	^2F	7/2	34 364.3	100		
		5/2	34 556.9	100		
$3d^3$	$^2D1$	5/2	52 976.4	78	22	$^2D2$
		3/2	53 143.8	77	23	
$3d^2(^3F)4s$	^4F	3/2	103 996.5	100		
		5/2	104 258.6	100		
		7/2	104 630.2	100		
		9/2	105 105.7	100		
$3d^2(^3F)4s$	^2F	5/2	109 941.5	100		
		7/2	110 691.8	100		
$3d^2(^1D)4s$	^2D	5/2	118 571.5	65	35	$(^3P)\ ^4P°$
		3/2	118 727.8	58	42	

(Continued)

Cr IV—Continued

Configuration	Term	J	Level (cm^{-1})	Leading percentages		
$3d^2(^3P)4s$	^4P	1/2	119 014.6	100		
		3/2	119 252.0	58	42	$(^1D)\ ^2D^\circ$
		5/2	119 671.3	65	35	$(^1D)\ ^2D^\circ$
$3d^2(^3P)4s$	^2P	1/2	124 410.5	100		
		3/2	124 734.4	99		
$3d^2(^1G)4s$	^2G	9/2	127 195.7	100		
		7/2	127 208.3	100		
$3d^2(^1S)4s$	^2S	1/2	155 354.4	100		
$3d^2(^3F)4p$	$^4G^\circ$	5/2	*157 361.4*	96		
		7/2	*157 933.3*	97		
		9/2	*158 629.5*	98		
		11/2	*159 449.2*	100		
$3d^2(^3F)4p$	$^4F^\circ$	3/2	*158 527.7*	97		
		5/2	*158 892.7*	98		
		7/2	*159 352.0*	98		
		9/2	*159 863.6*	98		
$3d^2(^3F)4p$	$^2F^\circ$	5/2	*160 305.4*	78	5	$(^1D)\ ^2F^\circ$
		7/2	*160 937.4*	79	14	$(^3F)\ ^4D^\circ$
$3d^2(^3F)4p$	$^2D^\circ$	3/2	*160 986.5*	45	42	$(^3F)\ ^4D^\circ$
		5/2	*162 301.4*	60	22	
$3d^2(^3F)4p$	$^4D^\circ$	1/2	*161 354.8*	94	6	$(^3P)\ ^4P^\circ$
		5/2	*161 495.3*	67	15	$(^3F)\ ^2D^\circ$
		3/2	*161 756.3*	52	36	$(^3F)\ ^2D^\circ$
		7/2	*162 064.9*	81	13	$(^3F)\ ^2F^\circ$
$3d^2(^3F)4p$	$^2G^\circ$	7/2	*164 909.7*	95		
		9/2	*165 430.0*	95		
$3d^2(^3P)4p$	$^2S^\circ$	1/2	*167 896.5*	99		
$3d^2(^3P)4p$	$^4S^\circ$	3/2	*171 081.3*	91	9	$(^1D)\ ^2P^\circ$
$3d^2(^1D)4p$	$^2P^\circ$	3/2	*172 184.0*	84	9	$(^3P)\ ^4S^\circ$
		1/2	*172 823.2*	96		
$3d^2(^1D)4p$	$^2F^\circ$	5/2	*172 636.4*	87	7	$(^3F)\ ^2F^\circ$
		7/2	*173 366.0*	84	7	$(^3P)\ ^4D^\circ$
$3d^2(^3P)4p$	$^4D^\circ$	1/2	*173 431.9*	93	5	$(^3F)\ ^4D^\circ$
		3/2	*173 659.1*	92	5	$(^3F)\ ^4D^\circ$
		5/2	*174 096.2*	89		
		7/2	*174 846.2*	88	7	$(^1D)\ ^2F^\circ$
$3d^2(^1D)4p$	$^2D^\circ$	3/2	*174 539.7*	83	6	$(^3F)\ ^2D^\circ$
		5/2	*174 968.6*	82	6	
$3d^2(^3P)4p$	$^4P^\circ$	1/2	*176 690.9*	99		
		3/2	*176 916.7*	99		
		5/2	*177 406.5*	95		

Cr IV—Continued

Configuration	Term	J	Level (cm^{-1})	Leading percentages		
$3d^2(^1G)4p$	$^2G^\circ$	7/2	*177 916.5*	95		
		9/2	*178 030.0*	95		
$3d^2(^3P)4p$	$^2D^\circ$	5/2	*181 243.1*	81	12	$(^3F)\ ^2D^\circ$
		3/2	*181 277.9*	79	10	
$3d^2(^1G)4p$	$^2H^\circ$	9/2	*182 679.2*	99		
		11/2	*183 444.2*	100		
$3d^2(^3P)4p$	$^2P^\circ$	1/2	*183 720.3*	99		
		3/2	*183 875.6*	97		
$3d^2(^1G)4p$	$^2F^\circ$	7/2	*186 980.8*	96		
		5/2	*187 519.0*	97		
$3d^2(^1S)4p$	$^2P^\circ$	1/2	*210 559.2*	98		
		3/2	*211 575.9*	98		
$3d^2(^3F)4d$	^4G	5/2	232 566.1	88	12	$(^3F)\ ^2F$
		7/2	232 896.6	87	11	$(^3F)\ ^4H$
		9/2	233 236.7	82	18	$(^3F)\ ^4H$
		11/2	233 647.0	80	19	$(^3F)\ ^4H$
$3d^2(^3F)4d$	^2F	5/2	233 117.3	82	12	$(^3F)\ ^4G$
		7/2	233 637.6	89		
$3d^2(^3F)4d$	^4H	7/2	233 358.8	89	11	$(^3F)\ ^4G$
		9/2	233 708.5	82	18	
		11/2	234 100.1	80	20	
		13/2	234 500.9	100		
$3d^2(^3F)4d$	^4D	1/2	233 618.4	97		
		3/2	233 798.9	98		
		5/2	234 085.2	98		
		7/2	234 502.1	95		
$3d^2(^3F)4d$	^2P	1/2	235 743.0	88	7	$(^1D)\ ^2P$
		3/2	236 491.4	89	6	
$3d^2(^3F)4d$	^4P	1/2	237 798.2	86	13	$(^3P)\ ^4P$
		3/2	238 100.4	85	13	
		5/2	238 447.1	84	14	
$3d^2(^3F)4d$	^2G	7/2	237 999.5	84	12	$(^1D)\ ^2G$
		9/2	238 561.3	84	12	
$3d^2(^3F)4d$	^2D	3/2	239 541.4	78	18	$(^1D)\ ^2D$
		5/2	239 822.2	75	19	
$3d^2(^3F)4d$	^2H	9/2	239 582.8	91	9	$(^1G)\ ^2H$
		11/2	240 268.8	90	9	
$3d^2(^3F)4d$	^4F	3/2	240 967.9	95		
		5/2	241 182.5	95		
		7/2	241 472.8	95		
		9/2	241 832.5	95		

(Continued)

Cr IV—Continued

Configuration	Term	J	Level (cm^{-1})	Leading percentages		
$3d^2(^3P)4d$	^2P	3/2	247 099.1	60	40	$(^1D)\ ^2P$
		1/2	247 765.0	59	38	
$3d^2(^1D)4d$	^2F	5/2	247 239.3	84	15	$(^3P)\ ^2F$
		7/2	247 546.7	86	11	
$3d^2(^1D)4d$	^2G	9/2	249 932.5	75	8	$(^1G)\ ^2G$
		7/2	250 007.7	79	8	
$3d^2(^3P)4d$	^4F	5/2	250 535.2	91		
		7/2	250 752.7	86	6	$(^3P)\ ^4D$
		9/2	251 093.1	86	8	$(^1G)\ ^2G$
$3d^2(^3P)4d$	^4D	3/2	251 017.3	97		
		1/2	251 095.6	99		
		5/2	251 137.8	94		
		7/2	251 342.7	93	5	$(^3P)\ ^4F$
$3d^2(^3F)5s$	^4F	3/2	251 563.9	100		
		5/2	251 807.0	97		
		7/2	252 188.3	98		
		9/2	252 694.5	100		
$3d^2(^3P)4d$	^2F	5/2	252 464.8	54	37	$(^1G)\ ^2F$
		7/2	252 532.8	56	36	
$3d^2(^3F)5s$	^2F	5/2	253 476.1	97		
		7/2	254 212.1	98		
$3d^2(^3P)4d$	^4P	5/2	254 534.6	81	13	$(^3F)\ ^4P$
$3d^2(^3P)4d$	^2D	3/2	255 662.3	35	27	$(^1D)\ ^2D$
		5/2	255 836.9	38	35	
$3d^2(^1D)4d$	^2P	1/2	255 803.6	51	34	$(^3P)\ ^2P$
		3/2	256 306.2	43	31	$(^3P)\ ^2D$
$3d^2(^1G)4d$	^2I	11/2	256 853.2	100		
		13/2	256 885.4	100		
$3d^2(^1G)4d$	^2G	7/2	257 588.6	88	7	$(^3F)\ ^2G$
		9/2	257 714.4	87	7	
$3d^2(^1G)4d$	^2H	9/2	259 755.4	91	9	$(^3F)\ ^2H$
		11/2	259 799.6	91	9	
$3d^2(^1G)4d$	^2F	7/2	263 875.9	62	27	$(^3P)\ ^2F$
		5/2	263 971.0	61	27	
$3d^2(^1D)5s$	^2D	5/2	265 031.6	96		
		3/2	265 048.6	98		
$3d^2(^1G)4d$	^2D	3/2	266 878.0	64	32	$(^3P)\ ^2D$
		5/2	266 931.7	64	33	
$3d^2(^3P)5s$	^4P	1/2	266 946.6	100		
		3/2	267 130.4	99		
		5/2	267 469.0	96		

Cr IV—Continued

Configuration	Term	*J*	Level (cm^{-1})	Leading percentages
$3d^2(^3P)5s$	^2P	$\frac{1}{2}$	268 458.1	100
		$\frac{3}{2}$	268 813.4	98
$3d^2(^1G)5s$	^2G	$\frac{9}{2}$	273 865.3	100
		$\frac{7}{2}$	273 866.6	100
Cr V ($^3F_2$)	***Limit***		**396 500**	

Cr v

$Z=24$

Ca I isoelectronic sequence

Ground state: $1s^2 2s^2 2p^6 3s^2 3p^6 3d^2\ ^3F_2$

Ionization energy = 560 200±300 cm^{-1} (69.46±0.04 eV)

The initial analysis is due to White (1929) who reported levels of the $3d^2$, $3d4s$, $3d4p$, and $3d4d$ configurations. Additions and revisions were made by Cady and Edlén and were communicated to Moore for inclusion in her AEL compilation (1952). The spectrum was completely reobserved in the range of 400–1800 Å by Ekberg (1973) whose results are quoted here. The uncertainty in his level values is given as ±0.5 cm^{-1}. He added the $^1S_0$ level of $3d^2$ and all the known levels of $3d4d$ and $3d5s$. The one term of $3d4d$ due to White was found to be false.

The ionization energy was derived by Ekberg from the series $3d4s$ and $3d5s$.

References

Ekberg, J. O. (1973), Phys. Scr. **7**, 59.

Moore, C. E. (1952), Atomic Energy Levels, Natl. Bur. Stand. (U.S.) Circ. 467, Vol. II (reissued in 1971 as Natl. Bur. Stand. (U.S.) Natl. Stand. Ref. Data Ser. 35, Vol. II).

White, H. E. (1929), Phys. Rev. **33**, 538.

Cr v

Configuration	Term	J	Level (cm^{-1})	Configuration	Term	J	Level (cm^{-1})
$3d^2$	^3F	2	0.0	$3d\,4d$	^1F	3	316 674.9
		3	508.2				
		4	1 141.7	$3d\,4d$	^3D	1	317 893.8
						2	318 227.6
$3d^2$	^1D	2	13 188.0			3	318 601.7
$3d^2$	^3P	0	15 491.8	$3d\,4d$	^3G	3	319 119.1
		1	15 676.6			4	319 516.8
		2	16 041.0			5	320 074.4
$3d^2$	^1G	4	22 019.2	$3d\,4d$	^1P	1	319 284.0
$3d^2$	^1S	0	51 146.4	$3d\,4d$	^3S	1	322 528.1
$3d\,4s$	^3D	1	167 176.4	$3d\,4d$	^3F	2	325 104.1
		2	167 491.0			3	325 472.5
		3	168 089.5			4	325 884.2
$3d\,4s$	^1D	2	171 698.1	$3d\,4d$	^1D	2	329 350.3
$3d\,4p$	$^1D°$	2	*226 119.8*	$3d\,4d$	^3P	0	330 084.8
						1	330 245.1
$3d\,4p$	$^3D°$	1	*228 001.8*			2	330 536.8
		2	*228 489.1*				
		3	*229 120.8*	$3d\,4d$	^1G	4	331 811.2
$3d\,4p$	$^3F°$	2	*229 551.7*	$3d\,5s$	^3D	1	356 744.8
		3	*230 316.3*			2	356 981.3
		4	*231 392.9*			3	357 675.9
$3d\,4p$	$^3P°$	1	*234 618.4*	$3d\,5s$	^1D	2	358 653.8
		0	*234 668.5*				
		2	*234 846.4*				
$3d\,4p$	$^1F°$	3	*237 529.5*	Cr VI ($^2D_{3/2}$)	*Limit*		**560 200**
$3d\,4p$	$^1P°$	1	*239 917.5*				

Cr VI

$Z=24$

K I isoelectronic sequence

Ground state: $1s^2 2s^2 2p^6 3s^2 3p^6 3d\ ^2D_{3/2}$

Ionization energy = 731 020±6 cm^{-1} (90.6356±0.0007 eV)

The first known terms were found by Gibbs and White (1926, 1929) who reported the positions of the $3p^6 3d\ ^2D$, $4s\ ^2S$, and $4p\ ^2P°$ terms. The $nf\ ^2F°$ series for $n=5$–10 was identified by Alexander, Feldman, and Fraenkel (1965) and the missing $4f\ ^2F°$ by Gabriel, Fawcett, and Jordan (1965) who replaced the false levels of this term given earlier by Kruger and Weissberg (1937). The $4p\ ^2P° - 4d\ ^2D$ multiplet was found by Fawcett (1970). Observations of open $3p^6$-core configurations were first reported by Feldman and Fraenkel (1966) who identified the $3p^6 3d - 3p^5 3d 4s$ transition array. Some of these lines were classified by Cowan (1967). Gabriel, Fawcett, and Jordan (1966) classified six lines of the $3p^6 3d - 3p^5 3d^2$ group.

The spectrum was reobserved in the range of 400–2500 Å by Ekberg (1973). He interpreted a considerable number of new lines and verified the earlier work. His revisions of the values for the known levels and his additions to the analysis are quoted here. Levels obtained from transitions to the ground term from 350 000 to 700 000 cm^{-1} are given to the units place and have an uncertainty varying from ±5 to ±20 cm^{-1}. The rest have an uncertainty of ±1 cm^{-1}. The ionization energy is derived by Ekberg from the nh series with an estimated uncertainty of ±6 cm^{-1}.

References

Alexander, E., Feldman, U., and Fraenkel, B. S. (1965), J. Opt. Soc. Am. **55**, 650.

Cowan, R. D. (1967), Astrophys. J. **147**, 377.

Ekberg, J. O. (1973), Phys. Scr. **8**, 35.

Fawcett, B. C. (1970), J. Phys. **B3**, 1732.

Feldman, U., and Fraenkel, B. S. (1966), Astrophys. J. **145**, 959.

Gabriel, A. H., Fawcett, B. C., and Jordan, C. (1965), Nature **206**, 390.

Gabriel, A. H., Fawcett, B. C., and Jordan, C. (1966), Proc. Phys. Soc. (London) **87**, 825.

Gibbs, R. C., and White, H. E. (1926), Proc. Natl. Acad. Sci. (U.S.) **12**, 675.

Gibbs, R. C., and White, H. E. (1929), Phys. Rev. **33**, 157.

Kruger, P. G., and Weissberg, S. G. (1937), Phys. Rev. **52**, 314.

Cr VI

Configuration	Term	J	Level (cm^{-1})	Configuration	Term	J	Level (cm^{-1})
$3p^6(^1S)3d$	^2D	3/2	0	$3p^5(^2P°)3d^2(^3P)$	$^2P°$	1/2	*493 247.1*
		5/2	940			3/2	*494 911.2*
$3p^6(^1S)4s$	^2S	1/2	227 857.9	$3p^5(^2P°)3d^2(^3F)$	$^2D°$	5/2	*496 958*
						3/2	*497 495*
$3p^6(^1S)4p$	$^2P°$	1/2	*296 573.2*				
		3/2	*298 396.7*	$3p^6(^1S)5d$	^2D	3/2	534 381.7
						5/2	534 489.7
$3p^5(^2P°)3d^2(^1G)$	$^2F°$	5/2	*356 962*				
		7/2	*359 165*	$3p^6(^1S)6s$	^2S	1/2	562 064.1
$3p^5(^2P°)3d^2(^1D)$	$^2F°$	7/2	*371 618*	$3p^6(^1S)5f$	$^2F°$	5/2	*568 957.4*
		5/2	*378 677*			7/2	*568 993.0*
$3p^6(^1S)4d$	^2D	3/2	402 661.7	$3p^6(^1S)5g$	^2G	7/2	572 272.3
		5/2	402 888.6			9/2	572 274.4
$3p^5(^2P°)3d^2(^3F)$	$^2F°$	5/2	*440 135.2*	$3p^6(^1S)6p$	$^2P°$	1/2	*574 135*
		7/2	*442 945.4*			3/2	*575 742*
$3p^6(^1S)5s$	^2S	1/2	461 253.0	$3p^5\,3d(^3P°)4s$	$^2P°$	1/2	*578 566*
$3p^6(^1S)4f$	$^2F°$	5/2	*481 956.0*			3/2	*580 697*
		7/2	*482 517.1*	$3p^5\,3d(^3F°)4s$	$^4F°$	7/2	*584 371*
						5/2	*586 273*
$3p^6(^1S)5p$	$^2P°$	1/2	*487 589.5*				
		3/2	*488 561.9*				

(Continued)

Cr VI—Continued

Configuration	Term	J	Level (cm^{-1})
$3p^5\,3d(^3F^\circ)4s$	$^2F^\circ$	7/2	591 137
		5/2	594 926
$3p^5\,3d(^3D^\circ)4s$	$^4D^\circ$	7/2	607 615
		5/2	608 631
		3/2	609 166
$3p^5\,3d(^1D^\circ)4s$	$^2D^\circ$	5/2	610 497
		3/2	611 568
$3p^5\,3d(^1F^\circ)4s$	$^2F^\circ$	5/2	614 385
		7/2	616 079
$3p^5\,3d(^3D^\circ)4s$	$^2D^\circ$	5/2	618 491
		3/2	619 419
$3p^6(^1S)6f$	$^2F^\circ$	5/2	618 583
		7/2	618 849
$3p^6(^1S)6g$	^2G	7/2	620 696.3
		9/2	620 700.5
$3p^6(^1S)6h$	$^2H^\circ$	9/2,11/2	621 162.9
$3p^6(^1S)7f$	$^2F^\circ$	5/2	648 521
		7/2	648 533
$3p^6(^1S)7h$	$^2H^\circ$	9/2,11/2	650 310.8
$3p^6(^1S)8f$	$^2F^\circ$	5/2,7/2	667 973
$3p^6(^1S)9f$	$^2F^\circ$	5/2,7/2	681 307
$3p^6(^1S)10f$	$^2F^\circ$	5/2,7/2	690 781
Cr VII ($^1S_0$)	***Limit***		**731 020**

Cr VII

$Z=24$

Ar I isoelectronic sequence

Ground state: $1s^2 2s^2 2p^6 3s^2 3p^6\ ^1S_0$

Ionization energy = 1 291 900±600 cm^{-1} (160.18±0.07 eV)

Most of the levels are taken from an extensive analysis by Ekberg (1976) who observed the spectrum in the range of 90–2000 Å with an accuracy of ±0.01 Å. The few earlier identifications are noted in his work. Relative to the ground state the level uncertainty is ±10 cm^{-1}. Among the excited levels the relative uncertainty is ±0.5 cm^{-1}. The designations and percentage compositions for the levels are from Ekberg. A value for the ionization energy of 1 291 900 (600) cm^{-1} was derived by him by extrapolation. The same value may be derived from the 3-member *ns* series.

The terms $3s3p^6 4p\ ^3P^\circ_1$, $5p\ ^1P^\circ_1$ and $^3P^\circ_1$, $6p\ ^1P^\circ_1$, and $7p\ ^1P^\circ_1$ were determined by Kastner, Crooker, Behring, and Cohen (1977) from observations of absorption in a spark between 70 and 100 Å. The level uncertainty obtained from these data is ±100 cm^{-1}.

References

Ekberg, J. O. (1976), Phys. Scr. **13**, 245.

Kastner, S. O., Crooker, A. M., Behring, W. E., and Cohen, L. (1977), Phys. Rev. **A16**, 577.

Cr VII

Configuration	Term	J	Level (cm^{-1})	Leading percentages		
$3s^2 3p^6$	^1S	0	0.0			
$3s^2 3p^5 3d$	$^3P^\circ$	0	*341 179.3*	100		
		1	*342 773.5*	100		
		2	*346 137.1*	99		
$3s^2 3p^5 3d$	$^3F^\circ$	4	*357 543.7*	100		
		3	*360 171.9*	97		
		2	*363 060.9*	96		
$3s^2 3p^5 3d$	$^1D^\circ$	2	*382 682.3*	77	19	$^3D^\circ$
$3s^2 3p^5 3d$	$^3D^\circ$	3	*382 737.4*	74	26	$^1F^\circ$
		1	*385 828.3*	99		
		2	*386 616.6*	78	20	$^1D^\circ$
$3s^2 3p^5 3d$	$^1F^\circ$	3	*389 226.2*	72	25	$^3D^\circ$
$3s^2 3p^5 3d$	$^1P^\circ$	1	*493 035.4*	100		
$3s3p^6 3d$	^3D	1	608 679.6	100		
		2	609 142.7	100		
		3	609 887.8	100		
$3s3p^6 3d$	^1D	2	627 826.7	100		
$3s^2 3p^5(^2P^\circ_{3/2})4s$	$^2[3/2]^\circ$	2	*668 858.6*	100		
		1	*672 427.7*	85	15	$(^2P^\circ_{1/2})\ ^2[1/2]^\circ$
$3s^2 3p^5(^2P^\circ_{1/2})4s$	$^2[1/2]^\circ$	0	*678 534.7*	100		
		1	*682 610.2*	85	15	$(^2P^\circ_{3/2})\ ^2[3/2]^\circ$
$3s^2 3p^5 4p$	^3S	1	734 605.3			

(Continued)

Cr VII—Continued

Configuration	Term	J	Level (cm^{-1})	Leading percentages		
$3s^2 3p^5 4p$	^{3}D	2	745 328.9			
		3	745 631.1			
		1	748 629.3			
$3s^2 3p^5 4p$	^{3}P	2	751 649.3			
		0	757 035.8			
		1	758 572.1			
$3s^2 3p^5 4p$	^{1}P	1	754 378.9			
$3s^2 3p^5 4p$	^{1}D	2	758 374.4			
$3s^2 3p^5(^2P^\circ_{3/2})4d$	$^2[1/2]^\circ$	0	*856 292.2*	100		
		1	*857 234.5*	75	20	$(^2P^\circ_{3/2})\ ^2[3/2]^\circ$
$3s^2 3p^5(^2P^\circ_{3/2})4d$	$^2[3/2]^\circ$	2	*859 407.1*	85	15	$(^2P^\circ_{1/2})\ ^2[3/2]^\circ$
		1	*866 502.8*	79	21	$(^2P^\circ_{3/2})\ ^2[1/2]^\circ$
$3s^2 3p^5(^2P^\circ_{3/2})4d$	$^2[7/2]^\circ$	4	*860 444.3*	100		
		3	*861 198.4*	93		
$3s^2 3p^5(^2P^\circ_{3/2})4d$	$^2[5/2]^\circ$	2	*864 129.5*	88	11	$(^2P^\circ_{1/2})\ ^2[5/2]^\circ$
		3	*865 155.8*	88	6	$(^2P^\circ_{3/2})\ ^2[7/2]^\circ$
$3s^2 3p^5(^2P^\circ_{1/2})4d$	$^2[5/2]^\circ$	2	*872 231.6*	88	12	$(^2P^\circ_{3/2})\ ^2[5/2]^\circ$
		3	*873 146.1*	92	7	$(^2P^\circ_{3/2})\ ^2[5/2]^\circ$
$3s^2 3p^5(^2P^\circ_{1/2})4d$	$^2[3/2]^\circ$	2	*873 565.5*	85	15	$(^2P^\circ_{3/2})\ ^2[3/2]^\circ$
		1	*875 380.5*	95		
$3s^2 3p^5(^2P^\circ_{3/2})4f$	$^2[3/2]$	1	941 811			
		2	943 149.1			
$3s^2 3p^5(^2P^\circ_{3/2})4f$	$^2[9/2]$	5	944 416.8			
		4	945 475.7			
$3s^2 3p^5(^2P^\circ_{3/2})4f$	$^2[5/2]$	3	944 866.7			
		2	954 623			
$3s^2 3p^5(^2P^\circ_{3/2})4f$	$^2[7/2]$	3	947 917.4			
		4	948 943.9			
$3s^2 3p^5(^2P^\circ_{3/2})5s$	$^2[3/2]^\circ$	1	*951 122*			
$3s^2 3p^5(^2P^\circ_{1/2})4f$	$^2[7/2]$	3	956 454			
		4	957 205.1			
$3s^2 3p^5(^2P^\circ_{1/2})4f$	$^2[5/2]$	3	957 004.6			
$3s^2 3p^5(^2P^\circ_{1/2})5s$	$^2[1/2]^\circ$	1	*960 366*			
$3s3p^6 4p$	$^3P^\circ$	1	*984 590*			
$3s3p^6 4p$	$^1P^\circ$	1	*994 105*			
$3s^2 3p^5(^2P^\circ_{3/2})5d$	$^2[3/2]^\circ$	1	*1 033 485*			
$3s^2 3p^5(^2P^\circ_{1/2})5d$	$^2[3/2]^\circ$	1	*1 042 568*			

Cr VII—Continued

Configuration	Term	J	Level (cm^{-1})	Leading percentages
$3s^2\,3p^5(^2P^\circ_{3/2})6s$	$^2[3/2]^\circ$	1	*1 075 627*	
$3s^2\,3p^5(^2P^\circ_{1/2})6s$	$^2[1/2]^\circ$	1	*1 085 446*	
$3s3p^6\,5p$	$^3P^\circ$	1	*1 219 810*	
$3s3p^6\,5p$	$^1P^\circ$	1	*1 227 130*	
Cr VIII ($^2P^\circ_{3/2}$)	***Limit***		**1 291 900**	
$3s3p^6\,6p$	$^1P^\circ$	1	*1 335 560*	
$3s3p^6\,7p$	$^1P^\circ$	1	*1 393 840*	

Cr VIII

$Z=24$

Cl I isoelectronic sequence

Ground state: $1s^22s^22p^63s^23p^5\,^2P^\circ_{3/2}$

Ionization energy = 1 490 000±3000 cm^{-1} (184.7±0.4 eV)

All the known levels are derived from transitions to the $3s^23p^5\,^2P^\circ$ ground term. Edlén (1937) identified lines originating from all levels of the $3s^23p^44s$ configuration except for the $^4P_{1/2}$. Earlier, Weissberg and Kruger (1936) had reported lines from the ^2P term of this configuration as well as from the ^2S term of $3s3p^6$. The transitions from $3s^23p^43d$ were identified by Gabriel, Fawcett, and Jordan (1966) and by Fawcett and Gabriel (1966). The parent states for the terms $3s^23p^43d$ ^2P and ^2D are changed to $3p^4$ (^3P) as indicated by the calculated states of Fe X by Bromage, Cowan, and Fawcett (1977). Transitions from $3s^23p^44d$ were given by Fawcett, Cowan, and Hayes (1972). Line identifications in the $3p^43d-3p^44f$ transition array were also given in this paper but cannot be used to derive energy levels because they do not combine with known levels.

The recent measurements of the $3s^23p^5-3s3p^6$ doublet by Smitt, Svensson, and Outred (1976) are used to determine the ground term interval with an uncertainty of ±2 cm^{-1} and the $3s3p^6\,^2S_{1/2}$ level with an uncertainty of ±10 cm^{-1}. The uncertainty in the values of the higher-lying levels is estimated to be ±100 cm^{-1}.

The ionization energy is an extrapolated value by Lotz (1967).

References

Bromage, G. E., Cowan, R. D., and Fawcett, B. C. (1977), Phys. Scr. **15**, 177.

Edlén, B. (1937), Z. Phys. **104**, 407.

Fawcett, B. C., and Gabriel, A. H. (1966), Proc. Phys. Soc. **88**, 262.

Fawcett, B. C., Cowan, R. D., and Hayes, R. W. (1972), J. Phys. **B5**, 2143.

Gabriel, A. H., Fawcett, B. C., and Jordan, C. (1965), Nature **206**, 390.

Lotz, W. J. (1967), J. Opt. Soc. Am. **57**, 873.

Smitt, R., Svensson, L. A., and Outred, M. (1976), Phys. Scr. **13**, 293.

Weissberg, S. G., and Kruger, P. G. (1936), Phys. Rev. **49**, 872.

Cr VIII

Configuration	Term	J	Level (cm^{-1})	Configuration	Term	J	Level (cm^{-1})
$3s^2\,3p^5$	$^2P^\circ$	3/2	*0*	$3s^2\,3p^4(^1D)4s$	^2D	5/2	769 240
		1/2	*9 892*			3/2	769 550
$3s3p^6$	^2S	1/2	242 065	$3s^2\,3p^4(^1S)4s$	^2S	1/2	805 260
$3s^2\,3p^4(^1D)3d$	^2S	1/2	461 540	$3s^2\,3p^4(^3P)4d$	^2D	5/2	946 200
						3/2	947 300
$3s^2\,3p^4(^3P)3d$	^2P	3/2	479 310				
		1/2	482 910	$3s^2\,3p^4(^1D)4d$	^2S	1/2	966 400
$3s^2\,3p^4(^3P)3d$	^2D	5/2	487 780	$3s^2\,3p^4(^1D)4d$	^2P	3/2	970 600
		3/2	496 170			1/2	972 200
$3s^2\,3p^4(^3P)4s$	^4P	5/2	735 880	$3s^2\,3p^4(^1D)4d$	^2D	5/2	976 100
		3/2	741 060			3/2	977 400
$3s^2\,3p^4(^3P)4s$	^2P	3/2	749 640	Cr IX ($^3P_2$)	*Limit*		**1 490 000**
		1/2	755 740				

Cr IX

$Z=24$

S I isoelectronic sequence

Ground state: $1s^2 2s^2 2p^6 3s^2 3p^4\ ^3P_2$

Ionization energy $=$ 1 688 000±3000 cm^{-1} (209.3±0.4 eV)

Edlén (1937) provided the initial spectral classifications by identifying the $3p^4-3p^3 4s$ transitions. Both singlet and triplet levels were detected but not the connection between them.

Classifications in the $3p^4-3p^3 3d$ transition array were made by Gabriel, Fawcett, and Jordan (1965) and by Fawcett and Gabriel (1966). Some new measurements were later provided by Fawcett (1971) as well as two newly classified lines of this group. The $3s3p^5$ configuration was found by Fawcett (1970). Classified lines of the $3p^4-3p^3 4d$ and $3p^3 3d-3p^3 4f$ arrays were reported by Fawcett, Cowan, and Hayes (1972). The latter array could not be used to derive $3p^3 4f$ levels because of the lack of transitions to known levels.

Smitt, Svensson, and Outred (1976) give improved measurements of the $3s^2 3p^4-3s3p^5$ transitions, obtaining an uncertainty of ±5 cm^{-1} for the levels, except for the $3s^2 3p^4\ ^1S_0$ determined from a blended line. They also found the intersystem line $^3P_2-{}^1P^{\circ}_1$ and identified the $^1S_0-{}^1P^{\circ}_1$ line. The uncertainty in the levels above the $3s3p^5$ configuration is ±100 cm^{-1}.

The ionization energy is an extrapolated value by Lotz (1976).

References

Edlén, B. (1937), Z. Phys. **104**, 188.

Fawcett, B. C., and Gabriel, A. H. (1966), Proc. Phys. Soc. **88**, 262.

Fawcett, B. C. (1970), J. Phys. **B3**, 1732.

Fawcett, B. C. (1971), J. Phys. **B4**, 1577.

Fawcett, B. C., Cowan, R. D., and Hayes, R. W. (1972), J. Phys. **B5**, 2143.

Gabriel, A. H., Fawcett, B. C., and Jordan, C. (1965), Proc. Phys. Soc. **87**, 825.

Lotz, W. J. (1967), J. Opt Soc. Am. **57**, 873.

Smitt, R., Svensson, L. A., and Outred, M. (1976), Phys. Scr. **13**, 293.

Cr IX

Configuration	Term	J	Level (cm^{-1})	Configuration	Term	J	Level (cm^{-1})
$3s^2\,3p^4$	^3P	2	0	$3s^2\,3p^3(^2P^{\circ})3d$	$^1P^{\circ}$	1	*531 880*
		1	7 821				
		0	9 549	$3s^2\,3p^3(^4S^{\circ})4s$	$^3S^{\circ}$	1	*821 100*
$3s^2\,3p^4$	^1D	2	30 284	$3s^2\,3p^3(^2D^{\circ})4s$	$^3D^{\circ}$	1	*845 900*
						2	*846 260*
$3s^2\,3p^4$	^1S	0	66 855			3	*847 870*
$3s3p^5$	$^3P^{\circ}$	2	*239 068*	$3s^2\,3p^3(^2D^{\circ})4s$	$^1D^{\circ}$	2	*854 730*
		1	*245 317*				
		0	*249 016*	$3s^2\,3p^3(^2P^{\circ})4s$	$^1P^{\circ}$	1	*881 810*
$3s3p^5$	$^1P^{\circ}$	1	*305 561*	$3s^2\,3p^3(^4S^{\circ})4d$	$^3D^{\circ}$	2	*1 028 500*
						3	*1 028 900*
$3s^2\,3p^3(^2D^{\circ})3d$	$^3P^{\circ}$	2	*454 510*			1	*1 029 100*
$3s^2\,3p^3(^2D^{\circ})3d$	$^1D^{\circ}$	2	*493 310*	$3s^2\,3p^3(^2D^{\circ})4d$	$^1D^{\circ}$	2	*1 066 800*
$3s^2\,3p^3(^4S^{\circ})3d$	$^3D^{\circ}$	3	*474 790*	$3s^2\,3p^3(^2D^{\circ})4d$	$^1F^{\circ}$	3	*1 070 100*
		2	*479 570*				
		1	*482 760*	$3s^2\,3p^3\,4d$	$^1P^{\circ}$	1	*1 102 600*
$3s^2\,3p^3(^2D^{\circ})3d$	$^1F^{\circ}$	3	*507 750*	Cr X ($^4S^{\circ}_{3/2}$)	*Limit*		**1 688 000**

Cr X

$Z = 24$

P I isoelectronic sequence

Ground state: $1s^2 2s^2 2p^6 3s^2 3p^3\ ^4S^\circ_{3/2}$

Ionization energy = 1 971 000±4000 cm^{-1} (244.4±0.5 eV)

The levels of $3s3p^4$ are due to Smitt, Svensson, and Outred (1976) who gave improved measurements and additional classifications in the $3s^2 3p^3 - 3s3p^4$ array, previously interpreted by Fawcett and Peacock (1967) and by Fawcett (1970). The important $^4S^\circ - ^2P^\circ$ forbidden lines of the $3s^2 3p^3$ configuration, which unify the term systems, were identified by Feldman and Doschek (1976) in solar coronal spectra. The uncertainty in these level values is ±5 cm^{-1}. Transitions from $3p^2 3d$ to the ground configuration were identified by Gabriel, Fawcett, and Jordan (1966), Fawcett, Gabriel, and Saunders (1967), and most completely by Fawcett (1970). The level value uncertainty is ±100 cm^{-1}. Lines of the $3s^2 3p^3 - 3s^2 3p^2 4s$ array were classified by Fawcett, Cowan, and Hayes (1972), giving a level uncertainty of ±300 cm^{-1}. These authors also identified lines arising from the $3p^2 4f$ configuration but the lower levels have not been determined.

The ionization energy was obtained by Lotz (1967) by extrapolation.

References

Fawcett, B. C., Gabriel, A. H., and Saunders, P. A. H. (1967), Proc. Phys. Soc. **90**, 863.

Fawcett, B. C., and Peacock, N. J. (1967), Proc. Phys. Soc. **91**, 973.

Fawcett, B. C. (1970), J. Phys. **B3**, 1732.

Fawcett, B. C., Cowan, R. D., and Hayes, R. W. (1972), J. Phys. **B5**, 2143.

Feldman, U., and Doschek, G. A. (1977), J. Opt. Soc. Am. **67**, 726.

Gabriel, A. H., Fawcett, B. C., and Jordan, C. (1966), Proc. Phys. Soc. **87**, 825.

Lotz, W. J. (1967), J. Opt. Soc. Am. **57**, 873.

Smitt, R., Svensson, L. A., and Outred, M. (1976), Phys. Scr. **13**, 293.

Cr X

Configuration	Term	J	Level (cm^{-1})	Configuration	Term	J	Level (cm^{-1})
$3s^2 3p^3$	$^4S^\circ$	3/2	*0*	$3s^2 3p^2(^1D)3d$	^2D	5/2	476 680
						3/2	476 820
$3s^2 3p^3$	$^2D^\circ$	3/2	*37 103*				
		5/2	*39 450*	$3s^2 3p^2(^1D)3d$	^2P	1/2	491 650
						3/2	496 430
$3s^2 3p^3$	$^2P^\circ$	1/2	*63 935*				
		3/2	*67 157*	$3s^2 3p^2(^3P)3d$	^2F	7/2	500 880
$3s3p^4$	^4P	5/2	233 890	$3s^2 3p^2(^3P)3d$	^2D	5/2	519 280
		3/2	239 987			3/2	520 820
		1/2	242 922				
				$3s^2 3p^2(^3P)4s$	^4P	1/2	928 500
$3s3p^4$	^2D	3/2	289 637			3/2	933 400
		5/2	290 606			5/2	939 100
$3s3p^4$	^2P	3/2	333 412	$3s^2 3p^2(^3P)4s$	^2P	1/2	943 300
						3/2	949 800
		1/2	337 370				
				$3s^2 3p^2(^1D)4s$	^2D	5/2	967 000
$3s3p^4$	^2S	1/2	348 760			3/2	967 800
$3s^2 3p^2(^3P)3d$	^2P	3/2	432 830				
		1/2	440 870				
				Cr XI ($^3P_0$)	*Limit*		**1 971 000**
$3s^2 3p^2(^3P)3d$	^4P	5/2	442 010				
		3/2	444 960				
		1/2	446 710				

Cr XI

$Z=24$

Si I isoelectronic sequence

Ground state: $1s^2 2s^2 2p^6 3s^2 3p^2\ ^3P_0$

Ionization energy = 2 184 000±4000 cm^{-1} (270.8±0.5 eV)

Fawcett (1970, 1971) gave the first interpretation of the $3s^2 3p^2 - 3s3p^3$ array. The level values were later revised by Smitt, Svensson, and Outred (1976) on the basis of their more accurate measurements with an estimated uncertainty of ±0.008 Å or better. The connection between the singlet and triplet systems results from this later work and from the solar coronal line at 3996.8 Å (air) identified by Jefferies (1969) as arising from the forbidden $^1D_2 - ^3P_2$ transition in $3s^2 3p^2$. The uncertainty in these level values is estimated to be ±5 cm^{-1}.

The levels of $3s^2 3p3d$ are from Fawcett (1971). Those of $3s^2 3p4s$ and the $^3D_3$ and $^1F_3$ of $3s^2 3p4d$ are due to Fawcett, Cowan, and Hayes (1972). In table I of this reference the line given as 98.48 Å must be changed to 99.48 Å to fit its classification. Kastner et al. (1978) added to the known levels of $3s^2 3p4d$ and the ^1G of $3s^2 3p4f$. The uncertainty in the $3d$ level values is ±100 cm^{-1} and for the others 300 cm^{-1}.

The ionization energy was obtained by Lotz (1967) by extrapolation.

References

Fawcett, B. C. (1970), J. Phys. **B3**, 1732.
Fawcett, B. C. (1971), J. Phys. **B4**, 1577.
Fawcett, B. C., Cowan, R. D., and Hayes, R. W. (1972), J. Phys. **B5**, 2143.
Jefferies, J. T. (1969), Mem. Soc. Roy. Sci. Liege, Collect. **17**, 213.
Kastner, S. O., Swartz, M., Bhatia, A. K., and Lapides, J. (1978), J. Opt. Soc. Am. **68**, 1558.
Lotz, W. J. (1967), J. Opt. Soc. Am. **57**, 873.
Smitt, R., Svensson, L. A., and Outred, M. (1976), Phys. Scr. **13**, 293.

Cr XI

Configuration	Term	J	Level (cm^{-1})	Configuration	Term	J	Level (cm^{-1})
$3s^2\,3p^2$	^3P	0	0	$3s^2\,3p3d$	$^3D^\circ$	1	434 240
		1	5 536			2	436 210
		2	11 980			3	436 550
$3s^2\,3p^2$	^1D	2	36 994	$3s^2\,3p3d$	$^1F^\circ$	3	478 590
$3s3p^3$	$^3D^\circ$	1	242 346	$3s^2\,3p4s$	$^3P^\circ$	0	1 008 800
		2	242 456			1	1 010 700
		3	243 916			2	1 021 100
$3s3p^3$	$^3P^\circ$	0	278 059	$3s^2\,3p4s$	$^1P^\circ$	1	1 028 100
		1	278 394				
		2	278 698	$3s^2\,3p4d$	$^3D^\circ$	1	1 234 300
						2	1 236 600
$3s3p^3$	$^1D^\circ$	2	306 570			3	1 237 900
$3s3p^3$	$^3S^\circ$	1	356 424	$3s^2\,3p4d$	$^3F^\circ$	3	1 243 800
$3s3p^3$	$^1P^\circ$	1	372 498	$3s^2\,3p4d$	$^1F^\circ$	3	1 255 500
$3s^2\,3p3d$	$^3P^\circ$	2	418 980	$3s^2\,3p4f$	^1G	4	1 347 200
		1	425 480				
				Cr XII ($^2P^\circ_{1/2}$)	*Limit*		**2 184 000**
$3s^2\,3p3d$	$^1D^\circ$	2	427 090				

Cr XII

$Z=24$

Al I isoelectronic sequence

Ground state: $1s^22s^22p^63s^23p\ ^2P^\circ_{1/2}$

Ionization energy = 2 404 000±5000 cm^{-1} (298.0±0.6 eV)

The $3p-4d$ doublet was identified by Edlén (1936). Thirty years later new observations of this spectrum by Gabriel, Fawcett, and Jordan (1966) resulted in the discovery of the $3p-3d$ doublet. The non-diagonal line of this multiplet was identified by Fawcett, Gabriel, and Saunders (1967) by means of a laser-produced plasma. The uncertainty in the $3d$ and $4d$ level values is ±100 cm^{-1}.

The ^2P and ^2D terms of the low-lying $3s3p^2$ configuration were first reported by Fawcett and Peacock (1967) again utilizing a laser plasma. Fawcett (1970) revised the classification of the $3s^23p\ ^2P^\circ-3s3p^2\ ^2D$ multiplet and added the $^2P^\circ-^2S$ lines. He also reported the $^4P-^4S^\circ$ lines of the $3s3p^2-3p^3$ array. The quartet term position relative to the doublet system is not observed or predicted and therefore cannot be included here.

Fawcett (1971) later revised the wavelengths of the $3s^23p-3s3p^2$ array from new plates taken with the theta pinch source. His value of 412.46 Å for the $^2P^\circ_{3/2}-^2D_{3/2}$ line is inconsistent with the $^2P^\circ_{1/2}-^2D_{3/2}$ principal line and was not used here. He also reported the $3p-3d$ doublet of the same array. Calculated wavelengths enabled Fawcett, Cowan, Kononov, and Hayes (1972) to identify more lines from the theta pinch spectrum. They classified the $3d-4f$ doublet and several quartet transitions unconnected with the doublets. The uncertainty of the level values of $3s3p^2$ is ±50 cm^{-1}.

The ground term $^2P^\circ$ splitting is obtained from the solar coronal line 8153.8 Å classified by Jefferies (1969) and has an uncertainty of ±1 cm^{-1}.

The value for the ionization energy was obtained by extrapolation by Lotz (1967).

References

Edlén, B. (1936), Z. Phys. **103**, 536.

Fawcett, B. C., Gabriel, A. H., and Saunders, P. A. H. (1967), Proc. Phys. Soc. **90**, 863.

Fawcett, B. C., and Peacock, N. J. (1967), Proc. Phys. Soc. **91**, 973.

Fawcett, B. C. (1970), J. Phys. **B3**, 1732.

Fawcett, B. C. (1971), J. Phys. **B4**, 1577.

Fawcett, B. C., Cowan, R. D., Kononov, E. Y., and Hayes, R. W. (1972), J. Phys. **B5**, 1255.

Gabriel, A. H., Fawcett, B. C., and Jordan, C. (1966), Proc. Phys. Soc. **87**, 825.

Jefferies, J. T. (1969), Mem. Soc. Roy. Sci. Liege, Collect. **17**, 213.

Lotz, W. J. (1967), J. Opt. Soc. Am. **57**, 873.

Cr XII

Configuration	Term	J	Level (cm^{-1})	Configuration	Term	J	Level (cm^{-1})
$3s^2\ 3p$	$^2P^\circ$	1/2	*0*	$3s^2\ 4d$	^2D	3/2	1 319 000
		3/2	*12 261*			5/2	1 319 660
$3s3p^2$	^2D	3/2	254 450	$3s^2\ 4f$	$^2F^\circ$	5/2	*1 395 000*
		5/2	255 620			7/2	*1 395 400*
$3s3p^2$	^2S	1/2	313 600				
$3s3p^2$	^2P	1/2	333 240	Cr XIII ($^1S_0$)	***Limit***		**2 404 000**
		3/2	339 250				
$3s^2\ 3d$	^2D	3/2	408 700				
		5/2	409 840				

Cr XIII

$Z=24$

Mg I isoelectronic sequence

Ground state: $1s^2 2s^2 2p^6 3s^2\ {}^1S_0$

Ionization energy = 2 862 000±6000 cm^{-1} (354.8±0.7 eV)

The $3s^2\ {}^1S_0 - 3s4p\ {}^1P^\circ_1$ resonance line was identified by Edlén (1936). He also reported two unconnected triplet systems, one containing levels of $3s3p$, $3s4s$, $3s4d$, and $3s5d$, and the other $3s3d$, $3s4f$, and $3s5f$. The analysis was resumed 30 years later by Fawcett, Gabriel, and Saunders (1967) who unified the triplets by discovering the $3s3p\ {}^3P^\circ - 3s3d\ {}^3D$ multiplet. They also identified the resonance line $3s^2\ {}^1S_0 - 3s3p\ {}^1P^\circ_1$ at 328.29±0.03 Å. The intersystem transition $3s^2\ {}^1S_0 - 3s3p\ {}^3P^\circ_1$ was observed at 482.2±0.2 Å in a tokamak plasma by Finkenthal, Bell, and Moos (1982). These two lines were measured in a tokamak plasma by Peacock, Stamp, and Silver (1984), who obtained the more accurate values of 328.267±0.004 Å and 482.17±0.02 Å, or level uncertainties of ±4 and ±9 cm^{-1}. The uncertainties in the rest of the levels of the above configurations is ±100 cm^{-1}.

The analysis was extended to $3p^2$ by Fawcett and Peacock (1967) who identified the $3s3p\ {}^3P^\circ - 3p^2\ {}^3P$ multiplet. Fawcett (1970) later found the ${}^1S_0$ of $3p^2$, the ${}^1D_2$ of $3s3d$ and all the levels of $3p3d$ presently known. The uncertainty of these level values is ±100 cm^{-1}.

The analysis was augmented by Fawcett, Cowan, Kononov, and Hayes (1972) with their identification of transitions from $3s4f\ {}^1F^\circ_3$, $3s4d\ {}^1D_2$, and levels of $3p4f$. Some of the last group are not connected to known levels and are therefore not used here.

Improved measurements of the $3s3p\ {}^3P^\circ - 3s3d\ {}^3D$ multiplet were made by Fawcett, Cowan, and Hayes (1972). In addition, they reported the ${}^1D_2$ of $3s3d$ and the ${}^1D_2$ of $3p^2$.

The ionization energy is an extrapolated value obtained by Lotz (1967).

References

Edlén, B. (1936), Z. Phys. **103**, 536.

Fawcett, B. C., Gabriel, A. H., and Saunders, P. A. H. (1967), Proc. Phys. Soc. **90**, 863.

Fawcett, B. C., and Peacock, N. J. (1967), Proc. Phys. Soc. **91**, 973.

Fawcett, B. C. (1970), J. Phys. **B3**, 1732.

Fawcett, B. C., Cowan, R. D., Kononov, E. Y., and Hayes, R. W. (1972), J. Phys. **B5**, 1255.

Fawcett, B. C., Cowan, R. D., and Hayes, R. W. (1972), J. Phys. **B5**, 2143.

Finkenthal, M., Bell, R. E., and Moos, H. W. (1982), Phys. Lett. **88A**, 165.

Lotz, W. J. (1967), J. Opt. Soc. Am. **57**, 873.

Peacock, N. J., Stamp, M. F., and Silver, J. D. (1984), Phys. Scr. **T8**, 10.

Cr XIII

Configuration	Term	J	Level (cm^{-1})	Configuration	Term	J	Level (cm^{-1})
$3s^2$	${}^1S$	0	0	$3p3d$	${}^3F^\circ$	2	*805 220*
						3	*811 550*
$3s3p$	${}^3P^\circ$	0	*203 470*			4	*818 910*
		1	*207 400*				
		2	*216 590*	$3p3d$	${}^3P^\circ$	2	*857 230*
$3s3p$	${}^1P^\circ$	1	*304 630*	$3p3d$	${}^3D^\circ$	3	*861 010*
$3p^2$	${}^3P$	0	482 160	$3p3d$	${}^1P^\circ$	1	*885 300*
		1	488 250				
		2	499 220	$3s4s$	${}^3S$	1	1 385 280
$3p^2$	${}^1D$	2	483 170	$3s4p$	${}^1P^\circ$	1	*1 492 920*
$3p^2$	${}^1S$	0	569 460	$3s4d$	${}^3D$	1	1 616 060
						2	1 616 460
$3s3d$	${}^3D$	1	588 580			3	1 617 190
		2	589 210				
		3	590 100	$3s4d$	${}^1D$	2	1 617 500
$3s3d$	${}^1D$	2	662 240	$3s4f$	${}^3F^\circ$	2	*1 678 240*
						3	*1 678 630*
						4	*1 678 770*

(Continued)

Cr XIII—Continued

Configuration	Term	J	Level (cm^{-1})	Configuration	Term	J	Level (cm^{-1})
$3s4f$	$^1F°$	3	*1 690 800*	$3s5d$	^3D	2	2 076 350
						3	2 076 540
$3p4f$	^1F	3	1 916 140				
				$3s5f$	$^3F°$	4	*2 105 990*
$3p4f$	^3G	4	1 920 560				
		5	1 930 140				
				Cr XIV ($^2S_{1/2}$)	***Limit***		**2 862 000**
$3p4f$	^3F	4	1 931 440				
$3p4f$	^3D	3	1 940 840				

Cr XIV

$Z=24$

Na I isoelectronic sequence

Ground state: $1s^2 2s^2 2p^6 3s\ ^2S_{1/2}$

Ionization energy = 3 098 520±200 (384.171±0.020 eV)

Edlén (1936) reported three independent systems of doublets: $3s-np$, $3p-nd$ (and $3p-4s$), and $3d-nf$. These were unified by Fawcett and Peacock (1967) who identified the $3s-3p$ and $3p-3d$ doublets. New observations by Peacock, Stamp, and Silver (1984) of the $3s-3p$ doublet in a tokamak plasma were reported at 389.864 and 412.051 Å. We use these and the improved values for the $3p-3d$ doublet by Edlén (1978). The $4s$, $4p$, $4d$, $4f$, $5d$, $5f$, and $6f(^2F_{7/2})$ terms are from the identifications and measurements of Edlén (1936). The uncertainty of the $n=3$ levels is ± 5 cm^{-1}, and the higher ones ± 200 cm^{-1}.

The additional series members $5s-6s$, $6p-9p$, $6d-9d$, and $7f-8f$ were identified by Fawcett, Cowan, and Hayes (1972). Improved measurements by Cohen and Behring (1976) in the range of 35–70 Å were used. They estimate their measurement uncertainty to be ±0.005 Å although their measurements differ from those of Edlén (1936) by about + 0.02 Å. The $2p^5 3s^2\ ^2P°$ term was found by Feldman and Cohen (1967).

The value for the ionization energy was derived by Edlén (1978) from core polarization theory applied to the nf series.

References

Cohen, L. and Behring, W. E. (1976), J. Opt. Soc. Am. **66**, 899.
Edlén, B. (1936), Z. Phys. **100**, 621.
Edlén, B. (1978), Phys. Scr. **17**, 565.
Fawcett, B. C., and Peacock, N. J. (1967), Proc. Phys. Soc. **91**, 973.
Fawcett, B. C., Cowan, R. D., and Hayes, R. W. (1972), J. Phys. **B5**, 2143.
Feldman, U., and Cohen, L. (1967), J. Opt. Soc. Am. **57**, 1128.
Peacock, N. J., Stamp, M. F., and Silver, J. D. (1984), Phys. Scr. **T8**, 10.

Cr XIV

Configuration	Term	J	Level (cm^{-1})
$2p^6 3s$	^2S	1/2	0
$2p^6 3p$	$^2P°$	1/2	*242 688*
		3/2	*256 500*
$2p^6 3d$	^2D	3/2	587 810
		5/2	589 505
$2p^6 4s$	^2S	1/2	1 478 580
$2p^6 4p$	$^2P°$	1/2	*1 574 180*
		3/2	*1 579 550*
$2p^6 4d$	^2D	3/2	1 701 150
		5/2	1 701 940
$2p^6 4f$	$^2F°$	5/2	*1 749 830*
		7/2	*1 750 080*
$2p^6 5s$	^2S	1/2	2 102 780
$2p^6 5p$	$^2P°$	1/2	*2 149 290*
		3/2	*2 152 020*
$2p^6 5d$	^2D	3/2	2 210 730
		5/2	2 211 080
$2p^6 5f$	$^2F°$	5/2	*2 235 280*
		7/2	*2 235 440*
$2p^6 6s$	^2S	1/2	2 424 470
$2p^6 6p$	$^2P°$	1/2	*2 450 990*
		3/2	*2 452 060*
$2p^6 6d$	^2D	3/2	2 484 990
		5/2	2 485 240
$2p^6 6f$	$^2F°$	5/2	*2 499 090*
		7/2	*2 499 260*
$2p^6 7s$	^2S	1/2	2 612 050
		5/2	2 649 530
$2p^6 7f$	$^2F°$	5/2	*2 658 200*
		7/2	*2 658 270*
$2p^6 8p$	$^2P°$	5/2,3/2	*2 742 280*
$2p^6 8d$	^2D	3/2	2 755 510
		5/2	2 755 380
$2p^6 8f$	$^2F°$	7/2	*2 761 580*
$2p^6 9p$	$^2P°$	5/2,3/2	*2 820 870*

(Continued)

Cr XIV—Continued

Configuration	Term	J	Level (cm^{-1})	Configuration	Term	J	Level (cm^{-1})
$2p^6\,9d$	^{2}D	$^5/_2$	2 827 260				
		$^3/_2$	2 828 070				
Cr XV (^{1}S$_0$)	***Limit***		3 098 520				
$2p^5\,3s^2$	^{2}P°	$^3/_2$	*4 593 500*				
		$^1/_2$	*4 658 300*				

Cr XV

$Z=24$

Ne I isoelectronic sequence

Ground state: $1s^2 2s^2 2p^6\ ^1S_0$

Ionization energy 8 151 000±5000 cm^{-1} (1010.6±0.6 eV)

Only resonance lines are classified for this ion. Tyrén (1938) identified the lines due to the $2s^2 2p^5 3s$, $3d$ and $4d$ as well as the $2s2p^6 3p$ levels. Swartz, Kastner, Rothe, and Neupert (1971) identified $5d$, and $6d$. The magnetic quadrupole transition $2p^6\ ^1S_0 - 2p^5 3s\ ^3P_2^\circ$ was identified in the spectrum of a tokamak plasma by Klapisch et al. (1978). They confirm the wavelengths of Tyrén for $2p^5 3s\ ^3P$, and ^1P, and $2p^5 3d\ ^3D$, to within ±0.003 Å, but give the value 18.488 Å for $2p^5 3d\ ^1P$, compared with Tyrén's value of 18.497 Å used here. The uncertainty in the level values is estimated to be ±2000 cm^{-1}.

Kastner, Behring, and Cohen (1975) identified transitions between $2p^5 3p$ and $2p^5 4d$ but there is no connection with known levels.

The percentage compositions were calculated by Bogdanovich et al. (1980).

We derived the ionization energy from the $2s^2 2p^5 nd\ ^3D_1^\circ$ series for $n=3, 4, 5$. The $n=6$ term does not fit well to a series calculation.

References

Bogdanovich, P. O., Bogdanovichene, M. I., Grudzinskas, I. I., Rudzikas, Z. B., Tutlis, V. I., and Shadzhjuvene, S. D. (1980), "Spectroscopy of Multiply Charged Ions," USSR Acad. of Sci., Sci. Counc. on Spectr.

Kastner, S. O., Behring, W. E., and Cohen, L. (1975), Astrophys. J. **199**, 777.

Klapisch, M., Bar Shalom, A., Schwob, J. L., Fraenkel, B. S., Breton, C., de Michelis, C., Finkenthal, M., and Mattioli, M. (1978), Phys. Lett. **69A**, 34.

Swartz, M., Kastner, S., Rothe, E., Neupert, W. (1971), J. Phys. **B4**, 1747.

Tyrén, F. (1938), Z. Phys. **111**, 314.

Cr XV

Configuration	Term	J	Level (cm^{-1})	Leading percentages		
$2s^2 2p^6$	^1S	0	0			
$2s^2 2p^5 3s$	$^3P^\circ$	2	*4 714 100*	100		
		1	*4 727 500*	50	50	$^1P^\circ$
$2s^2 2p^5 3s$	$^1P^\circ$	1	*4 793 200*	50	50	$^3P^\circ$
$2s^2 2p^5 3d$	$^3P^\circ$	1	*5 259 000*	94		
$2s^2 2p^5 3d$	$^3D^\circ$	1	*5 324 200*	83	12	$^1P^\circ$
$2s^2 2p^5 3d$	$^1P^\circ$	1	*5 406 300*	86		
$2s2p^6 3p$	$^3P^\circ$	1	*5 894 500*	92	8	$^1P^\circ$
$2s2p^6 3p$	$^1P^\circ$	1	*5 921 000*	92	8	$^3P^\circ$
$2s^2 2p^5 4d$	$^3D^\circ$	1	*6 576 000*	53	41	$^1P^\circ$
$2s^2 2p^5 4d$	$^1P^\circ$	1	*6 641 000*	59	34	$^3D^\circ$
$2s^2 2p^5 5d$	$^3D^\circ$	1	*7 148 000*			
$2s^2 2p^5 5d$	$^1P^\circ$	1	*7 215 000*			
$2s^2 2p^5 6d$	$^3D^\circ$	1	*7 452 000*			
$2s^2 2p^5 6d$	$^1P^\circ$	1	*7 524 000*			
Cr XVI ($^2P_{3/2}^\circ$)	***Limit***		8 151 000			

Cr XVI

$Z = 24$

F I isoelectronic sequence

Ground state: $1s^2 2s^2 2p^5\ ^2P^\circ_{3/2}$

Ionization energy = 8 850 000±18 000 cm^{-1} (1097±2 eV)

The ground term splitting was obtained by Fawcett (1971) from his identification of the $2s^22p^5-2s2p^6$ doublet. It was observed directly from a magnetic dipole transition at 1410.6±0.3 Å in a tokamak discharge by Hinnov et al. (1982). The present value is taken in the more accurate measurement of 1410.60±0.02 Å by Peacock, Stamp, and Silver (1984) in a similar light source, giving an uncertainty of 2 cm^{-1} for this interval. The $2s2p^6\ ^2S_{1/2}$ from the measurements of the resonance doublet by Doschek et al. (1974), and has an uncertainty of ±200 cm^{-1}.

The $2s^22p^5-2s^22p^43s$ and $3d$ arrays were first analyzed by Cohen, Feldman and Kastner (1968). This work was revised and extended by Feldman et al. (1973) from whose classified lines we derived the energy levels. Their reported wavelength accuracy is ±0.01 Å. The consequent level uncertainty is ±6000 cm^{-1}.

The $2s2p^6\ ^2S_{1/2}-2s2p^53s\ ^2P^\circ$ multiplet is from Feldman et al. (1973).

Bogdanovich et al. (1980) calculated the percentage compositions of the levels.

The ionization energy was obtained by extrapolation by Lotz (1967).

References

Bogdanovich, P. O., Bogdanovichene, M. I., Grudzinskas, I. I., Rudzikas, Z. B., Tutlis, V. I., and Shadzhjuvene S. D., "Spectroscopy of Multiply Charged Ions", USSR Acad. of Sci., Sci. Counc. on Spectr. (1980).

Cohen, L., Feldman, U., and Kastner, S. O. (1968), J. Opt. Soc. Am. **58**, 331.

Doschek, G. A., Feldman, U., Cowan, R. D., and Cohen, L. (1974), Astrophys. J. **188**, 417.

Fawcett, B. C. (1971), J. Phys. **B4**, 981.

Feldman, U., Doschek, G. A., Cowan, R. D., and Cohen, L. (1973), J. Opt. Soc. Am. **63**, 1445.

Hinnov, E., Suckewer, S., Cohen, S., and Sato, K. (1982), Phys. Rev. **A25**, 2293.

Lotz, W. J. (1967), J. Opt. Soc. Am. **57**, 873.

Peacock, N. J., Stamp, M. F., and Silver, J. D. (1984), Phys. Scr. **T8**, 10.

Cr XVI

Configuration	Term	J	Level (cm^{-1})	Leading percentages		
$2s^2\,2p^5$	$^2P^\circ$	3/2	*0*			
		1/2	*70 892*			
$2s2p^6$	^2S	1/2	937 940			
$2s^2\,2p^4(^3P)3s$	^4P	5/2	5 048 700	94		
		3/2	5 072 300	56	42	$(^3P)\ ^2P$
		1/2	5 109 300	94		
$2s^2\,2p^4(^3P)3s$	^2P	3/2	5 118 200	49	44	$(^3P)\ ^4P$
		1/2	5 143 400	94		
$2s^2\,2p^4(^1D)3s$	^2D	5/2	5 193 500	94		
		3/2	5 196 100	90		
$2s^2\,2p^4(^1S)3s$	^2S	1/2	5 323 500	90		
$2s^2\,2p^4(^3P)3d$	^4P	1/2	5 607 600	72	13	$(^3P)\ ^2P$
		3/2	5 620 600	64	18	$(^3P)\ ^2D$
		5/2	5 640 200	37	34	$(^3P)\ ^4F$
$2s^2\,2p^4(^3P)3d$	^4F	5/2	5 622 700	48	21	$(^3P)\ ^2F$
$2s^2\,2p^4(^3P)3d$	^2P	1/2	5 628 500	42	32	$(^3P)\ ^4D$
		3/2	5 671 200	52	23	$(^1D)\ ^2P$
$2s^2\,2p^4(^3P)3d$	^2D	3/2	5 648 100	29	17	$(^1D)\ ^2D$
		5/2	5 680 800	49	17	

Cr XVI—Continued

Configuration	Term	J	Level (cm^{-1})	Leading percentages		
$2s^2\,2p^4(^3P)3d$	^2F	$5/2$	5 659 000	46	36	$(^3P)\,^4P$
$2s^2\,2p^4(^1D)3d$	^2S	$1/2$	5 734 600	90		
$2s^2\,2p^4(^1D)3d$	^2P	$3/2$	5 756 200	67	24	$(^3P)\,^2P$
$2s^2\,2p^4(^1D)3d$	^2D	$5/2$	5 757 100	56	26	$(^3P)\,^2D$
		$3/2$	5 780 500	66	28	
$2s^2\,2p^4(^1S)3d$	^2D	$5/2$	5 857 200	90		
		$3/2$	5 870 600	83		
$2s2p^5(^3P^\circ)3s$	$^2P^\circ$	$3/2$	*5 950 200*	83		
		$1/2$	*5 986 600*	90		
Cr XVII ($^3P_2$)	***Limit***		8 850 000			

Cr XVII

$Z=24$

O I isoelectronic sequence

Ground state: $1s^2 2s^2 2p^4\ ^3P_2$

Ionization energy = 9 560 000±19 000 cm^{-1} (1185±2 eV)

Several magnetic dipole transitions within the $2s^2 2p^4$ ground configuration were observed by Hinnov et al. (1982) in a tokamak light source with an accuracy of ±0.3 Å. They are $^3P_2-^3P_1$ at 1656.3 Å, $^3P_2-^1D_2$ at 740.8 Å and $^3P_1-^1S_0$ at 493.8 Å. The $^3P_1$, $^1D_2$, and $^1S_0$ levels are derived from these data. Peacock, Stamp, and Silver (1984) obtained the value 740.75±0.03 Å for the $^3P_2-^1D_2$ in a similar light source. The $2s^2 2p^4-2s2p^5$ array was first interpreted by Fawcett (1971). It was reobserved by Lawson and Peacock (1980) who extended the analysis. The $2s2p^5\ ^1P_1^\circ-2p^6\ ^1S_0$ transition was identified by Doschek et al. (1975). Both lines of this array were observed by Lawson and Peacock. Their measurements are used here to derive the levels of $2s2p^5$, $2p^6$ and the $^3P_0$ of $2s^2 2p^4$ with an uncertainty of ±300 cm^{-1}. The percentage composition of these levels was provided by Kaufman and Sugar (1982). Their calculation includes configuration interaction between $2s^2 2p^4$ and $2p^6$.

The transition array $2s^2 2p^4-2s^2 2p^3 3s$ and 18 Å was analyzed by Doschek, Feldman, and Cohen (1973), and the $2s^2 2p^4-2s^2 2p^3 3d$ at 16 Å by Fawcett and Hayes (1975). Both report a measurement uncertainty of ±0.01 Å, resulting in level uncertainties of ±3000 cm^{-1}. Levels of $2p^2 3s$ with question marks were derived from doubly classified lines by Doschek et al. (1973). Some revisions of the $2p^3 3d$ levels due to Bromage and Fawcett (1977) are included.

The two levels of $2p^3 4d$ are from Spector et al. (1980) with an uncertainty of ±3000 cm^{-1}.

The value for the ionization energy was derived by Lotz (1967) by extrapolation.

References

Bromage, G. E. and Fawcett, B. C. (1977), Mon. Not. R. Astron. Soc. **178**, 591.

Doschek, G. A., Feldman, U., and Cohen, L. (1973), J. Opt. Soc. Am. **63**, 1463.

Doschek, G. A. Feldman, U., Davis, J., and Cowan, R. D. (1975), Phys. Rev. **A12**, 980.

Fawcett, B. C. (1971), J. Phys. **B4**, 981.

Fawcett, B. C., and Hayes, R. W. (1975), Mon. Not. R. Astron. Soc. **170**, 185.

Hinnov, E., Suckewer, S., Cohen, S. and Sato, K. (1982), Phys. Rev. **A25**, 2293.

Kaufman, V. and Sugar, J. (1982), private communication.

Lawson, K. D. and Peacock, N. J. (1980), J. Phys. **B13**, 3313.

Lotz, W. J. (1967), J. Opt. Soc. Am. **57**, 873.

Peacock, N. J., Stamp, M. F., and Silver, J. D. (1984), Phys. Scr. **T8**, 10.

Spector, N., Zigler, A., and Zmora, H. (1980), J. Opt. Soc. Am. **70**, 857.

Cr XVII

Configuration	Term	J	Level (cm^{-1})	Leading percentages		
$2s^2 2p^4$	^3P	2	0	93	7	^1D
		0	58 070	86	13	^1S
		1	60 376	100		
$2s^2 2p^4$	^1D	2	134 998	93	7	^3P
$2s^2 2p^4$	^1S	0	262 890	84	14	^3P
$2s2p^5$	$^3P^\circ$	2	*813 540*	100		
		1	*858 120*	98	2	$^1P^\circ$
		0	*887 920*	100		
$2s2p^5$	$^1P^\circ$	1	*1 116 380*	98	2	$^3P^\circ$
$2p^6$	^1S	0	1 886 920	98	2	$2s^2 2p^4\ ^1S$
$2s^2 2p^3(^4S^\circ)3s$	$^3S^\circ$	1	*5 455 000?*			
$2s^2 2p^3(^2D^\circ)3s$	$^3D^\circ$	1	*5 546 800?*			
		2	*5 549 400*			
		3	*5 568 900*			

Cr XVII—Continued

Configuration	Term	J	Level (cm^{-1})	Leading percentages
$2s^2\,2p^3(^2D^\circ)3s$	$^1D^\circ$	2	*5 588 700?*	
$2s^2\,2p^3(^2P^\circ)3s$	$^1P^\circ$	1	*5 700 700*	
$2s^2\,2p^3(^4S^\circ)3d$	$^3D^\circ$	3	*5 948 500*	
$2s^2\,2p^3(^2D^\circ)3d$	$^3D^\circ$	2	*6 070 000*	
		3	*6 074 000*	
$2s^2\,2p^3(^2D^\circ)3d$	$^1F^\circ$	3	*6 124 400*	
$2s^2\,2p^3(^2P^\circ)3d$	$^3P^\circ$	2	*6 131 000*	
$2s^2\,2p^3(^2P^\circ)3d$	$^3D^\circ$	3	*6 164 800*	
		1	*6 189 000*	
		2	*6 214 600*	
$2s^2\,2p^3(^2P^\circ)3d$	$^1F^\circ$	3	*6 338 000*	
$2s^2\,2p^3(^2P^\circ)4d$	$^1D^\circ$	2	*7 882 000*	
$2s^2\,2p^3(^2P^\circ)4d$	$^3F^\circ$	3	*7 960 000*	
Cr XVIII ($^4S^\circ_{3/2}$)	***Limit***		**9 560 000**	

Cr XVIII

$Z=27$

N I isoelectronic sequence

Ground state: $1s^2 2s^2 2p^3\ ^4S^\circ_{3/2}$

Ionization energy $= 10\,480\,000 \pm 21\,000$ cm^{-1} (1299 ± 3 eV)

Four lines of the $2s^2 2p^3\ ^2D^\circ - 2s2p^4(^2P, ^2D)$ multiplets were classified by Fawcett (1971). The analysis of this transition array was extended and partly revised by Doschek, Feldman, Cowan, and Cohen (1974). The $^2S_{1/2}$ of $2s2p^4$ was later reported by Feldman, Doschek, Cowan, and Cohen (1975). Fawcett and Hayes (1975) classified the resonance line arising from $2s^2 2p^2(^3P)3d\ ^4P_{5/2}$ and observed the $2s^2 2p^2(^1D)3d\ ^2F_{7/2}$ level of the doublet system. They also located the $2p^5$ configuration from transitions to $2s2p^4\ ^2D$. The $^2D_{3/2} - ^2P^\circ_{3/2}$ line of this multiplet was reported by Doschek, Feldman, Davis, and Cowan (1975).

The $2s^2 2p^3 - 2s2p^4$ and $2s2p^4 - 2p^5$ arrays were remeasured by Lawson and Peacock (1980) who found four intersystem lines. We used their wavelengths, accurate to ± 0.03 Å, to determine the energy levels with an uncertainty of ± 200 cm^{-1}. Magnetic dipole transitions within the ground configuration were observed in a tokamak plasma by Hinnov, Suckewer, Cohen, and Sato (1982). They reported privately the wavelengths 4038.6 Å for $^2D^\circ_{3/2} - ^2D^\circ_{5/2}$ and 2606.4 Å for $^2P^\circ_{1/2} - ^2P^\circ_{3/2}$, both in air with uncertainties of ± 0.3 Å. With these measurements and their published wavelength of 793.3 Å for the $^4S_{3/2} - ^2D_{3/2}$ transition, we obtained the position of the ^2D term with an uncertainty of ± 50 cm^{-1}, and the fine structure of the ^2P term with an uncertainty of ± 4 cm^{-1}.

The percentage compositions for the $2s^2 2p^3$, $2s2p^4$, and $2p^5$ configurations were provided by Kaufman and Sugar (1982). The calculation included configuration interaction between $2s^2 2p^3$ and $2p^5$.

The ionization energy was obtained by Lotz (1967) by extrapolation.

References

Doschek, G. A., Feldman, U., Cowan, R. D., and Cohen, L. (1974), Astrophys. J. **188**, 417.

Doschek, G. A., Feldman, U., Davis, J., and Cowan, R. D. (1975), Phys. Rev. **A12**, 980.

Fawcett, B. C. (1971), J. Phys. **B4**, 981.

Fawcett, B. C., and Hayes, R. W. (1975), Mon. Not. R. Astron. Soc. **170**, 185.

Feldman, U., Doschek, G. A., Cowan, R. D., and Cohen, L. (1975), Astrophys. J. **196**, 613.

Hinnov, E., Suckewer, S., Cohen, S., and Sato, K. (1982), Phys. Rev. **A25**, 2293.

Kaufman, V. and Sugar, J. (1982), private communication.

Lawson, K. D. and Peacock, N. J. (1980), J. Phys. **B13**, 3313.

Lotz, W. (1967), J. Opt. Soc. Am. **57**, 873.

Cr XVIII

Configuration	Term	J	Level (cm^{-1})	Leading percentages		
$2s^2 2p^3$	$^4S^\circ$	3/2	*0*	93	6	$^2P^\circ$
$2s^2 2p^3$	$^2D^\circ$	3/2	*126 060*	80	17	$^2P^\circ$
		5/2	*150 814*	100		
$2s^2 2p^3$	$^2P^\circ$	1/2	*226 100*	98	2	$2p^5\ ^2P^\circ$
		3/2	*264 456*	76	20	$2s^2 2p^3\ ^2D^\circ$
$2s2p^4$	^4P	5/2	667 560	98	2	^2D
		3/2	714 950	99	1	^2D
		1/2	732 490	97	3	^2S
$2s2p^4$	^2D	3/2	922 800	96	3	^2P
		5/2	931 420	98	2	^4P
$2s2p^4$	^2S	1/2	1 062 040	77	21	^2P
$2s2p^4$	^2P	3/2	1 103 370	96	3	^2D
		1/2	1 170 200	79	20	^2S
$2p^5$	$^2P^\circ$	3/2	*1 738 700*	98	2	$2s^2 2p^3\ ^2P^\circ$
		1/2	*1 813 490*	98	2	

Cr XVIII—Continued

Configuration	Term	J	Level (cm^{-1})	Leading percentages
$2s^2\,2p^2(^3P)3d$	^4P	$5/2$	6 443 000	
$2s^2\,2p^2(^1D)3d$	^2F	$7/2$	6 555 000	
Cr XIX ($^3P_0$)	***Limit***		**10 480 000**	

Cr XIX

$Z = 24$

C I isoelectronic sequence

Ground state: $1s^2 2s^2 2p^2\ ^3P_0$

Ionization energy = 11 260 000±22 000 cm^{-1} (1396±3 eV)

Transitions between the $2s^2 2p^2$ and $2s2p^3$ configurations were identified by Feldman et al. (1975). The spectrum was reobserved by Lawson and Peacock (1980) in the range of 95–203 Å with an accuracy of ±0.03 Å. They identified many more lines of this array, including several intersystem lines, and classified the $2s2p^3-2p^4$ array as well.

The levels of the $2s^2 2p^2$ ground configuration are determined from the magnetic dipole transitions observed in a tokamak plasma by Hinnov, Suckewer, Cohen, and Sato (1982). Their wavelengths, ranging from 398–2885 Å, have an uncertainty of ±0.3 Å. The uncertainty in the ^{3}P and ^{1}D levels is ±10, and for the ^{1}S ±100 cm^{-1}. With the exception of the $2s2p^3\ ^5S^\circ_2$ level, the rest of the levels are due to Lawson and Peacock with an uncertainty of ±200 cm^{-1}. Edlén (1984) has compared the known values of the $^5S^\circ_2$ level in the isoelectronic sequence with theoretical predictions. He concluded that the values given by Lawson and Peacock are inconsistent with the trend. We give Edlén's predicted value in brackets. The percentage compositions were provided by Kaufman and Sugar (1982). The calculation includes configuration interaction between $2s^2 2p^2$ and $2p^4$.

Bromage and Fawcett (1977) have given predicted wavelengths of the $2s^2 2p^2-2s^2 2p3d$ array.

The value for the ionization energy was obtained by extrapolation by Lotz (1967).

References

Bromage, G. E. and Fawcett, B. C. (1977), Mon. Not. R. Astron. Soc. **178**, 605.

Edlén, B. (1984), private communication.

Feldman, U., Doschek, G. A., Cowan, R. D., and Cohen, L. (1975), Astrophys. J. **196**, 613.

Hinnov, E., Suckewer, S., Cohen, S., and Sato, K. (1982), Phys. Rev. **A25**, 2293.

Kaufman, V. and Sugar, J. (1982), private communication.

Lawson, K. D. and Peacock, B. C. (1980), J. Phys. **B13**, 3313.

Lotz, W. (1967), J. Opt. Soc. Am. **57**, 873.

Cr XIX

Configuration	Term	J	Level (cm^{-1})	Leading percentages		
$2s^2\,2p^2$	^{3}P	0	0	92	7	^{1}S
		1	47 811	99	1	$2p^4\ ^3$P
		2	82 458	83	16	$2s^2 2p^2\ ^1$D
$2s^2\,2p^2$	^{1}D	2	184 597	83	16	^{3}P
$2s^2\,2p^2$	^{1}S	0	298 800	89	9	^{3}P
$2s2p^3$	^{5}S°	2	[*403 270*]	98	2	^{3}P°
$2s2p^3$	^{3}D°	2	*671 630*	88	12	^{3}P°
		1	*672 770*	90	8	
		3	*686 830*	100		
$2s2p^3$	^{3}P°	0	*789 160*	100		
		1	*794 130*	89	8	^{3}D°
		2	*804 750*	82	12	
$2s2p^3$	^{3}S°	1	*959 880*	86	12	^{1}P°
$2s2p^3$	^{1}D°	2	*976 220*	96	4	^{3}P°
$2s2p^3$	^{1}P°	1	*1 090 660*	87	12	^{3}S
$2p^4$	^{3}P	2	1 450 230	91	8	$2p^4\ ^1$D
		1	1 514 320	99	1	$2s^2 2p^2\ ^3$P
		0	1 517 990	90	8	$2p^4\ ^1$S

Cr XIX—Continued

Configuration	Term	J	Level (cm^{-1})	Leading percentages		
$2p^4$	^{1}D	2	1 586 230	91	8	^{3}P
$2p^4$	^{1}S	0	1 787 180	88	9	^{3}P
Cr XX (^{2}P$^\circ_{1/2}$)	***Limit***		**11 260 000**			

Cr XX

B I isoelectronic sequence

Ground state: $1s^2 2s^2 2p\ ^2P^\circ_{1/2}$

Ionization energy = 12 070 000±24 000 cm^{-1} (1496±3 eV)

The splitting of the ground $^2P^\circ$ term was determined with an uncertainty of 20 cm^{-1} from a magnetic dipole transition at 1205.9±0.3 Å observed in a tokamak plasma by Hinnov et al. (1982).

The transition arrays $2s^2 2p - 2s 2p^2$ and $2s 2p^2 - 2p^3$ were classified by Lawson and Peacock (1980). They observed the spectrum in the range of 116–271 Å with an accuracy of ±0.03 Å. The level values have an uncertainty of ±200 cm^{-1}. The percentage composition for these levels was provided by Kaufman and Sugar (1982). Their calculation includes configuration interaction between $2s^2 2p$ and $2p^3$.

The higher lying levels are from line identifications by Spector, Zigler, Zmora, and Schwob (1980) from laser-produced plasmas observed in the 10–14 Å range. The uncertainty in their values is ±3000 cm^{-1}.

The value for the ionization energy was obtained by extrapolation by Lotz (1967).

References

Hinnov, E., Suckewer, S., Cohen, S., and Sato, K. (1982), Phys. Rev. **A25**, 2293.

Kaufman, V. and Sugar, J. (1982) private communication.

Lawson, K. D. and Peacock, N. J. (1980), J. Phys. **B13**, 3313.

Lotz, W. (1967), J. Opt. Soc. Am. **57**, 873.

Spector, N., Zigler, A., Zmora, H., and Schwob, J. L. (1980), J. Opt. Soc. Am. **70**, 857.

Cr XX

Configuration	Term	J	Level (cm^{-1})	Leading percentages		
$2s^2 2p$	$^2P^\circ$	1/2	*0*	98	2	$2p^3\ ^2P^\circ$
		3/2	*82 926*	97	3	
$2s2p^2$	^4P	1/2	354 570	98	2	^2S
		3/2	391 360	100		
		5/2	430 650	97	3	^2D
$2s2p^2$	^2D	3/2	640 980	97	3	^2P
		5/2	652 990	97	3	^4P
$2s2p^2$	^2P	1/2	760 400	55	43	^2S
		3/2	861 660	97	3	^2D
$2s2p^2$	^2S	1/2	847 750	55	45	^2P
$2p^3$	$^4S^\circ$	3/2	*1 101 840*	95	4	$^2P^\circ$
$2p^3$	$^2D^\circ$	3/2	*1 229 660*	88	10	$^2P^\circ$
		5/2	*1 248 380*	100		
$2p^3$	$^2P^\circ$	1/2	*1 380 270*	98	2	$2s^2 2p\ ^2P^\circ$
		3/2	*1 414 590*	84	11	$^2D^\circ$
$2s2p(^3P^\circ)3d$	$^4F^\circ$	7/2	*7 443 000*			
$2s^2 4s$	^2S	1/2	9 145 000			
$2s^2 4d$	^2D	5/2	9 308 000			
		3/2	9 335 000			
Cr XXI ($^1S_0$)	*Limit*		**12 070 000**			

Cr XXI

$Z=24$

Be I isoelectronic sequence

Ground state: $1s^2 2s^2\ ^1S_0$

Ionization energy = 13 180 000±26 000 cm^{-1} (1634±3 eV)

Widing (1975) identified the $2s^2\ ^1S_0-2s2p\ ^3P^\circ_1$ intersystem line in a solar flare spectrum at 293.11±0.03 Å. The $^1S_0-^1P_1$ transition was found in a tokamak plasma at 149.90±0.03 Å by Hinnov (1979). The uncertainty of the $^3P_1$ is ±50 cm^{-1}, and of the $^1P_1$ ±150 cm^{-1}. The transition array $2s2p-2p^2$ was observed by Lawson and Peacock (1980) in a laser-generated plasma. They obtained a level uncertainty of ±200 cm^{-1}. They identified the intersystem line $2s2p\ ^3P^\circ_2-2p^2\ ^1D_2$, which confirms the solar identification of Widing.

The higher configurations are from the line classifications of Boiko et al. (1977) at 13 Å. A measurement uncertainty of ±0.003 Å is reported, giving a level uncertainty of ±2000 cm^{-1}. The two terms of $1s2s2p^2$ above the limit are from the observations of Boiko et al. (1978) at 2 Å, with a level uncertainty of ±10 000 cm^{-1}.

The ionization energy was obtained by extrapolation by Lotz (1967).

References

Boiko, V. A., Pikuz, S. A., Safronova, U. I., and Faenov, A. Y. (1977), J. Phys. **B10**, 1253.

Boiko, V. A., Chugunov, A. Y., Ivanova, T. G., Faenov, A. Y., Holin, I. V., Pikuz, S. A., Urnov, A. M., Vainstein, L. A., and Safronova, U. I. (1978), Mon. Not. R. Astron. Soc. **185**, 305.

Hinnov, E. (1979), Astrophys. J. **230**, L97.

Lawson, K. D. and Peacock, N. J. (1980), J. Phys. **B13**, 3313.

Lotz, W. (1967), J. Opt. Soc. Am. **57**, 873.

Widing, K. G. (1975), Astrophys. J. **197**, L33.

Cr XXI

Configuration	Term	J	Level (cm^{-1})	Configuration	Term	J	Level (cm^{-1})
$2s^2$	^1S	0	0	$2p3p$	^3D	3	8 087 000
$2s2p$	$^3P^\circ$	0	*318 080*	$2p3p$	^1P	1	8 022 000
		1	*341 170*				
		2	*405 070*	$2p3p$	^3P	2	8 109 000
$2s2p$	$^1P^\circ$	1	*667 110*	$2p3d$	$^3D^\circ$	2	*8 121 000*
						1	*8 134 000*
$2p^2$	^3P	0	864 780			3	*8 204 000*
		1	911 130				
		2	947 130	$2p3d$	$^3P^\circ$	1,2	*8 219 000*
$2p^2$	^1D	2	1 051 810	$2p3d$	$^1P^\circ$	1	*8 275 000*
$2p^2$	^1S	0	1 254 790	$2p3d$	$^1F^\circ$	3	*8 275 000*
$2s3s$	^3S	1	7 463 000	Cr XXII ($^2S_{1/2}$)	***Limit***		**13 180 000**
$2s3p$	$^1P^\circ$	1	*7 620 000*	$1s2s2p^2$	^3D	2	45 550 000
$2s3d$	^3D	2	7 721 000	$1s2s2p^2$	^1D	2	45 800 000
		3	7 733 000				

Cr XXII

$Z=24$

Li I isoelectronic sequence

Ground state: $1s^2 2s\ ^2S_{1/2}$

Ionization energy = 13 882 000±2900 cm^{-1} (1721.4±0.4 eV)

The resonance lines $2s\ ^2S-2p\ ^2P^\circ_{1/2}$, $^2P^\circ_{3/2}$ were observed in the solar corona at 222.99 Å and 279.69 Å by Sandlin, Brueckner, Scherrer, and Tousey (1976). We use the laboratory measurements of Lawson and Peacock (1980) with an uncertainty of ±0.03 Å, giving a level uncertainty of ±50 cm^{-1}. Hinnov (1979) reported observing these lines in a tokamak plasma.

The $2s-3p$ and $2p-3s$, $3d$ transitions were reported by Goldsmith, Feldman, Oren, and Cohen (1972). These series were remeasured and extended by Boiko, Faenov, and Pikuz (1978) to $4p$ and $4d$ by measurements in the range of 9–13 Å with an uncertainty of ±0.003 Å. The uncertainty of these high-lying levels is ±3000 cm^{-1}.

The ionization energy was determined by Edlén (1979) from the $2p-nd$ series.

References

Boiko, V. A., Faenov, A. Y., and Pikuz, S. A. (1978), J. Quant. Spectr. Rad. Trans. **19**, 11.

Edlén, B. (1979), Phys. Scr. **19**, 255.

Goldsmith, S., Feldman, U., Oren, L., and Cohen, L. (1972), Astrophys. J. **174**, 209.

Hinnov, E. (1979), Astrophys. J. **230**, L197.

Lawson, K. D. and Peacock, N. J. (1980), J. Phys. **B13**, 3313.

Sandlin, G. D., Brueckner, G. E., Scherrer, V. E., and Tousey, R. (1976), Astrophys. J. **205**, L47.

Cr XXII

Configuration	Term	J	Level (cm^{-1})	Configuration	Term	J	Level (cm^{-1})
$2s$	^2S	1/2	0	$3d$	^2D	3/2	7 963 000
						5/2	7 972 000
$2p$	$^2P^\circ$	1/2	*357 470*				
		3/2	*448 470*	$4p$	$^2P^\circ$	1/2,3/2	*10 534 000*
$3s$	^2S	1/2	7 826 000	$4d$	^2D	3/2	10 552 000
						5/2	10 585 000
$3p$	$^2P^\circ$	1/2	*7 896 000*				
		3/2	*7 922 000*	Cr XXIII ($^1S_0$)	*Limit*		13 882 000

Cr XXIII

$Z=24$

He I isoelectronic sequence

Ground state: $1s^2\,{}^1S_0$

Ionization energy $=$ 60 344 000$\pm$12 000 cm^{-1} (7481.8$\pm$0.6 eV)

Because of the excellent agreement of the calculated energies of the $n=2$ shell by Safronova (1981) with the few well-measured spectra in the He I sequence, we have compliled her results for the $n=2$ levels and for the ionization energy. Detailed comparisons are given in the Introduction. Levels of the $n=3$–5 shells are from the calculated binding energies by Ermolaev and Jones (1974) subtracted from Safronova's value for the binding energy of the ground state. We have assumed an uncertainty of 2 parts in 10^4 for the excited levels relative to the ground state, and for the ionization energy (see Introduction). For differences between excited levels where $\Delta n=0$, we assumed an uncertainty of 2 parts in 10^3.

Observations by Neupert (1971) of a solar flare spectrum place the $1s2p\,{}^3P^\circ_1$ level at 45 540 000 cm^{-1} and the $1s2p\,{}^1P^\circ_1$ at 45 890 000 cm^{-1} with an estimated uncertainty of $\pm$60000 cm^{-1}.

Percentage compositions are from Ermolaev and Jones.

References

Ermolaev, A. M., and Jones, M. (1974), J. Phys. **B7**, 199.
Neupert, W. (1971), Sol. Phys. **18**, 474.
Safronova, U. I., (1981), Phys. Scr. **23**, 241.

Cr XXIII

Configuration	Term	J	Level (cm^{-1})	Leading percentages		
$1s^2$	^1S	0	0			
$1s2s$	^3S	1	[45 383 500]			
$1s2p$	$^3P^\circ$	0	[*45 595 080*]			
		1	[*45 610 130*]	94	6	$^1P^\circ$
		2	[*45 691 370*]			
$1s2s$	^1S	0	[45 614 430]			
$1s2p$	$^1P^\circ$	1	[*45 827 110*]	94	6	$^3P^\circ$
$1s3s$	^3S	1	[53 759 820]			
$1s3p$	$^3P^\circ$	0	[*53 818 410*]			
		1	[*53 822 120*]	93	7	$^1P^\circ$
		2	[*53 846 400*]			
$1s3s$	^1S	0	[53 820 510]			
$1s3p$	$^1P^\circ$	1	[*53 883 700*]	93	7	$^3P^\circ$
$1s4s$	^3S	1	[56 658 460]			
$1s4p$	$^3P^\circ$	0	[*56 682 840*]			
		1	[*56 684 380*]	93	7	$^1P^\circ$
		2	[*56 694 650*]			
$1s4s$	^1S	0	[56 683 050]			
$1s4p$	$^1P^\circ$	1	[*56 709 860*]	93	7	$^3P^\circ$
$1s5s$	^3S	1	[57 992 520]			

(Continued)

Cr XXIII—Continued

Configuration	Term	J	Level (cm^{-1})	Leading percentages		
1s5s	^1S	0	[58 004 800]			
1s5p	$^3P^\circ$	0	[*58 004 840*]			
		1	[*58 005 640*]	92	8	$^1P^\circ$
		2	[*58 010 900*]			
1s5p	$^1P^\circ$	1	[*58 018 580*]	92	8	$^3P^\circ$
Cr XXIV ($^2S_{1/2}$)	***Limit***		**60 344 000**			

Cr XXIV

$Z=24$

H I isoelectronic sequence

Ground state: $1s\ ^2S_{1/2}$

Ionization energy = 63 675 900±200 cm^{-1} (7894.87±0.02 eV)

Swartz, Kastner, Rothe, and Neupert (1971) identified the $1s-2p$ unresolved pair of lines in a solar flare spectrum at 2.08 Å.

We give calculated values by Mohr (1983) for the $n=2$ shell and by Erickson (1977) for $n=3$–5 relative to the $2p\ ^2P^\circ_{3/2}$ level. Further details are given in the Introduction. Relative to the ground state, the level uncertainty is estimated to be 5 parts in 10^7. The uncertainty in the excited states relative to $2p\ ^2P^\circ_{3/2}$ is 1 part in 10^6.

References

Erickson, G. W. (1977), J. Phys. Chem. Ref. Data **6**, 831.
Mohr, P. J. (1983), At. Data Nucl. Data Tables **29**, 453.
Swartz, M., Kastner, S., Rothe, E., and Neupert, W. (1971), J. Phys. **B4**, 1747.

Cr XXIV

Configuration	Term	J	Level (cm^{-1})
$1s$	^2S	1/2	0
$2p$	$^2P^\circ$	1/2	*[47 719 790]*
		3/2	*[47 843 596]*
$2s$	^2S	1/2	[47 723 241]
$3p$	$^2P^\circ$	1/2	*[56 598 064]*
		3/2	*[56 634 766]*
$3s$	^2S	1/2	[56 599 135]
$3d$	^2D	3/2	[56 634 700]
		5/2	[56 646 759]
$4p$	$^2P^\circ$	1/2	*[59 700 397]*
		3/2	*[59 715 870]*
$4s$	^2S	1/2	[59 700 850]
$4d$	^2D	3/2	[59 715 842]
		5/2	[59 720 933]
$4f$	$^2F^\circ$	5/2	*[59 720 924]*
		7/2	*[59 723 460]*
$5p$	$^2P^\circ$	1/2	*[61 134 179]*
		3/2	*[61 142 095]*
$5s$	^2S	1/2	[61 134 411]
$5d$	^2D	3/2	[61 142 081]
		5/2	[61 144 687]
$5f$	$^2F^\circ$	5/2	*[61 144 683]*
		7/2	*[61 145 982]*
$5g$	^2G	7/2	[61 145 979]
		9/2	[61 146 758]
	Limit		**63 675 900**

Mn I

$Z=25$

Ground state: $1s^2 2s^2 2p^6 3s^2 3p^6 3d^5 4s^2\ ^6S_{5/2}$

Ionization energy = 59 959.4±0.1 cm^{-1} (7.43408±0.00002 eV)

The compact term structure characteristic of manganese atoms and ions produces obvious repeated patterns in manganese spectra which were first interpreted by Catalán in 1922. He called them multiplets and his work led to the general recognition of multiplet structure in complex spectra.

The analysis of Mn I initiated by Catalán was carried on by him and his associates over a period of 40 years. A final report of this effort was published by Catalán, Meggers, and Garcia-Riquelme (1964). The levels and g-values are reported here from that paper. From the differences of observed to calculated wavenumbers we estimate the level value uncertainty to be ±0.1 cm^{-1}. The g-value for the ground state was measured by Childs and Goodman (1965).

An absorption spectrum between 1305 Å and 2040 Å was observed by Brown and Ginter (1978). They identified two series converging to the $3d^5 4s\ ^7S_3$ state of Mn II: an intense series whose resolved lower members (n<15) are triplets was interpreted as $3d^5 4s^2\ ^6S_{5/2}-3d^5 4s(^7S)np\ ^6P^\circ_{3/2,5/2,7/2}$, for $n=6$ to 41, and a weak series with two resolved components in lower members ($n<10$) interpreted as $3d^5 4s^2\ ^6S_{5/2}-3d^5 4s(^7S)np\ ^8P^\circ_{5/2,7/2}$ for $n=6$ to 22. They estimate the uncertainty in these level values to be ±0.05 cm^{-1}. On the basis of the first of these they derived the value for the ionization energy quoted here. An intense series of Beutler-Fano profiles was interpreted as the Rydberg series $3d^5 4s^2\ ^6S_{5/2}-3d^5 4s(^5S)np\ ^6P^\circ$ through $n=55$. Two wavelengths are given for nearly all terms, representing the absorption and emission-like features of each profile. We used the average to represent the energy level position of the series members. Their estimated uncertainty is ±0.5 cm^{-1}.

Brown and Ginter also identified $3d^6 np$ series converging to the $3d^6\ ^5D$ levels of Mn II. They are labeled as to the parent ^5D level, but term names could not be assigned. These and the large number of unlabeled resonances are not tabulated here.

The percentage compositions for the levels of the $3d^5 4s^2$, $3d^6 4s$, and $3d^7$ configurations are taken from the fitted calculation of Dembczynski (1980), which include configuration interaction. The percentages for the odd configurations $3d^6 4p$, $3d^5 4s4p$, and $3d^4 4s^2 4p$ with configuration interaction were calculated by Roth (1980). He gave repeating terms of $3d^n$ alphabetic prefixes and assigned percentages equal to the sum of states differing in seniority only.

The alphabetic prefixing of final terms by Catalán et al., is retained except where the designations were revised by Roth.

References

Brown, C. M., and Ginter, M. I. (1978), J. Opt. Soc. Am. **68**, 1541.
Catalán, M. A., (1922), Phil. Trans. Roy. Soc. London, **A223**, 127.
Catalán, M. A., Meggers, W. F., and Garcia-Riquelme, O. (1964), J. Res. Natl. Bur. Stand. **68A**, 9.
Childs, W. J., and Goodman, L. S. (1965), Phys. Rev. **140**, 448.
Dembczynski, J. (1980), Physica **100C**, 105.
Roth, C. (1980), At. Data Nucl. Data Tables **25**, 91.

Mn I

Configuration	Term	J	Level (cm^{-1})	g	Leading percentages		
$3d^5 4s^2$	$a\ ^6S$	5/2	0.00	2.00152	100		
$3d^6(^5D)4s$	$a\ ^6D$	9/2	17 052.29	1.559	100		
		7/2	17 282.00	1.584	100		
		5/2	17 451.52	1.657	100		
		3/2	17 568.48	1.866	100		
		1/2	17 637.15	3.327	100		
$3d^5(^6S)4s4p(^3P^\circ)$	$z\ ^8P^\circ$	5/2	*18 402.46*	2.284	100		
		7/2	*18 531.64*	1.938	100		
		9/2	*18 705.37*	1.779	100		
$3d^6(^5D)4s$	$a\ ^4D$	7/2	23 296.67	1.427	98	2	$3d^5 4s^2\ ^4D$
		5/2	23 549.20	1.368	98	2	
		3/2	23 719.52	1.198	98	2	
		1/2	23 818.87	0.000	98	2	

Mn I—Continued

Configuration	Term	J	Level (cm^{-1})	g	Leading percentages		
$3d^5(^6S)4s4p(^3P^\circ)$	$z\ ^6P^\circ$	3/2	*24 779.32*	2.364	97		
		5/2	*24 788.05*	1.875	98		
		7/2	*24 802.25*	1.714	98		
$3d^5\ 4s^2$	$a\ ^4G$	11/2	25 265.74	1.270	100		
		5/2	25 281.04		100		
		9/2	25 285.43	1.173	100		
		7/2	25 287.74		100		
$3d^5\ 4s^2$	$a\ ^4P$	5/2	27 201.54	1.597	96	2	$3d^6(^3P2)4s\ ^4P$
		3/2	27 248.00	1.730	96	2	
		1/2	27 281.85	2.666	96	2	
$3d^5\ 4s^2$	$b\ ^4D$	7/2	30 354.21	1.425	96	2	$3d^6(^5D)4s\ ^4D$
		1/2	30 411.74	0.111	96	2	
		5/2	30 419.61	1.38	96	2	
		3/2	30 425.71		96	2	
$3d^5(^6S)4s4p(^3P^\circ)$	$z\ ^4P^\circ$	5/2	*31 001.15*	1.60	97		
		3/2	*31 076.42*	1.732	98		
		1/2	*31 124.95*	2.668	98		
$3d^6(^3P2)4s$	$b\ ^4P$	5/2	33 825.49	1.602	64	32	$(^3P1)\ ^4P$
		3/2	34 463.37	1.730	64	32	
		1/2	34 845.26	2.655	64	32	
$3d^6(^3H)4s$	$a\ ^4H$	13/2	34 138.88	1.231	100		
		11/2	34 250.52	1.135	100		
		9/2	34 343.90	0.971	100		
		7/2	34 423.27	0.665	100		
$3d^6(^3F2)4s$	$a\ ^4F$	9/2	34 938.70	1.328	77	18	$(^3F1)\ ^4F$
		7/2	35 041.37	1.238	79	18	
		5/2	35 114.98	1.024	79	18	
		3/2	35 165.05	0.430	79	17	
$3d^5(^6S)4s4p(^1P^\circ)$	$y\ ^6P^\circ$	3/2	*35 689.98*	2.400	87	11	$3d^6(^5D)4p\ ^6P^\circ$
		5/2	*35 725.85*	1.886	88	10	
		7/2	*35 769.97*	1.712	88	11	
$3d^5\ 4s^2$	$a\ ^2I$	11/2	37 148.66	0.94	98	1	^2H
		13/2	37 164.25		100		
$3d^6(^3G)4s$	$b\ ^4G$	11/2	37 420.24	1.263	88	10	$(^3H)\ ^2H$
		9/2	37 630.62	1.163	94	3	$(^3H)\ ^2H$
		7/2	37 737.22	0.989	96	2	$(^3F2)\ ^2F$
		5/2	37 789.93	0.59	96	2	$(^3F2)\ ^2F$
$3d^6(^3P2)4s$	$a\ ^2P$	3/2	37 586.03		64	33	$(^3P1)\ ^2P$
		1/2	38 351.78	0.675	65	33	
$3d^6(^3H)4s$	$a\ ^2H$	11/2	38 008.70	1.098	88	10	$(^3G)\ ^4G$
		9/2	38 120.18	0.914	94	3	
$3d^6(^3F2)4s$	$a\ ^2F$	7/2	38 669.60	1.128	74	16	$(^3F1)\ ^2F$
		5/2	38 934.94		72	15	
$3d^5\ 4s(^7S)5s$	$e\ ^8S$	7/2	39 431.31	2.000			

(Continued)

Mn I—Continued

Configuration	Term	J	Level (cm^{-1})	g	Leading percentages		
$3d^6(^3G)4s$	$a\ ^2G$	9/2	41 031.48	1.118	92	5	$3d^54s^2\ ^2G2$
		7/2	41 230.30	0.88	92	6	$(^3F2)\ ^2F$
$3d^5\ 4s(^7S)5s$	$e\ ^6S$	5/2	41 403.93	1.997			
$3d^6(^5D)4p$	$z\ ^6D°$	9/2	*41 789.48*	1.556	94	4	$3d^5(^4D)4s4p(^3P°)\ ^6D°$
		7/2	*41 932.64*	1.587	94	5	
		5/2	*42 053.73*	1.653	94	4	
		3/2	*42 143.57*	1.867	95	4	
		1/2	*42 198.56*	3.317	94	5	
$3d^6(^1I)4s$	$b\ ^2I$	13/2	43 053.30	1.07	100		
		11/2	43 139.27	0.924	100		
$3d^6(^5D)4p$	$z\ ^6F°$	11/2	*43 314.23*	1.464	82	16	$3d^5(^4G)4s4p(^3P°)\ ^6F$
		9/2	*43 428.58*	1.431	80	17	
		7/2	*43 524.08*	1.395	81	16	
		5/2	*43 595.50*	1.310	80	17	
		3/2	*43 644.45*	1.068	81	16	
		1/2	*43 672.66*	-0.602	80	16	
$3d^6(^5D)4p$	$z\ ^4F°$	9/2	*44 288.76*	1.317	90	6	$3d^5(^4G)4s4p(^3P°)\ ^4F°$
		7/2	*44 523.45*	1.240	88	6	
		5/2	*44 696.29*	1.030	89	7	
		3/2	*44 814.73*	0.400	90	7	
$3d^6(^5D)4p$	$x\ ^6P°$	7/2	*44 993.92*	1.717	64	20	$3d^5(^4P)4s4p(^3P°)\ ^6P°$
		5/2	*45 156.11*	1.885	67	21	
		3/2	*45 259.17*	2.399	68	20	
$3d^6(^5D)4p$	$z\ ^4D°$	7/2	*45 754.27*	1.427	81	9	$3d^6(^5D)4p\ ^6P°$
		5/2	*45 940.93*	1.372	92		
		3/2	*46 083.89*	1.200	92		
		1/2	*46 169.93*	0.000	93		
$3d^5\ 4s(^7S)5p$	$y\ ^8P°$	5/2	*45 981.44*				
		7/2	*46 000.77*				
		9/2	*46 026.75*				
$3d^5\ 4s(^7S)4d$	$e\ ^8D$	3/2	46 706.09				
		5/2	46 707.03				
		7/2	46 708.33				
		9/2	46 710.15				
		11/2	46 712.58				
$3d^6(^5D)4p$	$y\ ^4P°$	5/2	*46 901.13*	1.595	92		
		3/2	*47 154.51*	1.732	91		
		1/2	*47 299.29*	2.666	91		
$3d^5\ 4s(^7S)5p$	$w\ ^6P°$	7/2	*47 387.62*	1.713			
		5/2	*47 659.52*	1.952			
		3/2	*47 782.43*	2.666			
$3d^5\ 4s(^7S)4d$	$e\ ^6D$	9/2	47 207.28	1.554			
		7/2	47 212.06	1.581			
		5/2	47 215.61	1.634			
		3/2	47 218.15	1.759			
		1/2	47 219.64	3.934			

Mn I—Continued

Configuration	Term	J	Level (cm^{-1})	g	Leading percentages		
$3d^5(^4P)4s4p(^3P°)$	$y\ ^6D°$	1/2	*47 452.16*	3.174	82	16	$(^4D)(^3P°)\ ^6D°$
		3/2	*47 466.66*		83	15	$(^4D)(^3P°)\ ^6D°$
		5/2	*47 753.99*	1.820	82	16	$(^4D)(^3P°)\ ^6D°$
		7/2	*47 774.52*	1.594	74	13	$(^4D)(^3P°)\ ^6D°$
		9/2	*47 903.80*	1.540	65	17	$(^4G)(^3P°)\ ^6F°$
$3d^5(^4G)4s4p(^3P°)$	$y\ ^6F°$	11/2	*48 021.43*	1.460	81	14	$3d^6(^5D)4p\ ^6F°$
		9/2	*48 168.01*	1.432	61	20	$3d^5(^4P)4s4p(^3P°)\ ^6D°$
		7/2	*48 225.99*	1.043	72	17	$3d^6(^5D)4p\ ^6F°$
		5/2	*48 270.91*	1.319	76	20	$3d^6(^5D)4p\ ^6F°$
		3/2	*48 300.98*	1.068	76	19	$3d^6(^5D)4p\ ^6F°$
		1/2	*48 318.12*	−0.496	75	20	$3d^6(^5D)4p\ ^6F°$
$3d^5\ 4s(^5S)5s$	$f\ ^6S$	5/2	49 415.35	2.00			
$3d^5\ 4s(^5S)5s$	$e\ ^4S$	3/2	49 591.51	1.998			
$3d^5(^4P)4s4p(^3P°)$	$v\ ^6P°$	7/2	*49 888.01*	1.711	62	18	$3d^6(^5D)4p\ ^6P°$
		5/2	*50 012.50*	1.888	65	15	
		3/2	*50 099.03*	2.398	69	19	
$3d^5(^4G)4s4p(^3P°)$	$z\ ^4H°$	7/2	*50 065.46*		97		
		9/2	*50 072.59*		96		
		11/2	*50 081.31*		97		
		13/2	*50 094.60*	1.22	98		
$3d^5\ 4s(^7S)6s$	$f\ ^8S$	7/2	50 157.63	1.995			
$3d^5(^4G)4s4p(^3P°)$	$y\ ^4F°$	9/2	*50 341.30*	1.318	93		
		7/2	*50 359.28*	1.242	92		
		5/2	*50 373.23*	1.03	92		
		3/2	*50 383.27*		93		
$3d^5(^4D)4s4p(^3P°)$	$x\ ^6F°$	1/2	*50 818.64*	−0.62	91	6	$(^4G)(^3P°)\ ^6F°$
		3/2	*50 863.5*	1.07	92		
		5/2	*50 931.29*	1.316	93		
		7/2	*51 014.94*		93		
		9/2	*51 100.49*		93		
		11/2	*51 169.18*		94		
$3d^5\ 4s(^7S)6s$	$g\ ^6S$	5/2	50 904.68				
$3d^5(^4P)4s4p(^3P°)$	$x\ ^4P°$	5/2	*51 305.31*	1.591	75	19	$(^4D)(^3P°)\ ^4P°$
		3/2	*51 445.55*	1.728	72	16	
		1/2	*51 552.78*	2.664	73	14	
$3d^5(^4G)4s4p(^3P°)$	$z\ ^4G°$	5/2	*51 515.63*		93		
		7/2	*51 530.61*		92		
		9/2	*51 546.27*		93		
		11/2	*51 560.93*	1.273	93		
$3d^7$	$e\ ^4P$	5/2	51 638.17	1.601	90	7	$3d^6(^3P1)4s\ ^4P$
		3/2	51 718.22	1.733	86	10	
		1/2	51 787.92	2.65	83	13	
$3d^5(^4D)4s4p(^3P°)$	$u\ ^6P°$	3/2	*52 014.98*		86	8	$(^4P)(^3P°)\ ^6P°$
		5/2	*52 128.58*		69	20	$(^4P)(^3P°)\ ^6D°$
		7/2	*52 253.17*	1.71	60	21	$(^4P)(^3P°)\ ^6D°$

(Continued)

Mn I—Continued

Configuration	Term	J	Level (cm^{-1})	g	Leading percentages		
$3d^5\,4s(^7S)6p$	$^8P°$	5/2	52 490.0				
		7/2	52 497.21				
$3d^5\,4s(^7S)5d$	$f\,^8D$	3/2–7/2	52 702.48?				
		9/2	52 703.1				
		11/2	52 705.23				
$3d^5\,4s(^7S)5d$	$f\,^6D$	9/2	52 726.39				
		7/2	52 730.41				
		5/2	52 733.22				
		3/2	52 735.01				
		1/2	52 735.83				
$3d^5(^4D)4s4p(^3P°)$	$x\,^6D°$	9/2	52 758.11	1.552	81	15	$(^4P)(^3P°)\,^6D°$
		7/2	52 869.99	1.57	69	14	
		1/2	52 883.10		76	16	
		5/2	52 883.79		75	16	
		3/2	52 883.79		74	15	
$3d^5\,4s(^7S)4f$	$z\,^8F°$	1/2–13/2	52 974.5?				
$3d^5\,4s(^7S)4f$	$w\,^6F°$	9/2	52 977.75				
		7/2	52 977.82				
		11/2	52 977.89				
		3/2	52 977.93				
		5/2	52 978.03				
$3d^5(^4P)4s4p(^3P°)$	$y\,^4D°$	1/2	53 101.32		75	15	$(^4P)(^3P°)\,^6D°$
		3/2	53 103.09		80	12	
		5/2	53 109.14		75	12	
		7/2	53 124.00	1.423	77	13	
$3d^5\,4s(^7S)6p$	$t\,^6P°$	7/2	53 261.05				
		5/2	53 291.30				
		3/2	53 311.12				
$3d^5(^4P)4s4p(^3P°)$	$z\,^4S°$	3/2	54 218.62	1.770	70	15	$3d^6(a\,^3P)4p\,^4S°$
$3d^5\,4s(^7S)7s$	$h\,^6S$	5/2	54 460.30				
$3d^5\,4s(^5S)4d$	$g\,^6D$	9/2	54 938.94				
		7/2	54 946.55				
		1/2	54 949.60				
		5/2	54 950.81				
		3/2	54 953.21				
$3d^5(^4D)4s4p(^3P°)$	$x\,^4D°$	7/2	55 107.52	1.407	85	8	$3d^5(^4F)4s4p(^3P°)\,^4D°$
		5/2	55 186.1	1.365	87		
		3/2	55 279.91	0.826	62	21	$3d^5(^4P)4s4p(^3P°)\,^2P°$
$3d^5\,4s(^7S)7p$	$^8P°$	5/2	55 283.98				
		7/2	55 287.59				
$3d^5\,4s(^5S)5p$	$w\,^4P°$	3/2	55 368.66				
		5/2	55 406.0				
		1/2	55 457.20	2.28			

Mn I—Continued

Configuration	Term	J	Level (cm^{-1})	g	Leading percentages
$3d^5 4s(^7S)6d$	$g\ ^8D$	7/2	55 374.76		
		9/2	55 375.70		
		11/2	55 376.70		
$3d^5 4s(^7S)5f$	$v\ ^6F^\circ$	5/2	*55 491.57*		
		3/2	*55 491.95*		
		7/2	*55 492.08*		
		9/2	*55 492.52*		
		11/2	*55 492.74*		
$3d^5 4s(^7S)7p$	$^6P^\circ$	7/2	*55 492.41*		
		5/2	*55 493.51*		
$3d^5 4s(^7S)5f$	$y\ ^8F^\circ$	13/2	*55 498.5*		
		11/2	*55 499.09*		
		9/2	*55 499.09*		
		7/2	*55 499.5*		
		3/2	*55 499.75*		
		5/2	*55 499.90*		
$3d^5 4s(^7S)6d$	$h\ ^6D$	9/2	55 681.90		
		7/2	55 688.10		
		5/2	55 690.80		
		3/2	55 691.90		
		1/2	55 692.40		
$3d^5 4s(^5S)5p$	$^4P^\circ$	5/2	*55 923.98*		
		3/2	*55 939.27*		
$3d^5 4s(^5S)5p$	$s\ ^6P^\circ$	3/2	*55 996.90*		
		5/2	*56 008.18*		
		7/2	*56 012.61*		
$3d^5 4s(^7S)8s$	$g\ ^8S$	7/2	56 144.16		
$3d^6(^5D)5s$	$i\ ^6D$	9/2	56 189.45		
		7/2	56 356.21	1.57	
		5/2	56 490.79		
		3/2	56 567.93		
		1/2	56 666.06		
	$e\ ^4D$	7/2	56 462.08		
		5/2	56 561.95		
		3/2	56 601.63		
$3d^5 4s(^7S)8p$	$^8P^\circ$	7/2	*56 755.58*		
$3d^5 4s(^7S)7d$	$h\ ^8D$	3/2–11/2	56 801.4?		
$3d^5 4s(^7S)6f$	$u\ ^6F^\circ$	1/2–11/2	*56 867.1*		
$3d^5 4s(^7S)8p$	$^6P^\circ$	7/2	*56 926.32*		
		5/2	*56 931.14*		
		3/2	*56 934.17*		
$3d^5 4p^2$	$e\ ^8P$	5/2	57 086.33	2.27	
		7/2	57 218.15		
		9/2	57 388.90	1.767	

(Continued)

Mn I—Continued

Configuration	Term	J	Level (cm^{-1})	g	Leading percentages		
$3d^5(^4D)4s4p(^3P^\circ)$	$v\ ^4P^\circ$	1/2	*57 228.30*	2.671	62	27	$3d^5(^4P)4s4p(^3P^\circ)\ ^4P^\circ$
		3/2	*57 360.73*	1.736	49	17	$3d^6(a\ ^3P)4p\ ^4S^\circ$
		5/2	*57 487.05*	1.590	62	27	$3d^5(^4P)4s4p(^3P^\circ)\ ^4P^\circ$
$3d^6(^5D)5s$	$f\ ^4D$	7/2	57 305.62				
		5/2	57 485.97	1.372			
		3/2	57 621.90				
		1/2	57 705.83				
$3d^6(a\ ^3P)4p$	$y\ ^4S^\circ$	3/2	*57 512.08*	2.000	37	31	$3d^5(^4P)4s4p(^3P^\circ)\ ^4S^\circ$
$3d^5\ 4s(^7S)9p$	$^8P^\circ$	5/2	*57 623.33*				
		7/2	*57 624.65*				
$3d^5\ 4s(^7S)7f$	$t\ ^6F^\circ$	1/2–11/2	*57 697.2*				
$3d^5\ 4s(^7S)9p$	$^6P^\circ$	7/2	*57 716.70*				
		5/2	*57 719.70*				
		3/2	*57 722.04*				
$3d^5(^4G)4s4p(^1P^\circ)$	$y\ ^4G^\circ$	11/2	*58 075.06*	1.269	57	31	$3d^6(^3H)4p\ ^4G^\circ$
		9/2	*58 110.24*	1.168	58	30	
		7/2	*58 136.69*	0.980	59	29	
		5/2	*58 159.73*	0.578	60	28	
$3d^5\ 4s(^7S)10p$	$^8P^\circ$	5/2,7/2	*58 182.24*				
$3d^5\ 4s(^7S)10p$	$^6P^\circ$	7/2	*58 238.80*				
		5/2	*58 240.80*				
		3/2	*58 242.16*				
$3d^6(^3H)4p$	$y\ ^4H^\circ$	13/2	*58 338.67*	1.228	69	16	$3d^5(^4G)4s4p(^3P^\circ)\ ^4H^\circ$
		11/2	*58 427.30*	1.133	54	13	
		9/2	*58 485.52*	0.968	61	12	
		7/2	*58 519.90*	0.665	60	13	
$3d^5\ 4s(^7S)11p$	$^8P^\circ$	5/2,7/2	*58 561.18*				
$3d^5\ 4s(^7S)11p$	$^6P^\circ$	7/2	*58 598.73*				
		5/2	*58 600.33*				
		3/2	*58 600.92*				
$3d^5\ 4s(^7S)12p$	$^8P^\circ$		*58 830.70*				
$3d^6(^3H)4p$	$z\ ^4I^\circ$	13/2	*58 843.39*	1.09	51	35	$3d^5(^2I)4s4p(^3P^\circ)\ ^4I^\circ$
		11/2	*58 851.49*		52	46	
		15/2	*58 852.60*		54	35	
		9/2	*58 866.66*	0.73	51	46	
$3d^5\ 4s(^7S)12p$	$^6P^\circ$	7/2	*58 856.98*				
		5/2	*58 858.04*				
		3/2	*58 858.48*				
$3d^5\ 4s(^7S)13p$	$^8P^\circ$		*59 029.12*				
$3d^5\ 4s(^7S)13p$	$^6P^\circ$	7/2	*59 048.39*				
		3/2,5/2	*59 049.00*				

Mn I—Continued

Configuration	Term	J	Level (cm^{-1})	g	Leading percentages			
$3d^6(a\ ^3F)4p$	$^4D^\circ$	5/2	59 116.64	1.558	41		26	$3d^6(a\ ^3P)4p\ ^4P^\circ$
		7/2	59 339.49	1.362	56		23	$3d^6(a\ ^3P)4p\ ^4D^\circ$
		1/2	59 527.36		47		26	$3d^5(^4P)4s4p(^3P^\circ)\ ^4D^\circ$
		3/2	59 527.89		45		21	$3d^5(^4P)4s4p(^3P^\circ)\ ^4D^\circ$
$3d^5\ 4s(^7S)14p$	$^8P^\circ$		59 179.38					
$3d^5\ 4s(^7S)14p$	$^6P^\circ$	7/2	59 193.78					
		3/2,5/2	59 194.28					
$3d^6(a\ ^3F)4p$	$x\ ^4F^\circ$	9/2	59 257.44	1.327	55		21	$3d^5(^4G)4s4p(^1P^\circ)\ ^4F^\circ$
		7/2	59 290.11	1.325	53		19	
		5/2	59 360.67	1.11	52		22	
		3/2	59 416.15	0.39	49		19	
$3d^5\ 4s(^7S)15p$	$^8P^\circ$		59 296.00					
$3d^5\ 4s(^7S)15p$	$^6P^\circ$		59 307.46					
$3d^6(a\ ^3P)4p$	$u\ ^4P^\circ$	3/2	59 384.24	1.608	52		17	$3d^5(^4D)4s4p(^3P^\circ)\ ^4P^\circ$
		5/2	59 480.80	1.281	43		31	$3d^6(a\ ^3F)4p\ ^4D^\circ$
		1/2	59 568.29	1.94	61		18	$3d^5(^4D)4s4p(^3P^\circ)\ ^4P^\circ$
$3d^5\ 4s(^7S)16p$	$^8P^\circ$		59 388.47					
$3d^5\ 4s(^7S)16p$	$^6P^\circ$		59 397.35					
$3d^5\ 4s(^7S)17p$	$^8P^\circ$		59 462.48					
$3d^6(a\ ^3P)4p$	$^4D^\circ$	3/2	59 989.77	1.194	66		14	$3d^5(a\ ^2F)4s4p(^3P^\circ)\ ^4D^\circ$
		5/2	60 101.65	1.31	63		13	$3d^6(a\ ^3F)4p\ ^4D^\circ$
		1/2	60 141.98	0.17	67		13	$3d^5(a\ ^2F)4s4p(^3P^\circ)\ ^4D^\circ$
$3d^5\ 4s(^7S)17p$	$^6P^\circ$		59 470.31					
$3d^6(a\ ^3P)4p$		5/2	59 480.73	1.281	24	$^2D^\circ$	23	$3d^5(^4D)4s4p(^3P^\circ)\ ^2F^\circ$
$3d^5\ 4s(^7S)18p$	$^8P^\circ$		59 523.45					
$3d^5\ 4s(^7S)18p$	$^6P^\circ$		59 529.21					
$3d^5\ 4s(^7S)19p$	$^8P^\circ$		59 573.79					
$3d^5\ 4s(^7S)19p$	$^6P^\circ$		59 578.37					
$3d^6(a\ ^3P)4p$	$^2D^\circ$	5/2	59 600.35	1.277	48		21	$3d^5(^4D)4s4p(^3P^\circ)\ ^2F^\circ$
		3/2	60 395.64	0.91	53		12	$3d^6(a\ ^3F)4p\ ^4F^\circ$
$3d^5\ 4s(^7S)20p$	$^8P^\circ$		59 615.5					
$3d^6(^3H)4p$	$z\ ^2I^\circ$	13/2	59 617.12	1.074	87		11	$3d^5(^2I)4s4p(^3P^\circ)\ ^2I^\circ$
		11/2	59 827.88	0.93	86		10	
$3d^5\ 4s(^7S)20p$	$^6P^\circ$		59 619.60					
$3d^5\ 4s(^7S)21p$	$^8P^\circ$		59 650.9					

(Continued)

Mn I—Continued

Configuration	Term	J	Level (cm^{-1})	g	Leading percentages		
$3d^6(a\ ^3F)4p$	$x\ ^4G°$	11/2	*59 652.90*	1.239	62	25	$3d^6(^3H)4p\ ^4G°$
		9/2	*59 731.94*	1.169	51	19	$3d^6(a\ ^3F)4p\ ^4F°$
		7/2	*59 784.31*	0.990	52	20	$3d^6(^3H)4p\ ^4G°$
		5/2	*59 817.70*	0.584	55	21	$3d^6(^3H)4p\ ^4G°$
$3d^5\ 4s(^7S)21p$	$^6P°$		*59 654.46*				
$3d^5\ 4s(^7S)22p$	$^8P°$		*59 681.2*				
$3d^5\ 4s(^7S)22p$	$^6P°$		*59 684.25*				
$3d^5\ 4s(^7S)23p$	$^6P°$		*59 709.85*				
$3d^5\ 4s(^7S)24p$	$^6P°$		*59 732.03*				
$3d^5\ 4s(^7S)25p$	$^6P°$		*59 751.41*				
$3d^5\ 4s(^7S)26p$	$^6P°$		*59 768.39*				
$3d^5\ 4s(^7S)27p$	$^6P°$		*59 783.42*				
$3d^5\ 4s(^7S)28p$	$^6P°$		*59 796.72*				
$3d^5\ 4s(^7S)29p$	$^6P°$		*59 808.57*				
$3d^5\ 4s(^7S)30p$	$^6P°$		*59 819.14*				
$3d^5\ 4s(^7S)31p$	$^6P°$		*59 828.61*				
$3d^5\ 4s(^7S)32p$	$^6P°$		*59 837.27*				
$3d^5\ 4s(^7S)33p$	$^6P°$		*59 845.04*				
$3d^5\ 4s(^7S)34p$	$^6P°$		*59 851.99*				
$3d^5\ 4s(^7S)35p$	$^6P°$		*59 858.47*				
$3d^5\ 4s(^7S)36p$	$^6P°$		*59 864.44*				
$3d^5\ 4s(^7S)37p$	$^6P°$		*59 869.65*				
$3d^5\ 4s(^7S)38p$	$^6P°$		*59 874.55*				
$3d^5\ 4s(^7S)39p$	$^6P°$		*59 879.04*				
$3d^5\ 4s(^7S)40p$	$^6P°$		*59 883.26*				
$3d^5\ 4s(^7S)41p$	$^6P°$		*59 887.06*				
Mn II ($^7S_3$)	***Limit***		**59 959.4**				
$3d^6(^3H)4p$	$z\ ^2G°$	9/2	*60 668.49*	1.112	51	18	$3d^6(^3G)4p\ ^2G°$
		7/2	*60 739.42*		47	16	$3d^6(a\ ^3F)4p\ ^2F°$
$3d^5(a\ ^2D)4s4p(^3P°)$	$w\ ^4F°$	3/2	*60 760.87*		60	20	$3d^5(a\ ^2F)4s4p(^3P°)\ ^4F°$
		5/2	*60 820.35*		61	22	
		7/2	*60 902.80*		58	23	
		9/2	*60 938.97*		59	19	
$3d^5(^2I)4s4p(^3P°)$	$x\ ^4H°$	13/2	*60 891.48*	1.228	61	18	$(^4G)(^1P°)\ ^4H°$
		11/2	*60 933.73*	1.134	60	20	
		9/2	*60 955.88*		57	15	
		7/2	*60 957.21*		58	17	
$3d^5(^2I)4s4p(^3P°)$	$y\ ^4I°$	15/2	*61 204.54*	1.20	67	26	$3d^6(^3H)4p\ ^4I°$
		9/2	*61 211.43*	0.75	54	19	
		13/2	*61 225.55*		65	28	
		11/2	*61 225.77*		49	21	
$3d^6(^3G)4p$	$^4G°$	11/2	*61 469.21*	1.164	48	18	$3d^6(a\ ^3F)4p\ ^4G°$
		9/2	*61 714.52*		42	15	
		7/2	*62 034.04*		39	16	
		5/2	*62 075.55*	0.58	35	15	
$3d^6(a\ ^3F)4p$	$^2F°$	5/2	*61 469.7*		65	15	$3d^5(^4D)4s4p(^3P°)\ ^2F°$
		7/2	*61 480.6*	1.13	51	17	$3d^6(a\ ^3F)4p\ ^2G°$

Mn I—Continued

Configuration	Term	J	Level (cm^{-1})	g	Leading percentages		
$3d^6(a\ ^3F)4p$	$^2G°$	9/2	*61 485.34*	1.020	58	12	$3d^6(^3H)4p\ ^2G°$
		7/2	*61 785.94*	0.93	73	11	
$3d^5(^4F)4s4p(^3P°)$	$^6G°$	7/2	*61 710.98*		90		
		5/2	*61 727.28*		52	29	$3d^6(a\ ^3F)4p\ ^2F°$
		11/2	*61 744.04*		92		
$3d^6(^5D)4d$	$e\ ^6F$	11/2	61 713.62				
		9/2	62 030.18				
		7/2	62 294.66				
		5/2	62 905.81				
		3/2	63 083.24				
$3d^6(^3H)4p$	$^2H°$	11/2	*61 819.07*		59	14	$3d^5(^2I)4s4p(^3P°)\ ^2H°$
$3d^5\ 4s(^5S)6p$	$^6P°$		*61 870.6*				
$3d^5(^2I)4s4p(^3P°)$	$^2K°$	13/2	*61 912.57*		95		
$3d^6(^5D)4d$	$e\ ^6G$	13/2	62 001.09				
		11/2	62 134.45				
		9/2	62 300.63				
		7/2	62 426.48				
		5/2	62 514.59				
		3/2	62 573.11				
$3d^6(^5D)4d$	$e\ ^4G$	11/2	62 295.36				
		9/2	62 479.04				
		7/2	62 632.77				
		5/2	62 753.37				
$3d^6(a\ ^3P)4p$	$z\ ^2P°$	3/2	*62 354.76*	1.24	41	25	$3d^5(^4D)4s4p(^3P°)\ ^2P°$
		1/2	*62 391.05*	0.81	45	26	
$3d^6(^3G)4p$	$v\ ^4F°$	3/2	*62 390.20*		31	9	$3d^5(a\ ^2F)4s4p(^3P°)\ ^4F°$
		9/2	*62 392.82*	1.35	34	19	
		5/2	*62 486.99*		32	18	
		7/2	*62 505.14*		35	16	
$3d^5(^4F)4s4p(^3P°)$	$^6D°$	9/2	*62 670.81*		88		
		5/2	*62 760.73*		83		
		1/2	*62 768.16*		81		
		3/2	*62 787.11*		82		
		7/2	*62 851.24*		87		
$3d^6(a\ ^3F)4p$	$y\ ^2D°$	5/2	*63 081.28*	1.24	58	20	$(^3G)\ ^4F°$
		3/2	*63 114.11*	0.758	53	29	
$3d^5(^4G)4s4p(^1P°)$	$^4F°$	5/2	*63 139.70*		32	20	$3d^5(a\ ^2F)4s4p(^3P°)\ ^4F°$
$3d^6(^5D)4d$	$e\ ^4F$	9/2	63 231.43				
		7/2	63 424.00				
$3d^5(^4G)4s4p(^1P°)$		7/2	*63 288.78*		31 $^4F°$	19	$3d^6(^3G)4p\ ^4G°$
$3d^5(a\ ^2F)4s4p(^3P°)$	$^4G°$	9/2	*63 347.91*		67	12	$(^4G)(^1P°)\ ^4H°$
		11/2	*63 449.13*		68	11	

(Continued)

Mn I—Continued

Configuration	Term	J	Level (cm^{-1})	g	Leading percentages			
$3d^6(^3G)4p$	$^2H^\circ$	11/2	*63 288.78*	1.127	74		8	$(^3H)\ ^2H^\circ$
		9/2	*63 548.49*	0.92	73		9	
$3d^5(^4G)4s4p(^1P^\circ)$	$w\ ^4H^\circ$	13/2	*63 363.54*	1.231	51		22	$3d^6(^3G)4p\ ^4H^\circ$
		7/2	*63 395.45*	0.70	45		23	
		9/2	*63 444.61*		41		24	
		11/2	*63 457.85*	1.14	52		23	
		5/2	*63 371.58*					
$3d^5(^4F)4s4p(^3P^\circ)$	$^6F^\circ$	9/2	*63 374.53*		91			
		5/2	*63 522.98*		91			
		7/2	*63 546.30*		89			
		3/2	*63 583.04*		92			
$3d^5(a\ ^2D)4s4p(^3P^\circ)$	$x\ ^2D^\circ$	5/2	*63 764.64*		62		28	$(a\ ^2F)(^3P^\circ)\ ^2D^\circ$
		3/2	*63 845.32*		60		29	
$3d^5(^2I)4s4p(^3P^\circ)$	$^2I^\circ$	11/2	*64 051.91*		85		11	$3d^6(^1I)4p\ ^2I^\circ$
		9/2	*64 055.37*					
$3d^5(a\ ^2F)4s4p(^3P^\circ)$	$u\ ^4D^\circ$	7/2	*64 409.69*	1.42	71		15	$(a\ ^2D)(^3P^\circ)\ ^2F^\circ$
		1/2	*64 638.68*	0.22	62		19	$(a\ ^2D)(^3P^\circ)\ ^2P^\circ$
		3/2	*64 683.95*	1.22	68		12	$(a\ ^2D)(^3P^\circ)\ ^2P^\circ$
		5/2	*64 712.94*		65		18	$(a\ ^2D)(^3P^\circ)\ ^2F^\circ$
$3d^5(a\ ^2F)4s4p(^3P^\circ)$	$^4F^\circ$	9/2	*64 585.44*	1.307	42		23	$(a\ ^2D)(^3P^\circ)\ ^4F^\circ$
$3d^5(a\ ^2F)4s4p(^3P^\circ)$	$w\ ^2G^\circ$	7/2	*64 649.20*		62		21	$3d^6(^3G)4p\ ^2G^\circ$
		9/2	*65 262.28*	1.13	64		25	
$3d^6(^3G)4p$	$v\ ^4H^\circ$	13/2	*64 731.88*	1.236	55		19	$3d^5(^2I)4s4p(^3P^\circ)\ ^4H^\circ$
		11/2	*64 819.53*	1.137	54		8	
		9/2	*64 888.00*	0.974	53		19	
		7/2	*64 920.33*		41		18	
$3d^5\ 4s(^5S)7p$	$^6P^\circ$		*64 806.7*					
$3d^5(a\ ^2D)4s4p(^3P^\circ)$	$w\ ^2F^\circ$	5/2	*64 823.12*		62		25	$3d^5(a\ ^2F)4s4p(^3P^\circ)\ ^4D^\circ$
		7/2	*64 988.22*		61		18	$3d^6(^3G)4p\ ^4H^\circ$
		7/2	*65 305.13*					
$3d^6(^3G)4p$		7/2	*65 616.6*	1.015	35	$^2F^\circ$	32	$3d^6(^3G)4p\ ^2G^\circ$
		5/2	*65 649.13?*					
$3d^6(^3G)4p$		9/2	*65 768.81*	1.12	31	$^2G^\circ$	28	$(^3H)\ ^4G^\circ$
$3d^6(^3H)4p$		5/2	*65 873.40*		32	$^4G^\circ$	24	$(^3G)\ ^2F^\circ$
$3d^6(^3H)4p$		7/2	*65 876.34*		33	$^4G^\circ$	29	$(^3G)\ ^2F^\circ$
$3d^6(^3H)4p$	$v\ ^4G^\circ$	11/2	*65 887.31*	1.259	39		21	$3d^5(a\ ^2G)4s4p(^3P^\circ)\ ^4G^\circ$
		9/2	*65 908.92*	1.160	41		20	
$3d^5(^4P)4s4p(^1P^\circ)$	$^4D^\circ$	5/2	*65 946.59*	1.30	32		18	$3d^6(^3D)4p\ ^4D^\circ$

Mn I—Continued

Configuration	Term	J	Level (cm^{-1})	g	Leading percentages			
$3d^{5}(^{4}P)4s4p(^{1}P^{\circ})$		3/2	65 961.90		30	$^{4}D^{\circ}$	18	$3d^{5}(a\ ^{2}D)4s4p(^{3}P^{\circ})\ ^{2}P^{\circ}$
$3d^{6}(^{3}G)4p$	$v\ ^{2}F^{\circ}$	5/2	66 020.63		41		21	$3d^{5}(^{4}F)4s4p(^{3}P^{\circ})\ ^{4}G^{\circ}$
$3d^{6}(^{3}G)4p$		7/2	66 149.10	1.14	32	$^{2}G^{\circ}$	29	$(^{3}H)\ ^{4}G^{\circ}$
$3d^{5}(a\ ^{2}G)4s4p(^{3}P^{\circ})$	$u\ ^{4}H^{\circ}$	7/2	66 334.47	0.764	35		28	$(^{2}H)(^{3}P^{\circ})\ ^{4}H^{\circ}$
		9/2	66 356.40	1.022	34		27	
		11/2	66 418.55		36		25	
		13/2	66 568.58	1.23	39		26	
$3d^{5}(^{4}F)4s4p(^{3}P^{\circ})$		5/2	66 395.19	0.611	32	$^{4}G^{\circ}$	10	$3d^{6}(^{3}D)4p\ ^{4}F^{\circ}$
$3d^{5}(^{4}F)4s4p(^{3}P^{\circ})$		7/2	66 454.27	0.932	39	$^{4}G^{\circ}$	15	$3d^{6}(^{3}D)4p\ ^{4}F^{\circ}$
$3d^{5}(^{4}P)4s4p(^{1}P^{\circ})$	$^{4}S^{\circ}$	3/2	66 504.85		68		25	$3d^{6}(a\ ^{3}P)4p\ ^{4}S^{\circ}$
$3d^{5}(^{4}F)4s4p(^{3}P^{\circ})$	$u\ ^{4}G^{\circ}$	9/2	66 522.62	1.13	40		12	$3d^{6}(^{3}D)4p\ ^{4}F^{\circ}$
		11/2	66 573.60	1.24	59		16	$3d^{5}(^{2}H)4s4p(^{3}P^{\circ})\ ^{4}G^{\circ}$
$3d^{5}(a\ ^{2}F)4s4p(^{3}P^{\circ})$	$^{2}F^{\circ}$	7/2	66 600.17		39		18	$3d^{6}(a\ ^{1}G)4p\ ^{2}F^{\circ}$
		5/2	66 654.29		31		16	
$3d^{6}(a\ ^{1}G)4p$	$v\ ^{2}G^{\circ}$	9/2	66 630.92	1.13	48		16	$3d^{5}(^{4}F)4s4p(^{3}P^{\circ})\ ^{4}G^{\circ}$
		7/2	66 737.82	0.46	42		18	
$3d^{6}(^{3}D)4p$	$u\ ^{4}F^{\circ}$	7/2	66 783.50	1.21	34		21	$3d^{5}(^{4}D)4s4p(^{1}P^{\circ})\ ^{4}F^{\circ}$
		5/2	66 837.16		36		19	
		3/2	66 840.80	0.46	38		21	
		9/2	66 855.00	1.33	33		19	
$3d^{5}(^{4}D)4s4p(^{1}P^{\circ})$	$^{4}P^{\circ}$	5/2	66 910.54		41		20	$(^{4}D)(^{1}P^{\circ})\ ^{4}D^{\circ}$
$3d^{5}(^{4}D)4s4p(^{1}P^{\circ})$	$^{4}D^{\circ}$	7/2	66 981.30	1.33	59		18	$(a\ ^{2}G)(^{3}P^{\circ})\ ^{4}F^{\circ}$
$3d^{5}(a\ ^{2}F)4s4p(^{3}P^{\circ})$	$^{2}D^{\circ}$	5/2	67 008.29		52		21	$(a\ ^{2}D)(^{3}P^{\circ})\ ^{2}D^{\circ}$
$3d^{5}\ 4s(^{5}S)9p$	$^{6}P^{\circ}$		67 102.24					
$3d^{6}(^{1}I)4p$	$w\ ^{2}H^{\circ}$	11/2	67 504.90	1.09	51		32	$3d^{6}(a\ ^{1}G)4p\ ^{2}H^{\circ}$
		9/2	67 576.84	0.90	45		24	
$3d^{5}\ 4s(^{5}S)10p$	$^{6}P^{\circ}$		67 655.82					
$3d^{5}(^{2}H)4s4p(^{3}P^{\circ})$	$t\ ^{4}G^{\circ}$	11/2	67 752.84	1.266	42		22	$(a\ ^{2}G)(^{3}P^{\circ})\ ^{4}G^{\circ}$
		9/2	67 819.17		41		23	
		7/2	67 891.36		38		22	
		5/2	67 965.5		41		25	
$3d^{6}(^{5}D)6p$			67 890.					
$3d^{5}\ 4s(^{5}S)11p$	$^{6}P^{\circ}$		68 030.45					
$3d^{5}(^{4}F)4s4p(^{3}P^{\circ})$	$^{4}F^{\circ}$	9/2	68 286.44	1.320	52		18	$3d^{6}(^{3}D)4p\ ^{4}F^{\circ}$
		7/2	68 338.59		51		22	
$3d^{5}\ 4s(^{5}S)12p$	$^{6}P^{\circ}$		68 306.90					
$3d^{5}\ 4s(^{5}S)13p$	$^{6}P^{\circ}$		68 504.09					

(Continued)

Mn I—Continued

Configuration	Term	J	Level (cm^{-1})	g	Leading percentages
$3d^5\,4s(^5S)14p$	$^6P°$		*68 653.55*		
$3d^5\,4s(^5G)5s$	$f\,^4G$	11/2	68 693.02		
		9/2	68 716.22	1.17	
$3d^5\,4s(^5S)15p$	$^6P°$		*68 769.60*		
	$z\,^2K°$	15/2	*68 797.56?*	1.07	
		13/2	*68 842.52?*	0.93	
$3d^5\,4s(^5S)16p$	$^6P°$		*68 861.62*		
$3d^5\,4s(^5S)17p$	$^6P°$		*68 935.81*		
$3d^5\,4s(^5S)18p$	$^6P°$		*68 996.34*		
$3d^5\,4s(^5S)19p$	$^6P°$		*69 046.48*		
$3d^5\,4s(^5S)20p$	$^6P°$		*69 088.46*		
$3d^5\,4s(^5S)21p$	$^6P°$		*69 123.86*		
$3d^5\,4s(^5S)22p$	$^6P°$		*69 154.06*		
$3d^5\,4s(^5S)23p$	$^6P°$		*69 180.09*		
$3d^5\,4s(^5S)24p$	$^6P°$		*69 202.62*		
$3d^5\,4s(^5S)25p$	$^6P°$		*69 222.26*		
$3d^5\,4s(^5S)26p$	$^6P°$		*69 239.43*		
$3d^5\,4s(^5S)27p$	$^6P°$		*69 254.71*		
$3d^5\,4s(^5S)28p$	$^6P°$		*69 268.20*		
$3d^5\,4s(^5S)29p$	$^6P°$		*69 280.20*		
$3d^5\,4s(^5S)30p$	$^6P°$		*69 290.90*		
$3d^5\,4s(^5S)31p$	$^6P°$		*69 300.47*		
$3d^5\,4s(^5S)32p$	$^6P°$		*69 309.19*		
$3d^5\,4s(^5S)33p$	$^6P°$		*69 317.00*		
$3d^5\,4s(^5S)34p$	$^6P°$		*69 324.14*		
$3d^5\,4s(^5S)35p$	$^6P°$		*69 330.63*		
$3d^5\,4s(^5S)36p$	$^6P°$		*69 336.53*		
$3d^5\,4s(^5S)37p$	$^6P°$		*69 341.93*		
$3d^5\,4s(^5S)38p$	$^6P°$		*69 346.93*		
$3d^5\,4s(^5S)39p$	$^6P°$		*69 351.55*		
$3d^5\,4s(^5S)40p$	$^6P°$		*69 355.70*		
$3d^5\,4s(^5S)41p$	$^6P°$		*69 359.55*		
$3d^5\,4s(^5S)42p$	$^6P°$		*69 363.22*		
$3d^5\,4s(^5S)43p$	$^6P°$		*69 366.57*		
$3d^5\,4s(^5S)44p$	$^6P°$		*69 369.73*		
$3d^5\,4s(^5S)45p$	$^6P°$		*69 372.64*		
$3d^5\,4s(^5S)46p$	$^6P°$		*69 375.38*		
$3d^5\,4s(^5S)47p$	$^6P°$		*69 377.84*		
$3d^5\,4s(^5S)48p$	$^6P°$		*69 380.20*		
$3d^5\,4s(^5S)49p$	$^6P°$		*69 382.29*		
$3d^5\,4s(^5S)50p$	$^6P°$		*69 384.33*		
$3d^5\,4s(^5S)51p$	$^6P°$		*69 386.06*		
$3d^5\,4s(^5S)52p$	$^6P°$		*69 387.87*		
$3d^5\,4s(^5S)53p$	$^6P°$		*69 389.63*		
$3d^5\,4s(^5S)54p$	$^6P°$		*69 391.23*		
$3d^5\,4s(^5S)55p$	$^6P°$		*69 392.59*		
Mn II ($^5S_2$)	***Limit***		**69 432.37**		
	$x\,^2I°$	13/2	*69 560.88?*	1.07	
		11/2	*69 629.85?*	0.924	
	$v\,^2H°$	9/2	*69 663.20?*		
		11/2	*69 722.96*	1.10	

Mn II

$Z=25$

Cr I isoelectronic sequence

Ground state: $1s^22s^22p^63s^23p^63d^54s\ ^7S_3$

Ionization energy = 126 145.0±0.6 cm^{-1} (15.64011±0.00007 eV)

Work on the analysis of Mn II was initiated by Catalán in 1922 in conjunction with his work on Mn I. He discovered four multiplets, known in modern terminology as: the resonance septet $3d^54s\ a\ ^7S-3d^54p\ z\ ^7P°$, a higher septet $3d^54p\ z\ ^7P°-3d^54d\ e\ ^7D$, and two quintets, $3d^54s\ a\ ^5S-3d^54p\ z\ ^5P°$ and $3d^6\ a\ ^5D-3d^54p\ z\ ^5P°$. Russell (1927) interpreted these multiplets and added the higher $3d^55s\ e\ ^7S°$ term.

The analysis was continued by Curtis (1938), who found most of the quintet and septet terms and classified over 700 lines. In a later paper Curtis (1952) reported many new terms of the triplet system and classified about 500 more lines.

Subsequently, the work moved to Madrid where Iglesias and Velasco (1963) announced the discovery of about 200 new levels, including numerous singlets. In 1964 they published a monograph collecting all the data on wavelengths, energy levels, and classifications. From this publication the experimental data reported here are taken. The uncertainty in the level values is assumed to be ±0.05 cm^{-1} for levels given to two decimal places and ±0.5 cm^{-1} for those given to one decimal place.

From a theoretical study of the configurations of odd parity, Johansson, Litzén, Sinzelle, and Wyart (1980) have been able to give designations to 29 levels of the $3d^55p$ and $3d^44s4p$ configurations.

The Zeeman effect was observed at the Massachusetts Institute of Technology by Catalán for his work on Mn I. The analysis of the Mn II patterns was done by Catalán and Velasco, who determined the g-values.

The theoretical interpretation of the low even and odd configurations was started by Racah and Shadmi (1959) and Racah and Spector (1960). Their work had an important influence on the later stages of the analysis. Later Shadmi, Oreg, and Stein (1968) calculated the even configurations $3d^6$, $3d^54s$, and $3d^44s^2$ but gave percentage compositions for only $a\ ^3G$, $b\ ^3G$, $a\ ^5F$, $c\ ^3D$, and $b\ ^1F$, which are highly mixed terms. Calculations of percentage composition for $3d^54s$ terms, without configuration interaction, were made by Vizbaraite, Kupliauskis, and Tutlys (1968). Their results indicate that the percentages omitted by Shadmi et al. are greater than 70%.

Roth (1969) calculated the percentage compositions for $3d^54p$ used here. He labeled repeating terms of the $3d^5$ core by "a" and "b" rather than by seniority number. Percentages for these terms represent the sum of seniority states contributing to the same core term. Percentages for the $x\ ^3D°$, $y\ ^5G°$, $x\ ^5F°$, and $y\ ^3G°$ are omitted because of wrong identifications by Roth arising from misprints in the earlier published analysis by Iglesias and Velasco (1963). Velasco and Iglesias (1969) have commented on this matter.

The ionization energy was derived from numerous series in the papers of Curtis (1938), Garcia-Riquelme, Iglesias, and Velasco (1957), and by Iglesias and Velasco (1964). The value quoted here is from Iglesias and Velasco (1964). The uncertainty is given as 0.6 cm^{-1}.

References

Catalán, M. A. (1922), Phil. Trans. Roy. Soc. London **A223**, 127.
Curtis, C. W. (1938), Phys. Rev. **53**, 474.
Curtis, C. W. (1952), J. Opt. Soc. Am. **42**, 300.
Garcia-Riquelme, O., Iglesias, L., and Velasco, R. (1957), An. R. Soc. Esp. Fis. Quim. **53A**, 77.
Iglesias, L., and Velasco, R. (1963), An. R. Soc. Esp. Fis. Quim. **59A**, 227.
Iglesias, L., and Velasco, R. (1964), Publ. Inst. Opt. Madrid No. 23.
Johansson, S., Litzén, U., Sinzelle, J., and Wyart, J.-F. (1980), Phys. Scr. **21**, 40.
Racah, G., and Shadmi, Y. (1959), Bull. Research Council Israel **8F**, 15.
Racah, G., and Spector, N. (1960), Bull. Research Council Israel **9F**, 75.
Roth, C. (1969), J. Res. Natl. Bur. Stand. **73A**, 125.
Russell, H. N. (1927), Astrophys. J. **66**, 233.
Shadmi, Y., Oreg, J., and Stein, J. (1968), J. Opt. Soc. Am. **58**, 909.
Velasco, R., and Iglesias, L. (1969), Opt. Pur. Apl. **2**, 155.
Vizbaraite, J., Kupliauskis, Z., and Tutlys, V. (1968), Lietuvos Fizikos Rinkinys **8**, 497.

Mn II

Configuration	Term	J	Level (cm^{-1})	g	Leading percentages
$3d^5(^6S)4s$	$a\ ^7S$	3	0.00	1.992	
$3d^5(^6S)4s$	$a\ ^5S$	2	9 472.97	2.006	
$3d^6$	$a\ ^5D$	4	14 325.86	1.49	
		3	14 593.82	1.487	
		2	14 781.19	1.501	
		1	14 901.18	1.495	
		0	14 959.84		

(Continued)

Mn II—Continued

Configuration	Term	J	Level (cm^{-1})	g	Leading percentages		
$3d^5(^4G)4s$	$a\ ^5G$	6	27 547.25	1.349			
		5	27 571.25	1.274			
		4	27 583.57	1.16			
		3	27 588.50	0.98			
		2	27 589.28	0.34			
$3d^6$	$a\ ^3P$	2	29 869.48	1.497			
		1	30 685.07	1.48			
		0	31 022.05				
$3d^5(^4P)4s$	$a\ ^5P$	3	29 889.52	1.652			
		2	29 919.43	1.813			
		1	29 951.42	2.475			
$3d^6$	$a\ ^3H$	6	30 523.70	1.141			
		5	30 679.51	1.03			
		4	30 796.07	0.80			
$3d^6$	$a\ ^3F$	4	31 514.66	1.25			
		3	31 661.96	1.066			
		2	31 761.15	0.66			
$3d^5(^4D)4s$	$b\ ^5D$	4	32 787.87	1.497			
		0	32 818.33				
		1	32 836.66	1.507			
		3	32 857.19	1.515			
		2	32 859.09	1.497			
$3d^5(^4G)4s$	$a\ ^3G$	5	33 147.71	1.202	57	42	$3d^6\ ^3G$
		4	33 248.60	1.054	64	34	
		3	33 278.72	0.746	68	31	
$3d^6$	$b\ ^3G$	5	34 762.11	1.204	56	43	$3d^5(^4G)4s\ ^3G$
		4	34 910.75	1.054	63	36	
		3	35 004.77	0.745	67	32	
$3d^5(^4P)4s$	$b\ ^3P$	2	36 274.58	1.463			
		1	36 364.59	1.474			
		0	36 428.17				
$3d^6$	$a\ ^3D$	1	37 811.86	0.524			
		2	37 848.18	1.15			
		3	37 851.47	1.37			
$3d^5(^6S)4p$	$z\ ^7P^\circ$	2	*38 366.18*	2.315	100		
		3	*38 543.08*	1.923	100		
		4	*38 806.67*	1.740	100		
$3d^6$	$a\ ^1I$	6	38 720.02	1.022			
$3d^6$	$a\ ^1G$	4	38 901.8	0.96			
$3d^5(^4D)4s$	$b\ ^3D$	3	39 808.46	1.34			
		2	39 813.68	1.15			
		1	39 826.84	0.518			
$3d^6$	$a\ ^1S$	0	40 759.5				

Mn II—Continued

Configuration	Term	J	Level (cm^{-1})	g	Leading percentages		
$3d^5(^2I)4s$	$a\ ^3I$	7	41 182.53	1.133			
		6	41 190.47	1.025			
		5	41 204.34	0.849			
$3d^6$	$a\ ^1D$	2	43 131.4	1.01			
$3d^5(^4F)4s$	$a\ ^5F$	1	43 311.30	0.00	61	37	$(^2D)\ ^3D$
		5	43 528.64	1.44	99		
		4	43 537.18	1.39	99		
		3	43 696.19	1.251	70	12	$(^2D)\ ^3D$
		2	43 850.42	0.986	51	21	$(^2D)\ ^3D$
$3d^5(^2D)4s$	$c\ ^3D$	2	43 339.42	1.11	40	44	$(^4F)\ ^5F$
		3	43 395.38	1.23	54	28	
		1	44 138.96	0.347	58	39	
$3d^5(^6S)4p$	$z\ ^5P^\circ$	3	*43 370.51*	1.658	98		
		2	*43 484.64*	1.833	98		
		1	*43 557.14*	2.493	99		
$3d^5(^2I)4s$	$b\ ^1I$	6	44 315.17	1.009			
$3d^5(^2F)4s$	$b\ ^3F$	4	44 521.52	1.242			
		2	44 745.46	0.832			
		3	44 899.82	1.145			
$3d^5(^2D)4s$	$b\ ^1D$	2	46 903.26	0.971			
$3d^5(^2F)4s$	$a\ ^1F$	3	47 073.56	1.024			
$3d^5(^2H)4s$	$b\ ^3H$	4	48 317.85	0.807			
		5	48 333.06	1.030			
		6	48 435.96	1.144			
$3d^6$	$b\ ^1F$	3	49 291.31	0.90	49	30	$3d^5(^2G)4s\ ^3G$
$3d^5(^2G)4s$	$c\ ^3G$	4	49 427.32	1.051			
		3	49 465.13	0.874			
		5	49 517.58	1.186			
$3d^5(^4F)4s$	$c\ ^3F$	2	49 820.16	0.69			
		4	49 882.15	1.25			
		3	49 889.86	1.04			
$3d^5(^2H)4s$	$a\ ^1H$	5	51 553.06	1.002			
$3d^5(^2F)4s$	$d\ ^3F$	3	52 373.18				
		2	52 379.38	0.71			
		4	52 383.72	1.23			
$3d^5(^2G)4s$	$b\ ^1G$	4	52 653.27	1.004			
$3d^6$	$c\ ^3P$	0	52 718.8				
		1	53 017.16				
		2	53 597.13				
$3d^6$	$e\ ^3F$	2	53 698.7	0.69			
		3	53 781.71	1.07			
		4	53 805.8	1.23			

(Continued)

Mn II—Continued

Configuration	Term	J	Level (cm^{-1})	g	Leading percentages		
$3d^4\,4s^2$	$c\,^5$D	0	54 846.24				
		1	54 938.19				
		2	55 116.31				
		3	55 371.68				
		4	55 696.99				
$3d^5(^2$F$)4s$	$c\,^1$F	3	55 759.27	1.006			
$3d^5(^2$S$)4s$	$a\,^3$S	1	56 883.38				
$3d^6$	$c\,^1$G	4	59 537.4	0.98			
$3d^5(^2$D$)4s$	$d\,^3$D	1	62 564.7		100		
		2	62 572.2		100		
		3	62 587.5		100		
$3d^5(^4$G$)4p$	$z\,^5$G°	2	*64 456.69*		97		
		3	*64 473.39*		93		
		4	*64 494.14*	1.17	91	7	(^{4}G) ^{5}H°
		5	*64 518.87*	1.269	90	8	(^{4}G) ^{5}H°
		6	*64 550.04*	1.331	92	6	(^{4}G) ^{5}H°
$3d^5(^4$G$)4p$	$z\,^5$H°	3	*65 483.09*		95		
		4	*65 565.96*		92	7	(^{4}G) ^{5}G°
		5	*65 658.59*		92	8	
		6	*65 754.80*	1.23	94	6	
		7	*65 847.03*	1.31	100		
3 $^5(^2$D$)4s$	$c\,^1$D	2	65 567.07	1.04			
$3d^5(^4$G$)4p$	$z\,^5$F°	5	*66 542.53*	1.395	93		
		4	*66 643.31*	1.368	83	6	(^{4}P) ^{5}D°
		3	*66 686.70*	1.309	56	27	(^{4}P) ^{5}D°
		1	*66 894.09*	1.50	82	9	(^{4}P) ^{5}D°
		2	*66 901.44*	1.48	64	26	(^{4}P) ^{5}D°
$3d^5(^4$P$)4p$	$z\,^5$D°	0	*66 625.28*		83	16	(^{4}D) ^{5}D°
		1	*66 645.07*	1.36	73	14	(^{4}D) ^{5}D°
		2	*66 676.78*	1.312	54	26	(^{4}G) ^{5}F°
		3	*67 009.16*	1.44	52	35	(^{4}G) ^{5}F°
		4	*67 295.43*	1.479	78	11	(^{4}D) ^{5}D°
$3d^5(^4$P$)4p$	$z\,^5$S°	2	*66 929.52*	1.645	95		
$3d^5(^4$G$)4p$	$z\,^3$H°	6	*67 744.37*	1.160	97		
		5	*67 846.19*	1.028	98		
		4	*67 910.56*	0.789	98		
$3d^5(^4$G$)4p$	$z\,^3$F°	2	*67 766.76*	0.657	92		
		3	*67 812.05*	1.080	92		
		4	*67 865.85*	1.259	93		
$3d^5(^4$P$)4p$	$y\,^5$P°	3	*68 284.62*	1.655	74	20	(^{4}D) ^{5}P°
		2	*68 417.61*	1.80	74	12	
		1	*68 496.61*	2.41	83	8	
$3d^5(^4$P$)4p$	$z\,^3$P°	2	*69 044.90*	1.540	68	16	(^{4}D) ^{3}P°
		1	*69 216.02*	1.474	74	15	
		0	*69 319.26*		80	15	

Mn II—Continued

Configuration	Term	J	Level (cm^{-1})	g	Leading percentages		
3d^5(^{4}D)4p	y ^{5}F°	1	*70 150.74*	0.00	86	8	(^{4}G) ^{5}F°
		2	*70 231.47*	1.013	85	7	
		3	*70 342.93*	1.22	84	6	
		4	*70 497.80*	1.250	90	6	
		5	*70 657.58*	1.38	91	5	
3d^5(^{4}G)4p	z ^{3}G°	3	*70 518.05*	0.754	96		
		5	*70 527.62*	1.199	95		
		4	*70 546.37*	1.167	96		
3d^5(^{4}P)4p	z ^{3}D°	3	*70 745.38*	1.36	79	5	(^{4}D) ^{5}F°
		2	*70 940.54*	1.143	83	5	(^{4}D) ^{5}D°
		1	*71 078.44*	0.478	89		
3d^5(^{4}D)4p	x ^{5}P°	1	*71 264.42*	2.477	60	26	(^{4}D) ^{5}D°
		2	*71 323.62*	1.805	46	33	
		3	*71 390.48*	1.644	52	21	
3d^5(^{4}D)4p	y ^{5}D°	4	*72 011.02*	1.477	84	13	(^{4}P) ^{5}D°
		3	*72 247.71*	1.455	57	23	(^{4}D) ^{5}P°
		2	*72 307.21*		44	37	(^{4}D) ^{5}P°
		1	*72 321.06*	1.52	52	29	(^{4}D) ^{5}P°
		0	*72 322.49*		81	16	(^{4}P) ^{5}D°
3 ^{5}(^{2}G)4s	d ^{1}G	4	72 648.8	1.01			
3d^5(^{4}D)4p	y ^{3}D°	1	*73 385.46*	0.530	89	5	(^{4}F) ^{3}D°
		3	*73 395.44*	1.21	81	5	
		2	*73 396.26*		87	5	
3d^5(^{4}D)4p	y ^{3}F°	4	*73 683.44*	1.253	92		
		3	*73 781.11*	1.09	88		
		2	*73 785.53*	0.746	91		
3d^5(^{4}P)4p	z ^{3}S°	1	*73 911.6*	1.986	96		
3d^5(^{6}S)5s	e ^{7}S	3	74 560.01				
3d^5(^{4}D)4p	y ^{3}P°	0	*75 563.49*		81	16	(^{4}P) ^{3}P°
		1	*75 719.93*	1.490	78	16	
		2	*75 919.09*	1.50	77	19	
3d^5(^{6}S)5s	e ^{5}S	2	76 374.60				
3d^5(^{2}I)4p	z ^{3}K°	8	*77 820.28*	1.14	100		
		6	*77 842.09*	0.909	91	8	(^{2}I) ^{3}I°
		7	*77 945.71*	1.050	91	5	(^{2}I) ^{3}I°
3d^5(^{2}I)4p	z ^{3}I°	5	*78 085.2*	0.83	90	5	(^{2}I) ^{1}H°
		6	*78 340.72*	1.014	88	8	(^{2}I) ^{3}K°
		7	*78 475.43*	1.13	81	8	(^{2}I) ^{1}K°
3d^5(a ^{2}D)4p	z ^{1}D°	2	*78 913.17*	0.938	44	24	(a ^{2}F) ^{1}D°
3d^5(^{2}I)4p	z ^{1}H°	5	*79 112.56*	1.011	78	6	(^{2}I) ^{3}I°
3d^5(^{2}I)4p	z ^{1}K°	7	*79 147.10*	1.003	88	11	(^{2}I) ^{3}I°

(Continued)

Mn II—Continued

Configuration	Term	J	Level (cm^{-1})	g	Leading percentages		
$3d^5(a\ ^2D)4p$	$x\ ^3F^\circ$	2	*79 458.38*	0.734	44	29	$(a\ ^2F)\ ^3F^\circ$
		3	*79 512.73*	1.075	60	34	
		4	*79 913.35*	1.24	66	30	
$3d^5(^6S)4d$	$e\ ^7D$	1	79 540.87				
		2	79 544.68				
		3	79 550.44				
		4	79 558.54				
		5	79 569.27				
$3d^5(^2I)4p$	$y\ ^3H^\circ$	6	*79 592.14*	1.16	94		
		5	*79 739.88*	1.03	90		
		4	*79 800.52*	0.81	92		
$3d^5(a\ ^2D)4p$	$x\ ^3P^\circ$	2	*81 147.65*	1.372	77	12	$(a\ ^2D)\ ^3D^\circ$
		1	*81 322.33*		74	12	$(a\ ^2D)\ ^3D^\circ$
		0	*81 713.02*		92	4	$(^4F)\ ^5D^\circ$
$3d^5(a\ ^2D)4p$	$z\ ^1F^\circ$	3	*81 220.58*	0.92	45	35	$(a\ ^2F)\ ^3G^\circ$
$3d^5(a\ ^2F)4p$	$z\ ^1G^\circ$	4	*81 279.71*	1.010	53	24	$(a\ ^2F)\ ^3G^\circ$
$3d^5(^2D)4p$	$x\ ^3D^\circ$	3	*81 659.50*	1.32			
		1	*81 732.0*	0.44			
		2	*81 812.9*	0.90			
$3d^5(^4F)4p$	$y\ ^5G^\circ$	3	*81 780.70*	0.92			
		4	*81 862.84*	1.20			
		5	*82 117.16*	1.28			
		6	*82 142.46*	1.36			
		2	*82 193.04*				
$3d^5(^2I)4p$	$z\ ^1I^\circ$	6	*81 802.58*	1.00	94		
$3d^5(^4F)4p$	$x\ ^5F^\circ$	5	*81 886.33*				
		4	*81 993.85*	1.25			
		3	*82 051.62*	1.24			
		2	*82 054.35*	1.18			
		1	*82 236.89*				
$3d^5(^2F)4p$	$y\ ^3G^\circ$	4	*82 099.82*				
		5	*82 232.18*	1.29			
		3	*82 387.53*	0.90			
$3d^5(^6S)4d$	$e\ ^5D$	4	82 136.40				
		3	82 144.48				
		2	82 151.16				
		1	82 155.84				
		0	82 158.17				
$3d^5(a\ ^2F)4p$	$w\ ^3D^\circ$	3	*82 419.48*	1.33	57	21	$(a\ ^2F)\ ^3F^\circ$
		1	*82 939.00*	0.51	56	35	$(a\ ^2D)\ ^1P^\circ$
		2	*83 070.97*	1.21	80	6	$(^4F)\ ^5D^\circ$
$3d^5(^4F)4p$	$x\ ^5D^\circ$	4	*82 605.29*	1.47	86	9	$(^4F)\ ^5F^\circ$
		3	*82 712.55*	1.46	85	8	$(^4F)\ ^5F^\circ$
		2	*82 735.29*		82	5	$(a\ ^2F)\ ^3D^\circ$
		1	*82 774.73*		91	4	$(a\ ^2D)\ ^3P^\circ$
		0	*82 839.49*		94	4	$(a\ ^2D)\ ^3P^\circ$

Mn II—Continued

Configuration	Term	J	Level (cm^{-1})	g	Leading percentages		
$3d^5(a\ ^2F)4p$	$w\ ^3F°$	4	82 830.75	1.22	33	23	$(a\ ^2D)\ ^3F°$
		2	82 917.78	0.70	46	16	$(a\ ^2D)\ ^3F°$
		3	82 936.01	1.12	36	35	$(a\ ^2F)\ ^3D°$
$3d^4(^5D)4s4p(^3P°)$	$y\ ^7P°$	2	83 255.79				
		3	83 375.63				
		4	83 529.33				
$3d^5(^4F)4p$	$x\ ^3G°$	4	83 875.46	1.02	46	16	$(^2H)\ ^3H°$
		5	83 912.38	1.20			
		3	83 933.77	0.76	69	17	$(^2H)\ ^3G°$
$3d^5(a\ ^2D)4p$	$z\ ^1P°$	1	84 268.06	0.96	58	30	$(a\ ^2F)\ ^3D°$
$3d^5(a\ ^2G)4p$	$x\ ^3H°$	4	84 307.20	0.81	31	31	$(^2H)\ ^3H°$
		5	84 428.21	1.03	32	23	
		6	84 643.86	1.15	46	45	
$3d^5(a\ ^2F)4p$	$y\ ^1D°$	2	85 367.85	1.03	64	32	$(a\ ^2D)\ ^1D°$
$3d^5(^2H)4p$	$y\ ^3I°$	5	85 447.70	0.82	89	5	$(^2H)\ ^3H°$
		6	85 636.10	1.05	89	5	$(^2H)\ ^3H°$
		7	85 811.43	1.12	97		
$3d^5(^2H)4p$	$w\ ^3G°$	5	85 543.23	1.19	34	40	$(a\ ^2G)\ ^3G°$
		4	85 673.72	1.051	32	44	
		3	85 734.62	0.75	18	35	
$3d^5(a\ ^2G)4p$	$^1G°$	4	85 759.3	1.15	37	19	$(^2H)\ ^1G°$
$3d^5(^6S)5p$	$x\ ^7P°$	2	85 895.30				
		3	85 960.46				
		4	86 057.44				
$3d^5(^4F)4p$	$^3F°$	3	85 952.6	1.06	40	24	$(^4F)\ ^3D°$
		2	85 989.12	0.77	43	24	$(a\ ^2G)\ ^3F°$
		4	86 449.24	1.092	49	17	$(a\ ^2G)\ ^3F°$
$3d^5(a\ ^2F)4p$	$y\ ^1F°$	3	86 061.75		51	14	$(a\ ^2G)\ ^3G°$
$3d^5(^4F)4p$	$v\ ^3D°$	2	86 190.18		62	18	$(^4F)\ ^3F°$
		1	86 208.4	0.58	85	8	$(a\ ^2F)\ ^3D°$
		3	86 302.55	1.15	57	20	$(^4F)\ ^3F°$
$3d^5(^2H)4p$	$y\ ^1I°$	6	86 869.39	1.02	90	5	$(^2H)\ ^3H°$
$3d^4(^5D)4s4p(^3P°)$	$w\ ^5P°$	3	86 897.67				
		2	86 936.81				
		1	86 960.96				
$3d^5(a\ ^2G)4p$	$u\ ^3F°$	4	87 580.23	1.23	55	21	$(^4F)\ ^3F°$
		3	87 717.62	1.06	51	25	
		2	87 858.51		51	26	
$3d^5(^2H)4p$	$w\ ^3H°$	4	87 941.08	0.82	49	45	$(a\ ^2G)\ ^3H°$
		5	87 995.50	1.03	44	47	
		6	88 198.07	1.16	43	49	

(Continued)

Mn II—Continued

Configuration	Term	J	Level (cm^{-1})	g	Leading percentages		
$3d^5(a\ ^2G)4p$	$^3G°$	5	*88 772.30*		40	20	$(a\ ^2F)\ ^3G°$
		4	*89 096.56*	1.08	45	18	
		3	*89 126.0*	0.78	43	21	
$3d^5(^6S)5p$	$v\ ^5P°$	1	*88 839.7*				
		2	*89 079.2*				
		3	*89 429.0*				
$3d^5(a\ ^2G)4p$	$^1H°$	5	*89 062.6*	1.14	66	24	$(^2H)\ ^1H°$
$3d^5(b\ ^2F)4p$	$x\ ^1G°$	4	*89 465.3*		48	19	$(a\ ^2G)\ ^1G°$
$3d^5(b\ ^2F)4p$	$t\ ^3F°$	2	*89 519.25?*		75	13	$(a\ ^2G)\ ^3F°$
		3	*89 572.03*	1.14?	76	8	
		4	*89 800.34*		77	8	
$3d^5(^2H)4p$	$x\ ^1H°$	5	*89 760.33*	1.02	60	24	$(a\ ^2G)\ ^1H°$
$3d^5(a\ ^2G)4p$	$x\ ^1F°$	3	*89 950.40*		76	6	$(a\ ^2F)\ ^3F°$
$3d^4(^5D)4s4p(^3P°)$	$w\ ^5F°$	1	*90 459.9*				
		2	*90 582.2*	1.01			
		3	*90 785.7*				
		4	*91 037.8*				
		5	*91 385.1*				
$3d^5(b\ ^2F)4p$	$x\ ^1D°$	2	*90 596.8*	1.02	85	7	$(b\ ^2F)\ ^3F°$
$3d^5(b\ ^2F)4p$	$u\ ^3G°$	3	*91 018.3*	0.71	59	33	$(^2H)\ ^3G°$
		4	*91 178.8*		62	30	
		5	*91 301.8*	1.24	66	29	
$3d^5(b\ ^2F)4p$	$u\ ^3D°$	2	*92 039.8*	1.17	91	5	$(^4F)\ ^3D°$
		1	*92 060.8?*		93	4	
		3	*92 083.0*		83	8	
$3d^5(b\ ^2F)4p$	$w\ ^1G°$	4	*92 516.65*	1.01	41	37	$(^2H)\ ^1G°$
$3d^4(^5D)4s4p(^3P°)$	$w\ ^3P°$	0	*93 056.1*				
		1	*93 366.0*				
		2	*93 920.9*				
$3d^5(^2S)4p$	$^3P°$	0	*93 719.6*		89	9	$(b\ ^2D)\ ^3P°$
		1	*93 868.5*		88	9	
		2	*94 230.9*		88	10	
$3d^5(b\ ^2F)4p$	$w\ ^1F°$	3	*94 182.2*	1.00	95		
$3d^4(^5D)4s4p(^3P°)$	$w\ ^5D°$	0	*94 748.0?*				
		1	*94 795.5*				
		2	*94 886.5*				
		3	*94 989.7*				
		4	*95 206.3*				
$3d^5(^2S)4p$	$y\ ^1P°$	1	*95 080.76*		84	13	$(b\ ^2D)\ ^1P°$
$3d^4(^5D)4s4p(^3P°)$	$s\ ^3F°$	2	*96 614.1*				
		3	*96 799.9*				
		4	*97 049.9*				

Mn II—Continued

Configuration	Term	J	Level (cm^{-1})	g	Leading percentages		
$3d^5(^6S)6s$	^7S	3	97 728.54				
$3d^5(^6S)6s$	^5S	2	98 410.35				
$3d^5(^6S)4f$	$^7F°$	6	*98 423.3*				
		5	*98 423.4*				
		4	*98 423.6*				
		3	*98 423.7*				
		2	*98 423.8*				
$3d^5(^6S)4f$	$^5F°$	1	*98 461.77*				
		2	*98 462.41*				
		3	*98 463.23*				
		4	*98 464.20*				
		5	*98 465.21*				
$3d^5(^6S)5d$	^7D	1	99 890.51				
		2	99 892.32				
		3	99 895.13				
		4	99 898.98				
		5	99 904.10				
$3d^5(^6S)5d$	^5D	4	100 682.38				
		3	100 688.58				
		2	100 693.16				
		1	100 696.1				
$3d^4(^5D)4s4p(^3P°)$	$t\ ^3D°$	1	*100 928.2?*				
		2	*101 084.1*				
		3	*101 313.0*				
$3d^5(^4G)5s$	$e\ ^5G$	6	101 468.04				
		5	101 489.61				
		4	101 499.40				
		2	101 500.25				
		3	101 501.67				
$3d^5(b\ ^2D)4p$	$^1F°$	3	*101 588.3*	1.04	88	7	$(b\ ^2G)\ ^1F°$
$3d^5(^4G)5s$	$e\ ^3G$	5	102 680.35				
		4	102 703.67				
		3	102 705.66				
$3d^5(b\ ^2D)4p$	$w\ ^1D°$	2	*103 600.17*		94		
$3d^5(^4P)5s$	$e\ ^5P$	3	103 803.1				
		2	103 836.3				
		1	103 868.4				
$3d^5(^4P)5s$	$e\ ^3P$	0	104 914.48?				
		1	105 020.95				
		2	105 055.7				
$3d^5(^4G)4d$	$e\ ^5H$	3	106 157.83				
		4	106 164.38				
		7	106 167.98				
		5	106 169.03				
		6	106 170.25				

(Continued)

Mn II—Continued

Configuration	Term	J	Level (cm^{-1})	g	Leading percentages
$3d^4(b\,^3F)4s4p(^3P^\circ)$	$v\,^5F^\circ$	1	*106 265.7*		
		2	*106 374.3*		
		3	*106 526.2*		
		4	*106 707.9*		
		5	*106 894.4*		
$3d^4(b\,^3F)4s4p(^3P^\circ)$	$u\,^5P^\circ$	1	*106 479.2*		
		2	*106 750.0*		
		3	*107 172.9*		
$3d^5(^4G)4d$	$e\,^5I$	8	106 508.6		
		4	106 512.48		
		5	106 519.43		
		7	106 520.1		
		6	106 522.87		
$3d^5(^4D)5s$	$g\,^5D$	4	106 886.19		
		3	106 950.57		
		0	106 956.7		
		1	106 962.9		
		2	106 967.32		
$3d^5(^4D)5s$	$e\,^3D$	3	108 102.5		
		2	108 170.7		
		1	108 172.8		
$3d^5(^6S)7s$	^7S	3	108 126.18		
$3d^5(^6S)5f$	$^7F^\circ$	6	*108 409.9*		
		5	*108 410.0*		
		4	*108 410.1*		
		3	*108 410.2*		
		2	*108 410.3*		
$3d^5(^6S)5f$	$^5F^\circ$	1	*108 435.6*		
		2	*108 437.3*		
		3	*108 439.0*		
		4	*108 441.4*		
		5	*108 443.1*		
$3d^5(^6S)7s$	^5S	2	108 447.30		
$3d^4(^3H)4s4p(^3P^\circ)$	$x\,^5G^\circ$	2	*108 486.0*		
		3	*108 503.5*		
		4	*108 524.2*		
		5	*108 550.7*		
		6	*108 587.4*		
$3d^5(^6S)5g$	^7G	7–1	108 555.7		
$3d^5(^6S)5g$	^5G	6–2	108 556.1		
$3d^4(^5D)4s4p(^1P^\circ)$	$t\,^5P^\circ$	1	*108 727.1*		
		2	*108 975.0*		
		3	*109 379.5*		

Mn II—Continued

Configuration	Term	J	Level (cm^{-1})	g	Leading percentages
$3d^4(b\,^3F)4s4p(^1P°)$	$u\,^5F°$	1	*108 994.4*		
		2	*109 046.4*		
		3	*109 123.3*		
		4	*109 221.9*		
		5	*109 327.5*		
$3d^4(b\,^3F)4s4p(^3P°)$	$v\,^5D°$	0	*109 167.7*		
		1	*109 235.3*		
		2	*109 344.3*		
		3	*109 476.6*		
		4	*109 608.3*		
$3d^5(^6S)6d$	^7D	1	109 241.33		
		2	109 242.40		
		3	109 244.01		
		4	109 246.20		
		5	109 249.05		
$3d^5(b\,^2G)4p$	$v\,^1G°$	4	*109 473.9*		96
$3d^4(a\,^3P)4s4p(^3P°)$	$y\,^5S°$	2	*109 704.0*		
	$r\,^3F°$	2	*109 900.7*		
		3	*110 018.3*		
		4	*110 141.4*		
$3d^4\,4s4p$	$u\,^5D°$	0	*109 959.0*		
		1	*109 994.6*		
		2	*110 068.8*		
		3	*110 205.6*		
		4	*110 429.2*		
$3d^4(^3G)4s4p(^3P°)$	$y\,^5H°$	3	*110 547.7*		
		4	*110 602.3*		
		5	*110 691.9*		
		6	*110 795.3*		
		7	*110 926.3*		
$3d^4(a\,^3P)4s4p(^3P°)$	$y\,^3S°$	1	*110 671.6*		
$3d^4(^3G)4s4p(^3P°)$	$t\,^5F°$	1	*111 017.5*		
		2	*111 060.5*		
		3	*111 116.0*		
		4	*111 159.6*		
		5	*111 160.3*		
$3d^4(^3H)4s4p(^3P°)$	$x\,^3I°$	5	*111 083.8*		
		6	*111 339.2*		
		7	*111 816.5*		
	$s\,^5P°$	1	*111 162.3*		
		2	*111 178.5*		
		3	*111 213.3*		
$3d^4(^3H)4s4p(^3P°)$	$u\,^3H°$	4	*111 720.0*		
		5	*111 854.5*		
		6	*112 128.5*		

(Continued)

Mn II—Continued

Configuration	Term	J	Level (cm^{-1})	g	Leading percentages
	$s\ ^3D°$	1	*112 274.0?*		
		2	*112 717.6*		
		3	*113 251.4*		
$3d^5(^4G)5p$	$w\ ^5G°$	2	*112 424.2*		
		3	*112 453.2*		
		4	*112 463.2*		
		5	*112 468.8*		
		6	*112 472.4*		
	$q\ ^3F°$	2	*112 563.2*		
		3	*112 673.9*		
		4	*112 879.8*		
	$v\ ^5G°$	2	*113 092.5*		
		3	*113 109.7*		
		4	*113 181.7*		
		5	*113 251.4*		
		6	*113 322.7*		
$3d^5(^4G)5p$	$t\ ^3G°$	5	*113 515.0?*		
		3	*113 520.0*		
		4	*113 560.1*		
$3d^5(^4G)5p$	$s\ ^5F°$	3	*113 642.0*		
		2	*113 645.4*		
		1	*113 645.8*		
		4	*113 647.2*		
		5	*113 658.1*		
$3d^5(^6S)8s$	^7S	3	113 697.68		
$3d^4(b\ ^3F)4s4p(^3P°)$	$s\ ^3G°$	3	*113 773.4*		
		4	*113 926.2*		
		5	*114 090.6*		
$3d^5(^6S)6f$	$^7F°$	6	*113 840.0*		
		5	*113 840.2*		
		4	*113 840.3*		
		3	*113 840.4*		
$3d^5(^6S)8s$	^5S	2	113 895.25		
$3d^5(^6S)6g$	^7G	7–1	113 931.7		
$3d^5(^6S)6g$	^5G	6–2	113 932.1		
$3d^5(^6S)6f$	$^5F°$	1	*114 023.6*		
		2	*114 024.9*		
		3	*114 026.5*		
		5	*114 027.0*		
		4	*114 028.0*		
$3d^5(^6S)7d$	^7D	1	114 344.24		
		2	114 344.90		
		3	114 345.96		
		4	114 347.38		
		5	114 349.16		

Mn II—Continued

Configuration	Term	J	Level (cm^{-1})	g	Leading percentages
$3d^5(^6S)7d$	^5D	4	114 932.3		
		3	114 944.0		
		2	114 951.7		
		1	114 956.3		
		0	114 958.4		
$3d^5(^6S)9s$	^7S	3	117 031.23		
$3d^5(^6S)7f$	$^7F^\circ$	6	*117 113.1*		
		5–3	*117 113.2*		
$3d^5(^6S)7f$	$^5F^\circ$	1	*117 135*		
		2	*117 138*		
		4	*117 138*		
		3	*117 139*		
		5	*117 148*		
$3d^4(^3D)4s4p(^3P^\circ)$	$r\ ^5F^\circ$	1	*117 165.0*		
		2	*117 232.1*		
		3	*117 314.9*		
		4	*117 399.6*		
		5	*117 483.6*		
$3d^5(^6S)7g$	^7G	7–1	117 172.9		
$3d^5(^6S)7g$	^5G	6–2	117 173.8		
$3d^5(^6S)10s$	^7S	3	119 186.0		
$3d^5(^6S)8f$	$^5F^\circ$	1–5	*119 253*		
$3d^5(^6S)8g$	^7G	7–1	119 276.8		
$3d^5(^6S)8g$	^5G	6–2	119 278.3		
$3d^5(^6S)8h$	$^7H^\circ$	8–2	*119 282.3*		
$3d^5(^6S)9h$	$^7H^\circ$	8–2	*120 721.9*		
Mn III ($^6S_{5/2}$)	***Limit***		**126 145.0**		

Mn III

$Z=25$

V I isoelectronic sequence

Ground state: $1s^22s^22p^63s^23p^63d^5\ {}^6S_{5/2}$

Ionization energy = 271 550±100 cm^{-1} (33.668±0.010 eV)

The analysis is from Garcia-Riquelme (1968) with additional and corrected terms from Yarosewick, DaVia, and Moore (1971), and Garcia-Riquelme (1977). Level value uncertainties are presumed to be ±0.05 cm^{-1} for levels given with two decimal places and ±0.5 cm^{-1} for those with one decimal place.

The percentage compositions of the $3d^5$ levels are from Pasternak and Goldschmidt (1974). Percentages for the $3d^44s$ configuration are taken from an ab initio (Hartree-Fock) calculation by Vizbaraite, Kupliauskis and Tutlys (1968).

The compositions of the $3d^44p$ configuration are from Roth (1968). He distinguished repeating $3d^4$ core states by alphabetic prefixes and assigned percentages equal to the sum of states differing in seniority only.

The ionization energy was determined by Garcia-Riquelme from the three-member series of $3d^4ns\ {}^6D$.

References

Garcia-Riquelme, O. (1968), Opt. Pur. Apl. **1**, 53.
Garcia-Riquelme, O. (1977), Opt. Pur. Apl. **10**, 275.
Pasternak, A., and Goldschmidt, Z. B. (1974), Phys. Rev. **A9**, 1022.
Roth, C. (1968), J. Res. Natl. Bur. Stand. **72A**, 505.
Vizbaraite, J., Kupliauskis, Z., and Tutlys, V. (1968), Lietuvos Fizikos Rinkinys **8**, 497.
Yarosewick, S. J., DaVia, J. J., and Moore, F. L. (1971), J. Opt. Soc. Am. **61**, 732.

Mn III

Configuration	Term	J	Level (cm^{-1})	Leading percentages		
$3d^5$	$a\ {}^6S$	5/2	0.0	100		
$3d^5$	$a\ {}^4G$	11/2	26 824.40	100		
		9/2	26 851.10	100		
		7/2	26 859.90	100		
		5/2	26 857.80	100		
$3d^5$	$a\ {}^4P$	5/2	29 167.70	97		
		3/2	29 207.30	98		
		1/2	29 241.40	99		
$3d^5$	$a\ {}^4D$	7/2	32 307.30	100		
		1/2	32 368.9	99		
		5/2	32 383.70	97		
		3/2	32 384.70	98		
$3d^5$	$a\ {}^2I$	11/2	39 174.4	99		
		13/2	39 176.5	100		
$3d^5$	$a\ {}^2D3$	5/2	41 238.1	58	22	${}^2F1$
		3/2	41 569.8	73	24	${}^2D1$
$3d^5$	$a\ {}^2F1$	7/2	42 606.5	98		
		5/2	43 105.4	72	14	${}^2D3$
$3d^5$	$a\ {}^4F$	9/2	43 573.16	99		
		7/2	43 602.50	99		
		5/2	43 668.84	90		
		3/2	43 674.7	96		
$3d^5$	$a\ {}^2H$	9/2	46 515.9	92		
		11/2	46 670.7	99		

Mn III—Continued

Configuration	Term	J	Level (cm^{-1})	Leading percentages		
$3d^5$	$a\ ^2$G2	7/2	47 842.0	99		
		9/2	48 005.2	91		
$3d^5$	$b\ ^2$F2	5/2	51 002.7	99		
		7/2	51 059.7	99		
$3d^5$	$a\ ^2$S	1/2	55 677.7	100		
$3d^5$	$b\ ^2$D2	3/2	61 580.2	100		
		5/2	61 603.8	100		
$3d^4(^5$D$)4s$	$a\ ^6$D	1/2	62 456.99	100		
		3/2	62 568.08	100		
		5/2	62 747.50	100		
		7/2	62 988.92	100		
		9/2	63 285.37	100		
$3d^5$	$b\ ^2$G1	7/2	68 892.00	100		
		9/2	68 899.20	100		
$3d^4(^5$D$)4s$	$b\ ^4$D	1/2	71 395.27	100		
		3/2	71 564.21	100		
		5/2	71 831.98	100		
		7/2	72 183.33	100		
$3d^5$	$a\ ^2$P	3/2	83 176.0	100		
		1/2	83 229.0	100		
$3d^4(^3$P2$)4s$	$b\ ^4$P	1/2	84 610.53	61	38	$(^3$P1$)\ ^4$P
		3/2	85 173.88	61	38	
		5/2	86 051.50	62	38	
$3d^4(^3$H$)4s$	$a\ ^4$H	7/2	84 981.63	100		
		9/2	85 077.09	100		
		11/2	85 200.76	94	6	$(^3$G$)\ ^4$H
		13/2	85 346.72	100		
$3d^4(^3$F2$)4s$	$b\ ^4$F	3/2	86 486.77	79	20	$(^3$F1$)\ ^4$F
		5/2	86 520.94	79	20	
		7/2	86 578.24	79	19	
		9/2	86 654.07	79	19	
$3d^4(^3$G$)4s$	$b\ ^4$G	5/2	88 880.08	98		
		7/2	89 052.44	98		
		9/2	89 204.69	81	18	$(^3$H$)\ ^2$H
		11/2	89 307.22	94	6	$(^3$H$)\ ^4$H
$3d^5$	$c\ ^2$D1	5/2	89 496.3	76	24	^{2}D3
		3/2	89 543.4?	75	24	
$3d^4(^3$P2$)4s$	$b\ ^2$P	1/2	90 233.5	61	38	$(^3$P1$)\ ^2$P
		3/2	91 308.3	61	38	
$3d^4(^3$H$)4s$	$b\ ^2$H	9/2	90 440.50	81		
		11/2	90 746.06	98		
$3d^4(^3$F2$)4s$	$c\ ^2$F	5/2	91 906.10	79	20	$(^3$F1$)\ ^2$F
		7/2	91 948.30	79	19	

(Continued)

Mn III—Continued

Configuration	Term	J	Level (cm^{-1})	Leading percentages		
$3d^4(^3G)4s$	$c\ ^2G$	7/2	94 397.20	98		
		9/2	94 707.20	100		
$3d^4(^3D)4s$	$c\ ^4D$	7/2	94 697.85	100		
		5/2	94 771.47	100		
		3/2	94 850.66	100		
		1/2	94 906.45	100		
$3d^4(^1G2)4s$	$d\ ^2G$	9/2	96 430.4	66	34	$(^1G1)\ ^2G$
		7/2	96 487.5	66	34	
$3d^4(^1I)4s$	$b\ ^2I$	13/2	97 239.86	100		
		11/2	97 271.76	100		
$3d^4(^1S2)4s$	$b\ ^2S$	1/2	98 960.7	77	22	$(^1S1)\ ^2S$
$3d^4(^3D)4s$	$d\ ^2D$	5/2	100 001.30	100		
		3/2	100 085.20	100		
$3d^4(^1D2)4s$	$e\ ^2D$	5/2	104 470.8	77	22	$(^1D1)\ ^2D$
		3/2	104 517.9	77	21	
$3d^4(^1F)4s$	$d\ ^2F$	7/2	109 861.35	100		
		5/2	109 864.40	100		
$3d^4(^5D)4p$	$z\ ^6F^\circ$	1/2	*110 036.14*	100		
		3/2	*110 172.69*	100		
		5/2	*110 398.50*	100		
		7/2	*110 711.62*	100		
		9/2	*111 111.39*	100		
		11/2	*111 601.45*	100		
$3d^4(^5D)4p$	$z\ ^6P^\circ$	3/2	*111 776.60*	98		
		5/2	*111 883.60*	98		
		7/2	*112 057.80*	100		
$3d^4(^5D)4p$	$z\ ^4P^\circ$	1/2	*112 645.31*	73	24	$^6D^\circ$
		3/2	*113 078.34*	66	30	
		5/2	*113 676.53*	51	48	
$3d^4(^5D)4p$	$z\ ^6D^\circ$	1/2	*113 991.31*	75	24	$(^5D)\ ^4P^\circ$
		3/2	*114 094.97*	69	30	$(^5D)\ ^4P^\circ$
		7/2	*114 209.01*	99		
		5/2	*114 287.91*	52	46	$(^5D)\ ^4P^\circ$
		9/2	*114 501.18*	97		
$3d^4(^3F1)4s$	$c\ ^4F$	9/2	115 711.60	81	19	$(^3F2)\ ^4F$
		3/2	115 762.60	79	20	
		7/2	115 774.00	79	20	
		5/2	115 780.23	79	20	
$3d^4(^3P1)4s$	$c\ ^4P$	3/2	116 207.50	62	38	$(^3P2)\ ^4P$
		1/2	116 498.40	62	38	
		5/2	116 622.40	62	38	
$3d^4(^5D)4p$	$z\ ^4F^\circ$	3/2	*116 580.17*	96		
		5/2	*116 692.14*	96		
		7/2	*116 851.69*	95		
		9/2	*117 063.74*	94		

Mn III—Continued

Configuration	Term	J	Level (cm^{-1})	Leading percentages		
$3d^4(^3\mathrm{F}1)4s$	$e\ ^2\mathrm{F}$	7/2	120 955.45	81	19	$(^3\mathrm{F}2)\ ^2\mathrm{F}$
		5/2	121 003.10	79	20	
$3d^4(^5\mathrm{D})4p$	$z\ ^4\mathrm{D}°$	1/2	120 974.25	98		
		3/2	121 091.08	98		
		5/2	121 267.32	98		
		7/2	121 480.24	98		
$3d^4(^1\mathrm{G}1)4s$	$e\ ^2\mathrm{G}$	9/2	124 973.18	66	34	
		7/2	124 986.30	66	34	$(^1\mathrm{G}2)\ ^2\mathrm{G}$
$3d^4(^3\mathrm{H})4p$	$z\ ^4\mathrm{H}°$	7/2	130 731.31	80	18	$(^3\mathrm{G})\ ^4\mathrm{H}°$
		9/2	130 895.39	79	17	
		11/2	131 123.64	79	16	
		13/2	131 420.03	82	14	
$3d^4(a\ ^3\mathrm{P})4p$	$y\ ^4\mathrm{D}°$	1/2	131 287.18	83	14	$(a\ ^3\mathrm{F})\ ^4\mathrm{D}°$
		3/2	131 701.69	82	16	
		5/2	132 297.55	78	20	
		7/2	132 971.99	69	29	
$3d^4(a\ ^3\mathrm{F})4p$	$z\ ^4\mathrm{G}°$	5/2	132 897.74	72	18	$(^3\mathrm{G})\ ^4\mathrm{G}°$
		7/2	133 059.84	59	15	$(^3\mathrm{G})\ ^4\mathrm{G}°$
		9/2	133 276.40	45	13	$(^3\mathrm{G})\ ^4\mathrm{G}°$
		11/2	133 736.64	55	22	$(^3\mathrm{H})\ ^4\mathrm{G}°$
$3d^4(^3\mathrm{H})4p$	$z\ ^4\mathrm{I}°$	9/2	133 086.57	94		
		11/2	133 455.64	95		
		13/2	133 804.28	96		
		15/2	134 136.60	100		
$3d^4(a\ ^3\mathrm{P})4p$	$z\ ^2\mathrm{S}°$	1/2	133 662.6	50	42	$^4\mathrm{P}°$
$3d^4(^3\mathrm{H})4p$	$z\ ^2\mathrm{G}°$	7/2	133 751.50	49	28	$(a\ ^3\mathrm{F})\ ^2\mathrm{G}°$
		9/2	133 949.54	39	27	
$3d^4(a\ ^3\mathrm{P})4p$	$y\ ^4\mathrm{P}°$	3/2	134 638.39	91		
		1/2	134 789.10	52	32	$(a\ ^3\mathrm{P})\ ^2\mathrm{S}°$
		5/2	135 246.27	90		
$3d^4(a\ ^3\mathrm{F})4p$	$y\ ^4\mathrm{F}°$	3/2	135 116.80	50	30	$(a\ ^3\mathrm{F})\ ^2\mathrm{D}°$
		5/2	135 510.70	50	25	$(^3\mathrm{H})\ ^4\mathrm{G}°$
		7/2	135 869.76	84	8	$(^3\mathrm{H})\ ^4\mathrm{G}°$
		9/2	135 944.08	86	4	$(^3\mathrm{H})\ ^4\mathrm{G}°$
$3d^4(^3\mathrm{H})4p$	$y\ ^4\mathrm{G}°$	7/2	135 692.17	52	28	$(a\ ^3\mathrm{F})\ ^4\mathrm{G}°$
		5/2	135 709.90	34	24	$(a\ ^3\mathrm{F})\ ^4\mathrm{F}°$
		9/2	135 732.50	54	31	$(a\ ^3\mathrm{F})\ ^4\mathrm{G}°$
		11/2	135 781.09	51	40	$(a\ ^3\mathrm{F})\ ^4\mathrm{G}°$
$3d^4(a\ ^3\mathrm{P})4p$	$z\ ^2\mathrm{P}°$	3/2	135 859.0	44	28	$(a\ ^3\mathrm{P})\ ^4\mathrm{S}°$
		1/2	136 089.4	71	17	$(a\ ^3\mathrm{P})\ ^2\mathrm{S}°$
$3d^4(a\ ^3\mathrm{F})4p$		3/2	135 977.8	33 $^2\mathrm{D}°$	26	$(a\ ^3\mathrm{F})\ ^4\mathrm{F}°$
$3d^4(a\ ^3\mathrm{F})4p$	$z\ ^2\mathrm{D}°$	5/2	136 053.9	54	18	$(a\ ^3\mathrm{F})\ ^4\mathrm{F}°$
$3d^4(^3\mathrm{H})4p$	$z\ ^2\mathrm{I}°$	11/2	136 421.50	92	4	$(^1\mathrm{I})\ ^2\mathrm{I}°$
		13/2	136 500.63	93		

(Continued)

Mn III—Continued

Configuration	Term	J	Level (cm^{-1})	Leading percentages			
$3d^4(a\,^3F)4p$	$x\,^4D^\circ$	5/2	*136 796.61*	62		18	$(a\,^3P)\,^4D^\circ$
		7/2	*136 823.86*	51		27	
		3/2	*136 888.70*	61		15	
		1/2	*136 935.60*	78		15	
$3d^4(a\,^3P)4p$		3/2	*137 137.69*	39	$^4S^\circ$	30	$^2P^\circ$
$3d^4(a\,^3F)4p$	$z\,^2F^\circ$	5/2	*137 282.40*	35		28	$(^3G)\,^2F^\circ$
		7/2	*137 599.20*	38		23	
$3d^4(^3H)4p$	$z\,^2H^\circ$	9/2	*137 436.70*	71		14	$(a\,^1G)\,^2H^\circ$
		11/2	*137 908.20*	72		11	
$3d^4(^3G)4p$	$y\,^4H^\circ$	7/2	*137 660.24*	80		17	$(^3H)\,^4H^\circ$
		9/2	*137 935.53*	76		15	
		11/2	*138 245.09*	73		13	
		13/2	*138 605.80*	83		14	
$3d^4(^3G)4p$	$x\,^4F^\circ$	3/2	*138 234.79*	68		14	$(^3D)\,^4F^\circ$
		9/2	*138 312.70*	71		13	$(^3D)\,^4F^\circ$
		5/2	*138 340.62*	50		25	$(a\,^3P)\,^2D^\circ$
		7/2	*138 362.80*	69		14	$(^3D)\,^4F^\circ$
$3d^4(a\,^3P)4p$	$y\,^2D^\circ$	3/2	*138 692.0*	70		13	$(a\,^3F)\,^2D^\circ$
		5/2	*139 273.2*	55		21	$(^3G)\,^4F^\circ$
$3d^4(a\,^3F)4p$	$y\,^2G^\circ$	7/2	*139 643.6*	53		24	$(^3H)\,^2G^\circ$
		9/2	*139 857.40*	44		24	
$3d^4(^3G)4p$	$y\,^2H^\circ$	11/2	*139 902.06*	53		23	
		9/2	*139 970.90*	48		18	$(^3G)\,^4G^\circ$
$3d^4(^3G)4p$	$y\,^2F^\circ$	5/2	*139 944.4*	42		41	$(a\,^3F)\,^2F^\circ$
		7/2	*140 401.1*	48		40	
$3d^4(^3G)4p$	$x\,^4G^\circ$	5/2	*140 207.09*	57		25	$(^3H)\,^4G^\circ$
		7/2	*140 294.44*	62		24	$(^3H)\,^4G^\circ$
		9/2	*140 557.46*	46		17	$(^3G)\,^2H^\circ$
		11/2	*140 776.88*	47		27	$(^3G)\,^2H^\circ$
$3d^4(^3G)4p$	$x\,^2G^\circ$	7/2	*143 097.00*	74		15	$(^3H)\,^2G^\circ$
		9/2	*143 161.50*	67		17	
$3d^4(^3D)4p$	$w\,^4D^\circ$	1/2	*143 127.00*	92			
		3/2	*143 148.40*	87			
		5/2	*143 209.80*	79		12	$(^3D)\,^4P^\circ$
		7/2	*143 345.07*	88			
$3d^4(a\,^1G)4p$	$x\,^2F^\circ$	7/2	*144 129.8*	82		4	$(^3G)\,^2F^\circ$
		5/2	*144 718.4*	81		5	
$3d^4(^3D)4p$	$x\,^4P^\circ$	5/2	*144 146.50*	81		10	$(^3D)\,^4D^\circ$
		3/2	*144 619.00*	82		8	$(^3D)\,^4D^\circ$
		1/2	*144 832.90*	93			
$3d^4(^1I)4p$	$y\,^2I^\circ$	13/2	*144 258.5*	74		23	$(^1I)\,^2K^\circ$
		11/2	*144 311.0*	90		7	$(a\,^1G)\,^2H^\circ$

Mn III—Continued

Configuration	Term	J	Level (cm^{-1})	Leading percentages		
$3d^4(a\ ^1G)4p$	$x\ ^2H°$	9/2	144 522.30	80	10	$(^3H)\ ^2H°$
		11/2	144 880.10	76	10	
$3d^4(^3D)4p$	$w\ ^4F°$	3/2	144 525.50	70	16	$(^3G)\ ^4F°$
		5/2	144 672.38	72	16	
		7/2	144 817.68	77	16	
		9/2	144 959.74	83	16	
$3d^4(^3D)4p$	$y\ ^2P°$	1/2	144 879.1	45	47	$(a\ ^1S)\ ^2P°$
		3/2	145 120.8	56	38	
$3d^4(^1I)4p$	$z\ ^2K°$	13/2	145 008.36	77	23	$(^1I)\ ^2I°$
		15/2	145 455.10	100		
$3d^4(a\ ^1G)4p$	$w\ ^2G°$	7/2	146 591.9	80	10	$(^3G)\ ^2G°$
		9/2	146 788.4	78	15	
$3d^4(^1I)4p$	$w\ ^2H°$	11/2	148 206.90	85	9	$(^3G)\ ^2H°$
		9/2	148 560.50	87	9	
$3d^4(^3D)4p$	$w\ ^2F°$	7/2	148 355.60	73	9	$(^3G)\ ^2F°$
		5/2	148 568.40	73	11	
$3d^4(^3D)4p$	$x\ ^2D°$	5/2	149 234.40	58	30	$(a\ ^1D)\ ^2D°$
		3/2	149 540.5	64	15	
$3d^4(a\ ^1S)4p$	$x\ ^2P°$	3/2	149 512.90	43	24	$(^3D)\ ^2P°$
		1/2	149 621.00	41	42	
$3d^4(a\ ^1D)4p$	$w\ ^2D°$	3/2	151 628.8	67	15	$(^3D)\ ^2D°$
		5/2	151 871.3	58	23	
$3d^4(a\ ^1D)4p$	$v\ ^2F°$	7/2	152 370.65	83	10	$(^1F)\ ^2F°$
		5/2	153 067.70	81	5	
$3d^4(^1F)4p$	$u\ ^2F°$	5/2	156 621.7	81	7	$(a\ ^1D)\ ^2F°$
		7/2	156 809.6	78	10	
$3d^4(^1F)4p$	$v\ ^2G°$	9/2	158 016.3	95		
$3d^4(^1F)4p$	$v\ ^2D°$	5/2	160 039.7	66	18	$(b\ ^3P)\ ^2D°$
		3/2	160 758.5	70	16	
$3d^4(b\ ^3F)4p$	$v\ ^4F°$	3/2	162 999.8	96		
		5/2	163 022.3	92		
		7/2	163 078.5	92		
		9/2	163 194.6	98		
$3d^4(b\ ^3F)4p$	$w\ ^4G°$	5/2	165 382.0	79	16	$(b\ ^3F)\ ^2F°$
		7/2	165 428.7	80	15	$(b\ ^3F)\ ^2F°$
		9/2	165 740.0	97		
		11/2	165 947.0	99		
$3d^4(b\ ^3F)4p$	$t\ ^2F°$	7/2	165 702.9	73	17	$^4G°$
		5/2	165 922.9	73	16	
$3d^4(b\ ^3P)4p$	$y\ ^4S°$	3/2	167 808.4	97		

(Continued)

Mn III—Continued

Configuration	Term	J	Level (cm^{-1})	Leading percentages		
$3d^4(b\ ^3F)4p$	$u\ ^2G°$	9/2	*169 455.0*	97		
		11/2	*169 682.4*	97		
$3d^4(b\ ^3F)4p$	$u\ ^4D°$	7/2	*169 518.3*	68	31	$(b\ ^3P)\ ^4D°$
		5/2	*169 773.2*	69	30	
		3/2	*169 977.2?*	69	30	
$3d^4(^5D)4d$	$e\ ^6G$	3/2	172 451.23			
		5/2	172 547.85			
		7/2	172 682.64			
		9/2	172 855.02			
		11/2	173 063.55			
		13/2	173 306.45			
$3d^4(^5D)4d$	$e\ ^6P$	3/2	172 569.27			
		5/2	172 613.87			
		7/2	172 755.95			
$3d^4(^5D)4d$	$e\ ^6D$	1/2	173 835.07			
		3/2	173 920.69			
		5/2	174 048.89			
		7/2	174 243.33			
		9/2	174 464.07			
$3d^4(^5D)4d$	$e\ ^6F$	1/2	174 247.63			
		3/2	174 332.91			
		5/2	174 436.59			
		7/2	174 575.87			
		9/2	174 874.50			
		11/2	174 975.65			
$3d^4(^5D)4d$	$e\ ^4P$	1/2	175 918.50			
		3/2	176 265.78			
		5/2	176 683.76			
$3d^4(^5D)5s$	$f\ ^6D$	1/2	176 100.71			
		3/2	176 209.15			
		5/2	176 390.67			
		7/2	176 641.14			
		9/2	176 945.20			
$3d^4(^5D)4d$	$e\ ^4G$	5/2	176 270.25			
		7/2	176 464.78			
		9/2	176 695.06			
		11/2	176 970.01			
$3d^4(^5D)4d$	$e\ ^4D$	1/2	177 185.18			
		3/2	177 269.34			
		5/2	177 477.23			
		7/2	177 687.54			
$3d^4(^5D)5s$	$f\ ^4D$	1/2	178 386.00			
		3/2	178 558.30			
		5/2	178 816.96			
		7/2	179 152.09			
$3d^4(^5D)4d$	$e\ ^4F$	3/2	179 228.69			
		5/2	179 339.59			
		9/2	179 683.61			

Mn III—Continued

Configuration	Term	J	Level (cm^{-1})	Leading percentages
$3d^4(^5D)4d$	$e\ ^6S$	5/2	181 949.34	
$3d^4(^5D)5p$	$y\ ^6F°$	1/2	*191 915.43*	
		3/2	*192 012.75*	
		5/2	*192 172.89*	
		7/2	*192 392.20*	
		9/2	*192 665.80*	
		11/2	*192 989.05*	
$3d^4(^5D)5p$	$y\ ^6P°$	3/2	*192 365.52*	
		5/2	*192 576.42*	
		7/2	*192 880.16*	
$3d^4(^5D)5p$	$y\ ^6D°$	1/2	*192 741.67*	
		3/2	*192 948.29*	
		5/2	*193 235.43*	
		7/2	*193 498.07*	
		9/2	*193 756.79*	
$3d^4(^3H)4d$	$e\ ^4I$	9/2	194 446.1	
		11/2	194 893.0	
		13/2	195 141.8	
		15/2	195 455.5	
$3d^4(^5D)5p$	$w\ ^4P°$	1/2	*194 545.78*	
		3/2	*194 894.78*	
		5/2	*195 315.78*	
$3d^4(^3H)4d$	$e\ ^4H$	7/2	194 555.4	
		9/2	194 744.0	
		11/2	194 908.20	
		13/2	195 036.20	
$3d^4(^5D)5p$	$u\ ^4F°$	3/2	*194 849.10*	
		5/2	*194 971.15*	
		7/2	*195 112.81*	
		9/2	*195 308.79*	
$3d^4(^3H)4d$	$e\ ^4K$	11/2	194 900.40	
		13/2	195 036.20	
		15/2	195 186.90	
		17/2	195 343.10	
$3d^4(^3H)4d$	$f\ ^4G$	9/2	195 415.20	
		11/2	195 682.50	
$3d^4(^5D)5p$	$v\ ^4D°$	1/2	*195 462.26*	
		3/2	*195 592.09*	
		5/2	*195 752.48*	
		7/2	*195 948.44*	
$3d^4(^5D)4f$	$z\ ^6G°$	7/2	*208 650.54*	
		9/2	*208 751.60*	
		11/2	*209 032.30*	
		13/2	*209 323.03*	

(Continued)

Mn III—Continued

Configuration	Term	J	Level (cm^{-1})	Leading percentages
$3d^4(^5D)4f$	$z\ ^6H^{\circ}$	5/2	*208 743.10*	
		7/2	*208 819.64*	
		9/2	*208 935.88*	
		11/2	*209 091.41*	
		13/2	*209 289.76*	
		15/2	*209 530.38*	
$3d^4(^5D)5d$	$f\ ^6G$	3/2	215 473.20	
		5/2	215 520.22	
		7/2	215 663.50	
		9/2	215 917.52	
		11/2	216 194.05	
		13/2	216 533.06	
$3d^4(^5D)5d$	$f\ ^6F$	1/2	215 580.87	
		3/2	215 715.41	
		5/2	215 875.73	
		7/2	216 149.32	
		9/2	216 307.29	
		11/2	216 691.83	
$3d^4(^5D)5d$	$g\ ^6D$	3/2	215 911.73	
		5/2	216 062.13	
		7/2	216 283.58	
		9/2	216 670.58	
$3d^4(^5D)5d$	$f\ ^6P$	3/2	216 162.90	
		5/2	216 403.86	
		7/2	216 557.38	
$3d^4(^5D)6s$	$h\ ^6D$	1/2	216 423.25	
		3/2	216 531.99	
		5/2	216 712.73	
		7/2	216 961.62	
		9/2	217 271.59	
$3d^4(^5D)6s$	$g\ ^4D$	3/2	217 807.57?	
		5/2	218 067.19?	
		7/2	218 400.40?	
$3d^4(^5D)5d$	$h\ ^4G$	5/2	217 975.30	
		7/2	218 115.15	
		9/2	218 375.61	
		11/2	218 596.69	
$3d^4(^5D)5d$	$h\ ^4D$	1/2	218 085.88	
		3/2	218 180.90	
		5/2	218 307.98	
		7/2	218 506.39	
$3d^4(^5D)5d$	$f\ ^4P$	1/2	218 127.59	
		3/2	218 373.00	
		5/2	218 705.25	
$3d^4(^5D)5d$	$g\ ^4F$	3/2	218 669.26	
		5/2	218 816.97	
		7/2	218 977.34	
		9/2	219 195.95	

Mn III—Continued

Configuration	Term	J	Level (cm^{-1})	Leading percentages
$3d^4(^3H)4f$	$x\ ^4H^\circ$	13/2	*230 134.95*	
		11/2	*230 281.25*	
		9/2	*230 369.62*	
		7/2	*230 506.52*	
$3d^4(^5D)5f$	$y\ ^6H^\circ$	5/2	*231 361.28*	
		7/2	*231 387.66*	
		9/2	*231 487.34*	
		11/2	*231 752.76*	
		13/2	*232 075.44*	
		15/2	*232 433.03*	
$3d^4(^5D)5f$	$w\ ^6F^\circ$	5/2	*232 065.03*	
		7/2	*232 264.42*	
		9/2	*232 478.68*	
		11/2	*232 761.16*	
Mn IV ($^5D_0$)	***Limit***		**271 550**	

Mn IV

$Z=25$

Ti I isoelectronic sequence

Ground state: $1s^2 2s^2 2p^6 3s^2 3p^6 3d^4\ ^5D_0$

Ionization energy = 413 000±1000 cm^{-1} (51.2±0.1 eV)

White (1929) identified the $3d^3 4s\ ^5F-3d^3 4p\ ^5G°$ multiplet. Bowen (1937) found the low ^5D and four of the seven triplets in the $3d^4$ ground configuration. He also found $3d^3 4s\ ^3F$ and nine terms of $3d^3 4p$.

Yarosewick and Moore (1967) added two more triplets in $3d^4$, three more triplets in $3d^3 4s$ and nine new terms in $3d^3 4p$. They also found a ^5H term in $3d^3 4d$ and give new values for all levels. The estimated uncertainty is ±1.0 cm^{-1}. Yarosewick (1980) has subsequently made changes in some of the published level values of the $w\ ^3D°$, $w\ ^3F°$, $w\ ^3G°$, and $e\ ^5H$ terms.

The percentage compositions of the $3d^3 4p$ levels were calculated by Goldschmidt (1983). Some designations were changed to agree this calculation.

The ionization energy is from the isoelectronic extrapolation of Lotz (1967).

References

Bowen, I. S. (1937), Phys. Rev. **52**, 1153.
Goldschmidt, Z. B. (1983), private communication.
Lotz, W. (1967), J. Opt. Soc. Am. **57**, 873.
White, H. E. (1929), Phys. Rev. **33**, 914.
Yarosewick, S. J., and Moore, F. L. (1967), J. Opt. Soc. Am. **57**, 1381.
Yarosewick, S. J. (1980), private communication to R. L. Kelly.

Mn IV

Configuration	Term	J	Level (cm^{-1})	Leading percentages
$3d^4$	$a\,^5D$	0	0.0	
		1	99.0	
		2	286.8	
		3	552.2	
		4	885.8	
$3d^4$	$a\,^3P$	0	20 654.2	
		1	21 278.8	
		2	22 324.7	
$3d^4$	$a\,^3H$	4	21 280.7	
		5	21 474.8	
		6	21 679.3	
$3d^4$	$a\,^3F$	2	22 791.0	
		3	22 862.3	
		4	22 959.1	
$3d^4$	$a\,^3G$	3	25 440.4	
		4	25 670.8	
		5	25 877.7	
$3d^4$	$a\,^3D$	3	31 385.2	
		2	31 473.5	
		1	31 580.7	
$3d^4$	$b\,^3F$	4	53 212.3	
		2	53 260.7	
		3	53 286.7	

Mn IV—Continued

Configuration	Term	J	Level (cm^{-1})	Leading percentages		
$3d^3(^4F)4s$	$a\,^5F$	1	111 506.0			
		2	111 710.4			
		3	112 013.3			
		4	112 408.9			
		5	112 882.8			
$3d^3(^4F)4s$	$c\,^3F$	2	119 445.5			
		3	119 961.5			
		4	120 603.1			
$3d^3(^2G)4s$	$b\,^3G$	3	130 969.5			
		4	131 170.1			
		5	131 462.5			
$3d^3(^2H)4s$	$b\,^3H$	4	137 853.0			
		5	137 930.1			
		6	138 123.1			
$3d^3(^2F)4s$	$d\,^3F$	4	152 700.5			
		3	152 858.2			
		2	152 986.0			
$3d^3(^4F)4p$	$z\,^5G^\circ$	2	*167 890.1*	99		
		3	*168 300.3*	99		
		4	*168 839.7*	99		
		5	*169 499.2*	99		
		6	*170 285.8*	100		
$3d^3(^4F)4p$	$z\,^3D^\circ$	1	*170 577.8*	44	34	$^5F^\circ$
		2	*172 398.7*	57	29	$^5F^\circ$
		3	*172 952.6*	75	8	$^5D^\circ$
$3d^3(^4F)4p$	$z\,^5D^\circ$	2	*170 866.3*	43	28	$^3D^\circ$
		3	*171 276.9*	68	15	$^5F^\circ$
		1	*171 383.9*	77	16	$^5F^\circ$
		4	*171 765.0*	90	7	$^3F^\circ$
$3d^3(^4F)4p$	$z\,^5F^\circ$	2	*171 698.3*	46	47	$^5D^\circ$
		3	*172 081.9*	77	21	$^5D^\circ$
		1	*172 090.6*	49	43	$^3D^\circ$
		4	*172 477.2*	91	7	$^5D^\circ$
		5	*172 871.5*	96	3	$^3G^\circ$
$3d^3(^4F)4p$	$z\,^3G^\circ$	3	*175 436.0*	92	6	$(^2G)\ ^3G^\circ$
		4	*175 813.4*	91	6	
		5	*176 288.6*	89	5	
$3d^3(^4F)4p$	$z\,^3F^\circ$	2	*177 629.8*	95	2	$(^2D3)\ ^3F^\circ$
		3	*178 075.1*	94	2	
		4	*178 579.5*	94	2	
$3d^3(^4P)4p$	$z\,^5P^\circ$	1	*184 566.1*	98	1	$^3P^\circ$
		2	*184 901.2*	96	3	$^3P^\circ$
		3	*185 435.9*	99	1	$^5D^\circ$
$3d^3(^4P)4p$	$y\,^5D^\circ$	3	*187 004.0*	95	3	$(^4F)\ ^5D^\circ$
		4	*187 679.5*	97	3	

(Continued)

Mn IV—Continued

Configuration	Term	J	Level (cm^{-1})	Leading percentages		
$3d^3(^2G)4p$	$z^3H°$	4	*186 659.4*	80	17	$(^2H)\ ^2H°$
		5	*187 278.5*	77	18	
		6	*188 184.8*	79	20	
$3d^3(^2G)4p$	$y^3G°$	3	*188 756.2*	79	6	$(^4F)\ ^3G°$
		4	*189 210.9*	82	7	$(^2G)\ ^3F°$
		5	*189 553.9*	83	6	$(^4F)\ ^3G°$
$3d^3(^2G)4p$	$y^3F°$	4	*190 032.4*	48	36	$(^2G)\ ^1G°$
		2	*190 135.9*	68	15	$(^2D3)\ ^3F°$
		3	*190 282.8*	66	9	$(^2G)\ ^3G°$
$3d^3(^2P)4p$	$y^3D°$	1	*192 980.7*	77	13	$(^4P)\ ^3D°$
		2	*193 739.6*	66	12	
		3	*194 208.2*	54	20	
$3d^3(^2H)4p$	$y^3H°$	4	*193 761.7*	76	17	$(^2G)\ ^3H°$
		5	*193 879.4*	80	19	
		6	*194 155.7*	78	20	
$3d^3(^2H)4p$	$z^3I°$	5	*195 974.1*	96	3	$(^2G)\ ^1H°$
		6	*196 497.9*	99	1	$(^2H)\ ^3H°$
		7	*197 185.0*	100		
$3d^3(^2D3)4p$	$w^3D°$	1	*197 095.4*	62	16	$(^4P)\ ^3D°$
		2	*197 554.8*	68	14	$(^2D1)\ ^3D°$
		3	*197 879.5*	64	12	$(^2D1)\ ^3D°$
$3d^3(^2H)4p$	$x^3G°$	5	*200 192.2*	84	5	$(^2F)\ ^3G°$
		4	*200 290.4*	86	6	
		3	*200 333.2*	86	6	
$3d^3(^2F)4p$	$w^3F°$	2	*209 260.2*	94	2	$(^2D1)\ ^3F°$
		3	*209 315.5*	93	2	$(^2D1)\ ^3F°$
		4	*209 448.2*	93	2	$(^2F)\ ^3G°$
$3d^3(^2F)4p$	$w^3G°$	3	*212 399.3*	90	7	$(^2H)\ ^3G°$
		4	*212 666.1*	89	7	
		5	*212 965.7*	94	6	
$3d^3(^4F)4d$	e^5H	3	248 388.6			
		4	248 654.8			
		5	248 983.8			
		6	249 374.8			
		7	249 822.7			
Mn V $(^4F_{3/2})$	***Limit***		**413 000**			

Mn V

$Z=25$

Sc I isoelectronic sequence

Ground state: $1s^2 2s^2 2p^6 3s^2 3p^6 3d^3\ ^4F_{3/2}$

Ionization energy = 584 000±1200 cm^{-1} (72.4±0.1 eV)

The initial analysis was by White (1929) who found three terms in $3d^3$, two terms in $3d^2 4s$ and five in $3d^2 4p$. It was extended by Bowen (1935) who found seven new terms in $3d^2 4p$. Bowen's $3d^3\ ^2P$ and ^2D were published by Pasternack (1940).

More precise values for the $3d^2 4s$ levels were provided in advance of publication by Yarosewick and Moore (1976). They are based on Bowen's value for $3d^2 4p\ ^4G^\circ_{5/2}$ taken as 241 907.1 cm^{-1}. Yarosewick and Moore also found a ^4H term in $3d^2 4d$.

Kovalev, Ramonas, and Ryabtsev (1976) have extended the analysis of the $3d^3-3d^2 4p$ transition array with new observations in the range 382–548 Å with a reported accuracy of ±0.002 Å. We used their level values for $3d^3$ and $3d^2 4p$ and adjusted Yarosewick and Moore's values for the ^4H term of $3d^2 4d$ accordingly. The percentage compositions are from parametric calculations of Kovalev, Ramonas, and Ryabtsev for $3d^3$ and $3d^2 4p$.

The transition array $3d^2 4s-3d^2 4p$ was remeasured by Podobedova, Ramonas, and Ryabtsev (1978) in the range of 1086–1914 Å with an accuracy of ±0.007 Å. With the complete set of levels of $3d^2 4p$ provided by Kovalev et al. (1977), they were able to identify all the missing terms of the $3d^2 4s$ configuration. Their results are given here along with their calculated percentage compositions of these levels.

The uncertainty in the energy level values is ±1.0 cm^{-1}.

The ionization energy is from an isoelectronic extrapolation by Lotz (1967).

References

Bowen, I. S. (1935), Phys. Rev. **47**, 924.

Kovalev, V. I., Ramonas, A. A., and Ryabtsev, A. N. (1977), Opt. Spectrosc. (USSR) **43**, 4.

Lotz, W. (1967), J. Opt. Soc. Am. **57**, 873.

Pasternack, S. (1940), Astrophys. J. **92**, 129.

Podobedova, L. I., Ramonas, A. A., and Ryabtsev, A. N. (1978), Opt. Spectrosc. (USSR) **45**, 237.

White, H. E. (1929), Phys. Rev. **33**, 672.

Yarosewick, S. J., and Moore, F. L. (1976), private communication.

Mn V

Configuration	Term	J	Level (cm^{-1})	Leading percentages		
$3d^3$	^4F	3/2	0.0	100		
		5/2	359.0	100		
		7/2	835.1	100		
		9/2	1 412.2	100		
$3d^3$	^4P	1/2	16 434.0	100		
		3/2	16 594.6	98		
		5/2	17 048.6	100		
$3d^3$	^2G	7/2	17 892.4	100		
		9/2	18 398.8	98		
$3d^3$	^2P	3/2	22 918.8	64	27	$^2D2$
		1/2	23 081.6	100		
$3d^3$	$^2D2$	5/2	24 630.0	79	20	$^2D1$
		3/2	24 670.5	50	35	^2P
$3d^3$	^2H	9/2	24 974.8	98		
		11/2	25 333.8	100		
$3d^3$	^2F	7/2	40 423.3	100		
		5/2	40 707.1	100		
$3d^3$	$^2D1$	5/2	62 608.2	79	20	$^2D2$
		3/2	62 853.5	77	22	

(Continued)

Mn v—Continued

Configuration	Term	J	Level (cm^{-1})	Leading percentages		
$3d^2(^3F)4s$	^4F	3/2	176 946.3	100		
		5/2	177 326.9	100		
		7/2	177 874.2	100		
		9/2	178 572.5	100		
$3d^2(^3F)4s$	^2F	5/2	183 538.8	100		
		7/2	184 635.6	100		
$3d^2(^1D)4s$	^2D	5/2	193 781.6	75	25	$(^3P)\ ^4P$
		3/2	194 011.4	83	17	
$3d^2(^3P)4s$	^4P	1/2	194 288.8	100		
		3/2	194 664.4	83	17	$(^1D)\ ^2D$
		5/2	195 302.7	75	25	$(^1D)\ ^2D$
$3d^2(^3P)4s$	^2P	1/2	200 176.9	100		
		3/2	200 670.1	100		
$3\ ^2(^1G)4s$	^2G	9/2	203 811.8	100		
		7/2	203 827.3	100		
$3d^2(^1S)4s$	^2S	1/2	236 634.7	100		
$3d^2(^3F)4p$	$^4G°$	5/2	*241 907.1*	92	5	$(^3F)\ ^2F°$
		7/2	*242 766.9*	94	4	$(^3F)\ ^4F°$
		9/2	*243 803.1*	94	5	$(^3F)\ ^4F°$
		11/2	*245 046.7*	100		
$3d^2(^3F)4p$	$^4F°$	3/2	*243 120.9*	94	4	$(^3F)\ ^2D°$
		5/2	*243 678.5*	96		
		7/2	*244 382.1*	96	4	$(^3F)\ ^4G°$
		9/2	*245 135.6*	94	5	$(^3F)\ ^4G°$
$3d^2(^3F)4p$	$^2F°$	5/2	*245 497.8*	67	10	$(^3F)\ ^2D°$
		7/2	*246 324.7*	67	26	$(^3F)\ ^4D°$
$3d^2(^3F)4p$	$^2D°$	3/2	*246 161.2*	44	41	$(^3F)\ ^4D°$
		5/2	*248 088.4*	59	21	
$3d^2(^3F)4p$	$^4D°$	1/2	*246 539.1*	94	6	$(^3P)\ ^4D°$
		5/2	*246 897.6*	62	18	$(^3F)\ ^2F°$
		3/2	*247 222.1*	53	35	$(^3F)\ ^2D°$
		7/2	*247 705.2*	69	24	$(^3F)\ ^2F°$
$3d^2(^3F)4p$	$^2G°$	7/2	*250 966.5*	94	5	$(^1G)\ ^2G°$
		9/2	*251 719.6*	94	4	
$3d^2(^3P)4p$	$^2S°$	1/2	*253 893.8*	98		
$3d^2(^3P)4p$	$^4S°$	3/2	*257 436.8*	90	9	$(^1D)\ ^2P°$
$3d^2(^1D)4p$	$^2P°$	3/2	*259 043.8*	83	10	$(^3P)\ ^4S°$
		1/2	*260 042.8*	92		
$3d^2(^1D)4p$	$^2F°$	5/2	*259 583.3*	85	7	$(^3F)\ ^2F°$
		7/2	*260 680.8*	79	10	$(^3P)\ ^4D°$

Mn V—Continued

Configuration	Term	J	Level (cm^{-1})	Leading percentages		
$3d^2(^3P)4p$	$^4D°$	1/2	*260 460.8*	90	6	$(^3F)\ ^4D°$
		3/2	*260 806.2*	90	6	$(^3F)\ ^4D°$
		5/2	*261 464.4*	86	5	$(^3F)\ ^4D°$
		7/2	*262 574.9*	85	10	$(^1D)\ ^2F°$
$3d^2(^1D)4p$	$^2D°$	3/2	*262 252.0*	81	6	$(^3F)\ ^2D°$
		5/2	*262 836.1*	79	7	$(^3P)\ ^4P°$
$3d^2(^3P)4p$	$^4P°$	1/2	*264 400.6*	98		
		3/2	*264 732.4*	98		
		5/2	*265 488.4*	92	8	$(^1D)\ ^2D°$
$3d^2(^1G)4p$	$^2G°$	7/2	*265 579.3*	94	5	$(^3F)\ ^2G°$
		9/2	*265 734.0*	94	5	
$3d^2(^3P)4p$	$^2D°$	5/2	*269 955.4*	81	12	$(^3F)\ ^2D°$
		3/2	*269 968.3*	79	11	
$3d^2(^1G)4p$	$^2H°$	9/2	*271 494.6*	98		
		11/2	*272 640.4*	100		
$3d^2(^3P)4p$	$^2P°$	1/2	*272 982.9*	98		
		3/2	*273 212.3*	96		
$3d^2(^1G)4p$	$^2F°$	7/2	*276 592.0*	96	4	$(^1D)\ ^2F°$
		5/2	*277 398.1*	96		
$3d^2(^1S)4p$	$^2P°$	1/2	*303 705*	98		
		3/2	*305 247.4*	98		
$3d^2(^3F)4d$	^4H	7/2	338 057.2			
		9/2	338 584.9			
		11/2	339 165.5			
		13/2	339 745.1			
Mn VI $(^3F_2)$	*Limit*		584 000			

Mn VI

$Z=25$

Ca I isoelectronic sequence

Ground state: $1s^2 2s^2 2p^6 3s^2 3p^6 3d^2\ ^3F_2$

Ionization energy = 771 100±200 cm^{-1} (95.60±0.02 eV)

The $3d^2$ and $3d4p$ configurations were identified by Cady (1933) from observations he made in the range 307–330 Å. Revised level values were communicated to C. E. Moore by B. Edlén in 1949.

Observations in the range of 800–1550 Å were made by King (1977) with an accuracy of ±0.007 Å, including the transition arrays $3d4s-3d4p$ and $3d4p-3d4d$. King improved the accuracy of the $3d4p$ levels, setting them relative to the value 329 992 cm^{-1} for the $3d4p\ ^3P_2^\circ$, the number given by Moore (1952). He also reported the four levels of $3d4s$ and 15 of the 18 levels of $3d4d$.

Ryabtsev (1982) continued the analysis of this spectrum with observations from 140–220 Å with a reported uncertainty of ±0.003 Å. These lines were classified as transitions between $3d^2$ and the excited configurations $3dnf$ ($n=4$–8), $3d5p$, and $3p^5 3d^3$. Five transitions to the $3d^2\ ^1S_0$ level were found. The uncertainty in the values of the levels of $3d^2$ and these excited configurations is ±10 cm^{-1}. Relative to the lowest level of $3d4s$, the levels of $3d4s$, $3d4p$, and $3d4d$ have an uncertainty of ±1 cm^{-1}.

We fit a linear Rydberg formula to the $3dnf\ ^1H_5$ series ($n=5$-7) to obtain the value for the ionization energy, assuming the series limit is $3p^6 3d\ ^2D_{5/2}$.

References

Cady, W. M. (1933), Phys. Rev, **43**, 322.

King, W. H. (1977), J. Phys. **B10**, 3381.

Moore, C. E. (1952), Atomic Energy Levels, Natl. Bur. Stand. (U.S.) Circ. 467, Vol. II (reissued in 1971 as Natl. Bur. Stand. (U.S.) Natl. Stand. Ref. Data Ser. 35, Vol. II).

Ryabtsev, A. N. (1982), Opt. Spectrosc. (USSR) **53**, 13.

Mn VI

Configuration	Term	J	Level (cm^{-1})
$3d^2$	^3F	2	0
		3	746
		4	1 669
$3d^2$	^1D	2	15 336
$3d^2$	^3P	0	17 782
		1	18 057
		2	18 628
$3d^2$	^1G	4	25 511
$3d^2$	^1S	0	59 265
3d 4s	^3D	1	250 096.6
		2	250 527.0
		3	251 403.0
3d 4s	^1D	2	255 239.7
3d 4p	$^1D^\circ$	2	*319 821.2*
3d 4p	$^3D^\circ$	1	*321 693.5*
		2	*322 409.6*
		3	*323 282.5*
3d 4p	$^3F^\circ$	2	*323 796.1*
		3	*324 849.1*
		4	*326 372.6*
3d 4p	$^3P^\circ$	1	*329 634.5*
		0	*329 729.3*
		2	*329 992.0*
3d 4p	$^1F^\circ$	3	*333 054.5*
3d 4p	$^1P^\circ$	1	*336 130.8*
3d 4d	^1F	3	429 104.8
3d 4d	^3D	2	431 059.4
		3	431 606.6
3d 4d	^1P	1	432 313.3
3d 4d	^3G	3	432 090.5
		4	432 652.6
		5	433 463.6
3d 4d	^3S	1	436 451.0
3d 4d	^3F	2	439 105.0
		3	439 643.4
		4	440 234.1
3d 4d	^1D	2	444 637.1
3d 4d	^3P	1	445 590.9
		2	446 044.2
3d 4d	^1G	4	447 701.8

Mn VI—Continued

Configuration	Term	J	Level (cm^{-1})
$3p^6\,3d4f$	$^3F^\circ$	2	*501 976*
		3	*502 639*
		4	*503 432*
$3p^6\,3d4f$	$^1G^\circ$	4	*518 905*
$3p^6\,3d4f$	$^3G^\circ$	3	*520 698*
		4	*521 305*
		5	*521 892*
$3p^6\,3d\,5p$	$^3D^\circ$	1	*523 203*
		2	*524 146*
		3	*524 608*
$3p^6\,3d\,5p$	$^1D^\circ$	2	*523 443*
$3p^5(^2P^\circ)3d^3(^2D1)$	^1D	2	524 162
$3p^6\,3d\,5p$	$^3F^\circ$	2	*524 590*
		3	*524 985*
		4	*526 054*
$3p^5(^2P^\circ)3d^3(^2G)$	^1F	3	526 055
$3p^6\,3d4f$	$^1H^\circ$	5	*526 092*
$3p^5(^2P^\circ)3d^3(^4F)$	^3F	2	527 514
		3	529 488
		4	530 550
$3p^6\,3d\,5p$	$^1F^\circ$	3	*528 532*
$3p^6\,3d\,5p$	$^1P^\circ$	1	*531 252*
$3p^5(^2P^\circ)3d^3(^2P)$	^1P	1	549 303
$3p^5(^2P^\circ)3d^3(^4F)$	^3D	3	550 258
		2	550 654
		1	551 400
$3p^5(^2P^\circ)3d^3(^2H)$	^1G	4	556 973
$3p^5(^2P^\circ)3d^3(^4P)$	^3S	1	568 390
$3p^5(^2P^\circ)3d^3(^2F)$	^1F	3	568 974
$3p^5(^2P^\circ)3d^3(^2D1)$	^1P	1	575 512
$3p^6\,3d5f$	$^1G^\circ$	4	*608 125*
$3p^6\,3d5f$	$^3F^\circ$	2	*608 193*
		3	*608 407*
		4	*609 095*
$3p^6\,3d5f$	$^3G^\circ$	3	*609 568*
		4	*610 314*
		5	*610 595*
$3p^6\,3d5f$	$^1D^\circ$	2	*610 051*
$3p^6\,3d5f$	$^3D^\circ$	1	*610 678*
		2	*611 340*
		3	*611 405*
$3p^6\,3d5f$	$^1F^\circ$	3	*610 723*
$3p^6\,3d5f$	$^3P^\circ$	2	*612 044*
		1	*612 161*
$3p^6\,3d5f$	$^1H^\circ$	5	*612 678*
$3p^6\,3d5f$	$^1P^\circ$	1	*613 361*
$3p^6\,3d6f$	$^3F^\circ$	2	*657 696*
		3	*658 238*
		4	*658 777*
$3p^6\,3d6f$	$^1G^\circ$	4	*657 867*
$3p^6\,3d6f$	$^3G^\circ$	3	*658 896*
		4	*659 783*
		5	*660 000*
$3p^6\,3d6f$	$^1D^\circ$	2	*659 181*
$3p^6\,3d6f$	$^3D^\circ$	1	*659 849*
		2	*659 951*
		3	*660 007*
$3p^6\,3d6f$	$^1F^\circ$	3	*660 522*
$3p^6\,3d6f$	$^1H^\circ$	5	*661 233*
$3p^6\,3d6f$	$^1P^\circ$	1	*661 258*
$3p^6\,3d7f$	$^3F^\circ$	2	*688 212*
		3	*688 438*
		4	*689 181*
$3p^6\,3d7f$	$^1D^\circ$	2	*688 844*
$3p^6\,3d7f$	$^3G^\circ$	3	*688 846*
		4	*689 844*
		5	*689 995*
$3p^6\,3d7f$	$^3D^\circ$	1	*689 333*
		3	*689 778*
		2	*689 892*
$3p^6\,3d7f$	$^1H^\circ$	5	*690 741*
$3p^6\,3d8f$	$^3G^\circ$	3	*708 250*
		4	*709 370*
		5	*709 500*
$3p^6\,3d8f$	$^1H^\circ$	5	*709 890*
Mn VII ($^2D_{3/2}$)	*Limit*		**771 100**

Mn VII

$Z=25$

K I isoelectronic sequence

Ground state: $1s^2 2s^2 2p^6 3s^2 3p^6 3d$ $^2D_{3/2}$

Ionization energy = 961 440±100 cm^{-1} (119.204±0.010 eV)

The three lines of the $3p^6 3d$ $^2D-3p^6 4p$ $^2P°$ multiplet were reported by Kruger and Weissberg (1937). His identifications were based on the assumption of regular isoelectronic behavior of the $3p^6 4p$ $^2P°$ term. The recent analysis of this spectrum reported by Ramonas and Ryabtsev (1980) showed that the strong interaction of this term with the $3p^5 3d^2(^1D)$ $^2P°$ term greatly distorts both term splittings. They found that the identification of the line $3p^6 3d$ $^2D_{3/2}-3p^6 4p$ $^2P°_{1/2}$ by Kruger and Weissberg was wrong, but confirmed the $3p^6 3d$ ^2D splitting and the $3p^6 4p$ $^2P°_{3/2}$ level. They also confirmed the $4f$, $6f$, $7f$, and $8f$ identifications. The $4s$ ^2S term identified by Gibbs and White (1926) and the $5s$ and $6s$ ^2S terms by Kruger and Weissberg are based on the incorrect $4p$ $^2P°$ splitting.

Analyses of the transition arrays $3p^6 3d-3p^5 3d 4s$ by Cowan (1967), using the measurements of Feldman and Fraenkel (1966), and $3p^6 3d-3p^5 3d^2$ by Gabriel, Fawcett, and Jordan (1966) were greatly expanded by Ramonas and Ryabtsev. Their results, obtained with a reported wavelength accuracy of ±0.003 Å, are quoted here. The level value uncertainty is ±20 cm^{-1}.

The ionization energy was derived from the $4p^6 nf$ series by Ramonas and Ryabtsev.

References

Cowan, R. D. (1967), Astrophys. J. **147**, 377.

Feldman, U., and Fraenkel, B. S. (1966), Astrophys. J. **145**, 959.

Gabriel, A. H., Fawcett, B. C., and Jordan, C. (1966), Proc. Phys. Soc. **87**, 825.

Gibbs, R. C., and White, H. E. (1926), Proc. Natl. Acad. Sci. **12**, 675.

Kruger, P. G., and Weissberg, S. G. (1937), Phys. Rev. **52**, 314.

Ramonas, A. A., and Ryabtsev, A. N. (1980), Opt. Spectrosc. (USSR) **48**, 348.

Mn VII

Configuration	Term	J	Level (cm^{-1})	Configuration	Term	J	Level (cm^{-1})
$3p^6 3d$	^2D	3/2	0	$3p^5 3d(^3F°)4s$	$^4F°$	7/2	*710 337*
		5/2	1 338			5/2	*712 642*
						3/2	*714 990*
$3p^5(^2P°)3d^2(^1G)$	$^2F°$	5/2	*394 238*				
		7/2	*396 618*	$3p^5 3d(^3F°)4s$	$^2F°$	7/2	*717 696*
						5/2	*722 331*
$3p^6 4p$	$^2P°$	1/2	*395 633*				
		3/2	*399 795*	$3p^5 3d(^3D°)4s$	$^4D°$	7/2	*735 676*
						5/2	*737 173*
$3p^5(^2P°)3d^2(^1D)$	$^2P°$	1/2	*404 085*			3/2	*737 833*
		3/2	*408 273*			1/2	*738 585*
$3p^5(^2P°)3d^2(^1D)$	$^2F°$	7/2	*409 891*	$3p^6 5f$	$^2F°$	5/2	*738 765*
		5/2	*419 081*			7/2	*739 482*
$3p^5(^2P°)3d^2(^3F)$	$^2F°$	5/2	*489 916*	$3p^5 3d(^1F°)4s$	$^2F°$	5/2	*739 930*
		7/2	*494 337*			7/2	*746 550*
$3p^5(^2P°)3d^2(^3P)$	$^2P°$	1/2	*541 894*	$3p^5 3d(^1D°)4s$	$^2D°$	3/2	*740 894*
		3/2	*544 342*			5/2	*744 126*
$3p^5(^2P°)3d^2(^3F)$	$^2D°$	5/2	*547 367*	$3p^5 3d(^3D°)4s$	$^2D°$	3/2	*748 302*
		3/2	*547 949*			5/2	*749 532*
$3p^6 4f$	$^2F°$	5/2	*616 007*	$3p^6 6f$	$^2F°$	5/2	*807 820*
		7/2	*616 132*			7/2	*807 835*
$3p^5 3d(^3P°)4s$	$^2P°$	1/2	*701 189*	$3p^6 7f$	$^2F°$	5/2	*848 947*
		3/2	*705 425*			7/2	*848 954*

Mn VII—Continued

Configuration	Term	J	Level (cm^{-1})	Configuration	Term	J	Level (cm^{-1})
$3p^6\,8f$	$^2F^\circ$	5/2,7/2	*875 620*				
Mn VIII ($^1S_0$)	*Limit*		**961 440**				

Mn VIII

$Z=25$

Ar I isoelectronic sequence

Ground state: $1s^2 2s^2 2p^6 3s^2 3p^6\ ^1S_0$

Ionization energy = 1 569 000±3000 cm^{-1} (194.5±0.4 eV)

The resonance lines arising from the two $J=1$ levels of the $3p^5 4s$ configuration were observed and identified by Kruger, Weissberg, and Phillips (1937). The levels of $3p^5 3d$ and $3s3p^6 3d$ were determined by Smitt and Svensson (1983) from observations in the range of 185–668 Å. Their level accuracy was estimated to be ±6 cm^{-1}.

The levels of $3p^5 4d$ were identified by Alexander, Feldman, and Fraenkel (1965), who designated them in *jj* notation. The suitability of the $j_1 l$-coupling designations for this configuration and for $3p^5 4s$ is indicated by the calculations of Ekberg (1976) in Cr VII for the corresponding configurations.

The $3d-4f$ transition array was identified by Wagner and House (1971). Their classifications are combined with the $3d$ levels of Smitt and Svensson.

The uncertainty in the levels of $3p^5 4s$, $4d$, and $4f$ is ±400 cm^{-1}.

We derived the ionization energy from $3p^5 3d$ and $4d\ ^1P^\circ_1$ levels, assuming a change in effective quantum number Δn^* between them of 0.9986 obtained from the same terms in Cr VII.

References

Alexander, E., Feldman, U., and Fraenkel, B. S. (1965), J. Opt. Soc. Am. **55**, 650.

Ekberg, J. O. (1976), Phys. Scr. **13**, 245.

Kruger, P. G., Weissberg, S. G., and Phillips, L. W. (1937), Phys. Rev. **51**, 1090.

Smitt, R., and Svensson, L. A. (1983), Phys. Scr. **27**, 364.

Wagner, W. J., and House, L. L. (1971), Astrophys. J. **166**, 683.

Mn VIII

Configuration	Term	J	Level (cm^{-1})	Configuration	Term	J	Level (cm^{-1})
$3p^6$	^1S	0	0	$3p^5(^2P^\circ_{3/2})4s$	$^2[3/2]^\circ$	1	*806 100*
$3p^5 3d$	$^3P^\circ$	0	*373 658*	$3p^5(^2P^\circ_{1/2})4s$	$^2[1/2]^\circ$	1	*818 500*
		1	*375 710*				
		2	*379 993*	$3p^5(^2P^\circ_{3/2})4d$	$^2[3/2]^\circ$	1	*1 026 600*
$3p^5 3d$	$^3F^\circ$	4	*391 836*	$3p^5(^2P^\circ_{1/2})4d$	$^2[3/2]^\circ$	1	*1 038 100*
		3	*394 921*				
		2	*398 564*	$3p^5 4f$	$^3D^\circ$	1	*1 116 100*
						2	*1 117 600*
$3p^5 3d$	$^3D^\circ$	3	*419 374*			3	*1 119 900*
		1	*423 337*				
		2	*424 641*	$3p^5 4f$	$^3G^\circ$	5	*1 119 100*
						4	*1 120 500*
$3p^5 3d$	$^1D^\circ$	2	*419 817*			3	*1 123 600*
$3p^5 3d$	$^1F^\circ$	3	*427 531*	$3p^5 4f$	$^1G^\circ$	4	*1 124 800*
$3p^5 3d$	$^1P^\circ$	1	*539 214*	$3p^5 4f$	$^1F^\circ$	3	*1 134 500*
$3s3p^6 3d$	^3D	1	667 677	$3p^5 4f$	$^3F^\circ$	3	*1 135 200*
		2	668 308			4	*1 135 300*
		3	669 326				
$3s3p^6 3d$	^1D	2	688 850	Mn IX ($^2P^\circ_{3/2}$)	***Limit***		**1 569 000**

Mn IX

$Z=25$

Cl I isoelectronic sequence

Ground state: $1s^22s^22p^63s^23p^5\,^2P^\circ_{3/2}$

Ionization energy = 1 789 000±4000 cm^{-1} (221.8±0.5 eV)

The transition array $3s^23p^5-3s^23p^44s$ was observed by Edlén (1937). The separation of the $^2P^\circ$ ground term is from Smitt, Svensson, and Outred (1976) who observed the $3s^23p^5\,^2P^\circ-3s3p^6\,^2S$ doublet at 376.778 and 395.473 Å with an uncertainty of 2 cm^{-1}.

The three terms of $3p^43d$ are from Fawcett and Gabriel (1966). Their parent state designations are given according to the percentages in Fe X. The level values for $3p^44d$ are from Fawcett, Cowan, Kononov, and Hayes (1972). The latter authors give six lines of the array $3p^43d-3p^44f$, none of which are connected with the present system of levels.

The uncertainty in the levels of $3p^43d$ is ±100 cm^{-1}. For levels of $3p^44d$ the uncertainty is ±500 cm^{-1}.

The ionization energy is an extrapolated value by Lotz (1967).

References

Edlén, B. (1937), Z. Phys. **104**, 407.

Fawcett, B. C., Cowan, R. D., Kononov, E. Y., and Hayes, R. W. (1972), J. Phys. **B5**, 1255.

Fawcett, B. C., and Gabriel, A. H. (1966), Proc. Phys. Soc. **88**, 262.

Lotz, W. (1967), J. Opt. Soc. Am. **57**, 873.

Smitt, R., Svensson, L. A., and Outred, M (1976), Phys. Scr. **13**, 293.

Mn IX

Configuration	Term	J	Level (cm^{-1})	Configuration	Term	J	Level (cm^{-1})
$3s^2 3p^5$	$^2P^\circ$	3/2	*0*	$3s^2 3p^4(^3P)4d$	^2D	5/2	1 109 500
		1/2	*12 546*			3/2	1 110 700
$3s3p^6$	^2S	1/2	265 408	$3s^2 3p^4(^3P)4d$	^4F	5/2	1 112 200
$3s^2 3p^4(^1D)3d$	^2S	1/2	501 710	$3s^2 3p^4(^3P)4d$	^2F	5/2	1 113 800
$3s^2 3p^4(^3P)3d$	^2P	3/2	521 840	$3s^2 3p^4(^3P)4d$	^2P	3/2	1 116 300
		1/2	526 380	$3s^2 3p^4(^1D)4d$	^2S	1/2	1 130 700
$3s^2 3p^4(^3P)3d$	^2D	5/2	530 560				
		3/2	541 160	$3s^2 3p^4(^1D)4d$	^2P	3/2	1 137 000
						1/2	1 139 000
$3s^2 3p^4(^3P)4s$	^4P	5/2	873 580				
		3/2	880 070	$3s^2 3p^4(^1D)4d$	^2D	5/2	1 142 200
						3/2	1 145 700
$3s^2 3p^4(^3P)4s$	^2P	3/2	889 560				
		1/2	896 860	Mn X ($^3P_2$)	*Limit*		**1 789 000**
$3s^2 3p^4(^1D)4s$	^2D	5/2	910 890				
		3/2	911 310				
$3s^2 3p^4(^1S)4s$	^2S	1/2	950 060				

Mn X

S I isoelectronic sequence

Ground state: $1s^2 2s^2 2p^6 3s^2 3p^4\ ^3P_2$

Ionization energy = 2 003 000±4000 cm^{-1} (248.3±0.5 eV)

Edlén (1937) observed and identified the $3p^4-3p^3 4s$ array which occurs at 100 Å. He identified singlets and triplets but no intercombinations. The present values for the ^{3}P ground term were measured by Smitt, Svensson and Outred (1976) who also measured the $3s3p^5\ ^3P°$ and the ^{1}P transitions to $3p^4\ ^1$D and ^{1}S. The uncertainty in the triplet levels is ±4 cm^{-1}.

The $3p^3 3d$ configuration was first observed by Gabriel, Fawcett, and Jordan (1966) and by Fawcett and Gabriel (1966) except for the ^{1}P° term observed by Fawcett (1971). Fawcett also reported the $3s3p^5$ configuration and two intersystem combinations between $3p^4\ ^1$D and $3p^3 3d\ ^3P°$. These two combinations are used to establish the values for the singlet levels relative to the triplets with an uncertainty of ±30 cm^{-1}. The present values for $3p^3 3d\ ^3P°$ and ^{3}D° are also from Fawcett (1971).

The $3p^3 4d$ terms were first identified by Fawcett, Peacock, and Cowan (1968). The present values are from Fawcett, Cowan, Kononov, and Hayes (1972). The latter group also observed nine transitions in the $3d-4f$ array but they are not connected with the present systems of levels.

The uncertainty in the level values of $3p^3 3d$, $4s$, and $4d$ is ±100 cm^{-1}.

The ionization energy is an extrapolated value by Lotz (1967).

References

Edlén, B. (1937), Z. Phys. **104**, 188.

Fawcett, B. C., and Gabriel, A. H. (1966), Proc. Phys. Soc. **88**, 262.

Fawcett, B. C., Peacock, N. J., and Cowan, R. D. (1968), J. Phys. **B1**, 295.

Fawcett, B. C. (1971), J. Phys. **B4**, 1577.

Fawcett, B. C., Cowan, R. D., Kononov, E. Y., and Hayes, R. W. (1972), J. Phys. **B5**, 1255.

Gabriel, A. H., Fawcett, B. C., and Jordan, C. (1966), Proc. Phys. Soc. **87**, 825.

Lotz, W. (1967), J. Opt. Soc. Am. **57**, 873.

Smitt, R., Svensson, L. A., and Outred, M. (1976), Phys. Scr. **13**, 293.

Mn X

Configuration	Term	J	Level (cm^{-1})	Configuration	Term	J	Level (cm^{-1})
$3s^2 3p^4$	^{3}P	2	0	$3s^2 3p^3(^2D°)3d$	^{1}P°	1	*577 530*
		1	10 019				
		0	11 797	$3s^2 3p^3(^4S°)4s$	^{3}S°	1	*965 970*
$3s^2 3p^4$	^{1}D	2	33 820	$3s^2 3p^3(^2D°)4s$	^{3}D°	1	*991 860*
						2	*992 220*
$3s^2 3p^4$	^{1}S	0	73 545			3	*994 180*
$3s3p^5$	^{3}P°	2	*261 072*	$3s^2 3p^3(^2D°)4s$	^{1}D°	2	*1 002 160*
		1	*268 874*				
		0	*273 615*	$3s^2 3p^3(^2P°)4s$	^{1}P°	1	*1 032 090*
$3s3p^5$	^{1}P°	1	*333 402*	$3s^2 3p^3(^4S°)4d$	^{3}D°	2	*1 196 370*
						3	*1 197 350*
$3s^2 3p^3(^2D°)3d$	^{3}P°	1	*492 320*				
		2	*492 770*	$3s^2 3p^3(^2D°)4d$	^{1}D°	2	*1 237 650*
				$3s^2 3p^3(^2D°)4d$	^{1}F°	3	*1 241 140*
$3s^2 3p^3(^4S°)3d$	^{3}D°	3	*514 670*				
		2	*520 620*				
		1	*524 520*	Mn XI ($^4S°_1$)	***Limit***		**2 003 000**
$3s^2 3p^3(^2D°)3d$	^{1}D°	2	*536 130*				
$3s^2 3p^3(^2D°)3d$	^{1}F°	3	*550 800*				

Mn XI

$Z=25$

P I isoelectronic sequence

Ground state: $1s^2s^22p^63s^23p^3\,{}^4S^\circ_{3/2}$

Ionization energy = 2 307 000±5000 cm^{-1} (286.0±0.6 eV)

The levels of the $3s^23p^3-3s3p^4$ array are from Smitt, Svensson, and Outred (1976) who improved the measurements and extended the classifications of the earlier work of Fawcett and Peacock (1967) and Fawcett (1970, 1971). The coronal line at 1359.59 Å observed on the NRL-Skylab spectra by Feldman and Doschek (1977) is classified as the forbidden transition $3p^3\,{}^4S^\circ_{3/2}-3p^3\,{}^2P^\circ_{3/2}$, in exact agreement with the calculated value of Svensson (1971). The uncertainty in the levels of $3s^23p^3$ and $3s3p^4$ is ±5 cm^{-1}.

The $3p^23d$ configuration is from Fawcett (1971) who greatly extended the earlier identifications of Gabriel, Fawcett and Jordan (1966) and Fawcett, Gabriel and Saunders (1967).

The $3p^24s$, and $4d$ configurations are from the paper by Fawcett, Cowan, Kononov, and Hayes (1972) who also identified two unconnected multiplets in the $3d-4f$ array. The uncertainty in the levels of $3p^23d$, $4s$, and $4d$ is ±100 cm^{-1}. The ionization energy was obtained by Lotz (1967) by extrapolation.

References

Fawcett, B. C., and Peacock, N. J. (1967), Proc. Phys. Soc. **91**, 973.

Fawcett, B. C., Gabriel, A. H., and Saunders, P. A. H. (1967) Proc. Phys. Soc. **90**, 863.

Fawcett, B. C. (1970), J. Phys. **B3**, 1732.

Fawcett, B. C. (1971), J. Phys. **B4**, 1577.

Fawcett, B. C., Cowan, R. D., Kononov, E. Y., and Hayes, R. W. (1972), J. Phys. **B5**, 1255.

Feldman, U., and Doschek, G. A. (1977), J. Opt. Soc. Am. **67**, 726.

Gabriel, A. H., Fawcett, B. C., and Jordan, C. (1966), Proc. Phys. Soc. **87**, 825.

Lotz, W. (1967), J. Opt. Soc. Am. **57**, 873.

Smitt, R., Svensson, L. A., and Outred, M. (1976), Phys. Scr. **13**, 293.

Svensson, L. A. (1971), Sol. Phys. **18**, 232.

Mn XI

Configuration	Term	J	Level (cm^{-1})	Configuration	Term	J	Level (cm^{-1})
$3s^2\,3p^3$	$^4S^\circ$	3/2	*0*	$3s^2\,3p^2(^1D)3d$	^2P	1/2	530 620
						3/2	536 800
$3s^2\,3p^3$	$^2D^\circ$	3/2	*39 384*				
		5/2	*42 702*	$3s^2\,3p^2(^3P)3d$	^2F	7/2	541 030
$3s^2\,3p^3$	$^2P^\circ$	1/2	*68 945*	$3s^2\,3p^2(^3P)3d$	^2D	5/2	561 400
		3/2	*73 552*			3/2	563 060
$3s3p^4$	^4P	5/2	253 974	$3s^2\,3p^2(^3P)4s$	^4P	1/2	1 078 200
		3/2	261 683			3/2	1 084 130
		1/2	265 144			5/2	1 091 160
$3s3p^4$	^2D	3/2	314 532	$3s^2\,3p^2(^3P)4s$	^2P	3/2	1 102 840
		5/2	315 881				
				$3s^2\,3p^2(^1D)4s$	^2D	5/2	1 120 870
$3s3p^4$	^2P	3/2	361 400			3/2	1 121 880
		1/2	365 689				
				$3s^2\,3p^2(^3P)4d$	^4P	5/2	1 324 910
$3s3p^4$	^2S	1/2	379 093			3/2	1 332 280
$3s^2\,3p^2(^3P)3d$	^2P	3/2	467 240	$3s^2\,3p^2(^3P)4d$	^4F	5/2	1 329 310
		1/2	476 980				
				$3s^2\,3p^2(^3P)4d$	^2F	5/2	1 331 340
$3s^2\,3p^2(^3P)3d$	^4P	5/2	477 170			7/2	1 336 860
		3/2	480 720				
		1/2	483 040	$3s^2\,3p^2(^3P)4d$	^4D	7/2	1 345 410
$3s^2\,3p^2(^1D)3d$	^2D	3/2	515 210	$3s^2\,3p^2(^3P)4d$	^2D	5/2	1 348 630
		5/2	515 430			3/2	1 350 080

(Continued)

Mn XI—Continued

Configuration	Term	J	Level (cm^{-1})	Configuration	Term	J	Level (cm^{-1})
$3s^2 3p^2(^1D)4d$	^2F	$7/2$ $5/2$	1 360 590 1 361 630	$3s^2 3p^2(^1D)4d$	^2S	$1/2$	1 374 650
$3s^2 3p^2(^1D)4d$	^2D	$5/2$	1 362 940	Mn XII ($^3P_0$)	***Limit***		**2 307 000**

Mn XII

$Z=25$

Si I isoelectronic sequence

Ground state: $1s^2 2s^2 2p^6 3s^2 3p^2\ {}^3P_0$

Ionization energy = 2 536 000±5000 cm^{-1} (314.4±0.6 eV)

With one exception, all of the terms below $3p4s$ are derived from the observations and identifications of Fawcett (1971) with an uncertainty of ±100 cm^{-1}. Compared with Cr XI and Fe XIII the $3s3p^3\ {}^3D^\circ_1$ level seems to be about 5500 cm^{-1} too low. The $3p^2\ {}^1S_0$ level value was determined by interpolation by Smitt, Svensson, and Outred (1976).

The $3p4s$ levels are from Fawcett, Cowan, and Hayes (1972) and the $3p4d$ are from the classifications of Kastner et al. (1978) with an uncertainty of ±500 cm^{-1}.

The ionization energy was obtained by Lotz (1967) by extrapolation.

References

Fawcett, B. C. (1971), J. Phys. **B4**, 1577.

Fawcett, B. C., Cowan, R. D., and Hayes, R. W. (1972), J. Phys. **B5**, 2143.

Kastner, S. O., Swartz, M., Bhatia, A. K., and Lapides, J. (1978), J. Opt. Soc. Am. **68**, 1558.

Lotz, W. (1967), J. Opt. Soc. Am. **57**, 873.

Smitt, R., Svensson, L. A., and Outred, M. (1976), Phys. Scr. **13**, 293.

Mn XII

Configuration	Term	J	Level (cm^{-1})	Configuration	Term	J	Level (cm^{-1})
$3s^2\,3p^2$	^3P	0	0	$3s^2\,3p3d$	$^1D^\circ$	2	462 700
		1	7 200				
		2	15 000	$3s^2\,3p3d$	$^3D^\circ$	1	469 910
						2	472 250
$3s^2\,3p^2$	^1D	2	42 150			3	472 540
$3s^2\,3p^2$	^1S	0	82 860+x	$3s^2\,3p3d$	$^1F^\circ$	3	517 360
$3s3p^3$	$^3D^\circ$	1	258 890?	$3s^2\,3p3d$	$^1P^\circ$	1	530 170+x
		2	264 550				
		3	266 600	$3s^2\,3p4s$	$^3P^\circ$	1	1 168 300
						2	1 181 300
$3s3p^3$	$^3P^\circ$	1	303 690				
		2	303 980	$3s^2\,3p4d$	$^3D^\circ$	1	1 414 000
						2	1 414 900
$3s3p^3$	$^1D^\circ$	2	333 970			3	1 417 100
$3s3p^3$	$^3S^\circ$	1	385 620	$3s^2\,3p4d$	$^3F^\circ$	3	1 425 600
$3s3p^3$	$^1P^\circ$	1	404 750	$3s^2\,3p4d$	$^1F^\circ$	3	1 437 000
$3s^2\,3p3d$	$^3P^\circ$	2	452 420	Mn XIII ($^2P^\circ_{1/2}$)	*Limit*		2 536 000
		1	460 000				

Mn XIII

$Z=25$

Al I isoelectronic sequence

Ground state: $1s^22s^22p^63s^23p\ ^2P^\circ_{1/2}$

Ionization energy = 2 771 000±6000 cm^{-1} (343.6±0.7 eV)

The splitting of the $3s^23p\ ^2P^\circ$ ground term of 15320 cm^{-1} is the average of five observations by Fawcett (1971). This separation has also been observed (Jefferies (1969)) as a forbidden line of the solar corona at 6536.3 Å (15 295 cm^{-1}), which we use here. The uncertainty in the value of the term interval is ±1 cm^{-1}.

The doublet terms of the $3s3p^2$ configuration and $3s^23d$ are determined from the combinations with the ground term observed by Fawcett (1971). The $4d\ ^2D$ term is from two lines identified by Edlén (1936) at 67 Å. The $4f\ ^2F^\circ$ term is from Fawcett, Cowan, Kononov, and Hayes (1972). The uncertainty in these levels is ±100 cm^{-1}.

The quartet system is based on the $3s3p^2\ ^4P_{5/2}$ level, the position of which was obtained by extrapolation in the sequence Al I–Ar VI . The uncertainty in our extrapolation is about ±2000 cm^{-1}. The values for the $3p^3\ ^4S^\circ$ and $3s3p3d\ ^4D^\circ$ levels are from Fawcett (1971). The $3s3p4s\ ^4P^\circ$ level is from Fawcett, Cowan, Kononov, and Hayes (1972). They also observed the $3s3p3d\ ^4F^\circ-3s3p3f\ ^4G$ multiplet but the terms cannot be connected to the present system.

The ionization energy was obtained by Lotz (1967) by extrapolation.

References

Edlén, B. (1936), Z. Phys. **103**, 536.

Fawcett, B. C. (1971), J. Phys. **B4**, 1577.

Fawcett, B. C., Cowan, R. D., Kononov, E. Y., and Hayes, R. W. (1972), J. Phys. **B5**, 1255.

Jefferies, J. T. (1969), Mem. Soc. Roy. Sci. Liege. Collect. **17**, 213.

Lotz, W. (1967), J. Opt. Soc. Am. **57**, 873.

Mn XIII

Configuration	Term	J	Level (cm^{-1})	Configuration	Term	J	Level (cm^{-1})
$3s^2\ 3p$	$^2P^\circ$	1/2	*0*	$3s3p3d$	$^4D^\circ$	7/2	*628 840+x*
		3/2	*15 295*	$3s3p4s$	$^4P^\circ$	5/2	*1 464 300+x*
$3s3p^2$	^4P	5/2	201 000+x	$3s^2\ 4d$	^2D	3/2	1 502 090
$3s3p^2$	^2D	3/2	276 560			5/2	1 503 060
		5/2	278 160	$3s^2\ 4f$	$^2F^\circ$	5/2	*1 586 200*
$3s3p^2$	^2S	1/2	339 020			7/2	*1 586 400*
$3s3p^2$	^2P	1/2	360 450				
		3/2	367 520	Mn XIV ($^1S_0$)	*Limit*		2 771 000
$3s^2\ 3d$	^2D	3/2	440 690				
		5/2	442 210				
$3p^3$	$^4S^\circ$	3/2	*524 890+x*				

Mn XIV

$Z=25$

Mg I isoelectronic sequence

Ground state: $1s^22s^22p^63s^2\ ^1S_0$

Ionization energy = 3 250 000±2000 cm^{-1} (403.0±0.2 eV)

Edlén (1936) reported three unconnected systems of levels for this ion: the resonance line $3s^2\ ^1S_0-3s4p\ ^1P_1^\circ$; the triplet system of $3s3p-3s4s$, $3s4d$, and the triplets of $3s3d-3s4f$, $3s5f$. The triplets were unified by the work of Fawcett, Gabriel, and Saunders (1967) who identified the $3s3p\ ^3P^\circ-3s3d\ ^3D$ multiplet. They also reported the $3s^2\ ^1S_0-3s3p\ ^1P_1^\circ$ line. The triplet system remains unconnected to the ground state but its position is estimated from a linearly interpolated value for $3s3p\ ^3P_1^\circ$ along the isoelectronic sequence with an uncertainty of ±2000 cm^{-1}.

The $3p^2\ ^3P$ and $3s3p\ ^1P^\circ$ terms were reported by Fawcett and Peacock (1967) and later remeasured by Fawcett (1971). Fawcett (1970) also provided the ^1S and ^1D of $3p^2$ and the known terms of $3p3d$, the $^3D^\circ$ and $^3F^\circ$. He classified the line at 260.41 as $3p^2\ ^3P_2-3p3d\ ^3P_2^\circ$ but this identification was later changed to $3s3p\ ^1P^\circ-3s3d\ ^1D$ by Fawcett, Cowan, and Hayes (1972). These authors also found the singlets of $3s4d$ and $3s4f$. Within the singlet or triplet system the uncertainty in these levels is ±100 cm^{-1}. Their publication was accompanied by a supplementary report, referenced in the same paper, which provides some extensions of the analysis. Here they identify the $^3S_1$ term of $3s5s$ and $3s6s$, the $^3F_4^\circ$ of $3s6f$, the $^1P_1^\circ$ of $3s5p$ and $3s6p$, the ^3D term of $3s5d$ and $3s6d$, the ^1D term of $3s4d$ and $3s5d$, and the $^3D_3$ of $3p4f$. Fawcett, Cowan, Kononov, and Hayes (1972) measured the $3p3d-3p4f$ array and identified the ^1F and ^3G terms of $3p4f$. The uncertainty in these level values is ±500 cm^{-1}.

We derived the ionization energy from the $3snp\ ^1P_1^\circ$ $n=4$, 5, 6 series and the $3snf\ ^3F_4^\circ$ $n=4$, 5, 6 series and obtained the values of 3 249 000 cm^{-1} and 3 251 000, respectively. We use the average of these values. Lotz obtained the value 404.1 eV by extrapolation. Several high-lying levels of other series reported by Fawcett, Cowan, and Hayes (1972) do not exhibit quantum defects that agree well with the isoelectronic sequence members analyzed by Ekberg (1971), K VIII through Ti XI. On this basis we find that the $3s5d\ ^3D$ and $^1D_1$, the $3s6s\ ^3S$, and the $3s6d\ ^3D$ terms need further confirmation.

References

Edlén, B. (1936), Z. Phys. **103**, 536.

Ekberg, J. O. (1971), Phys. Scr. **4**, 101.

Fawcett, B. C., Gabriel, A. H., Saunders, P. A. H. (1967), Proc. Phys. Soc. **90**, 863.

Fawcett, B. C., and Peacock, N. J. (1967), Proc. Phys. Soc. **91**, 973.

Fawcett, B. C. (1970), J. Phys. **B3**, 1732.

Fawcett, B. C. (1971), J. Phys. **B4**, 1577.

Fawcett, B. C., Cowan, R. D., Kononov, E. Y., and Hayes, R. W. (1972), J. Phys. **B5**, 1255.

Fawcett, B. C., Cowan, R. D., and Hayes, R. W. (1972), J. Phys. **B5**, 2143.

Lotz, W. J. (1967), J. Opt. Soc. Am. **57**, 873.

Mn XIV

Configuration	Term	J	Level (cm^{-1})	Configuration	Term	J	Level (cm^{-1})
$3s^2$	^1S	0	0	$3s3d$	^1D	2	712 040
$3s3p$	$^3P^\circ$	0	*218 720+x*	$3p3d$	$^3F^\circ$	2	*866 510+x*
		1	*223 500+x*			3	*874 400+x*
		2	*234 900+x*			4	*883 630+x*
$3s3p$	$^1P^\circ$	1	*328 030*	$3p3d$	$^3P^\circ$	2	*924 010+x*
$3p^2$	^3P	0	518 320+x	$3p3d$	$^3D^\circ$	3	*927 400+x*
		1	526 090+x				
		2	540 000+x	$3s4s$	^3S	1	1 568 920+x
$3p^2$	^1D	2	521 020	$3s4p$	$^1P^\circ$	1	*1 685 630*
$3p^2$	^1S	0	614 040	$3s4d$	^3D	1	1 818 070+x
						2	1 818 540+x
$3s3d$	^3D	1	633 540+x			3	1 819 460+x
		2	634 260+x				
		3	635 480+x	$3s4d$	^1D	2	1 820 100

(Continued)

Mn XIV—Continued

Configuration	Term	J	Level (cm^{-1})	Configuration	Term	J	Level (cm^{-1})
$3p4s$	$^3P°$	2	*1 856 810+x*	$3s5p$	$^1P°$	1	*2 286 000*
$3s4f$	$^3F°$	2	*1 887 930+x*	$3s5d$	^3D	3	2 320 100+x
		3	*1 888 010+x*			2	2 321 100+x
		4	*1 888 210+x*			1	2 328 100+x
$3s4f$	$^1F°$	3	*1 901 260*	$3s5d$	^1D	2	2 351 500
$3p4d$	$^3D°$	2	*2 087 900+x*	$3s5f$	$^3F°$	4	*2 383 000+x*
		3	*2 096 900+x*				
				$3s6s$	^3S	1	2 560 000+x
$3p4d$	$^3P°$	2	*2 121 500+x*				
				$3s6p$	$^1P°$	1	*2 595 000*
$3p4f$	^3G	3	2 141 700+x				
		4	2 147 600+x	$3s6d$	^3D	3	2 631 000+x
		5	2 159 900+x			2	2 632 000+x
$3p4f$	^3F	4	2 161 200+x	$3s6f$	$^3F°$	4	*2 650 000+x*
$3p4f$	^3D	3	2 170 200+x				
				Mn XV ($^2S_{1/2}$)	***Limit***		**3 250 000**
$3s5s$	^3S	1	2 223 000+x				

Mn XV

$Z=25$

Na I isoelectronic sequence

Ground state: $1s^2 2s^2 2p^6 3s\ ^2S_{1/2}$

Ionization energy = 3 509 820±300 cm^{-1} (435.166±0.030 eV)

Edlén (1936) reported three independent systems of doublets: $3s-4p$; $3p-4s$, $4d$, and $5d$; and $3d-4f$ and $5f$. They were unified by the observation by Fawcett, Gabriel, and Saunders (1967) of the $3s-3p$ and $3p-3d$ transitions. The series were extended to include $5s$ to $7s$, $5p$ to $10p$, $6d$ to $10d$, and $6f$ to $11f$ by Fawcett, Cowan, and Hayes (1972).

A new set of measurements by Cohen and Behring (1976) in the limited wavelength range of 30–55 Å, containing wavelengths given to the thousandths place, are used to derive the level values. We note from their paper, where all the known measurements are compiled, that the consistency is ±300 cm^{-1} for the $3p\ ^2P°$ term interval. Therefore we have rounded off the level values accordingly that are based on transitions to this term. These include all levels through $8d$. The rest are from the measurements of Fawcett, Cowan, and Hayes with an uncertainty of ±3000 cm^{-1}.

The value for the ionization energy was derived by Edlén (1978) from core polarization theory applied to the nf series.

References

Cohen, L., and Behring, W. E. (1976), J. Opt. Soc. Am. **66**, 899.

Edlén , B. (1936), Z. Phys. **100**, 621.

Edlén, B. (1978), Phys. Scr. **17**, 565.

Fawcett, B. C., Cowan, R. D., and Hayes, R. W. (1972), J. Phys. **B5**, 2143.

Fawcett, B. C., Gabriel, A. H., and Saunders, P. A. H. (1967), Proc. Phys. Soc. **90**, 863.

Mn XV

Configuration	Term	J	Level (cm^{-1})
$3s$	^2S	1/2	0
$3p$	$^2P°$	1/2	*259 900*
		3/2	*277 000*
$3d$	^2D	3/2	631 500
		5/2	633 700
$4s$	^2S	1/2	1 667 500
$4p$	$^2P°$	1/2	*1 770 400*
		3/2	*1 777 100*
$4d$	^2D	3/2	1 906 800
		5/2	1 907 800
$4f$	$^2F°$	5/2	*1 961 600*
		7/2	*1 962 000*
$5s$	^2S	1/2	2 375 200
$5p$	$^2P°$	1/2	*2 424 600*
		3/2	*2 428 100*
$5d$	^2D	3/2	2 491 100
		5/2	2 491 700
$5f$	$^2F°$	5/2	*2 519 100*
		7/2	*2 519 400*
$6s$	^2S	1/2	2 742 000
$6p$	$^2P°$	1/2	*2 768 600*
		3/2	*2 770 200*
$6d$	^2D	3/2	2 805 300
		5/2	2 805 700
$6f$	$^2F°$	5/2	*2 821 700*
		7/2	*2 821 900*
$7s$	^2S	1/2	2 950 000
$7p$	$^2P°$	1/2,3/2	*2 973 100*
$7d$	^2D	3/2	2 993 900
		5/2	2 994 200
$7f$	$^2F°$	5/2	*3 003 900*
		7/2	*3 004 200*
$8p$	$^2P°$	1/2,3/2	*3 102 700*
$8d$	^2D	5/2	3 115 600
$8f$	$^2F°$	7/2	*3 125 000*
$9p$	$^2P°$	3/2	*3 188 000*
$9d$	^2D	5/2	3 199 000
$9f$	$^2F°$	7/2	*3 205 000*

(Continued)

Mn XV—Continued

Configuration	Term	J	Level (cm^{-1})	Configuration	Term	J	Level (cm^{-1})
10p	$^2P^\circ$	3/2	*3 246 000*	11f	$^2F^\circ$	7/2	*3 306 000*
10d	^2D	5/2	3 258 000	Mn XVI ($^1S_0$)	***Limit***		3 509 820
10f	$^2F^\circ$	7/2	*3 264 000*				

Mn XVI

$Z=25$

Ne I isoelectronic sequence

Ground state: $1s^2 2s^2 2p^6\ {}^1S_0$

Ionization energy = 9 152 000±5000 cm^{-1} (1134.7±0.6 eV)

Only resonance lines are classified by this system of energy levels. Tyrén (1938) identified resonance lines from the $2s^2 2p^5 3s$, $3d$, and $4d$ as well as the $2s 2p^6 3p$ levels. Swartz, Kastner, Rothe, and Neupert (1971) made identifications of $2s^2 2p^5 4s$, $4d$, $5d$, and $6d$ levels and confirmed Tyrén's $2s 2p^6 3p$ levels. The uncertainty in the level values is ±2000 cm^{-1}. The percentage compositions were calculated by Bogdanovich, et al. (1980).

Kastner, Behring, and Cohen (1975) identified transitions between $2p^5 3p$ and $2p^5 4d$, but there is no connection with the levels given here.

We derived the ionization energy from the $2s^2 2p^5 nd\ {}^3D_1^\circ$ series for $n=4$, 5, and 6 assuming a linear quantum defect formula.

References

Bogdanovich, P. O., Bogdanovichene, M. I., Grudzinskas, I. I., Rudzikas, Z. B., Tutlis, V. I., and Shadzhjuvene, S. D., "Spectroscopy of Multiply Charged Ions," USSR Acad. of Sci., Sci. Council on Spectr. (1980).

Kastner, S. O., Behring, W. E., and Cohen, L. (1975), Astrophys. J. **199**, 777.

Swartz, M., Kastner, S., Rothe, E., and Neupert, W. (1971), J. Phys. **B4**, 1747.

Tyrén, F. (1938), Z. Phys. **111**, 314.

Mn XVI

Configuration	Term	J	Level (cm^{-1})	Leading percentages		
$2s^2 2p^6$	^1S	0	0			
$2s^2 2p^5(^2P^\circ_{3/2})3s$	$^3P^\circ$	1	*5 281 200*	52	48	$^1P^\circ$
$2s^2 2p^5(^2P^\circ_{1/2})3s$	$^1P^\circ$	1	*5 360 800*	52	48	$^3P^\circ$
$2s^2 2p^5(^2P^\circ_{3/2})3d$	$^3P^\circ$	1	*5 849 700*	92		
$2s^2 2p^5(^2P^\circ_{3/2})3d$	$^3D^\circ$	1	*5 923 500*	79	14	$^1P^\circ$
$2s^2 2p^5(^2P^\circ_{1/2})3d$	$^1P^\circ$	1	*6 018 300*	85	13	$^3D^\circ$
$2s2p^6\ 3p$	$^3P^\circ$	1	*6 530 800*	92	8	$^1P^\circ$
$2s2p^6\ 3p$	$^1P^\circ$	1	*6 562 500*	92	8	$^3P^\circ$
$2s^2 2p^5\ 4s$	$^3P^\circ$	1	*7 092 000*	62	37	$^1P^\circ$
$2s^2 2p^5\ 4d$	$^3D^\circ$	1	*7 348 000*	52	44	$^1P^\circ$
$2s^2 2p^5\ 4d$	$^1P^\circ$	1	*7 429 000*	56	35	$^3D^\circ$
$2s^2 2p^5\ 5d$	$^3D^\circ$	1	*7 994 000*			
$2s^2 2p^5\ 5d$	$^1P^\circ$	1	*8 084 000*			
$2s^2 2p^5\ 6d$	$^3D^\circ$	1	*8 354 000*			
$2s^2 2p^5\ 6d$	$^1P^\circ$	1	*8 439 000*			
Mn XVII ($^2P^\circ_{3/2}$)	*Limit*		**9 152 000**			

Mn XVII

$Z = 25$

F I isoelectronic sequence

Ground state: $1s^2 2s^2 2p^5\ ^2P^\circ_{3/2}$

Ionization energy = 9 872 000±20 000 cm^{-1} (1224±2.5 eV)

The ground term interval is obtained from the doublet $2s^2 2p^5\ ^2P^\circ - 2s2p^6\ ^2S$ measured by Fawcett (1971) at 100.00 Å and 109.35 Å with an uncertainty of ±0.05 Å, giving a level uncertainty of ±500 cm^{-1}.

The transition arrays $2p^5 - 2p^4 3s$ and $2p^5 - 2p^4 3d$ at 15–18 Å were analyzed by Fawcett, Gabriel, and Saunders (1967), and later remeasured by Feldman, Doschek, Cowan, and Cohen (1973) with a wavelength accuracy of ±0.01 Å. The later analysis is quoted here. Their level uncertainty is ±4000 cm^{-1}.

Bogdanovich et al. (1980) calculated the percentage compositions of the levels.

The ionization energy was obtained by extrapolation by Lotz (1967).

References

Bogdanovich, P. O., Bogdanovichene, M. I., Grudzinskas, I. I., Rudzikas, Z. B., Tutlis, V. I., and Shadzhjuvene S. D., "Spectroscopy of Multiply Charged Ions", USSR Acad. of Sci., Sci. Counc. on Spectr. (1980).

Fawcett, B. C. (1971), J. Phys. **B4**, 981.

Fawcett, B. C., Gabriel, A. H., and Saunders, P. A. H. (1967), Proc. Phys. Soc. **90**, 863.

Feldman, U., Doschek, G. A., Cowan, R. D., and Cohen, L. (1973), J. Opt. Soc. Am. **63**, 1445.

Lotz, W. (1967), J. Opt. Soc. Am. **57**, 873.

Mn XVII

Configuration	Term	J	Level (cm^{-1})	Leading percentages			
$2s^2 2p^5$	$^2P^\circ$	3/2	*0*	100			
		1/2	*85 500*	100			
$2s2p^6$	^2S	1/2	1 000 000				
$2s^2 2p^4(^3P)3s$	^4P	5/2	5 619 900	92			
		3/2	5 644 800	62		36	$(^3P)\ ^2P$
$2s^2 2p^4(^3P)3s$	^2P	3/2	5 701 100	53		37	$(^3P)\ ^4P$
		1/2	5 725 800	92			
$2s^2 2p^4(^1D)3s$	^2D	5/2	5 780 000	92			
		3/2	5 783 500	88			
$2s^2 2p^4(^1S)3s$	^2S	1/2	5 922 900	86			
$2s^2 2p^4(^3P)3d$	^4P	1/2	6 215 000	69		17	$(^3P)\ ^2P$
		3/2	6 228 900	58		21	$(^3P)\ ^2D$
$2s^2 2p^4(^3P)3d$		5/2	6 234 000	38	$(^3P)\ ^4F$	24	$(^3P)\ ^2F$
$2s^2 2p^4(^3P)3d$	^4F	5/2	6 255 100	40		30	$(^3P)\ ^4P$
$2s^2 2p^4(^3P)3d$		3/2	6 266 400	38	$(^3P)\ ^4F$	24	$(^3P)\ ^2F$
$2s^2 2p^4(^3P)3d$	^2F	5/2	6 279 000	42		38	$(^3P)\ ^4P$
$2s^2 2p^4(^3P)3d$	^2D	5/2	6 300 800	46		24	$(^3P)\ ^2F$
$2s^2 2p^4(^1D)3d$	^2S	1/2	6 356 600	88			
$2s^2 2p^4(^1D)3d$	^2P	3/2	6 379 200	67		23	$(^3P)\ ^2P$
$2s^2 2p^4(^1D)3d$	^2D	5/2	6 381 600	53		27	$(^3P)\ ^2D$
		3/2	6 404 200	65		28	

Mn XVII—Continued

Configuration	Term	J	Level (cm^{-1})	Leading percentages
$2s^2\,2p^4(^1S)3d$	^2D	$5/2$	6 491 800	88
		$3/2$	6 508 100	79
Mn XVIII ($^3P_2$)	***Limit***		**9 872 000**	

Mn XVIII

$Z=25$

O I isoelectronic sequence

Ground state: $1s^2 2s^2 2p^4\ ^3P_2$

Ionization energy $=$ 10 620 000$\pm$20 000 cm^{-1} (1317$\pm$3 eV)

Doschek, Feldman, Cowan, and Cohen (1974) observed and classified the $2s^2 2p^4 - 2s 2p^5$ array. The $2p^6\ ^1S$ level was reported by Fawcett, Galanti, and Peacock (1974). The spectrum was reobserved by Lawson and Peacock (1980) who found six intersystem lines. The levels given below were derived from their data with an uncertainty of $\pm$300 cm^{-1}. The percentage compositions for these levels were provided by Kaufman and Sugar (1982).

Lines of the $2p^4 - 2p^3 3s$ array at ~16 Å were classified by Doschek, Feldman, and Cohen (1973). Tentative assignments in the $2p^4 - 2p^3 3d$ array were made by Fawcett and Hayes (1975). The uncertainty in the $2p^3 3s$ levels is $\pm$3000 cm^{-1}.

The ionization energy was obtained by extrapolation by Lotz (1967).

References

Doschek, G. A., Feldman, U., and Cohen, L. (1973), J. Opt. Soc. Am. **63**, 1463.

Doschek, G. A., Feldman, U., and Cowan, R. D., and Cohen, L. (1974), Astrophys. J. **188**, 417.

Fawcett, B. C., Galanti, M., and Peacock, N. J. (1974), J. Phys. **B7**, 1149.

Fawcett, B. C., and Hayes, R.W. (1975), Mon. Not. R. Astron. Soc. **170**, 185.

Kaufman, V. and Sugar, J. (1982), private communication.

Lawson, K. D. and Peacock, N. J. (1980), J. Phys. **B13**, 3313.

Lotz, W.J. (1967), J. Opt. Soc. Am. **57**, 873.

Mn XVIII

Configuration	Term	J	Level (cm^{-1})	Leading percentages		
$2s^2 2p^4$	^3P	2	0	91	9	^1D
		0	66 560	83	16	^1S
		1	73 590	100		
$2s^2 2p^4$	^1D	2	150 610	91	9	^3P
$2s^2 2p^4$	^1S	0	292 020	81	17	^3P
$2s2p^5$	$^3P^\circ$	2	*866 690*	100		
		1	*919 450*	98	2	^1P
		0	*956 200*	100		
$2s2p^5$	$^1P^\circ$	1	*1 189 770*	98	2	^3P
$2p^6$	^1S	0	2 007 500	98	2	$2s^2 2p^4\ ^1S$
$2s^2 2p^3(^4S^\circ)3s$	$^3S^\circ$	1	*6 052 900*			
$2s^2 2p^3(^2D^\circ)3s$	$^3D^\circ$	2	*6 152 100*			
		1	*6 154 800*			
		3	*6 178 600*			
$2s^2 2p^3(^2D^\circ)3s$	$^1D^\circ$	2	*6 196 600*			
$2s^2 2p^3(^2P^\circ)3s$	$^1P^\circ$	1	*6 324 500*			
Mn XIX ($^4S^\circ_{3/2}$)	*Limit*		**10 620 000**			

Mn XIX

$Z = 25$

N I isoelectronic sequence

Ground state: $1s^2 2s^2 2p^3\ ^4S^\circ_{3/2}$

Ionization energy = 11 590 000±20 000 cm^{-1} (1437±3 eV)

The transition array $2s^2 2p^3 - 2s2p^4$ was classified by Doschek et al. (1974) and the term $2p^5\ ^2P^\circ$ was found by Doschek et al. (1975). These arrays were remeasured by Lawson and Peacock (1980) in the range of 85–148 Å with an uncertainty of ±0.03 Å. They identified five intersystem lines connecting the doublet and quartet terms. The uncertainty in the level values is ±300 cm^{-1}.

The percentage compositions of these levels were provided by Kaufman and Sugar (1982).

The ionization energy was obtained by extrapolation by Lotz (1967).

References

Doschek, G. A., Feldman, U., Cowan, R. D., and Cohen, L. (1974), Astrophys. J. **188**, 417.

Doschek, G. A., Feldman, U., Davis, J., and Cowan, R. D. (1975), Phys. Rev. **A12**, 980.

Kaufman, V. and Sugar, J. (1982), private communication.

Lawson, K. D., and Peacock, N. J. (1980), J. Phys. **B13**, 3313.

Lotz, W. (1967), J. Opt. Soc. Am. **57**, 873.

Mn XIX

Configuration	Term	J	Level (cm^{-1})	Leading percentages		
$2s^2 2p^3$	$^4S^\circ$	3/2	*0*	91	7	$^2P^\circ$
$2s^2 2p^3$	$^2D^\circ$	3/2	*131 770*	77	17	$^2P^\circ$
		5/2	*162 450*	100		
$2s^2 2p^3$	$^2P^\circ$	1/2	*242 110*	98	2	$2p^5\ ^2P^\circ$
		3/2	*291 730*	73	21	$2s^2 2p^3\ ^2D^\circ$
$2s2p^4$	^4P	5/2	709 070	98	2	^2D
		3/2	765 760	99	1	^2D
		1/2	785 670	96	4	^2S
$2s2p^4$	^2D	3/2	981 160	95	4	^2P
		5/2	992 600	98	2	^4P
$2s2p^4$	^2S	1/2	1 126 740	74	23	^2P
$2s2p^4$	^2P	3/2	1 170 820	95	4	^2D
		1/2	1 252 090	77	22	^2S
$2p^5$	$^2P^\circ$	3/2	*1 844 360*	98	2	$2s^2 2p^3\ ^2P^\circ$
		1/2	*1 934 390*	98	2	
Mn XX ($^3P_0$)	*Limit*		**11 590 000**			

Mn XX

$Z = 25$

C I isoelectronic sequence

Ground state: $1s^2 2s^2 2p^2\ ^3P_0$

Ionization energy = 12 410 000±20 000 cm^{-1} (1539±3 eV)

The transition arrays $2s^2 2p^2 - 2s2p^3$ was identified by Feldman et al. (1975). It was reobserved by Lawson and Peacock (1980) in the range of 89–192 Å with an accuracy of ±0.03 Å. They identified many more lines of this array, including several intersystem lines, and classified the $2s2p^3 - 2p^4$ array as well. The level uncertainty is ±300 cm^{-1}. Edlén (1984) has compared the known values of the $2s2p^3\ ^5S_2^\circ$ level in the isoelectronic sequence with the theoretical predictions. He concluded that the values given by Lawson and Peacock are inconsistent with the trend. We give Edlén's predicted value in brackets. The percentage compositions with configuration interaction were provided by Kaufman and Sugar (1982).

Bromage and Fawcett (1977) have given predicted wavelengths for the $2s^2 2p^2 - 2s^2 2p3d$ array.

The ionization energy is an extrapolated value by Lotz (1967).

References

Bromage, G. E. and Fawcett, B. C. (1977), Mon. Not. R. Astron. Soc. **178**, 605.

Edlén, B. (1984), private communication.

Feldman, U., Doschek, G. A., Cowan, R. D., and Cohen, L. (1975), Astrophys. J. **196**, 613.

Kaufman, V. and Sugar, J. (1982), private communication.

Lawson, K. D. and Peacock, B. C. (1980), J. Phys. **B13**, 3313.

Lotz, W. (1967), J. Opt. Soc. Am. **57**, 873.

Mn XX

Configuration	Term	J	Level (cm^{-1})	Leading percentages		
$2s^2 2p^2$	^3P	0	0	91	8	$2s^2 2p^2\ ^1S$
		1	59 580	99	1	$2p^4\ ^3P$
		2	98 650	79	20	$2s^2 2p^2\ ^1D$
$2s^2 2p^2$	^1D	2	212 260	79	19	^3P
$2s^2 2p^2$	^1S	0	333 080	88	8	^3P
$2s2p^3$	$^5S^\circ$	2	[*443 060*]	97	2	$^3P^\circ$
$2s2p^3$	$^3D^\circ$	2	*722 710*	85	14	$^3P^\circ$
		1	*723 090*	88	10	$^3P^\circ$
		3	*742 940*	100		
$2s2p^3$	$^3P^\circ$	0	*850 340*	100		
		1	*856 900*	87	10	$^3D^\circ$
		2	*870 580*	78	14	$^3D^\circ$
$2s2p^3$	$^3S^\circ$	1	*1 025 510*	83	14	$^1P^\circ$
$2s2p^3$	$^1D^\circ$	2	*1 048 880*	94	6	$^3P^\circ$
$2s2p^3$	$^1P^\circ$	1	*1 172 570*	84	14	$^3S^\circ$
$2p^4$	^3P	2	1 545 800	89	9	$2p^4\ ^1D$
		1	1 623 650	99	1	$2s^2 2p^2\ ^3P$
		0	1 623 890	88	11	$2p^4\ ^1S$
$2p^4$	^1D	2	1 698 290	90	9	^3P
$2p^4$	^1S	0	1 912 980	85	11	^3P
Mn XXI ($^2P^\circ_{1/2}$)	***Limit***		**12 410 000**			

Mn XXI

$Z=25$

B I isoelectronic sequence

Ground state: $1s^2 2s^2 2p$ $^2P^\circ_{1/2}$

Ionization energy = 13 260 000±27 000 cm^{-1} (1644±3 eV)

Doschek, Feldman, and Cohen (1975) reported two lines of the $2s^2 2p - 2s2p^2$ array, giving the ground term interval. Several more lines of this array were identified in a tokamak discharge by the TFR Group (1979). The spectrum was reobserved by Lawson and Peacock (1980) in the range of 108–258 Å with an accuracy of ±0.03 Å. Their measurements included the $2s^2 2p - 2s2p^2$ and $2s2p^2 - 2p^3$ arrays and the intersystem line $2s2p^2$ $^4P_{5/2} - 2p^3$ $^2D^\circ_{5/2}$ at 115.69 Å. Their level uncertainty is ±300 cm^{-1}. The percentage compositions of these levels were provided by Kaufman and Sugar (1982). Their calculation includes configuration interaction between $2s^2 2p$ and $2p^3$.

The ionization energy was obtained by Lotz (1967) by extrapolation.

References

Doschek, G. A., Feldman, U., and Cohen, L. (1975), J. Opt. Soc. Am. **65**, 463.

Kaufman, V. and Sugar, J. (1982), unpublished.

Lawson, K. D., and Peacock, N. J. (1980), J. Phys. **B13**, 3313.

TFR Group (1979), Phys. Lett. **74A**, 57.

Lotz, W. (1967), J. Opt. Soc. Am. **57**, 873.

Mn XXI

Configuration	Term	J	Level (cm^{-1})	Leading percentages		
$2s^2\,2p$	$^2P^\circ$	1/2	*0*	98	2	$2p^3$ $^2P^\circ$
		3/2	*99 360*	97	3	
$2s2p^2$	^4P	1/2	379 660	97	3	^2S
		3/2	424 980	100		
		5/2	470 670	96	4	^2D
$2s2p^2$	^2D	3/2	687 540	96	4	^2D
		5/2	704 190	96	4	^4P
$2s2p^2$	^2P	1/2	805 930	59	39	^2S
		3/2	924 710	96	4	^2D
$2s2p^2$	^2S	1/2	910 880	58	41	^2P
$2p^3$	$^4S^\circ$	3/2	*1 177 430*	93	5	$^2P^\circ$
$2p^3$	$^2D^\circ$	3/2	*1 310 890*	86	11	$^2P^\circ$
		5/2	*1 335 070*	100		
$2p^3$	$^2P^\circ$	1/2	*1 472 710*	98	2	$2s^2 2p$ $^2P^\circ$
		3/2	*1 517 410*	81	13	$2p^3$ $^2D^\circ$
Mn XXII ($^1S_0$)	***Limit***		**13 260 000**			

Mn XXII

$Z=25$

Be I isoelectronic sequence

Ground state: $1s^22s^2\,{}^1S_0$

Ionization energy = 14 420 000±29 000 cm^{-1} (1788±4 eV)

The intersystem resonance line, $2s^2\,{}^1S_0-2s2p\,{}^3P^\circ_1$ at 277.80±0.03 Å, was identified by Sandlin, Brueckner, Scherrer, and Tousey (1976) from Skylab data. The transition from the $2s2p\,{}^1P$ was outside their range.

Lawson and Peacock (1980) observed the $2s^2\,{}^1S_0-2s2p\,{}^1P^\circ_1$ line at 141.09 Å and classified the $2s2p-2p^2$ array, including the intersystem line $2s2p\,{}^3P^\circ_2-2p^2\,{}^1D_2$. Their measurements from 141–239 Å have a reported uncertainty of ±0.03 Å below 180 Å and ±0.06 Å above. The level uncertainty is ±300 cm^{-1}.

The higher-lying configurations are from the observations of Boiko et al. (1977) at 11–13 Å with a measurement uncertainty of ±0.003 Å, giving a level uncertainty of ±3000 cm^{-1}.

The ionization energy was obtained by Lotz (1967) by extrapolation.

References

Boiko, V. A., Pikuz, S. A., Safronova, U. I., and Faenov, A. Y. (1977), J. Phys. **B10**, 1253.

Lawson, K. D., and Peacock, N. J. (1980), J. Phys. **B13**, 3313.

Lotz, W. (1967), J. Opt. Soc. Am. **57**, 873.

Sandlin, G. D., Brueckner, G. E., Scherrer, V. E., and Tousey, R. (1976), Astrophys. J. **205**, L47.

Mn XXII

Configuration	Term	J	Level (cm^{-1})	Configuration	Term	J	Level (cm^{-1})
$2s^2$	^{1}S	0	0	$2s3d$	^{1}D	2	8 512 000
$2s2p$	^{3}P°	0	*333 350*	$2p3s$	^{1}P°	1	*8 702 000*
		1	*359 970*				
		2	*437 300*	$2p3p$	^{1}P	1	8 756 000
$2s2p$	^{1}P°	1	*708 770*	$2p3p$	^{3}P	2	8 858 000
$2p^2$	^{3}P	0	910 360	$2p3d$	^{3}D°	2	*8 860 000*
		1	967 950			1	*8 878 000*
		2	1 008 140			3	*8 957 000*
$2p^2$	^{1}D	2	1 125 660	$2p3p$	^{1}D	2	8 924 000
$2p^2$	^{1}S	0	1 336 400	$2p3d$	^{1}D°	2	*8 928 000*
$2s3s$	^{3}S	1	8 168 000	$2p3d$	^{3}P°	1	*8 975 000*
						2	*8 976 000*
$2s3p$	^{3}P°	1	*8 335 000*				
				$2p3d$	^{1}F°	3	*9 027 000*
$2s3p$	^{1}P°	1	*8 354 000*				
$2s3d$	^{3}D	1	8 433 000	Mn XXIII (^{2}S$_{1/2}$)	***Limit***		**14 420 000**
		2,3	8 445 000				

Mn XXIII

$Z=25$

Li I isoelectronic sequence

Ground state: $1s^22s\ ^2S_{1/2}$

Ionization energy = 15 162 000±35 000 cm^{-1} (1879.9±0.4 eV)

The $2s-2p$ resonance lines were measured in solar flare spectra at 206.90 Å and 266.88 Å with an uncertainty of ±0.02 Å by Widing and Purcell (1976), and were also reported by Sandlin, Brueckner, Scherrer, and Tousey (1976). The level uncertainty for $2p\ ^2P°$ is ±50 cm^{-1}. Laboratory observations of the $2p-3s$ and $2p-3d$ lines at ~12 Å were given by Goldsmith, Feldman, Oren, and Cohen (1972). These lines were remeasured with an uncertainty of ±0.003 Å and the additional transitions $2s-3p$, $2p-4d$, and $2s-4p$ were identified by Boiko, Faenov, and Pikuz (1978). Present results are from this paper. Their level uncertainty is ±3000 cm^{-1}.

The ionization energy was derived by Edlén (1979) from the *nd* series.

References

Boiko, V. A., Faenov, A. Y., and Pikuz, S. A. (1978), J. Quant. Spectr. Rad. Trans. **19**, 11.

Edlén, B. (1979), Phys. Scr. **19**, 255.

Goldsmith, S., Feldman, U., Oren, L., and Cohen, L. (1972), Astrophys. J. **174**, 209.

Sandlin, G. D., Brueckner, G. E., Scherrer, V. E., and Tousey, R. (1976), Astrophys. J. **205**, L47.

Widing, K. G., and Purcell, J. D. (1976), Astrophys. J. **204**, L151.

Mn XXIII

Configuration	Term	J	Level (cm^{-1})	Configuration	Term	J	Level (cm^{-1})
$1s^2\ 2s$	^2S	1/2	0	$1s^2\ 4p$	$^2P°$	1/2,3/2	*11 510 000*
$1s^2\ 2p$	$^2P°$	1/2	*374 700*	$1s^2\ 4d$	^2D	3/2,5/2	11 530 000
		3/2	*483 320*				
$1s^2\ 3s$	^2S	1/2	8 518 000	Mn XXIV ($^1S_0$)	***Limit***		**15 162 000**
$1s^2\ 3p$	$^2P°$	1/2	*8 618 000*				
		3/2	*8 648 000*				
$1s^2\ 3d$	^2D	3/2	8 691 000				
		5/2	8 708 000				

Mn XXIV

$Z=25$

He I isoelectronic sequence

Ground state: $1s^2\,{}^1S_0$

Ionization energy $=$ 65 660 000$\pm$13 000 cm^{-1} (8140.7$\pm$2.0 eV)

Because of the excellent agreement of the calculated energies of the $n=2$ shell by Safronova (1981) with the few well-measured spectra in the He I sequence, we have compiled her results for the $n=2$ levels and for the ionization energy. Detailed comparisons are given in the Introduction. Levels of the $n=3$–5 shells are from the calculated binding energies by Ermolaev and Jones (1974) subtracted from Safronova's value for the binding energy of the ground state. We have assumed an uncertainty of 2 parts in 10^4 for the excited levels relative to the ground state, and for the ionization energy. For differences between excited levels where $\Delta n=0$, we assumed an uncertainty of 2 parts in 10^3.

Percentage compositions are from Ermolaev and Jones.

References

Ermolaev, A. M., and Jones, M. (1974), J. Phys. **B7**, 199.
Safronova, U. I. (1980), Phys. Scr. **23**, 241.

Mn XXIV

Configuration	Term	J	Level (cm^{-1})	Leading percentages		
$1s^2$	^1S	0	0			
$1s2s$	^3S	1	[49 369 590]			
$1s2p$	$^3P°$	0	*[49 591 860]*			
		1	*[49 608 800]*	93	7	$^1P°$
		2	*[49 706 640]*			
$1s2s$	^1S	0	[49 612 060]			
$1s2p$	$^1P°$	1	*[49 846 540]*	93	7	$^3P°$
$1s3s$	^3S	1	[58 488 650]			
$1s3p$	$^3P°$	0	*[58 550 250]*			
		1	*[58 554 320]*	92	8	$^1P°$
		2	*[58 583 550]*			
$1s3s$	^1S	0	[58 552 360]			
$1s3p$	$^1P°$	1	*[58 622 130]*	92	8	$^3P°$
$1s4s$	^3S	1	[61 645 020]			
$1s4p$	$^3P°$	0	*[61 670 660]*			
		1	*[61 672 360]*	91	9	$^1P°$
		2	*[61 684 720]*			
$1s4s$	^1S	0	[61 670 830]			
$1s4p$	$^1P°$	1	*[61 700 440]*	91	9	$^3P°$
$1s5s$	^3S	1	[63 097 780]			
$1s5s$	^1S	0	[63 110 670]			

Mn XXIV—Continued

Configuration	Term	J	Level (cm^{-1})	Leading percentages		
1s5p	$^3P^\circ$	0	[*63 110 750*]			
		1	[*63 111 610*]	91	9	$^1P^\circ$
		2	[*63 117 950*]			
1s5p	$^1P^\circ$	1	[*63 125 890*]	91	9	$^3P^\circ$
Mn XXV ($^2S_{1/2}$)	***Limit***		**65 660 000**			

Mn XXV

$Z=25$

H I isoelectronic sequence

Ground state: $1s\ ^2S_{1/2}$

Ionization energy $=$ 69 137 400$\pm$300 cm^{-1} (8572.01$\pm$0.04 eV)

No observations of this spectrum are reported. We give calculated values by Mohr (1983) for the $n=2$ shell and by Erickson (1977) for $n=3$–5 relative to the $2p\ ^2P^{\circ}_{3/2}$ level. Further details are given in the Introduction. Relative to the ground state, the level uncertainty is estimated to be 5 parts in 10^7. The uncertainty in the excited states relative to $2p\ ^2P^{\circ}_{3/2}$ is 1 part in 10^6.

References

Erickson, G. W. (1977), J. Phys. Chem. Ref. Data **6**, 831.

Mohr, P. J. (1983), At. Data Nucl. Data Tables **29**, 453.

Mn XXV

Configuration	Term	J	Level (cm^{-1})
$1s$	^2S	1/2	0
$2p$	$^2P^{\circ}$	1/2	*[51 809 520]*
		3/2	*[51 955 528]*
$2s$	^2S	1/2	[51 813 492]
$3p$	$^2P^{\circ}$	1/2	*[61 452 412]*
		3/2	*[61 495 697]*
$3s$	^2S	1/2	[61 453 650]
$3d$	^2D	3/2	[61 495 620]
		5/2	[61 509 824]
$4p$	$^2P^{\circ}$	1/2	*[64 821 451]*
		3/2	*[64 839 699]*
$4s$	^2S	1/2	[64 821 974]
$4d$	^2D	3/2	[64 839 666]
		5/2	[64 845 662]
$4f$	$^2F^{\circ}$	5/2	*[64 845 652]*
		7/2	*[64 848 639]*
$5p$	$^2P^{\circ}$	1/2	*[66 378 293]*
		3/2	*[66 387 628]*
$5s$	^2S	1/2	[66 378 561]
$5d$	^2D	3/2	[66 387 611]
		5/2	[66 390 682]
$5f$	$^2F^{\circ}$	5/2	*[66 390 676]*
		7/2	*[66 392 206]*
$5g$	^2G	7/2	[66 392 203]
		9/2	[66 393 120]
	Limit		**69 137 400**

Fe I

$Z=26$

Ground state: $1s^2 2s^2 2p^6 3s^2 3p^6 3d^6 4s^2\ {}^5D_4$

Ionization energy = 63 737±1 cm^{-1} (7.9024±0.0001 eV)

The principal contributors to the analysis of Fe I are Walters, Laporte, Burns, and Catalán, who together provided 404 energy levels. Following their work and utilizing new Zeeman effect data from the Massachusetts Institute of Technology, Russell, Moore, and Weeks (1944) were able to confirm the older analysis. A few high levels were later found by Kiess, Rubin, and Moore (1961). The five-place g-values are from Childs and Goodman (1965), who give uncertainties of about ±0.00005 with the values. The rest, from Russell et al., have an uncertainty of ±0.001 to 0.009 depending on the determination.

Revised values for many of the levels have been provided by Crosswhite (1975). These result from a new set of observations made with a low pressure hollow cathode discharge. The values given below to three decimal places are due to Crosswhite. He ascribes an accuracy of ±0.002 cm^{-1} to these levels. A comparison of his results with the earlier data, which were derived from arc sources at atmospheric pressure, shows that the earlier values of levels belonging to the $3d^8$, $3d^7 4s$, $3d^7 4p$, $3d^7 4d$, $3d^6 4s 4p$, $3d^6 4s 4d$, and $3d^5 4s^2 4p$ configurations should be reduced by 0.04 cm^{-1} to obtain values consistent with observations from low pressure sources. This correction has been applied by us to all levels of these configurations whose values were not already revised by Crosswhite. These are given below to two decimal places. Insufficient information is available to establish corresponding corrections for levels of the $3d^6 4s 5s$, $3d^6 4s 6s$, $3d^6 4s 7s$, $3d^6 4s 5d$, $3d^7 5s$, and $3d^6 4s 5p$ configurations. The uncorrected values rounded off to two decimal places are given below. The two decimal place values for energy levels are assumed to have an uncertainty of ±0.05 cm^{-1}.

Litzén (1976) observed the spectrum from 13 350–24 924 Å and identified new terms in $3d^6 4s 5p$ and $3d^7 5p$. He also determined revised level values for a few high even terms. His results are included here.

A fitted calculation of the even configurations $3d^6 4s^2$, $3d^7 4s$, and $3d^8$ in intermediate coupling by Dembczynski (1980) provided leading percentages for these levels.

The leading percentages for the levels of odd parity are from Roth (1981). He has calculated the $3d^7 4p$, $3d^6 4s 4p$, and $3d^5 4s^2 4p$ groups of levels with configuration interaction. Roth distinguished repeating terms of the $3d^n$ core by the letters a, b ... rather than by seniority. His percentage composition for a given level is the sum of the percentages of states that differ only in the seniority of the core term.

The alphabetic prefixing of final terms with lower case letters, which served to distinguish final terms of the same type, has been repeated here from the literature except for levels that have been redesignated as a result of a new theoretical interpretation. Similarly, the authors' numerical designations for uninterpreted levels have been retained.

The ionization energy was obtained by laser-scanning of Rydberg series by Worden et al. (1983).

References

Burns, K., and Walters, Jr., F. M. (1929), Publ. Allegheny Observ. **6**, 159.

Catalán, M. A. (1930), An. R. Soc. Esp. Fis. Quim. **28**, 1239.

Catalán, M. A., and Velasco, R. (1952), An. R. Soc. Esp. Fis. Quim. **48A**, 247.

Childs, W. J., and Goodman, L. S. (1965), Phys. Rev. **140 A**, 447.

Crosswhite, H. M. (1975), J. Res. Natl. Bur. Stand. (U.S.) **79A**, 17.

Dembczynski, J. (1980), Physica **100 C**, 105.

Kiess, C. C., Rubin, V. C., and Moore, C. E. (1961), J. Res. Natl. Bur. Stand. (U.S.) **65A**, 1.

Laporte, O. (1924), Z. Physik **23**, 135; **26**, 1.

Litzén, U. (1976), Phys. Scr. **14**, 165.

Roth, C. (1980), At. Data Nucl. Data Tables **25**, 91.

Russell, H. N., Moore, C. E., and Weeks, D. W. (1944), Trans. Am. Phil. Soc. **34-II**, 111.

Walters, F. M. (1924), J. Opt. Soc. Am. **8**, 245.

Walters, F. M. (1923), J. Wash. Acad. Sci. **13**, 243.

Worden, E. F., Comaskey, B., Densberger, J., Christensen, J., McAfee, J. M., and Paisner, J. A. (1984), J. Opt. Soc. Am. **B1**, 314

Fe I

Configuration	Term	J	Level (cm^{-1})	g	Leading percentages		
$3d^6\,4s^2$	$a\ ^5$D	4	0.000	1.50020	100		
		3	415.932	1.50034	100		
		2	704.004	1.50041	100		
		1	888.129	1.50022	100		
		0	978.072		100		
$3d^7(^4$F$)4s$	$a\ ^5$F	5	6 928.266	1.40021	100		
		4	7 376.760	1.35004	100		
		3	7 728.056	1.24988	100		
		2	7 985.780	0.99953	100		
		1	8 154.710	-0.014	100		
$3d^7(^4$F$)4s$	$a\ ^3$F	4	11 976.234	1.254	98	1	$3d^64s^2\ ^3$F2
		3	12 560.930	1.086	98	1	
		2	12 968.549	0.670	98	1	
$3d^7(^4$P$)4s$	$a\ ^5$P	3	17 550.175	1.666	99		
		2	17 726.981	1.820	99		
		1	17 927.376	2.499	99		
$3d^6\,4s^2$	$a\ ^3$P2	2	18 378.181	1.506	55	32	^{3}P1
		1	19 552.473	1.500	55	32	
		0	20 037.813		55	32	
$3d^6(^5$D$)4s4p(^3$P°$)$	$z\ ^7$D°	5	*19 350.892*	1.597	99		
		4	*19 562.440*	1.642	98		
		3	*19 757.033*	1.746	99		
		2	*19 912.494*	2.008	99		
		1	*20 019.635*	2.999	100		
$3d^6\,4s^2$	$a\ ^3$H	6	19 390.164	1.163	100		
		5	19 621.005	1.038	100		
		4	19 788.245	0.811	100		
$3d^6\,4s^2$	$b\ ^3$F2	4	20 641.109	1.235	71	21	^{3}F1
		3	20 874.484	1.073	71	21	
		2	21 038.985	0.663	71	21	
$3d^7(^2$G$)4s$	$a\ ^3$G	5	21 715.730	1.197	88	10	$3d^64s^2\ ^3$G
		4	21 999.127	1.051	88	10	
		3	22 249.428	0.756	88	10	
$3d^6(^5$D$)4s4p(^3$P°$)$	$z\ ^7$F°	6	*22 650.421*	1.498	100		
		5	*22 845.868*	1.498	99		
		4	*22 996.676*	1.493	99		
		3	*23 110.937*	1.513	99		
		2	*23 192.497*	1.504	99		
		1	*23 244.834*	1.549	100		
		0	*23 270.374*		100		
$3d^7(^4$P$)4s$	$b\ ^3$P	2	22 838.318	1.498	92	4	$3d^64s^2\ ^3$P1
		1	22 946.808	1.489	79	10	$3d^7(^2$P$)4s\ ^3$P
		0	23 051.742		79	12	$3d^7(^2$P$)4s\ ^3$P
$3d^6(^5$D$)4s4p(^3$P°$)$	$z\ ^7$P°	4	*23 711.457*	1.747	98		
		3	*24 180.864*	1.908	99		
		2	*24 506.919*	2.333	98		

Fe I—Continued

Configuration	Term	J	Level (cm^{-1})	g	Leading percentages		
$3d^6\,4s^2$	$b\ ^3G$	5	23 783.614	1.200	88	10	$3d^7(^2G)4s\ ^3G$
		4	24 118.814	1.048	88	10	
		3	24 338.762	0.761	88	10	
$3d^7(^2P)4s$	$c\ ^3P$	2	24 335.759	1.484	90	4	$(^2D2)\ ^3D$
		1	24 772.017	1.466	81	7	$(^4P)\ ^3P$
		0	25 091.597		79	12	$(^4P)\ ^3P$
$3d^7(^2G)4s$	$a\ ^1G$	4	24 574.650	1.001	90	3	$(^2H)\ ^3H$
$3d^6(^5D)4s4p(^3P^\circ)$	$z\ ^5D^\circ$	4	*25 899.987*	1.502	91	6	$3d^7(^4F)4p\ ^5D^\circ$
		3	*26 140.177*	1.500	91	6	
		2	*26 339.691*	1.503	92	6	
		1	*26 479.376*	1.495	92	6	
		0	*26 550.476*		93	5	
$3d^7(^2H)4s$	$b\ ^3H$	6	26 105.904	1.165	100		
		5	26 351.039	1.032	98	2	$(^2H)\ ^1H$
		4	26 627.604	0.811	98	2	$(^2G)\ ^1G$
$3d^7(^2D2)4s$	$a\ ^3D$	3	26 224.966	1.335	74	3	$(^2D1)\ ^3D$
		1	26 406.470	0.731	45	35	$(^2P)\ ^1P$
		2	26 623.730	1.178	67	18	$(^2D1)\ ^3D$
$3d^6(^5D)4s4p(^3P^\circ)$	$z\ ^5F^\circ$	5	*26 874.549*	1.399	95	4	$3d^7(^4F)4p\ ^5F^\circ$
		4	*27 166.819*	1.355	94	4	
		3	*27 394.688*	1.250	94	4	
		2	*27 559.581*	1.004	95	4	
		1	*27 666.346*	−0.012	95	4	
$3d^7(^2P)4s$	$a\ ^1P$	1	27 543.004	0.817	62	23	$(^2D2)\ ^3D$
$3d^7(^2D2)4s$	$a\ ^1D$	2	28 604.606	1.028	64	16	$(^2D1)\ ^1D$
$3d^7(^2H)4s$	$a\ ^1H$	5	28 819.946	1.000	98		
$3d^6(^5D)4s4p(^3P^\circ)$	$z\ ^5P^\circ$	3	*29 056.321*	1.657	98		
		2	*29 469.020*	1.835	97		
		1	*29 732.733*	2.487	97		
$3d^6\,4s^2$	$a\ ^1I$	6	29 313.003	1.014	100		
$3d^6\,4s^2$	$b\ ^3D$	1	29 320.028		88	8	$3d^7(^2D2)4s\ ^3D$
		2	29 356.740		81	7	
		3	29 371.811	1.326	94	4	
$3d^6\,4s^2$	$b\ ^1G2$	4	29 798.933	0.979	62	35	$^1G1$
$3d^6(^5D)4s4p(^3P^\circ)$	$z\ ^3F^\circ$	4	*31 307.243*	1.250	94	5	$3d^7(^4F)4p\ ^3F^\circ$
		3	*31 805.067*	1.086	97		
		2	*32 133.986*	0.682	93	5	$3d^7(^4F)4p\ ^3F^\circ$
$3d^6(^5D)4s4p(^3P^\circ)$	$z\ ^3D^\circ$	3	*31 322.611*	1.321	90	8	$3d^7(^4F)4p\ ^3D^\circ$
		2	*31 686.346*	1.168	90	8	
		1	*31 937.316*	0.513	91	8	
$3d^8$	$c\ ^3F$	4	32 873.619	1.264	92	3	$3d^7(^2F)4s\ ^3F$
		3	33 412.713	1.066	92	5	
		2	33 765.304	0.677	86	6	

(Continued)

Fe I—Continued

Configuration	Term	J	Level (cm^{-1})	g	Leading percentages		
$3d^7(^4F)4p$	$y\,^5D^\circ$	4	*33 095.937*	1.496	61	34	$3d^6(^5D)4s4p(^1P^\circ)\ ^5D^\circ$
		3	*33 507.120*	1.492	60	34	$3d^6(^5D)4s4p(^1P^\circ)\ ^5D^\circ$
		2	*33 801.567*	1.495	56	34	$3d^6(^5D)4s4p(^1P^\circ)\ ^5D^\circ$
		1	*34 017.098*	1.492	47	30	$3d^6(^5D)4s4p(^3P^\circ)\ ^3P^\circ$
		0	*34 121.58*		42	28	$3d^6(^5D)4s4p(^3P^\circ)\ ^3P^\circ$
$3d^7(^4F)4p$	$y\,^5F^\circ$	5	*33 695.394*	1.417	84	11	$3d^6(^5D)4s4p(^1P^\circ)\ ^5F^\circ$
		4	*34 039.513*	1.344	81	12	
		3	*34 328.749*	1.244	81	11	
		2	*34 547.206*	0.998	82	13	
		1	*34 692.144*	-0.016	84	12	
$3d^6(^5D)4s4p(^3P^\circ)$	$z\,^3P^\circ$	2	*33 946.929*	1.493	91	4	$3d^7(^4F)4p\ ^5D^\circ$
		1	*34 362.871*	1.496	50	30	
		0	*34 555.60*		69	18	
$3d^6\,4s^2$	$b\,^1D2$	2	34 636.78		67	20	$^1D1$
$3d^7(^4F)4p$	$z\,^5G^\circ$	5	*34 782.416*	1.218	58	35	$3d^7(^4F)4p\ ^3G^\circ$
		6	*34 843.94*	1.332	94	4	$3d^6(^3H)4s4p(^3P^\circ)\ ^5G^\circ$
		4	*35 257.319*	1.103	75	16	$3d^7(^4F)4p\ ^3G^\circ$
		3	*35 611.619*	0.887	86	6	$3d^7(^4F)4p\ ^3G^\circ$
		2	*35 856.400*	0.335	92	5	$3d^6(^3H)4s4p(^3P^\circ)\ ^5G^\circ$
$3d^7(^4F)4p$	$z\,^3G^\circ$	5	*35 379.206*	1.248	61	33	$^5G^\circ$
		4	*35 767.561*	1.100	78	16	
		3	*36 079.366*	0.791	89	6	
$3d^7(^4F)4p$	$y\,^3F^\circ$	4	*36 686.164*	1.246	86	5	$3d^6(^5D)4s4p(^3P^\circ)\ ^3F^\circ$
		3	*37 162.740*	1.086	84	5	
		2	*37 521.157*	0.688	87	5	
$3d^6(^5D)4s4p(^1P^\circ)$	$y\,^5P^\circ$	3	*36 766.962*	1.661	60	34	$3d^5(^6S)4s^24p\ ^5P^\circ$
		2	*37 157.557*	1.836	60	35	
		1	*37 409.542*	2.502	59	36	
$3d^7(^2F)4s$	$d\,^3F$	2	36 940.56		92	6	$3d^8\ ^3F$
		3	36 975.60		94	5	
		4	37 045.96		96	3	
$3d^7(^4F)4p$	$y\,^3D^\circ$	3	*38 175.350*	1.324	84	8	$3d^6(^5D)4s4p(^3P^\circ)\ ^3D^\circ$
		2	*38 678.032*	1.151	85	7	
		1	*38 995.730*	0.493	86	7	
$3d^6(^5D)4s4p(^1P^\circ)$	$x\,^5D^\circ$	4	*39 625.800*	1.489	55	18	$3d^7(^4F)4p\ ^5D^\circ$
		3	*39 969.844*	1.504	54	19	
		2	*40 231.332*	1.501	53	19	
		1	*40 404.506*	1.498	53	20	
		0	*40 491.274*		53	20	
$3d^5(^6S)4s^2\,4p$	$y\,^7P^\circ$	2	*40 052.030*	2.340	97		
		3	*40 207.086*	1.908	98		
		4	*40 421.85*	1.75?	91	7	$3d^6(^5D)4s4p(^1P^\circ)\ ^5F^\circ$
$3d^6(^5D)4s4p(^1P^\circ)$	$x\,^5F^\circ$	5	*40 257.307*	1.390	90	5	$3d^7(^4F)4p\ ^5F^\circ$
		4	*40 594.429*	1.328	82	5	
		3	*40 842.151*	1.254	88	5	
		2	*41 018.050*	0.998	88	5	
		1	*41 130.627*	-0.006	88	5	

Fe I—Continued

Configuration	Term	J	Level (cm^{-1})	g	Leading percentages			
$3d^8$	^{3}P	2	40 871.410		85		6	$3d^64s^2\ ^3$P2
		1	41 178.406		85		8	
		0	41 234.498		83		8	
$3d^6(a\ ^3$P$)4s4p(^3$P$^\circ)$	$z\ ^5$S$^\circ$	2	*40 894.986*	1.985	59		36	$3d^7(^4$P$)4p\ ^5$S$^\circ$
$3d^6(a\ ^3$P$)4s4p(^3$P$^\circ)$	$x\ ^5$P$^\circ$	3	*42 532.736*	1.650	86		5	$3d^7(^4$P$)4p\ ^5$P$^\circ$
		2	*42 859.770*	1.822	76		10	$3d^7(^4$P$)4p\ ^5$S$^\circ$
		1	*43 079.026*	2.464	85		7	$3d^7(^4$P$)4p\ ^5$P$^\circ$
$3d^6(^3$H$)4s4p(^3$P$^\circ)$	$y\ ^5$G$^\circ$	6	*42 784.35*	1.342	60		30	$(a\ ^3$F$)(^3$P$^\circ)\ ^5$G$^\circ$
		5	*42 911.908*	1.203	53		33	
		4	*43 022.975*	1.024	44		37	
		3	*43 137.479*	0.905	38		22	
		2	*43 210.021*	0.331	46		45	
$3d^6(^5$D$)4s\ (^6$D$)5s$	$e\ ^7$D	5	42 815.858	1.585				
		4	43 163.327	1.655				
		3	43 434.629	1.755				
		2	43 633.534	2.009				
		1	43 763.980	3.002				
$3d^6(^3$H$)4s4p(^3$P$^\circ)$	$z\ ^5$H$^\circ$	5	*42 991.62*	1.054	65		27	^{5}I$^\circ$
		4	*43 108.90*	0.871	67		17	^{5}I$^\circ$
		6	*43 321.08?*		64		30	^{5}I$^\circ$
		3	*43 325.958*	0.509	48		26	^{5}G$^\circ$
$3d^6(a\ ^3$P$)4s4p(^3$P$^\circ)$	$w\ ^5$D$^\circ$	4	*43 499.496*	1.492	51		34	$(a\ ^3$F$)(^3$P$^\circ)\ ^5$D$^\circ$
		3	*43 922.664*	1.481	35		28	
		2	*44 183.620*	1.533	44		22	
		1	*44 411.151*	1.315	48		19	
		0	*44 458.92*		45		19	
$3d^6(a\ ^3$F$)4s4p(^3$P$^\circ)$	^{5}D$^\circ$	4	*44 022.535*	1.444	42		23	$(a\ ^3$P$)(^3$P$^\circ)\ ^5$D$^\circ$
		3	*44 166.203*	1.351	39		28	
		2	*44 664.068*	1.378	41		29	
		1	*44 760.75*	1.389	40		28	
		0	*44 826.88*		60		25	
$3d^6(a\ ^3$F$)4s4p(^3$P$^\circ)$	^{5}F$^\circ$	5	*44 243.673*	1.382	84		4	$(^3$D$)(^3$P$^\circ)\ ^5$F$^\circ$
		2	*44 285.443*	1.117	59		10	$(a\ ^3$F$)(^3$P$^\circ)\ ^5$D$^\circ$
		1	*44 378.38?*	0.283	81		6	$(^3$D$)(^3$P$^\circ)\ ^5$F$^\circ$
		4	*44 415.070*	1.401	62		18	$(a\ ^3$F$)(^3$P$^\circ)\ ^5$D$^\circ$
		3	*44 551.330*	1.386	45		26	$(a\ ^3$F$)(^3$P$^\circ)\ ^5$D$^\circ$
$3d^7(^4$P$)4p$	$y\ ^5$S$^\circ$	2	*44 511.806*	1.888	38		32	$3d^6(a\ ^3$P$)4s4p(^3$P$^\circ)\ ^5$S$^\circ$
$3d^6(^5$D$)4s\ (^6$D$)5s$	$e\ ^5$D	4	44 677.004	1.502				
		3	45 061.327	1.508				
		2	45 333.874	1.503				
		1	45 509.150	1.518				
		0	45 595.08					
$3d^6(a\ ^3$P$)4s4p(^3$P$^\circ)$		3	*45 220.676*	1.352	29	^{3}D$^\circ$	29	^{5}D$^\circ$
$3d^6(a\ ^3$P$)4s4p(^3$P$^\circ)$		2	*45 281.831*	1.200	33	^{5}D$^\circ$	31	^{3}D$^\circ$

(Continued)

Fe I—Continued

Configuration	Term	J	Level (cm^{-1})	g	Leading percentages			
$3d^6(^3H)4s4p(^3P^\circ)$	$y\ ^3G^\circ$	5	*45 294.846*	1.207	57		21	$3d^7(^2G)4p\ ^3G^\circ$
		4	*45 428.397*	1.053	55		20	
		3	*45 562.970*	0.765	54		19	
$3d^6(a\ ^3P)4s4p(^3P^\circ)$		1	*45 551.763*	0.556	32	$^5D^\circ$	30	$^3D^\circ$
$3d^6(a\ ^3F)4s4p(^3P^\circ)$	$x\ ^5G^\circ$	6	*45 608.31?*	1.336	65		30	$(^3H)(^3P^\circ)\ ^5G^\circ$
		5	*45 726.117*	1.269	60		32	
		4	*45 833.20*	1.158	55		36	
		3	*45 913.488*	0.928	53		40	
		2	*45 964.959*	0.323	52		45	
$3d^6(^3H)4s4p(^3P^\circ)$	$z\ ^3I^\circ$	7	*45 978.00?*	1.149	93		3	$3d^7(^2H)4p\ ^3I^\circ$
		6	*46 026.94*	1.040	91		3	
		5	*46 135.88*	0.833	94		4	
$3d^7(^4P)4p$	$w\ ^5P^\circ$	3	*46 137.10*	1.658	45		35	$3d^5(^6S)4s^24p\ ^5P^\circ$
		2	*46 313.57*	1.822	41		31	
		1	*46 410.40*	2.436	38		20	
$3d^6(a\ ^3P)4s4p(^3P^\circ)$	$z\ ^3S^\circ$	1	*46 600.814*	1.888	49		14	$3d^7(^2P)4p\ ^3S^\circ$
$3d^6(a\ ^3P)4s4p(^3P^\circ)$	$y\ ^3P^\circ$	0	*46 672.527*		37		24	$3d^7(^4P)4p\ ^3P^\circ$
		2	*46 727.068*	1.444	32		36	
		1	*46 901.820*	1.600	31		21	
$3d^7(^4P)4p$	$^5D^\circ$	4	*46 720.836*	1.341	50		17	$3d^6(a\ ^3F)4s4p(^3P^\circ)\ ^5D^\circ$
		3	*46 744.988*	1.397	54		14	
		2	*47 136.072*	1.216	53		19	
		0	*47 171.48?*		52		20	
		1	*47 177.225*	1.410	55		21	
$3d^7(^4P)4p$	$^3D^\circ$	2	*46 888.510*	1.260	51		15	$3d^6(a\ ^3F)4s4p(^3P^\circ)\ ^3D^\circ$
		3	*47 017.188*	1.346	54		18	
		1	*47 272.016*	0.767	47		20	
$3d^6(a\ ^3F)4s4p(^3P^\circ)$	$^3F^\circ$	4	*46 889.143*	1.344	38		17	$3d^7(^2G)4p\ ^3F^\circ$
		3	*47 092.707*	1.159	51		25	$3d^7(^2G)4p\ ^3G^\circ$
		2	*47 197.014*	0.743	41		24	$3d^6(^3G)4s4p(^3P^\circ)\ ^5G^\circ$
$3d^6(^3H)4s4p(^3P^\circ)$	$z\ ^3H^\circ$	6	*46 982.34*	1.200	36		37	$3d^7(^2G)4p\ ^3H^\circ$
		5	*47 008.366*	1.060	34		36	
		4	*47 106.477*	0.880	31		18	
$3d^7(^4F)5s$	$e\ ^5F$	5	47 005.508	1.421				
		4	47 377.962	1.331				
		3	47 755.539	1.236				
		2	48 036.666	0.991				
		1	48 221.314	0.007				
$3d^6(^3G)4s4p(^3P^\circ)$	$w\ ^5G^\circ$	6	*47 363.369*	1.306	78		11	$3d^7(^2G)4p\ ^3H^\circ$
		5	*47 420.229*	1.305	73		7	$3d^6(^3G)4s4p(^3P^\circ)\ ^5F^\circ$
		4	*47 590.047*	1.145	73		10	$3d^6(a\ ^3F)4s4p(^1P^\circ)\ ^3G^\circ$
		3	*47 693.227*	0.931	42		18	$3d^6(a\ ^3F)4s4p(^1P^\circ)\ ^3G^\circ$
		2	*47 831.150*	0.472	65		16	$3d^7(^2G)4p\ ^3F^\circ$
$3d^6(a\ ^3P)4s4p(^3P^\circ)$	$^1D^\circ$	2	*47 419.674*	1.137	36		12	$3d^7(^2P)4p\ ^1D^\circ$
$3d^7(^2G)4p$	$z\ ^1G^\circ$	4	*47 452.716*	1.025	31		23	$3d^6(^3H)4s4p(^3P^\circ)\ ^1G^\circ$

Fe I—Continued

Configuration	Term	J	Level (cm^{-1})	g	Leading percentages			
$3d^7(^4P)4p$	$y\,^3S^\circ$	1	47 555.598	1.884	60		9	$^3D^\circ$
$3d^6(^3G)4s4p(^3P^\circ)$	$v\,^5F^\circ$	5	47 606.094	1.317	61		11	$(^3G)(^3P^\circ)\,^5H^\circ$
		4	47 929.999	1.264	55		20	$(^3G)(^3P^\circ)\,^5H^\circ$
		3	48 122.928	1.236	70		7	$(^3D)(^3P^\circ)\,^5F^\circ$
		2	48 238.844	1.267	74		7	$(^3D)(^3P^\circ)\,^5F^\circ$
		1	48 350.601	0.230	70		7	$(^3D)(^3P^\circ)\,^5F^\circ$
$3d^6(a\,^3F)4s4p(^3P^\circ)$	$x\,^3G^\circ$	4	47 812.118	1.061	51		22	$(^3G)(^3P^\circ)\,^5H^\circ$
		3	47 834.218	0.668	39		37	$(^3G)(^3P^\circ)\,^5G^\circ$
		5	47 834.542	1.203	49		18	$(^3G)(^3P^\circ)\,^5H^\circ$
$3d^7(^4F)5s$	$e\,^3F$	4	47 960.941	1.288				
		3	48 531.864	1.107				
		2	48 928.389	0.622				
	$v\,^5P^\circ$	3	47 966.59	1.646				
		2	48 163.438	1.740				
		1	48 289.865	2.213				
$3d^6(^3G)4s4p(^3P^\circ)$	$^5H^\circ$	5	48 231.270	1.27?	67		10	$(a\,^3F)(^3P^\circ)\,^3G^\circ$
		4	48 361.878	0.934	44		18	
		3	48 475.668	0.584	54		31	
$3d^7(^4P)4p$	$x\,^3P^\circ$	2	48 304.638	1.263	36		21	$(^2P)\,^3P^\circ$
		0	48 460.098		42		23	
		1	48 516.135	1.547	39		18	
$3d^7(^2G)4p$	$z\,^1H^\circ$	5	48 382.597	1.018	68		10	$3d^6(^3H)4s4p(^3P^\circ)\,^1H^\circ$
$3d^6(^3H)4s4p(^3P^\circ)$	$y\,^1G^\circ$	4	48 702.526	1.063	36		20	$3d^7(^2G)4p\,^1G^\circ$
$3d^7(^2G)4p$	$w\,^3F^\circ$	4	49 108.890	1.181	39		26	$3d^6(a\,^3F)4s4p(^3P^\circ)\,^3F^\circ$
		3	49 242.881	1.165	37		25	
		2	49 433.121	0.677	50		21	
$3d^6(a\,^3F)4s4p(^3P^\circ)$		3	49 135.022	1.211	31	$^3D^\circ$	19	$3d^7(^2G)4p\,^3F^\circ$
$3d^6(a\,^3F)4s4p(^3P^\circ)$	$^1F^\circ$	3	49 227.12		40		39	$3d^7(^2G)4p\,^1F^\circ$
$3d^6(a\,^3F)4s4p(^3P^\circ)$	$v\,^3D^\circ$	2	49 242.593	0.954	52		12	$3d^7(^2P)4p\,^3D^\circ$
		1	49 297.620	0.562	47		13	$3d^7(^2P)4p\,^3D^\circ$
$3d^6(^5D)4s\,(^6D)5p$	$^7D^\circ$	5	49 352.335					
		4	49 558.724					
		3	49 805.249					
		2	50 008.515					
		1	50 152.609					
$3d^7(^2G)4p$	$y\,^3H^\circ$	6	49 434.156	1.17?	43		43	$3d^6(^3H)4s4p(^3P^\circ)\,^3H^\circ$
		5	49 604.415	1.075	38		26	
		4	49 726.977	0.929	42		29	
$3d^7(^2G)4p$	$v\,^3G^\circ$	5	49 460.890	1.163	38		24	$3d^6(a\,^3F)4s4p(^3P^\circ)\,^3G^\circ$
		4	49 627.877	0.914	41		18	$3d^6(^3H)4s4p(^3P^\circ)\,^3G^\circ$
		3	49 850.581	0.763	43		25	$3d^6(a\,^3F)4s4p(^3P^\circ)\,^3G^\circ$
	$z\,^1D^\circ$	2	49 477.10	0.92?				

(Continued)

Fe I—Continued

Configuration	Term	J	Level (cm^{-1})	g	Leading percentages		
$3d^6(^5D)4s\ (^6D)5p$	$^7F^\circ$	6	*49 758.133*				
		5	*50 052.184*				
		4	*50 303.216*				
		3	*50 433.015*				
		2	*50 555.750*				
		1	*50 627.429*				
		0	*50 659.672*				
$3d^7(^2P)4p$	$w\ ^3P^\circ$	0	*49 951.341*		52	24	$3d^6(a\ ^3P)4s4p(^1P^\circ)\ ^3P^\circ$
		1	*50 043.205*	1.389	50	11	$3d^5(^6S)4s^24p\ ^5P^\circ$
		2	*50 186.830*	1.469	46	10	$3d^7(^4P)4p\ ^5P^\circ$
		2	*50 045.9*				
$3d^6(^5D)4s\ (^6D)5p$	$^7P^\circ$	4	*50 185.730*				
		3	*50 628.360*				
		2	*50 901.157*				
$3d^6(^5D)4s\ (^6D)4d$	$e\ ^7F$	6	50 342.14	1.490			
		5	50 833.428	1.505			
		3	51 148.859	1.499			
		4	51 192.270	1.617			
		1	51 207.991	2.490			
		2	51 331.044				
$3d^6(^5D)4s\ (^6D)4d$	$f\ ^7D$	5	50 377.913	1.510			
		4	50 807.991	1.574			
		3	50 861.816				
		2	50 998.641	1.844			
		1	51 048.113				
$3d^6(^5D)4s\ (^6D)4d$	$f\ ^5D$	4	50 423.136	1.514			
		3	50 534.391	1.615			
		2	50 698.624	1.614			
		1	50 880.098	1.662			
		0	50 980.98				
$3d^6(^5D)4s\ (^6D)4d$	$e\ ^7P$	4	50 475.287	1.585			
		3	50 611.260	1.687			
		2	50 861.321				
$3d^6(^5D)4s\ (^6D)4d$	$e\ ^5G$	6	50 522.946	1.351			
		5	50 703.866	1.360			
		4	50 979.578	1.238			
		3	51 219.017	1.294			
		2	51 370.130	0.953			
$3d^7(^2G)4p$	$z\ ^1F^\circ$	3	*50 586.874*	1.018	36	23	$(a\ ^2D)\ ^1F^\circ$
$3d^6(a\ ^3F)4s4p(^3P^\circ)$	$x\ ^1G^\circ$	4	*50 613.972*	0.978	64	9	$3d^7(^2H)4p\ ^1G^\circ$
$3d^6(^5D)4s\ (^6D)4d$	$e\ ^7G$	7	50 651.72?				
		6	50 967.826	1.415			
		5	51 228.555	1.379			
		4	51 334.909	1.338			
		3	51 460.516	1.244			
		2	51 539.712				
		1	51 566.82	-0.374			

Fe I—Continued

Configuration	Term	J	Level (cm^{-1})	g	Leading percentages		
$3d^6(^5D)4s\,(^6D)5p$	$u\,^5F^\circ$	5	*51 016.660*				
		4	*51 381.460*				
		3	*51 619.078*				
		2	*51 827.413*				
		1	*51 945.819*				
$3d^6(^3G)4s4p(^3P^\circ)$	$x\,^3H^\circ$	6	*51 023.152*	1.161	80	15	$(^3H)(^3P^\circ)\,^3H^\circ$
		5	*51 068.710*	1.038	74	13	$(^3H)(^3P^\circ)\,^3H^\circ$
		4	*51 409.117*	0.953	67	15	$(a\,^3F)(^3P^\circ)\,^1G^\circ$
$3d^6(^5D)4s\,(^6D)5p$	$t\,^5D^\circ$	4	*51 076.622*	1.486			
		3	*51 361.390*				
		2	*51 629.998*				
		1	*51 827.854*				
		0	*51 941.531*				
$3d^6(^5D)4s\,(^6D)4d$	$f\,^5F$	5	51 103.187	1.384			
		4	51 461.672	1.355?			
		3	51 604.102				
		2	51 705.007	0.967			
		1	51 754.490				
$3d^6(^5D)4s\,(^6D)4d$	$e\,^5S$	2	51 148.883	1.952			
$3d^7(a\,^2D)4p$	$v\,^3F^\circ$	2	*51 201.284*	0.803	31	23	$3d^6(^3G)4s4p(^3P^\circ)\,^3F^\circ$
		4	*51 304.603*	1.122	34	30	$3d^7(a\,^2D)4p\,^3F^\circ$
		3	*51 365.308*	1.096	36	26	$3d^6(^3G)4s4p(^3P^\circ)\,^3F^\circ$
$3d^6(^5D)4s\,(^4D)5s$	$e\,^3D$	3	51 294.222	1.345			
		2	51 739.920	1.125			
		1	52 039.886	0.801			
$3d^6(^5D)4s\,(^4D)5s$	$g\,^5D$	4	51 350.491	1.487			
		3	51 770.554	1.492			
		2	52 049.814	1.57?			
		1	52 214.336				
		0	52 257.33				
$3d^7(^2H)4p$	$u\,^3G^\circ$	5	*51 373.909*	1.140	30	17	$3d^6(^3G)4s4p(^3P^\circ)\,^3G^\circ$
		4	*51 668.189*	1.067	35	32	
		3	*51 825.773*	0.801	35	32	
$3d^6(^5D)4s\,(^6D)4d$	$e\,^7S$	3	51 570.084	1.92?			
$3d^6(^3H)4s4p(^3P^\circ)$	$^1H^\circ$	5	*51 630.172*	1.061	39	10	$3d^7(^2G)4p\,^1H^\circ$
$3d^6(^5D)4s\,(^6D)5p$	$u\,^5P^\circ$	3	*51 692.007*				
		2	*51 944.784*				
		1	*52 110.607*	2.633			
$3d^7(^2P)4p$	$y\,^1D^\circ$	2	*51 708.309*	1.025	49	18	$(a\,^2D)\,^1D^\circ$
$3d^6(a\,^3F)4s4p(^3P^\circ)$	$x\,^1D^\circ$	2	*51 762.067*	0.883	56	15	$3d^7(a\,^2D)4p\,^1D^\circ$
$3d^6(^5D)4s\,(^6D)4d$	$e\,^5P$	3	51 837.24	1.664			
		1	52 019.67	2.432			
		2	52 067.460				

(Continued)

Fe I—Continued

Configuration	Term	J	Level (cm^{-1})	g	Leading percentages		
$3d^7(^2P)4p$	$u\,^3D^\circ$	3	*51 969.079*	1.306	61	19	$3d^6(a\,^3F)4s4p(^1P^\circ)\,^3D^\circ$
		2	*52 296.899*	1.156	53	21	
		1	*52 512.445*	0.700	48	22	
$3d^7(^2P)4p$	$^1P^\circ$	1	*52 180.804*	0.801	47	20	$(a\,^2D)\,^3D^\circ$
$3d^7(a\,^2D)4p$	$^3D^\circ$	3	*52 213.226*	1.317	75	19	$3d^6(a\,^3F)4s4p(^1P^\circ)\,^3D^\circ$
		2	*52 682.915*	1.145	60	11	$3d^7(^2P)4p\,^3D^\circ$
		1	*52 857.790*	1.246	45	10	$3d^7(^2P)4p\,^3D^\circ$
$3d^7(^2H)4p$	$w\,^3H^\circ$	6	*52 431.418*	1.177	63	17	$3d^6(^3H)4s4p(^1P^\circ)\,^3H^\circ$
		5	*52 613.084*	1.033	60	14	
		4	*52 768.721*	0.810	61	18	
$3d^7(^2H)4p$	$y\,^3I^\circ$	6	*52 513.549*	1.019	65	22	$3d^7(^2H)4p\,^1I^\circ$
		7	*52 655.00?*	1.147	88	8	$3d^6(^1I)4s4p(^3P^\circ)\,^3I^\circ$
		5	*52 898.971*	0.830	85	9	$3d^6(^1I)4s4p(^3P^\circ)\,^3I^\circ$
$3d^7(a\,^2D)4p$	$^3P^\circ$	2	*52 916.292*	1.495	55	19	$3d^6(a\,^3P)4s4p(^1P^\circ)\,^3P^\circ$
		1	*53 229.942*	1.266	41	13	$3d^7(^2P)4p\,^1P^\circ$
$3d^7(^4F)5p$	$^5F^\circ$	4	*52 953.625*				
		5	*53 084.791*				
		3	*53 357.508*				
		2	*53 749.405*				
$3d^7(^4F)4d$	$g\,^5F$	5	53 061.24				
		4	53 393.68				
		3	53 830.973				
		2	54 257.505				
		1	54 386.189				
$3d^7(^4F)5p$	$^5G^\circ$	6	*53 069.350*				
		5	*53 586.501*				
		4	*53 610.414*				
		3	*53 852.108*				
$3d^7(^2H)4p$	$z\,^1I^\circ$	6	*53 093.521*	1.010	65	21	$^3I^\circ$
$3d^7(^4F)4d$	$h\,^5D$	4	53 155.09	1.435			
		3	53 545.847				
		2	53 966.68				
		1	54 132.550				
$3d^7(^4F)4d$	$f\,^5P$	3	53 160.49				
		2	53 568.68				
		1	53 925.22				
$3d^7(^4F)4d$	$f\,^5G$	6	53 169.17	1.323			
		5	53 281.70	1.221			
		4	53 768.969				
		3	54 161.132	1.142			
		2	54 375.68				
$3d^7(^4F)4d$	$e\,^5H$	7	53 275.16?	1.30?			
		6	53 352.98?	1.191			
		5	53 874.26?	1.102			
		4	54 237.16	0.90?			
		3	54 491.04	0.484			

Fe I—Continued

Configuration	Term	J	Level (cm^{-1})	g	Leading percentages		
$3d^6(^3D)4s4p(^3P^\circ)$	$^5F^\circ$	2	*53 275.23*		84	11	$(^3G)(^3P^\circ)\ ^5F^\circ$
		3	*53 661.09*	1.21?	84	10	
		5	*54 013.747*	1.356	88	8	
$3d^7(^4F)5p$	$^5D^\circ$	4	*53 328.827*				
		3	*53 733.583*				
		2	*54 042.516*				
		1	*54 224.402*				
$3d^6(^3G)4s4p(^3P^\circ)$	$y\ ^1H^\circ$	5	*53 722.40*	1.03?	77	15	$(^3H)(^3P^\circ)\ ^1H^\circ$
$3d^7(^4F)4d$	$e\ ^3G$	5	53 739.433	1.248			
		4	54 066.53	1.096			
		3	54 379.40	0.842			
$3d^7(^4F)4d$	$f\ ^3D$	3	53 747.51	1.258			
		2	54 066.758				
		1	54 449.29				
$3d^6(^3G)4s4p(^3P^\circ)$	$x\ ^3F^\circ$	3	*53 763.272*	1.079	34	30	$3d^7(a\ ^2D)4p\ ^3F^\circ$
$3d^6(^5D)4s\ (^6D)6s$	$g\ ^7D$	5	53 800.841	1.586			
		4	54 124.724	1.65?			
		3	54 404.765				
		2	54 611.691				
		1	54 747.581				
$3d^6(^3D)4s4p(^3P^\circ)$	$^5D^\circ$	1	*53 808.353*	1.418	72	14	$(^3D)(^3P^\circ)\ ^5P^\circ$
		3	*53 891.520*	1.476	74	13	$(^3D)(^3P^\circ)\ ^5P^\circ$
		4	*54 301.334*		89	6	$(b\ ^3F)(^3P^\circ)\ ^5D^\circ$
$3d^7(^4F)5p$	$^3D^\circ$	3	*53 837.847*				
		2	*54 342.762*				
$3d^7(^4F)4d$	$e\ ^3H$	6	53 840.64?	1.225			
		5	54 266.72?	1.109			
		4	54 555.41?	0.871			
$3d^6(^3G)4s4p(^3P^\circ)$	$t\ ^3G^\circ$	5	*53 983.284*	1.234	39	35	$3d^7(^2H)4p\ ^3G^\circ$
		4	*54 237.415*	1.183	34	36	
		3	*54 600.346*	0.922	32	31	
$3d^6(^3D)4s4p(^3P^\circ)$	$^5P^\circ$	3	*54 004.78*		68	13	$^5D^\circ$
		2	*54 112.218*	1.70?	47	31	$^5D^\circ$
		1	*54 271.057*		57	9	$3d^7(^2P)4p\ ^3S^\circ$
$3d^7(^4F)5p$	$^3G^\circ$	4	*54 017.573*				
		3	*54 357.398*				
		3	*54 289.09*				
$3d^7(^4F)4d$	$f\ ^3F$	4	54 683.35	1.141			
		3	55 124.93	1.071			
		2	55 378.80	0.676			
$3d^6(^3G)4s4p(^3P^\circ)$	$w\ ^1G^\circ$	4	*54 810.841*	1.001	44	22	$3d^7(^2G)4p\ ^1G^\circ$

(Continued)

Fe I—Continued

Configuration	Term	J	Level (cm^{-1})	g	Leading percentages			
$3d^7(^4\mathrm{F})4d$	$e\ ^3\mathrm{P}$	2	54 879.68	1.459				
		1	55 376.08	1.459				
		0	55 726.50?					
$3d^6(a\ ^1\mathrm{G})4s4p(^3\mathrm{P}^\circ)$	$^3\mathrm{G}^\circ$	5	*55 429.815*	1.057	46		23	$3d^7(^2\mathrm{H})4p\ ^1\mathrm{H}^\circ$
		3	*55 790.673*	0.908	53		21	$3d^6(^3\mathrm{G})4s4p(^3\mathrm{P}^\circ)\ ^1\mathrm{F}^\circ$
		4	*55 905.538*		61		17	$3d^6(a\ ^1\mathrm{G})4s4p(^3\mathrm{P}^\circ)\ ^3\mathrm{H}^\circ$
$3d^6(a\ ^1\mathrm{G})4s4p(^3\mathrm{P}^\circ)$	$^3\mathrm{H}^\circ$	4	*55 446.000*	0.804	59		11	$3d^6(^1\mathrm{I})4s4p(^3\mathrm{P}^\circ)\ ^3\mathrm{H}^\circ$
		6	*55 489.77*	1.169	48		33	$3d^6(^1\mathrm{I})4s4p(^3\mathrm{P}^\circ)\ ^3\mathrm{H}^\circ$
		5	*55 525.54*	1.018	47		23	$3d^7(^2\mathrm{H})4p\ ^1\mathrm{H}^\circ$
$3d^7(a\ ^2\mathrm{D})4p$	$w\ ^1\mathrm{D}^\circ$	2	*55 754.239*	0.990	62		15	$(^2\mathrm{P})\ ^1\mathrm{D}^\circ$
$3d^7(^2\mathrm{H})4p$		5	*55 907.171*	1.145	33	$^1\mathrm{H}^\circ$	31	$3d^6(a\ ^1\mathrm{G})4s4p(^3\mathrm{P}^\circ)\ ^3\mathrm{G}^\circ$
$3d^6(^3\mathrm{G})4s4p(^3\mathrm{P}^\circ)$	$^1\mathrm{F}^\circ$	3	*56 097.829*	0.857	45		25	$(a\ ^1\mathrm{G})(^3\mathrm{P}^\circ)\ ^3\mathrm{G}^\circ$
$3d^6(^1\mathrm{I})4s4p(^3\mathrm{P}^\circ)$	$u\ ^3\mathrm{H}^\circ$	6	*56 333.957*	1.166	44		47	$(a\ ^1\mathrm{G})(^3\mathrm{P}^\circ)\ ^3\mathrm{H}^\circ$
		5	*56 382.662*	1.029	46		26	$(a\ ^1\mathrm{G})(^3\mathrm{P}^\circ)\ ^3\mathrm{H}^\circ$
		4	*56 423.279*	0.859	50		18	$(^3\mathrm{H})(^1\mathrm{P}^\circ)\ ^3\mathrm{H}^\circ$
$3d^6\ 4s5d$	1	5	56 428.06					
$3d^6\ 4s5d$	2	4,5	56 452.04					
$3d^6(a\ ^1\mathrm{G})4s4p(^3\mathrm{P}^\circ)$	$u\ ^3\mathrm{F}^\circ$	4	*56 592.699*	1.148	47		17	$3d^7(a\ ^2\mathrm{D})4p\ ^3\mathrm{F}^\circ$
		3	*56 783.317*	1.077	54		20	
		2	*56 858.659*	0.687	47		26	
$3d^6\ 4s5d$	3	4	56 842.70					
$3d^7(^2\mathrm{H})4p$	$v\ ^1\mathrm{G}^\circ$	4	*56 951.286*	1.053	39		23	$3d^6(^3\mathrm{G})4s4p(^3\mathrm{P}^\circ)\ ^1\mathrm{G}^\circ$
$3d^6(^1\mathrm{I})4s4p(^3\mathrm{P}^\circ)$	$x\ ^3\mathrm{I}^\circ$	7	*57 027.52?*	1.145	86		6	$3d^7(^2\mathrm{H})4p\ ^3\mathrm{I}^\circ$
		6	*57 070.21*	1.028	85		7	
		5	*57 104.22*	0.832	84		7	
$3d^6(^3\mathrm{D})4s4p(^3\mathrm{P}^\circ)$	$t\ ^3\mathrm{F}^\circ$	4	*57 550.000*	1.235	67		15	$(a\ ^3\mathrm{F})(^1\mathrm{P}^\circ)\ ^3\mathrm{F}^\circ$
		3	*57 641.000*		60		17	
		2	*57 708.747*	0.698	61		16	
$3d^6(^5\mathrm{D})4s\ (^4\mathrm{D})4d$	$i\ ^5\mathrm{D}$	4	57 697.55	1.384				
		3	57 813.940	1.415				
		2	57 974.129					
$3d^6(^5\mathrm{D})4s\ (^6\mathrm{D})7s$	$h\ ^7\mathrm{D}$	5	57 897.17					
$3d^6(^5\mathrm{D})4s\ (^4\mathrm{D})4d$	$g\ ^5\mathrm{G}$	6	58 001.84	1.40?				
		5	58 271.46?					
		4	58 520.14?					
		3	58 710.05?					
		2	58 824.77	0.343				
$3d^6\ 4s5d$	4	2	58 213.121					
$3d^6(a\ ^3\mathrm{F})4s4p(^1\mathrm{P}^\circ)$	$r\ ^3\mathrm{G}^\circ$	5	*59 926.62?*	1.190	81		8	$(^3\mathrm{H})(^1\mathrm{P}^\circ)\ ^3\mathrm{H}^\circ$
		4	*60 172.06*	1.030	65		25	$(a\ ^1\mathrm{D})(^3\mathrm{P}^\circ)\ ^3\mathrm{F}^\circ$
		3	*60 364.76?*	0.780	63		17	$(a\ ^1\mathrm{D})(^3\mathrm{P}^\circ)\ ^3\mathrm{F}^\circ$

Fe I—Continued

Configuration	Term	J	Level (cm^{-1})	g	Leading percentages		
$3d^6(^3H)4s4p(^1P°)$	$t\ ^3H°$	6	*60 365.70?*	1.163	59	25	$3d^7(^2H)4p\ ^3H°$
		5	*60 549.18*	1.040	49	22	
		4	*60 757.68*	0.805	50	22	
		3	*60 563.61*				
$3d^6(a\ ^3F)4s4p(^1P°)$	$^3F°$	4	*60 754.71?*		32	33	$(a\ ^1D)(^3P°)\ ^3F°$
		3	*60 806.654*		29	28	
Fe II ($^6D_{9/2}$)	***Limit***		**63 737**				

Fe II

$Z=26$

Mn I isoelectronic sequence

Ground state: $1s^22s^22p^63s^23p^63d^64s\ ^6D_{9/2}$

Ionization energy = 130 563±10 cm^{-1} (16.1879±0.0012 eV)

The earlier work on this spectrum was mainly by Dobbie (1938), Green (1939), and unpublished material of Edlén. Johansson and Litzén (1974) found the complete set of $3d^6(^5D)4f$ levels as well as many new $3d^64d$ levels.

The spectrum has now been reobserved in the regions 900–2200 Å and 4800–11 200 Å by Johansson (1978) by using a pulsed hollow cathode discharge. With the new measurements and Dobbie's list in the region 2200–4800 Å, Johansson has contributed some 250 new levels to the presently known 576 levels. He has redetermined all the level values and discarded 23 earlier levels. The accuracy of levels given to three decimal places is about ±0.01 cm^{-1}, those with two places are about 0.1 cm^{-1}, and with one place, ±0.5 cm^{-1}.

The $3d^7$, $3d^64s$, and $3d^54s^2$ configurations have been treated theoretically by Shadmi, Oreg, and Stein (1968), whose results confirm the assignments given by Johansson. The configurations $3d^64p$, $3d^65p$, and $3d^54s4p$ were calculated by Sinzelle and Wyart (1978, unpublished) with configuration interaction. Since $3d^54s4p$ was not recoupled in the scheme which exhibits the highest percentages, we give only its admixture with the other two configurations. Experience has shown that $3d^54s4p$ should be coupled $3d^5(S_1,L_1)4s4p\ (S_2,L_2)\ S_3,L_3$. A discussion of their calculations is given by Johansson, Litzén, Sinzelle, and Wyart (1980). Johansson's designations for the $3d^54s4p$ levels are quoted here.

The g-values were derived by Weeks in 1949 and are taken from Moore (1952). The observations were made at M.I.T. The uncertainty in the g-value determinations varies from ±0.002 to ±0.009.

The ionization energy was determined by Johansson from the $3d^6(^5D)ns\ ^6D_{9/2}$ series.

References

Dobbie, J. C. (1938), Ann. Solar Phys. Obs. Cambridge **5**, 1.

Green, L. C. (1939), Phys. Rev. **55**, 1211.

Johansson, S. (1978), Phys. Scr. **18**, 217.

Johansson, S., and Litzén, U. (1974), Phys. Scr. **10**, 121.

Johansson, S., Litzén, U., Sinzelle, J., and Wyart, J.-F. (1980), Phys. Scr. **21**, 40.

Moore, C. E. (1952), Atomic Energy Levels, Natl. Bur. Stand. (U.S.) Circ. 467, Vol. II (reissued in 1971 as Natl. Bur. Stand. (U.S.) Natl. Stand. Ref. Data Ser. 35, Vol. II).

Shadmi, Y., Oreg, J., and Stein, J. (1968), J. Opt. Soc. Am. **58**, 909.

Fe II

Configuration	Term	J	Level (cm^{-1})	g	Leading percentages
$3d^6(^5D)4s$	$a\ ^6D$	9/2	0.000	1.58	
		7/2	384.790	1.58	
		5/2	667.683	1.655	
		3/2	862.613	1.862	
		1/2	977.053	3.31	
$3d^7$	$a\ ^4F$	9/2	1 872.567	1.33	
		7/2	2 430.097	1.223	
		5/2	2 837.950	1.02	
		3/2	3 117.461	0.385	
$3d^6(^5D)4s$	$a\ ^4D$	7/2	7 955.299	1.419	
		5/2	8 391.938	1.365	
		3/2	8 680.454	1.200	
		1/2	8 846.768	-0.05	
$3d^7$	$a\ ^4P$	5/2	13 474.411	1.609	
		3/2	13 673.185	1.737	
		1/2	13 904.824	2.67	
$3d^7$	$a\ ^2G$	9/2	15 844.65		
		7/2	16 369.36		

Fe II—Continued

Configuration	Term	J	Level (cm^{-1})	g	Leading percentages
$3d^7$	$a\ ^2P$	3/2	18 360.646	1.28	
		1/2	18 886.780		
$3d^7$	$a\ ^2H$	11/2	20 340.30		
		9/2	20 805.77	0.92	
$3d^7$	$a\ ^2D2$	5/2	20 516.960	1.22	
		3/2	21 308.04		
$3d^6(^3P2)4s$	$b\ ^4P$	5/2	20 830.582	1.583	
		3/2	21 812.055	1.720	
		1/2	22 409.852	2.68	
$3d^6(^3H)4s$	$a\ ^4H$	13/2	21 251.608	1.20	
		11/2	21 430.359	1.119	
		9/2	21 581.638	0.951	
		7/2	21 711.917	0.661	
$3d^6(^3F2)4s$	$b\ ^4F$	9/2	22 637.205	1.307	
		7/2	22 810.357	1.210	
		5/2	22 939.358	1.019	
		3/2	23 031.300	0.398	
$3d^5\ 4s^2$	$a\ ^6S$	5/2	23 317.633	1.996	
$3d^6(^3G)4s$	$a\ ^4G$	11/2	25 428.784	1.237	
		9/2	25 805.328	1.15	
		7/2	25 981.629	0.98	
		5/2	26 055.423	0.574	
$3d^6(^3P2)4s$	$b\ ^2P$	3/2	25 787.598	1.33	
		1/2	26 932.748	0.67	
$3d^6(^3H)4s$	$b\ ^2H$	11/2	26 170.181	1.09	
		9/2	26 352.766	0.927	
$3d^6(^3F2)4s$	$a\ ^2F$	7/2	27 314.922	1.129	
		5/2	27 620.412	0.851	
$3d^6(^3G)4s$	$b\ ^2G$	9/2	30 388.542	1.10	
		7/2	30 764.485	0.898	
$3d^6(^3D)4s$	$b\ ^4D$	3/2	31 364.440		
		1/2	31 368.450		
		5/2	31 387.948	1.327	
		7/2	31 483.176	1.41	
$3d^7$	$b\ ^2F$	5/2	31 811.822	0.86	
		7/2	31 999.048	1.124	
$3d^6(^1I)4s$	$a\ ^2I$	13/2	32 875.646	1.062	
		11/2	32 909.905	0.92	
$3d^6(^1G2)4s$	$c\ ^2G$	9/2	33 466.463	1.099	
		7/2	33 501.253	0.88	
$3d^6(^3D)4s$	$b\ ^2D$	3/2	36 126.387	0.799	
		5/2	36 252.918	1.179	

(Continued)

Fe II—Continued

Configuration	Term	J	Level (cm^{-1})	g	Leading percentages		
$3d^6(^1S2)4s$	$a\ ^2S$	1/2	37 227.326	2.06			
$3d^6(^1D2)4s$	$c\ ^2D$	5/2	38 164.194	1.176			
		3/2	38 214.507	0.79			
$3d^6(^5D)4p$	$z\ ^6D^{\circ}$	9/2	*38 458.981*	1.542	99		
		7/2	*38 660.043*	1.584	98		
		5/2	*38 858.958*	1.653	98		
		3/2	*39 013.206*	1.86	99		
		1/2	*39 109.307*	3.35	99		
$3d^6(^5D)4p$	$z\ ^6F^{\circ}$	11/2	*41 968.046*		99		
		9/2	*42 114.818*	1.43	96		
		7/2	*42 237.033*	1.399	96		
		5/2	*42 334.822*	1.304	97		
		3/2	*42 401.302*	1.04	98		
		1/2	*42 439.822*	-0.647	98		
$3d^6(^5D)4p$	$z\ ^6P^{\circ}$	7/2	*42 658.224*	1.702	93	3	$3d^54s4p$
		5/2	*43 238.586*	1.869	95	3	
		3/2	*43 620.957*	2.398	96	3	
$3d^6(^5D)4p$	$z\ ^4F^{\circ}$	9/2	*44 232.512*	1.32	96	2	$(^5D)\ ^6F^{\circ}$
		7/2	*44 753.799*	1.29	91	5	$(^5D)\ ^4D^{\circ}$
		5/2	*45 079.879*	1.069	93	4	$(^5D)\ ^4D^{\circ}$
		3/2	*45 289.801*	0.445	96	2	$(^5D)\ ^4D^{\circ}$
$3d^6(^5D)4p$	$z\ ^4D^{\circ}$	7/2	*44 446.878*	1.40	90	5	$^4F^{\circ}$
		5/2	*44 784.761*	1.35	91	4	$^4F^{\circ}$
		3/2	*45 044.168*	1.15	93	2	$^4F^{\circ}$
		1/2	*45 206.450*	-0.021	95		
$3d^6(^1F)4s$	$c\ ^2F$	7/2	44 915.046				
		5/2	44 929.55				
$3d^6(^5D)4p$	$z\ ^4P^{\circ}$	5/2	*46 967.444*	1.592	96	2	$3d^54s4p$
		3/2	*47 389.779*	1.717	96	2	
		1/2	*47 626.076*	2.70	96	2	
$3d^7$	$d\ ^2D1$	3/2	47 674.721				
		5/2	48 039.090				
$3d^6(^3P1)4s$	$c\ ^4P$	1/2	49 100.976				
		3/2	49 506.934				
		5/2	50 212.826				
$3d^6(^3F1)4s$	$c\ ^4F$	3/2	50 075.910				
		5/2	50 142.786				
		9/2	50 157.452				
		7/2	50 187.813				
$3d^5(^6S)4s4p(^3P^{\circ})$	$z\ ^8P^{\circ}$	5/2	*52 299.39*				
		7/2	*52 582.51*				
		9/2	*52 965.82*				
$3d^6(^3P1)4s$	$c\ ^2P$	1/2	54 063.459				
		3/2	54 902.315				

Fe II—Continued

Configuration	Term	J	Level (cm^{-1})	g	Leading percentages			
$3d^5\,4s^2$	$b\ ^4$G	11/2	54 232.195					
		9/2	54 273.641					
		5/2	54 275.637					
		7/2	54 283.220					
$3d^6(^3$F1$)4s$	$d\ ^2$F	5/2	54 870.528					
		7/2	54 904.222					
$3d^5\,4s^2$	$d\ ^4$P	5/2	57 411.065					
		3/2	57 493.321					
		1/2	57 578.484					
$3d^6(^1$G1$)4s$	$d\ ^2$G	9/2	58 631.531					
		7/2	58 666.258					
$3d^6(^3$P2$)4p$	$z\ ^4$S°	3/2	*59 663.456*	1.89	46		19	$(^3$P1$)\ ^4$S°
$3d^5\,4s^2$	$c\ ^4$D	7/2	60 270.339					
		1/2	60 384.370					
		3/2	60 441.033					
		5/2	60 445.275					
$3d^6(^3$P2$)4p$	$y\ ^4$P°	5/2	*60 402.342*	1.58	46		36	$(^3$P1$)\ ^4$P°
		1/2	*61 035.287*	2.613	52		42	
		3/2	*61 332.764*	1.74	28		22	
$3d^6(^3$H$)4p$	$z\ ^4$G°	11/2	*60 625.449*	1.24	66		24	$(^3$F2$)\ ^4$G°
		9/2	*60 807.230*	1.155	53		29	
		7/2	*60 956.781*	0.969	50		33	
		5/2	*61 041.748*	0.799	48		35	
$3d^6(^3$H$)4p$	$z\ ^4$H°	13/2	*60 837.569*		47		42	$(^3$H$)\ ^4$I°
		11/2	*60 887.598*		44		44	$(^3$H$)\ ^4$I°
		9/2	*60 989.444*		45		37	$(^3$H$)\ ^4$I°
		7/2	*61 156.835*	0.720	66		12	$(^3$H$)\ ^2$G°
$3d^6(^3$P2$)4p$	$z\ ^2$D°	5/2	*61 093.413*	1.01	42		27	$(^3$P1$)\ ^2$D°
		3/2	*62 125.600*	1.019	35		23	
$3d^6(^3$H$)4p$	$z\ ^4$I°	15/2	*61 347.614*		100			
		9/2	*61 512.634*		56		26	$(^3$H$)\ ^4$H°
		13/2	*61 527.616*		49		43	
		11/2	*61 587.214*		51		42	
$3d^6(^3$P2$)4p$	$y\ ^4$D°	7/2	*61 726.077*	1.411	56		33	$(^3$P1$)\ ^4$D°
		5/2	*62 689.880*	1.349	49		27	
		1/2	*62 829.075*		60		32	
		3/2	*62 962.205*	1.14	45		25	
$3d^5(^6$S$)4s4p(^3$P°$)$	$y\ ^6$P°	3/2	*61 974.933*					
		5/2	*62 049.025*					
		7/2	*62 171.615*	1.68				
$3d^6(^3$F2$)4p$		7/2	*62 065.521*	1.198	31	^{4}F°	13	$(^3$H$)\ ^2$G°
$3d^6(^3$H$)4p$	$z\ ^2$G°	9/2	*62 083.108*	1.097	61		15	$(^3$G$)\ ^2$G°
		7/2	*62 322.431*		38		23	$(^3$F2$)\ ^4$F°

(Continued)

Fe II—Continued

Configuration	Term	J	Level (cm^{-1})	g	Leading percentages			
$3d^6(^3F2)4p$	$y\ ^4F^\circ$	5/2	62 151.561	1.025	61		19	$(^3F1)\ ^4F^\circ$
		9/2	62 158.110	1.33	63		21	
		3/2	62 244.520	0.43	66		21	
$3d^6(^3H)4p$	$z\ ^2I^\circ$	13/2	62 293.164	1.069	90		8	$(^3H)\ ^4I^\circ$
		11/2	62 662.244	0.910	93		4	
$3d^6(^3F2)4p$	$x\ ^4D^\circ$	7/2	62 945.038	1.385	60		14	$(^3F1)\ ^4D^\circ$
		5/2	63 272.976	1.351	67		15	
		3/2	63 465.109	1.21	71		16	
		1/2	63 559.488	0.013	71		15	
$3d^6(^3F2)4p$	$y\ ^4G^\circ$	11/2	63 876.317	1.24	50		28	$(^3H)\ ^4G^\circ$
		9/2	63 948.790	1.15	41		30	
		7/2	64 040.886	0.975	29		25	
		5/2	64 087.418	0.617	26		29	
$3d^6(^3F2)4p$	$z\ ^2F^\circ$	7/2	64 286.345	1.135	30		13	$(^3F1)\ ^2F^\circ$
		5/2	64 425.408	0.82	32		13	
$3d^6(^3P2)4p$	$z\ ^2P^\circ$	1/2	64 806.487		45		31	$(^3P1)\ ^2P^\circ$
		3/2	64 834.073	1.329	54		34	
$3d^6(^3F2)4p$	$y\ ^2G^\circ$	9/2	64 831.943	1.101	62		16	$(^3F1)\ ^2G^\circ$
		7/2	65 109.679	0.896	67		16	
$3d^6(^3H)4p$		11/2	65 363.595	1.066	39	$^2H^\circ$	37	$(^3G)\ ^4G^\circ$
$3d^6(^3H)4p$	$z\ ^2H^\circ$	9/2	65 556.280	0.913	51		20	$(^3G)\ ^2H^\circ$
$3d^6(^3G)4p$	$x\ ^4G^\circ$	11/2	65 580.041		53		20	$(^3G)\ ^2H^\circ$
		9/2	65 696.038		48		27	$(^3G)\ ^4F^\circ$
		7/2	65 931.334	1.00	76		10	$(^3G)\ ^4F^\circ$
		5/2	66 078.269	0.62	83		6	$(^3F2)\ ^4G^\circ$
$3d^6(^3G)4p$	$x\ ^4F^\circ$	9/2	66 012.750		53		33	$(^3G)\ ^4G^\circ$
		7/2	66 377.283	1.21	71		11	$(^3G)\ ^4G^\circ$
		5/2	66 522.304	1.02	67		9	$(^3F2)\ ^2D^\circ$
		3/2	66 612.656		67		12	$(^3F2)\ ^2D^\circ$
$3d^6(^3P2)4p$	$z\ ^2S^\circ$	1/2	66 248.66		58		28	$(^3P1)\ ^2S^\circ$
$3d^6(^3G)4p$	$y\ ^4H^\circ$	13/2	66 411.686		89		9	$(^3H)\ ^4H^\circ$
		11/2	66 463.528	1.13	79		8	
		9/2	66 589.008	0.959	83		9	
		7/2	66 672.334	0.69	85		11	
$3d^6(^3F2)4p$	$y\ ^2D^\circ$	5/2	67 000.517	1.16	64		10	$(^3G)\ ^4F^\circ$
		3/2	67 273.826	0.719	59		14	
$3d^6(^3G)4p$	$y\ ^2H^\circ$	11/2	67 516.332	1.07	55		32	$(^3H)\ ^2H^\circ$
		9/2	68 000.788	0.907	59		31	
$3d^5(^6S)4s4p(^3P^\circ)$	$x\ ^4P^\circ$	5/2	69 102.38					
		3/2	69 302.09					
		1/2	69 426.98					
$3d^6(^3G)4p$	$y\ ^2F^\circ$	7/2	69 606.552	1.13	62		12	$(^3F2)\ ^2F^\circ$
		5/2	69 650.484	0.857	63		12	$(^3D)\ ^2F^\circ$

Fe II—Continued

Configuration	Term	J	Level (cm^{-1})	g	Leading percentages			
$3d^6(^3G)4p$	$x\ ^2G^\circ$	9/2	*70 314.604*	1.11	78		17	$(^3H)\ ^2G^\circ$
		7/2	*70 523.706*	0.87	73		14	
$3d^6(^1I)4p$	$z\ ^2K^\circ$	13/2	*70 986.677*		99			
		15/2	*71 432.680*	1.05	100			
$3d^6(^3D)4p$	$^4P^\circ$	5/2	*71 964.710*		86		5	$^4D^\circ$
		3/2	*72 043.026*	1.66	54		23	$^4F^\circ$
		1/2	*72 429.711*		51		35	$^4D^\circ$
$3d^6(^1G2)4p$	$x\ ^2H^\circ$	9/2	*72 130.39*	0.91	51		22	$(^1G1)\ ^2H^\circ$
		11/2	*72 261.729*	1.08	42		35	$(^1I)\ ^2H^\circ$
$3d^6(^3D)4p$	$w\ ^4F^\circ$	3/2	*72 168.998*		58		16	$(^3D)\ ^4P^\circ$
		5/2	*72 238.513*		77		11	$(^3G)\ ^4F^\circ$
		7/2	*72 352.024*		67		9	$(^3G)\ ^4F^\circ$
		9/2	*72 650.658*		86		9	$(^3G)\ ^4F^\circ$
$3d^6(^3D)4p$		1/2	*72 212.978*		41	$^4P^\circ$	38	$(^3D)\ ^4D^\circ$
$3d^6(^3D)4p$	$w\ ^4D^\circ$	3/2	*72 524.566*		63		20	$^4P^\circ$
		5/2	*72 619.490*		79		5	$^4P^\circ$
		7/2	*72 651.876*		49		16	$^4F^\circ$
$3d^6(^3D)4p$		7/2	*73 016.147*		31	$^4D^\circ$	20	$(^1G2)\ ^2F^\circ$
$3d^6(^3D)4p$		5/2	*73 054.881*		32	$^2F^\circ$	30	$(^1G2)\ ^2F^\circ$
$3d^6(^1G2)4p$	$w\ ^2G^\circ$	9/2	*73 091.590*		58		26	$(^1G1)\ ^2G^\circ$
		7/2	*73 143.288*	0.91	55		25	
$3d^6(^3D)4p$	$y\ ^2P^\circ$	1/2	*73 187.280*		66		15	$(^3D)\ ^4D^\circ$
		3/2	*73 189.11*		70		10	
$3d^5\ 4s^2$	^4F	9/2	73 393.745					
		5/2	73 395.93					
		7/2	73 492.215					
		3/2	73 637.34					
$3d^6(^1I)4p$	$w\ ^2H^\circ$	11/2	*73 603.50*		44		21	$(^1G2)\ ^2H^\circ$
		9/2	*73 751.282*		68		8	
$3d^6(^1I)4p$	$y\ ^2I^\circ$	13/2	*73 966.832*		98			
		11/2	*73 969.767*		89		7	$(^1I)\ ^2H^\circ$
$3d^6(^3D)4p$	$x\ ^2D^\circ$	3/2	*74 498.057*		91		3	$(^1F)\ ^2D^\circ$
		5/2	*74 606.841*		91		2	
$3d^6(^3D)4p$	$w\ ^2F^\circ$	7/2	*75 600.931*	1.125	57		16	$(^1G2)\ ^2F^\circ$
		5/2	*75 915.215*	0.844	50		22	
$3d^6(^1S2)4p$	$x\ ^2P^\circ$	3/2	*76 129.446*	1.34	37		31	$(^1D2)\ ^2P^\circ$
		1/2	*76 577.482*		41		21	$(^3D)\ ^2P^\circ$
$3d^5\ 4s^2$	^2H	9/2	77 230.90					
$3d^6(^1D2)4p$	$v\ ^2F^\circ$	5/2	*77 742.730*		60		16	$(^1D1)\ ^2F^\circ$
		7/2	*78 137.364*	1.13	65		18	

(Continued)

Fe II—Continued

Configuration	Term	*J*	Level (cm^{-1})	*g*	Leading percentages		
$3d^6(^5D)5s$	$e\ ^6D$	9/2	77 861.625				
		7/2	78 237.685				
		5/2	78 525.407				
		3/2	78 725.790				
		1/2	78 843.992				
$3d^5\ 4s^2$	^2G	7/2	78 185.03				
		9/2	78 577.28				
$3d^6(^1D2)4p$	$w\ ^2D^\circ$	3/2	*78 487.153*		68	15	$(^1D1)\ ^2D^\circ$
		5/2	*78 690.846*		64	15	
$3d^6(^1D2)4p$	$w\ ^2P^\circ$	1/2	*78 841.96*		48	20	$(^1S2)\ ^2P^\circ$
		3/2	*79 243.60*		36	25	
$3d^5(^6S)4s4p(^1P^\circ)$	$x\ ^6P^\circ$	3/2	*79 246.17*				
		5/2	*79 285.11*				
		7/2	*79 331.50*				
$3d^6(^5D)5s$	$e\ ^4D$	7/2	79 439.467				
		5/2	79 885.493				
		3/2	80 177.975				
		1/2	80 346.016				
$3d^5\ 4s^2$	$^2F2$	5/2	81 639.26				
		7/2	81 734.75				
$3d^6(^5D)4d$	$e\ ^6F$	11/2	82 853.658				
		9/2	82 978.677				
		7/2	83 136.487				
		5/2	83 308.194				
		3/2	83 459.67				
		1/2	83 558.54				
$3d^6(^1F)4p$	$v\ ^2G^\circ$	7/2	*83 305.251*		92		
		9/2	*83 871.184*		94		
$3d^6(^5D)4d$	^6D	7/2	83 713.536				
		9/2	83 726.364				
		5/2	83 812.316				
		3/2	83 990.063				
		1/2	84 131.563				
$3d^6(^1F)4p$	$v\ ^2D^\circ$	5/2	*83 868.45*		80	6	$(^1D2)\ ^2D^\circ$
		3/2	*84 359.80*		85	4	
$3d^6(^5D)4d$	$e\ ^6G$	13/2	84 035.14				
		11/2	84 296.83	1.33			
		9/2	84 527.778				
		7/2	84 710.685				
		5/2	84 844.834				
		3/2	84 938.18				
$3d^6(^5D)4d$	^6P	7/2	84 266.556				
		5/2	84 326.912				
		3/2	84 424.37				

Fe II—Continued

Configuration	Term	J	Level (cm^{-1})	g	Leading percentages			
$3d^6(^5D)4d$	$f\,^4D$	7/2	84 685.198					
		5/2	84 870.863					
		3/2	85 048.602					
		1/2	85 172.809					
$3d^6(^5D)4d$	$e\,^4G$	11/2	84 863.351	1.27				
		9/2	85 184.734					
		7/2	85 462.862					
		5/2	85 679.698					
$3d^6(^5D)4d$	^6S	5/2	85 495.304					
$3d^6(^5D)4d$	^4S	3/2	85 728.806					
$3d^6(^5D)4d$	$e\,^4F$	9/2	86 124.301	1.29				
		7/2	86 416.333					
		5/2	86 599.738					
		3/2	86 710.837					
$3d^6(^3P1)4p$	$v\,^4D°$	1/2	*86 388.82*		35		36	$(^3F1)\ ^4D°$
		3/2	*86 543.974*		33		38	
		5/2	*86 767.577*		30		40	
		7/2	*86 929.649*		25		45	
$3d^6(^1F)4p$	$u\,^2F°$	7/2	*86 482.75*		88		2	$(^1G2)\ ^2F°$
		5/2	*86 547.49*		90		2	
$3d^5(^4G)4s4p(^3P°)$	$y\,^6F°$	11/2	*87 340.983*					
		9/2	*87 471.765*					
		7/2	*87 537.652*					
		5/2	*87 572.431*					
		3/2	*87 602.25*					
		1/2	*87 898.12*					
$3d^5(^4P)4s4p(^3P°)$	$^6D°$	1/2	*87 635.92*					
		3/2	*87 964.65*					
		5/2	*88 059.38*					
		7/2	*88 209.45*					
		9/2	*88 614.52*					
$3d^6(^5D)4d$	^4P	5/2	87 985.628					
		3/2	88 157.116					
		1/2	88 189.030					
$3d^6(^5D)5p$	$^6D°$	9/2	*88 723.400*		90		8	$3d^54s4p$
		7/2	*88 853.533*		53		19	$(^5D)\ ^6P°$
		5/2	*89 119.457*		71		12	$(^5D)\ ^6P°$
		3/2	*89 331.195*		84		6	$(^5D)\ ^6P°$
		1/2	*89 471.365*		97		2	$3d^54s4p$
$3d^6(^3P1)4p$	$^2S°$	1/2	*89 003.46*		61		32	$(^3P2)\ ^2S°$
$3d^6(^5D)5p$		7/2	*89 128.561*		40	$^6D°$	22	$^6P°$
$3d^5(^4P)4s4p(^3P°)$	$^6P°$	5/2	*89 444.458*					
		3/2	*89 625.940*					

(Continued)

Fe II—Continued

Configuration	Term	J	Level (cm^{-1})	g	Leading percentages			
$3d^6(^3F1)4p$	$^4G°$	5/2	89 727.342		77		17	$(^3F2)\ ^4G°$
		7/2	89 890.373		76		17	
		9/2	90 042.779		70		16	
		11/2	90 211.70		75		18	
$3d^6(^5D)5p$	$^6F°$	11/2	89 924.175		87		12	$3d^54s4p$
		7/2	90 300.625		59		26	$(^5D)\ ^4F°$
		5/2	90 487.810		77		9	$(^5D)\ ^4F°$
		3/2	90 593.497		82		13	$3d^54s4p$
		1/2	90 648.217		83		15	$3d^54s4p$
$3d^6(^5D)5p$		9/2	90 067.347		45	$^4F°$	39	$(^5D)\ ^6F°$
$3d^6(^5D)5p$	$^4F°$	9/2	90 386.528		48		44	$(^5D)\ ^6F°$
		7/2	90 780.621		63		18	$(^5D)\ ^4D°$
		5/2	91 070.547		67		13	$(^5D)\ ^4D°$
		3/2	91 208.887		64		28	$(^5D)\ ^4D°$
$3d^6(^5D)5p$	$^4D°$	7/2	90 397.868		77		12	$^6F°$
		5/2	90 638.822		69		19	$^4F°$
		3/2	91 048.256		65		30	$^4F°$
		1/2	91 199.746		97		1	$^4P°$
$3d^6(^3P1)4p$	$^4S°$	3/2	90 629.902		61		27	$(^3P2)\ ^4S°$
$3d^6(^3P1)4p$	$^4P°$	1/2	90 839.486		52		37	$(^3P2)\ ^4P°$
		3/2	90 898.873		28		20	
		5/2	92 274.12		45		31	
$3d^6(^5D)5p$	$^4P°$	5/2	90 901.124		79		16	$^4D°$
		3/2	92 225.538		91		5	
		1/2	92 314.758		95		1	
$3d^6(^5D)5p$	$w\ ^6P°$	7/2	91 167.937		53		39	$3d^54s4p$
		5/2	91 575.139		55		38	
$3d^6(^3P1)4p$		3/2	91 843.470		28	$^4P°$	24	$3d^6(^5D)5p\ ^6P°$
$3d^5(^4D)4s4p(^3P°)$	$^6F°$	1/2	91 850.722					
		3/2	91 915.95					
		5/2	92 018.729					
		7/2	92 154.165					
		9/2	92 300.277					
		11/2	92 432.136					
$3d^5(^4G)4s4p(^3P°)$	$x\ ^4H°$	7/2	92 089.26					
		9/2	92 116.78					
		11/2	92 166.60					
		13/2	92 250.21					
$3d^6(^3F1)4p$	$u\ ^2G°$	9/2	92 171.716		50		13	$(^3F2)\ ^2G°$
		7/2	92 602.703		62		15	
$3d^6(^3F1)4p$	$u\ ^2D°$	3/2	92 216.32		46		20	$(^3P1)\ ^2D°$
		5/2	92 695.374		50		17	

Fe II—Continued

Configuration	Term	J	Level (cm⁻¹)	g	Leading percentages		
$3d^5(^4G)4s4p(^3P^\circ)$	$v\ ^4F^\circ$	7/2	92 282.46				
		5/2	92 329.89				
		3/2	92 358.61				
		9/2	92 426.98				
$3d^6(^3F1)4p$	$^4D^\circ$	1/2	92 453.46		42	25	$(^3P1)\ ^4D^\circ$
		3/2	92 647.51		36	25	
		5/2	92 899.20		26	23	
		7/2	93 129.90		18	22	
$3d^6(^3F1)4p$	$u\ ^4F^\circ$	3/2	93 328.48		46	17	$(^3F2)\ ^4F^\circ$
		5/2	93 395.36		45	17	
		9/2	93 484.58		54	21	
		7/2	93 487.65		35	13	
$3d^5(^4D)4s4p(^3P^\circ)$	$^6D^\circ$	5/2	93 830.979				
		3/2	93 840.34				
		7/2	93 987.457				
		1/2	94 031.378				
		9/2	94 057.773				
$3d^5(^4G)4s4p(^3P^\circ)$	$w\ ^4G^\circ$	5/2	93 988.17				
		7/2	94 073.24				
		9/2	94 148.51				
		11/2	94 189.88				
$3d^5(^4P)4s4p(^3P^\circ)$	$^4P^\circ$	5/2	94 211.739				
		3/2	94 739.17				
		1/2	94 880.74				
$3d^5(^4D)4s4p(^3P^\circ)$	$^6P^\circ$	5/2	94 685.09				
		7/2	94 763.219				
$3d^6(^3P1)4p$	$^2D^\circ$	5/2	94 700.66		37	27	$(^3P2)\ ^2D^\circ$
$3d^6(^3P1)4p$	$^2P^\circ$	3/2	95 039.2		43	26	$(^3P2)\ ^2P^\circ$
$3d^6(^3F1)4p$	$^2F^\circ$	5/2	95 046.10		56	20	$3d^54s4p$
		7/2	95 079.64		54	20	
$3d^5(^4P)4s4p(^3P^\circ)$	$^4D^\circ$	1/2	95 767.70				
		3/2	95 858.05				
		5/2	95 995.69				
		7/2	96 217.42				
$3d^5(^4G)4s4p(^3P^\circ)$	$v\ ^2H^\circ$	11/2	96 062.06				
		9/2	96 239.20				
$3d^5(^4G)4s4p(^3P^\circ)$	$t\ ^2F^\circ$	5/2	96 279.49				
		7/2	96 356.96				
$3d^5(^4P)4s4p(^3P^\circ)$	$^2P^\circ$	3/2	97 326.27				
$3d^6(^1G1)4p$	$^2H^\circ$	9/2	97 851.35		62	33	$(^1G2)\ ^2H^\circ$
		11/2	98 278.77		63	34	

(Continued)

Fe II—Continued

Configuration	Term	J	Level (cm^{-1})	g	Leading percentages		
$3d^6(^3\mathrm{H})5s$	$e\ ^4\mathrm{H}$	13/2	98 130.131				
		11/2	98 294.401				
		9/2	98 445.400				
		7/2	98 568.912				
$3d^5(^4\mathrm{D})4s4p(^3\mathrm{P}^\circ)$	$^4\mathrm{F}^\circ$	3/2	*98 196.00*				
		5/2	*98 354.66*				
		7/2	*98 535.85*				
		9/2	*98 596.65*				
$3d^5(^4\mathrm{P})4s4p(^3\mathrm{P}^\circ)$	$^4\mathrm{S}^\circ$	3/2	*98 338.28*				
$3d^5(^4\mathrm{D})4s4p(^3\mathrm{P}^\circ)$	$^4\mathrm{D}^\circ$	7/2	*98 391.33*				
		5/2	*98 505.10*				
		3/2	*98 770.14*				
		1/2	*99 007.70*				
$3d^6(^1\mathrm{G}1)4p$	$^2\mathrm{F}^\circ$	7/2	*98 898.71*		43	22	$(^1\mathrm{G}2)\ ^2\mathrm{F}^\circ$
$3d^6(^3\mathrm{H})5s$	$e\ ^2\mathrm{H}$	11/2	99 093.452				
		9/2	99 332.102				
$3d^6(^3\mathrm{F}2)5s$	$f\ ^4\mathrm{F}$	9/2	99 573.225				
		7/2	99 688.337				
		5/2	99 824.045				
		3/2	99 918.569				
$3d^5(^4\mathrm{G})4s4p(^3\mathrm{P}^\circ)$	$^2\mathrm{G}^\circ$	7/2	*99 635.52*				
		9/2	*99 653.22*				
$3d^6(^1\mathrm{G}1)4p$	$^2\mathrm{G}^\circ$	9/2	*99 757.12*		42	31	$3d^54s4p$
		7/2	*99 808.40*		44	20	$3d^6(^1\mathrm{G}2)4p\ ^2\mathrm{G}^\circ$
$3d^5(^4\mathrm{P})4s4p(^3\mathrm{P}^\circ)$	$^2\mathrm{D}^\circ$	5/2	*100 400.36*				
$3d^6(^3\mathrm{F}2)5s$	$e\ ^2\mathrm{F}$	7/2	100 492.02				
		5/2	100 749.81				
$3d^5(^4\mathrm{D})4s4p(^3\mathrm{P}^\circ)$	$^4\mathrm{P}^\circ$	1/2	*101 402.38*				
		3/2	*101 573.90*				
$3d^6(^5\mathrm{D})6s$	$^6\mathrm{D}$	9/2	101 698.489				
		7/2	102 030.912				
		5/2	102 334.112				
		3/2	102 543.648				
		1/2	102 666.694				
$3d^5(^2\mathrm{I})4s4p(^3\mathrm{P}^\circ)$	$^4\mathrm{K}^\circ$	11/2	*102 340.3*				
		13/2	*102 489.9*				
		15/2	*102 851.2?*				
$3d^6(^5\mathrm{D})6s$	$^4\mathrm{D}$	7/2	102 394.718				
		5/2	102 802.312				
		3/2	103 118.400				
		1/2	103 265.694				
$3d^5(^4\mathrm{D})4s4p(^3\mathrm{P}^\circ)$	$^2\mathrm{D}^\circ$	5/2	*102 449.10*				
		3/2	*102 503.81*				

Fe II—Continued

Configuration	Term	J	Level (cm^{-1})	g	Leading percentages
$3d^6(^3G)5s$	$f\,^4G$	11/2	102 584.963		
		9/2	102 842.119		
$3d^6(^5D_4)4f$	$^2[5]°$	11/2	*102 831.32*		
		9/2	*102 851.36*		
$3d^6(^5D_4)4f$	$^2[6]°$	13/2	*102 840.25*		
		11/2	*102 893.38*		
$3d^6(^5D_4)4f$	$^2[4]°$	9/2	*102 882.37*		
		7/2	*102 887.12*		
$3d^6(^5D_4)4f$	$^2[3]°$	7/2	*102 942.20*		
		5/2	*102 952.12*		
$3d^5(^2I)4s4p(^3P°)$	$^4I°$	9/2	*102 951.5*		
		11/2	*102 980.3*		
		13/2	*103 120.9*		
		15/2	*103 232.1*		
$3d^6(^5D_4)4f$	$^2[7]°$	13/2	*103 019.67*		
		15/2	*103 040.32*		
$3d^6(^5D_4)4f$	$^2[2]°$	5/2	*103 024.29*		
		3/2	*103 034.76*		
$3d^5(^6S)4s\,(^7S)5s$	^8S	7/2	103 094.73		
$3d^6(^5D_4)4f$	$^2[1]°$	3/2	*103 110.79*		
		1/2	*103 125.65*		
$3d^5(^4D)4s4p(^3P°)$	$^2F°$	7/2	*103 183.7*		
		5/2	*103 334.1*		
$3d^6(^5D_3)4f$	$^2[5]°$	11/2	*103 325.95*		
		9/2	*103 352.68*		
$3d^6(^5D_3)4f$	$^2[4]°$	9/2	*103 326.41*		
		7/2	*103 340.64*		
$3d^6(^5D_3)4f$	$^2[3]°$	5/2	*103 364.84*		
		7/2	*103 385.73*		
$3d^6(^5D_3)4f$	$^2[2]°$	3/2	*103 391.29*		
		5/2	*103 406.25*		
$3d^6(^5D_3)4f$	$^2[1]°$	3/2	*103 417.91*		
		1/2	*103 437.28*		
$3d^6(^5D_3)4f$	$^2[0]°$	1/2	*103 418.08*		
$3d^6(^5D_3)4f$	$^2[6]°$	11/2	*103 420.16*		
		13/2	*103 421.18*		
$3d^6(^3H)4d$	^4H	13/2	103 600.44		
		11/2	103 751.66		
$3d^6(^3G)5s$	^2G	9/2	103 608.909		
		7/2	103 983.51		

(Continued)

Fe II—Continued

Configuration	Term	J	Level (cm^{-1})	g	Leading percentages
$3d^6(^5D_2)4f$	$^2[2]°$	3/2	*103 645.22*		
		5/2	*103 660.98*		
$3d^6(^5D_2)4f$	$^2[1]°$	3/2	*103 668.69*		
		1/2	*103 676.22*		
$3d^6(^5D_2)4f$	$^2[3]°$	7/2	*103 676.78*		
		5/2	*103 698.44*		
$3d^6(^5D_2)4f$	$^2[4]°$	9/2	*103 680.64*		
		7/2	*103 711.57*		
$3d^6(^5D_2)4f$	$^2[5]°$	11/2	*103 691.05*		
		9/2	*103 701.72*		
$3d^6(^5D_1)4f$	$^2[2]°$	5/2	*103 857.74*		
		3/2	*103 869.02*		
$3d^6(^5D_1)4f$	$^2[4]°$	9/2	*103 873.99*		
		7/2	*103 882.68*		
$3d^6(^3H)4d$	^4I	15/2	103 878.34		
		13/2	104 064.67		
		11/2	104 174.27		
$3d^6(^5D)5d$	^6F	11/2	103 936.60		
		9/2	103 950.59		
		7/2	104 380.94		
		5/2	104 426.46		
$3d^5(^4P)4s4p(^3P°)$	$^2S°$	1/2	*103 967.49*		
$3d^6(^5D_1)4f$	$^2[3]°$	7/2	*103 969.76*		
		5/2	*103 987.93*		
$3d^6(^5D)5d$	^6P	7/2	104 000.81		
		5/2	104 120.27		
		3/2	104 630.43		
$3d^6(^5D_0)4f$	$^2[3]°$	7/2	*104 022.89*		
		5/2	*104 046.35*		
$3d^6(^3H)4d$	^2K	15/2	104 119.71		
		13/2	104 315.37		
$3d^6(^3H)4d$	^4F	3/2	104 189.38		
$3d^6(^5D)5d$	^6G	13/2	104 366.82		
		11/2	104 593.27		
		9/2	104 868.50		
		7/2	105 065.63		
		5/2	105 205.79		
		3/2	105 288.53		
$3d^6(^5D)5d$	^6D	9/2	104 411.69		
		3/2	104 588.71		
		7/2	104 705.42		
		1/2	104 757.11		
		5/2	104 828.16		

Fe II—Continued

Configuration	Term	J	Level (cm^{-1})	g	Leading percentages
		5/2	104 569.23		
$3d^5(^2I)4s4p(^3P°)$	$^4H°$	13/2	*104 659.26*		
		11/2	*104 816.80*		
		9/2	*104 937.8*		
		7/2	*105 028.6*		
		5/2	104 761.10		
		3/2	104 840.02		
$3d^6(^5D)5d$	^4G	11/2	104 863.43		
		9/2	105 211.14		
		7/2	105 449.54		
		5/2	105 630.75		
$3d^6(^5D)5d$	^4D	7/2	104 873.23		
		5/2	105 127.77		
		1/2	105 230.29		
$3d^6(^3F2)4d$	^4G	11/2	105 063.55		
		9/2	105 155.09		
		7/2	105 291.01		
		5/2	105 414.18		
		5/2	105 234.06		
		5/2	105 238.77		
$3d^6(^3F2)4d$	^4H	13/2	105 288.847		
		11/2	105 398.852		
		9/2	105 524.461		
		7/2	105 589.42		
$3d^6(^5D)5d$	^6S	5/2	105 711.73		
$3d^6(^3F2)4d$	^2H	11/2	105 763.270		
		9/2	106 018.643		
$3d^5(^2I)4s4p(^3P°)$	$^2K°$	13/2	*106 183.1*		
		15/2	*106 524.4*		
$3d^5(^2I)4s4p(^3P°)$	$^2H°$	11/2	*106 690.17*		
		9/2	*107 006.35*		
$3d^5(^6S)4p^2(^3P)$	^8P	7/2	106 836.0		
		9/2	107 219.5		
$3d^5(^4G)4s4p(^1P°)$	$^4G°$	11/2	*108 483.87*		
		9/2	*108 570.56*		
		5/2	*108 629.25*		
		7/2	*108 631.09*		
$3d^6(^1I)5s$	$e\ ^2I$	11/2	108 630.429		
		13/2	108 648.695		
$3d^5(^4G)4s4p(^1P°)$	$^4H°$	13/2	*108 729.16*		
		11/2	*108 809.31*		
		9/2	*108 868.98*		
		7/2	*108 906.64*		

(Continued)

Fe II—Continued

Configuration	Term	J	Level (cm^{-1})	g	Leading percentages
$3d^6(^3D)5s$	^4D	$7/2$	108 804.667		
$3d^5(^2I)4s4p(^3P°)$	$^2I°$	$13/2$	*109 149.68*		
		$11/2$	*109 271.71*		
$3d^5(^6S)4s\ (^7S)4d$	^8D	$3/2$	109 449.53		
		$5/2$	109 455.25		
		$7/2$	109 463.22		
		$9/2$	109 473.65		
		$11/2$	109 486.15		
Fe III ($^5D_4$)	*Limit*		**130 563**		

Fe III

$Z=26$

Cr I isoelectronic sequence

Ground state: $1s^22s^22p^63s^23p^63d^6\ {}^5D_4$

Ionization energy = 247 220±100 cm^{-1} (30.652±0.010 eV)

The present list of energy levels for Fe III is a combination of the results of Edlén and Swings (1942), who observed the spectrum from 500–6500 Å, and those of Glad (1956), who reobserved the long wavelength portion from 2600–8600 Å. A correction of 0.8 cm^{-1} has been added to the published level values to place the ground state at zero. No discussion of the level accuracy was given. We assume an uncertainty of ± 0.05 for levels given to two decimal places and ±0.5 for those given to one decimal place. The ^{7}P term of $3d^44s4p$ was found by Johansson and Ekberg (1982).

The percentage compositions for levels of the $3d^6$ configuration were taken from the theoretical work of Pasternak and Goldschmidt (1972). For the $3d^54s$ configuration, we have used the percentages given by Shadmi, Caspi, and Oreg (1969), who listed compositions only for highly mixed states. Although no statement was made concerning the percentage compositions of the remaining levels, it appears that their purity is at least 90%. For the $3d^54p$ configuration we have used the results of Roth (1968). Roth distinguished repeating terms of $3d^n$ by the letters a, b ... rather than by seniority. Each of his percentages is the sum of LS term contributions differing only in the seniority of the core term.

Transitions among levels of the $3d^6$ configuration observed in nebular spectra have been given by Bowen (1960).

The ionization energy was determined by Glad from the $3d^5(^6S)ns\ ^7S$ levels ($n=5$, 6, 7).

References

Bowen, I. S. (1960), Astrophys. J. **132**, 1.

Edlén, B., and Swings, P. (1942), Astrophys. J. **95**, 532.

Glad, S. (1956), Ark. Fys. **10**, 21.

Johansson, S., and Ekberg, J. O. (1982), Atomic Spectroscopy Annual Report, University of Lund.

Pasternak, A., and Goldschmidt, Z. B. (1972), Phys. Rev. **A6**, 55.

Roth, C. (1968), J. Res. Natl. Bur. Stand (U.S.) **72A**, 505.

Shadmi, Y., Caspi, E., and Oreg, J. (1969), J. Res. Natl. Bur. Stand. (U.S.) **73A**, 173.

Fe III

Configuration	Term	J	Level (cm^{-1})	Leading percentages		
$3d^6$	^{5}D	4	0.0	100		
		3	436.2	100		
		2	738.9	100		
		1	932.4	100		
		0	1 027.3	100		
$3d^6$	^{3}P2	2	19 404.8	61	38	^{3}P1
		1	20 688.4	62	38	
		0	21 208.5	62	37	
$3d^6$	^{3}H	6	20 051.1	100		
		5	20 300.8	99		
		4	20 481.9	97		
$3d^6$	^{3}F2	4	21 462.2	74	21	^{3}F1
		3	21 699.9	77	21	
		2	21 857.2	79	20	
$3d^6$	^{3}G	5	24 558.8	99		
		4	24 940.9	97		
		3	25 142.4	98		
$3d^5(^6S)4s$	^{7}S	3	30 088.84			
$3d^6$	^{1}I	6	30 356.2	100		

(Continued)

Fe III—Continued

Configuration	Term	J	Level (cm^{-1})	Leading percentages		
$3d^6$	^{3}D	2	30 716.2	99		
		1	30 725.8	100		
		3	30 857.8	100		
$3d^6$	^{1}G2	4	30 886.4	65	34	^{1}G1
$3d^6$	^{1}S2	0	34 812.4	76	23	^{1}S1
$3d^6$	^{1}D2	2	35 803.7	77	22	^{1}D1
$3d^5(^6$S$)4s$	^{5}S	2	40 999.87			
$3d^6$	^{1}F	3	42 896.9	99		
$3d^6$	^{3}P1	0	49 148	62	38	^{3}P2
		1	49 576.9	62	38	
		2	50 412.3	61	39	
$3d^6$	^{3}F1	2	50 184.9	80	20	^{3}F2
		4	50 276.1	78	22	
		3	50 295.2	78	21	
$3d^6$	^{1}G1	4	57 221.7	65	35	^{1}G2
$3d^5(^4$G$)4s$	^{5}G	6	63 425.17			
		5	63 466.39			
		4	63 486.78			
		3	63 494.00			
		2	63 494.56			
$3d^5(^4$P$)4s$	^{5}P	3	66 464.64			
		2	66 522.95			
		1	66 591.68			
$3d^5(^4$D$)4s$	^{5}D	4	69 695.73			
		0	69 747.40			
		1	69 788.19			
		3	69 836.83			
		2	69 837.76			
$3d^5(^4$G$)4s$	^{3}G	5	70 694.03			
		3	70 725.01			
		4	70 728.75			
$3d^5(^4$P$)4s$	^{3}P	2	73 727.64			
		1	73 849.10			
		0	73 935.96			
$3d^5(^4$D$)4s$	^{3}D	3	76 956.79			
		1	77 075.30			
		2	77 102.43			
$3d^5(^2$I$)4s$	^{3}I	7	79 840.12			
		6	79 844.74			
		5	79 860.42			
$3d^5(^6$S$)4p$	^{7}P°	2	*82 001.73*	100		
		3	*82 333.92*	99		
		4	*82 846.59*	100		

Fe III—Continued

Configuration	Term	J	Level (cm^{-1})	Leading percentages		
$3d^5(^2D3)4s$	^3D	3	82 382.87	76	16	$(^2F2)\ ^3F$
		2	82 410.94	69	17	$(^4F)\ ^5F$
		1	82 494.88	66	34	$(^4F)\ ^5F$
$3d^5(^4F)4s$	^5F	5	83 138.23			
		4	83 161.48			
		3	83 237.86			
		2	83 358.88	77	15	$(^2F2)\ ^3F$
		1	83 646.98	66	34	$(^2D3)\ ^3D$
$3d^5(^2I)4s$	^1I	6	83 429.61			
$3d^5(^2F2)4s$	^3F	4	84 159.55			
		2	84 369.92	60	17	$(^2D3)\ ^3D$
		3	84 671.87	77	18	$(^2D3)\ ^3D$
$3d^5(^2D3)4s$	^1D	2	86 847.11			
$3d^5(^2F2)4s$	^1F	3	87 901.87			
$3d^5(^2H)4s$	^3H	4	88 663.87			
		5	88 694.67			
		6	88 923.07			
$3d^5(^6S)4p$	$^5P°$	3	*89 084.79*	98		
		2	*89 334.51*	98		
		1	*89 491.39*	98		
$3d^5(^2G2)4s$	^3G	3	89 697.52			
		4	89 783.59			
		5	89 907.85			
$3d^5(^4F)4s$	^3F	2	90 423.68			
		4	90 472.53			
		3	90 483.94			
$3d^5(^2H)4s$	^1H	5	92 523.91			
$3d^5(^2F1)4s$	^3F	4	93 388.75	58	41	$(^2G2)\ ^1G$
		3	93 392.45			
		2	93 412.93			
$3d^5(^2G2)4s$	^1G	4	93 512.64	55	40	$(^2F1)\ ^3F$
$3d^5(^2F1)4s$	^1F	3	97 041.38			
$3d^5(^2S)4s$	^3S	1	98 662.68			
$3d^5(^2D2)4s$	^3D	1	105 895.35			
		2	105 906.23			
		3	105 929.16			
$3d^5(^2D2)4s$	^1D	2	109 570.84			
$3d^5(^4G)4p$	$^5G°$	2	*113 584.20*	96		
		3	*113 605.37*	91	5	$(^4G)\ ^5H°$
		4	*113 635.34*	89	8	
		5	*113 677.01*	88	9	
		6	*113 739.62*	90	7	

(Continued)

Fe III—Continued

Configuration	Term	J	Level (cm^{-1})	Leading percentages			
$3d^5(^2G1)4s$	^3G	5	114 325.35				
		4	114 339.95				
		3	114 351.92				
$3d^5(^4G)4p$	$^5H^\circ$	3	114 948.55	94		5	$(^4G)\ ^5G^\circ$
		4	115 110.92	90		8	
		5	115 289.91	90		9	
		6	115 474.25	92		7	
		7	115 642.23	100			
$3d^5(^4G)4p$	$^5F^\circ$	5	116 316.63	90		5	$(^4D)\ ^5F^\circ$
		4	116 467.41	81		7	$(^4D)\ ^5F^\circ$
		3	116 475.44	55		22	$(^4P)\ ^5D^\circ$
		1	116 937.57	76		12	$(^4P)\ ^5D^\circ$
		2	116 975.05	57		28	$(^4P)\ ^5D^\circ$
$3d^5(^4P)4p$	$^5D^\circ$	0	116 364.76	80		16	$(^4D)\ ^5D^\circ$
		1	116 380.07	67		16	$(^4D)\ ^5D^\circ$
		2	116 419.39	46		29	$(^4G)\ ^5F^\circ$
		3	117 068.56	49		32	$(^4G)\ ^5F^\circ$
		4	117 521.91	75		14	$(^4D)\ ^5D^\circ$
$3d^5(^4P)4p$	$^5S^\circ$	2	116 898.22	92			
$3d^5(^2G1)4s$	^1G	4	117 950.32				
$3d^5(^4G)4p$	$^3F^\circ$	2	118 163.56	90			
		3	118 246.52	75		10	$(^4P)\ ^5P^\circ$
		4	118 350.24	89			
$3d^5(^4G)4p$	$^3H^\circ$	6	118 355.01	96			
		5	118 557.25	97			
		4	118 686.25	95			
$3d^5(^4P)4p$	$^5P^\circ$	3	118 442.92	53		22	$(^4D)\ ^5P^\circ$
		2	118 721.60	69		19	
		1	118 867.87	78		14	
$3d^5(^4P)4p$	$^3P^\circ$	2	119 697.64	66		18	$(^4D)\ ^3P^\circ$
		1	119 982.26	71		18	
		0	120 179.95	76		17	
$3d^5(^4D)4p$	$^5F^\circ$	1	120 697.10	85		11	$(^4G)\ ^5F^\circ$
		2	120 826.17	84		10	
		3	121 008.78	84		8	
		4	121 241.67	87		7	
		5	121 468.82	92		6	
$3d^5(^4G)4p$	$^3G^\circ$	3	121 919.74	94			
		4	121 941.29	95			
		5	121 949.62	95			
$3d^5(^4P)4p$	$^3D^\circ$	3	122 346.61	53		29	$(^4D)\ ^5D^\circ$
		2	122 628.34	46		36	
		1	122 843.03	46		35	
$3d^5(^4D)4p$		3	122 829.55	36	$^5D^\circ$	31	$(^4P)\ ^3D^\circ$
$3d^5(^4P)4p$		2	122 898.84	40	$^3D^\circ$	25	$(^4D)\ ^5D^\circ$

Fe III—Continued

Configuration	Term	J	Level (cm^{-1})	Leading percentages			
$3d^5(^4P)4p$		1	*122 921.37*	41	$^3D^\circ$	22	$(^4D)\ ^5P^\circ$
$3d^5(^4D)4p$	$^5D^\circ$	4	*122 944.15*	78		16	$(^4P)\ ^5D^\circ$
		0	*123 455.92*	75		19	$(^4P)\ ^5D^\circ$
$3d^5(^4D)4p$	$^5P^\circ$	1	*123 552.95*	56		20	$(^4D)\ ^5D^\circ$
		2	*123 697.18*	55		18	$(^4P)\ ^5P^\circ$
		3	*123 750.39*	45		23	$(^4P)\ ^5P^\circ$
$3d^5(^4D)4p$	$^3D^\circ$	3	*124 854.04*	71		12	$(^4D)\ ^5P^\circ$
		2	*124 903.92*	84		7	$(^4F)\ ^3D^\circ$
		1	*124 954.88*	84		8	$(^4F)\ ^3D^\circ$
$3d^5(^4D)4p$	$^3F^\circ$	4	*125 443.58*	90		6	$(a\ ^2G)\ ^3F^\circ$
		3	*125 637.98*	86		6	
		2	*125 672.83*	88		6	
$3d^5(^4P)4p$	$^3S^\circ$	1	*126 390.57*	95			
$3d^5(^4D)4p$	$^3P^\circ$	0	*128 371.53*	77		18	$(^4P)\ ^3P^\circ$
		1	*128 605.65*	74		19	
		2	*128 917.51*	72		21	
$3d^5(^2I)4p$	$^3K^\circ$	6	*129 854.80*	83		15	$(^2I)\ ^3I^\circ$
		7	*130 040.56*	76		17	
		8	*130 852.25*	100			
$3d^5(^2I)4p$	$^3I^\circ$	5	*130 256.27*	82		9	(^2I) [illegible]H°
		6	*130 756.84*	78		16	$(^2I)\ ^3K^\circ$
		7	*131 035.07*	71		21	$(^2I)\ ^3K^\circ$
$3d^5(a\ ^2D)4p$		2	*131 445.03*	32	$^1D^\circ$	26	$(a\ ^2F)\ ^3F^\circ$
$3d^5(^2I)4p$	$^1H^\circ$	5	*131 710.79*	69		13	$(^2I)\ ^3I^\circ$
$3d^5(^2I)4p$	$^1K^\circ$	7	*131 991.58*	89		9	$(^2I)\ ^3I^\circ$
$3d^5(a\ ^2D)4p$	$^3F^\circ$	3	*132 079.91*	58		25	$(a\ ^2F)\ ^3F^\circ$
		2	*132 104.94*	42		26	$(a\ ^2D)\ ^1D^\circ$
		4	*132 785.36*	58		22	$(a\ ^2F)\ ^3F^\circ$
$3d^5(^2I)4p$	$^3H^\circ$	6	*132 262.66*	90			
		5	*132 564.71*	86		6	$(^2I)\ ^1H^\circ$
		4	*132 659.17*	84			
$3d^5(a\ ^2D)4p$	$^3P^\circ$	2	*134 265.42*	67		25	$(a\ ^2F)\ ^3D^\circ$
		1	*134 549.38*	59		20	$(a\ ^2D)\ ^3D^\circ$
		0	*135 088.60*	90		6	$(^4F)\ ^5D^\circ$
$3d^5(a\ ^2F)4p$	$^1G^\circ$	4	*134 360.40*	57		17	$(a\ ^2F)\ ^3G^\circ$
$3d^5(a\ ^2F)4p$	$^3G^\circ$	3	*134 549.00*	53		25	$(a\ ^2D)\ ^1F^\circ$
		5	*135 316.42*	55		35	$(^4F)\ ^5G^\circ$
		4	*135 554.41*	54		36	$(a\ ^2F)\ ^3F^\circ$
$3d^5(^4F)4p$	$^5G^\circ$	2	*134 937.84*	75		10	$(a\ ^2F)\ ^3F^\circ$
		3	*135 096.84*	54		27	$(a\ ^2D)\ ^3D^\circ$
		4	*135 239.74*	81		8	$(a\ ^2F)\ ^1G^\circ$
		6	*135 582.08*	50		44	$(^2I)\ ^1I^\circ$
		5	*135 735.31*	58		39	$(a\ ^2F)\ ^3G^\circ$

(Continued)

Fe III—Continued

Configuration	Term	J	Level (cm^{-1})	Leading percentages			
$3d^5(a\,^2D)4p$	$^3D^\circ$	3	134 976.22	32		25	$(a\,^2F)\,^3D^\circ$
		1	135 217.17	60		22	$(a\,^2D)\,^3P^\circ$
		2	135 279.04	62		12	$(^4F)\,^5G^\circ$
$3d^5(a\,^2F)4p$	$^3D^\circ$	3	135 705.57	65		11	$(a\,^2D)\,^3D^\circ$
		1	136 464.9	66		19	$(a\,^2D)\,^1P^\circ$
		2	136 793.82	36		37	$(^4F)\,^5F^\circ$
$3d^5(^2I)4p$	$^1I^\circ$	6	135 739.47	50		46	$(^4F)\,^5G^\circ$
$3d^5(^4F)4p$	$^5F^\circ$	4	135 990.62	74		17	$(^4F)\,^5D^\circ$
		3	136 008.74	65		13	$(^4F)\,^5D^\circ$
		2	136 117.94	38		36	$(a\,^2F)\,^3D^\circ$
		5	136 185.17	88			
		1	136 235.84	76		10	$(a\,^2D)\,^3D^\circ$
$3d^5(a\,^2D)4p$		3	136 200.13	31	$^1F^\circ$	24	$(a\,^2F)\,^3G^\circ$
$3d^5(a\,^2F)4p$	$^3F^\circ$	2	136 532.45	46		19	$(a\,^2D)\,^3F^\circ$
		4	136 612.78	42		28	
		3	136 797.05	41		14	
$3d^5(^4F)4p$	$^5D^\circ$	4	137 209.73	75		16	$(^4F)\,^5F^\circ$
		3	137 423.00	74		14	$(^4F)\,^5F^\circ$
		2	137 544.60	77		9	$(^4F)\,^5F^\circ$
		1	137 561.1	85		6	$(a\,^2D)\,^3P^\circ$
		0	137 573.2	91		6	$(a\,^2D)\,^3P^\circ$
$3d^5(^2H)4p$	$^3H^\circ$	4	137 527.92	46		44	$(a\,^2G)\,^3H^\circ$
		5	137 763.70	43		42	
		6	138 264.47	46		41	
$3d^5(^2H)4p$	$^3G^\circ$	5	138 054.59	47		29	$(^4F)\,^3G^\circ$
		4	138 103.12	43		30	
		3	138 187.93	41		28	
$3d^5(a\,^2D)4p$	$^1P^\circ$	1	138 691.81	71		17	$(a\,^2F)\,^3D^\circ$
$3d^5(^4F)4p$	$^3G^\circ$	5	139 463.36	43		25	$(a\,^2G)\,^3G^\circ$
		4	139 625.17	42		36	
		3	139 680.47	42		41	
$3d^5(^2H)4p$	$^3I^\circ$	5	139 509.44	79		8	$(^2H)\,^3H^\circ$
		6	139 846.18	87		5	
		7	140 196.33	96			
$3d^5(a\,^2F)4p$	$^1D^\circ$	2	139 764.48	56		38	$(a\,^2D)\,^1D^\circ$
$3d^5(a\,^2G)4p$	$^1G^\circ$	4	139 827.17	40		19	$(a\,^2F)\,^1G^\circ$
$3d^5(a\,^2F)4p$	$^1F^\circ$	3	140 453.10	72		8	$(a\,^2D)\,^1F^\circ$
$3d^5(a\,^2G)4p$	$^3F^\circ$	3	140 693.36	42		26	$(^4F)\,^3F^\circ$
		2	140 750.98	42		31	
		4	141 002.99	45		26	
$3d^5(^4F)4p$	$^3D^\circ$	2	141 399.04	68		8	$(a\,^2G)\,^3F^\circ$
		3	141 466.53	64		7	$(a\,^2G)\,^3F^\circ$
		1	141 469.45	84		6	$(^4D)\,^3D^\circ$

Fe III—Continued

Configuration	Term	J	Level (cm^{-1})	Leading percentages			
$3d^5(^2H)4p$	$^1I°$	6	*141 539.55*	88		5	$(^2H)\ ^3H°$
$3d^5(^4F)4p$	$^3F°$	4	*142 047.0*	50		25	$(a\ ^2G)\ ^3F°$
		3	*142 312.90*	50		24	
		2	*142 535.07*	48		24	
$3d^5(a\ ^2G)4p$	$^3H°$	4	*142 855.59*	45		47	$(^2H)\ ^3H°$
		5	*142 908.48*	46		38	
		6	*143 320.85*	50		40	
$3d^5(a\ ^2G)4p$	$^3G°$	5	*143 883.74*	40		20	$(a\ ^2F)\ ^3G°$
		4	*144 085.97*	42		23	
		3	*144 116.64*	43		24	
$3d^5(b\ ^2F)4p$		4	*144 332.21*	35	$^1G°$	30	$(a\ ^2F)\ ^3F°$
$3d^5(b\ ^2F)4p$	$^3F°$	2	*144 501.74*	66		19	$(a\ ^2G)\ ^3F°$
		3	*144 570.53*	73		11	$(a\ ^2G)\ ^3F°$
		4	*144 968.50*	48		20	$(b\ ^2F)\ ^1G°$
$3d^5(a\ ^2G)4p$	$^1H°$	5	*144 586.83*	66		18	$(^2H)\ ^1H°$
$3d^5(^2H)4p$	$^1H°$	5	*144 843.24*	70		23	$(a\ ^2G)\ ^1H°$
$3d^5(a\ ^2G)4p$	$^1F°$	3	*145 038.61*	76		5	$(b\ ^2F)\ ^1F°$
$3d^5(b\ ^2F)4p$	$^1D°$	2	*145 618.39*	82		7	$(b\ ^2F)\ ^3F°$
$3d^5(b\ ^2F)4p$	$^3G°$	3	*146 891.04*	55		36	$(^2H)\ ^3G°$
		4	*147 161.36*	59		32	
		5	*147 406.14*	66		28	
$3d^5(^6S)4d$	^7D	1	147 281.69				
		2	147 291.21				
		3	147 305.97				
		4	147 326.85				
		5	147 354.70				
$3d^5(b\ ^2F)4p$	$^3D°$	1	*147 556.45*	90			
		2	*147 614.65*	89			
		3	*147 635.95*	86		7	$(^4F)\ ^3D°$
$3d^5(^2S)4p$	$^3P°$	0	*148 655*	85		12	$(b\ ^2D)\ ^3P°$
		1	*148 915.3*	82		13	
		2	*149 525.63*	82		14	
$3d^5(b\ ^2F)4p$	$^1G°$	4	*149 013.36*	44		34	$(^2H)\ ^1G°$
$3d^5(^6S)5s$	^7S	3	149 285.00				
$3d^5(b\ ^2F)4p$	$^1F°$	3	*150 654.9*	93			
$3d^5(^6S)4d$	^5D	3	151 534.13				
		2	151 534.90				
		1	151 536.68				
		4	151 537.80				
		0	151 537.91				
$3d^5(^2S)4p$	$^1P°$	1	*151 637.3?*	78		19	$(b\ ^2D)\ ^1P°$

(Continued)

Fe III—Continued

Configuration	Term	J	Level (cm^{-1})	Leading percentages		
$3d^5(^6S)5s$	^5S	2	151 757.67			
$3d^5(b\ ^2D)4p$	$^3F^\circ$	2	*157 684.3*	75	18	$(b\ ^2D)\ ^3D^\circ$
		3	*157 982.0*	61	27	
		4	*158 562.7*	94		
$3d^5(b\ ^2D)4p$	$^3D^\circ$	1	*158 257.37*	95		
		2	*158 417.31*	76	18	$(b\ ^2D)\ ^3F^\circ$
		3	*158 729.89*	67	29	
$3d^5(b\ ^2D)4p$	$^1F^\circ$	3	*159 493.0*	82	12	$(b\ ^2G)\ ^1F^\circ$
$3d^5(b\ ^2D)4p$	$^3P^\circ$	2	*160 037.9*	81	14	$(^2S)\ ^3P^\circ$
$3d^5(b\ ^2D)4p$	$^1D^\circ$	2	*162 084.8?*	92	6	$(b\ ^2F)\ ^1D^\circ$
$3d^5(b\ ^2G)4p$	$^3H^\circ$	4	*165 719.20*	93	5	$(b\ ^2G)\ ^3G^\circ$
		5	*165 939.47*	90	6	
		6	*166 187.50*	98		
$3d^5(^6S)5p$	$^7P^\circ$	2	*166 144.63*			
		3	*166 252.74*			
		4	*166 421.33*			
$3d^5(b\ ^2G)4p$	$^3F^\circ$	4	*166 222.2*	81	11	$(b\ ^2G)\ ^3G^\circ$
		3	*166 498*	50	46	$(b\ ^2G)\ ^3G^\circ$
		2	*167 002*	93	5	$(c\ ^2D)\ ^3F^\circ$
$3d^5(b\ ^2G)4p$	$^3G^\circ$	3	*167 085.12*	53	44	$(b\ ^2G)\ ^3F^\circ$
		4	*167 207.30*	85	11	$(b\ ^2G)\ ^3F^\circ$
		5	*167 299.60*	91	7	$(b\ ^2G)\ ^3H^\circ$
$3d^5(^6S)5p$	$^5P^\circ$	3	*168 329.67*			
		2	*168 420.99*			
		1	*168 477.36*			
$3d^5(b\ ^2G)4p$	$^1H^\circ$	5	*168 780.1*	95		
$3d^5(b\ ^2G)4p$	$^1G^\circ$	4	*169 277.6?*	96		
$3d^5(b\ ^2G)4p$	$^1F^\circ$	3	*170 310.6?*	87	12	$(c\ ^2D)\ ^1F^\circ$
$3d^5(^4G)4d$	^5H	3	179 178.62			
		4	179 194.22			
		5	179 207.57			
		6	179 216.47			
		7	179 221.45			
$3d^5(^4G)4d$	^5F	5	179 579.83			
		4	179 630.77			
		3	179 661.48			
		2	179 676.89			
		1	179 682.94			
$3d^5(^4G)4d$	^5G	6	179 725.31			
		5	179 748.17			
		4	179 757.98			
		2	179 759.49			
		3	179 760.72			

Fe III—Continued

Configuration	Term	J	Level (cm^{-1})	Leading percentages
$3d^5(^4G)4d$	^5I	4	179 876.71	
		8	179 889.03	
		5	179 893.56	
		7	179 904.56	
		6	179 904.56	
$3d^5(^4G)5s$	^5G	6	181 772.59	
		5	181 808.70	
		4	181 825.67	
		2	181 828.66	
		3	181 830.02	
$3d^5(^4P)4d$	^5F	5	182 379.86	
		4	182 412.65	
		3	182 444.70	
		2	182 480.72	
		1	182 486.40	
$3d^5(^4G)4d$	^3F	2	182 392.55	
		3	182 408.91	
		4	182 418.70	
$3d^5(^4G)4d$	^3I	5	182 810.66	
		6	182 830.76	
		7	182 852.05	
$3d^5(^4G)5s$	^3G	5	183 431.28	
		3	183 456.69	
		4	183 457.15	
$3d^5(^6S)4f$	$^7F^\circ$	1	*184 181.39*	
		2	*184 247.16*	
		3	*184 316.58*	
		4	*184 374.59*	
		5	*184 417.27*	
		6	*184 447.38*	
$3d^5(^6S)4f$	$^5F^\circ$	1	*184 777.3*	
		2	*184 777.6*	
		3	*184 778.5*	
		4	*184 779.5*	
		5	*184 780.8*	
$3d^5(^4P)5s$	^5P	3	184 951.62	
		2	185 003.35	
		1	185 061.35	
$3d^5(^4D)4d$	^5G	2	186 268.69	
		3	186 303.44	
		4	186 378.94	
		5	186 454.09	
		6	186 597.30	
$3d^5(^4D)4d$	^5D	1	186 712.02	
		2	186 791.78	
		3	186 882.98	
		4	186 998.60	

(Continued)

Fe III—Continued

Configuration	Term	J	Level (cm^{-1})	Leading percentages
$3d^4(^5D)4s4p(^3P^\circ)$	$^7P^\circ$	2	*186 740.6*	
		3	*186 891.5*	
		4	*187 090.4*	
$3d^5(^4D)5s$	^5D	4	188 013.40	
		0	188 109.32	
		3	188 109.58	
		1	188 131.70	
		2	188 142.64	
$3d^5(^4D)4d$	^3G	3	188 955.56	
		4	189 011.84	
		5	189 024.53	
$3d^5(^4D)5s$	^3D	3	189 679.07	
		2	189 784.52	
		1	189 796.03	
$3d^5(^6S)5d$	^7D	1	190 393.27	
		2	190 397.71	
		3	190 404.31	
		4	190 413.57	
		5	190 425.72	
$3d^5(^6S)6s$	^7S	3	190 918.17	
$3d^5(^6S)6s$	^5S	2	192 006.94	
$3d^5(^6S)5d$	^5D	0	193 595.30	
		1	193 599.54	
		2	193 605.99	
		3	193 610.92	
		4	193 611.37	
$3d^5(^2I)5s$	^3I	7	196 881.47	
		6	196 886.01	
		5	196 901.27	
$3d^5(^4G)5p$	$^5G^\circ$	2	*198 333.56*	
		6	*198 333.76*	
		5	*198 336.58*	
		3	*198 337.06*	
		4	*198 338.62*	
$3d^5(^6S)6p$	$^7P^\circ$	2	*198 606.37*	
		3	*198 655.66*	
		4	*198 737.05*	
$3d^5(^4G)5p$	$^5H^\circ$	3	*198 658.80*	
		4	*198 717.60*	
		5	*198 773.95*	
		6	*198 821.39*	
		7	*198 848.38*	
$3d^5(^4G)5p$	$^5F^\circ$	5	*199 139.76*	
		4	*199 212.72*	
		3	*199 262.44*	
		2	*199 300.15*	
		1	*199 327.95*	

Fe III—Continued

Configuration	Term	J	Level (cm^{-1})	Leading percentages
$3d^5(^4G)5p$	$^3F°$	2	*199 577.71*	
		3	*199 595.30*	
		4	*199 603.61*	
$3d^5(^4G)5p$	$^3H°$	6	*199 634.92*	
		5	*199 660.84*	
		4	*199 700.83*	
$3d^5(^4F)4d$	^5H	3	199 701.82	
		4	199 804.81	
		5	199 884.39	
		6	199 906.03	
		7	200 003.70	
$3d^5(^4F)4d$	^5G	4	200 325.56	
		3	200 384.28	
		5	200 395.33	
		2	200 437.94	
		6	200 656.02	
$3d^5(^4G)5p$	$^3G°$	3	*200 504.99*	
		4	*200 514.46*	
		5	*200 524.12*	
$3d^5(^4P)5p$	$^5D°$	2	*201 164.21*	
		3	*201 166.35*	
		0	*201 170.10*	
		1	*201 178.01*	
		4	*201 207.29*	
$3d^5(^4P)5p$	$^5S°$	2	*201 293.75*	
$3d^5(^4F)5s$	^5F	5	201 892.44	
		4	201 919.53	
		3	202 030.38	
		2	202 156.13	
		1	202 429.04	
$3d^5(^4P)5p$	$^5P°$	3	*202 200.51*	
		2	*202 282.65*	
		1	*202 334.39*	
$3d^5(^4D)5p$	$^5F°$	1	*204 907.13*	
		2	*204 943.26*	
		3	*205 002.47*	
		4	*205 092.53*	
		5	*205 195.15*	
$3d^5(^4D)5p$	$^5D°$	4	*205 672.01*	
		1	*205 694.09*	
		3	*205 732.37*	
		2	*205 737.51*	
$3d^5(^4D)5p$	$^3D°$	3	*206 180.41*	
		2	*206 233.31*	
		1	*206 295.81*	
$3d^5(^4D)5p$	$^3F°$	4	*206 261.33*	
		2	*206 324.89*	
		3	*206 328.22*	

(Continued)

Fe III—Continued

Configuration	Term	J	Level (cm^{-1})	Leading percentages
$3d^5(^6S)5f$	$^7F°$	6	*207 118.1*	
		5	*207 118.6*	
		4	*207 119.1*	
		3	*207 119.6*	
		2	*207 120.1*	
$3d^5(^6S)5f$	$^5F°$	1	*207 252.5*	
		2	*207 257.8*	
		3	*207 263.0*	
		4	*207 268.2*	
		5	*207 273.23*	
$3d^5(^6S)5g$	^7G	7	207 640.8	
		6	207 640.8	
		5	207 640.9	
		4	207 640.9	
		3	207 641.1	
		2	207 641.3	
$3d^5(^6S)5g$	^5G	6	207 642.9	
		5	207 643.1	
		4	207 643.3	
		3	207 643.3	
		2	207 643.5	
$3d^5(^6S)6d$	^7D	1	210 393.67	
		2	210 396.00	
		3	210 399.57	
		4	210 404.61	
		5	210 411.32	
$3d^5(^6S)7s$	^7S	3	210 615.21	
$3d^5(^2I)5p$	$^3I°$	7	*213 457.82*	
		6	*213 505.73*	
		5	*213 563.08*	
$3d^5(^2I)5p$	$^3H°$	6	*213 974.42*	
		5	*214 010.32*	
		4	*214 047.38*	
$3d^5(^4F)5p$	$^5G°$	2	*218 860.43*	
		3	*218 923.08*	
		4	*219 004.53*	
		5	*219 092.86*	
		6	*219 162.42*	
$3d^5(^4F)5p$	$^5F°$	5	*219 415.61*	
		4	*219 471.97*	
		3	*219 566.08*	
		2	*219 655.55*	
		1	*219 743.04*	
$3d^5(^6S)6g$	^7G	1–7	219 740	
$3d^5(^6S)6g$	^5G	6	219 741.9	
		5	219 741.9	
		4	219 742.0	
		3	219 742.1	
		2	219 742.1	

Fe III—Continued

Configuration	Term	J	Level (cm^{-1})	Leading percentages
$3d^5(^6S)6h$	$^7H°$	2–8	*219 780.2*	
$3d^5(^6S)6h$	$^5H°$	3–7	*219 780.6*	
$3d^5(^4G)5d$	^5H	7	222 590.86	
		6	222 602.50	
		3	222 605.24	
		4	222 605.82	
		5	222 611.16	
$3d^5(^4G)5d$	^5F	5	222 699.09	
		4	222 734.33	
		3	222 750.23	
		2	222 774.22	
		1	222 776.89	
$3d^5(^4G)5d$	^5G	6	222 714.30	
		5	222 744.69	
		2	222 758.28	
		4	222 765.97	
		3	222 766.04	
$3d^5(^4G)5d$	^5I	8	222 797.97	
		4	222 823.33	
		7	222 824.71	
		5	222 832.48	
		6	222 834.77	
$3d^5(^4G)6s$	^5G	6	223 272.06	
		5	223 309.37	
		4	223 326.76	
		2	223 327.87	
		3	223 330.71	
$3d^5(^4G)6s$	^3G	5	224 038.73	
		3	224 051.63	
		4	224 058.70	
$3d^5(^4P)6s$	^5P	3	226 381.91	
		2	226 447.88	
		1	226 506.54	
$3d^5(^4D)6s$	^5D	4	229 421.73	
		3	229 509.56	
		1	229 530.67	
		2	229 570.36	
$3d^5(^4D)6s$	^3D	3	230 192.86	
		1	230 248.26	
		2	230 257.15	
Fe IV ($^6S_{5/2}$)	***Limit***		**247 220**	

Fe IV

$Z=26$

V I isoelectronic sequence

Ground state: $1s^2 2s^2 2p^6 3s^2 3p^6 3d^5\ ^6S_{5/2}$

Ionization energy = 442 000±1000 cm^{-1} (54.8±0.1 eV)

The early work of Kruger and Gilroy (1935) and Edlén (1969) has now been superseded by that of Ekberg and Edlén (1978), who have made a nearly complete analysis of the three lowest configurations. They have classified 706 lines from the transition array $3d^5-3d^4 4p$ in the region 446–789 Å and 560 lines of the $3d^4 4s-3d^4 4p$ array in the region 1247–2028 Å. Only four of the 280 possible levels are undiscovered. The uncertainty of the $3d^5$ level values is ±0.4 cm^{-1} and of the $3d^4 4s$ and $3d^4 4p$ levels relative to each other is ±0.1 cm^{-1}.

The leading percentages for $3d^5$ were provided to Ekberg and Edlén by R. Poppe, A. J. J. Raassen, and Th. A. M. van Kleef. The rest were calculated by the authors.

Transitions among levels of the $3d^5$ configuration observed in nebular spectra have been identified by Bowen (1960).

The ionization energy is taken from an isoelectronic extrapolation by Lotz (1967).

References

Bowen, I. S. (1960), Astrophys. J. **132**, 1.
Edlén, B. (1969), Mon. Not. R. Astron. Soc. **144**, 391.
Ekberg, J. O., and Edlén, B. (1978), Phys. Scr. **18**, 107.
Kruger, P. G., and Gilroy, H. T. (1935), Phys. Rev. **48**, 720.
Lotz, W. (1967), J. Opt. Soc. Am. **57**, 873.

Fe IV

Configuration	Term	J	Level (cm^{-1})	Leading percentages		
$3d^5$	^6S	5/2	0.0	100		
$3d^5$	^4G	11/2	32 245.5	100		
		9/2	32 292.8	100		
		5/2	32 301.2	100		
		7/2	32 305.7	100		
$3d^5$	^4P	5/2	35 253.8	95		
		3/2	35 333.3	97		
		1/2	35 406.6	99		
$3d^5$	^4D	7/2	38 779.4	100		
		1/2	38 896.7	99		
		5/2	38 935.1	96		
		3/2	38 938.2	97		
$3d^5$	^2I	11/2	47 078.6	99		
		13/2	47 090.5	100		
$3d^5$	$^2D3$	5/2	49 541.5	57	24	$^2F2$
		3/2	50 051.4	73	23	$^2D1$
$3d^5$	$^2F2$	7/2	51 394.2	97		
		5/2	52 166.7	70	15	$^2D3$
$3d^5$	^4F	9/2	52 620.7	98		
		7/2	52 695.4	98		
		3/2	52 837.1	96		
		5/2	52 838.0	89	5	$^2F2$
$3d^5$	^2H	9/2	56 058.3	86	14	$^2G2$
		11/2	56 368.8	99		

Fe IV—Continued

Configuration	Term	J	Level (cm^{-1})	Leading percentages		
$3d^5$	^{2}G2	7/2	57 408.0	99		
		9/2	57 721.2	84	14	^{2}H
$3d^5$	^{2}F1	5/2	61 156.5	99		
		7/2	61 254.4	98		
$3d^5$	^{2}S	1/2	66 720.1	100		
$3d^5$	^{2}D2	3/2	74 096.6	100		
		5/2	74 133.1	100		
$3d^5$	^{2}G1	9/2	82 894.9	100		
		7/2	82 897.3	100		
$3d^5$	^{2}P	3/2	100 118.0	100		
		1/2	100 126.0	100		
$3d^5$	^{2}D1	5/2	108 242.1	76	24	^{2}D3
		3/2	108 258.3	76	24	
$3d^4(^5D)4s$	^{6}D	1/2	127 766.15	100		
		3/2	127 929.12	100		
		5/2	128 191.54	100		
		7/2	128 541.85	100		
		9/2	128 967.67	100		
$3d^4(^5D)4s$	^{4}D	1/2	137 700.81	100		
		3/2	137 949.29	100		
		5/2	138 338.83	100		
		7/2	138 844.03	100		
$3d^4(^3P2)4s$	^{4}P	1/2	153 651.74	60	39	(^{3}P1) ^{4}P
		3/2	154 474.85	60	39	
		5/2	155 744.87	61	39	
$3d^4(^3H)4s$	^{4}H	7/2	154 185.85	98		
		9/2	154 325.96	98		
		11/2	154 512.67	99		
		13/2	154 731.29	100		
$3d^4(^3F2)4s$	^{4}F	3/2	156 012.29	78	21	(^{3}F1) ^{4}F
		5/2	156 049.32	77	21	
		7/2	156 123.77	76	20	
		9/2	156 224.88	76	19	
$3d^4(^3G)4s$	^{4}G	5/2	158 738.69	96		
		7/2	159 010.39	95		
		9/2	159 227.90	93		
		11/2	159 342.88	92	7	(^{3}H) ^{2}H
$3d^4(^3P2)4s$	^{2}P	1/2	160 015.88	59	39	(^{3}P1) ^{2}P
		3/2	161 571.59	60	38	
$3d^4(^3H)4s$	^{2}H	9/2	160 311.64	96		
		11/2	160 778.60	93	7	(^{3}G) ^{4}G
$3d^4(^3F2)4s$	^{2}F	5/2	162 074.42	77	21	(^{3}F1) ^{2}F
		7/2	162 087.81	74	19	

(Continued)

Fe IV—Continued

Configuration	Term	J	Level (cm^{-1})	Leading percentages		
$3d^4(^3G)4s$	^2G	7/2	164 950.50	94		
		9/2	165 392.58	98		
$3d^4(^3D)4s$	^4D	7/2	165 493.10	100		
		5/2	165 600.96	99		
		3/2	165 720.94	99		
		1/2	165 804.47	99		
$3d^4(^1G2)4s$	^2G	9/2	167 712.50	65	32	$(^1G1)\ ^2G$
		7/2	167 795.92	64	32	
$3d^4(^1I)4s$	^2I	13/2	168 526.37	100		
		11/2	168 566.43	99		
$3d^4(^1S2)4s$	^2S	1/2	170 729.49	78	20	$(^1S1)\ ^2S$
$3d^4(^3D)4s$	^2D	5/2	171 345.33	99		
		3/2	171 476.39	99		
$3d^4(^1D2)4s$	^2D	5/2	177 005.97	78	21	$(^1D1)\ ^2D$
		3/2	177 066.72	78	21	
$3d^4(^1F)4s$	^2F	5/2	183 159.61	99		
		7/2	183 164.49	99		
$3d^4(^5D)4p$	$^6F°$	1/2	*187 878.81*	99		
		3/2	*188 086.05*	99		
		5/2	*188 428.78*	99		
		7/2	*188 904.55*	99		
		9/2	*189 515.88*	99		
		11/2	*190 276.85*	100		
$3d^4(^5D)4p$	$^6P°$	3/2	*189 885.11*	96		
		5/2	*190 008.28*	97		
		7/2	*190 226.87*	99		
$3d^4(^3P1)4s$	^4P	5/2	189 975.01	61	39	$(^3P2)\ ^4P$
		3/2	190 811.79	60	39	
		1/2	191 337.82	60	39	
$3d^4(^3F1)4s$	^4F	9/2	190 318.34	80	20	$(^3F2)\ ^4F$
		3/2	190 406.45	78	22	
		7/2	190 424.14	79	20	
		5/2	190 435.47	78	21	
$3d^4(^5D)4p$	$^4P°$	1/2	*191 021.18*	70	28	$(^5D)\ ^6D°$
		3/2	*191 694.11*	61	33	
		5/2	*193 549.25*	54	44	
$3d^4(^5D)4p$	$^6D°$	5/2	*192 595.28*	55	41	$(^5D)\ ^4P°$
		1/2	*193 120.34*	72	27	$(^5D)\ ^4P°$
		3/2	*193 271.27*	66	33	$(^5D)\ ^4P°$
		7/2	*193 386.17*	97		
		9/2	*193 789.19*	94	5	$(^5D)\ ^4F°$
$3d^4(^3P1)4s$	^2P	3/2	195 864.15	61	39	$(^3P2)\ ^2P$
		1/2	196 875.62	60	39	

Fe IV—Continued

Configuration	Term	J	Level (cm^{-1})	Leading percentages			
$3d^4(^3F1)4s$	^2F	7/2	196 131.19	80		20	$(^3F2)\ ^2F$
		5/2	196 220.71	79		21	
$3d^4(^5D)4p$	$^4F°$	3/2	*196 186.88*	96			
		5/2	*196 334.63*	95			
		7/2	*196 549.59*	94			
		9/2	*196 846.82*	91		6	$(^5D)\ ^6D°$
$3d^4(^1G1)4s$	^2G	9/2	201 178.05	66		33	$(^1G2)\ ^2G$
		7/2	201 212.22	66		33	
$3d^4(^5D)4p$	$^4D°$	1/2	*201 919.38*	98			
		3/2	*202 085.22*	98			
		5/2	*202 328.53*	97			
		7/2	*202 608.33*	97			
$3d^4(^3H)4p$	$^4H°$	7/2	*212 135.79*	77		20	$(^3G)\ ^4H°$
		9/2	*212 374.04*	75		19	
		11/2	*212 714.37*	76		17	
		13/2	*213 162.59*	80		15	
$3d^4(^3P2)4p$	$^4D°$	1/2	*212 812.66*	48		33	$(^3P1)\ ^4D°$
		3/2	*213 445.05*	47		32	$(^3P1)\ ^4D°$
		5/2	*214 317.21*	44		30	$(^3P1)\ ^4D°$
		7/2	*215 220.67*	26		19	$(^3F2)\ ^4D°$
$3d^4(^3F2)4p$	$^4G°$	5/2	*214 821.74*	47		21	$(^3F1)\ ^4G°$
		7/2	*215 033.81*	27		13	$(^3G)\ ^4G°$
		9/2	*215 385.23*	24		14	$(^3G)\ ^4G°$
		11/2	*216 002.72*	26		31	$(^3H)\ ^4G°$
$3d^4(^3H)4p$	$^4I°$	9/2	*215 155.69*	87			
		11/2	*215 808.91*	92		6	$(^3H)\ ^4H°$
		13/2	*216 367.80*	94		5	$(^3H)\ ^4H°$
		15/2	*216 877.83*	100			
$3d^4(^3P2)4p$	$^4P°$	1/2	*215 860.50*	32		19	$(^3P1)\ ^4P°$
		3/2	*217 031.89*	56		33	
		5/2	*218 023.81*	26		15	
$3d^4(^3H)4p$	$^2G°$	7/2	*216 111.69*	46		22	$(^3F2)\ ^2G°$
		9/2	*216 428.44*	36		22	
$3d^4(^3F2)4p$	$^4F°$	3/2	*217 466.19*	42		23	$(^3F2)\ ^2D°$
		7/2	*218 478.36*	71		13	$(^3F1)\ ^4F°$
		9/2	*218 601.03*	70		12	$(^3F1)\ ^4F°$
$3d^4(^3P2)4p$		1/2	*217 607.71*	26	$^4P°$	21	$(^3P1)\ ^2S°$
$3d^4(^3H)4p$	$^4G°$	5/2	*217 845.29*	31		20	$(^3F2)\ ^4G°$
		7/2	*218 159.77*	51		28	
		9/2	*218 238.51*	50		33	
		11/2	*218 375.12*	31		26	
$3d^4(^3F2)4p$		5/2	*218 195.66*	30	$^4F°$	28	$(^3P2)\ ^4P°$
$3d^4(^3P2)4p$		3/2	*218 613.88*	25	$^2P°$	15	$(^3P2)\ ^4S°$
$3d^4(^3F2)4p$	$^2D°$	5/2	*218 871.21*	37		13	$(^3D)\ ^2D°$

(Continued)

Fe IV—Continued

Configuration	Term	J	Level (cm^{-1})	Leading percentages			
$3d^4(^3F2)4p$		3/2	*219 091.67*	21	$^2D°$	19	$(^3F2)\ ^4F°$
$3d^4(^3P2)4p$	$^2P°$	1/2	*219 333.90*	33		15	$(^3P1)\ ^2P°$
		3/2	*220 360.44*	27		11	
$3d^4(^3H)4p$	$^2I°$	11/2	*219 564.46*	89			
		13/2	*219 640.76*	89		6	$(^3G)\ ^4H°$
$3d^4(^3F2)4p$	$^4D°$	5/2	*219 590.84*	30		10	$(^3P2)\ ^4D°$
		7/2	*219 700.53*	23		18	$(^3P2)\ ^4D°$
		3/2	*219 826.61*	39		13	$(^3F1)\ ^4D°$
		1/2	*220 059.33*	43		14	$(^3F1)\ ^4D°$
$3d^4(^3G)4p$	$^2F°$	5/2	*220 197.25*	26		14	$(^3F2)\ ^2F°$
		7/2	*220 649.22*	21		20	
$3d^4(^3H)4p$	$^2H°$	9/2	*220 461.33*	51		16	$(^3G)\ ^4H°$
		11/2	*221 161.02*	40		33	
$3d^4(^3G)4p$	$^4H°$	7/2	*220 658.04*	77		19	$(^3H)\ ^4H°$
		9/2	*221 104.17*	56		16	$(^3H)\ ^2H°$
		11/2	*221 647.49*	46		35	$(^3H)\ ^2H°$
		13/2	*222 154.99*	79		15	$(^3H)\ ^4H°$
$3d^4(^3G)4p$	$^4F°$	3/2	*221 219.29*	44		13	$(^3D)\ ^4F°$
		9/2	*221 239.21*	60		12	
		5/2	*221 320.54*	58		16	
		7/2	*221 346.06*	56		14	
$3d^4(^3P2)4p$	$^2D°$	3/2	*222 020.09*	34		23	$(^3P1)\ ^2D°$
		5/2	*222 880.23*	47		32	
$3d^4(^1D1)4s$	^2D	5/2	222 840.58	79		21	$(^1D2)\ ^2D$
		3/2	222 851.68	79		21	
$3d^4(^3F2)4p$	$^2G°$	7/2	*223 398.62*	32		19	$(^3H)\ ^2G°$
$3d^4(^3F2)4p$	$^2F°$	5/2	*223 478.96*	48		35	$(^3G)\ ^2F°$
		7/2	*224 046.02*	44		42	
$3d^4(^3G)4p$	$^2H°$	11/2	*223 550.14*	50		19	$(^3G)\ ^4G°$
		9/2	*223 745.82*	41		15	$(^3H)\ ^2G°$
$3d^4(^3G)4p$		9/2	*223 629.59*	18	$^4G°$	15	$(^3F2)\ ^2G°$
$3d^4(^3G)4p$	$^4G°$	5/2	*224 045.96*	55		29	$(^3H)\ ^4G°$
		7/2	*224 230.60*	55		25	$(^3H)\ ^4G°$
		9/2	*224 576.43*	42		17	$(^3H)\ ^4G°$
		11/2	*224 870.85*	47		24	$(^3G)\ ^2H°$
$3d^4(^3D)4p$	$^4D°$	1/2	*226 851.93*	86			
		3/2	*226 892.10*	81		7	$(^3D)\ ^4P°$
		5/2	*226 983.58*	60		29	$(^3D)\ ^4P°$
		7/2	*227 258.80*	81			
$3d^4(^3G)4p$	$^2G°$	7/2	*227 604.73*	64		14	$(^3H)\ ^2G°$
		9/2	*227 660.25*	60		20	

Fe IV—Continued

Configuration	Term	J	Level (cm^{-1})	Leading percentages		
$3d^4(^3D)4p$	$^4P°$	$5/2$	*227 919.05*	63	25	$(^3D)\ ^4D°$
		$3/2$	*228 589.67*	82	5	$(^3D)\ ^4D°$
		$1/2$	*229 037.02*	90	5	$(^3P2)\ ^4P°$
$3d^4(^1G2)4p$	$^2F°$	$7/2$	*228 193.67*	53	24	$(^1G1)\ ^2F°$
		$5/2$	*229 138.90*	53	25	
$3d^4(^1I)4p$	$^2I°$	$13/2$	*228 204.33*	69	27	$(^1I)\ ^2K°$
		$11/2$	*228 315.03*	89		
$3d^4(^1G2)4p$	$^2H°$	$9/2$	*228 793.86*	51	25	$(^1G1)\ ^2H°$
		$11/2$	*229 306.61*	52	22	
$3d^4(^3D)4p$	$^4F°$	$3/2$	*228 862.61*	69	20	$(^3G)\ ^4F°$
		$5/2$	*229 062.58*	68	18	
		$7/2$	*229 288.02*	71	19	
		$9/2$	*229 494.74*	80	19	
$3d^4(^1S2)4p$	$^2P°$	$1/2$	*228 946.59*	39	37	$(^3D)\ ^2P°$
		$3/2$	*233 786.72*	33	27	
$3d^4(^3D)4p$	$^2P°$	$3/2$	*229 428.91*	52	32	$(^1S2)\ ^2P°$
		$1/2$	*233 927.12*	48	29	
$3d^4(^1I)4p$	$^2K°$	$13/2$	*229 472.83*	72	27	$(^1I)\ ^2I°$
		$15/2$	*230 195.02*	100		
$3d^4(^1G2)4p$	$^2G°$	$7/2$	*231 473.32*	47	30	$(^1G1)\ ^2G°$
		$9/2$	*231 804.00*	44	31	
$3d^4(^1I)4p$	$^2H°$	$11/2$	*233 272.84*	80	9	$(^3G)\ ^2H°$
		$9/2$	*233 802.12*	86	9	
$3d^4(^3D)4p$	$^2F°$	$7/2$	*233 780.86*	70	11	$(^3G)\ ^2F°$
		$5/2$	*234 106.72*	70	12	
$3d^4(^3D)4p$	$^2D°$	$5/2$	*234 472.00*	60	19	$(^1D2)\ ^2D°$
		$3/2$	*234 984.35*	68	9	$(^3F2)\ ^2D°$
$3d^4(^1D2)4p$	$^2D°$	$3/2$	*236 918.79*	54	18	$(^1D1)\ ^2D°$
		$5/2$	*237 283.09*	44	18	$(^3D)\ ^2D°$
$3d^4(^1D2)4p$	$^2F°$	$5/2$	*238 512.84*	59	14	$(^1D1)\ ^2F°$
		$7/2$	*239 071.40*	59	20	$(^1F)\ ^2F°$
$3d^4(^1D2)4p$	$^2P°$	$3/2$	*242 259.25*	68	14	$(^1D1)\ ^2P°$
$3d^4(^1F)4p$	$^2F°$	$5/2$	*242 614.60*	74	10	$(^1D2)\ ^2F°$
		$7/2$	*242 965.62*	66	16	
$3d^4(^1F)4p$	$^2G°$	$7/2$	*244 759.25*	91		
		$9/2$	*245 742.29*	94		
$3d^4(^1F)4p$	$^2D°$	$5/2$	*246 990.80*	61	15	$(^3P1)\ ^2D°$
		$3/2$	*248 077.97*	64	13	

(Continued)

Fe IV—Continued

Configuration	Term	J	Level (cm^{-1})	Leading percentages			
$3d^4(^3F1)4p$	$^4F^\circ$	5/2	250 195.07	73		11	$(^3F2)\ ^4F^\circ$
		3/2	250 249.39	78		13	
		7/2	250 279.06	74		11	
		9/2	250 502.29	84		12	
$3d^4(^3P1)4p$	$^4P^\circ$	3/2	250 891.05	38		19	$(^3P2)\ ^4P^\circ$
		1/2	251 156.94	43		23	
		5/2	251 958.94	39		20	
$3d^4(^3P1)4p$		5/2	251 014.02	22	$^4D^\circ$	22	$(^3P1)\ ^4P^\circ$
$3d^4(^3P1)4p$	$^4D^\circ$	7/2	251 658.34	37		21	$(^3P2)\ ^4D^\circ$
		1/2	251 984.20	30		18	
$3d^4(^3P1)4p$		3/2	251 944.03	25	$^4D^\circ$	21	$(^3P1)\ ^4P^\circ$
$3d^4(^3F1)4p$	$^4G^\circ$	5/2	252 884.48	54		18	$(^3F2)\ ^4G^\circ$
		7/2	253 254.37	52		20	$(^3F1)\ ^2F^\circ$
		9/2	253 827.43	73		22	$(^3F2)\ ^4G^\circ$
		11/2	254 164.84	76		22	$(^3F2)\ ^4G^\circ$
$3d^4(^3P1)4p$	$^2D^\circ$	5/2	253 575.65	23		25	$(^1F)\ ^2D^\circ$
		3/2	253 868.43	25		26	
$3d^4(^3F1)4p$	$^2F^\circ$	7/2	253 923.59	52		21	$(^3F1)\ ^4G^\circ$
		5/2	254 169.13	55		16	
$3d^4(^3P1)4p$	$^4S^\circ$	3/2	257 503.26	50		45	$(^3P2)\ ^4S^\circ$
$3d^4(^3F1)4p$	$^2G^\circ$	9/2	258 566.02	75		21	$(^3F2)\ ^2G^\circ$
		7/2	259 039.72	75		22	
$3d^4(^3F1)4p$	$^4D^\circ$	7/2	258 591.92	50		17	$(^3F2)\ ^4D^\circ$
		5/2	258 986.64	50		18	
		3/2	259 183.82	49		18	
		1/2	259 254.34	52		20	
$3d^4(^3P1)4p$	$^2P^\circ$	3/2	259 011.48	61		29	$(^3P2)\ ^2P^\circ$
		1/2	259 581.64	61		29	
$3d^4(^3P2)4p$	$^2S^\circ$	1/2	262 348.36	55		43	$(^3P1)\ ^2S^\circ$
$3d^4(^1G1)4p$	$^2H^\circ$	9/2	262 557.77	42		21	$(^1G2)\ ^2H^\circ$
		11/2	264 011.54	65		32	
$3d^4(^1G1)4p$	$^2G^\circ$	7/2	262 995.01	57		35	$(^1G2)\ ^2G^\circ$
		9/2	263 876.91	39		23	
$3d^4(^1G1)4p$	$^2F^\circ$	7/2	265 084.77	58		21	$(^1G2)\ ^2F^\circ$
		5/2	265 369.48	49		18	
$3d^4(^3F1)4p$	$^2D^\circ$	5/2	266 181.09	41		15	$(^3F2)\ ^2D^\circ$
		3/2	266 335.24	47		18	
$3d^4(^1D1)4p$	$^2P^\circ$	3/2	280 758.37	76		16	$(^1D2)\ ^2P^\circ$
		1/2	281 446.30	76		17	
$3d^4(^1D1)4p$	$^2F^\circ$	5/2	285 052.77	72		19	$(^1D2)\ ^2F^\circ$
		7/2	286 084.72	74		19	

Fe IV—Continued

Configuration	Term	J	Level (cm^{-1})	Leading percentages		
$3d^4(^1D1)4p$	$^2D°$	$3/2$	*289 400.72*	73	26	$(^1D2)\ ^2D°$
		$5/2$	*289 818.77*	73	25	
Fe V ($^5D_0$)	***Limit***		**442 000**			

Fe V

$Z = 26$

Ti I isoelectronic sequence

Ground state: $1s^2 2s^2 2p^6 3s^2 3p^6 3d^4\ ^5D_0$

Ionization energy = 605 000±1200 cm^{-1} (75.0±0.2 eV)

Bowen's contribution in 1937 established terms of $3d^4$, $3d^3 4s$, and $3d^3 4p$, greatly expanding the start made by White (1929). Additions to all three configurations have been made by Fawcett and Henrichs (1974). The analysis of these configurations has been greatly extended by Ekberg (1975), who reobserved the spectrum from 302–1715 Å. He improved the level uncertainty to ±0.4 cm^{-1}. The leading percentages given below are also due to Ekberg.

Bowen (1960) has observed lines in nebular spectra due to transitions among levels of the $3d^4$ configuration.

The ionization energy is from the isoelectronic extrapolation of Lotz (1967).

References

Bowen, I. S. (1937), Phys. Rev. **52**, 1153.
Bowen, I. S. (1960), Astrophys. J. **132**, 1.
Ekberg, J. O. (1975), Phys. Scr. **12**, 42.
Fawcett, B. C., and Henrichs, H. F. (1974), Astron. Astrophys. **18**, 157.
Lotz, W. (1967), J. Opt. Soc. Am. **57**, 873.
White, H. E. (1929), Phys. Rev. **33**, 914.

Fe V

Configuration	Term	J	Level (cm^{-1})	Leading percentages		
$3d^4$	^5D	0	0.0	100		
		1	142.1	100		
		2	417.3	100		
		3	803.1	100		
		4	1 282.8	100		
$3d^4$	$^3P2$	0	24 055.4	59	40	$^3P1$
		1	24 972.9	60	40	
		2	26 468.3	60	39	
$3d^4$	^3H	4	24 932.5	97		
		5	25 225.9	99		
		6	25 528.5	100		
$3d^4$	$^3F2$	2	26 760.7	78	22	$^3F1$
		3	26 842.3	75	20	
		4	26 974.0	75	19	
$3d^4$	^3G	3	29 817.1	96		
		4	30 147.0	94		
		5	30 430.1	99		
$3d^4$	$^1G2$	4	36 586.3	65	33	$^1G1$
$3d^4$	^3D	3	36 630.1	100		
		2	36 758.5	99		
		1	36 925.4	100		
$3d^4$	^1I	6	37 511.7	100		
$3d^4$	$^1S2$	0	39 633.4	78	21	$^1S1$
$3d^4$	$^1D2$	2	46 291.2	78	21	$^1D1$
$3d^4$	^1F	3	52 732.7	99		

Fe v—Continued

Configuration	Term	J	Level (cm^{-1})	Leading percentages		
$3d^4$	^{3}P1	2	61 854.4	61	39	^{3}P2
		1	62 914.2	60	40	
		0	63 420.0	60	40	
$3d^4$	^{3}F1	4	62 238.1	80	20	^{3}F2
		2	62 321.1	78	22	
		3	62 364.4	78	21	
$3d^4$	^{1}G1	4	71 280.3	66	34	^{1}G2
$3d^4$	^{1}D1	2	93 832.3	78	22	^{1}D2
$3d^4$	^{1}S1	0	121 130.2	79	21	^{1}S2
$3d^3(^4F)4s$	^{5}F	1	186 433.6	100		
		2	186 725.5	100		
		3	187 157.5	100		
		4	187 719.0	100		
		5	188 395.3	100		
$3d^3(^4F)4s$	^{3}F	2	195 196.3	100		
		3	195 933.0	100		
		4	196 838.6	100		
$3d^3(^4P)4s$	^{5}P	1	204 729.9	99		
		2	204 975.4	99		
		3	205 536.4	100		
$3d^3(^2G)4s$	^{3}G	3	208 838.2	100		
		4	209 110.1	99		
		5	209 523.9	98		
$3d^3(^4P)4s$	^{3}P	0	212 542.1	85	15	(^{2}P) ^{3}P
		1	212 818.1	88	6	
		2	213 649.2	91	8	
$3d^3(^2G)4s$	^{1}G	4	213 534.1	94	5	(^{2}H) ^{3}H
$3d^3(^2P)4s$	^{3}P	2	214 525.8	61	23	(^{2}D2) ^{3}D
		1	214 611.4	72	14	
$3d^3(^2D2)4s$	^{3}D	1	215 782.6	56	20	(^{2}P) ^{3}P
		3	216 538.1	80	20	(^{2}D1) ^{3}D
		2	216 592.7	55	28	(^{2}P) ^{3}P
$3d^3(^2H)4s$	^{3}H	4	216 779.1	94	5	(^{2}G) ^{1}G
		5	216 860.4	99		
		6	217 122.5	100		
$3d^3(^2P)4s$	^{1}P	1	219 486.9	90	5	(^{2}D2) ^{3}D
$3d^3(^2D2)4s$	^{1}D	2	220 621.0	77	20	(^{2}D1) ^{1}D
$3d^3(^2H)4s$	^{1}H	5	221 305.2	99		
$3d^3(^2F)4s$	^{3}F	4	233 633.6	100		
		3	233 848.9	100		
		2	234 027.4	100		
$3d^3(^2F)4s$	^{1}F	3	237 729.6	100		

(Continued)

Fe v—Continued

Configuration	Term	J	Level (cm^{-1})	Leading percentages		
$3d^3(^4F)4p$	$^5G°$	2	*254 803.3*	99		
		3	*255 399.2*	99		
		4	*256 177.9*	99		
		5	*257 138.0*	99		
		6	*258 297.4*	100		
$3d^3(^4F)4p$		1	*257 742.3*	38 $^5F°$	36	$(^4F)\ ^3D°$
$3d^3(^4F)4p$	$^5D°$	2	*258 128.5*	48	25	$(^4F)\ ^5F°$
		0	*258 619.5*	96		
		3	*258 680.0*	71	15	$(^4F)\ ^5F°$
		1	*258 891.5*	72	20	$(^4F)\ ^5F°$
		4	*259 344.8*	89	7	$(^4F)\ ^5F°$
$3d^3(^2D1)4s$	^3D	3	258 434.1	80	20	$(^2D2)\ ^3D$
		2	258 628.5	79	21	
		1	258 769.5	78	22	
$3d^3(^4F)4p$	$^5F°$	2	*259 376.1*	51	42	$(^4F)\ ^5D°$
		3	*259 954.7*	78	19	
		4	*260 521.0*	90	6	
		5	*261 051.9*	94		
$3d^3(^4F)4p$	$^3D°$	1	*259 995.2*	49	42	$(^4F)\ ^5F°$
		2	*260 411.4*	62	23	$(^4F)\ ^5F°$
		3	*261 179.6*	76	8	$(^4P)\ ^3D°$
$3d^3(^2D1)4s$	^1D	2	262 509.3	79	21	$(^2D2)\ ^1D$
$3d^3(^4F)4p$	$^3G°$	3	*263 898.6*	92	5	$(^2G)\ ^3G°$
		4	*264 434.2*	91		
		5	*265 112.6*	88	6	$(^4F)\ ^5F°$
$3d^3(^4F)4p$	$^3F°$	2	*266 612.8*	94		
		3	*267 240.1*	94		
		4	*267 928.6*	94		
$3d^3(^4P)4p$	$^5P°$	1	*273 643.1*	98		
		2	*274 136.1*	96		
		3	*274 930.3*	98		
$3d^3(^4P)4p$	$^5D°$	0	*274 753.3*	54	36	$(^4P)\ ^3P°$
		1	*275 146.6*	59	34	
		2	*276 759.2*	58	32	
		3	*277 068.5*	94		
		4	*278 075.8*	96		
$3d^3(^4P)4p$	$^3P°$	2	*275 374.3*	52	36	$(^4P)\ ^5D°$
		0	*276 434.9*	43	40	
		1	*276 765.9*	54	35	
$3d^3(^2G)4p$	$^3H°$	4	*276 429.7*	79	16	$(^2H)\ ^3H°$
		5	*277 292.7*	73	18	
		6	*278 650.7*	78	21	
$3d^3(^2G)4p$	$^3G°$	3	*278 794.2*	77	7	$(^2G)\ ^1F°$
		4	*279 502.6*	78	9	$(^2G)\ ^3F°$
		5	*280 039.6*	79	7	$(^2G)\ ^3H°$

Fe v—Continued

Configuration	Term	J	Level (cm^{-1})	Leading percentages			
$3d^3(^2G)4p$	$^3F^\circ$	4	280 367.2	46		35	$(^2G)\ ^1G^\circ$
		2	280 539.7	60		17	$(^2D2)\ ^3F^\circ$
		3	280 832.2	64		11	$(^2G)\ ^3G^\circ$
$3d^3(^2P)4p$	$^3P^\circ$	1	281 944.9	52		22	$(^2D2)\ ^3P^\circ$
		0	282 234.5	50		17	$(^2D2)\ ^3P^\circ$
		2	290 407.7	42		38	$(^2D2)\ ^3P^\circ$
$3d^3(^2G)4p$	$^1G^\circ$	4	282 038.1	50		30	$(^2G)\ ^3F^\circ$
$3d^3(^4P)4p$		2	282 423.5	27	$^5S^\circ$	12	$(^2P)\ ^1D^\circ$
$3d^3(^2G)4p$	$^1F^\circ$	3	282 571.6	67		13	$(^2D2)\ ^1F^\circ$
$3d^3(^4P)4p$	$^5S^\circ$	2	282 604.8	49		21	$(^2P)\ ^1D^\circ$
$3d^3(^2G)4p$	$^1H^\circ$	5	282 871.9	72		18	$(^2H)\ ^1H^\circ$
$3d^3(^2P)4p$		2	283 686.3	27	$^3P^\circ$	22	$(^4P)\ ^5S^\circ$
$3d^3(^2P)4p$	$^3D^\circ$	1	283 754.0	81		8	$(^4P)\ ^3D^\circ$
		2	284 911.2	66		9	$(^2P)\ ^1D^\circ$
		3	285 474.0	54		15	$(^2D2)\ ^3F^\circ$
$3d^3(^2H)4p$	$^3H^\circ$	4	284 690.3	69		15	$(^2G)\ ^3H^\circ$
		5	284 790.8	78		19	
		6	285 196.1	77		21	
$3d^3(^2D2)4p$	$^1P^\circ$	1	285 961.7	40		21	$(^2P)\ ^1P^\circ$
$3d^3(^2D2)4p$	$^3F^\circ$	2	286 154.9	45		15	$(^2G)\ ^3F^\circ$
		4	287 620.2	70		16	$(^2D1)\ ^3F^\circ$
$3d^3(^2P)4p$	$^3S^\circ$	1	286 187.7	83		6	$(^2P)\ ^3P^\circ$
$3d^3(^4P)4p$	$^3D^\circ$	3	286 431.3	41		24	$(^2D2)\ ^3F^\circ$
		1	286 855.3	48		15	$(^2D2)\ ^3D^\circ$
		2	286 862.7	52		20	$(^2P)\ ^3D^\circ$
$3d^3(^2P)4p$		3	287 109.6	33	$^3D^\circ$	23	$(^2D2)\ ^3F^\circ$
$3d^3(^2H)4p$	$^3I^\circ$	5	287 440.5	93		5	$(^2G)\ ^1H^\circ$
		6	288 167.2	98			
		7	289 171.9	100			
$3d^3(^2D2)4p$	$^3D^\circ$	1	288 669.8	58		20	$(^4P)\ ^3D^\circ$
		2	289 389.7	65		15	
		3	289 913.0	57		10	
$3d^3(^2H)4p$	$^1G^\circ$	4	289 545.9	75		17	$(^2F)\ ^1G^\circ$
$3d^3(^2H)4p$	$^1H^\circ$	5	290 099.1	75		17	$(^2G)\ ^1H^\circ$
$3d^3(^2D2)4p$	$^3P^\circ$	1	290 583.7	43		33	
		0	290 903.4	45		35	
$3d^3(^2D2)4p$	$^1F^\circ$	3	291 231.4	53		16	$(^2D1)\ ^1F^\circ$

(Continued)

Fe V—Continued

Configuration	Term	J	Level (cm^{-1})	Leading percentages		
$3d^3(^2H)4p$	$^3G°$	5	*292 287.6*	83	6	$(^2F)\ ^3G°$
		4	*292 430.7*	82	7	
		3	*292 513.2*	82	7	
$3d^3(^2H)4p$	$^1I°$	6	*292 365.9*	98		
$3d^3(^4P)4p$	$^3S°$	1	*294 644.0*	83	8	$(^2P)\ ^1P°$
$3d^3(^2D2)4p$	$^1D°$	2	*295 716.4*	46	41	$(^2P)\ ^1D°$
$3d^3(^2P)4p$	$^1P°$	1	*295 973.2*	62	18	$(^2D2)\ ^1P°$
$3d^3(^2F)4p$	$^3F°$	2	*302 292.7*	92		
		3	*302 377.1*	90		
		4	*302 602.5*	90		
$3d^3(^2F)4p$	$^3G°$	3	*306 193.9*	86	8	$(^2H)\ ^3G°$
		4	*306 622.8*	86	8	
		5	*307 064.4*	93	7	
$3d^3(^2F)4p$	$^3D°$	3	*307 288.7*	85	8	$(^2D1)\ ^3D°$
		2	*308 165.0*	75	12	$(^2F)\ ^1D°$
		1	*308 671.5*	90	8	$(^2D1)\ ^3D°$
$3d^3(^2F)4p$	$^1D°$	2	*307 644.4*	62	18	$(^2D1)\ ^1D°$
$3d^3(^2F)4p$	$^1G°$	4	*311 180.9*	80	18	$(^2H)\ ^1G°$
$3d^3(^2F)4p$	$^1F°$	3	*311 538.7*	92		
$3d^3(^2D1)4p$	$^3D°$	1	*327 533.8*	76	18	$(^2D2)\ ^3D°$
		2	*327 605.4*	75	16	
		3	*327 924.4*	76	15	
$3d^3(^2D1)4p$	$^1D°$	2	*329 848.6*	47	18	$(^2D2)\ ^1D°$
$3d^3(^2D1)4p$	$^3F°$	2	*331 333.8*	57	18	$(^2D2)\ ^3F°$
		3	*331 367.0*	70	21	
		4	*332 017.3*	76	22	
$3d^3(^2D1)4p$	$^3P°$	2	*334 509.1*	75	22	$(^2D2)\ ^3P°$
		1	*335 267.8*	75	24	
		0	*335 642.7*	75	24	
$3d^3(^2D1)4p$	$^1F°$	3	*335 947.4*	75	19	$(^2D2)\ ^1F°$
$3d^3(^2D1)4p$	$^1P°$	1	*342 462.2*	76	23	$(^2D2)\ ^1P°$
Fe VI ($^4F_{3/2}$)	***Limit***		**605 000**			

Fe VI

$Z = 26$

Sc I isoelectronic sequence

Ground state: $1s^22s^22p^63s^23p^63d^3\ ^4F_{3/2}$

Ionization energy = 799 000±2000 cm^{-1} (99.1±0.2 eV)

The original analysis was by Bowen (1935), whose observations yielded levels of the $3d^3$ and $3d^24p$ configurations. Several levels due to Bowen were published later in a paper by Pasternak (1940). Fawcett and Cowan (1973) observed the $3p^63d^3-3p^53d^4$ transition array between 162 and 180 Å. Fawcett and Henrichs (1974) have classified a number of lines of the $3d^24s-3d^24p$ array. Ekberg (1975) has observed the spectrum from 250–1580 Å. He has found all the terms of $3d^3$, $3d^24s$ and $3d^24p$ except $3d^2(^1S)4s\ ^2S$.

The present list of levels and leading percentages is compiled from Ekberg, except the configuration $3p^53d^4$, the levels of which are from Fawcett and Cowan. Ekberg's levels are stated to be uncertain by ±0.4 cm^{-1} and those of Fawcett and Cowan by ±100 cm^{-1}.

Bowen (1960) has observed lines in nebular spectra due to transitions within the $3d^3$ configuration.

The ionization energy is from an isoelectronic extrapolation by Lotz (1967).

References

Bowen, I. S. (1960), Astrophys. J. **132**, 1.
Bowen, I. S. (1935), Phys. Rev. **47**, 924.
Ekberg, J. O. (1975), Phys. Scr. **11**, 23.
Fawcett, B. C., and Cowan, R. D. (1973), Solar Physics **31**, 339.
Fawcett, B. C., and Henrichs, H. F. (1974), Astron. Astrophys. **18**, 157.
Lotz, W. (1967), J. Opt. Soc. Am. **57**, 873.
Pasternak, S. (1940), Astrophys. J. **92**, 140.

Fe VI

Configuration	Term	J	Level (cm^{-1})	Leading percentages		
$3d^3$	^4F	3/2	0.0	100		
		5/2	511.3	100		
		7/2	1 188.3	100		
		9/2	2 000.6	100		
$3d^3$	^4P	1/2	18 738.3	99		
		3/2	18 942.0	98		
		5/2	19 610.8	100		
$3d^3$	^2G	7/2	20 616.4	100		
		9/2	21 315.0	98		
$3d^3$	^2P	3/2	26 214.9	58	31	$^2D2$
		1/2	26 495.5	99		
$3d^3$	$^2D2$	5/2	28 484.3	80	20	$^2D1$
		3/2	28 627.9	46	40	^2P
$3d^3$	^2H	9/2	28 724.3	98		
		11/2	29 202.9	100		
$3d^3$	^2F	7/2	46 217.3	100		
		5/2	46 603.7	100		
$3d^3$	$^2D1$	5/2	71 707.6	80	20	$^2D2$
		3/2	72 048.9	78	22	
$3d^2(^3F)4s$	^4F	3/2	261 841.4	100		
		5/2	262 368.4	99		
		7/2	263 135.9	99		
		9/2	264 118.3	100		

Fe VI—Continued

Configuration	Term	J	Level (cm^{-1})	Leading percentages		
$3d^2(^3F)4s$	^2F	5/2	269 140.2	99		
		7/2	270 672.6	99		
$3d^2(^1D)4s$	^2D	5/2	280 901.5	61	38	$(^3P)\ ^4P$
		3/2	281 217.8	51	49	
$3d^2(^3P)4s$	^4P	1/2	281 477.0	100		
		3/2	282 035.0	51	47	$(^1D)\ ^2D$
		5/2	282 951.9	62	38	$(^1D)\ ^2D$
$3d^2(^3P)4s$	^2P	1/2	287 919.2	100		
		3/2	288 638.3	98		
$3d^2(^1G)4s$	^2G	9/2	292 313.0	100		
		7/2	292 330.1	100		
$3d^2(^3F)4p$	$^4G°$	5/2	*338 256.4*	91	6	$(^3F)\ ^2F°$
		7/2	*339 477.0*	92	5	$(^3F)\ ^4F°$
		9/2	*340 935.0*	93	7	$(^3F)\ ^4F°$
		11/2	*342 730.6*	100		
$3d^2(^3F)4p$	$^4F°$	3/2	*339 539.8*	94		
		5/2	*340 344.0*	94		
		7/2	*341 365.3*	94	5	$(^3F)\ ^4G°$
		9/2	*342 434.4*	91	7	$(^3F)\ ^4G°$
$3d^2(^3F)4p$	$^2F°$	5/2	*342 571.5*	58	16	$(^3F)\ ^4D°$
		7/2	*343 608.2*	54	40	
$3d^2(^3F)4p$	$^4D°$	3/2	*343 210.9*	55	30	$(^3F)\ ^2D°$
		1/2	*343 619.3*	92	7	$(^3P)\ ^4D°$
		5/2	*344 273.3*	63	23	$(^3F)\ ^2F°$
		7/2	*345 422.6*	53	37	$(^3F)\ ^2F°$
$3d^2(^3F)4p$	$^2D°$	3/2	*344 652.6*	47	36	$(^3F)\ ^4D°$
		5/2	*345 907.1*	62	15	$(^3P)\ ^2D°$
$3d^2(^3F)4p$	$^2G°$	7/2	*348 962.1*	94	4	$(^1G)\ ^2G°$
		9/2	*350 017.8*	94	4	
$3d^2(^3P)4p$	$^2S°$	1/2	*351 805.8*	98		
$3d^2(^3P)4p$	$^4S°$	3/2	*355 657.1*	90	9	$(^1D)\ ^2P°$
$3d^2(^1D)4p$	$^2P°$	3/2	*357 755.2*	80	9	$(^3P)\ ^4S°$
		1/2	*359 099.3*	51	40	$(^3P)\ ^4D°$
$3d^2(^1D)4p$	$^2F°$	5/2	*358 334.6*	83	6	$(^3F)\ ^2F°$
		7/2	*359 884.0*	77	12	$(^3P)\ ^4D°$
$3d^2(^3P)4p$	$^4D°$	1/2	*359 395.9*	52	44	$(^1D)\ ^2P°$
		3/2	*359 781.3*	90	7	$(^3F)\ ^4D°$
		5/2	*360 707.1*	85	6	$(^3F)\ ^4D°$
		7/2	*362 270.0*	82	12	$(^1D)\ ^2F°$
$3d^2(^1D)4p$	$^2D°$	3/2	*361 858.2*	80	6	$(^3F)\ ^2D°$
		5/2	*362 602.9*	76	11	$(^3P)\ ^4P°$

Fe VI—Continued

Configuration	Term	J	Level (cm^{-1})	Leading percentages		
$3d^2(^3P)4p$	$^4P°$	1/2	*363 945.7*	98		
		3/2	*364 392.9*	97		
		5/2	*365 494.0*	87	13	$(^1D)\ ^2D°$
$3d^2(^1G)4p$	$^2G°$	7/2	*365 077.0*	93	5	$(^3F)\ ^2G°$
		9/2	*365 266.6*	94	4	
$3d^2(^3P)4p$	$^2D°$	3/2	*370 538.1*	78	12	$(^3F)\ ^2D°$
		5/2	*370 579.6*	80	15	
$3d^2(^1G)4p$	$^2H°$	9/2	*372 095.6*	98		
		11/2	*373 706.1*	100		
$3d^2(^3P)4p$	$^2P°$	1/2	*374 088.3*	98		
		3/2	*374 425.6*	95		
$3d^2(^1G)4p$	$^2F°$	7/2	*377 951.8*	95	4	$(^1D)\ ^2F°$
		5/2	*379 077.6*	97		
$3d^2(^1S)4p$	$^2P°$	1/2	*408 207.4*	97		
		3/2	*410 389.5*	98		
$3p^5(^2P°)3d^4(^1G)$	$^2G°$	7/2	*575 930*			
		9/2	*576 990*			
$3p^5(^2P°)3d^4(^5D)$	$^4D°$	1/2	*603 210*			
		3/2	*604 230*			
		5/2	*605 420*			
		7/2	*606 230*			
$3p^5(^2P°)3d^4(^3H)$	$^2H°$	9/2	*603 340*			
		11/2	*605 740*			
$3p^5(^2P°)3d^4(^3F)$	$^2D°$	5/2	*617 520*			
		3/2	*618 290*			
$3p^5(^2P°)3d^4(^3H)$	$^2G°$	9/2	*630 240*			
		7/2	*631 240*			
$3p^5(^2P°)3d^4(^3G)$	$^2F°$	7/2	*634 170*			
		5/2	*635 430*			
Fe VII $(^3F_2)$	***Limit***		**799 000**			

Fe VII

$Z = 26$

Ca I isoelectronic sequence

Ground state: $1s^2 2s^2 2p^6 3s^2 3p^6 3d^2\ {}^3F_2$

Ionization energy = 1 008 000±100 cm^{-1} (124.98±0.01 eV)

The initial work by Cady (1933) on this spectrum was extended by numerous contributions. Ekberg (1981) has completely reobserved the spectrum and greatly extended the analysis. His paper gives more than 400 lines in the region 104–270 Å classified as transitions to the lowest configuration, $3p^6 3d^2$, and 20 lines between 1010 and 1362 Å in the $3d4s - 3d4p$ transition array. He states that the uncertainty of the levels of $3d^2$ is ± 1 cm^{-1}, while for the excited configurations the uncertainty increases from ± 4 to ± 20 cm^{-1} as the level value rises. He has made parametric calculations for $3d^2$, $3d4s$, $3dnf\,(n = 4\text{–}6)$, and $3p^5 3d^2 4s$. In the calculations for $3d4f$, configuration interaction with $3p^5 3d^3$ was included. The repeating ^{2}D terms of $3d^3$ were distinguished by the letters *a* and *b* by Ekberg. No indication of seniority contributions was given.

The $3d4p$ configuration has been calculated by Warner and Kirkpatrick (1969). Their percentage compositions as communicated privately to us are given here.

A number of forbidden transitions among levels of the $3d^2$ configuration have been listed by Bowen (1960). Their wavelengths are improved by Ekberg.

The ionization energy was determined by Ekberg from five different $3dnf$ series.

References

Bowen, I. S. (1960), Astrophys. J. **132**, 1.

Cady, W. M. (1933), Phys. Rev. **43**, 324.

Ekberg, J. O. (1981), Phys. Scr. **23**, 7.

Warner, B., and Kirkpatrick, R. C. (1969), Mon. Not. R. Astron. Soc. **144**, 397.

Fe VII

Configuration	Term	J	Level (cm^{-1})	Leading percentages		
$3p^6\,3d^2$	^{3}F	2	0.0	100		
		3	1 051.5	100		
		4	2 331.5	100		
$3p^6\,3d^2$	^{1}D	2	17 475.5	93	6	^{3}P
$3p^6\,3d^2$	^{3}P	0	20 040.3	100		
		1	20 430.1	100		
		2	21 278.6	94	6	^{1}D
$3p^6\,3d^2$	^{1}G	4	28 927.3	100		
$3p^6\,3d^2$	^{1}S	0	67 078.3	100		
$3p^6\,3d\,4s$	^{3}D	1	344 463.3	100		
		2	345 028.7	97		
		3	346 262.2	100		
$3p^6\,3d\,4s$	^{1}D	2	350 332.6	97		
$3p^6\,3d\,4p$	^{3}D°	1	*425 128.6*	97	2	^{1}P°
		2	*427 784.7*	83	7	^{1}D°
		3	*430 948.6*	75	23	^{3}F°
$3p^6\,3d\,4p$	^{1}D°	2	*425 386.1*	76	21	^{3}F°
$3p^6\,3d\,4p$	^{3}F°	2	*430 213.4*	72	15	^{1}D°
		3	*431 609.5*	77	23	^{3}D°
		4	*433 871.2*	100		

Fe VII—Continued

Configuration	Term	J	Level (cm^{-1})	Leading percentages		
$3p^6\,3d\,4p$	$^3P^\circ$	1	*436 952.2*	92	6	$^1P^\circ$
		0	*437 001.3*	100		
		2	*437 558.0*	96		
$3p^6\,3d\,4p$	$^1F^\circ$	3	*439 811.6*	98		
$3p^6\,3d\,4p$	$^1P^\circ$	1	*443 447.0*	92	6	$^3P^\circ$
$3p^5(^2P^\circ)3d^3(^2G)$	$^1H^\circ$	5	*464 034*			
$3p^5(^2P^\circ)3d^3(^2F)$	$^3G^\circ$	5	*472 559*			
		4	*472 903*			
		3	*481 435*			
$3p^5(^2P^\circ)3d^3(^2F)$	$^1G^\circ$	4	*496 454*			
$3p^5(^2P^\circ)3d^3(^2H)$	$^3G^\circ$	3	*510 086*			
		4	*510 158*			
		5	*514 133*			
$3p^5(^2P^\circ)3d^3(a^2D)$	$^1D^\circ$	2	*538 290*			
$3p^5(^2P^\circ)3d^3(^2F)$	$^3D^\circ$	2	*548 274*			
		3	*551 568*			
$3p^5(^2P^\circ)3d^3(^2F)$	$^1D^\circ$	2	*553 220*			
$3p^5(^2P^\circ)3d^3(^2G)$	$^1F^\circ$	3	*556 422*			
$3p^5(^2P^\circ)3d^3(^4P)$	$^3P^\circ$	1	*561 303*			
		2	*565 275*			
$3p^5(^2P^\circ)3d^3(^4F)$	$^3F^\circ$	2	*564 425*			
		3	*566 256*			
		4	*568 118*			
$3p^5(^2P^\circ)3d^3(^2P)$	$^1P^\circ$	1	*598 638*			
$3p^5(^2P^\circ)3d^3(^4F)$	$^3D^\circ$	3	*603 419*			
		2	*603 757*			
		1	*604 270*			
$3p^5(^2P^\circ)3d^3(^2H)$	$^1G^\circ$	4	*605 489*			
$3p^5(^2P^\circ)3d^3(^4P)$	$^3S^\circ$	1	*623 699*			
$3p^5(^2P^\circ)3d^3(b^2D)$	$^1P^\circ$	1	*630 283*			
$3p^6\,3d4f$	$^1G^\circ$	4	*659 917*	56	25	$^3F^\circ$
$3p^6\,3d4f$	$^3F^\circ$	2	*660 015*	94		
		3	*660 358*	94		
		4	*661 169*	70	26	$^1G^\circ$
$3p^6\,3d4f$	$^3G^\circ$	3	*663 097*	87	9	$^1F^\circ$
		4	*663 950*	94		
		5	*664 482*	97		
$3p^6\,3d4f$	$^1D^\circ$	2	*663 871*	90		

(Continued)

Fe VII—Continued

Configuration	Term	J	Level (cm^{-1})	Leading percentages		
$3p^6\,3d4f$	$^1F^\circ$	3	665 417	58	30	$^3D^\circ$
$3p^6\,3d4f$	$^3D^\circ$	1	665 832	92	5	$^3P^\circ$
		2	665 923	85	10	$^3P^\circ$
		3	666 651	66	31	$^1F^\circ$
$3p^6\,3d4f$	$^3P^\circ$	2	667 899	85	11	$^3D^\circ$
		1	668 253	92	6	$^3D^\circ$
		0	668 489	99		
$3p^6\,3d4f$	$^1H^\circ$	5	669 978	98		
$3p^6\,3d4f$	$^1P^\circ$	1	672 820	90	8	$3p^5(^2P^\circ)3d^3(b\,^2D)\ ^1P^\circ$
$3p^5(^2P^\circ)\,3d^2(^3F)4s\,(^2F)$	$^3D^\circ$	1	745 556	78	20	$(^2P^\circ)(^2P)\ ^3D^\circ$
		2	746 965	73	24	
		3	749 166	67	29	
$3p^5(^2P^\circ)\,3d^2(^3F)4s\,(^2F)$	$^3G^\circ$	5	766 991	94	5	$(^2P^\circ)(^4F)\ ^5F^\circ$
		4	768 813	83	7	
		3	769 991	50	16	
$3p^5(^2P^\circ)\,3d^2(^3P)4s\,(^2P)$	$^3P^\circ$	2	768 425	95		
		1	771 612	95		
		0	773 488	97		
$3p^5(^2P^\circ)\,3d^2(^1G)4s\,(^2G)$	$^3F^\circ$	3	778 420	45	36	$(^2P^\circ)(^4F)\ ^3F^\circ$
		4	779 575	36	32	
$3p^5(^2P^\circ)\,3d^2(^1D)4s\,(^2D)$	$^1D^\circ$	2	779 009	45	21	$(^2P^\circ)(^2F)\ ^3F^\circ$
$3p^5(^2P^\circ)\,3d^2(^3F)4s\,(^2F)$	$^3F^\circ$	4	782 690	72	9	$(^2P^\circ)(^2G)\ ^3F^\circ$
		3	783 119	52	13	$(^2P^\circ)(^2F)\ ^3G^\circ$
$3p^6\,3d5f$	$^1G^\circ$	4	784 174	48	46	$^3H^\circ$
$3p^6\,3d5f$	$^3H^\circ$	4	784 477	50	29	$^1G^\circ$
$3p^6\,3d5f$	$^3F^\circ$	2	784 733	86	12	$^1D^\circ$
		3	785 012	90		
		4	785 809	76	21	$^1G^\circ$
$3p^6\,3d5f$	$^3G^\circ$	3	786 732	78	16	$^1F^\circ$
		4	787 737	91	5	$^3F^\circ$
		5	788 146	97		
$3p^6\,3d5f$	$^1D^\circ$	2	786 830	74	12	$^3P^\circ$
$3p^6\,3d5f$	$^3D^\circ$	1	787 945	75	18	$^3P^\circ$
		2	788 030	51	32	$^3P^\circ$
		3	788 303	60	20	$^1F^\circ$
$3p^6\,3d5f$	$^3P^\circ$	2	788 995	56	41	$^3D^\circ$
		1	789 172	78	21	$^3D^\circ$
		0	789 365	100		
$3p^6\,3d5f$	$^1F^\circ$	3	789 215	64	31	$^3D^\circ$
$3p^6\,3d5f$	$^1P^\circ$	1	790 708	93		

Fe VII—Continued

Configuration	Term	J	Level (cm^{-1})	Leading percentages			
$3p^6\,3d5f$	^{1}H°	5	791 168	98			
$3p^5(^2$P°$)\,3d^2(^3$F$)4s\,(^4$F$)$	^{3}G°	3	794 149	45		30	$(^2$P°$)(^2$D$)\ ^3$F°
		4	797 712	65		9	$(^2$P°$)(^2$F$)\ ^1$G°
		5	800 633	69		12	$(^2$P°$)(^2$G$)\ ^1$H°
$3p^5(^2$P°$)\,3d^2(^3$F$)4s\,(^4$F$)$		3	797 257	38	^{3}G°	33	$(^2$P°$)(^2$D$)\ ^3$F°
$3p^5(^2$P°$)\,3d^2(^3$F$)4s\,(^2$F$)$	^{1}G°	4	802 462	67		13	$(^2$P°$)(^4$F$)\ ^3$G°
$3p^5(^2$P°$)\,3d^2(^1$G$)4s\,(^2$G$)$	^{1}H°	5	806 033	61		21	$(^2$P°$)(^2$G$)\ ^3$H°
$3p^5(^2$P°$)\,3d^2(^1$D$)4s\,(^2$D$)$	^{1}F°	3	807 627	64		30	$(^2$P°$)(^2$D$)\ ^3$F°
$3p^5(^2$P°$)\,3d^2(^3$P$)4s\,(^2$P$)$	^{3}D°	3	812 086	63		30	$(^2$P°$)(^2$F$)\ ^3$D°
		2	813 877	65		23	
		1	817 195	71		19	
$3p^5(^2$P°$)\,3d^2(^3$P$)4s\,(^4$P$)$	^{3}D°	1	822 689	79		8	$(^2$P°$)(^2$P$)\ ^3$D°
		2	824 184	75		7	$(^2$P°$)(^2$P$)\ ^1$D°
		3	827 533	81		7	$(^2$P°$)(^4$F$)\ ^3$D°
$3p^5(^2$P°$)\,3d^2(^3$P$)4s\,(^4$P$)$	^{3}S°	1	826 106	97			
$3p^5(^2$P°$)\,3d^2(^3$P$)4s\,(^2$P$)$	^{1}D°	2	829 626	72		8	$(^2$P°$)(^4$P$)\ ^3$D°
$3p^5(^2$P°$)\,3d^2(^1$G$)4s\,(^2$G$)$	^{3}G°	5	832 889	88		8	$(^2$P°$)(^4$F$)\ ^3$G°
		4	832 893	91		6	
		3	833 128	93		5	
$3p^5(^2$P°$)\,3d^2(^3$P$)4s\,(^2$P$)$	^{3}S°	1	837 472	97			
$3p^6\,3d6f$	^{3}H°	4	852 601	61		34	^{1}G°
$3p^6\,3d6f$		4	853 307	33	^{3}H°	30	^{1}G°
$3p^6\,3d6f$	^{3}F°	2	853 433	77		19	^{1}D°
		3	853 697	79		11	^{3}G°
		4	854 767	63		31	^{1}G°
$3p^6\,3d6f$	^{3}G°	3	854 760	65		24	^{1}F°
		4	855 969	82		11	^{3}F°
		5	856 260	93		6	^{3}H°
$3p^6\,3d6f$	^{1}D°	2	854 838	40		34	^{3}P°
$3p^6\,3d6f$	^{3}D°	1	855 346	64		22	^{3}P°
		3	856 109	49		19	^{3}F°
$3p^6\,3d6f$		2	855 903	38	^{1}D°	29	^{3}D°
$3p^6\,3d6f$	^{1}F°	3	856 797	61		32	^{3}D°
$3p^6\,3d6f$	^{3}P°	2	856 811	48		48	^{3}D°
		1	856 975	71		28	^{3}D°
		0	857 082	100			
$3p^6\,3d6f$	^{1}H°	5	857 881	94			

(Continued)

Fe VII—Continued

Configuration	Term	J	Level (cm^{-1})	Leading percentages		
$3p^6\,3d7f$	$^3F°$	2	*894 718*			
		3	*894 944*			
		4	*896 382*			
$3p^6\,3d7f$	$^3G°$	3	*895 744*			
		4	*897 077*			
		5	*897 254*			
$3p^6\,3d7f$	$^1H°$	5	*898 243*			
$3p^6\,3d8f$	$^3F°$	3	*921 694*			
		4	*923 282*			
$3p^6\,3d8f$	$^3G°$	4	*923 716*			
		5	*923 838*			
$3p^6\,3d8f$	$^1H°$	5	*924 479*			
$3p^5(^2P°)\,3d^2(^3P)4s\,(^4P)$	$^3P°$	2	*928 684*	75	16	$(^2P°)(^2D)\,^3P°$
$3p^6\,3d9f$	$^3G°$	4	*941 918*			
		5	*942 022*			
$3p^6\,3d9f$	$^1H°$	5	*942 477*			
$3p^6\,3d\,10f$	$^3G°$	4	*954 904*			
		5	*954 966*			
$3p^6\,3d\,10f$	$^1H°$	5	*955 307*			
Fe VIII ($^2D_{3/2}$)	***Limit***		**1 008 000**			

Fe VIII

$Z=26$

K I isoelectronic sequence

Ground state: $1s^2 2s^2 2p^6 3s^2 3p^6 3d\ ^2D_{3/2}$

Ionization energy = 1 218 380±100 cm^{-1} (151.061±0.012 eV)

The ground-term splitting was determined by Cowan and Peacock (1965) by means of four pairs of lines arising from the $3p^5 3d^2$ configuration. This upper configuration was interpreted by the same authors. Earlier, Kruger and Weissberg (1937) reported the one-electron terms $3p^6(^1S)$ $5s$, $6s$, $4p$, $4f$, $5f$, $6f$, and $7f$. With light sources allowing differentiation among highly ionized species Alexander, Feldman, and Fraenkel (1965) determined that the lines used by Kruger et al. to establish $5s$, $6s$, and $4p$ were erroneously assigned to Fe VIII. This finding has been confirmed privately by Ekberg.

The levels of $3p^5 3d4s$ were deduced by Cowan (1967) from lines reported by Feldman and Fraenkel (1966).

The spectrum has now been remeasured between 93 and 233 Å with an uncertainty of ±0.003 Å by Ramonas and Ryabtsev (1980). They have redetermined all the known level values and extended the analysis. Their levels given below, have an uncertainty of ±50 cm^{-1}. The leading percentages were supplied privately by Cowan.

We have determined the ionization energy from the new measurements of the nf series. The $7f$ term is predicted to be 590 cm^{-1} below the observed value.

References

Alexander, E., Feldman, U., and Fraenkel, B. S. (1965), J. Opt. Soc. Am. **55**, 650.

Cowan, R. D., and Peacock, N. J. (1965), Astrophys. J. **142**, 390.

Cowan, R. D. (1967), Astrophys. J. **147**, 377.

Feldman, U., and Fraenkel, B. S. (1966), Astrophys. J. **145**, 959.

Kruger, P. G., and Weissberg, S. G. (1937), Phys. Rev. **52**, 314.

Ramonas, A. A., and Ryabtsev, A. N. (1980), Opt. Spectrosc. **48**, 348.

Fe VIII

Configuration	Term	J	Level (cm^{-1})	Leading percentages		
$3p^6(^1S)3d$	^2D	3/2	0			
		5/2	1 836			
$3p^5(^2P^\circ)3d^2(^1G)$	$^2F^\circ$	5/2	*431 250*	44	37	$(^2P^\circ)(^3F)\ ^2F^\circ$
		7/2	*434 555*	45	41	
$3p^5(^2P^\circ)3d^2(^1D)$	$^2F^\circ$	7/2	*447 658*	72	23	$(^2P^\circ)(^3F)\ ^2G^\circ$
		5/2	*459 367*	91	4	$(^2P^\circ)(^3P)\ ^4D^\circ$
$3p^5(^2P^\circ)3d^2(^1S)$	$^2P^\circ$	3/2	*508 518*	61	23	$3p^6 4p\ ^2P^\circ$
		1/2	*520 822?*	77	14	$3p^5(^2P^\circ)3d^2(^1D)\ ^2P^\circ$
$3p^6(^1S)4p$	$^2P^\circ$	1/2	*510 277*	94	4	$3p^5(^2P^\circ)3d^2(^1S)\ ^2P^\circ$
		3/2	*515 550*	74	18	
$3p^5(^2P^\circ)3d^2(^3F)$	$^2F^\circ$	5/2	*535 909*	53	45	$(^2P^\circ)(^1G)\ ^2F^\circ$
		7/2	*541 755*	50	48	
$3p^5(^2P^\circ)3d^2(^3P)$	$^2P^\circ$	1/2	*591 964*	72	14	$(^2P^\circ)(^1D)\ ^2P^\circ$
		3/2	*595 152*	73	16	
$3p^5(^2P^\circ)3d^2(^3F)$	$^2D^\circ$	5/2	*596 463*	71	17	$(^2P^\circ)(^1D)\ ^2D^\circ$
		3/2	*597 065*	71	17	
$3p^6(^1S)4f$	$^2F^\circ$	5/2	*763 703*	98		
		7/2	*763 799*	98		
$3p^5\ 3d(^3P^\circ)4s$	$^2P^\circ$	1/2	*837 661*	98		
		3/2	*842 829*	95		

(Continued)

Fe VIII—Continued

Configuration	Term	J	Level (cm^{-1})	Leading percentages		
$3p^5\,3d(^3F°)4s$	$^4F°$	7/2	*847 145*	95		
		5/2	*849 899*	92		
		3/2	*852 849?*	96		
$3p^5\,3d(^3F°)4s$	$^2F°$	7/2	*855 100*	94		
		5/2	*860 615*	89	4	$(^1D°)\ ^2D°$
$3p^5\,3d(^3D°)4s$	$^4D°$	7/2	*874 711*	81	18	$(^1F°)\ ^2F°$
		5/2	*876 765*	74	10	$(^1D°)\ ^2D°$
		3/2	*877 476*	76	19	$(^1D°)\ ^2D°$
		1/2	*878 264?*	98		
$3p^5\,3d(^1D°)4s$	$^2D°$	5/2	*879 021*	48	25	$(^3D°)\ ^2D°$
		3/2	*881 345*	77	20	$(^3D°)\ ^4D°$
$3p^5\,3d(^1F°)4s$	$^2F°$	5/2	*884 331*	46	27	$(^1D°)\ ^2D°$
		7/2	*887 325*	78	16	$(^3D°)\ ^4D°$
$3p^5\,3d(^3D°)4s$	$^2D°$	3/2	*889 113*	98		
		5/2	*890 845*	68	20	$(^1F°)\ ^2F°$
$3p^6(^1S)5f$	$^2F°$	5/2	*927 059*			
		7/2	*927 102*			
$3p^6(^1S)6f$	$^2F°$	5/2	*1 016 560*			
		7/2	*1 016 570*			
$3p^6(^1S)7f$	$^2F°$	5/2	*1 069 873*			
		7/2	*1 070 029*			
Fe IX ($^1S_0$)	*Limit*		1 218 380			

Fe IX

$Z=26$

Ar I isoelectronic sequence

Ground state: $1s^2 2s^2 2p^6 3s^2 3p^6\ ^1S_0$

Ionization energy = 1 884 000±3000 cm^{-1} (233.6±0.4 eV)

This spectrum was first investigated by Kruger, Weissberg, and Phillips (1937), who identified the resonance lines arising from the two $J=1$ levels of the $3p^5 4s$ configuration. The present values of these levels are taken from the paper of Fawcett, Cowan, Kononov, and Hayes (1972). The level uncertainty is ±100 cm^{-1}.

All levels of the $3s^2 3p^5 3d$ configuration were determined from combinations with $3s3p^6 3d$ by Svensson, Ekberg and Edlén (1974). Using these levels to predict forbidden transitions within $3s^2 3p^5 3d$, Edlén and Smitt (1978) identified a number of well-measured solar lines obtained with Skylab and redetermined most of the level values with an uncertainty of ± 3 cm^{-1}. Their results are quoted here. The $3s3p^6 3d$ levels are from Smitt and Svensson (1983) with an uncertainty of ±6 cm^{-1}. The uncertainty of the connection of these two configurations to the ground state is ±6 cm^{-1}.

The $3p^5 4d$ and $3p^5 5s$ levels were found by Alexander, Feldman, and Fraenkel (1965). The present values for the two $3p^5 4d$ levels are obtained from the measurements of Fawcett et al. The uncertainty is ±100 cm^1.

The $3p^5 4f$ level values given here were derived by combining the $3p^5 3d - 3p^5 4f$ line identifications of Wagner and House (1971) and of Fawcett et al. with the level values of $3p^5 3d$ of Edlén and Smitt, giving a level uncertainty of ±300 cm^{-1}. The $3p^5 4f$ levels clearly follow a $J_1 l$ coupling scheme, the designations having been assigned by comparison with isoelectronic spectra.

The $3s3p^6 4p$ term is from Kastner, Crooker, Behring, and Cohen (1977) with an uncertainty of ±200 cm^{-1}.

We have derived the ionization energy from the $3p^5 4s$ and $3p^5 5s$ configurations under the assumption of a change in effective quantum number $\Delta n^* = n^*(5s) - n^*(4s) = 1.024$, as observed in similar spectra. The stated uncertainty in the ionization energy is based on an estimated uncertainty of ± 0.005 in the value of Δn^*.

References

Alexander, E., Feldman, U., and Fraenkel, B. S. (1965), J. Opt. Soc. Am. **55**, 650.

Edlén, B., and Smitt, R. (1978), Sol. Phys. **57**, 329.

Fawcett, B. C., Cowan, R. D., Kononov, E. Y., and Hayes, R. W. (1972), J. Phys. **B5**, 1255.

Kastner, S. O., Crooker, A. M., Behring, W. E., and Cohen, L. (1977), Phys. Rev. **A16**, 577.

Kruger, P. G., Weissberg, S. G., and Phillips, L. W. (1937), Phys. Rev. **51**, 1090.

Smitt, R., and Svensson, L. A. (1983), Phys. Scr. **27**, 364.

Svensson, L. A., Ekberg, J. O., and Edlén, B., (1974), Sol. Phys. **34**, 173.

Wagner, W. J., and House, L. L. (1971), Astrophys. J. **166**, 683.

Fe IX

Configuration	Term	J	Level (cm^{-1})	Configuration	Term	J	Level (cm^{-1})
$3p^6$	^1S	0	0	$3s3p^6\,3d$	^3D	1	726 734
						2	727 560
$3p^5\,3d$	$^3P°$	0	*405 772*			3	728 935
		1	*408 315.1*				
		2	*413 669.2*	$3s3p^6\,3d$	^1D	2	749 871
$3p^5\,3d$	$^3F°$	4	*425 809.8*	$3p^5(^2P°_{3/2})4s$	$(3/2,1/2)°$	1	*950 500*
		3	*429 310.9*				
		2	*433 818.8*	$3p^5(^2P°_{1/2})4s$	$(1/2,1/2)°$	1	*965 570*
$3p^5\,3d$	$^3D°$	3	*455 612.2*	$3p^5\,4d$	$^3P°$	1	*1 198 220*
		1	*460 616*				
		2	*462 616.6*	$3p^5\,4d$	$^1P°$	1	*1 213 150*
$3p^5\,3d$	$^1D°$	2	*456 752.7*	$3p^5(^2P°_{3/2})4f$	$^2[3/2]$	1	1 300 920
						2	1 302 840
$3p^5\,3d$	$^1F°$	3	*465 828.4*				
				$3p^5(^2P°_{3/2})4f$	$^2[9/2]$	5	1 304 600
$3p^5\,3d$	$^1P°$	1	*584 546*			4	1 306 320

(Continued)

Fe IX—Continued

Configuration	Term	J	Level (cm^{-1})	Configuration	Term	J	Level (cm^{-1})
$3p^5(^2P^\circ_{3/2})4f$	$^2[5/2]$	3	1 305 760	$3p^5(^2P^\circ_{3/2})5s$	$(3/2,1/2)^\circ$	1	*1 358 140*
$3p^5(^2P^\circ_{3/2})4f$	$^2[7/2]$	3	1 310 160	$3s3p^6\,4p$	$^1P^\circ$	1	*1 371 910*
		4	1 311 750	$3p^5(^2P^\circ_{1/2})5s$	$(1/2,1/2)^\circ$	1	*1 372 670*
$3p^5(^2P^\circ_{1/2})4f$	$^2[5/2]$	3	1 323 660	Fe X ($^2P^\circ_{3/2}$)	***Limit***		**1 884 000**
$3p^5(^2P^\circ_{1/2})4f$	$^2[7/2]$	3	1 324 720				
		4	1 324 800				

Fe X

$Z=26$

Cl I isoelectronic sequence

Ground state: $1s^2 2s^2 2p^6 3s^2 3p^5\ {}^2P^\circ_{3/2}$

Ionization energy = 2 114 000±1000 cm^{-1} (262.1±0.1 eV)

The $3s^2 3p^4 4s$ levels and the $3s^2 3p^5\ {}^2P^\circ$ term interval were established with an uncertainty of ±50 cm^{-1} by the identification of a group of eight lines at 94–98 Å by Edlén in 1937. The value of the $3s^2 3p^5\ {}^2P^\circ$ interval was later more precisely determined by Grotrian (1939) to ±0.1 cm^{-1} through his identification of the solar coronal line at 6374.51 Å as the $3s^2 3p^5$ magnetic dipole transition ${}^2P^\circ_{3/2} - {}^2P^\circ_{1/2}$ in Fe X, which was confirmed by Edlén (1942). The $3s3p^5 3d\ {}^4F^\circ$ term is from Smitt (1977). The $3s3p^6\ {}^2S$ term was first located by Fawcett (1971). We use the improved measurements of Smitt, Svensson, and Outred (1976), which provide a level of uncertainty of ±5 cm^{-1}.

Edlén and Smitt (1978) derived levels of the $3s^2 3p^4 3d$ configuration relative to its lowest level (${}^4D_{7/2}$) with an uncertainty of ±5 cm^{-1} using forbidden transitions within this configuration observed in the solar corona. The ${}^4D_{7/2}$ and ${}^4D_{5/2}$ levels are unresolved. Bromage, Cowan, and Fawcett (1977) give the connection of ${}^4D_{5/2}$ to the $3s^2 3p^5\ {}^2P_{3/2}$ ground state with an uncertainty of ±10 cm^{-1} and report additional levels of $3s^2 3p^4 3d$. They have also provided the leading percentages for the levels. The $3s^2 3p^4 4p$, $4d$, and $4f$ level values are derived from the identifications of Fawcett, Cowan, Kononov, and Hayes (1972), who also give the percentage composition of the levels. Their level uncertainty is ±100 cm^{-1}.

Edlén (1937) derived the ionization energy by isoelectronic extrapolation.

References

Bromage, G. E., Cowan, R. D., and Fawcett, B. C. (1977), Phys. Scr. **15**, 177.

Edlén, B. (1937), Z. Phys. **104**, 407.

Edlén, B. (1942), Z. Astrophys. **22**, 30.

Edlén, B., and Smitt, R. (1978), Sol. Phys. **57**, 329.

Fawcett, B. C. (1971), J. Phys. **B4**, 1577.

Fawcett, B. C., Cowan, R. D., Kononov, E. Y., and Hayes, R. W. (1972), J. Phys. **B5**, 1255.

Grotrian, W. (1939), Naturwiss. **27**, 214.

Smitt, R. (1977), Sol. Phys. **51**, 113.

Smitt, R., Svensson, L. A., and Outred, M. (1976), Phys. Scr. **13**, 293.

Fe X

Configuration	Term	J	Level (cm^{-1})	Leading percentages		
$3s^2 3p^5$	${}^2P^\circ$	3/2	*0.0*			
		1/2	*15 683.1*			
$3s3p^6$	${}^2S$	1/2	289 249	77	23	$3s^2 3p^4(^1D)3d\ {}^2S$
$3s^2 3p^4(^3P)3d$	${}^4D$	7/2	388 709	97		
		5/2	388 709	95		
		3/2	390 050	94		
		1/2	391 555	96		
$3s^2 3p^4(^3P)3d$	${}^4F$	9/2	417 653	94		
		7/2	422 795	89		
		5/2	426 763	94		
		3/2	428 298	90		
$3s^2 3p^4(^1D)3d$	${}^2P$	3/2	431 928	45	33	$(^3P)\ {}^2P$
$3s^2 3p^4(^1D)3d$	${}^2D$	3/2	434 614	43	27	$(^3P)\ {}^2D$
$3s^2 3p^4(^3P)3d$	${}^4P$	1/2	434 800	96		
		5/2	441 853	45	23	$(^1D)\ {}^2D$
$3s^2 3p^4(^3P)3d$	${}^2F$	7/2	440 840	55	28	$(^1D)\ {}^2G$
		5/2	452 730?	78	15	$(^1D)\ {}^2F$
$3s^2 3p^4(^1D)3d$	${}^2G$	9/2	450 751	94		
		7/2	451 084	67	25	$(^3P)\ {}^2F$

(Continued)

Fe x—Continued

Configuration	Term	J	Level (cm^{-1})	Leading percentages		
$3s^2 3p^4(^1D)3d$	^2F	7/2	485 983	84		
$3s^2 3p^4(^1S)3d$	^2D	3/2	511 800	69	26	$(^1D)\ ^2D$
$3s^2 3p^4(^1D)3d$	^2S	1/2	541 879	73	23	$3s3p^6\ ^2S$
$3s^2 3p^4(^3P)3d$	^2P	3/2	564 198	52	40	$(^1D)\ ^2P$
		1/2	569 985	54	45	
$3s^2 3p^4(^3P)3d$	^2D	5/2	572 954	66	21	$(^1D)\ ^2D$
		3/2	586 244	61	17	
$3s3p^5(^3P^\circ)3d$	$^4F^\circ$	9/2	*696 661*			
		7/2	*699 492*			
		5/2	*702 585*			
		3/2	*705 430*			
$3s^2 3p^4(^3P)4s$	^4P	5/2	1 022 100			
		3/2	1 029 630			
$3s^2 3p^4(^3P)4s$	^2P	3/2	1 040 350			
		1/2	1 048 890			
$3s^2 3p^4(^1D)4s$	^2D	5/2	1 063 690			
		3/2	1 064 190			
$3s^2 3p^4(^3P)4p$	$^4P^\circ$	5/2	*1 118 490?*	84	14	$(^3P)\ 4d$
$3s^2 3p^4(^3P)4p$	$^4D^\circ$	7/2	*1 130 430*			
$3s^2 3p^4(^1D)4p$	$^2F^\circ$	5/2	*1 161 930*			
		7/2	*1 165 710*			
$3s^2 3p^4(^1D)4p$	$^2D^\circ$	5/2	*1 178 850?*			
$3s^2 3p^4(^3P)4d$	^2D	5/2	1 284 270	65	14	^2F
		3/2	1 285 180	58	20	^4P
$3s^2 3p^4(^3P)4d$	^4F	5/2	1 286 540	77	14	$(^3P)\ ^2D$
$3s^2 3p^4(^3P)4d$	^2F	5/2	1 288 210			
$3s^2 3p^4(^3P)4d$	^2P	3/2	1 295 260	82	10	$(^3P)\ ^2D$
$3s^2 3p^4(^1D)4d$	^2P	3/2	1 315 690			
		1/2	1 317 390	79	18	$(^3P)\ ^2P$
$3s^2 3p^4(^1D)4d$	^2D	5/2	1 321 270			
		3/2	1 322 960			
$3s^2 3p^4(^3P)4f$	$^4F^\circ$	9/2	*1 388 450*			
$3s^2 3p^4(^3P)4f$	$^4G^\circ$	11/2	*1 397 130*			
		9/2	*1 399 850*	47	40	$(^3P)\ ^2G^\circ$
		7/2	*1 409 730*	61	14	$(^3P)\ ^2F^\circ$
$3s^2 3p^4(^3P)4f$	$^2G^\circ$	9/2	*1 408 650*	49	42	$(^3P)\ ^4G^\circ$
$3s^2 3p^4(^1D)4f$	$^2H^\circ$	11/2	*1 429 300*			

Fe X—Continued

Configuration	Term	J	Level (cm^{-1})	Leading percentages
$3s^2 3p^4(^1D)4f$	$^2G°$	9/2	*1 441 660*	
$3s^2 3p^4(^1S)4f$	$^2F°$	5/2	*1 484 290*	
Fe XI ($^3P_2$)	***Limit***		**2 114 000**	

Fe XI

$Z = 26$

S I isoelectronic sequence

Ground state: $1s^2 2s^2 2p^6 3s^2 3p^4\ ^3P_2$

Ionization energy = 2 341 000±5000 cm^{-1} (290.3±0.6 eV)

This spectrum was first investigated by Edlén (1937), who observed and identified the group of $3s^2 3p^4 - 3s^2 3p^3 4s$ transitions occurring at about 90 Å. He established most of the levels of these two configurations. Grotrian (1939) identified a solar coronal line at 7891 Å as the transition between the $3s^2 3p^4\ ^3P_2$ and $^3P_1$ levels. This was subsequently confirmed by Edlén (1942), who also identified a coronal line at 3986.9 Å as the $3s^2 3p^4$ transition $^3P_1 - ^1D_2$.

The separations of these levels quoted here are derived from wavelengths from the coronal observations of Jefferies (1969). He assigns an estimated accuracy of ±0.4 Å, which corresponds to about ±1 cm^{-1} in the levels. The $3s^2 3p^4\ ^1S_0$ level is derived from the solar line at 1467.08 Å observed by Doschek et al. (1976), which was identified by both Svensson (1971) and Jordan (1971) as the $3s^2 3p^4$ transition $^3P_1 - ^1S_0$.

The $3s^2 3p^4 - 3s3p^5$ array was analyzed by Fawcett (1971). The more accurate measurements of Smitt, Svensson, and Outred (1976) are used here to obtain the levels of $3s3p^5$ and the $^3P_0$ level of $3s^2 3p^4$ with an uncertainty of ±5 cm^{-1}.

The $3s^2 3p^3 4s$ levels are derived from the 1937 observations of Edlén, with the dropping of the identification of the original singlet-triplet intercombination lines at 86.149 and 89.771 Å, as noted by Edlén in 1942. The level uncertainty for this configuration is ±100 cm^{-1}. The line at 89.771 Å has been given by Fawcett, Cowan, Kononov, and Hayes (1972) as the $3s^2 3p^4\ ^1S_0 - 3s^2 3p^3 4s\ ^1P^\circ_1$ transition in Fe XI. However, this identification is inconsistent with Edlén's identification of the line at 86.513 Å as the $3s^2 3p^4\ ^1D_2 - 3s^2 3p^3 4s\ ^1P^\circ_1$ transition, which fixes the position of the $3s^2 3p^3 4s\ ^1P^\circ_1$ level. The $3s^2 3p^3 4s\ ^5S^\circ$ and $^3P^\circ$ terms have not yet been located.

The classifications for the $3p^4 - 3p^3 3d$ array are from Bromage, Cowan, and Fawcett (1977) and Fawcett (1971). The leading percentages for the $3p^3 3d$ were calculated including configuration interaction among $3s3p^5$, $3p^5 3d$ and $3s3p^3 3d^2$ by Bromage et al., but only the results for $3p^3 3d$ were published. Levels are derived with an uncertainty of ±20 cm^{-1} from the more accurate solar wavelengths (±0.008 Å) of Behring, Cohen, and Feldman (1972).

The $3s^2 3p^3 4d$ levels are taken from the work of Fawcett, Cowan, Kononov, and Hayes and are reliable to ±100 cm^{-1}. These authors have also observed a number of lines identified as $3s^2 3p^3 3d - 3s^2 3p^3 4f$ and $3s^2 3p^3 3d - 3s^2 3p^3 4p$ transitions of Fe XI. However, it is not possible to derive level values from these identifications inasmuch as none of the levels involved is part of the system of levels given here. The leading percentages for the $3s^2 3p^3 4d\ ^1D^\circ_2$ level are from Fawcett et al.

The ionization energy is from an isoelectronic extrapolation by Lotz (1967).

References

Behring, W. E., Cohen, L., and Feldman, U. (1972), Astrophys. J. **175**, 493.

Bromage, G. E., Cowan, R. D., and Fawcett, B. C. (1977), Phys. Scr. **15**, 177.

Doschek, G. A., Feldman, U., Van Hoosier, M. E., and Bartoe, J.-D. F. (1976), Astrophys. J Suppl. Ser. **31**, 417.

Edlén, B. (1937), Z. Phys. **104**, 188.

Edlén, B. (1942), Z. Astrophys. **22**, 30.

Fawcett, B. C. (1971), J. Phys. **B4**, 1577.

Fawcett, B. C., Cowan, R. D., Kononov, E. Y., and Hayes, R. W. (1972), J. Phys. **B5**, 1255.

Grotrian, W. (1939), Naturwiss. **27**, 214.

Jefferies, J. T. (1969), Mem. Soc. Roy. Sci. Liege, Collect. **17**, 213.

Jordan, C. (1971), Solar Phys. **21**, 381.

Lotz, W. (1967), J. Opt. Soc. Am. **57**, 873.

Smitt, R., Svensson, L. A., and Outred, M. (1976), Phys. Scr. **13**, 293.

Svensson, L. A. (1971), Solar Phys. **18**, 232.

Configuration	Term	J	Level (cm^{-1})	Leading percentages
$3s^2\,3p^4$	^3P	2	0.0	
		1	12 667.9	
		0	14 312	
$3s^2\,3p^4$	^1D	2	37 743.6	
$3s^2\,3p^4$	^1S	0	80 814.7	

Fe XI—Continued

Configuration	Term	*J*	Level (cm^{-1})	Leading percentages			
$3s3p^5$	$^3P^\circ$	2	*283 558*				
		1	*293 158*				
		0	*299 163*				
$3s3p^5$	$^1P^\circ$	1	*361 842*				
$3s^2\,3p^3(^2P^\circ)3d$	$^3P^\circ$	2	*496 090*	80			
$3s^2\,3p^3(^2D^\circ)3d$		1	*526 480*	38	$^3S^\circ$	23	$^3P^\circ$
$3s^2\,3p^3(^2D^\circ)3d$	$^3P^\circ$	2	*531 290*	74		15	$3s3p^5\ ^3P^\circ$
		1	*541 390*	42		16	$3s^23p^3(^2P^\circ)3d\ ^3P^\circ$
		0	*541 720*	62		19	$3s^23p^3(^2P^\circ)3d\ ^3P^\circ$
$3s^2\,3p^3(^2D^\circ)3d$	$^3S^\circ$	1	*533 450*	53		28	$(^2D^\circ)\ ^1P^\circ$
$3s^2\,3p^3(^4S^\circ)3d$	$^3D^\circ$	3	*554 300*	46		31	$(^2P^\circ)\ ^3D^\circ$
		2	*561 610*	41		31	
		1	*566 380*	42		36	
$3s^2\,3p^3(^2D^\circ)3d$	$^1D^\circ$	2	*578 860*	70		20	$(^2P^\circ)\ ^1D^\circ$
$3s^2\,3p^3(^2D^\circ)3d$	$^1F^\circ$	3	*594 030*	62		33	$(^2P^\circ)\ ^1S^\circ$
$3s^2\,3p^3(^2P^\circ)3d$	$^1P^\circ$	1	*623 080*	90			
$3s^2\,3p^3(^4S^\circ)4s$	$^3S^\circ$	1	*1 121 230*				
$3s^2\,3p^3(^2D^\circ)4s$	$^3D^\circ$	1	*1 148 590*				
		2	*1 149 100*				
		3	*1 152 450*				
$3s^2\,3p^3(^2D^\circ)4s$	$^1D^\circ$	2	*1 160 030*				
$3s^2\,3p^3(^2P^\circ)4s$	$^1P^\circ$	1	*1 193 640*				
$3s^2\,3p^3(^4S^\circ)4d$	$^3D^\circ$	3	*1 376 750*				
$3s^2\,3p^3(^2D^\circ)4d$	$^1D^\circ$	2	*1 420 680*	72		10	$(^2D^\circ)\ ^3D^\circ$
$3s^2\,3p^3(^2D^\circ)4d$	$^1F^\circ$	3	*1 423 440*				
Fe XII ($^4S^\circ_{3/2}$)	*Limit*		2 341 000				

Fe XII

$Z=26$

P I isoelectronic sequence

Ground state: $1s^22s^22p^63s^23p^3\,{}^4S^\circ_{3/2}$

Ionization energy = 2 668 000±5000 cm^{-1} (330.8±0.6 eV)

The level values of the $3s^23p^3$ ground configuration are determined from four solar coronal lines. The lines at 1242 and 1349 Å were identified by Burton, Ridgeley, and Wilson (1967) as $3p^3\,{}^4S^\circ_{3/2}-3p^3\,{}^2P^\circ_{3/2,1/2}$. New measurements for these lines by Doschek et al. (1976) with an uncertainty of ±0.05 Å are used here. The line at 2169.7 Å measured by Gabriel et al. (1971) was identified as $3p^3\,{}^4S^\circ_{3/2}-3p^3\,{}^2D^\circ_{5/2}$. The transition $3p^3\,{}^2D_{3/2}-3p^3\,{}^2P^\circ_{1/2}$ was assigned by Svensson (1971) to the coronal line reported at 3072.0 Å by Jefferies (1969). The level uncertainty in the ground configuration is ±5 cm^{-1}.

The classifications of the $3s^23p^3-3s3p^4$ and $3s^23p^3-3s^23p^23d$ transition arrays are due to Fawcett (1971), who points out that they are strong in the solar spectrum between 180 and 390 Å, and to Bromage, Cowan, and Fawcett (1978), who also provided the leading percentages for $3s3p^4$ and $3s^23p^23d$. Configuration interaction between them was included in the calculation. Improved measurements of these wavelengths by Behring, Cohen, Feldman, and Doschek (1976) with an uncertainty of ±0.004 Å made from solar observations are used to obtain the levels with an uncertainty of ±10 cm^{-1}.

Lines classified as transitions between the $3s^23p^3$ configuration and the $3s^23p^24s$ and $4d$ configurations in the range of 66–81 Å are given by Fawcett, Cowan, Kononov, and Hayes (1972) with an uncertainty of 0.01 Å. The upper levels are derived with these data with an uncertainty of ±200 cm^{-1}. Classified lines from $3s^23p^24p$ and $4f$ are also given but are not connected with known lower levels. The percentage compositions for $3p^24s$ and $3p^24d$ are given in the same paper.

The ionization energy is an extrapolated value by Lotz (1967).

References

Behring, W. E., Cohen, L., Feldman, U., and Doschek, G. A. (1976), Astrophys. J. **203**, 521.

Bromage, G. E., Cowan, R. D., and Fawcett,B. C. (1978), Mon. Not. R. Astron. Soc. **183**, 19.

Burton, W. M., Ridgeley, A., and Wilson, R. (1967), Mon. Not. R. Astron. Soc. **135**, 207.

Doschek, G. A., Feldman, U., Van Hoosier, M. E., and Bartoe, J.-D. F. (1976), Astrophys. J. Suppl. Ser. **31**, 417.

Fawcett, B. C. (1971), J. Phys. **B4**, 1577.

Fawcett, B. C., Cowan, R. D., Kononov, E. Y., and Hayes, R. W. (1972), J. Phys. **B5**, 1255.

Gabriel, A. H., Garton, W. R. S., Goldberg, L., Jones, T. J. L., Jordan, C., Morgan, F. J., Nicholls, R. W., Parkinson, W. J., Paxton, H. J. B., Reeves, E. M., Shenton, C.B., Speer, R. J., and Wilson, R. (1971), Astrophys. J. **169**, 595.

Jefferies, J. T. (1969), Mem. Soc. Roy. Sci. Liege, Collect. **17**, 213.

Lotz, W., (1967), J. Opt. Soc. Am. **57**, 873.

Svensson, L. A. (1971), Solar Phys. **18**, 232.

Fe XII

Configuration	Term	J	Level (cm^{-1})	Leading percentages			
$3s^2\,3p^3$	$^4S^\circ$	3/2	*0*				
$3s^2\,3p^3$	$^2D^\circ$	3/2	*41 566*				
		5/2	*46 075*				
$3s^2\,3p^3$	$^2P^\circ$	1/2	*74 109*				
		3/2	*80 515*				
$3s3p^4$	^4P	5/2	274 373	89		9	$3p^2(^3P)3d\ ^4P$
		3/2	284 005	89		9	
		1/2	288 307	88		9	
$3s3p^4$	^2D	3/2	340 020	78		16	$3p^2(^1D)3d\ ^2D$
		5/2	341 703	79		16	
$3s3p^4$	^2P	3/2	389 706	48		44	$3p^2(^3P)3d\ ^2P$
$3s3p^4$	^2S	1/2	394 120	41		27	^2P
$3s3p^4$		3/2	501 800	41	^2P	26	$3p^2(^1D)3d\ ^2P$

Fe XII—Continued

Configuration	Term	J	Level (cm^{-1})	Leading percentages			
$3s^2 3p^2(^3P)3d$	^4P	5/2	512 510	86		8	$3s3p^4\ ^4P$
		3/2	516 740	82		8	
		1/2	519 770	80		8	
$3s3p^4$		1/2	513 850	35	^2P	32	$3s^2 3p^2(^1D)3d\ ^2P$
$3s^2 3p^2(^1S)3d$	^2D	3/2	526 120	46		40	$(^3P)\ ^2D$
		5/2	538 040	41		43	
$3s^2 3p^2(^1D)3d$	^2D	3/2	554 030	78		14	$3s3p^4\ ^2D$
		5/2	554 610	71		17	$3p^2(^1S)3d\ ^2D$
$3s^2 3p^2(^1D)3d$	^2P	1/2	568 940	58		24	$(^3P)\ ^2P$
		3/2	577 740	61		26	
$3s^2 3p^2(^3P)3d$	^2F	5/2	576 740	49		35	$(^1D)\ ^2F$
		7/2	581 180	61		37	
$3s^2 3p^2(^1D)3d$	^2S	1/2	579 630	72		14	$3s3p^4\ ^2S$
$3s^2 3p^2(^3P)3d$	^2D	5/2	603 930	47		31	$(^1S)\ ^2D$
		3/2	605 480	55		41	
$3s^2 3p^2(^3P)4s$	^4P	1/2	1 242 000				
		3/2	1 249 660				
		5/2	1 258 050				
$3s^2 3p^2(^3P)4s$	^2P	1/2	1 257 730				
		3/2	1 266 360	81		17	$(^1D)\ ^2D$
$3s^2 3p^2(^1D)4s$	^2D	5/2	1 287 700				
		3/2	1 289 060	82		16	$(^3P)\ ^2P$
$3s^2 3p^2(^3P)4d$	^4P	5/2	1 508 360	35		35	$(^3P)\ ^4F$
		3/2	1 517 340	65		19	$(^3P)\ ^2P$
$3s^2 3p^2(^3P)4d$	^4F	5/2	1 514 070	49		48	$(^3P)\ ^4P$
$3s^2 3p^2(^3P)4d$	^2F	5/2	1 516 030	77		9	$(^3P)\ ^4F$
		7/2	1 523 140	39		12	$(^1D)\ ^2F$
$3s^2 3p^2(^3P)4d$	^4D	7/2	1 532 160	48		48	$(^3P)\ ^2F$
$3s^2 3p^2(^3P)4d$	^2D	5/2	1 534 990	67		26	$(^1D)\ ^2F$
		3/2	1 536 480				
$3s^2 3p^2(^1D)4d$	^2F	7/2	1 549 250	81		14	$(^3P)\ ^4D$
		5/2	1 551 400?	40		39	$(^1D)\ ^2D$
$3s^2 3p^2(^1D)4d$	^2D	5/2	1 551 640	40		24	$(^3P)\ ^2D$
$3s^2 3p^2(^1D)4d$	^2P	3/2	1 565 720				
$3s^2 3p^2(^1D)4d$	^2S	1/2	1 569 410				
Fe XIII ($^3P_0$)	*Limit*		**2 668 000**				

Fe XIII

$Z=26$

Si I isoelectronic sequence

Ground state: $1s^22s^22p^63s^23p^2\,{}^3P_0$

Ionization energy = 2 912 000±6000 cm^{-1} (361.0±0.7 eV)

The ^{3}P and ^{1}D terms of the $3s^23p^2$ configurations are from solar coronal line identifications by Edlén (1942). We used the improved wavelength measurements given by Jefferies (1969). The level uncertainty is ±0.4 cm^{-1}. The ^{1}S term is determined by Doschek et al. (1976) with an uncertainty of ±1 cm^{-1}. From laboratory observations, Fawcett (1971) interpreted the $3s^23p^2-3s3p^3$ array except for the $^3P_0-{}^3D_1^\circ$ transition, which was established through a solar identification by Widing, Sandlin, and Cowan (1971). Fawcett also identified the $3s^23p3d$ configuration.

Behring, Cohen, Feldman, and Doschek (1976) observed the $3s^23p^2-3s3p^3$ and $3s^23p3d$ transition arrays with improved accuracy (±0.004 Å) in the spectrum of the quiet sun in the region from 200–360 Å by using a rocket-borne spectrograph. The level values are mostly from their observations and have an uncertainty of ±10 cm^{-1}. They also identified the $3s3p^3\,{}^3P_1^\circ$ level. The leading percentages for $3s3p^3$ and $3s^23p3d$ are taken from the calculations of Bromage, Cowan, and Fawcett (1978), who also identified the $3p3d\,{}^3P_0^\circ$ level. Configuration interaction between them was included.

The level values for $3p4d$ and the level values and leading percentages for $3p4f$ are from Kastner, Swartz, Bhatia, and Lapides (1978). The uncertainty of the level values is ±100 cm^{-1}.

The configurations $3s^23p4s$, $4p$, $4d$, and $4f$ were found also by Fawcett, Cowan, Kononov, and Hayes (1972) from observations in the region 60–100 Å. Some of the lines they identified involve transitions to the unknown ^{3}F° term of $3s^23p3d$ from $3s^23p4f$ levels, and therefore cannot be used to establish connected levels of the latter configuration. In the same paper percentage compositions are given for levels whose first component is not "high," generally less than 90%. The uncertainty of the level values is ±200 cm^{-1}.

The ionization energy is an extrapolated value by Lotz (1967).

References

Behring, W. E., Cohen, L., Feldman, U., and Doschek, G. A. (1976), Astrophys. J. **203**, 521.

Bromage, G. E., Cowan, R. D., and Fawcett, B. C. (1978), Mon. Not. R. Astron. Soc. **183**, 19.

Doschek, G. A., Feldman, U., Van Hoosier, M. E., and Bartoe, J.-D. F. (1976), Astrophys. J Suppl. Ser. **31**, 417.

Edlén, B. (1942), Z. Astrophys. **22**, 30.

Fawcett, B. C. (1971), J. Phys. **B4**, 1577.

Fawcett, B. C., Cowan, R. D., Kononov, E. Y., and Hayes, R. W. (1972), J. Phys. **B5**, 1255.

Jefferies, J. T. (1969), Mem. Soc. Roy. Sci. Liege, Collect. **17**, 213.

Kastner, S. O., Swartz, M., Bhatia, A. K., and Lapides, J. (1978), J. Opt. Soc. Am. **68**, 1558.

Lotz, W. (1967), J. Opt. Soc. Am. **57**, 873.

Widing, K. G., Sandlin, G. D., and Cowan, R. D. (1971), Astrophys. J. **169**, 405.

Fe XIII

Configuration	Term	J	Level (cm^{-1})	Leading percentages		
$3s^2\,3p^2$	^{3}P	0	0.0			
		1	9 302.5			
		2	18 561.0			
$3s^2\,3p^2$	^{1}D	2	48 068			
$3s^2\,3p^2$	^{1}S	0	91 508			
$3s3p^3$	^{3}D°	1	*287 205*	85	10	$3p3d\,{}^3$D°
		2	*287 360*	83	10	
		3	*290 210*	89	10	
$3s3p^3$	^{3}P°	1	*329 647*	86	9	$3p3d\,{}^3$P°
		2	*330 279*	78	9	
$3s3p^3$	^{1}D°	2	*362 330*	54	39	$3p3d\,{}^1$D°
$3s3p^3$	^{3}S°	1	*415 462*	78	17	^{1}P°

Fe XIII—Continued

Configuration	Term	J	Level (cm^{-1})	Leading percentages		
$3s3p^3$	$^1P°$	1	*438 050*	69	19	$^3S°$
$3s^2\,3p3d$	$^3P°$	2	*486 358*	43	23	$3p3d\ ^1D°$
		1	*494 942*	49	38	$3p3d\ ^3D°$
		0	*503 340*	89	9	$3s3p^3\ ^3P°$
$3s^2\,3p3d$	$^1D°$	2	*498 870*	32	21	$3s3p^3\ ^1D°$
$3s^2\,3p3d$	$^3D°$	1	*506 502*	48	39	$3p3d\ ^3P°$
		3	*509 176*	86	10	$3s3p^3\ ^3D°$
		2	*509 250*	61	24	$3p3d\ ^3P°$
$3s^2\,3p3d$	$^1F°$	3	*556 870*	97		
$3s^2\,3p3d$	$^1P°$	1	*570 690*	86	10	$3s3p^3\ ^1P°$
$3s^2\,3p4s$	$^3P°$	1	*1 336 220*	84	16	$^1P°$
		2	*1 354 680*			
$3s^2\,3p4s$	$^1P°$	1	*1 361 830*	84	16	$^3P°$
$3s^2\,3p4p$	^1D	2	1 488 110	86	13	^3P
$3s^2\,3p4p$	^1P	1	1 515 260?	39	35	^3D
$3s^2\,3p4p$	^3D	1	1 603 770	59	39	^1P
$3s^2\,3p4d$	$^3D°$	2	*1 604 220*	61	17	$^3F°$
		3	*1 606 800*	45	41	$^3F°$
$3s^2\,3p4d$	$^3F°$	3	*1 619 600*			
$3s^2\,3p4d$	$^1F°$	3	*1 630 650*	84	12	$^3F°$
$3s^2\,3p4d$	$^1P°$	1	*1 650 620*			
$3s^2\,3p4f$	^1D	2	*1 740 800*			
$3s^2\,3p4f$	^3F	4	1 741 290	56	44	^3G
$3s^2\,3p4f$	^1G	4	1 743 460			
Fe XIV ($^2P°_{1/2}$)	***Limit***		**2 912 000**			

Fe XIV

$Z=26$

Al I isoelectronic sequence

Ground state: $1s^2 2s^2 2p^6 3s^2 3p\ ^2P^\circ_{1/2}$

Ionization energy = 3 163 000±6000 cm^{-1} (392.2±0.8 eV)

The early laboratory work on this one-electron spectrum consisted of Edlén's (1936) identification of the $3p-4d$ resonance lines at about 59 Å. Later Edlén (1942) identified the strong solar coronal line at 5302.86 Å as the transition between the $3s^2 3p\ ^2P^\circ_{1/2}$ and $^2P^\circ_{3/2}$ levels. This identification serves as the basis for the ground term splitting in this ion. The uncertainty is ±0.2 cm^{-1}.

The first excited configuration in Fe XIV, $3s3p^2$, was identified from the laboratory observations of Fawcett and Peacock (1967) and by Fawcett (1970). The $3s^2 3d$ levels were found by Peacock, Cowan, and Sawyer (1967). The level values for these two configurations are derived from the more accurate solar observations of Behring, Cohen, Feldman, and Doschek (1976) in the region 210–290 Å, which give an uncertainty of ±20 cm^{-1}. The $3s^2 4s$, $3s^2 4p$, and $3s^2 4f$ levels are derived from the work of Fawcett, Cowan, Konovov, and Hayes (1972) with a level uncertainty of ±200 cm^{-1}.

The $3s^2 4d$ levels are obtained from Edlén (1936) with an uncertainty of ±100 cm^{-1}.

Fawcett (1970) has classified the $3s3p^2\ ^4P-3p^3\ ^4S^\circ$ group and Fawcett et al. (1972) reported the observation of $3s3p^2\ ^4P-3s3p4s\ ^4P^\circ$. The quartet level uncertainty is ±200 relative to the calculated position of $3s3p^2\ ^4P_{1/2}$ by Fischer (1978).

We calculated the percentage composition of the levels of $3s3p^2$ and $3s^2 3d$ with configuration interaction. Unlike the analogous configuration $2s2p^2$ of Fe XXII, the $^2S^\circ$ state of $3s3p^2$ remains below ^2P throughout the iron period.

The ionization energy is the value given by Lotz (1967) from his isoelectronic extrapolation.

References

Behring, W. E., Cohen, L., Feldman, U., and Doschek, G. A. (1976), Astrophys. J. **203**, 521.

Edlén, B. (1936), Z. Phys. **103**, 536.

Edlén, B. (1942), Z. Astrophys. **22**, 30.

Fawcett, B. C., and Peacock, N. J. (1967), Proc. Phys. Soc. (London) **91**, 973.

Fawcett, B. C. (1970), J. Phys. **B3**, 1732.

Fawcett, B. C., Cowan, R. D., Kononov, E. Y., and Hayes, R. W. (1972), J. Phys. **B5**, 1255.

Fischer, C. F. (1978), Penn. State Univ. Dept. Computer Sci., CS-78-24.

Lotz, W. (1967), J. Opt. Soc. Am. **57**, 873.

Peacock, N. J., Cowan, R. D., and Sawyer, G. A., Proc. 7th Int. Conf. on Ionization Phenomena in Gases, Belgrade, 1967 (Belgrade: Gradevinska Knjiga).

Fe XIV

Configuration	Term	J	Level (cm^{-1})	Leading percentages		
$3s^2 3p$	$^2P^\circ$	1/2	*0.0*			
		3/2	*18 852.5*			
$3s3p^2$	^4P	1/2	221 700+x	99	1	^2S
		3/2	222 986+x	100		
		5/2	239 450+x	98	2	^2D
$3s3p^2$	^2D	3/2	299 248	88	11	$3s^2 3d\ ^2D$
		5/2	301 472	87	11	
$3s3p^2$	^2S	1/2	364 693	75	24	^2P
$3s3p^2$	^2P	1/2	388 510	76	24	^2S
		3/2	396 515	99		
$3s^2 3d$	^2D	3/2	473 227	89	11	$3s3p^2\ ^2D$
		5/2	475 217	89	11	
$3p^3$	$^4S^\circ$	3/2	*586 130+x*			
$3s^2 4s$	^2S	1/2	1 435 020			
$3s^2 4p$	$^2P^\circ$	1/2	*1 568 840*			
		3/2	*1 574 010*			

Fe XIV—Continued

Configuration	Term	J	Level (cm^{-1})	Leading percentages
$2s3p(^3P^\circ)4s$	$^4P^\circ$	$3/2$	*1 662 920+x*	
		$5/2$	*1 675 450+x*	
$3s^2\,4d$	^2D	$3/2$	1 695 980	
		$5/2$	1 697 290	
$3s^2\,4f$	$^2F^\circ$	$7/2$	*1 788 380*	
		$5/2$	*1 788 640*	
Fe XV ($^1S_0$)	***Limit***		**3 163 000**	

Fe XV

$Z = 26$

Mg I isoelectronic sequence

Ground state: $1s^2 2s^2 2p^6 3s^2\ ^1S_0$

Ionization energy = 3 686 000±20 000 cm^{-1} (457.0±2.5 eV)

Edlén's (1936) work on this spectrum consisted of the observation of the multiplets $3s^2\ ^1S - 3s4p\ ^1P^\circ$, $3s3p\ ^3P^\circ - 3s4d\ ^3D$, $3s3d\ ^3D - 3s4f\ ^3F^\circ$ and $3s3d\ ^3D - 3s5f\ ^3F^\circ$. These groups were unconnected and the relative positions of the terms were estimated by isoelectronic extrapolation.

The absolute energy of the system of excited triplet levels is based on the measurement of the transition $3s^2\ ^1S_0 - 3s3p\ ^3P_1^\circ$ at 417.258 Å in the solar spectrum by Behring, Cohen, Feldman, and Doschek (1976). The error in this wavelength is stated to be < 0.01 Å, giving an error in the intersystem connection of ±5 cm^{-1}. The error in the levels due to Edlén is about ±100 cm^{-1}.

Peacock, Cowan, and Sawyer (1967) gave the classification of the $3s3p - 3s3d$ array and the $3s^2\ ^1S_0 - 3s3p\ ^1P_1^\circ$ resonance line measured in the range of 224–284 Å with an accuracy of ±0.05 Å. Several of these lines are found in the list of solar lines of Behring et al. (1976) with improved accuracy and are used here. They are $3s^2\ ^1S_0 - 3s3p\ ^1P_1^\circ$ and $3s3p - 3s3d\ ^1P_1^\circ - ^1D_2$, $^3P_1^\circ - ^3D_2$, and $^3P_2^\circ - ^3D_3$ at 284.160 Å, 243.790 Å, 227.208 Å, and 233.857 Å, respectively. The resulting level values have an uncertainty of ±20 cm^{-1}.

The values of the levels of the $3p^2$ configuration are derived with an uncertainty of ±20 cm^{-1} from the classifications of Fawcett (1971) and Fawcett, Cowan, and Hayes (1972). The $3p^2\ ^1S_0$ level has been tentatively located by Cowan and Widing (1973). This identification is disputed by Kastner and Bhatia (1979) on the basis of an isoelectronic study. The observations by Fawcett (1970) provide values for the three levels of the $3p3d\ ^3F^\circ$ term.

The measurements of Edlén were used to obtain the values of the $3s4f\ ^3F^\circ$, $3s5d\ ^3D$, and $3s5f\ ^3F^\circ$ levels.

The levels of $3s4s\ ^3S$ and $3s4d\ ^3D$ given here are derived from the work of Feldman, Katz, Behring, and Cohen (1971). The $3s4d\ ^1D$ and $3s4f\ ^1F^\circ$ levels are taken from the paper of Fawcett, Cowan, Kononov, and Hayes (1972).

The $3s5s\ ^3S$, $3s5p\ ^1D^\circ$, $3s5f\ ^1F^\circ$, and $3s6f\ ^{1,3}F^\circ$ levels are from Fawcett, Gabriel, Irons, Peacock, and Saunders (1966). Ekberg (1971) has noted that the $3s5f\ ^1F^\circ$ and $3s6f\ ^1F^\circ$ levels are questionable. We have not included them here.

The transitions $3p3d - 3p4f$ are classified by Kastner, Swartz, Bhatia, and Lapides (1978), who gave percentages for $3p4f$.

With observations of L-series satellite spectra, Burkhalter, Cohen, Cowan, and Feldman (1979) have made some tentative assignments to $2p^6 3s3p - 2p^5 3s^2 3p$ transitions. They are not quoted here.

We derived the ionization energy from the $3snf\ ^3F_4^\circ$ ($n = 4$–6) levels.

References

Behring, W. E., Cohen, L., Feldman, U., and Doschek, G. A. (1976), Astrophys. J. **203**, 521.

Burkhalter, P.G., Cohen, L., Cowan, R. D., and Feldman, U. (1979), J. Opt. Soc. Am. **69**, 1133.

Cowan, R. D., and Widing, K. C. (1973), Astrophys. J. **180**, 285.

Edlén, B. (1936), Z. Phys. **103**, 536.

Ekberg, J. O. (1971), Phys. Scr. **4**, 101.

Fawcett, B. C. (1970), J. Phys. **B3**, 1732.

Fawcett, B. C. (1971), J. Phys. **B4**, 1577.

Fawcett, B. C., Cowan, R. D., and Hayes, R. W. (1972), J. Phys. **B5**, 2143.

Fawcett, B. C., Cowan, R. D., Kononov, E. Y., and Hayes, R. W. (1972), J. Phys. **B5**, 1255.

Fawcett, B. C., Gabriel, A. H., Irons, F. E., Peacock, N. J., and Saunders, P. A. H. (1966), Proc. Phys. Soc. **88**, 1051.

Feldman, U., Katz, L.,Behring, W., and Cohen L. (1971), J. Opt. Soc. Am. **61**, 91.

Kastner, S. O., Swartz, M., Bhatia, A. K., and Lapides, J. (1978), J. Opt. Soc. Am. **68**, 1558.

Kastner,S. O., and Bhatia, A. K. (1979), J. Opt. Soc. Am. **69**, 1391.

Peacock, N. J., Cowan, R. D., and Sawyer, G. A., Proc. 7th Int. Conf. on Ionization Phenomena in Gases, Belgrade, 1967 (Belgrade: Gradevinska Knjiga).

Fe XV

Configuration	Term	J	Level (cm^{-1})	Leading percentages
$3s^2$	^1S	0	0	
$3s3p$	$^3P^\circ$	0	*233 910*	
		1	*239 660*	
		2	*253 820*	
$3s3p$	$^1P^\circ$	1	*351 914*	

Fe XV—Continued

Configuration	Term	J	Level (cm^{-1})	Leading percentages		
$3p^2$	^{3}P	0	554 500			
		1	564 570			
		2	581 690			
$3p^2$	^{1}D	2	559 590			
$3p^2$	^{1}S	0	660 970?			
$3s3d$	^{3}D	1	678 830			
		2	679 785			
		3	681 410			
$3s3d$	^{1}D	2	762 103			
$3p3d$	^{3}F°	2	*928 420*			
		3	*938 180*			
		4	*949 660*			
$3s4s$	^{3}S	1	1 763 700			
$3s4p$	^{1}P°	1	*1 889 970*			
$3s4d$	^{3}D	1	2 031 310			
		2	2 032 020			
		3	2 033 180			
$3s4d$	^{1}D	2	2 035 280			
$3s4f$	^{3}F°	2	*2 108 520*			
		3	*2 108 620*			
		4	*2 108 880*			
$3s4f$	^{1}F°	3	*2 123 150*			
$3p4f$	^{3}G	3	2 380 160	46	51	^{1}F
		4	2 386 700	57	39	^{3}F
		5	2 402 100			
$3s5s$	^{3}S	1	2 544 800			
$3s5p$	^{1}P°	1	*2 567 000*			
$3s5d$	^{3}D	2	2 639 900			
		1	2 640 100			
		3	2 640 300			
$3s5f$	^{3}F°	2	*2 676 400*			
		3	*2 676 400*			
		4	*2 676 600*			
$3s5f$	^{1}F°	3	*2 782 700?*			
$3s6f$	^{3}F°	4	*2 986 100*			
$3s6f$	^{1}F°	3	*3 091 500?*			
Fe XVI ($^2S_{1/2}$)	***Limit***		**3 686 000**			

Fe XVI

$Z=26$

Na I isoelectronic sequence

Ground state: $1s^2 2s^2 2p^6 3s$ $^2S_{1/2}$

Ionization energy = 3 946 280±300 cm^{-1} (489.262±0.040 eV)

The original work on this spectrum by Edlén (1936) in the region 40–66 Å gave the position of only the $4p$ $^2P°$ term relative to the ground state, but included transitions from $4s$, $4d$, and $5d$ to the $3p$ $^2P°$ term, and transitions from $4f$ and $5f$ to $3d$ ^2D. Peacock, Cowan, and Sawyer (1967) brought these unconnected systems together by identifying the $3s-3p$ and $3p-3d$ transitions in the region 250–360 Å. New observations by Peacock, Stamp, and Silver (1984) of the $3s-3p$ doublet in a tokamak plasma were reported at 335.410 and 360.761 Å. We used these and the measurements of the $3p-3d$ doublet by Behring, Cohen, Feldman, and Doschek (1976) in a solar spectrum. The $4s$, $4p$, $4d$, $4f$, $5d$, and $5f$ terms are from the identifications and measurements of Edlén (1936). The uncertainty of the $n=3$ levels is ±5 cm^{-1} and the higher ones ±200 cm^{-1}.

Additional series members of ns, np, nd, and nf were identified by Fawcett, Gabriel, Irons, Peacock, and Saunders (1966) between 27 and 64 Å. The values of the $5p$, $6p$, $6d$, $7d$, and $8d$ terms given below were derived by using the wavelengths of Feldman et al. (1971). Their uncertainty is ±800 cm^{-1}. The $5g$ term is from Kononov, Kovalev, Ryabtsev, and Churilov (1977).

The $2p^5 3s\,nl$ configurations are from classifications of L-series satellite spectra by Burkhalter, Cohen, Cowan, and Feldman (1979). The leading percentages, in two alternate coupling schemes, are taken from this paper. Other inner shell transitions have been tentatively identified by Connerade, Peacock, and Speer (1970).

The ionization energy was determined from a polarization formula applied to the nf series by Edlén (1978).

References

Behring, W. E., Cohen, L., Feldman, U., and Doschek, G. A. (1976), Astrophys. J. **203**, 521.

Burkhalter, P. G., Cohen, L., Cowan, R. D., and Feldman, U. (1979), J. Opt. Soc. Am. **69**, 1133.

Connerade, J. P., Peacock, N. J., and Speer, R. J. (1970), Solar Phys. **14**, 159.

Edlén, B. (1936), Z. Phys. **100**, 621.

Edlén, B. (1978), Phys. Scr. **17**, 565.

Fawcett, B. C., Gabriel, A. H., Irons, F. E., Peacock, N. J., and Saunders, P. A. H. (1966), Proc. Phys. Soc. **88**, 1051.

Feldman, U., Katz, L., Behring, W., and Cohen, L. (1971), J. Opt. Soc. Am. **61**, 91.

Kononov, E. Y., Kovalev, V. I., Ryabtsev, A. N., and Churilov, S. S. (1977), Sov. J. Quan. Elec. **7**, 111.

Peacock, N. J., Cowan, R. D., and Sawyer, G. A., Proc. 7th Int. Conf. on Ionization Phenomena in Gases, Belgrade, 1967 (Belgrade: Gradevinska Knjiga).

Peacock, N. J., Stamp, M. F., and Silver, J. D. (1984), Phys. Scr. **T8**, 10.

Fe XVI

Configuration	Term	J	Level (cm^{-1})	Leading percentages
$2p^6(^1S)3s$	^2S	1/2	0	
$2p^6(^1S)3p$	$^2P°$	1/2	*277 192*	
		3/2	*298 143*	
$2p^6(^1S)3d$	^2D	3/2	675 481	
		5/2	678 394	
$2p^6(^1S)4s$	^2S	1/2	1 867 540	
$2p^6(^1S)4p$	$^2P°$	1/2	*1 978 040*	
		3/2	*1 986 100*	
$2p^6(^1S)4d$	^2D	3/2	2 124 190	
		5/2	2 125 360	
$2p^6(^1S)4f$	$^2F°$	5/2	*2 184 620*	
		7/2	*2 185 150*	
$2p^6(^1S)5s$	^2S	1/2	2 662 000	

Fe XVI—Continued

Configuration	Term	J	Level (cm^{-1})	Leading percentages
$2p^6(^1S)5p$	$^2P^{\circ}$	1/2	*2 717 170*	
		3/2	*2 721 160*	
$2p^6(^1S)5d$	^2D	3/2	2 788 050	
		5/2	2 788 610	
$2p^6(^1S)5f$	$^2F^{\circ}$	5/2	*2 818 600*	
		7/2	*2 818 900*	
$2p^6(^1S)5g$	^2G	7/2,9/2	2 822 100	
$2p^6(^1S)6s$	^2S	1/2	3 076 000	
$2p^6(^1S)6p$	$^2P^{\circ}$	1/2	*3 106 400*	
		3/2	*3 108 900*	
$2p^6(^1S)6d$	^2D	3/2	3 146 070	
		5/2	3 146 670	
$2p^6(^1S)6f$	$^2F^{\circ}$	5/2	*3 163 130*	
		7/2	*3 163 190*	
$2p^6(^1S)7s$	^2S	1/2	3 323 000	
$2p^6(^1S)7p$	$^2P^{\circ}$	3/2	*3 341 000*	
$2p^6(^1S)7d$	^2D	3/2	3 360 500	
		5/2	3 360 800	
$2p^6(^1S)7f$	$^2F^{\circ}$	7/2	*3 371 070*	
		5/2	*3 371 210*	
$2p^6(^1S)8p$	$^2P^{\circ}$	3/2	*3 488 000*	
$2p^6(^1S)8d$	^2D	3/2	3 498 800	
		5/2	3 499 000	
$2p^6(^1S)8f$	$^2F^{\circ}$	5/2	*3 505 700*	
		7/2	*3 505 800*	
$2p^6(^1S)9p$	$^2P^{\circ}$	3/2	*3 587 000*	
$2p^6(^1S)9d$	^2D	5/2	3 595 000	
		3/2	3 599 000	
$2p^6(^1S)9f$	$^2F^{\circ}$	5/2,7/2	*3 600 000*	
Fe XVII ($^1S_0$)	***Limit***		**3 946 280**	
$2p^5(^2P^{\circ})3s^2$	$^2P^{\circ}$	3/2	*5 773 000*	
		1/2	*5 873 000*	
$2p^5(^2P^{\circ}_{3/2})3s3p(^3P_1)$	(3/2,1)	5/2	5 982 000	98 or 83 $2p^53s(^3P^{\circ})3p\ ^4D$
		1/2	6 001 000	92 or 35 $2p^53s(^1P^{\circ})3p\ ^2P$
$2p^5(^2P^{\circ}_{3/2})3s3p(^3P_2)$	(3/2,2)	3/2	6 013 000	69 or 30 $2p^53s(^1P^{\circ})3p\ ^2D$
		5/2	6 013 000	99 or 65 $2p^53s(^3P^{\circ})3p\ ^4P$
		1/2	6 042 000	71 or 40 $2p^53s(^1P^{\circ})3p\ ^2S$

(Continued)

Fe XVI—Continued

Configuration	Term	J	Level (cm^{-1})	Leading percentages
$2p^5(^2P^\circ_{1/2})3s3p(^3P_0)$	($\frac{1}{2}$,0)	$\frac{1}{2}$	6 075 000	68 or 65 $2p^53s(^3P^\circ)3p\ ^4D$
$2p^5(^2P^\circ_{1/2})3s3p(^3P_1)$	($\frac{1}{2}$,1)	$\frac{3}{2}$	6 089 000	88 or 50 $2p^53s(^3P^\circ)3p\ ^4D$
$2p^5(^2P^\circ_{3/2})3s3p(^1P_1)$	($\frac{3}{2}$,1)	$\frac{1}{2}$	6 089 000	44 or 39 $2p^53s(^3P^\circ)3p\ ^4P$
		$\frac{5}{2}$	6 110 000	57 or 69 $2p^53s(^1P^\circ)3p\ ^2D$
		$\frac{3}{2}$	6 129 000	94 or 40 $2p^53s(^3P^\circ)3p\ ^2P$
$2p^5(^2P^\circ_{1/2})3s3p(^3P_2)$	($\frac{1}{2}$,2)	$\frac{3}{2}$	6 096 000	86 or 45 $2p^53s(^3P^\circ)3p\ ^4P$
$2p^5(^2P^\circ_{1/2})3s3p(^1P_1)$	($\frac{1}{2}$,1)	$\frac{1}{2}$	6 197 000	47 or 55 $2p^53s(^3P^\circ)3p\ ^2P$
$2p^5(^2P^\circ_{1/2})3s3p(^1P_1)$	($\frac{1}{2}$,1)	$\frac{3}{2}$	6 217 000	98 or 59 $2p^53s(^3P^\circ)3p\ ^2D$
		$\frac{1}{2}$	6 267 000	50 or 80 $2p^53s(^3P^\circ)3p\ ^2S$
$2p^5(^2P^\circ_{3/2})3s3d(^3D_3)$	($\frac{3}{2}$,3)°	$\frac{5}{2}$	*6 393 000*	50 or 67 $2p^53s(^3P^\circ)3d\ ^4P^\circ$
		$\frac{7}{2}$	*6 422 000*	100 or 66 $2p^53s(^3P^\circ)3d\ ^4D^\circ$
		$\frac{3}{2}$	*6 436 000*	45 or 57 $2p^53s(^1P^\circ)3d\ ^2P^\circ$
$2p^5(^2P^\circ_{3/2})3s3d(^3D_1)$	($\frac{3}{2}$,1)°	$\frac{5}{2}$	*6 406 000*	83 or 54 $2p^53s(^3P^\circ)3d\ ^4F^\circ$
		$\frac{3}{2}$	*6 419 000*	60 or 40 $2p^53s(^3P^\circ)3d\ ^4F^\circ$
$2p^5(^2P^\circ_{3/2})3s3d(^3D_2)$	($\frac{3}{2}$,2)°	$\frac{1}{2}$	*6 423 000*	45 or 61 $2p^53s(^1P^\circ)3d\ ^2P^\circ$
		$\frac{5}{2}$	*6 425 000*	64 or 39 $2p^53s(^1P^\circ)3d\ ^2F^\circ$
$2p^5(^2P^\circ_{3/2})3s3d(^1D_2)$	($\frac{3}{2}$,2)°	$\frac{7}{2}$	*6 445 000*	100 or 72 $2p^53s(^3P^\circ)3d\ ^2F^\circ$
		$\frac{5}{2}$	*6 464 000*	95 or 50 $2p^53s(^3P^\circ)3d\ ^2D^\circ$
		$\frac{3}{2}$	*6 473 000*	44 or 35 $2p^53s(^3P^\circ)3d\ ^4D^\circ$
$2p^5(^2P^\circ_{1/2})3s3d(^3D_3)$	($\frac{1}{2}$,3)°	$\frac{5}{2}$	*6 502 000*	54 or 41 $2p^53s(^3P^\circ)3d\ ^4D^\circ$
		$\frac{7}{2}$	*6 517 000*	98 or 41 $2p^53s(^1P^\circ)3d\ ^2F^\circ$
$2p^5(^2P^\circ_{1/2})3s3d(^3D_1)$	($\frac{1}{2}$,1)°	$\frac{3}{2}$	*6 502 000*	88 or 54 $2p^53s(^3P^\circ)3d\ ^4F^\circ$
		$\frac{1}{2}$	*6 574 000*	74 or 52 $2p^53s(^3P^\circ)3d\ ^2P^\circ$
$2p^5(^2P^\circ_{1/2})3s3d(^3D_2)$	($\frac{1}{2}$,2)°	$\frac{5}{2}$	*6 516 000*	51 or 45 $2p^53s(^1P^\circ)3d\ ^2D^\circ$
		$\frac{3}{2}$	*6 530 000*	46 or 38 $2p^53s(^1P^\circ)3d\ ^2P^\circ$
$2p^5(^2P^\circ_{1/2})3s3d(^1D_2)$	($\frac{1}{2}$,2)°	$\frac{5}{2}$	*6 556 000*	97 or 51 $2p^53s(^3P^\circ)3d\ ^2F^\circ$
		$\frac{3}{2}$	*6 595 000*	61 or 68 $2p^53s(^3P^\circ)3d\ ^2P^\circ$

Fe XVII

$Z = 26$

Ne I isoelectronic sequence

Ground state: $1s^2 2s^2 2p^6\ {}^1S_0$

Ionization energy = 10 180 000±8000 cm^{-1} (1262.2±1.0 eV)

Tyrén (1938) identified resonance lines in the region of 12–17 Å arising from the $2p^5 3s$, $2p^5 3d$, $2p^5 4d$, and $2s2p^6 3p$ configurations. The magnetic quadrupole transition from the $2p^5 3s\ {}^3P^\circ_2$ level to the ground state was first observed by Parkinson (1973). New laboratory observations in the region 10–17 Å were reported by Gordon, Hobby, and Peacock (1980), who identified resonance transitions from $2p^5 3s - 4s$, $2p^5 3d - 6d$, and $2s2p^6 4p$. Their results, with a reported wavelength accuracy of ±0.005 Å are used to obtain the odd parity $J = 1$ levels with an uncertainty of ±3000 cm^{-1}. From solar coronal observations Hutcheon, Pye, and Evans (1976) identified resonance lines from $2p^5 5s$, $6s$, $7s$, $5p$, and $8d$ and obtained the value for the ionization energy quoted here. Their measurement uncertainty is given as ±0.003 Å, giving a level of uncertainty of ±2000 cm^{-1}.

The $2p^5 3s - 2p^5 3p$ and $3d$ transition arrays were identified in published solar flare spectra by Jupén and Litzén (1984). Improved measurements of some of the lines and a value for the $2p^5 3s\ {}^3P^\circ_1 - {}^3P^\circ_0$ magnetic-dipole transition were obtained by Feldman, Doschek, and Seely (1985) by remeasuring the spectroheliograms. A self-consistent set of levels for these three configurations are derived from these identifications relative to the lowest level of $2p^5 3s$ with an uncertainty of 50 cm^{-1}. Their positions relative to the $2p^6\ {}^1S_0$ ground state is determined from the $2p^6 - 2p^5 3s$ resonance line measurements of Gordon, Hobby, and Peacock, giving an uncertainty relative to the ground state of ±3000 cm^{-1}. The $J = 1$ levels of $2p^5 3d$ are also obtained from this paper. The percentage compositions of these levels are from Jupén and Litzén, with some reordering of level designations.

Classifications in the $2p^5 3d - 2p^5 4f$ array at 59 Å by Fawcett, Bromage, and Hayes (1979) could not be used to derive additional levels because they are not connected with the known system.

References

Fawcett, B. C., Bromage, G. E., and Hayes, R. W. (1979), Mon. Not. R. Astron. Soc. **186**, 113.

Feldman, U., Doschek, G. A., and Seely, J. F. (1985), Mon. Not. R. Astron. Soc. **212**, 41.

Gordon, H., Hobby, M. G., and Peacock, N. J. (1980), J. Phys. **B13**, 1985.

Hutcheon, R. J., Pye, J. P., and Evans, K. D. (1976), Astron. Astrophys. **51**, 451.

Jupén, C., and Litzén, U. (1984), Phys. Scr. **30**, 112.

Parkinson, J. H. (1973), Astron. Astrophys. **24**, 213.

Tyrén, F. (1938), Z. Phys. **111**, 314.

Fe XVII

Configuration	Term	J	Level (cm^{-1})	Leading percentages			
$2s^2 2p^6$	^1S	0	0				
$2s^2 2p^5(^2P^\circ_{3/2})3s$	$(3/2,1/2)^\circ$	2	*5 848 620*	100			
		1	*5 864 000*	98		2	$(1/2,1/2)^\circ$
$2s^2 2p^6(^2P^\circ_{1/2})3s$	$(1/2,1/2)^\circ$	0	*5 950 710*	100			
		1	*5 960 270*	98		2	$(3/2,1/2)^\circ$
$2s^2 2p^5 3p$	^3S	1	6 092 810	82		17	^3P
$2s^2 2p^5 3p$	^3D	2	6 121 010	59		28	^1D
		3	6 134 030	100			
		1	6 218 510	67		31	^1P
$2s^2 2p^5 3p$	^1P	1	6 143 140	50		26	^3D
$2s^2 2p^5 3p$	^3P	2	6 157 770	67		33	^1D
		0	6 201 860	94		6	^1S
		1	6 244 720	63		18	^1P
$2s^2 2p^5 3p$		2	6 247 750	41	^3D	39	^1D

(Continued)

Fe XVII—Continued

Configuration	Term	*J*	Level (cm^{-1})	Leading percentages			
$2s^2\,2p^5\,3p$	^{1}S	0	6 352 630	94		6	^{3}P
$2s^2\,2p^5\,3d$	^{3}P°	0	*6 463 340*	100			
		1	*6 472 500*	92		8	^{3}D°
		2	*6 485 220*	72		20	^{3}D°
$2s^2\,2p^5\,3d$	^{3}F°	4	*6 485 930*	100			
		3	*6 492 190*	65		29	^{1}F°
		2	*6 593 860*	64		27	^{1}D°
$2s^2\,2p^5\,3d$	^{1}D°	2	*6 506 060*	41		35	^{3}F°
$2s^2\,2p^5\,3d$	^{3}D°	3	*6 514 730*	64		34	^{1}F°
		1	*6 552 200*	73		22	^{1}P°
		2	*6 601 990*	48		27	^{3}P°
$2s^2\,2p^5\,3d$		3	*6 605 900*	37	^{1}F°	33	^{3}F°
$2s^2\,2p^5\,3d$	^{1}P°	1	*6 660 000*	77		19	^{3}D°
$2s2p^6\,3p$	^{3}P°	1	*7 198 900*				
$2s2p^6\,3p$	^{1}P°	1	*7 234 300*				
$2s^2\,2p^5\,(^2P^\circ_{3/2})\,4s$	$(\frac{3}{2},\frac{1}{2})^\circ$	1	*7 885 800*				
$2s^2\,2p^5\,(^2P^\circ_{1/2})\,4s$	$(\frac{1}{2},\frac{1}{2})^\circ$	1	*7 983 000*				
$2s^2\,2p^5\,4d$	^{3}P°	1	*8 116 000*				
$2s^2\,2p^5\,4d$	^{3}D°	1	*8 154 000*				
$2s^2\,2p^5\,4d$	^{1}P°	1	*8 249 000*				
$2s^2\,2p^5\,(^2P^\circ_{3/2})\,5s$	$(\frac{3}{2},\frac{1}{2})^\circ$	1	*8 757 000*				
$2s^2\,2p^5\,(^2P^\circ_{1/2})\,5s$	$(\frac{1}{2},\frac{1}{2})^\circ$	1	*8 860 000*				
$2s^2\,2p^5\,5d$	^{3}P°	1	*8 860 000*				
$2s^2\,2p^5\,5d$	^{3}D°	1	*8 887 000*				
$2s^2\,2p^5\,5d$	^{1}P°	1	*8 982 000*				
$2s2p^6\,4p$	^{3}P°	1	*9 056 000*				
$2s2p^6\,4p$	^{1}P°	1	*9 072 000*				
$2s^2\,2p^5(^2P^\circ_{3/2})6s$	$(\frac{3}{2},\frac{1}{2})^\circ$	1	*9 216 000*				
$2s^2\,2p^5(^2P^\circ_{3/2})6d$	$^2[\frac{3}{2}]^\circ$	1	*9 285 000*				
$2s^2\,2p^5(^2P^\circ_{1/2})6d$	$^2[\frac{3}{2}]^\circ$	1	*9 383 000*				
$2s^2\,2p^5(^2P^\circ_{3/2})7s$	$(\frac{3}{2},\frac{1}{2})^\circ$	1	*9 479 000*				
$2s^2\,2p^5(^2P^\circ_{3/2})7d$	$^2[\frac{3}{2}]^\circ$	1	*9 524 000*				
$2s^2\,2p^5(^2P^\circ_{1/2})7d$	$^2[\frac{3}{2}]^\circ$	1	*9 628 000*				

Fe XVII—Continued

Configuration	Term	J	Level (cm^{-1})	Leading percentages
$2s^2\,2p^5(^2P^\circ_{3/2})8d$	$^2[^3/_2]^\circ$	1	*9 690 000*	
$2s^2\,2p^5(^2P^\circ_{1/2})8d$	$^2[^3/_2]^\circ$	1	*9 784 000*	
$2s2p^6\,5p$	$^3P^\circ$	1	*9 878 000*	
$2s2p^6\,5p$	$^1P^\circ$	1	*9 878 000*	
Fe XVIII ($^2P^\circ_{3/2}$)	***Limit***		10 180 000	

Fe XVIII

$Z=26$

F I isoelectronic sequence

Ground state: $1s^2 2s^2 2p^5\ ^2P^\circ_{3/2}$

Ionization energy = 10 985 000±30 000 cm^{-1} (1362±4 eV)

The $2s^2 2p^5\ ^2P^\circ - 2s2p^6\ ^2S$ doublet was observed in a laser-produced plasma by Feldman, Doschek, Nagel, Behring, and Cohen (1973). Kovalev (1981) remeasured these wavelengths and obtained the values 93.923 Å and 103.937 Å with an accuracy of ±0.004 Å. They are published by Kononov (1983), who found a systematic shift in the measurements of Feldman et al. of up to 0.02 Å. We use this result to obtain the $2s2p^6\ ^2S$ level with an uncertainty of ±50 cm^{-1}. The $^2P^\circ$ ground term splitting was measured directly from a magnetic dipole transition in a tokamak discharge by Hinnov et al. (1982). An improved value of 974.86±0.02 was obtained by Peacock, Stamp, and Silver (1984) giving an uncertainty for the ^2P interval of ±2 cm^{-1}.

The $2p^4 3s$ and $2p^4 3d$ levels were derived by Feldman, Doschek, Cowan, and Cohen (1973) from measurements at 14–16 Å. The range of observation was extended to 10 Å by Gordon, Hobby, and Peacock (1980), who also obtained the $2p^4 4s$, $2p^4 4d$, and $2s2p^5 3p$ levels. Their measurements, with a reported accuracy of ±0.005 Å, are used to determine these configurations. The uncertainty in the levels values is about ±3000 cm^{-1}. All of the leading percentages are taken from Gordon et al.

The $2p^4 5d$ and $6d$ levels are from observations of exploding wires at 10 Å by Burkhalter, Dozier, Stallings, and Cowan (1978) with a wavenumber uncertainty of ±5000 cm^{-1}. These levels are designated by the authors in *LS* coupling, although *Jj* coupling is probably more appropriate. We include designations from both coupling schemes.

The $1s2s^2 2p^6\ ^2S$ level was obtained from the *x*-ray observations of Fraenkel and Schwob (1972). An uncertainty of ±30 000 cm^{-1} is estimated for this level.

We derived the value for the ionization energy from the $2p^4(^1D)nd\ ^2D_{3/2}$ series for n = 3–5.

Reference

Burkhalter, P. G., Dozier, C. M., Stallings, E., and Cowan, R. D. (1978), J. Appl. Phys. **49**, 1092.

Feldman, U., Doschek, G. A., Cowan, R. D., and Cohen L. (1973), J. Opt. Soc. Am. **63**, 1445.

Feldman, U., Doschek, G. A., Nagel, D. J., Behring, W. E., and Cohen, L. (1973), Astrophys. J. **183**, L43.

Fraenkel, B. S., and Schwob, J. L. (1972), Phys. Lett. **40A**, 83.

Gordon, H., Hobby, M. G., and Peacock, N. J. (1980), J. Phys. **B13**, 1985.

Hinnov, E., Suckewer, S., Cohen, S., and Sato, K. (1982), Phys. Rev. **A25**, 2293.

Kononov, E. Ya. (1983), Phys. Scr. **27**, 117.

Kovalev, V. I. (1981), Inst. for Spectrosc., Troitsk, U.S.S.R., preprint No. 8.

Peacock, N. J., Stamp, M. F., and Silver, J. D. (1984), Phys. Scr. **T8**, 10.

Fe XVIII

Configuration	Term	J	Level (cm^{-1})	Leading percentages		
$2s^2 2p^5$	$^2P^\circ$	3/2	*0*	100		
		1/2	*102 579*	100		
$2s2p^6$	^2S	1/2	1 064 702	100		
$2s^2 2p^4(^3P)3s$	^4P	5/2	6 222 000	90	9	$(^1D)\ ^2D$
		1/2	6 301 200	84	15	$(^1S)\ ^2S$
		3/2	6 317 900	68	30	$(^3P)\ ^2P$
$2s^2 2p^4(^3P)3s$	^2P	3/2	6 248 100	57	31	$(^3P)\ ^4P$
		1/2	6 342 600	90	5	$(^1S)\ ^2S$
$2s^2 2p^4(^1D)3s$	^2D	5/2	6 400 000	90	9	$(^3P)\ ^4P$
		3/2	6 403 800	85	13	$(^3P)\ ^2P$
$2s^2 2p^4(^1S)3s$	^2S	1/2	6 575 100	80	11	$(^3P)\ ^4P$
$2s^2 2p^4(^3P)3d$	^4P	1/2	6 858 200	61	22	$(^3P)\ ^2P$
		3/2	6 872 400	48	27	$(^3P)\ ^2D$
		5/2	6 903 700	42	36	$(^3P)\ ^2F$

Fe XVIII—Continued

Configuration	Term	J	Level (cm^{-1})	Leading percentages			
$2s^2\,2p^4(^3P)3d$		5/2	6 880 400	26	^2F	23	$(^3P)\ ^4P$
$2s^2\,2p^4(^3P)3d$	^4D	1/2	6 903 200	53		26	$(^3P)\ ^2P$
$2s^2\,2p^4(^3P)3d$		3/2	6 919 000	29	^4D	19	$(^3P)\ ^2P$
$2s^2\,2p^4(^3P)3d$	^2P	3/2	6 947 300	49		21	$(^1D)\ ^2P$
$2s^2\,2p^4(^3P)3d$	^2D	5/2	6 963 800	42		29	$(^3P)\ ^2F$
$2s^2\,2p^4(^1D)3d$	^2S	1/2	7 014 300	83		10	$(^3P)\ ^4P$
$2s^2\,2p^4(^1D)3d$	^2P	3/2	7 038 400	66		20	$(^3P)\ ^2P$
		1/2	7 074 200	61		34	
$2s^2\,2p^4(^1D)3d$	^2D	5/2	7 040 800	44		26	$(^3P)\ ^2D$
		3/2	7 066 100	65		28	
$2s^2\,2p^4(^1S)3d$	^2D	5/2	7 166 400	72		13	$(^3P)\ ^2D$
		3/2	7 184 300	72		13	
$2s2p^5(^3P^\circ)3p$		3/2	7 464 400	38	^4D	29	$(^3P^\circ)\ ^2P$
$2s2p^5(^3P^\circ)3p$	^2D	5/2	7 477 200	50		49	$(^3P^\circ)\ ^4D$
		3/2	7 567 000	45		22	
$2s2p^5(^3P^\circ)3p$	^2P	3/2	7 487 800	59		37	$(^3P^\circ)\ ^4D$
		1/2	7 508 100	58		16	
$2s2p^5(^3P^\circ)3p$	^4P	5/2	7 508 100				
		3/2	7 529 900	44		41	$(^3P^\circ)\ ^2D$
$2s2p^5(^3P^\circ)3p$	^2S	1/2	7 599 400	62		21	$(^3P^\circ)\ ^2P$
$2s2p^5(^1P^\circ)3p$	^2D	3/2	7 763 400	88		7	$(^1P^\circ)\ ^2P$
		5/2	7 783 900	96		2	$(^3P^\circ)\ ^4D$
$2s2p^5(^1P^\circ)3p$	^2P	1/2	7 786 000	93		2	$(^3P^\circ)\ ^2S$
		3/2	7 794 400	89		7	$(^1P^\circ)\ ^2D$
$2s^2\,2p^4(^3P_2)4s$	(2,1/2)	3/2	8 428 200	88	or	70	$(^3P)\ ^2P$
$2s^2\,2p^4(^3P_1)4s$	(1,1/2)	3/2	8 517 200	99	or	80	$(^3P)\ ^4P$
$2s^2\,2p^4(^1D_2)4s$	(2,1/2)	5/2	8 591 100	90	or	90	$(^1D)\ ^2D$
		3/2	8 593 000	88	or	88	
$2s^2\,2p^4(^3P_2)4d$	(2,5/2)	5/2	8 676 000	73	or	40	$(^3P)\ ^2D$
		3/2	8 676 000	86	or	32	
$2s^2\,2p^4(^3P_0)4d$	(0,3/2)	3/2	8 727 500	73	or	55	$(^3P)\ ^4F$
$2s^2\,2p^4(^3P_0)4d$	(0,5/2)	5/2	8 727 500	66	or	44	$(^3P)\ ^4F$
$2s^2\,2p^4(^3P_1)4d$	(1,5/2)	5/2	8 756 600	94	or	44	$(^3P)\ ^4P$
		3/2	8 759 900	68	or	51	$(^3P)\ ^2P$
$2s^2\,2p^4(^1D_2)4d$	(2,3/2)	5/2	8 829 200	82	or	50	$(^1D)\ ^2F$
		1/2	8 829 200	68	or	78	$(^1D)\ ^2S$
		3/2	8 843 900	49	or	81	$(^1D)\ ^2D$

(Continued)

Fe XVIII—Continued

Configuration	Term	J	Level (cm^{-1})	Leading percentages			
$2s^2 2p^4(^1D_2)4d$	(2,5/2)	5/2	8 829 200	86	or	54	$(^1D)\ ^2D$
		3/2	8 829 200	52	or	87	$(^1D)\ ^2P$
		1/2	8 843 900	61	or	71	$(^1D)\ ^2P$
$2s^2 2p^4(^1S_0)4d$	(0,3/2)	3/2	8 989 200	79	or	79	$(^1S)\ ^2D$
$2s^2 2p^4(^3P)5d$	^2D	5/2	9 510 000				
$2s^2 2p^4(^3P)5d$	^2F	5/2	9 610 000				
$2s^2 2p^4(^3P)5d$	^2P	3/2	9 640 000				
$2s^2 2p^4(^1D)5d$	^2P	1/2,3/2	9 680 000				
$2s^2 2p^4(^1D)5d$	^2D	3/2	9 680 000				
$2s^2 2p^4(^1D)5d$	^2F	5/2	9 680 000				
$2s^2 2p^4(^3P)6d$	^2D	5/2	9 970 000				
$2s^2 2p^4(^1D)6d$	^2P	1/2	10 120 000				
$2s^2 2p^4(^1D)6d$	^2D	3/2	10 120 000				
Fe XIX $(^3P_2)$	***Limit***		**10 985 000**				
$1s2s^2 2p^6$	^2S	1/2	51 902 000				

Fe XIX

$Z=26$

O I isoelectronic sequence

Ground state: $1s^2 2s^2 2p^4\ ^3P_2$

Ionization energy = 11 850 000±30 000 cm^{-1} (1469±4 eV)

All levels of the ground configuration except $^3P_0$ are determined from forbidden lines in solar flares. The transition $^3P_2 - ^3P_1$ at 1118.1±0.1 Å was reported by Doschek et al. (1975). The lines at 592.16 Å and 424.26 Å are classified as $^3P_2 - ^1D_2$ and $^3P_1 - ^1S_0$, respectively, by Widing (1978). From observations of a tokamak plasma Peacock, Stamp, and Silver (1984) obtained for two of these lines the more accurate values 592.234±0.006 Å and 1118.060±0.010 Å.

The transition array $2s^2 2p^4 - 2s2p^5$ in the range of 80–120 Å was observed in a laser-produced iron plasma and classified by Feldman, Doschek, Nagel, Behring, and Cohen (1973). Fawcett, Galanti, and Peacock (1974) identified the transition from $2p^6\ ^1S_0$ to $2s2p^5\ ^1P^\circ_1$ at 115.42 Å. Kovalev (1981) remeasured these lines with an accuracy of ±0.004 Å. Most of them are published by Kononov (1983), who found a systematic shift in the measurements of Feldman et al. of up to 0.02 Å. The levels of $2s^2 2p^4$ (except for $^3P_2$ and $^1D_2$), $2s2p^5$, and $2p^6$ have an uncertainty of ±40 cm^{-1}. With the magnetic dipole lines reported by Peacock, Stamp, and Silver, the $2s^2 2p^4\ ^3P_2$ and $^1D_2$ levels are determined with an uncertainty of ±2 cm^{-1}. Percentage compositions for the $2s^2 2p^4$ and $2p^6$ configurations with configuration interaction, and the $2s2p^5$ configuration were provided by Kaufman and Sugar (1982).

Wavelengths in the range of 10–14 Å observed in a laser-produced plasma by Gordon, Hobby, and Peacock (1980a) were classified in the transition arrays $2p^4 - 2p^3 3s$, $2p^3 3d$, and $2p^3 4d$. A wavelength accuracy of ±0.005 Å is reported, permitting an energy level accuracy of ±3000 cm^{-1}. Their calculated leading percentages in $J_1 j$ and LS coupling appear in a Culham Laboratory Report (1980b).

The $2p^3 5d$ and $6d$ levels are from spectra of exploding wires between 9 and 10 Å reported by Burkhalter et al. (1978). The uncertainty in their level values is ±30 000 cm^{-1}. Designations for these configurations are given in LS coupling.

A position for the $1s2s^2 2p^5$ configuration of 52 138 000 cm^{-1} was obtained by Lie and Elton (1971).

We derived the value for the ionization energy from the $2p^3(^4S)nd\ ^3D_3$ series for $n=3$–5.

References

Burkhalter, P. G., Dozier, C. M., Stallings, E., and Cowan, R. D. (1978), J. Appl. Phys. **49**, 1092.

Doschek, G. A., Feldman, U., Dere, K. P., Sandlin, G. D., Van Hoosier, M. E., Brueckner, G. E., Purcell, J. D., and Tousey, R. (1975), Astrophys. J. **196**, L83.

Fawcett, B.C., Galanti, M., and Peacock, N. J. (1974), J. Phys. **B7**, 1149.

Feldman, U., Doschek, G. A., Nagel, D. J., Behring, W. E., and Cohen, L. (1973), Astrophys. J. **183**, L43.

Gordon, H., Hobby, M. G., and Peacock, N. J. (1980a), J. Phys. **B13**, 1985.

Gordon, H., Hobby, M. G., and Peacock, N J. (1980b), Culham Laboratory Report No. CLM-P592, Abingdon England.

Kaufman, V., and Sugar J. (1982), private communication.

Kononov, E. Y. (1983), Phys. Scr. **27**, 117.

Kovalev, V. I. (1981), Inst. Spectrosc., Troitsk, U.S.S.R., preprint No. 8.

Lie, T. N., and Elton, R. C. (1971), Phys. Rev. **A3**, 865.

Peacock, N. J., Stamp, M F., and Silver, J. D (1984), Phys. Scr. **T8**, 10.

Widing, K. G. (1978), Astrophys. J. **222**, 735.

Fe XIX

Configuration	Term	J	Level (cm^{-1})	Leading percentages		
$2s^2 2p^4$	^3P	2	0	90	10	^1D
		0	75 250	80	20	^1S
		1	89 441	100		
$2s^2 2p^4$	^1D	2	168 852	90	10	^3P
$2s^2 2p^4$	^1S	0	325 140	78	20	^3P
$2s2p^5$	$^3P^\circ$	2	*922 890*	100		
		1	*984 740*	97	3	^1P
		0	*1 030 020*	100		
$2s2p^5$	$^1P^\circ$	1	*1 267 600*	97	3	^3P
$2p^6$	^1S	0	2 134 180	98	2	$2s^2 2p^4\ ^1S$

(Continued)

Fe XIX—Continued

Configuration	Term	J	Level (cm^{-1})	Leading percentages			
$2s^2\,2p^3(^4S^\circ)3s$	$^3S^\circ$	1	6 680 000	81		8	$(^2P^\circ)\ ^1P^\circ$
$2s^2\,2p^3(^2D^\circ)3s$	$^3D^\circ$	2	6 787 000	64		17	$(^2P^\circ)\ ^3P^\circ$
		1	6 788 000	76		13	$(^4S^\circ)\ ^3S^\circ$
		3	6 818 000	99			
$2s^2\,2p^3(^2D^\circ)3s$	$^1D^\circ$	2	6 834 000	75		22	$(^2D^\circ)\ ^3D^\circ$
$2s^2\,2p^3(^2P^\circ)3s$	$^3P^\circ$	0	6 907 000	99			
		1	6 923 000	76		22	$(^2P^\circ)\ ^1P^\circ$
		2	6 970 000	72		12	$(^2D^\circ)\ ^3D^\circ$
$2s^2\,2p^3(^2P^\circ)3s$	$^1P^\circ$	1	6 985 000	63		17	$(^2D^\circ)\ ^3D^\circ$
$2s^2\,2p^3(^4S^\circ_{3/2})3d$	$(^3/_2,^5/_2)^\circ$	3	7 249 000	52	or	69	$(^4S^\circ)\ ^3D^\circ$
$2s^2\,2p^3(^2D^\circ_{3/2})3d$	$(^3/_2,^5/_2)^\circ$	2	7 370 000	55	or	27	$(^2D^\circ)\ ^3P^\circ$
$2s^2\,2p^3(^2D^\circ_{5/2})3d$	$(^5/_2,^5/_2)^\circ$	3	7 396 000	44	or	74	$(^2D^\circ)\ ^3D^\circ$
$2s^2\,2p^3(^2D^\circ_{5/2})3d$	$(^5/_2,^3/_2)^\circ$	2	7 405 000	73	or	43	$(^2D^\circ)\ ^3D^\circ$
$2s^2\,2p^3(^2D^\circ_{5/2})3d$	$(^5/_2,^5/_2)^\circ$	3	7 449 000	37	or	59	$(^2D^\circ)\ ^1F^\circ$
$2s^2\,2p^3(^2P^\circ_{1/2})3d$	$(^1/_2,^5/_2)^\circ$	3	7 450 000	69	or	68	$(^2P^\circ)\ ^3F^\circ$
		2	7 468 000	62	or	31	$(^2P^\circ)\ ^3P^\circ$
$2s^2\,2p^3(^2P^\circ_{3/2})3d$	$(^3/_2,^3/_2)^\circ$	2	7 554 000	57	or	32	$(^2P^\circ)\ ^3D^\circ$
		3	7 565 000	51	or	44	$(^2P^\circ)\ ^1F^\circ$
		1	7 567 000	48	or	33	$(^2P^\circ)\ ^3P^\circ$
$2s^2\,2p^3(^2P^\circ_{3/2})3d$	$(^3/_2,^5/_2)^\circ$	1	7 606 000	55	or	68	$(^2P^\circ)\ ^1P^\circ$
$2s^2\,2p^3(^4S^\circ_{3/2})4d$	$(^3/_2,^5/_2)^\circ$	2	9 242 000	68	or	76	$(^4S^\circ)\ ^3D^\circ$
		1	9 244 000	81	or	87	
		3	9 248 000	53	or	82	
$2s^2\,2p^3(^2D^\circ_{3/2})4d$	$(^3/_2,^3/_2)^\circ$	3	9 359 000	37	or	36	$(^2D^\circ)\ ^3F^\circ$
$2s^2\,2p^3(^2D^\circ_{3/2})4d$	$(^3/_2,^5/_2)^\circ$	2	9 374 000	70	or	41	$(^2D^\circ)\ ^3D^\circ$
$2s^2\,2p^3(^2D^\circ_{5/2})4d$	$(^5/_2,^3/_2)^\circ$	3	9 383 000	62	or	71	$(^2D^\circ)\ ^3D^\circ$
		2	9 395 000	93	or	44	$(^2D^\circ)\ ^3D^\circ$
		1	9 403 000	57	or	62	$(^2D^\circ)\ ^3S^\circ$
$2s^2\,2p^3(^2D^\circ_{5/2})4d$	$(^5/_2,^5/_2)^\circ$	3	9 417 000	64	or	77	$(^2D^\circ)\ ^1F^\circ$
		2	9 417 000	93	or	65	$(^2D^\circ)\ ^1D^\circ$
$2s^2\,2p^3(^2P^\circ_{1/2})4d$	$(^1/_2,^5/_2)^\circ$	3	9 483 000	97	or	55	$(^2P^\circ)\ ^3F^\circ$
		2	9 492 000	99	or	45	$(^2P^\circ)\ ^3P^\circ$
$2s^2\,2p^3(^2P^\circ_{1/2})4d$	$(^1/_2,^3/_2)^\circ$	1	9 494 000	97	or	56	$(^2P^\circ)\ ^3D^\circ$
$2s^2\,2p^3(^2P^\circ_{3/2})4d$	$(^3/_2,^3/_2)^\circ$	3	9 552 000	72	or	39	$(^2P^\circ)\ ^1F^\circ$
		1	9 556 000	56	or	57	$(^2P^\circ)\ ^3P^\circ$
$2s^2\,2p^3(^2P^\circ_{3/2})4d$	$(^3/_2,^5/_2)^\circ$	1	9 573 000	57	or	59	$(^2P^\circ)\ ^1P^\circ$
$2s^2\,2p^3(^4S^\circ)5d$	$^3D^\circ$	3	10 190 000				

Fe XIX—Continued

Configuration	Term	J	Level (cm^{-1})	Leading percentages
$2s^2\,2p^3(^2D°)5d$	$^3D°$	2	*10 330 000*	
		3	*10 330 000*	
$2s^2\,2p^3(^2D°)5d$	$^1D°$	2	*10 360 000*	
$2s^2\,2p^3(^2D°)5d$	$^1F°$	3	*10 390 000*	
$2s^2\,2p^3(^2D°)5d$	$^3F°$	3	*10 420 000*	
$2s^2\,2p^3(^2P°)5d$	$^3D°$	1	*10 450 000*	
		3	*10 500 000*	
$2s^2\,2p^3(^2P°)5d$	$^3F°$	3	*10 500 000*	
$2s^2\,2p^3(^2P°)5d$	$^3P°$	1	*10 500 000*	
$2s^2\,2p^3(^2P°)5d$	$^1F°$	3	*10 500 000*	
$2s^2\,2p^3(^2P°)5d$	$^1P°$	1	*10 510 000*	
$2s^2\,2p^3(^4S°)6d$	$^5D°$	2	*10 680 000*	
$2s^2\,2p^3(^4S°)6d$	$^3D°$	3	*10 710 000*	
$2s^2\,2p^3(^2P°)6d$	$^3D°$	1	*10 760 000*	
$2s^2\,2p^3(^2D°)6d$	$^1F°$	3	*11 030 000*	
Fe XX ($^4S°_{3/2}$)	***Limit***		**11 850 000**	

Fe XX

$Z=26$

N I isoelectronic sequence

Ground state: $1s^22s^22p^3\ {}^4S^{\circ}_{3/2}$

Ionization energy = 12 708 000±4000 cm^{-1} (1582.0±0.5 eV)

The array $2s^22p^3-2s2p^4$ was identified in the range of 90–132 Å from spectra of a laser-produced plasma by Doschek, Feldman, Cowan, and Cohen (1974) and by Feldman, Doschek, Cowan, and Cohen (1975). Further classifications in the same wavelength region by Doschek, Feldman, Davis, and Cowan (1975) provided the $2p^5\ {}^2P^{\circ}$ term. These arrays were reobserved by Lawson and Peacock (1980) with an accuracy of ±0.03 Å. They also identified intersystem transitions connecting the quartet and doublet levels. New measurments by Kovalev (1981) with an accuracy of ±0.004 Å are used to obtain the doublet levels of $2s^22p^3$, $2s2p^4$, and $2p^5$ with an uncertainty of ±40 cm^{-1}. The percentage compositions for these levels were provided by Kaufman and Sugar (1982).

A measurement of the $2s^22p^3\ {}^2D^{\circ}_{3/2}-{}^2D^{\circ}_{5/2}$ magnetic dipole transition at 2665.1 Å was obtained by Suckewer and Hinnov (1979) from a tokamak discharge. New predictions of the magnetic dipole transitions in the N I sequence by Edlén (1984) indicate that the solar flare line at 567.76(5) Å reported by Widing (1978) is the $2s^22p^3\ {}^4S_{3/2}-{}^2D_{5/2}$ transition. Edlén confirms Widing's classification of 541.35(5) Å as $2s^22p^3\ {}^2D_{3/2}-{}^2P_{3/2}$. We have used these lines and the line at 2665.1(3) Å classified by Suckewer and Hinnov in determining the levels of $2s^22p^3$.

Transition arrays between the ground configuration and $2p^23d$, $4d$, and $5d$ are analyzed in the report of Bromage et al. (1977) who used laser-produced plasma spectra observed in the range of 8–18 Å. The estimated wavelength accuracy is about ±0.005 Å. The levels derived from their measurements have an uncertainty of ±3000 cm^{-1}. Bromage and Fawcett (1977) have given the leading percentages for the $2p^23d$ levels.

The identification of the inner shell transitions giving the position of the $1s2s^22p^4$ configuration at 52 470 000 cm^{-1} was made by Lie and Elton (1971).

We derived the value for the ionization energy from the $2p^2({}^1D)nd\ {}^2F$ series.

References

Bromage, G. E., Cowan, R. D., Fawcett, B. C., Gordon, H., Hobby, M. G., Peacock, N. J., and Ridgley, A. (1977), UKAEA Report CLMR 170, Culham Lab.

Bromage, G. E., and Fawcett, B. C. (1977), Mon. Not. R. Astron. Soc. **179**, 683.

Doschek, G. A., Feldman, U., Cowan, R. D., and Cohen, L. (1974), Astrophys. J. **188**, 417.

Doschek, G. A., Feldman, U., Davis, J., and Cowan, R. D. (1975), Phys. Rev. **A12**, 980.

Edlén, B. (1984) private communication.

Feldman, U., Doschek, G. A., Cowan, R. D., and Cohen, L. (1975), Astrophys. J. **196**, 613.

Kaufman, V. and Sugar, J. (1982), private communication.

Kovalev, V. I. (1981), Inst. Spectrosc., Troitsk, U.S.S.R., preprint No. 8.

Lawson, K. D., and Peacock, N. J. (1980), J. Phys. **B13**, 3313.

Lie, T. N., and Elton, R. C. (1971), Phys. Rev. **A3**, 865.

Suckewer, S., and Hinnov, E. (1979), Phys. Rev. **A20**, 578.

Widing, K. G. (1978), Astrophys. J. **222**, 735.

Fe XX

Configuration	Term	J	Level (cm^{-1})	Leading percentages		
$2s^2\ 2p^3$	$^4S^{\circ}$	3/2	*0*	88	9	$^2P^{\circ}$
$2s^2\ 2p^3$	$^2D^{\circ}$	3/2	*138 620*	75	17	$^2P^{\circ}$
		5/2	*176 130*	100		
$2s^2\ 2p^3$	$^2P^{\circ}$	1/2	*260 270*	98	2	$2p^5\ {}^2P^{\circ}$
		3/2	*323 340*	71	23	$^2D^{\circ}$
$2s\ 2p^4$	^4P	5/2	752 730	97	3	^2D
		3/2	820 820	98	1	^2D
		1/2	842 740	95	5	^2S
$2s\ 2p^4$	^2D	3/2	1 042 570	93	5	^2P
		5/2	1 058 360	97	3	^4P

Fe XX—Continued

Configuration	Term	J	Level (cm^{-1})	Leading percentages		
$2s\,2p^4$	^{2}S	1/2	1 195 260	71	25	^{2}P
$2s\,2p^4$	^{2}P	3/2	1 242 430	94	6	^{2}D
		1/2	1 340 040	75	24	^{2}S
$2p^5$	^{2}P°	3/2	*1 954 520*	98	2	$2s^22p^3$ ^{2}P°
		1/2	*2 062 200*	98	2	
$2s^2\,2p^2(^3P)3d$	^{4}P	5/2	7 802 000	54	29	^{4}D
		3/2	7 802 000	77		
$2s^2\,2p^2(^3P)3d$	^{2}F	7/2	7 820 000	40	28	^{4}D
$2s^2\,2p^2(^3P)3d$	^{2}D	5/2	7 843 000	65	14	(^{1}D) ^{2}D
		3/2	7 859 000	72		
$2s^2\,2p^2(^1D)3d$	^{2}D	5/2	7 913 000	54	22	^{2}F
		3/2	7 919 000	80		
$2s^2\,2p^2(^1D)3d$	^{2}F	7/2	7 935 000	41	41	(^{3}P) ^{2}F
		5/2	7 983 000	31	30	(^{3}P) ^{2}D
$2s^2\,2p^2(^1D)3d$	^{2}P	3/2	7 967 000	73	11	(^{3}P) ^{2}P
$2s^2\,2p^2(^1S)3d$	^{2}D	5/2	8 047 000	86		
$2s^2\,2p^2(^3P)4d$	^{4}P	5/2	9 880 000			
		3/2	10 009 000			
		1/2	10 009 000			
$2s^2\,2p^2(^3P)4d$	^{4}F	5/2	9 942 000			
$2s^2\,2p^2(^3P)4d$	^{2}F	5/2	9 964 000			
		7/2	10 019 000			
$2s^2\,2p^2(^3P)4d$	^{4}D	5/2	9 992 000			
$2s^2\,2p^2(^3P)4d$	^{2}D	5/2	10 019 000			
		3/2	10 043 000			
$2s^2\,2p^2(^1D)4d$	^{2}D	3/2	10 130 000			
		5/2	10 142 000			
$2s^2\,2p^2(^1D)4d$	^{2}G	7/2	10 142 000			
$2s^2\,2p^2(^1D)4d$	^{2}F	5/2	10 149 000			
$2s^2\,2p^2(^1D)4d$	^{2}P	3/2	10 149 000			
$2s^2\,2p^2(^1S)4d$	^{2}D	3/2	10 269 000			
		5/2	10 289 000			
$2s^2\,2p^2(^3P)5d$	^{4}P	5/2	10 930 000			
		3/2	11 048 000			
		1/2	11 048 000			
$2s^2\,2p^2(^3P)5d$	^{4}F	5/2	10 994 000			

(Continued)

Fe XX—Continued

Configuration	Term	J	Level (cm^{-1})	Leading percentages
$2s^2\,2p^2(^3P)5d$	^2F	5/2 7/2	10 998 000 11 047 000	
$2s^2\,2p^2(^3P)5d$	^2D	5/2	11 036 000	
$2s^2\,2p^2(^3P)5d$	^4D	5/2	11 048 000	
$2s^2\,2p^2(^1D)5d$	^2G	7/2	11 153 000	
$2s^2\,2p^2(^1D)5d$	^2D	5/2 3/2	11 153 000 11 160 000	
$2s^2\,2p^2(^1D)5d$	^2F	5/2	11 169 000	
$2s^2\,2p^2(^1D)5d$	^2P	3/2	11 169 000	
Fe XXI ($^3P_0$)	*Limit*		**12 708 000**	

Fe XXI

$Z=26$

C I isoelectronic sequence

Ground state: $1s^2 2s^2 2p^2\ {}^3P_0$

Ionization energy = 13 620 000±30 000 cm^{-1} (1689±4 eV)

Identifications of transitions in the $2s^2 2p^2 - 2s2p^3$ array were reported by Feldman et al. (1975) in the region 91–121 Å. Kononov et al. (1976) extended the analysis and found transitions from $2p^4$, but reported no intersystem lines. Lawson and Peacock (1980) remeasured this spectrum with an accuracy of ±0.03 Å and found the connection among all the terms. The wavelengths of these groups fall in the range of 84–182 Å. The levels are determined from this work with an uncertainty of ±300 cm^{-1} except as noted below.

The $2s^2 2p^2\ {}^3P$ and ${}^1D$ terms are obtained from forbidden transitions. The ${}^3P_0 - {}^3P_1$ interval is derived from a solar line at 1354.1±0.1 Å observed by Doschek et al. (1975) from Skylab. Hinnov et al. (1982) reported the observations of the ${}^3P_1 - {}^3P_2$ transition at 2298.0±0.3 Å (in air) using the PLT tokamak. They also observed the transitions ${}^3P_2 - {}^1D_2$ and ${}^3P_1 - {}^1D_2$ within the ground configuration at 786.1 Å and 585.8 Å with an uncertainty of ±0.3 Å. Dere (1978) has identified the two transitions $2s^2 2p^2\ {}^3P_{1,2} - 2s2p^3\ {}^5S_2^\circ$ in solar flare spectra at 242.07 Å and 270.52 Å ±0.03 Å.

Percentage compositions of $2s^2 2p^2$ and $2p^4$ with configuration interaction between them, and of $2s2p^3$ were provided by Kaufman and Sugar (1982).

The levels of the $2p3d$ configuration are from Bromage and Fawcett (1977). The ${}^3F_4$ level of $2s2p^2 3d$ is from Boiko et al. (1978).

The transition $2p^2 - 2p4d$, $2p5d$ in the range of 8.5–9.5 Å were observed by Bromage et al. (1977). Their classifications provide the levels of $2p4d$ and $2p5d$ included here with an uncertainty of ±6000 cm^{-1}.

The resonance line reported by Lie and Elton (1971) at 1.896 Å arising from the $1s2s^2 2p^3$ configuration was resolved into three components by Feldman, Doschek, and Kreplin (1980), each classified as a blend of three lines. A center of gravity value for the configuration is about 52 910 000 cm^{-1}.

The ionization energy was obtained by extrapolation by Lotz (1967).

References

Boiko, V. A., Faenov, A. Y. and Pikuz, S. A. (1978), J. Quant. Spect. Rad. Transf. **19**, 11.

Bromage, G. E., Cowan, R. D., Fawcett, B. C., Gordon, H., Hobby, M. G., Peacock, N. J., and Ridgeley, A. (1977), UKAEA Report CLMR170, Culham Lab.

Bromage, G. E., and Fawcett, B. C. (1977), Mon. Not. R. Astron. Soc. **178**, 605.

Dere, K. P. (1978), Astrophys. J. **221**, 1062.

Doschek, G. A., Feldman, U., Dere, K. P., Sandlin, G. D., Van Hoosier, M. E., Brueckner, G. E., Purcell, J. D., and Tousey, R. (1975), Astrophys. J. **196**, L83.

Feldman, U., Doschek, G. A., Cowan, R. D., and Cohen, L. (1975), Astrophys. J. **196**, 613.

Feldman, U., Doschek, G. A., and Kreplin, R. W. (1980), Astrophys. J. **238**, 365.

Hinnov, E., Suckewer, S., Cohen, S., and Sato, K. (1982), Phys. Rev. **A25**, 2293.

Kaufman, V. and Sugar, J. (1982), private communication.

Kononov, E. Y., Koshelev, K. N., Podobedova, L. I., Chekalin, S. V., and Churlov, S. S. (1976), J. Phys. **B9**, 565.

Lawson, K. D., and Peacock, N. J. (1980), J. Phys. **B13**, 3313.

Lie, F. N., and Elton, R. C. (1971), Phys. Rev. **3**, 865.

Lotz, W. (1967), J. Opt. Soc. Am. **57**, 873.

Fe XXI

Configuration	Term	J	Level (cm^{-1})	Leading percentages		
$2s^2 2p^2$	${}^3P$	0	0	90	7	$2s^2 2p^2\ {}^1S$
		1	73 850	99	1	$2p^4\ {}^3P$
		2	117 353	75	24	$2s^2 2p^2\ {}^1D$
$2s^2 2p^2$	${}^1D$	2	244 560	75	24	${}^3P$
$2s^2 2p^2$	${}^1S$	0	371 900	87	9	${}^3P$
$2s2p^3$	${}^5S^\circ$	2	*486 950*	97	3	${}^3P^\circ$
$2s2p^3$	${}^3D^\circ$	1	*776 780*	86	11	${}^3P^\circ$
		2	*777 350*	83	15	${}^3P^\circ$
		3	*803 930*	100		

(Continued)

Fe XXI—Continued

Configuration	Term	J	Level (cm^{-1})	Leading percentages		
$2s2p^3$	$^3P°$	0	*916 380*	100		
		1	*924 880*	84	12	$^3D°$
		2	*942 320*	74	16	$^3D°$
$2s2p^3$	$^3S°$	1	*1 095 600*	80	16	$^1P°$
$2s2p^3$	$^1D°$	2	*1 126 800*	92	7	$^3P°$
$2s2p^3$	$^1P°$	1	*1 261 000*	82	16	$^3S°$
$2p^4$	^3P	2	1 646 300	88	11	$2s^22p^2\ ^1D$
		0	1 735 700	85	13	$2p^4\ ^1S$
		1	1 740 500	99	1	$2s^22p^2\ ^3P$
$2p^4$	^1D	2	1 817 300	88	11	^3P
$2p^4$	^1S	0	2 048 200	83		
$2s^2\,2p3d$	$^1D°$	2	*8 098 000*			
$2p3d$	$^3F°$	3	*8 101 400*			
$2s^2\,2p3d$	$^3D°$	2	*8 187 400*			
		3	*8 195 000*			
$2s^2\,2p3d$	$^3P°$	2	*8 230 900*			
$2s^2\,2p3d$	$^1P°$	1	*8 293 900*			
$2s^2\,2p3d$	$^1F°$	3	*8 313 600*			
$2s^2\,2p4d$	$^3F°$	3	*10 554 000*			
$2s^2\,2p4d$	$^3P°$	2	*10 580 000*			
		1	*10 688 000*			
$2s^2\,2p4d$	$^3D°$	1	*10 581 000*			
		2	*10 655 000*			
		3	*10 688 000*			
$2s^2\,2p4d$	$^1D°$	2	*10 675 000*			
$2s^2\,2p4d$	$^1F°$	3	*10 681 000*			
$2s^2\,2p5d$	$^3D°$	3	*11 802 000*			
$2s^2\,2p5d$	$^3P°$	1	*11 810 000*			
$2s^2\,2p5d$	$^1D°$	2	*11 810 000*			
$2s^2\,2p5d$	$^1F°$	3	*11 814 000*			
Fe XXII ($^2P°_{1/2}$)	***Limit***		**13 620 000**			

Fe XXII

$Z=26$

B I isoelectronic sequence

Ground state: $1s^22s^22p\ ^2P^\circ_{1/2}$

Ionization energy = 14 510 000±30 000 cm^{-1} (1799±4 eV)

Spectra in the range of 100–160 Å arising from transitions among configurations $2s^22p$, $2s2p^2$, and $2p^3$ were analyzed by Kononov et al. (1976). New measurements and some corrections to $2p^3$ were made by Lawson and Peacock (1980), whose wavelength uncertainty is ±0.03 Å. The results of Lawson and Peacock, with a level uncertainty of ±300 cm^{-1}, are given below. They have identified the intersystem line $2s2p^2\,^4P_{5/2}-2p^3\,^2D^\circ_{5/2}$ at 109.53 Å. Sandlin, Brueckner, Scherrer, and Tousey (1976) identified the intersystem multiplet $2s^22p\ ^2P^\circ-2s2p^2\,^4P$ from solar flare data which predicts the value 109.45±0.03 Å for the line of Lawson and Peacock. We use the solar measurements for the intersystem connection. The ^2P ground term splitting is from the magnetic dipole transition at 845.5±0.3 Å observed by Hinnov et al. (1982) in a tokamak discharge. The percentage compositions are provided by Kaufman and Sugar (1982) with configuration interaction between $2s^22p$ and $2p^3$.

Bromage, Cowan, Fawcett, and Ridgeley (1978), using the wavelength measurements of Boiko, Faenov, and Pikuz (1978) in the region of 9–12 Å and Hartree-Fock calculations, classified spectra arising from the transition arrays $2s^22p-2s^23d$, $2s^22p-2s2p3p$, $2s2p^2-2s2p3d$, and $2s2p^2-2s2p4d$. The uncertainty in the level values derived from these data is ±4000 cm^{-1}.

Exploding wire spectra were analyzed by Burkhalter et al. (1978), who reported the observation of the $2s^22p\ ^2P^\circ-2s^24d\ ^2D$ and $2s^22p\ ^2P^\circ-2s^24s\ ^2S$ lines at 9 Å measured with an uncertainty of ±0.03 Å. The resulting level uncertainty is ±40 000 cm^{-1}.

The transition array $1s^22s^22p-1s2s^22p^2$ was measured in solar flare spectra at ~1.8 Å by Feldman, Doschek, and Kreplin (1980) with an uncertainty of ±0.0005 Å. The level accuracy is ±14 000 cm^{-1}.

The ionization energy is from the isoelectronic extrapolation by Lotz (1967).

References

Boiko, V. A., Faenov, A. Y., and Pikuz, S. A. (1978), J. Quant. Spectrosc. Radiat. Transfer **19**, 11.

Bromage, G. E., Cowan, R. D., Fawcett, B. C., and Ridgeley, A. (1978), J. Opt. Soc. Am. **68**, 48.

Burkhalter, P. G.,Dozier, C. M., Stallings, C., and Cowan, R. D. (1978), J. Appl. Phys. **49**, 1092.

Feldman, U.,Doschek, G. A., and Kreplin, R. W. (1980), Astrophys. J. **238**, 365.

Hinnov,E., Suckewer, S., Cohen, S., and Sato, K. (1982) Phys. Rev. **A25**, 2293.

Kaufman, V. and Sugar, J. (1982), private communication.

Kononov, E. Y., Koshelev, K. N., Podobedova, L. I., Chekalin, S. V., and Churilov, S. S. (1976), J. Phys. **B9**, 565.

Lawson, K. D., and Peacock, N. J. (1980), J. Phys. **B13**, 3313.

Lotz, W. (1967), J. Opt. Soc. Am. **57**, 873.

Sandlin, G. D., Brueckner, G. E., Scherrer, V. E., and Tousey, R. (1976), Astrophys. J. **205**, L 47.

Fe XXII

Configuration	Term	J	Level (cm^{-1})	Leading percentages		
$1s^2\,2s^2\,2p$	$^2P^\circ$	1/2	*0*	98	2	$2p^3\ ^2P^\circ$
		3/2	*118 270*	97	3	
$1s^2\,2s2p^2$	^4P	1/2	404 550	96	3	^2S
		3/2	460 200	99	1	^2D
		5/2	513 260	94	6	^2D
$1s^2\,2s2p^2$	^2D	3/2	736 520	95	4	^2P
		5/2	759 620	94	6	^4P
$1s^2\,2s2p^2$	^2P	1/2	853 480	61	36	^2S
		3/2	992 290	95	4	^2D
$1s^2\,2s2p^2$	^2S	1/2	978 220	60	38	^2P
$1s^2\,2p^3$	$^4S^\circ$	3/2	*1 255 700*	91	7	$^2P^\circ$
$1s^2\,2p^3$	$^2D^\circ$	3/2	*1 396 420*	83	11	$^2P^\circ$
		5/2	*1 426 880*	100		

(Continued)

Fe XXII—Continued

Configuration	Term	J	Level (cm^{-1})	Leading percentages		
$1s^2\,2p^3$	$^2P°$	1/2	*1 569 630*	98	2	$2s^22p^2\ ^2P$°
		3/2	*1 627 720*	79	15	$2p^3\ ^2D$°
$1s^2\,2s^2\,3d$	^2D	3/2	8 498 000			
		5/2	8 507 000			
$1s^2\,2s2p3p$	^2P	1/2	8 584 000			
		3/2	8 688 000			
$1s^2\,2s2p3p$	^2D	3/2	8 740 000			
		5/2	8 845 000			
$1s^2\,2s2p(^3P°)3d$	$^4F°$	7/2	*8 864 000*			
$1s^2\,2s2p(^3P°)3d$	$^4P°$	5/2	*8 874 000*			
		3/2	*8 972 000*			
		1/2	*8 973 000*			
$1s^2\,2s2p(^3P°)3d$	$^4D°$	3/2	*8 882 000*			
		1/2	*8 888 000*			
		7/2	*8 962 000*			
		5/2	*8 973 000*			
$1s^2\,2s2p(^3P°)3d$	$^2D°$	5/2	*8 938 000*			
$1s^2\,2s2p(^3P°)3d$	$^2P°$	1/2	*8 967 000*			
		3/2	*9 180 000*			
$1s^2\,2s2p(^3P°)3d$	$^2F°$	5/2	*9 030 000*			
		7/2	*9 062 000*			
$1s^2\,2s2p(^1P°)3d$	$^2D°$	3/2	*9 134 000*			
		5/2	*9 272 000*			
$1s^2\,2s2p(^1P°)3d$	$^2P°$	3/2	*9 168 000*			
$1s^2\,2s2p(^1P°)3d$	$^2F°$	7/2	*9 242 000*			
		5/2	9 249 000			
$1s^2\,2s^2\,4s$	^2S	1/2	11 050 000			
$1s^2\,2s^2\,4d$	^2D	5/2	11 149 000			
		3/2	11 161 000			
$1s^2\,2s2p(^3P°)4d$	$^4F°$	5/2	*11 492 000*			
$1s^2\,2s2p(^3P°)4d$	$^4D°$	3/2	*11 526 000*			
		5/2	*11 618 000*			
		7/2	*11 618 000*			
$1s^2\,2s2p(^3P°)4d$	$^2F°$	5/2	*11 558 000*			
		7/2	*11 900 000*			
$1s^2\,2s2p(^3P°)4d$	$^2D°$	5/2	*11 611 000*			
$1s^2\,2s2p(^1P°)4d$	$^2F°$	7/2	*11 649 000*			
		5/2	*11 897 000*			

Fe XXII—Continued

Configuration	Term	J	Level (cm^{-1})	Leading percentages
$1s^2 2s2p(^1P°)4d$	$^2D°$	5/2	*11 906 000*	
Fe XXIII ($^1S_0$)	***Limit***		**14 510 000**	
$1s2s^2 2p^2$	^2P	1/2	53 122 000	
		3/2	53 242 000	
$1s2s^2 2p^2$	^2D	3/2	53 124 000	
		5/2	53 166 000	
$1s2s^2 2p^2$	^2S	1/2	53 327 000	

Fe XXIII

$Z=26$

Be I isoelectronic sequence

Ground state: $1s^22s^2\,^1S_0$

Ionization energy = 15 797 000±30 000 cm^{-1} (1958.6±3.7 eV)

Widing (1975) identified the resonance transitions $2s^2\,^1S_0-2s2p\,^1P_1^\circ$ and $^3P_1^\circ$ at 132.83 Å and 263.76 Å with an uncertainty of ±0.03 Å in spectra of solar flares. The magnetic dipole transition $2s2p\,^3P_1-^3P_2$ was observed by Hinnov, Suckewer, Cohen, and Sato (1982) at 1079.3±0.3 Å in a tokamak plasma.

These data were combined with the laser plasma observations by Lawson and Peacock (1980) of the $2s2p-2p^2$ array in the range of 136–221 Å, with a measurement uncertainty of ±0.03 Å. Edlén (1983) has concluded that some of the identifications in this array are doubtful. The percentage compositions of the $2s^2$, $2s2p$, and $2p^2$ configurations with mixing of $2s^2$ and $2p^2$ were calculated by Scott and Burke (1980). We give the two leading percentages.

Laser produced spectra of iron in the range of 6–17 Å arising from L-shell excitations were reported by Boiko, Faenov, and Pikuz (1978) with an accuracy of ±0.003 Å. Their classifications of these lines was revised and extended by Bromage, Cowan, Fawcett, and Ridgeley (1978), who obtained the spectra with improved ionization discrimination. The classifications by Bromage et al. are used to determine the levels of the $2snp$, $2snd$, $2pnp$, and $2pnd$ configurations ($n=3$–5) with an uncertainty of ±5000 cm^{-1}. They also gave percentage compositions for only the highly mixed levels of this group. The $2s3s$ and $2p3s$ levels are from Boiko et al.

Bhatia and Mason (1981) calculated the $2s3s$, $2s3p$, and $2s3d$ configurations and showed that the identification of $2s3s\,^1S$ by Boiko et al. is not correct. We have omitted this level pending clarification.

Spectral lines of the $1s^22s2p-1s2s2p^2$ array at ~1.8 Å were identified by Kononov, Koshelev, and Sidelnikov (1977). The complete designations are given by Safronova and Lisina (1979). The line at 1.8704 Å was assigned to $1s^22s^2\,^1S_0-1s2s^22p\,^1P_1^\circ$ by Feldman, Doschek, and Kreplin (1980). The uncertainty of these level values is about ±10 000 cm^{-1}.

We obtained an average value of 15 797 000±30 000 cm^{-1} for the ionization energy from the $2snp\,^1P_1^\circ$ and the $2snd\,^1D_2$ series.

References

Bhatia, A. K., and Mason, H. E. (1981), Astron. Astrophys. **103**, 324.

Boiko, V. A., Faenov, A. Y., and Pikuz, S. A. (1978), J. Quant. Spectrosc. Radiat. Transfer **19**, 11.

Bromage, G. E., Cowan, R. D., Fawcett, B. C., and Ridgeley, A. (1978), J. Opt. Soc. Am. **68**, 48.

Edlén, B. (1983), Phys. Scr. **28**, 51.

Feldman, U., Doschek, G. A., and Kreplin, R. W. (1980), Astrophys. J. **238**, 365.

Hinnov, E., Suckewer, S., Cohen, S. and Sato, K. (1982), Phys. Rev. **A25**, 2293.

Kononov, E. Y., Koshelev, K. N., and Sidelnikov, Y. V. (1977), Sov. J. Plasma Phys. **3**, 375.

Lawson, K. D., and Peacock, N. J. (1980), J. Phys. **B13**, 3313.

Safronova, U. I., and Lisina, T. G. (1979), At. Data Nucl. Data Tables **24**, 49.

Scott, N. S., and Burke, P. G. (1980), J. Phys. **B13**, 4299.

Widing, K. G.(1975), Astrophys. J. **197**, L33.

Fe XXIII

Configuration	Term	J	Level (cm^{-1})	Leading percentages		
$1s^2\,2s^2$	^1S	0	0	96	4	$2p^2\,^1S$
$1s^2\,2s2p$	$^3P^\circ$	0	*348 180*	100		
		1	*379 130*	96	2	$^1P^\circ$
		2	*471 780*	100		
$1s^2\,2s2p$	$^1P^\circ$	1	*752 840*	96	2	$^3P^\circ$
$1s^2\,2p^2$	^3P	0	956 100	94	6	^1S
		1	1 027 200	100		
		2	1 071 700	76	24	^1D
$1s^2\,2p^2$	^1D	2	1 204 200	76	24	^3P
$1s^2\,2p^2$	^1S	0	1 423 000	90	6	^3P

Fe XXIII—Continued

Configuration	Term	*J*	Level (cm^{-1})	Leading percentages			
1s^2 2s3s	^{3}S	1	8 894 000				
1s^2 2s3p	^{3}P°	1	*9 076 000*	70		29	^{1}P°
1s^2 2s3p	^{1}P°	1	*9 107 000*	64		30	^{3}P°
1s^2 2s3d	^{3}D	1	9 199 000				
		2	9 209 000				
		3	9 212 000				
1s^2 2p3s	^{3}P°	0	*9 295 000*				
1s^2 2s3d	^{1}D	2	9 273 000				
1s^2 2p3p	^{3}D	1	9 455 000				
		2	9 524 000				
		3	9 624 000				
1s^2 2p3s	^{1}P°	1	*9 470 000*				
1s^2 2p3d	^{3}F°	3	*9 625 000*				
1s^2 2p3d	^{3}D°	1	*9 637 000*				
		3	*9 749 000*				
1s^2 2p3d		2	*9 728 000*	35	^{3}P°	29	^{3}D°
1s^2 2p3d	^{1}D°	2	*9 638 000*	46		29	^{3}D°
1s^2 2p3p	^{3}P	2	9 644 000				
1s^2 2p3p	^{1}D	2	9 709 000				
1s^2 2p3d	^{3}P°	2	*9 753 000*	54		40	^{3}D°
1s^2 2p3d	^{1}P°	1	*9 828 000*				
1s^2 2p3d	^{1}F°	3	*9 830 000*				
1s^2 2s4p	^{1}P°	1	*12 044 000*				
1s^2 2s4d	^{3}D	1	12 073 000				
		2	12 075 000				
		3	12 081 000				
1s^2 2s4d	^{1}D	2	12 098 000				
1s^2 2p4d		2	*12 481 000*	44	^{3}P°	35	^{3}D°
1s^2 2p4d	^{3}D°	1	*12 488 000*				
		3	*12 603 000*				
1s^2 2p4d	^{3}F°	3	*12 484 000*				
1s^2 2p4p	^{3}D	3	12 560 000				
1s^2 2p4d	^{1}D°	2	*12 597 000*	56		21	^{3}D°

(Continued)

Fe XXIII—Continued

Configuration	Term	J	Level (cm^{-1})	Leading percentages
$1s^2\,2p4d$	^{3}P°	2	*12 614 000*	50 42 ^{3}D°
		1	*12 615 000*	
$1s^2\,2p4d$	^{1}F°	3	*12 631 000*	
$1s^2\,2s5d$	^{3}D	1	13 369 000	
		2	13 400 000	
		3	13 404 000	
$1s^2\,2s5p$	^{1}P°	1	*13 383 000*	
$1s^2\,2s5d$	^{1}D	2	13 438 000	
$1s^2\,2p5d$	^{3}F°	3	*13 804 000*	
$1s^2\,2p5d$	^{3}D°	2	*13 805 000*	
		3	*13 929 000*	
$1s^2\,2p5p$	^{3}D	3	13 904 000	
$1s^2\,2p5d$	^{1}D°	2	*13 922 000*	
$1s^2\,2p5d$	^{1}F°	3	*13 945 000*	
Fe XXIV (^{2}S$_{1/2}$)	***Limit***		**15 797 000**	
$1s2s^2\,2p$	^{1}P°	1	*53 464 000*	
$1s(^2S)2s2p^2(^4P)$	^{3}P	0	53 707 000	
$1s(^2S)2s2p^2(^2D)$	^{3}D	1	53 800 000	
$1s(^2S)2s2p^2(^2S)$	^{3}S	1	53 925 000	
$1s(^2S)2s2p^2(^2D)$	^{1}D	2	54 045 000	
$1s(^2S)2s2p^2(^2P)$	^{1}P	1	54 182 000	
$1s(^2S)2s2p^2(^2S)$	^{1}S	0	54 252 000	

Fe XXIV

$Z=26$

Li I isoelectronic sequence

Ground state: $1s^2 2s\ ^2S_{1/2}$

Ionization energy $=$ 16 500 000±4000 cm^{-1} (2045.8±0.5 eV)

The $2s-2p$ transitions have been observed with an uncertainty of ±0.02 Å at 192.04 and 255.10 Å in solar flares from Skylab as reported by Widing and Purcell (1976).

The $1s^2nl$ levels with $n>2$ are from the observations of Boiko, Faenov, and Pikuz (1978) from 6–11 Å with a laser-produced plasma. They report a measurement uncertainty of ±0.003 Å.

The levels above the ionization energy are from the analysis by Kononov, Koshelev, and Sidelnikov (1977). They obtained the spectrum at ~1.8 Å from the x-ray emitting hot spot in a low inductance spark discharge with a measurement uncertainty of ±0.0003 Å. The designations are obtained from Vainstein and Safronova (1978). Klapisch et al. (1977) reported two resonance lines from the $1s2s3p$ configuration.

The ionization energy was calculated by Edlén (1979).

References

Boiko, V. A., Faenov, A. Y., and Pikuz, S. A. (1978), J. Quant. Spectrosc. Radiat. Transfer **19**, 11.

Edlén, B. (1979), Phys. Scr. **19**, 255.

Klapisch, M., Schwob, J. L., Fraenkel, B. S., and Oreg. J. (1977), J. Opt. Soc. Am. **67**, 148.

Kononov, E. Y., Koshelev, K. N., and Sidelnikov, Y. V. (1977), Sov. J. Plasma Phys. **3**, 375.

Vainstein, L. A., and Safronova, U. I. (1978), At. Data Nucl. Data Tables **21**, 49.

Widing, K. G., and Purcell, J. D. (1976), Astrophys. J. **204**, L151.

Fe XXIV

Configuration	Term	J	Level (cm^{-1})
$1s^2\,2s$	^2S	1/2	0
$1s^2\,2p$	$^2P°$	1/2	*392 000*
		3/2	*520 720*
$1s^2\,3s$	^2S	1/2	9 272 400
$1s^2\,3p$	$^2P°$	1/2	*9 378 000*
		3/2	*9 417 000*
$1s^2\,3d$	^2D	3/2	9 459 000
		5/2	9 472 000
$1s^2\,4s$	^2S	1/2	12 464 000
$1s^2\,4p$	$^2P°$	1/2	*12 511 000*
		3/2	*12 527 000*
$1s^2\,4d$	^2D	3/2	12 541 000
		5/2	12 546 000
$1s^2\,5p$	$^2P°$	1/2,3/2	*13 949 000*
$1s^2\,5d$	^2D	3/2	13 961 000
		5/2	13 965 000
$1s^2\,6p$	$^2P°$	1/2,3/2	*14 734 000*
$1s^2\,6d$	^2D	3/2	14 735 000
		5/2	14 739 000
$1s^2\,7d$	^2D	3/2,5/2	15 209 000
Fe XXV ($^1S_0$)	*Limit*		**16 500 000**
$1s(^2S)2s2p(^3P°)$	$^4P°$	3/2	*53 390 000*
$1s(^2S)2s2p(^3P°)$	$^2P°$	1/2	*53 657 000*
		3/2	*53 752 000*
$1s2p^2$	^4P	1/2	53 806 000
		3/2	53 877 000
		5/2	53 937 000
$1s(^2S)2s2p(^1P°)$	$^2P°$	1/2	*53 844 000*
		3/2	*53 903 000*
$1s2p^2$	^2D	3/2	54 070 000
		5/2	54 126 000
$1s2p^2$	^2P	1/2	54 077 000
		3/2	54 244 000
$1s2p^2$	^2S	1/2	54 385 000
$1s2s(^3S_1)3p$	$(1,1/2)°$	1/2	*62 790 000*
$1s2s(^1S_0)3p$	$(0,1/2)°$	1/2	*62 970 000*
$1s2p3s$	$^2P°$	3/2	*63 209 000*
$1s2p(^3P°)3d$	$^4D°$	3/2	*63 281 000*
$1s2p3p$	^2D	5/2	63 543 000
$1s2p3p$	^2S	1/2	63 572 000
$1s2p3d$	$^2F°$	7/2	*63 618 000*

Fe XXV

$Z=26$

He I isoelectronic sequence

Ground state: $1s^2\,{}^1S_0$

Ionization energy $=$ 71 203 000±14 000 cm^{-1} (8828.1±2.0 eV)

The $1s^2-1s2p$ and $1s^2-1s3p$ lines were first observed in a solar flare spectrum by Neupert et al. (1967). The first laboratory observations of these lines were obtained by Cohen et al. (1968) using a spark discharge. They reported the wavelengths 2.86 and 1.59 Å.

Because of the excellent agreement of the calculated energies of the $n=2$ shell by Safronova (1981) with the few well-measured spectra in the He I sequence, we have compiled her results for the $n=2$ levels and for the ionization energy. Detailed comparisons are given in the Introduction. Kononov, Koshelev, and Sidelnikov (1977) have measured the wavelengths of the $1s^2-1s2p$ $^1P^\circ_1$ and $^3P^\circ_1$ transitions with an estimated uncertainty of $\pm3\times10^{-4}$ Å. Their values are 1.8510 Å and 1.8592 Å respectively, compared with calculated values of 1.85048 Å and 1.85945 Å. A beam foil observation of the $1s2s\,{}^3S_1-1s2p\,{}^3P^\circ_2$ line by Buchet et al. (1981) gave a wavelength of 271.02±0.09 Å. The calculated value by Safronova is 270.929 Å. A new calculation of $1s2s\,{}^3S_1-1s2p\,{}^3P^\circ_{0,2}$ by Hata and Grant (1983) gives the values 427.982 Å and 271.153 Å for these transitions. We use the experimental value for the $^3S_1-{}^3P^\circ_2$ transition by Buchet et al. to set the value of $1s2p\,{}^3P^\circ_2$ relative to $1s2s\,{}^3S_1$.

For $n=3$ to 5 we give the calculated levels by Ermolaev and Jones (1974). These are obtained by subtracting their binding energies for the excited states from the binding energy of the ground state obtained by Safronova. The radiative corrections are reduced considerably with increasing n, which should bring the two calculations into closer agreement for $n<2$. Resonance transitions from $1s3p$ $^{1,3}P^\circ_1$ were observed by Klapisch et al. (1977) at 1.5738 Å and 1.5755 Å. The calculated levels predict the wavelengths 1.5732 Å and 1.5750 Å.

We have assumed an uncertainty of 2 parts in 10^4 for the excited levels relative to the ground state, and for the ionization energy. For the differences between excited levels where $\Delta n=0$, we assumed an uncertainty of 2 parts in 10^3.

The mixing coefficients for the $1snp$ $^{1,3}P$ levels were obtained from Ermolaev and Jones.

References

Buchet, J. P., Buchet-Poulizac, M. C., Denis, A., Desequelles, J., and Druetta, M. (1981), Phys. Rev. **A23**, 3354.

Cohen, L., Feldman, U., Swartz, M., and Underwood, J. H. (1968), J. Opt. Soc. Am. **58**, 843.

Ermolaev, A. M., and Jones, M. (1974), J. Phys. **B7**, 199.

Hata, J., and Grant, I. P. (1983), J. Phys. **B16**, 523.

Kononov, E. Y., Koshelev, K. N., and Sidelnikov, Y.V. (1977), Sov. J. Plasma Phys. **3**, 375.

Klapisch, M., Schwob, J. L., Fraenkel, B. S., and Oreg, J. (1977), J. Opt. Soc. Am. **67**, 148.

Neupert, W. M., Gates, W., Swartz, M., and Young, R. (1967), Astrophys. J. **149**, L49.

Safronova, U. I. (1981), Phys. Scr. **23**, 241.

Fe XXV

Configuration	Term	J	Level (cm^{-1})	Leading percentages		
$1s^2$	^1S	0	0			
$1s2s$	^3S	1	[53 527 090]			
$1s2p$	$^3P^\circ$	0	*[53 760 280]*			
		1	*[53 779 140]*	91	9	$^1P^\circ$
		2	*53 895 550+x*			
$1s2s$	^1S	0	[53 781 300]			
$1s2p$	$^1P^\circ$	1	*[54 040 000]*	91	9	$^3P^\circ$
$1s3s$	^3S	1	[63 421 610]			
$1s3p$	$^3P^\circ$	0	*[63 486 290]*			
		1	*[63 490 690]*	89	11	$^1P^\circ$
		2	*[63 525 620]*			

Fe XXV—Continued

Configuration	Term	*J*	Level (cm^{-1})	Leading percentages		
1*s*3*s*	^{1}S	0	[63 488 390]			
1*s*3*p*	^{1}P°	1	[*63 565 470*]	89	11	^{3}P°
1*s*4*s*	^{3}S	1	[66 847 000]			
1*s*4*p*	^{3}P°	0	[*66 873 940*]			
		1	[*66 875 780*]	89	11	^{1}P°
		2	[*66 890 550*]			
1*s*4*s*	^{1}S	0	[66 874 060]			
1*s*4*p*	^{1}P°	1	[*66 906 790*]	89	11	^{3}P°
1*s*5*s*	^{3}S	1	[68 423 650]			
1*s*5*s*	^{1}S	0	[68 437 160]			
1*s*5*p*	^{3}P°	0	[*68 437 270*]			
		1	[*68 438 210*]	89	11	^{1}P°
		2	[*68 445 780*]			
1*s*5*p*	^{1}P°	1	[*68 453 990*]	89	11	^{3}P°
Fe XXVI ($^{2}S_{1/2}$)	***Limit***		**71 203 000**			

Fe XXVI

$Z=26$

H I isoelectronic sequence

Ground state: $1s\ ^2S_{1/2}$

Ionization energy $= 74\,829\,600 \pm 300\ \text{cm}^{-1}$ (9277.76 ± 0.04 eV)

Briand, Tavernier, and Indelicato (1983) have measured values for the $1s\ ^2S_{1/2}-2p\ ^2P_{1/2,3/2}$ energies of 56 070 000 cm^{-1} and 56 247 000 cm^{-1} with an uncertainty of $\pm 5000\ \text{cm}^{-1}$.

We give calculated values by Mohr (1983) for the $n=2$ shell and by Erickson (1977) for $n=3$ to 5 relative to the $2p\ ^2P^\circ_{3/2}$ level. Further details are given in the Introduction. Relative to ground state, the level uncertainty is estimated to be 5 parts in 10^7. The uncertainty in the excited states relative to $2p\ ^2P^\circ_{3/2}$ is 1 part in 10^6.

References

Briand, J. P., Tavernier, M., and Indelicato, P. (1983), Phys. Rev. Lett. **50**, 832.

Erickson, G. W. (1977), J. Phys. Chem. Ref. Data **6**, 831.

Mohr, P. J. (1983), At. Data Nucl. Data Tables **29**, 453.

Fe XXVI

Configuration	Term	J	Level (cm^{-1})
$1s$	^2S	$1/2$	0
$2p$	$^2P^\circ$	$1/2$	*[56 071 350]*
		$3/2$	*[56 242 454]*
$2s$	^2S	$1/2$	[56 075 896]
$3p$	$^2P^\circ$	$1/2$	*[66 511 661]*
		$3/2$	*[66 562 389]*
$3s$	^2S	$1/2$	[66 513 080]
$3d$	^2D	$3/2$	[66 562 298]
		$5/2$	[66 578 923]
$4p$	$^2P^\circ$	$1/2$	*[70 158 764]*
		$3/2$	*[70 180 148]*
$4s$	^2S	$1/2$	[70 159 365]
$4d$	^2D	$3/2$	[70 180 109]
		$5/2$	[70 187 128]
$4f$	$^2F^\circ$	$5/2$	*[70 187 116]*
		$7/2$	*[70 190 611]*
$5p$	$^2P^\circ$	$1/2$	*[71 843 872]*
		$3/2$	*[71 854 812]*
$5s$	^2S	$1/2$	[71 844 180]
$5d$	^2D	$3/2$	[71 854 791]
		$5/2$	[71 850 386]
$5f$	$^2F^\circ$	$5/2$	*[71 858 379]*
		$7/2$	*[71 860 169]*
$5g$	^2G	$7/2$	[71 860 166]
		$9/2$	[71 861 238]
	Limit		**74 829 600**

Co I

$Z=27$

Ground state: $1s^2 2s^2 2p^6 3s^2 3p^6 3d^7 4s^2\ {}^4F_{9/2}$

Ionization energy = 63 400±500 cm^{-1} (7.86±0.06 eV)

The first regularities in the arc spectrum of cobalt were found by Walters (1924), who classified 88 lines in 12 multiplets of the quartet system. The work was then taken up by Catalán, who published a series of papers culminating in a rather complete analysis by Catalán and Antunes (1936). Russell, King, and Moore (1940) continued the analysis using wavelengths assembled from various sources covering the range of 1814–11 895 Å, including the very precise interferometric determinations of Burns and Sullivan (1942). Their results are reported here. All classified lines fit the level system with a deviation of less than ±0.1 cm^{-1}.

The five-place g-values for the seven low levels were measured by Childs and Goodman (1968) in an atomic beam with an accuracy of 2 parts in the fifth place. The three-place g-values were calculated by D. Weeks in 1951 from measurements of plates made at the Massachusetts Institute of Technology, and were tabulated in the energy-level compilation by C. E. Moore (1952). The two-place g-values were measured by R. B. King for the analysis by Russell et al. (1940).

The configuration assignments for the terms of $3d^8 4s$ and $3d^7 4s^2$ are those given by Racah (1942). The leading percentages of the $3d^8 4s$ terms are from the Hartree-Fock calculation of Vizbaraite, Rudzikas, and Grabauskas (1968).

The $3d^7 4s 4p$ and $3d^8 4p$ configurations have been calculated by Roth (1970), with configuration interaction. We give his identifications and leading percentages. Roth distinguished the two ${}^2D$ terms of the $3d^7$ core by the letters a and b, rather than by seniority. The percentages include the sum of seniority states contributing to the term.

The alphabetic prefixing of terms with lower case letters for distinguishing terms of the same type has been retained from Russell et al. except where the levels were reinterpreted by Roth on the basis of his theoretical treatment. When the leading component is less than 40% the level name is usually omitted, but the prefix is included in the percentage column.

The ionization energy given here was obtained by Catalán and Velasco (1952) from a study of regularities along the iron period of the elements. It agrees well with the mean of the values found by Catalán and Antunes and by Russell et al.

References

Burns, K., and Sullivan, F. (1942), Science Studies St. Bonaventure College **10**, 10; ibid. **11**, 4.

Catalán, M. A., and Antunes, M. T. (1936), An. Soc. Esp. Fis. Quim. **34**, 103 and 207.

Catalán, M. A., and Velasco, R. (1952), An. R. Soc. Esp. Fis. Quim. **48**, 247.

Childs, W. J., and Goodman, L. S. (1968), Phys. Rev. **170**, 50.

Moore, C. E. (1952), Atomic Energy Levels, Natl. Bur. Stand. (U. S.) Circ. 467, Vol. II (reissued in 1971 as Natl. Bur. Stand. (U.S.) Natl. Stand. Ref. Data Ser. 35, Vol. II).

Racah, G. (1942), Phys. Rev. **61**, 537.

Roth, C. (1970), J. Res. Natl. Bur. Stand. (U.S.) **74A**, 703.

Russell, H. N., King, R. B., and Moore, C. E. (1940), Phys. Rev. **58**, 407.

Vizbaraite, J., Rudzikas, A., and Grabauskas, D. (1968), Leituvos Fizikos Rinkinys **8**, 481.

Walters, F. M. (1924), J. Wash. Acad. Sci. **14**, 407.

Co I

Configuration	Term	J	Level (cm^{-1})	g	Leading percentages
$3d^7\ 4s^2$	$a\ {}^4F$	9/2	0.00	1.33289	
		7/2	816.00	1.23778	
		5/2	1 406.84	1.02826	
		3/2	1 809.33	0.39940	
$3d^8({}^3F)4s$	$b\ {}^4F$	9/2	3 482.82	1.33343	100
		7/2	4 142.66	1.23661	98
		5/2	4 690.18	1.02709	100
		3/2	5 075.83	0.404	100
$3d^8({}^3F)4s$	$a\ {}^2F$	7/2	7 442.41	1.147	98
		5/2	8 460.81	0.862	100
$3d^7\ 4s^2$	$a\ {}^4P$	5/2	13 795.52	1.604	
		3/2	14 036.28	1.722	
		1/2	14 399.28	2.651	

(Continued)

Co I—Continued

Configuration	Term	J	Level (cm^{-1})	g	Leading percentages		
$3d^8(^3P)4s$	$b\ ^4P$	5/2	15 184.04	1.515	84	16	$(^1D)\ ^2D$
		3/2	15 744.04	1.476	98		
		1/2	16 195.68	2.682	100		
$3d^7\ 4s^2$	$a\ ^2G$	9/2	16 467.90	1.109			
		7/2	17 233.68	0.883			
$3d^8(^1D)4s$	$a\ ^2D$	3/2	16 470.60	1.101	96		
		5/2	16 778.16	1.296	83	16	$(^3P)\ ^4P$
$3d^8(^3P)4s$	$a\ ^2P$	3/2	18 389.57	1.300	98		
		1/2	18 775.01	0.695	100		
$3d^7\ 4s^2$	$b\ ^2P$	3/2	20 500.71	1.284			
		1/2	21 215.90	0.680			
$3d^7\ 4s^2$	$a\ ^2H$	11/2	21 780.47	1.100			
		9/2	22 475.36	0.921			
$3d^7\ 4s^2$	$b\ ^2D$	5/2	21 920.09	1.180			
		3/2	23 152.57	0.955			
$3d^8(^1G)4s$	$b\ ^2G$	9/2	23 184.23	1.098	100		
		7/2	23 207.76	0.883	100		
$3d^7(^4F)4s4p(^3P^\circ)$	$z\ ^6F^\circ$	11/2	*23 611.78*	1.466	97		
		9/2	*23 855.62*	1.481	78	18	$^6D^\circ$
		7/2	*24 326.11*	1.436	84	12	$^6D^\circ$
		5/2	*24 733.28*	1.336	90	7	$^6D^\circ$
		3/2	*25 041.16*	1.118	95		
		1/2	*25 232.79*	-0.622	99		
$3d^7(^4F)4s4p(^3P^\circ)$	$z\ ^6D^\circ$	9/2	*24 627.79*	1.569	78	17	$^6F^\circ$
		7/2	*25 269.25*	1.550	83	11	$^6F^\circ$
		5/2	*25 739.93*	1.612	87	6	$^6F^\circ$
		3/2	*26 063.11*	1.812	91		
		1/2	*26 250.49*	3.286	94		
$3d^7(^4F)4s4p(^3P^\circ)$	$z\ ^6G^\circ$	13/2	*25 138.88*	1.40	100		
		11/2	*25 568.68*	1.354	96		
		9/2	*25 937.59*	1.281	94		
		7/2	*26 232.05*	1.150	94		
		5/2	*26 450.02*	0.876	95		
		3/2	*26 597.64*	0.006	97		
$3d^9$	$c\ ^2D$	5/2	27 497.06	1.200			
		3/2	28 470.51	0.907			
$3d^7(^4F)4s4p(^3P^\circ)$	$z\ ^4F^\circ$	9/2	*28 345.86*	1.330	82	15	$3d^8(^3F)4p\ ^4F^\circ$
		7/2	*28 777.27*	1.247	80	13	
		5/2	*29 216.37*	1.033	82	13	
		3/2	*29 563.17*	0.410	85	13	
$3d^7(^4F)4s4p(^3P^\circ)$	$z\ ^4G^\circ$	11/2	*28 845.22*	1.276	95		
		9/2	*29 269.73*	1.175	90		
		7/2	*29 735.18*	0.995	91		
		5/2	*30 102.96*	0.577	94		

Co I—Continued

Configuration	Term	J	Level (cm^{-1})	g	Leading percentages			
$3d^7(^4F)4s4p(^3P^\circ)$	$z\ ^4D^\circ$	7/2	*29 294.52*	1.425	53		42	$3d^8(^3F)4p\ ^4D^\circ$
		5/2	*29 948.76*	1.359	57		36	
		3/2	*30 443.63*	1.192	62		33	
		1/2	*30 742.65*	-0.006	65		31	
$3d^7(^4F)4s4p(^3P^\circ)$	$z\ ^2G^\circ$	9/2	*31 699.69*	1.126	47		29	$3d^8(^3F)4p\ ^4G^\circ$
		7/2	*32 733.07*	0.899	41		29	$3d^8(^3F)4p\ ^4F^\circ$
$3d^7(^4F)4s4p(^3P^\circ)$	$z\ ^2F^\circ$	7/2	*31 871.15*	1.777	66		28	$3d^8(^3F)4p\ ^2F^\circ$
		5/2	*32 781.71*	0.870	71		26	
$3d^8(^3F)4p$	$y\ ^4D^\circ$	7/2	*32 027.50*	1.395	46		45	$3d^7(^4F)4s4p(^3P^\circ)\ ^4D^\circ$
		5/2	*32 654.50*	1.366	51		41	
		3/2	*33 150.68*	1.195	56		37	
		1/2	*33 449.18*	0.012	59		36	
$3d^8(^3F)4p$	$y\ ^4G^\circ$	11/2	*32 430.59*	1.287	94			
		9/2	*32 464.73*	1.154	53		28	$^2G^\circ$
		5/2	*33 674.38*	0.704	52		30	$^4F^\circ$
$3d^8(^3F)4p$	$y\ ^4F^\circ$	9/2	*32 841.99*	1.313	51		22	$3d^7(^4F)4s4p(^1P^\circ)\ ^4F^\circ$
		3/2	*34 196.21*	0.430	67		24	
$3d^8(^3F)4p$		7/2	*33 173.36*	1.039	34	$^4G^\circ$	31	$^2G^\circ$
$3d^8(^3F)4p$	$y\ ^2G^\circ$	9/2	*33 439.72*	1.165	56		27	$3d^7(^4F)4s4p(^3P^\circ)\ ^2G^\circ$
		7/2	*34 133.59*	0.917	50		21	
$3d^7(^4F)4s4p(^3P^\circ)$	$z\ ^2D^\circ$	5/2	*33 462.83*	1.186	53		42	$3d^8(^3F)4p\ ^2D^\circ$
		3/2	*34 352.42*	0.787	53		37	
$3d^8(^3F)4p$		7/2	*33 466.87*	1.155	33	$^4G^\circ$	25	$3d^8(^3F)4p\ ^4F^\circ$
$3d^8(^3F)4p$		5/2	*33 945.90*	0.900	42	$^4G^\circ$	38	$3d^8(^3F)4p\ ^4F^\circ$
$3d^8(^3F)4p$	$y\ ^2F^\circ$	7/2	*35 450.56*	1.145	61		25	$3d^7(^4F)4s4p(^3P^\circ)\ ^2F^\circ$
		5/2	*36 329.86*	0.892	53		18	$3d^8(^3F)4p\ ^2D^\circ$
$3d^8(^3F)4p$	$y\ ^2D^\circ$	5/2	*36 092.44*	1.186	35		36	$3d^7(^4F)4s4p(^3P^\circ)\ ^2D^\circ$
		3/2	*36 875.13*	0.794	53		40	
$3d^7(^4F)4s4p(^1P^\circ)$	$x\ ^4D^\circ$	7/2	*39 649.16*	1.428	65		18	$3d^8(^3P)4p\ ^4D^\circ$
		5/2	*40 345.95*	1.370	63		21	
		3/2	*40 827.77*	1.240	62		23	
		1/2	*41 101.80*	0.026	62		24	
$3d^7(^4P)4s4p(^3P^\circ)$	$z\ ^4S^\circ$	3/2	*40 621.62*	2.017	74		10	$3d^8(^3P)4p\ ^4S^\circ$
$3d^7(^4P)4s4p(^3P^\circ)$	$^6P^\circ$	7/2	*41 041.43*	1.40	97			
		5/2	*41 104.96*	1.863	93			
$3d^7(^4F)4s4p(^1P^\circ)$	$x\ ^4F^\circ$	9/2	*41 225.76*	1.319	66		19	$(^2G)(^3P^\circ)\ ^4F^\circ$
		7/2	*41 918.41*	1.248	63		23	
		5/2	*42 434.23*	1.024	59		24	
		3/2	*42 796.67*	0.406	61		27	

(Continued)

Co I—Continued

Configuration	Term	J	Level (cm^{-1})	g	Leading percentages			
$3d^7(^4F)4s4p(^1P^\circ)$	$x\ ^4G^\circ$	11/2	*41 528.53*	1.291	95			
		9/2	*42 269.32*	1.169	93			
		7/2	*42 811.44*	1.004	92			
		5/2	*43 199.65*	0.649	93			
$3d^8(^3P)4p$	$z\ ^4P^\circ$	5/2	*41 968.89*	1.627	65		12	$3d^7(^4P)4s4p(^3P^\circ)\ ^4P^\circ$
		1/2	*41 969.90*	2.51	66		14	$3d^7(^2P)4s4p(^3P^\circ)\ ^4P^\circ$
		3/2	*41 982.66*	1.732	64		12	$3d^7(^4P)4s4p(^3P^\circ)\ ^4P^\circ$
$3d^7(^2G)4s4p(^3P^\circ)$	$^4H^\circ$	7/2	*42 988.12*		95			
$3d^8(^1D)4p$	$z\ ^2P^\circ$	1/2	*43 130.24*	0.727	42		25	$3d^8(^3P)4p\ ^2P^\circ$
$3d^7(^4P)4s4p(^3P^\circ)$	$w\ ^4D^\circ$	5/2	*43 242.95*	1.101	46		28	$3d^8(^1D)4p\ ^2F^\circ$
		3/2	*43 263.57*	1.191	58		9	$3d^8(^1D)4p\ ^2D^\circ$
		7/2	*43 398.62*	1.334	88		3	$3d^8(^1D)4p\ ^2F^\circ$
		1/2	*43 435.58*	0.169	75		9	$3d^7(^2P)4s4p(^3P^\circ)\ ^4D^\circ$
$3d^7(^2G)4s4p(^3P^\circ)$	$w\ ^4F^\circ$	9/2	*43 295.32*	1.295	71		15	$3d^7(^4F)4s4p(^1P^\circ)\ ^4F^\circ$
		3/2	*44 555.71*	0.415	68		18	$3d^7(^4F)4s4p(^1P^\circ)\ ^4F^\circ$
$3d^8(^1D)4p$		5/2	*43 425.71*	1.119	37	$^2F^\circ$	36	$3d^7(^4P)4s4p(^3P^\circ)\ ^4D^\circ$
$3d^8(^1D)4p$		3/2	*43 537.71*	1.120	31	$^2P^\circ$	23	$3d^8(^1D)4p\ ^2D^\circ$
$3d^8(^1D)4p$	$x\ ^2F^\circ$	7/2	*43 555.22*	1.229	45		20	$3d^7(^2G)4s4p(^3P^\circ)\ ^4F^\circ$
$3d^7(^2G)4s4p(^3P^\circ)$		7/2	*43 847.98*	1.197	39	$^4F^\circ$	23	$3d^8(^1D)4p\ ^2F^\circ$
$3d^8(^1D)4p$	$x\ ^2D^\circ$	3/2	*43 911.36*	1.127	40		33	$3d^8(^1D)4p\ ^2P^\circ$
		5/2	*43 921.89*	1.230	52		20	$3d^7(^2G)4s4p(^3P^\circ)\ ^4F^\circ$
$3d^7(^2G)4s4p(^3P^\circ)$	$w\ ^4G^\circ$	11/2	*43 952.06?*	1.279	93			
		9/2	*44 183.34*	1.163	85		5	$3d^7(^2G)4s4p(^3P^\circ)\ ^4F^\circ$
		7/2	*44 394.47*	1.004	77		11	
		5/2	*44 568.47*	0.676	69		19	
	4°	7/2	*43 969.90*	1.081				
$3d^7(^2G)4s4p(^3P^\circ)$		5/2	*44 201.92*	0.950	29	$^4F^\circ$	23	$3d^8(^1D)4p\ ^2D^\circ$
	5°	3/2	*44 381.32?*					
$3d^7(^4P)4s4p(^3P^\circ)$	$z\ ^2S^\circ$	1/2	*44 454.51*	2.10	61		21	$3d^8(^3P)4p\ ^2S^\circ$
$3d^7(^4P)4s4p(^1P^\circ)$	$y\ ^4P^\circ$	5/2	*44 480.14*	1.557	42		33	$(^2P)(^3P^\circ)\ ^4P^\circ$
		3/2	*44 658.03*	1.674	40		37	$(^4P)(^3P^\circ)\ ^4P^\circ$
$3d^7(^2P)4s4p(^3P^\circ)$	$^4P^\circ$	1/2	*44 857.57*	2.371	44		40	$(^4P)(^1P^\circ)\ ^4P^\circ$
$3d^8(^3F)5s$	$e\ ^4F$	9/2	44 782.13	1.33				
		7/2	45 105.59	1.21				
		5/2	45 876.58	1.01				
		3/2	46 375.17	0.44				
$3d^7(^2G)4s4p(^3P^\circ)$	$z\ ^2H^\circ$	9/2	*45 111.48*	0.897	87			
		11/2	*45 540.28*	1.097	90			

Co I—Continued

Configuration	Term	J	Level (cm^{-1})	g	Leading percentages			
$3d^7\,4s(^5F)5s$	$e\,^6F$	$11/2$	45 676.00	1.475				
		$9/2$	46 223.01	1.443				
		$7/2$	46 706.83	1.396				
		$5/2$	47 090.65	1.301				
		$3/2$	47 364.73	1.054				
		$1/2$	47 528.44	-0.666				
$3d^8(^3P)4p$	$^2D^\circ$	$5/2$	45 688.15	1.219	70		8	$3d^7(^4P)4s4p(^3P^\circ)\ ^2D^\circ$
		$3/2$	46 186.41	1.218	72		7	$3d^8(^1D)4p\ ^2D^\circ$
$3d^7(^2G)4s4p(^3P^\circ)$	$x\,^2G^\circ$	$7/2$	45 766.63	0.898	79		10	$(^2H)(^3P^\circ)\ ^2G^\circ$
		$9/2$	46 032.10	1.131	80		10	
$3d^8(^3P)4p$	$^4S^\circ$	$3/2$	45 904.68	1.674	32		25	$3d^7(^4P)4s4p(^3P^\circ)\ ^4S^\circ$
$3d^8(^3F)5s$	$e\,^2F$	$7/2$	45 924.98	1.14				
		$5/2$	46 746.00	0.49				
$3d^7(^4P)4s4p(^3P^\circ)$	$x\,^4P^\circ$	$1/2$	45 957.29	2.522	44		22	$3d^8(^3P)4p\ ^4P^\circ$
		$5/2$	46 002.83	1.543	30		24	$3d^7(^2P)4s4p(^3P^\circ)\ ^4P^\circ$
		$3/2$	46 260.02	1.508	30		18	$3d^8(^3P)4p\ ^4P^\circ$
$3d^8(^3P)4p$	$^4D^\circ$	$7/2$	45 971.19	1.424	44		22	$3d^7(^2P)4s4p(^3P^\circ)\ ^4D^\circ$
		$5/2$	46 329.63	1.365	28		25	$3d^7(^2P)4s4p(^3P^\circ)\ ^4D^\circ$
		$1/2$	46 502.15	0.161	27		21	$3d^7(a\ ^2D)4s4p(^3P^\circ)\ ^4D^\circ$
$3d^7(^4P)4s4p(^3P^\circ)$	$^2D^\circ$	$3/2$	46 454.95	0.869	61		14	$(^2P)(^3P^\circ)\ ^2D^\circ$
		$5/2$	46 671.94	1.233	45		20	$(^2G)(^3P^\circ)\ ^2F^\circ$
$3d^8(^3P)4p$		$3/2$	46 562.87	1.273	16	$^2P^\circ$	14	$3d^8(^3P)4p\ ^4D^\circ$
$3d^8(^3P)4p$		$3/2$	46 685.43	1.352	27	$^2P^\circ$	12	$3d^8(^1D)4p\ ^2P^\circ$
$3d^7(^2P)4s4p(^3P^\circ)$		$7/2$	46 872.74	1.332	26	$^4D^\circ$	23	$3d^7(a\ ^2D)4s4p(^3P^\circ)\ ^4D^\circ$
$3d^8(^3P)4p$	$y\,^2P^\circ$	$1/2$	47 091.14	0.656	46		30	$(^1D)\ ^2P^\circ$
$3d^7(^2G)4s4p(^3P^\circ)$		$5/2$	47 128.96	0.858	33	$^2F^\circ$	22	$(^4P)(^3P^\circ)\ ^2D^\circ$
$3d^7(^2G)4s4p(^3P^\circ)$		$7/2$	47 225.11	1.229	36	$^2F^\circ$	31	$(^2P)(^3P^\circ)\ ^4D^\circ$
$3d^7(^2P)4s4p(^3P^\circ)$	$u\,^4D^\circ$	$5/2$	47 393.93	1.324	45		13	$3d^7(^4P)4s4p(^3P^\circ)\ ^2D^\circ$
		$3/2$	47 612.18	1.122	47		18	$3d^8(^3P)4p\ ^4D^\circ$
$3d^7\,4s(^5F)5s$	$f\,^4F$	$9/2$	47 524.47	1.328				
		$7/2$	48 201.60	1.226				
		$5/2$	48 718.57	1.041				
		$3/2$	49 078.43	0.401				
	6°	$7/2$	47 839.15					
$3d^7(^2P)4s4p(^3P^\circ)$		$1/2$	47 905.26	0.016	37	$^4D^\circ$	21	$3d^8(^3P)4p\ ^2S^\circ$
	$y\,^2S^\circ$	$1/2$	47 977.94	2.093				
$3d^8(^3P)4p$		$1/2$	48 026.34	1.699	35	$^2S^\circ$	26	$3d^7(^2P)4s4p(^3P^\circ)\ ^4D^\circ$

(Continued)

Co I—Continued

Configuration	Term	J	Level (cm^{-1})	g	Leading percentages			
$3d^7(a\ ^2D)4s4p(^3P^\circ)$	$t\ ^4D^\circ$	7/2	48 217.32	1.211	51		18	$3d^8(^3P)4p\ ^4D^\circ$
		5/2	48 443.76	1.340	52		19	
		1/2	48 571.77	0.452	52		16	
$3d^7(^2H)4s4p(^3P^\circ)$	$^4G^\circ$	7/2	48 317.17	1.173	84		4	$(^2G)(^3P^\circ)\ ^2F^\circ$
		5/2	48 615.56?	0.619	81		7	
$3d^7(^4P)4s4p(^3P^\circ)$	$^2P^\circ$	3/2	48 334.37	1.436	40		15	$(^2P)(^3P^\circ)\ ^4S^\circ$
		1/2	48 837.72	1.50	60		12	$(^2P)(^3P^\circ)\ ^2S^\circ$
$3d^7(a\ ^2D)4s4p(^3P^\circ)$		3/2	48 546.07	1.050	21	$^4D^\circ$	22	$(^2P)(^3P^\circ)\ ^4S^\circ$
$3d^7(^2P)4s4p(^3P^\circ)$	$x\ ^4S^\circ$	3/2	48 753.72	1.728	48		26	$(a\ ^2D)(^3P^\circ)\ ^4D^\circ$
$3d^7(a\ ^2D)4s4p(^3P^\circ)$		5/2	48 828.87		34	$^4P^\circ$	32	$(^2P)(^3P^\circ)\ ^2D^\circ$
	8°	3/2	48 851.58?					
$3d^7(^2P)4s4p(^3P^\circ)$		3/2	49 025.42	1.099	39	$^2D^\circ$	34	$3d^7(^4P)4s4p(^3P^\circ)\ ^2P^\circ$
$3d^7(a\ ^2D)4s4p(^3P^\circ)$	$^4F^\circ$	9/2	49 197.74?		98			
		7/2	49 484.05	1.260	75		6	$(^2P)(^3P^\circ)\ ^4D^\circ$
		5/2	49 847.08	1.079	43		24	$(^2P)(^3P^\circ)\ ^2D^\circ$
		3/2	50 105.05	0.569	50		15	$(^2P)(^3P^\circ)\ ^2D^\circ$
	$w\ ^2P^\circ$	1/2	49 754.73	1.365				
$3d^8(^1G)4p$	$y\ ^2H^\circ$	9/2	50 210.80	0.899	81		9	$3d^7(^2H)4s4p(^3P^\circ)\ ^2H^\circ$
		11/2	50 375.91	1.091	69		17	
$3d^8(^1G)4p$	$u\ ^2F^\circ$	7/2	50 578.73	1.125	60		12	$3d^7(a\ ^2D)4s4p(^3P^\circ)\ ^2F^\circ$
		5/2	50 712.45	0.905	57		17	
$3d^8(^1G)4p$	$w\ ^2G^\circ$	9/2	50 593.38	1.10	40		38	$3d^7(^2H)4s4p(^3P^\circ)\ ^2G^\circ$
		7/2	50 611.22	0.82	38		37	
$3d^7(^2H)4s4p(^3P^\circ)$	$^4H^\circ$	11/2	50 703.08	1.110	86		11	$3d^8(^1G)4p\ ^2H^\circ$
		9/2	50 902.61	0.941	86		5	$3d^8(^1G)4p\ ^2G^\circ$
		7/2	51 184.63		89		5	$3d^8(^1G)4p\ ^2G^\circ$
	13°	7/2,9/2	50 738.20					
	$s\ ^4D^\circ$	7/2	50 741.66	1.458				
		5/2	51 139.38					
		3/2	51 847.27					
		1/2	52 264.01?					
	14°	5/2	50 806.55	1.136				
$3d^7(^2P)4s4p(^3P^\circ)$	$v\ ^2P^\circ$	3/2	50 925.11	1.340	53		15	$(a\ ^2D)(^3P^\circ)\ ^4D^\circ$
		1/2	50 945.47	0.732	71		13	
$3d^8(^3F)4d$	$e\ ^4P$	5/2	51 042.26	1.59				
		3/2	52 033.26	1.40				
		1/2	52 915.92?					

Co I—Continued

Configuration	Term	J	Level (cm^{-1})	g	Leading percentages		
$3d^8(^3F)4d$	$e\ ^4D$	7/2	51 052.98	1.402			
		5/2	51 560.76	1.354			
		3/2	52 264.49?				
		1/2	52 634.62	1.58			
$3d^8(^3F)4d$	$e\ ^4H$	13/2	51 142.53	1.224			
		11/2	51 174.28	1.147			
		9/2	52 121.21	0.96			
		7/2	52 716.70	0.896			
$3d^7(a\ ^2D)4s4p(^3P^\circ)$	$w\ ^4P^\circ$	5/2	*51 160.03*	1.578	73	13	$(^2P)(^3P^\circ)\ ^4P^\circ$
		3/2	*52 014.45*	1.616	57	17	$(^2P)(^3P^\circ)\ ^2P^\circ$
		1/2	*52 355.12*	2.304	66	12	$(^2P)(^3P^\circ)\ ^4P^\circ$
$3d^8(^3F)4d$	$g\ ^4F$	9/2	51 170.14	1.337			
		7/2	51 199.58	1.16			
		5/2	52 070.00	1.08			
		3/2	52 702.76?	0.76			
$3d^8(^3F)4d$	$e\ ^2P$	3/2	51 200.60	1.368			
		1/2	52 041.14	0.48			
$3d^8(^3F)4d$	$e\ ^4G$	11/2	51 203.75	1.218			
		9/2	51 267.93	1.083			
		7/2	52 162.02	1.13			
		5/2	52 772.30	0.74			
	16°	5/2	*51 863.18*				
	17°	5/2	*51 989.31?*				
$3d^8(^3F)4d$	$f\ ^2F$	7/2	52 095.00	1.118			
		5/2	52 970.62	1.13			
$3d^8(^3F)4d$	$e\ ^2H$	11/2	52 113.91	1.13			
		9/2	52 775.47	0.97			
$3d^8(^3F)4d$	$e\ ^2G$	9/2	52 156.46	1.12			
		7/2	52 856.68	0.92			
$3d^8(^3F)4d$	$e\ ^2D$	5/2	52 460.10	0.92			
		3/2	53 343.27	0.80			
	18°	7/2	*52 476.64*				
	19°	3/2	*52 498.17*				
	20°	3/2,5/2	*52 526.04*				
$3d^7\ 4s(^3F)5s$	$g\ ^2F$	7/2	52 763.68	0.933			
		5/2	53 704.14	0.923			
$3d^7(a\ ^2D)4s4p(^3P^\circ)$	$^2F^\circ$	5/2	*52 796.13*	0.883	58	17	$3d^8(^1G)4p\ ^2F^\circ$
		7/2	*53 103.78*	1.136	75	11	$3d^7(^2G)4s4p(^3P^\circ)\ ^2F^\circ$
$3d^7\ 4s(^3F)5s$	$h\ ^4F$	9/2	52 864.41	1.307			
		7/2	53 694.57?	1.28			
		5/2	54 258.75?	0.986			
		3/2	54 426.64	0.422			

(Continued)

Co I—Continued

Configuration	Term	J	Level (cm^{-1})	g	Leading percentages		
	21°	7/2,9/2	*53 065.96*				
$3d^7(a\ ^2D)4s4p(^3P°)$	$u\ ^2D°$	3/2	*53 074.92*	0.823	51	21	$(^2P)(^3P°)\ ^2D°$
		5/2	*53 195.98*	1.206	49	18	
$3d^7(^2H)4s4p(^3P°)$	$v\ ^2G°$	9/2	*53 276.02*	1.124	44	28	$3d^8(^1G)4p\ ^2G°$
		7/2	*53 373.53*	0.888	44	30	
	22°	7/2	*53 463.10*				
$3d^7\ 4s(^5F)4d$	$f\ ^4G$	11/2	53 511.83	1.274			
		9/2	54 158.17	1.25			
		7/2	54 514.67	1.23			
		5/2	55 165.63?				
$3d^7\ 4s(^5F)4d$	$f\ ^4H$	13/2	53 618.08	1.227			
		11/2	54 315.67	1.168			
		9/2	54 860.93	1.10			
		7/2	55 268.75	0.857			
$3d^7\ 4s(^5F)4d$	$f\ ^6F$	11/2	53 660.37	1.421			
		9/2	54 356.45	1.403			
		7/2	54 896.57	1.27			
		5/2	55 283.02	1.17			
		3/2	55 577.28?	1.07			
$3d^7\ 4s(^5F)4d$	$f\ ^4D$	7/2	53 702.13	1.377			
		5/2	54 282.73?				
$3d^7\ 4s(^5F)4d$	$e\ ^6D$	9/2	53 725.20	1.387			
		7/2	54 352.30	1.48			
		5/2	54 946.90	1.47			
		3/2	55 407.10?	2.14			
$3d^7\ 4s(^5F)4d$	$e\ ^6G$	13/2	53 728.36?	1.35			
		11/2	54 367.43	1.320			
		9/2	54 682.91	1.23			
		7/2	54 989.62?	1.23			
		3/2	55 389.73				
		5/2	55 449.97	1.199			
$3d^7\ 4s(^5F)4d$	$i\ ^4F$	9/2	53 788.78	1.316			
		7/2	54 477.07				
		5/2	54 904.99	0.85			
$3d^7\ 4s(^5F)4d$	$e\ ^6P$	7/2	53 789.12	1.635			
		5/2	54 445.61	1.594			
		3/2	54 949.97	2.134			
$3d^7\ 4s(^5F)4d$	$e\ ^6H$	15/2	53 822.08	1.34			
		13/2	54 452.38	1.289			
		11/2	54 947.68	1.22			
		9/2	55 312.96	1.108			
		7/2	55 520.64	0.922			
		5/2	55 555.34				
$3d^7\ 4s(^5F)4d$	$f\ ^4P$	5/2	53 936.68	1.46			
$3d^7(a\ ^2D)4s4p(^3P°)$	$^2P°$	3/2	*54 165.35*	1.353	87		

Co I—Continued

Configuration	Term	J	Level (cm^{-1})	g	Leading percentages
	24°	7/2	*54 398.60*		
	1	5/2	54 561.74	1.323	
	2	3/2	55 078.76	1.46	
	3	7/2	55 223.14	1.14	
	4	5/2	55 598.74		
	5	3/2, 5/2	55 721.01		
	6	1/2	55 826.81		
	v ^{4}F°	9/2	*54 791.2*		
		7/2	*55 314.04?*		
		3/2	*55 622.84*		
		5/2	*55 684.7*		
	25°	9/2	*54 874.08*		
	26°	7/2	*54 932.32*		
	27°	5/2	*55 061.49*	1.539	
	28°	7/2	*55 120.30?*		
	29°	5/2	*55 387.11?*	1.170	
	30°	5/2	*55 508.78*		
	31°	3/2	*55 737.87*		
	32°	3/2	*55 818.91*	1.31	
	33°	3/2, 5/2	*55 922.3*		
$3d^8$(^{3}P)$5s$	g ^{4}P	5/2	56 545.51?		
	g ^{4}H	13/2	57 922.06		
		11/2	58 441.03		
		9/2	58 673.73		
		7/2	59 314.82?		
	36°	7/2, 9/2	*58 187.39*		
	37°	5/2, 7/2	*59 388.89*		
Co II (^{3}F$_4$)	***Limit***		**63 400**		

Co II

$Z=27$

Fe I isoelectronic sequence

Ground state: $1s^22s^22p^63s^23p^63d^8\ ^3F_4$

Ionization energy = 137 795±10 cm^{-1} (17.083±0.001 eV)

The first analysis of Co II was by Meggers (1928), who measured the spectrum between 2150 and 5000 Å, found eight multiplets and identified 14 lines of Co II in the solar spectrum. The $3d^8\ ^3$F ground state could not be found until Findlay (1930) extended the observations to 1940 Å. Findlay corrected some of Meggers' assignments on the basis of Zeeman effects he observed in the $3d^74s-3d^74p$ transition array.

The analysis was extended to include terms from the configurations $3d^64s^2$, $3d^64s4p$, and $3d^75s$ by Hagar (1951) and by Velasco and Adames (1966). The latter paper includes a table of about 1500 observed lines and a table of about 450 classified lines.

The present compilation is taken from the papers of Iglesias (1972, 1979). She has extended the analysis to include terms of the configurations $3d^76s$, $7s$, $5p$, $4d$, $5d$, $6d$ and $7d$. She has measured most of the g-values from Zeeman patterns obtained by Catalán at the Massachusetts Institute of Technology in 1949. A few of the g-values are from Findlay. The uncertainty in the level values given with two decimal places is probably ±0.05 cm^{-1}, and those given with on decimal place ±0.1 cm^{-1}.

Roth (1969) has calculated the odd-parity configuration $3d^74p$. His leading percentages for the experimental levels are listed here. Roth distinguished repeating terms of the $3d^7$ core by the letters, a, b,... rather than by seniority. The percentages include the sum of seniority states contributing to the term.

The compositions of levels of the even configurations $3d^8$, $3d^74s$, and $3d^64s^2$ were calculated by Shadmi, Oreg, and Stein (1968). Percentages are given for only two levels, where the mixing is large. Roth's designation of repeating core terms are used in this work is well.

The ionization energy was calculated by Iglesias from ns and nd series ($n=5$–7). She obtained an average value of 137 780 cm^{-1}. The two limits differ by 23 cm^{-1}. We have adopted a value we calculated from the 4 members of the $ns\ ^5F_5$ series ($n=4$–7).

References

Findlay, J. H. (1930), Phys. Rev. **36**, 5.
Hagar, N. E. (1951), Dissertation (Princeton Univ.).
Iglesias, L. (1972), Opt. Pur. Apl. **5**, 195.
Iglesias, L. (1979), Opt. Pur. Apl. **12**, 63.
Meggers, W. F. (1928), J. Wash. Acad. Sci. **18**, 325.
Roth, C. (1969), J. Res. Natl. Bur. Stand. **73A**, 125.
Shadmi, T., Oreg. J., and Stein, J. (1968), J. Opt. Soc. Am. **58**, 909.
Velasco, R., and Adames, J. (1966), Publ. Inst. Opt. Madrid., No. 26.

Co II

Configuration	Term	J	Level (cm^{-1})	g	Leading percentages
$3d^8$	$a\ ^3$F	4	0.00		
		3	950.51		
		2	1 597.32		
$3d^7(^4F)4s$	$a\ ^5$F	5	3 350.58	1.413	
		4	4 029.00	1.354	
		3	4 560.81	1.258	
		2	4 950.20	0.997	
		1	5 204.82	0.00	
$3d^7(^4F)4s$	$b\ ^3$F	4	9 812.96	1.243	
		3	10 708.47	1.082	
		2	11 321.96	0.68	
$3d^8$	$a\ ^1$D	2	11 651.48	1.111	
$3d^8$	$a\ ^3$P	2	13 260.77	1.415	
		1	13 404.48	1.484	
		0	13 593.32		
$3d^7(^4P)4s$	$a\ ^5$P	3	17 771.71	1.68	
		2	18 031.65	1.839	
		1	18 338.80	2.510	

Co II—Continued

Configuration	Term	J	Level (cm^{-1})	g	Leading percentages		
$3d^8$	$a\ ^1\mathrm{G}$	4	19 190.11				
$3d^7(^2\mathrm{G})4s$	$a\ ^3\mathrm{G}$	5	21 624.64	1.186			
		4	22 009.47	1.066			
		3	22 414.60	0.75			
$3d^7(^4\mathrm{P})4s$	$b\ ^3\mathrm{P}$	2	24 074.60	1.500			
		1	24 267.57	1.498			
		0	24 411.36				
$3d^7(^2\mathrm{P})4s$	$c\ ^3\mathrm{P}$	2	24 886.57	1.49			
		1	25 317.65	1.47			
		0	25 861.60				
$3d^7(^2\mathrm{G})4s$	$b\ ^1\mathrm{G}$	4	25 147.37	0.997			
$3d^7(^2\mathrm{H})4s$	$a\ ^3\mathrm{H}$	6	27 105.96	1.172			
		5	27 469.13	1.041			
		4	27 902.37	0.803			
$3d^7(a\ ^2\mathrm{D})4s$	$a\ ^3\mathrm{D}$	3	27 484.61	1.36			
		2	28 112.06	1.18			
		1	29 268.97	0.77	52	45	$(^2\mathrm{P})\ ^1\mathrm{P}$
$3d^7(^2\mathrm{P})4s$	$a\ ^1\mathrm{P}$	1	27 585.31	0.83	48	43	$(a\ ^2\mathrm{D})\ ^3\mathrm{D}$
$3d^7(^2\mathrm{H})4s$	$a\ ^1\mathrm{H}$	5	30 567.36	1.027			
$3d^7(a\ ^2\mathrm{D})4s$	$b\ ^1\mathrm{D}$	2	31 199.52	1.02			
$3d^6\ 4s^2$	$a\ ^5\mathrm{D}$	4	40 695.60				
		3	41 314.20				
		2	41 738.3				
		1	42 008.9				
$3d^7(^2\mathrm{F})4s$	$c\ ^3\mathrm{F}$	2	40 771.11				
		3	40 879.44				
		4	41 047.06				
$3d^7(^2\mathrm{F})4s$	$a\ ^1\mathrm{F}$	3	44 090.94?				
$3d^7(^4\mathrm{F})4p$	$z\ ^5\mathrm{F}^\circ$	5	*45 197.78*	1.386	96		
		4	*45 378.85*	1.42	73	22	$(^4\mathrm{F})\ ^5\mathrm{D}^\circ$
		3	*45 972.17*	1.30	83	13	$(^4\mathrm{F})\ ^5\mathrm{D}^\circ$
		2	*46 452.82*	1.062	91	7	$(^4\mathrm{F})\ ^5\mathrm{D}^\circ$
		1	*46 786.53*	0.00	97		
$3d^7(^4\mathrm{F})4p$	$z\ ^5\mathrm{D}^\circ$	4	*46 320.96*	1.447	72	20	$(^4\mathrm{F})\ ^3\mathrm{F}^\circ$
		3	*47 039.27*	1.442	80	11	$(^4\mathrm{F})\ ^5\mathrm{F}^\circ$
		2	*47 537.48*	1.468	85	5	$(^4\mathrm{P})\ ^5\mathrm{D}^\circ$
		1	*47 848.89*	1.436	92	6	$(^4\mathrm{P})\ ^5\mathrm{D}^\circ$
		0	*47 995.74*		94	6	$(^4\mathrm{P})\ ^5\mathrm{D}^\circ$
$3d^7(^4\mathrm{F})4p$	$z\ ^5\mathrm{G}^\circ$	6	*47 078.74*	1.33	100		
		5	*47 345.94*	1.260	84	11	$(^4\mathrm{F})\ ^3\mathrm{G}^\circ$
		4	*47 807.58*	1.154	87	6	$(^4\mathrm{F})\ ^3\mathrm{F}^\circ$
		3	*48 151.07*	0.927	90	5	$(^4\mathrm{F})\ ^5\mathrm{F}^\circ$
		2	*48 388.62*	0.341	93	3	$(^4\mathrm{F})\ ^5\mathrm{F}^\circ$

(Continued)

Co II—Continued

Configuration	Term	J	Level (cm^{-1})	g	Leading percentages			
$3d^7(^4F)4p$	$z\ ^3G^\circ$	5	48 556.16	1.194	88		12	$(^4F)\ ^5G^\circ$
		4	49 348.43	1.11	59		36	$(^4F)\ ^3F^\circ$
		3	50 036.55	0.80	75		22	$(^4F)\ ^3F^\circ$
$3d^7(^4F)4p$	$z\ ^3F^\circ$	4	49 697.81	1.19	60		34	$(^4F)\ ^3G^\circ$
		3	50 381.86	1.055	69		22	$(^4F)\ ^3G^\circ$
		2	50 914.51	0.688	94		3	$(^2G)\ ^3F^\circ$
$3d^7(^4F)4p$	$z\ ^3D^\circ$	3	51 512.41	1.32	92		4	$(^4F)\ ^3F^\circ$
		2	52 229.92	1.161	94			
		1	52 684.77	0.50	95			
$3d^7(^4P)4p$	$z\ ^5S^\circ$	2	56 010.62	2.01	99			
$3d^7(^4P)4p$	$y\ ^5D^\circ$	3	61 240.96	1.505	89		5	$(^4F)\ ^5D^\circ$
		2	61 260.55	1.502	88		5	
		1	61 350.42	1.51	87		6	
		4	61 388.43	1.51	95		4	
		0	61 458.22		92		6	
$3d^7(^4P)4p$	$z\ ^3S^\circ$	1	62 440.62	2.06	60		16	$(^4P)\ ^5P^\circ$
$3d^7(^4P)4p$	$z\ ^5P^\circ$	3	63 344.50	1.67	82		13	$(^4P)\ ^3D^\circ$
		2	63 367.43	1.697	67		21	$(^4P)\ ^3D^\circ$
		1	63 665.32	2.377	77		19	$(^4P)\ ^3S^\circ$
$3d^7(^2G)4p$	$y\ ^3F^\circ$	4	63 510.40	1.152	66		12	$(^2G)\ ^3H^\circ$
		3	64 360.26	1.06	83		10	$(^2G)\ ^3G^\circ$
		2	65 017.06	0.78	95			
$3d^7(^4P)4p$	$y\ ^3D^\circ$	3	63 587.01	1.35	77		14	$(^4P)\ ^5P^\circ$
		2	63 616.12	1.32	57		27	$(^4P)\ ^5P^\circ$
		1	63 865.30	0.58	81		7	$(^2P)\ ^3D^\circ$
$3d^7(^2G)4p$	$z\ ^3H^\circ$	5	63 306.66	1.03	78		15	$(^2G)\ ^1H^\circ$
		6	63 597.59	1.16	97			
		4	63 792.90	0.86	83		8	$(^2G)\ ^3G^\circ$
$3d^7(^2G)4p$	$z\ ^1G^\circ$	4	64 401.46	1.07	46		23	$(^2G)\ ^3F^\circ$
$3d^7(^2G)4p$	$y\ ^3G^\circ$	5	64 601.72	1.16	62		35	$(^2G)\ ^1H^\circ$
		4	65 154.18	1.06	67		22	$(^2G)\ ^1G^\circ$
		3	65 174.89	0.78	85		9	$(^2G)\ ^3F^\circ$
$3d^7(^2P)4p$	$z\ ^3P^\circ$	0	64 605.65		57		20	$(a\ ^2D)\ ^3P^\circ$
		2	64 914.81	1.430	45		35	$(^4P)\ ^3P^\circ$
		1	65 028.63	1.49	59		17	$(a\ ^2D)\ ^3P^\circ$
$3d^7(^2G)4p$	$z\ ^1H^\circ$	5	64 957.50	1.05	48		32	$(^2G)\ ^3G^\circ$
$3d^7(^4P)4p$	$y\ ^3P^\circ$	2	65 408.59	1.50	53		23	$(^2P)\ ^3P^\circ$
		0	65 512.12		63		24	
		1	65 657.58	1.50	83		8	
$3d^7(^2G)4p$	$z\ ^1F^\circ$	3	66 017.79	1.01	72		18	$(a\ ^2D)\ ^1F^\circ$
$3d^7(^2P)4p$		2	67 209.42	1.05	37	$^1D^\circ$	19	$(^2P)\ ^3D^\circ$

Co II—Continued

Configuration	Term	J	Level (cm⁻¹)	g	Leading percentages			
$3d^7(^2P)4p$	$x\ ^3D^\circ$	3	*67 524.20*	1.33	81		10	$(a\ ^2D)\ ^3F^\circ$
		1	*67 824.65*	0.56	68		18	$(a\ ^2D)\ ^3D^\circ$
		2	*67 939.51*	1.15	54		21	$(^2P)\ ^1D^\circ$
$3d^7(^2H)4p$	$x\ ^3G^\circ$	5	*68 203.39*	1.18	95			
		4	*68 843.67*	1.05	92			
		3	*69 356.64*	0.78	85		4	$(a\ ^2D)\ ^3F^\circ$
$3d^7(^2H)4p$	$z\ ^3I^\circ$	6	*68 311.47*	1.02	75		24	$(^2H)\ ^1I^\circ$
		7	*68 614.25*	1.14	100			
		5	*68 829.56*	0.81	97			
$3d^7(a\ ^2D)4p$	$w\ ^3D^\circ$	3	*69 060.26*	1.31	85		8	$(a\ ^2D)\ ^3F^\circ$
		1	*69 317.45*	0.75	50		25	$(^2P)\ ^1P^\circ$
		2	*69 637.27*	1.16	67		11	$(^2P)\ ^3D^\circ$
$3d^7(^2H)4p$	$z\ ^1I^\circ$	6	*69 617.60*	1.02	75		24	$(^2H)\ ^3I^\circ$
$3d^7(a\ ^2D)4p$	$x\ ^3F^\circ$	4	*70 186.30*	1.26	98			
		3	*70 457.93*	1.09	70		8	$(^2P)\ ^3D^\circ$
		2	*70 775.25*	0.78	75		9	$(a\ ^2D)\ ^3D^\circ$
$3d^7(^2P)4p$	$y\ ^3S^\circ$	1	*70 266.20*		53		16	$(^2P)\ ^1P^\circ$
$3d^7(^2P)4p$		1	*70 857.1*	0.90	38	$^1P^\circ$	27	$(^2P)\ ^3S^\circ$
$3d^7(^2H)4p$	$y\ ^3H^\circ$	6	*71 404.0*	1.16	99			
		5	*71 703.2*		96			
		4	*72 009.0*	0.70	96			
$3d^7(a\ ^2D)4p$	$x\ ^3P^\circ$	2	*71 846.75*		46		30	$(a\ ^2D)\ ^1D^\circ$
		1	*73 144.56*		68		12	$(^2P)\ ^3P^\circ$
		0	*73 440.00*		79		17	$(^2P)\ ^3P^\circ$
$3d^7(a\ ^2D)4p$	$y\ ^1F^\circ$	3	*72 535.35*	1.04	72		16	$(^2G)\ ^1F^\circ$
$3d^7(a\ ^2D)4p$		2	*72 654.32?*		39	$^1D^\circ$	32	$(a\ ^2D)\ ^3P^\circ$
$3d^7(^2H)4p$	$y\ ^1G^\circ$	4	*73 147.23*	1.04	76		22	$(^2G)\ ^1G^\circ$
$3d^7(^2H)4p$	$y\ ^1H^\circ$	5	*74 377.07*	1.03	97			
$3d^7(a\ ^2D)4p$	$y\ ^1P^\circ$	1	*74 522.96*	1.03	86		8	$(^2P)\ ^1P^\circ$
$3d^6(^5D)4s4p$	$x\ ^5D^\circ$	4	*80 299.5*					
		3	*80 543.3*					
		2	*80 788.7*					
		1	*80 971.5*					
		0	*81 066.6*					
$3d^6(^5D)4s4p$	$y\ ^5F^\circ$	5	*81 970.7*					
		4	*82 343.5*					
		3	*82 627.5*					
		2	*82 830.0*					
		1	*82 959.9*					
$3d^7(^2F)4p$	$w\ ^3G^\circ$	3	*83 110.53*					
		4	*83 398.6*					
		5	*83 877.6*					

(Continued)

Co II—Continued

Configuration	Term	J	Level (cm^{-1})	g	Leading percentages
$3d^7(^2F)4p$	$w\ ^3F°$	3	*83 862.4 ?*		
		4	*84 140.38?*		
$3d^7(^4F)5s$	$e\ ^5F$	5	84 014.26		
		4	84 586.96	1.32	
		3	85 167.78	1.29	
		2	85 596.28	1.00	
		1	85 876.40		
$3d^7(^2F)4p$	$v\ ^3D°$	3	*84 216.4*		
		2	*84 394.6*		
		1	*84 419.6*		
$3d^6(^5D)4s4p$	$y\ ^5P°$	3	*85 044.3*		
		2	*85 712.6*		
		1	*86 133.8 ?*		
$3d^7(^4F)5s$	$e\ ^3F$	4	85 481.80	1.26	
		3	86 346.85	1.09	
		2	86 940.29	0.68	
$3d^6(^5D)4s4p$	$u\ ^3D°$	3	*86 981.3*		
		2	*87 490.0*		
		1	*87 847.3*		
$3d^6(^5D)4s4p$	$v\ ^3F°$	4	*87 112.5*		
		3	*87 896.2*		
		2	*88 417.5*		
$3d^7(^2F)4p$	$x\ ^1F°$	3	*87 633.0 ?*	1.00	98
$3d^7(^4F)4d$	$f\ ^5F$	5	89 931.64		
		4	90 199.41		
		3	90 753.43		
		2	90 967.82		
		1	91 359.20		
$3d^7(^4F)4d$	$e\ ^5G$	6	90 303.00		
		5	90 697.69		
		4	91 049.62		
		3	91 667.63		
		2	91 929.21		
$3d^7(^4F)4d$	$e\ ^5P$	3	90 464.80		
		2	91 425.12		
		1	91 932.31		
$3d^7(^4F)4d$	$e\ ^5H$	7	90 668.10		
		6	90 975.76		
		5	91 646.58		
		4	92 121.10		
		3	92 406.25		
$3d^7(^4F)4d$	$e\ ^3G$	5	91 327.06		
		4	91 918.35		
		3	92 520.29		

Co II—Continued

Configuration	Term	J	Level (cm^{-1})	g	Leading percentages
$3d^7(^4F)4d$	$e\ ^5D$	4	91 352.96		
		3	91 958.58		
		2	92 317.51		
$3d^7(^4F)4d$	$e\ ^3D$	3	91 408.7		
		2	92 045.1		
		1	92 639.95		
$3d^7(^4F)4d$	$e\ ^3H$	6	91 623.54		
		5	92 452.30		
		4	93 019.60		
$3d^7(^4F)4d$	$e\ ^3P$	2	93 678.7		
		1	94 405.7		
		0	94 785.4		
$3d^7(^4F)4d$	$f\ ^3F$	4	93 739.04		
		3	94 368.83		
		2	94 744.0		
$3d^7(^4F)5p$	$^5D°$	4	*95 036.19*		
		3	*95 683.50*		
		2	*96 206.0*		
$3d^7(^4F)5p$	$^5F°$	5	*95 456.10*		
		4	*95 976.63*		
		3	*96 496.16*		
		2	*96 891.07*		
		1	*97 153.84*		
$3d^7(^4F)5p$	$^5G°$	6	*95 998.72*		
		5	*96 236.66*		
		4	*96 882.43*		
		3	*97 335.81*		
		2	*97 628.56*		
$3d^7(^4F)5p$	$^3G°$	5	*96 942.21*		
		4	*97 705.16*		
		3	*98 278.64*		
$3d^7(^4F)5p$	$^3F°$	4	*97 062.95*		
		3	*97 967.79*		
		2	*98 648.67*		
$3d^7(^4F)5p$	$^3D°$	3	*97 496.53*		
		2	*98 287.75?*		
$3d^7(^4P)5s$	$e\ ^5P$	3	99 022.66		
		2	99 235.28		
		1	99 552.67		
$3d^7(^2G)5s$	$f\ ^3G$	5	101 233.65		
		4	101 451.12		
		3	102 020.46		
$3d^7(^2G)5s$	$e\ ^1G$	4	103 787.43		

(Continued)

Co II—Continued

Configuration	Term	J	Level (cm^{-1})	g	Leading percentages
$3d^7(^4F)6s$	^5F	5	108 474.38		
		4	108 819.90		
		3	109 528.3		
$3d^7(^4F)6s$	^3F	4	109 442.2		
$3d^7(^4F)5d$	^5F	5	111 013.85?		
		4	111 145.96		
$3d^7(^4F)5d$	^5G	6	111 141.00?		
		5	111 350.14		
$3d^7(^4F)5d$	^5H	7	111 288.03		
		6	111 438.81		
		5	112 210.70		
$3d^7(^4F)5d$	^3H	6	112 155.67		
$3d^7(^4F)7s$	^5F	5	119 311.7		
		4	119 497.4 ?		
		3	120 228.5 ?		
$3d^7(^4F)7s$	^3F	4	120 201.3		
		3	120 867.0 ?		
$3d^7(^4F)6d$	^5G	6	120 681.76		
$3d^7(^4F)6d$	^5H	7	120 760.3		
		6	120 843.7		
		5	121 651.6		
		4	122 169.2 ?		
$3d^7(^4F)6d$	^3H	6	121 612.00		
$3d^7(^4F)7d$	^5H	7	125 920.7 ?		
		6	125 969.5		
Co III ($^4F_{9/2}$)	*Limit*		**137 795**		

Co III

$Z=27$

Mn I isoelectronic sequence

Ground state: $1s^2 2s^2 2p^6 3s^2 3p^6 3d^7\ ^4F_{9/2}$

Ionization energy = 270 200±500 cm^{-1} (33.50±0.06 eV)

The analysis is by Shenstone (1960) who observed the spectrum from 650–3800 Å. The uncertainty in the energy level values is probably ±0.5 cm^{-1}. The leading percentages of the levels of the $3d^7$ configuration calculated by Pasternak and Goldschmidt (1972) and of $3d^6 4s$ by Vizbaraite, Kupliauskis, and Tutlys (1968) are reported here. These two configurations have also been calculated by Shadmi, Caspi, and Oreg (1969) but no percentages are given.

Roth (1968) calculated the percentages for the $3d^6 4p$ levels. His alphabetic prefixing of $3d^6$ core states denotes the order of similar terms. The percentages include the sum of the seniority states contributing to the core term.

The ionization energy was estimated by Catalán and Velasco (1952) by means of regularities in spectra of the iron group of elements.

References

Catalán, M. A., and Velasco, R. (1952), An. R. Soc. Esp. Fis. Quim. **48A**, 247.

Pasternak, A., and Goldschmidt, Z. B. (1972), Phys. Rev. **6A**, 55.

Roth, C. (1968), J. Res. Natl. Bur. Stand. **72A**, 505.

Shadmi, Y., Caspi, E., and Oreg, J. (1969), J. Res. Natl. Bur. Stand. **73A**, 173.

Shenstone, A. G. (1960), Can. J. Phys. **38**, 677.

Vizbaraite, J., Kupliauskis, Z., and Tutlys. V. (1968), Lietuvos Fizikos Rinkinys **8**, 497.

Co III

Configuration	Term	J	Level (cm^{-1})	Leading percentages		
$3d^7$	a ^{4}F	9/2	0.0	100		
		7/2	841.2	100		
		5/2	1 451.3	100		
		3/2	1 866.8	100		
$3d^7$	a ^{4}P	5/2	15 201.9	100		
		3/2	15 428.2	95	5	^{2}P°
		1/2	15 811.4	98		
$3d^7$	a ^{2}G	9/2	16 977.6	98		
		7/2	17 766.2	100		
$3d^7$	a ^{2}P	3/2	20 194.9	87	6	^{2}D°2
		1/2	20 918.5?	98		
$3d^7$	a ^{2}H	11/2	22 720.3	100		
		9/2	23 434.3	98		
$3d^7$	a ^{2}D2	5/2	23 058.8	76	23	^{2}D°1
		3/2	24 236.8	72	19	
$3d^7$	a ^{2}F	5/2	37 021.0	100		
		7/2	37 316.5	100		
$3d^6(^5D)4s$	a ^{6}D	9/2	46 438.3	100		
		7/2	47 003.1	100		
		5/2	47 415.4	100		
		3/2	47 698.6	100		
		1/2	47 864.8	100		
$3d^6(^5D)4s$	a ^{4}D	7/2	55 729.2	100		
		5/2	56 373.8	100		
		3/2	56 794.8	100		
		1/2	57 036.8	100		

(Continued)

Co III—Continued

Configuration	Term	J	Level (cm^{-1})	Leading percentages		
$3d^6(^3P2)4s$	$b\ ^4P$	5/2	70 934.1	61	38	$(^3P1)\ ^4P^\circ$
		3/2	72 341.9	61	37	
		1/2	73 214.5	61	37	
$3d^6(^3H)4s$	$a\ ^4H$	13/2	71 623.1	100		
		11/2	71 873.7	94	6	$(^3G)\ ^4G^\circ$
		9/2	72 083.3	92	7	$(^3G)\ ^4G^\circ$
		7/2	72 270.5	96	4	$(^3G)\ ^4G^\circ$
$3d^6(^3F2)4s$	$b\ ^4F$	9/2	73 286.0	59	16	$(^3F1)\ ^4F^\circ$
		7/2	73 540.2	61	16	
		5/2	73 726.6	67	17	
		3/2	73 861.8	79	20	
$3d^6(^3G)4s$	$a\ ^4G$	11/2	76 518.9	94	6	$(^3H)\ ^4H^\circ$
		9/2	77 121.1	71	17	$(^3F2)\ ^4F^\circ$
		7/2	77 383.1	74	17	$(^3F2)\ ^4F^\circ$
		5/2	77 472.3	83	13	$(^3F2)\ ^4F^\circ$
$3d^6(^3P2)4s$	$b\ ^2P$	3/2	76 791.1	61	37	$(^3P1)\ ^2P^\circ$
		1/2	78 434.3	62	37	
$3d^6(^3H)4s$	$b\ ^2H$	11/2	77 411.6	98	1	$(^3G)\ ^4G^\circ$
		9/2	77 622.9	92	7	$(^3G)\ ^2G^\circ$
$3d^6(^3F2)4s$	$b\ ^2F$	7/2	78 927.8	56	28	$(^3G)\ ^2G^\circ$
		5/2	79 425.3	79	20	$(^3F1)\ ^2F^\circ$
$3d^6(^3G)4s$	$b\ ^2G$	9/2	82 363.3	92	7	$(^3H)\ ^2H^\circ$
		7/2	82 920.7	71	21	$(^3F2)\ ^2F^\circ$
$3d^6(^3D)4s$	$b\ ^4D$	3/2	83 773.4	100		
		1/2	83 789.3	100		
		5/2	83 799.6	100		
		7/2	83 938.9	100		
$3d^6(^1I)4s$	$a\ ^2I$	13/2	85 474.1	100		
		11/2	85 517.3	100		
$3d^6(^1G2)4s$	$c\ ^2G$	9/2	86 283.8	66	34	$(^1G1)\ ^2G^\circ$
		7/2	86 327.1	66	34	
$3d^6(^1D2)4s$	$b\ ^2D$	5/2	91 715.1?	77	21	$(^1D1)\ ^2D^\circ$
$3d^6(^5D)4p$	$z\ ^6D^\circ$	9/2	*98 290.3*	99		
		7/2	*98 545.6*	97		
		5/2	*98 823.2*	98		
		3/2	*99 043.5*	99		
		1/2	*99 182.2*	99		
$3d^6(^5D)4p$	$z\ ^6F^\circ$	11/2	*103 245.0*	100		
		9/2	*103 386.7*	96		
		7/2	*103 501.7*	96		
		5/2	*103 593.7*	97		
		3/2	*103 655.7*	98		
		1/2	*103 690.8*	98		
$3d^6(^5D)4p$	$z\ ^6P^\circ$	7/2	*105 008.7*	91	7	$(^5D)\ ^4D^\circ$
		5/2	*105 965.1*	94		
		3/2	*106 591.9*	98		

Co III—Continued

Configuration	Term	J	Level (cm^{-1})	Leading percentages			
$3d^6(^5D)4p$	$z\ ^4D^\circ$	7/2	106 489.2	89		7	$(^5D)\ ^6P^\circ$
		5/2	106 954.7	91		5	$(^5D)\ ^6P^\circ$
		3/2	107 297.0	94			
		1/2	107 507.6	95			
$3d^6(^5D)4p$	$z\ ^4F^\circ$	9/2	106 764.9	96			
		7/2	107 530.1	96			
		5/2	108 052.9	97			
		3/2	108 403.4	98			
$3d^6(^5D)4p$	$z\ ^4P^\circ$	5/2	110 371.2	98			
		3/2	110 961.5	98			
		1/2	111 283.1	98			
$3d^6(a\ ^3P)4p$	$z\ ^4S^\circ$	3/2	123 123.1	61		36	$(a\ ^3P)\ ^4P^\circ$
$3d^6(^3H)4p$	$z\ ^4G^\circ$	11/2	124 765.9	64		27	$(a\ ^3F)\ ^4G^\circ$
		9/2	125 012.4	47		33	
		7/2	125 227.0	46		40	
		5/2	125 369.1	44		44	
$3d^6(a\ ^3P)4p$	$y\ ^4P^\circ$	5/2	125 003.3	60		25	$(a\ ^3P)\ ^4D^\circ$
$3d^6(^3H)4p$	$^4I^\circ$	13/2	125 276.2	58		31	$(^3H)\ ^4H^\circ$
		11/2	125 296.2	59		30	$(^3H)\ ^4H^\circ$
		9/2	125 421.6	57		26	$(^3H)\ ^4H^\circ$
		15/2	126 119.0	100			
$3d^6(^3H)4p$	$^4H^\circ$	7/2	125 690.5	60		14	$(^3H)\ ^2G^\circ$
		9/2	126 239.4	41		35	$(^3H)\ ^4I^\circ$
		13/2	126 475.5	59		34	$(^3H)\ ^4I^\circ$
		11/2	126 501.3	55		36	$(^3H)\ ^4I^\circ$
$3d^6(a\ ^3P)4p$	$z\ ^2D^\circ$	5/2	125 992.7	58		25	$(a\ ^3P)\ ^4P^\circ$
		3/2	127 458.1	32	$^2D^\circ$	23	$(a\ ^3P)\ ^4P^\circ$
$3d^6(a\ ^3P)4p$		3/2	126 381.8?	30	$^4P^\circ$	28	$(a\ ^3P)\ ^4D^\circ$
$3d^6(a\ ^3P)4p$	$y\ ^4D^\circ$	7/2	126 549.3	80		7	$(a\ ^3F)\ ^4F^\circ$
		5/2	128 085.2	60		15	$(a\ ^3P)\ ^2D^\circ$
		3/2	128 423.1	42		31	$(a\ ^3P)\ ^2D^\circ$
		1/2	128 536.3	90		6	$(a\ ^3P)\ ^4P^\circ$
$3d^6(a\ ^3F)4p$	$y\ ^4F^\circ$	5/2	126 870.9	78		6	$(^3D)\ ^4F^\circ$
		7/2	126 892.1	48		16	$(a\ ^3P)\ ^4D^\circ$
		3/2	126 987.6	86		7	$(^3D)\ ^4F^\circ$
		9/2	126 998.1	82			
$3d^6(^3H)4p$	$z\ ^2G^\circ$	9/2	127 050.8	59		15	$(^3G)\ ^2G^\circ$
		7/2	127 317.7	43		18	$(a\ ^3F)\ ^4F^\circ$
		1/2	127 224.7?				
$3d^6(^3H)4p$	$z\ ^2I^\circ$	13/2	127 672.9	90		7	$(^3H)\ ^4I^\circ$
		11/2	128 258.9	92			

(Continued)

Co III—Continued

Configuration	Term	J	Level (cm^{-1})	Leading percentages			
$3d^6(a\,^3F)4p$	$x\,^4D°$	7/2	128 017.7	69		8	$(^3D)\,^4D°$
		5/2	128 524.9	73		9	
		3/2	128 804.9	80		10	
		1/2	128 937.4	84		11	
$3d^6(a\,^3F)4p$	$y\,^4G°$	11/2	129 556.0	68		22	$(^3H)\,^4G°$
		9/2	129 592.3	52		25	
		7/2	129 706.9	44		26	
		5/2	129 747.3	35		27	
$3d^6(a\,^3F)4p$	$z\,^2F°$	7/2	130 183.8	49		12	$(^3G)\,^2F°$
		5/2	130 407.8	41		19	$(^3H)\,^4G°$
$3d^6(a\,^3P)4p$	$z\,^2P°$	1/2	130 799.5	55		34	$(a\,^3P)\,^2S°$
		3/2	130 909.0	87		6	$(a\,^1D)\,^2P°$
$3d^6(a\,^3F)4p$	$y\,^2G°$	9/2	130 801.9	60		13	$(^3G)\,^2H°$
		7/2	131 279.9	62		17	$(^3G)\,^4G°$
$3d^6(^3G)4p$		11/2	131 054.2	34	$^2H°$	23	$(^3H)\,^2H°$
$3d^6(^3G)4p$	$^4F°$	9/2	131 098.2	52		32	$(^3G)\,^4G°$
		7/2	132 277.1	44		28	$(^3G)\,^4G°$
		5/2	132 489.0	56		18	$(^3G)\,^4G°$
		3/2	132 592.4	71		14	$(^3D)\,^4F°$
$3d^6(^3G)4p$	$^4G°$	11/2	131 371.1	59		19	$(^3H)\,^2H°$
		5/2	131 883.9	63		17	$(^3G)\,^4F°$
		9/2	131 887.1	49		24	$(^3G)\,^4F°$
$3d^6(^3H)4p$	$z\,^2H°$	9/2	131 538.3	37		23	$(^3G)\,^2H°$
$3d^6(^3G)4p$		7/2	131 581.6	38	$^4G°$	24	$(^3G)\,^4F°$
$3d^6(a\,^3P)4p$	$z\,^2S°$	1/2	132 314.9	65		29	$(a\,^3P)\,^2P°$
$3d^6(^3G)4p$	$y\,^4H°$	13/2	132 376.8	88		10	$(^3H)\,^4H°$
		11/2	132 506.2	74		7	
		9/2	132 587.0	76		9	
		7/2	132 623.8	74		11	
$3d^6(a\,^3F)4p$	$y\,^2D°$	5/2	133 711.4	78		13	$(a\,^3P)\,^2D°$
		3/2	134 079.8	72		19	
$3d^6(^3G)4p$	$y\,^2H°$	11/2	134 696.1	47		45	$(^3H)\,^2H°$
		9/2	135 404.0	50		39	
$3d^6(^3G)4p$	$y\,^2F°$	5/2	136 129.1	54		18	$(^3D)\,^2F°$
		7/2	136 290.1	57		13	
$3d^6(^1I)4p$	$z\,^2K°$	13/2	137 363.0	98			
		15/2	138 234.3	100			
$3d^6(^3G)4p$	$x\,^2G°$	9/2	137 661.4	74		18	$(^3H)\,^2G°$
		7/2	137 812.4	72		14	
$3d^6(^3D)4p$	$x\,^4P°$	5/2	138 774.7	90			
		3/2	138 970.5	82		5	$(^3D)\,^4D°$
		1/2	139 314.2	82		8	

Co III—Continued

Configuration	Term	J	Level (cm^{-1})	Leading percentages			
$3d^6(a\ ^1G)4p$	$x\ ^2H°$	9/2	*138 921.0*	69		17	$(^1I)\ ^2H°$
$3d^6(a\ ^1G)4p$		11/2	*139 137.5*	49	$^2H°$	44	$(^1I)\ ^2H°$
$3d^6(^3D)4p$	$w\ ^4D°$	3/2	*139 627.2*	55		18	$(^3D)\ ^2P°$
		1/2	*139 731.0*	53		20	$(^3D)\ ^2P°$
		5/2	*139 743.4*	77		9	$(a\ ^3F)\ ^4D°$
		7/2	*140 645.4*	44		30	$(a\ ^1G)\ ^2F°$
$3d^6(^3D)4p$		7/2	*139 818.7*	27	$^4D°$	22	$(^3D)\ ^4F°$
$3d^6(^3D)4p$	$w\ ^4F°$	3/2	*140 200.1*	72		19	$(^3G)\ ^4F°$
		5/2	*140 211.6*	68		16	$(^3G)\ ^4F°$
		7/2	*140 225.8*	46		14	$(^3D)\ ^4D°$
		9/2	*140 492.4*	54		30	$(a\ ^1G)\ ^2G°$
$3d^6(a\ ^1G)4p$	$w\ ^2G°$	9/2	*140 358.4*	54		29	$(^3D)\ ^4F°$
		7/2	*140 382.9*	61		13	
$3d^6(a\ ^1G)4p$	$x\ ^2F°$	5/2	*140 787.3*	48		21	$(^3D)\ ^2F°$
$3d^6(a\ ^1G)4p$	$w\ ^2H°$	11/2	*141 190.5*	50		33	$(^1I)\ ^2H°$
		9/2	*141 347.3*	68		21	
$3d^6(^1I)4p$	$y\ ^2I°$	13/2	*141 868.8*	99			
		11/2	*141 873.8*	87		9	$(^1I)\ ^2H°$
$3d^6(^3D)4p$	$w\ ^2F°$	7/2	*143 677.1*	61		17	$(a\ ^1G)\ ^2F°$
$3d^6(a\ ^1D)4p$	$v\ ^2F°$	7/2	*146 815.6*	72		15	$(a\ ^1G)\ ^2F°$
$3d^6(a\ ^1D)4p$	$x\ ^2D°$	5/2	*146 950.7*	62		17	$(a\ ^1D)\ ^2F°$
$3d^6(b\ ^3P)4p$	$^4D°$	3/2	*155 702.4*	50		45	$(b\ ^3F)\ ^4D°$
		3/2	*156 047.8?*				
$3d^6(b\ ^3F)4p$	$^4D°$	5/2	*156 291.4*	45		38	$(b\ ^3P)\ ^4D°$
	$v\ ^4P°$	5/2	*161 811.3*				
$3d^6(^5D)4d$	$e\ ^6F$	11/2	168 012.7				
		9/2	168 211.5				
		7/2	168 449.6				
		5/2	168 688.9				
		3/2	168 885.5				
		1/2	169 011.2				
$3d^6(^5D)4d$	$e\ ^6D$	9/2	169 527.7				
		7/2	169 595.9				
		5/2	169 770.3				
		3/2	169 996.0				
		1/2	170 194.7				
$3d^6(^5D)4d$	$e\ ^6P$	7/2	169 591.0				
		5/2	170 355.5				
		3/2	170 804.6				

(Continued)

Co III—Continued

Configuration	Term	J	Level (cm^{-1})	Leading percentages
$3d^6(^5D)5s$	$f\,^6D$	$9/2$	170 535.2	
		$7/2$	171 079.3	
		$5/2$	171 466.6	
		$3/2$	171 740.4	
		$1/2$	171 941.5	
$3d^6(^5D)4d$	$e\,^6G$	$13/2$	170 725.7	
		$11/2$	171 028.9	
		$9/2$	171 269.1	
		$7/2$	171 527.6	
		$5/2$	171 729.6	
		$3/2$	171 825.8	
$3d^6(^5D)4d$	$e\,^4F$	$9/2$	171 942.6	
		$7/2$	172 377.0	
		$5/2$	172 802.7	
		$3/2$	173 133.5	
$3d^6(^5D)5s$	$f\,^4D$	$7/2$	172 583.6	
		$5/2$	173 194.9	
		$3/2$	173 609.3	
		$1/2$	173 851.1	
Co IV ($^5D_4$)	***Limit***		**270 200**	

Co IV

$Z=27$

Cr I isoelectronic sequence

Ground state: $1s^2 2s^2 2p^6 3s^2 3p^6 3d^6\ {}^5D_4$

Ionization energy = 413 500±800 cm^{-1} (51.3±0.1 eV)

In 1974, Poppe, van Kleef, and Raassen reported an observation at 600 Å of the $3d^6\ {}^5D - 3d^5({}^6S)4p\ {}^5P°$ multiplet of Co IV. Poppe has extended the analysis to include all but the high ^1S term of $3d^6$, and most of the $3d^5 4s$ and $3d^5 4p$ configurations. The results are confirmed by parametric calculations of the level structure. Poppe has provided the energy levels and percentage composition to us in advance of publication. The level uncertainty is presumably ±0.5 cm^{-1}.

The ionization energy was obtained by extrapolation by Lotz (1967).

References

Lotz, W. (1967), J. Opt. Soc. Am. **57**, 873.

Poppe, R., van Kleef, Th. A. M., and Raassen, A. J. J. (1974), Physica **77**, 165.

Co IV

Configuration	Term	J	Level (cm^{-1})	Leading percentages		
$3d^6$	^5D	4	0.0	100		
		3	639.1	100		
		2	1 077.7	100		
		1	1 357.3	100		
		0	1 493.6	100		
$3d^6$	$^3P2$	2	22 883.3	62	38	$^3P1$
		1	24 729.2	62	37	
		0	25 448.7	62	37	
$3d^6$	^3H	6	23 679.5	100		
		5	24 031.8	98	2	^3G
		4	24 272.0	93	3	^3G
$3d^6$	$^3F2$	4	25 396.0	71	21	$^3F1$
		3	25 735.9	76	21	
		2	25 969.0	79	20	
$3d^6$	^3G	5	29 021.8	98	2	^3H
		4	29 592.2	95	3	$^3F2$
		3	29 867.5	97	3	$^3F2$
$3d^6$	^1I	6	35 942.7	100		
$3d^6$	^3D	2	36 348.0	98	1	$^1D2$
		1	36 382.0	100		
		3	36 554.5	100		
$3d^6$	$^1G2$	4	36 683.3	65	33	$^1G1$
$3d^6$	$^1S2$	0	41 441.9	76	23	$^1S1$
$3d^6$	$^1D2$	2	42 341.8	76	22	$^1D1$
$3d^6$	^1F	3	50 630.1	99	1	$^3F1$
$3d^6$	$^3P1$	0	58 320.6	63	37	$^3P2$
		1	58 919.8	63	37	
		2	60 098.4	62	38	

(Continued)

Co IV—Continued

Configuration	Term	J	Level (cm^{-1})	Leading percentages		
$3d^6$	^{3}F1	2	59 748.8	80	20	^{3}F2
		4	59 838.2	77	22	
		3	59 902.8	78	21	
$3d^6$	^{1}G1	4	67 907.0	66	34	^{1}G2
$3d^5$(^{6}S)4s	^{7}S	3	90 554.4	100		
$3d^6$	^{1}D1	2	91 218.4	78	22	^{1}D2
$3d^5$(^{6}S)4s	^{5}S	2	102 773.9	100		
$3d^5$(^{4}G)4s	^{5}G	6	129 134.0	100		
		5	129 195.2	100		
		2	129 222.2	100		
		4	129 223.7	100		
		3	129 228.7	100		
$3d^5$(^{4}P)4s	^{5}P	3	132 729.1	93	6	(^{4}D) ^{5}D
		2	132 828.7	94	6	
		1	132 952.6	97	3	
$3d^5$(^{4}D)4s	^{5}D	4	136 362.9	99		
		0	136 399.9	99	1	(^{4}P) ^{3}P
		1	136 519.1	96	3	(^{4}P) ^{5}P
		2	136 611.2	93	6	(^{4}P) ^{5}P
		3	136 612.5	93	6	(^{4}P) ^{5}P
$3d^5$(^{4}G)4s	^{3}G	5	137 280.9	100		
		3	137 325.6	99		
		4	137 336.4	100		
$3d^5$(^{4}P)4s	^{3}P	2	140 863.5	92	7	(^{4}D) ^{3}D
		1	141 076.0	95	4	
$3d^5$(^{4}D)4s	^{3}D	3	144 509.6	99		
		1	144 720.3	96	4	(^{4}P) ^{3}P
		2	144 764.2	93	7	(^{4}P) ^{3}P
$3d^5$(^{2}I)4s	^{3}I	6	148 098.1	99	1	(^{2}H) ^{3}H
		5	148 105.5	99	1	(^{2}H) ^{3}H
		7	148 106.0	100		
$3d^5$(^{2}D3)4s	^{3}D	3	150 970.9	57	18	(^{2}F1) ^{3}F
		2	151 040.6	52	16	(^{2}D1) ^{3}D
$3d^5$(^{2}I)4s	^{1}I	6	152 113.3	99	1	(^{2}H) ^{3}H
$3d^5$(^{4}F)4s	^{5}F	5	152 126.2	98	1	(^{2}G2) ^{3}G
		4	152 161.1	95	3	(^{2}F1) ^{3}F
		3	152 264.2	90	7	(^{2}F1) ^{3}F
		2	152 351.4	74	19	(^{2}F1) ^{3}F
$3d^5$(^{2}F1)4s	^{3}F	4	153 265.3	94	4	(^{4}F) ^{5}F
		2	153 524.9	48	20	(^{2}D3) ^{3}D
		3	154 012.2	73	15	(^{2}D3) ^{3}D

Co IV—Continued

Configuration	Term	J	Level (cm^{-1})	Leading percentages		
$3d^5(^6S)4p$	$^7P^\circ$	2	*155 743.6*	99	1	$(^6S)\ ^5P^\circ$
		3	*156 272.8*	99	1	$(^6S)\ ^5P^\circ$
		4	*157 123.7*	100		
$3d^5(^2D3)4s$	^1D	2	156 395.4	60	18	$(^2D1)\ ^1D$
$3d^5(^2F1)4s$	^1F	3	157 513.1	89	4	$(^2G2)\ ^3G$
$3d^5(^2H)4s$	^3H	4	158 359.9	78	19	$(^2G2)\ ^3G$
		5	158 423.4	76	23	$(^2G2)\ ^3G$
		6	158 846.7	98	1	$(^2I)\ ^1I$
$3d^5(^2G2)4s$	^3G	3	159 482.3	85	9	$(^4F)\ ^3F$
		4	159 638.1	60	20	$(^2H)\ ^3H$
		5	159 851.5	73	24	$(^2H)\ ^3H$
$3d^5(^4F)4s$	^3F	2	160 351.4	95	2	$(^2F2)\ ^3F$
		4	160 404.3	74	20	$(^2G2)\ ^3G$
$3d^5(^2H)4s$	^1H	5	162 889.0	97	3	$(^2G2)\ ^3G$
$3d^5(^2G2)4s$	^1G	4	163 717.9	62	36	$(^2F2)\ ^3F$
$3d^5(^2F2)4s$	^3F	3	163 752.9	98	1	$(^2F1)\ ^3F$
		4	163 870.6	61	32	$(^2G2)\ ^1G$
$3d^5(^6S)4p$	$^5P^\circ$	3	*164 803.2*	97	1	$(^4D)\ ^5P^\circ$
		2	*165 226.2*	97	1	
		1	*165 488.5*	98	1	
$3d^5(^2F2)4s$	^1F	3	167 856.5	97	1	$(^4F)\ ^3F$
$3d^5(^2S)4s$	^3S	1	169 800.9	100		
$3d^5(^2D2)4s$	^3D	2	178 396.5	100		
		3	178 439.4	99		
$3d^5(^2D2)4s$	^1D	2	182 501.4	100		
$3d^5(^2G1)4s$	^3G	5	188 138.4	100		
		4	188 164.4	100		
		3	188 179.4	100		
$3d^5(^2G1)4s$	^1G	4	192 204.9	100		
$3d^5(^4G)4p$	$^5G^\circ$	2	*192 352.8*	94	3	$(^4G)\ ^3F^\circ$
		3	*192 382.0*	88	7	$(^4G)\ ^5H^\circ$
		4	*192 424.3*	85	10	$(^4G)\ ^5H^\circ$
		5	*192 490.2*	85	11	$(^4G)\ ^5H^\circ$
		6	*192 600.3*	88	8	$(^4G)\ ^5H^\circ$
$3d^5(^4G)4p$	$^5H^\circ$	4	*194 351.6*	87	9	$(^4G)\ ^5G^\circ$
		5	*194 644.3*	85	10	
		6	*194 958.0*	89	9	
		7	*195 222.8*	100		

(Continued)

Co IV—Continued

Configuration	Term	J	Level (cm^{-1})	Leading percentages			
$3d^5(^4G)4p$	$^5F°$	5	*195 558.9*	85		5	$(^4D)\ ^5F°$
		3	*195 712.4*	50		21	$(^4P)\ ^5D°$
		4	*195 770.5*	77		8	$(^4D)\ ^5F°$
		1	*196 490.2*	73		11	$(^4P)\ ^5D°$
		2	*196 550.2*	55		23	$(^4P)\ ^5D°$
$3d^5(^4P)4p$	$^5D°$	2	*195 606.5*	44		25	$(^4G)\ ^5F°$
		1	*195 703.1*	63		18	$(^4D)\ ^5D°$
		3	*196 636.2*	42		33	$(^4G)\ ^5F°$
		4	*197 322.8*	69		18	$(^4D)\ ^5D°$
$3d^5(^4P)4p$	$^5S°$	2	*196 846.6*	83		6	$(^4P)\ ^3P°$
$3d^5(^4P)4p$		3	*198 048.9*	33	$^5P°$	33	$(^4G)\ ^3F°$
$3d^5(^4G)4p$	$^3F°$	2	*198 084.5*	87		3	$(^4F)\ ^3F°$
		3	*198 242.2*	55		21	$(^4P)\ ^5P°$
		4	*198 390.5*	89		4	$(^4F)\ ^3F°$
$3d^5(^4P)4p$	$^5P°$	2	*198 499.6*	62		23	$(^4D)\ ^5P°$
		1	*198 670.1*	73		15	$(^4D)\ ^5P°$
$3d^5(^4G)4p$	$^3H°$	6	*198 567.8*	94		2	$(^4G)\ ^5H°$
		5	*198 890.9*	95		2	$(^2I)\ ^3H°$
		4	*199 092.4*	96		2	$(^2I)\ ^3H°$
$3d^5(^4P)4p$	$^3P°$	2	*199 817.3*	58		19	$(^4D)\ ^3P°$
		1	*200 193.8*	62		19	
		0	*200 506.1*	70		20	
$3d^5(^4D)4p$	$^5F°$	1	*200 723.5*	81		14	$(^4G)\ ^5F°$
		2	*200 911.1*	80		12	
		3	*201 176.2*	80		9	
		4	*201 524.1*	83		8	
		5	*201 874.2*	91		6	
$3d^5(^4G)4p$	$^3G°$	3	*202 849.1*	49		22	$(^4D)\ ^5D°$
		4	*202 883.7*	93		2	$(^2F1)\ ^3G°$
		5	*202 898.8*	94		1	$(^2F1)\ ^3G°$
$3d^5(^4G)4p$		3	*203 059.0*	44	$^3G°$	25	$(^4D)\ ^5D°$
$3d^5(^4D)4p$	$^5D°$	2	*203 370.3*	53		17	$(^4P)\ ^3D°$
		4	*203 398.0*	72		20	$(^4P)\ ^5D°$
		1	*203 633.3*	52		16	$(^4P)\ ^5D°$
		0	*204 177.8*	69		21	$(^4P)\ ^5D°$
$3d^5(^4P)4p$	$^3D°$	3	*203 560.6*	54		15	$(^4D)\ ^5P°$
		2	*203 833.4*	60		18	
		1	*203 972.8*	49		34	
$3d^5(^4D)4p$		1	*204 485.3*	34	$^5P°$	26	$(^4P)\ ^3D°$
$3d^5(^4D)4p$	$^5P°$	2	*204 792.8*	43		18	$(^4P)\ ^5P°$
$3d^5(^4D)4p$		3	*204 845.1*	30	$^3D°$	24	$(^4D)\ ^5P°$

Co IV—Continued

Configuration	Term	J	Level (cm^{-1})	Leading percentages			
$3d^5(^4D)4p$	$^3D°$	3	*205 928.7*	51		23	$(^4D)\ ^5P°$
		2	*205 999.7*	80		8	$(^4F)\ ^3D°$
		1	*206 167.2*	77		9	$(^4F)\ ^3D°$
$3d^5(^4D)4p$	$^3F°$	4	*206 648.3*	86		5	$(^2G2)\ ^3F°$
		3	*206 993.6*	81		6	$(^4P)\ ^3D°$
		2	*207 052.6*	84		5	$(^4P)\ ^3D°$
$3d^5(^4P)4p$	$^3S°$	1	*208 310.4*	94		2	$(^4D)\ ^3P°$
$3d^5(^4D)4p$	$^3P°$	0	*210 392.2*	74		21	$(^4P)\ ^3P°$
		1	*210 724.9*	69		20	
		2	*211 151.7*	67		24	
$3d^5(^2I)4p$	$^3K°$	6	*211 275.3*	71		25	$(^2I)\ ^3I°$
		7	*211 556.6*	59		33	$(^2I)\ ^3I°$
		8	*212 946.5*	100			
$3d^5(^2I)4p$	$^3I°$	5	*211 722.0*	70		16	$(^2I)\ ^1H°$
		6	*212 602.7*	61		26	$(^2I)\ ^3K°$
		7	*213 131.5*	63		36	$(^2I)\ ^3K°$
$3d^5(^2F1)4p$	$^3F°$	2	*213 158.6*	25		21	$(^2D3)\ ^3F°$
$3d^5(^2I)4p$	$^1H°$	5	*213 735.1*	59		24	$(^2I)\ ^3I°$
$3d^5(^2D3)4p$	$^3F°$	3	*213 897.9*	40		25	$(^2F1)\ ^3F°$
		4	*215 006.4*	42		26	$(^2F1)\ ^3F°$
$3d^5(^2D3)4p$		2	*214 059.3*	29	$^3F°$	22	$(^2D3)\ ^1D°$
$3d^5(^2I)4p$	$^3H°$	6	*214 260.5*	82		10	$(^2I)\ ^3I°$
		5	*214 739.1*	82		8	$(^2I)\ ^1H°$
		4	*214 831.6*	86		4	$(^2G2)\ ^3H°$
$3d^5(^2I)4p$	$^1K°$	7	*214 352.1*	93		4	$(^2I)\ ^3K°$
$3d^5(^2D3)4p$	$^3P°$	2	*216 640.4*	46		14	$(^2D1)\ ^3P°$
$3d^5(^2D1)4s$	^3D	3	216 651.1	77		23	$(^2D3)\ ^3D$
		2	216 686.5	77		23	
		1	216 721.1	76		23	
$3d^5(^2F1)4p$	$^1G°$	4	*216 717.8*	47		13	$(^2H)\ ^1G°$
$3d^5(^2D3)4p$		1	*216 974.3*	38	$^3P°$	23	$(^2D3)\ ^3D°$
$3d^5(^2F1)4p$	$^3G°$	3	*217 060.7*	45		21	$(^2D3)\ ^1F°$
		4	*218 414.0*	47		35	$(^2F1)\ ^3F°$
		5	*218 703.5*	67		27	$(^4F)\ ^5G°$
$3d^5(^4F)4p$	$^5G°$	3	*217 508.9*	48		23	$(^2D3)\ ^3D°$
		4	*217 830.7*	79		3	$(^4F)\ ^5F°$
		5	*217 963.0*	58		20	$(^2F1)\ ^3G°$
		6	*218 733.8*	81		15	$(^2I)\ ^1I°$
$3d^5(^4F)4p$		3	*217 585.4*	28	$^5G°$	23	$(^2F1)\ ^3D°$
$3d^5(^2D3)4p$		1	*217 951.4*	36	$^3D°$	24	$(^2D3)\ ^3P°$

(Continued)

Co IV—Continued

Configuration	Term	J	Level (cm^{-1})	Leading percentages			
$3d^5(^2D3)4p$	$^3D°$	2	*217 993.0*	46		15	$(^2D1)\ ^3D°$
$3d^5(^2F1)4p$	$^3D°$	3	*218 321.2*	51		10	$(^2F2)\ ^3D°$
		1	*219 260.1*	56		19	$(^2D3)\ ^1P°$
$3d^5(^2I)4p$	$^1I°$	6	*218 560.2*	76		13	$(^4F)\ ^5G°$
$3d^5(^4F)4p$	$^5F°$	4	*218 868.5*	57		25	$(^4F)\ ^5D°$
		3	*218 876.3*	65		22	$(^4F)\ ^5D°$
		2	*219 052.1*	50		20	$(^4F)\ ^5D°$
		5	*219 231.6*	78		6	$(^4F)\ ^5G°$
		1	*219 388.2*	76		7	$(^4F)\ ^5D°$
$3d^5(^2F1)4p$		2	*219 453.9*	35	$^3D°$	16	$(^2F1)\ ^3F°$
$3d^5(^2F1)4p$		4	*219 694.2*	23	$^3F°$	22	$(^2D3)\ ^3F°$
$3d^5(^2F1)4p$		2	*219 969.0*	31	$^3F°$	21	$(^2F1)\ ^3D°$
$3d^5(^2F1)4p$		3	*220 052.2*	31	$^3F°$	14	$(^2D3)\ ^1F°$
$3d^5(^2G2)4p$	$^3H°$	4	*220 145.6*	37		36	$(^2H)\ ^3H°$
		5	*226 983.3*	41		34	
		6	*227 737.1*	47		38	
$3d^5(^2H)4p$	$^3H°$	5	*220 423.7*	37		40	$(^2G2)\ ^3H°$
		6	*221 274.1*	43		38	$(^2G2)\ ^3H°$
		4	*226 950.3*	46		40	$(^2G2)\ ^3H°$
$3d^5(^4F)4p$	$^5D°$	4	*220 535.2*	61		27	$(^4F)\ ^5F°$
		3	*220 887.1*	66		21	
		2	*221 058.4*	70		13	
		1	*221 062.2*	80		7	
$3d^5(^2D1)4s$	^1D	2	220 548.5	77		23	$(^2D3)\ ^1D$
$3d^5(^2H)4p$	$^3G°$	5	*220 950.8*	51		14	$(^2G2)\ ^3G°$
		4	*221 068.1*	46		15	$(^2F2)\ ^3G°$
		3	*221 287.7*	43		17	$(^2F2)\ ^3G°$
$3d^5(^2D3)4p$	$^1P°$	1	*222 336.1*	48		21	$(^2F1)\ ^3D°$
$3d^5(^2H)4p$	$^3I°$	5	*222 848.1*	87		4	$(^2H)\ ^3H°$
		6	*223 382.3*	84		6	$(^2H)\ ^1I°$
		7	*223 981.3*	96		2	$(^2I)\ ^3I°$
$3d^5(^2G2)4p$		4	*222 994.2*	24	$^1G°$	18	$(^2G2)\ ^3G°$
$3d^5(^4F)4p$	$^3G°$	5	*223 041.1*	61		19	$(^2G2)\ ^3G°$
		3	*223 192.4*	45		38	
		4	*223 260.7*	47		14	
$3d^5(^2F1)4p$	$^1D°$	2	*223 381.4*	56		29	$(^2D3)\ ^1D°$
$3d^5(^2F1)4p$	$^1F°$	3	*224 038.7*	43		21	$(^2G2)\ ^3F°$
$3d^5(^2G2)4p$		3	*224 411.0*	38	$^3F°$	25	$(^2F1)\ ^1F°$

Co IV—Continued

Configuration	Term	J	Level (cm^{-1})	Leading percentages		
$3d^5(^2G2)4p$	$^3F°$	2	224 463.9	56	18	$(^4F)\ ^3F°$
		4	224 611.0	56	11	$(^2G2)\ ^1G°$
$3d^5(^2H)4p$	$^1I°$	6	225 479.8	81	8	$(^2H)\ ^3H°$
$3d^5(^4F)4p$	$^3D°$	2	225 553.4	64	7	$(^2F2)\ ^3D°$
		3	225 666.3	61	10	$(^2F2)\ ^3D°$
		1	225 780.8	81	7	$(^4D)\ ^3D°$
$3d^5(^4F)4p$	$^3F°$	4	225 938.7	63	25	$(^2F2)\ ^3F°$
		3	226 367.6	63	20	
		2	226 646.1	57	24	
$3d^5(^2G2)4p$	$^3G°$	5	227 971.6	31	22	$(^2F2)\ ^3G°$
		4	228 208.5	34	26	
		3	228 214.7	34	28	
$3d^5(^2F2)4p$	$^1G°$	4	228 448.1	35	15	$(^2G2)\ ^1G°$
$3d^5(^2F2)4p$	$^3F°$	2	228 736.2	34	18	$(^2G2)\ ^3F°$
		3	228 809.0	55	11	$(^2G2)\ ^1F°$
		4	229 414.0	51	12	$(^4F)\ ^3F°$
$3d^5(^2G2)4p$	$^1H°$	5	229 056.6	62	20	$(^2H)\ ^1H°$
$3d^5(^2G2)4p$	$^1F°$	3	229 388.2	63	6	$(^2F2)\ ^3F°$
$3d^5(^2H)4p$	$^1H°$	5	229 583.3	70	23	$(^2G2)\ ^1H°$
$3d^5(^2F2)4p$	$^1D°$	2	229 890.3	63	23	$(^2F2)\ ^3F°$
$3d^5(^2F2)4p$	$^3G°$	3	231 876.6	42	40	$(^2H)\ ^3G°$
		4	232 288.1	51	35	
		5	232 615.6	60	30	
$3d^5(^2F2)4p$	$^3D°$	1	232 133.7	75	9	$(^2F1)\ ^3D°$
		2	232 257.0	75	10	
		3	232 265.7	66	11	
$3d^5(^2S)4p$	$^3P°$	1	233 158.8	77	12	$(^2D2)\ ^3P°$
		2	234 109.4	79	15	
$3d^5(^2F2)4p$	$^1G°$	4	234 584.4	39	38	$(^2H)\ ^1G°$
$3d^5(^2F2)4p$	$^1F°$	3	236 240.4	90	4	$(^2F1)\ ^1F°$
$3d^5(^2S)4p$	$^1P°$	1	236 588.9	74	17	$(^2D2)\ ^1P°$
$3d^5(^2D2)4p$	$^3F°$	2	243 999.4	68	22	$(^2D2)\ ^3D°$
		3	244 285.6	53	29	$(^2D2)\ ^3D°$
		4	245 323.9	92	5	$(^2G2)\ ^3F°$
$3d^5(^2D2)4p$	$^3D°$	2	244 801.0	67	22	$(^2D2)\ ^3F°$
		3	245 361.7	61	33	
$3d^5(^2D2)4p$	$^1F°$	3	246 316.4	76	11	$(^2G1)\ ^1F°$

(Continued)

Co IV—Continued

Configuration	Term	J	Level (cm^{-1})	Leading percentages		
$3d^5(^2D2)4p$	$^3P°$	2	*247 091.4*	76	15	$(^2S)\ ^3P°$
		1	*247 130.4*	78	14	
		0	*247 138.4*	85	13	
$3d^5(^2D2)4p$	$^1P°$	1	*248 390.4*	74	16	$(^2S)\ ^1P°$
$3d^5(^2D2)4p$	$^1D°$	2	*249 444.1*	89	5	$(^2F2)\ ^1D°$
$3d^5(^2G1)4p$	$^3H°$	4	*253 119.4*	79	11	$(^2G1)\ ^3G°$
		5	*253 446.4*	81	12	$(^2G1)\ ^3G°$
		6	*254 347.1*	98	2	$(^2I)\ ^3H°$
$3d^5(^2G1)4p$	$^3F°$	4	*253 472.8*	75	11	$(^2G1)\ ^3H°$
		3	*253 921.8*	56	37	$(^2G1)\ ^3G°$
		2	*254 722.1*	92	5	$(^2D1)\ ^3F°$
$3d^5(^2G1)4p$	$^3G°$	3	*254 903.0*	60	36	$(^2G1)\ ^3F°$
		4	*255 138.9*	82	9	$(^2G1)\ ^3F°$
		5	*255 307.3*	82	14	$(^2G1)\ ^3H°$
$3d^5(^2G1)4p$	$^1H°$	5	*257 289.6*	92	3	$(^2G1)\ ^3G°$
$3d^5(^2G1)4p$	$^1G°$	4	*257 680.7*	94	2	$(^2G1)\ ^3F°$
$3d^5(^2G1)4p$	$^1F°$	3	*258 840.7*	81	10	$(^2D2)\ ^1F°$
$3d^5(^2P)4p$	$^3P°$	1	*267 508.7*	75	20	$(^2D1)\ ^3P°$
		2	*268 151.4*	74	21	
$3d^5(^2P)4p$	$^3D°$	2	*272 522.3*	59	27	$(^2P)\ ^1D°$
		1	*272 523.8*	91	5	$(^2D1)\ ^3D°$
		3	*273 632.6*	90	6	$(^2D1)\ ^3D°$
$3d^5(^2P)4p$	$^1D°$	2	*274 104.5*	51	32	$(^2P)\ ^3D°$
$3d^5(^2P)4p$	$^3S°$	1	*275 594.1*	95	4	$(^2P)\ ^1P°$
$3d^5(^2D1)4p$	$^3F°$	2	*282 925.2*	66	21	$(^2D3)\ ^3F°$
		3	*283 211.1*	63	20	
		4	*284 089.0*	72	23	
$3d^5(^2D1)4p$	$^3D°$	1	*284 496.9*	71	22	$(^2D3)\ ^3D°$
		2	*285 045.7*	64	19	
		3	*285 642.1*	63	19	
$3d^5(^2D1)4p$	$^1D°$	2	*285 928.4*	47	16	$(^2P)\ ^1D°$
$3d^5(^2D1)4p$	$^3P°$	2	*287 239.6*	45	20	$(^2P)\ ^3P°$
$3d^5(^2D1)4p$	$^1F°$	3	*287 698.3*	70	22	$(^2D3)\ ^1F°$
Co V $(^6S_{5/2})$	***Limit***		**413 500**			

Co V

$Z=27$

V I isoelectronic sequence

Ground state: $1s^2 2s^2 2p^6 3s^2 3p^6 3d^5\ ^6S_{5/2}$

Ionization energy = 641 000±1600 cm^{-1} (79.5±0.2 eV)

Kruger and Gilroy (1935) observed the $3d^5\,^6P-3d^44p\,^6P°$ multiplet of Co V near 356 Å. No further analysis of this spectrum appeared until 1979, when Raassen and van Kleef (1979) published all terms of $3d^5$ and nearly all of $3d^44s$ and $3d^44p$ along with their calculations of the level values and percentage composition. The $3d^5-3d^44p$ array lies in the region of 300–450 Å and is measured with an accuracy of ±0.002 Å. The $3d^44s-3d^44p$ array appears from 1100–1500 Å and has been measured with an accuracy of ±0.005 Å. The level values and leading percentages are taken from their paper. The level uncertainty is ±2 cm^{-1}.

The ionization energy was obtained by extrapolation by Lotz (1967).

References

Kruger, P. G., and Gilroy, H. T. (1935), Phys. Rev. **48**, 720.
Lotz, W. (1967), J. Opt. Soc. Am. **57**, 873.
Raassen, A. J. J., and van Kleef, Th. A. M. (1979), Physica **96C**, 367.

Co V

Configuration	Term	J	Level (cm^{-1})	Leading percentages		
$3d^5$	^6S	5/2	0.0	100		
$3d^5$	^4G	11/2	37 217.5	100		
		9/2	37 288.8	100		
		5/2	37 289.5	99		
		7/2	37 304.0	100		
$3d^5$	^4P	5/2	40 753.2	93	6	^4D
		3/2	40 890.9	95	5	^4D
		1/2	41 023.8	98	2	^4D
$3d^5$	^4D	7/2	44 709.1	100		
		1/2	44 907.5	98	2	^4P
		5/2	44 984.1	93	6	^4P
		3/2	44 986.7	95	5	^4P
$3d^5$	^2I	11/2	54 339.2	99	1	^2H
		13/2	54 376.6	100		
$3d^5$	$^2D3$	5/2	57 082.6	54	27	$^2F1$
		3/2	57 823.2	72	22	$^2D2$
$3d^5$	$^2F1$	7/2	59 454.9	96	2	^4F
		5/2	60 532.2	58	25	^4F
$3d^5$	^4F	9/2	60 830.3	97	3	$^2G2$
		7/2	60 973.6	96	2	$^2F1$
		3/2	61 213.2	94	4	$^2D3$
		5/2	61 284.5	73	15	$^2F1$
$3d^5$	^2H	9/2	64 742.3	79	20	$^2G2$
		11/2	65 283.8	99	1	^2I
$3d^5$	$^2G2$	7/2	66 228.7	99	1	^4F
		9/2	66 760.4	77	21	^2H
$3d^5$	$^2F2$	5/2	70 502.5	99	1	^4F
		7/2	70 652.8	98	1	$^2F1$

(Continued)

Co V—Continued

Configuration	Term	*J*	Level (cm^{-1})	Leading percentages		
$3d^5$	^{2}S	1/2	76 864.5	100		
$3d^5$	^{2}D2	3/2	85 573.5	100		
		5/2	85 636.2	99		
$3d^5$	^{2}G1	9/2	95 708.7	100		
		7/2	95 726.5	100		
$3d^5$	^{2}P	3/2	115 437.1	99		
		1/2	115 468.5	100		
$3d^5$	^{2}D1	5/2	125 022.7	77	23	^{2}D3
		3/2	125 068.8	76	23	^{2}D3
$3d^4(^5D)4s$	^{6}D	1/2	205 878.9	100		
		3/2	206 108.5	100		
		5/2	206 476.0	100		
		7/2	206 962.2	100		
		9/2	207 548.5	100		
$3d^4(^5D)4s$	^{4}D	1/2	216 820.5	100		
		3/2	217 171.0	100		
		5/2	217 713.2	100		
		7/2	218 407.2	100		
$3d^4(^3P2)4s$	^{4}P	1/2	235 199.9	59	39	(^{3}P1) ^{4}P
		3/2	236 353.1	60	39	
		5/2	238 106.2	59	37	
$3d^4(^3H)4s$	^{4}H	7/2	235 961.1	98		
		9/2	236 158.0	97		
		11/2	236 431.2	98		
		13/2	236 746.0	100		
$3d^4(^3F2)4s$	^{4}F	3/2	238 129.6	78	21	(^{3}F1) ^{4}F
		5/2	238 164.6	73	20	
		7/2	238 259.1	74	19	
		9/2	238 390.3	75	18	
$3d^4(^3G)4s$	^{4}G	5/2	241 174.1	94		
		7/2	241 582.4	93	5	(^{3}F2) ^{4}F
		9/2	241 875.4	87	6	(^{3}H) ^{2}H
		11/2	241 985.3	86	12	(^{3}H) ^{2}H
$3d^4(^3H)4s$	^{2}H	9/2	242 765.0	91	6	(^{3}G) ^{4}G
		11/2	243 455.2	87	12	(^{3}G) ^{4}G
$3d^4(^3P2)4s$	^{2}P	1/2	242 340.3	59	38	(^{3}P1) ^{2}P
		3/2	244 485.4	60	38	(^{3}P1) ^{2}P
$3d^4(^3F2)4s$	^{2}F	7/2	244 821.2	72	18	(^{3}F1) ^{2}F
		5/2	244 866.7	77	20	(^{3}F1) ^{2}F
$3d^4(^3G)4s$	^{2}G	7/2	248 098.9	92		
		9/2	248 698.2	97		
$3d^4(^3D)4s$	^{4}D	7/2	248 855.9	100		
		5/2	249 006.8	99		
		3/2	249 179.7	99		
		1/2	249 300.0	99		

Co v—Continued

Configuration	Term	J	Level (cm^{-1})	Leading percentages		
$3d^4(^1\mathrm{G}2)4s$	^{2}G	9/2	251 582.2	65	32	(^{1}G1) ^{2}G
		7/2	251 701.5	63	31	(^{1}G1) ^{2}G
$3d^4(^1\mathrm{I})4s$	^{2}I	13/2	252 331.9	100		
		11/2	252 385.0	99		
$3d^4(^1\mathrm{S}2)4s$	^{2}S	1/2	255 002.9	78	20	(^{1}S1) ^{2}S
$3d^4(^3\mathrm{D})4s$	^{2}D	5/2	255 292.5	99		
		3/2	255 484.1	98		
$3d^4(^1\mathrm{D}2)4s$	^{2}D	5/2	261 994.0	78	21	(^{1}D1) ^{2}D
		3/2	262 077.3	77	21	(^{1}D1) ^{2}D
$3d^4(^1\mathrm{F})4s$	^{2}F	5/2	268 877.0	99		
		7/2	268 885.8	99		
$3d^4(^3\mathrm{P}1)4s$	^{4}P	5/2	276 608.8	61	38	(^{3}P2) ^{4}P
		3/2	277 747.9	60	39	(^{3}P2) ^{4}P
$3d^4(^3\mathrm{F}1)4s$	^{4}F	9/2	277 192.4	80	19	(^{3}F2) ^{4}F
		3/2	277 330.2	78	22	(^{3}F2) ^{4}F
		7/2	277 360.7	79	20	(^{3}F2) ^{4}F
		5/2	277 376.8	78	21	(^{3}F2) ^{4}F
$3d^4(^5\mathrm{D})4p$	^{6}F°	1/2	*278 407.0*	99		
		3/2	*278 703.5*	99		
		5/2	*279 192.4*	99		
		7/2	*279 871.1*	99		
		9/2	*280 748.2*	99		
		11/2	*281 860.6*	100		
$3d^4(^5\mathrm{D})4p$	^{6}P°	3/2	*280 778.7*	94	5	(^{5}D) ^{4}P°
		5/2	*280 923.5*	95		
		7/2	*281 201.4*	99		
$3d^4(^5\mathrm{D})4p$	^{4}P°	1/2	*282 002.8*	71	26	(^{5}D) ^{6}D°
		3/2	*282 995.6*	61	30	
		5/2	*284 259.9*	46	48	
$3d^4(^3\mathrm{F}1)4s$	^{2}F	7/2	283 592.1	80	19	(^{3}F2) ^{4}F
		5/2	283 736.7	79	21	(^{3}F2) ^{4}F
$3d^4(^5\mathrm{D})4p$	^{6}D°	1/2	*285 033.7*	74	26	(^{5}D) ^{4}P°
		3/2	*285 248.1*	68	31	(^{5}D) ^{4}P°
		7/2	*285 335.0*	96		
		5/2	*285 643.0*	50	48	(^{5}D) ^{4}P°
		9/2	*285 855.7*	91	7	(^{5}D) ^{4}F°
$3d^4(^5\mathrm{D})4p$	^{4}F°	3/2	*288 539.7*	95		
		5/2	*288 738.4*	94		
		7/2	*289 029.0*	92		
		9/2	*289 447.5*	88	8	(^{5}D) ^{6}D°
$3d^4(^1\mathrm{G}1)4s$	^{2}G	9/2	289 449.2	67	33	(^{1}G2) ^{2}G
		7/2	289 495.1	66	33	(^{1}G2) ^{2}G

(Continued)

Co v—Continued

Configuration	Term	J	Level (cm^{-1})	Leading percentages			
3d^4(^{5}D)4p	^{4}D°	1/2	*295 510.0*	97			
		3/2	*295 741.8*	97			
		5/2	*296 071.4*	97			
		7/2	*296 430.2*	97			
3d^4(^{3}H)4p	^{4}H°	7/2	*306 046.4*	76		20	(^{3}G) ^{4}H°
		9/2	*306 372.8*	74		19	
		11/2	*306 854.6*	75		17	
		13/2	*307 496.4*	79		14	
3d^4(^{3}P2)4p	^{4}D°	1/2	*306 725.2*	48		32	(^{3}P1) ^{4}D°
		3/2	*307 649.2*	47		32	(^{3}P1) ^{4}D°
		5/2	*308 863.5*	41		28	(^{3}P1) ^{4}D°
		7/2	*309 993.7*	29		28	(^{3}F2) ^{4}D°
3d^4(^{3}F2)4p	^{4}G°	5/2	*309 225.4*	42		23	(^{3}G) ^{4}G°
		11/2	*313 577.1*	54		28	(^{3}H) ^{4}G°
3d^4(^{3}F2)4p		7/2	*309 481.8*	27	^{4}G°	16	(^{3}G) ^{4}G°
3d^4(^{3}H)4p	^{4}I°	9/2	*309 625.8*	58		9	(^{3}H) ^{2}G°
		11/2	*310 649.4*	85		7	(^{3}H) ^{4}H°
		13/2	*311 534.5*	93		6	(^{3}H) ^{4}H°
		15/2	*312 260.6*	100			
3d^4(^{3}H)4p		9/2	*310 031.1*	33	^{4}I°	14	(^{3}F2) ^{4}G°
3d^4(^{3}P2)4p		1/2	*310 357.9*	31	^{4}P°	22	(^{3}P1) ^{2}S°
3d^4(^{3}H)4p	^{4}G°	11/2	*310 842.1*	39		27	(^{3}G) ^{4}G°
		5/2	*312 792.9*	31		24	(^{3}F2) ^{4}G°
		7/2	*313 142.7*	50		31	(^{3}F2) ^{4}G°
		9/2	*313 278.7*	45		38	(^{3}F2) ^{4}G°
3d^4(^{3}H)4p	^{2}G°	7/2	*311 005.7*	40		18	(^{3}F2) ^{2}G°
		9/2	*311 479.6*	33		18	(^{3}F2) ^{2}G°
3d^4(^{3}P2)4p	^{4}P°	3/2	*311 912.5*	43		25	(^{3}P1) ^{4}P°
		5/2	*313 311.0*	50		28	(^{3}P1) ^{4}P°
3d^4(^{3}F2)4p		3/2	*312 235.6*	29	^{4}F°	20	(^{3}F2) ^{2}D°
3d^4(^{3}P2)4p		1/2	*312 518.0*	27	^{4}P°	18	(^{3}P1) ^{2}S°
3d^4(^{3}F2)4p		5/2	*313 083.1*	34	^{4}F°	17	(^{3}H) ^{4}G°
3d^4(^{3}F2)4p	^{4}F°	7/2	*313 626.7*	68		12	(^{3}F1) ^{4}F°
		9/2	*313 817.8*	65		11	(^{3}F1) ^{4}F°
3d^4(^{3}P2)4p		3/2	*313 825.8*	27	^{2}P°	15	(^{3}P2) ^{4}S°
3d^4(^{3}F2)4p		5/2	*314 219.1*	28	^{2}D°	21	(^{3}F2) ^{4}F°
3d^4(^{3}F2)4p		3/2	*314 454.8*	24	^{4}F°	19	(^{3}F2) ^{2}D°
3d^4(^{3}F2)4p		5/2	*314 845.5*	18	^{4}D°	13	(^{3}G) ^{2}F°
3d^4(^{3}P2)4p		1/2	*314 997.3*	24	^{2}P°	28	(^{3}F2) ^{4}D°

Co v—Continued

Configuration	Term	J	Level (cm^{-1})	Leading percentages			
$3d^4(^3P2)4p$		7/2	*315 092.2*	19	$^4D^\circ$	13	$(^3F2)\ ^4D^\circ$
$3d^4(^3H)4p$	$^2I^\circ$	11/2	*315 216.1*	84		6	$(^1I)\ ^2I^\circ$
		13/2	*315 289.8*	85		8	$(^3G)\ ^4H^\circ$
$3d^4(^3F2)4p$	$^4D^\circ$	3/2	*315 255.4*	40		13	$(^3F1)\ ^4D^\circ$
$3d^4(^3G)4p$		5/2	*315 657.5*	22	$^2F^\circ$	20	$(^3F2)\ ^4D^\circ$
$3d^4(^3F2)4p$		1/2	*315 868.7*	33	$^4D^\circ$	20	$(^3P2)\ ^2P^\circ$
$3d^4(^3H)4p$	$^2H^\circ$	9/2	*315 916.7*	46		16	$(^3G)\ ^4H^\circ$
		11/2	*316 884.4*	40		30	$(^3G)\ ^4H^\circ$
$3d^4(^3G)4p$	$^4H^\circ$	7/2	*316 132.0*	74		19	$(^3H)\ ^4H^\circ$
		9/2	*316 936.2*	60		14	$(^3H)\ ^2H^\circ$
		11/2	*317 669.7*	47		33	$(^3H)\ ^2H^\circ$
		13/2	*318 344.6*	77		15	$(^3H)\ ^4H^\circ$
$3d^4(^3P2)4p$		3/2	*316 178.9*	24	$^2P^\circ$	13	$(^3P2)\ ^4S^\circ$
$3d^4(^3G)4p$		7/2	*316 254.0*	18	$^2F^\circ$	15	$(^3D)\ ^2F^\circ$
$3d^4(^3G)4p$	$^4F^\circ$	9/2	*316 581.0*	60		12	$(^3F2)\ ^4F^\circ$
		3/2	*316 761.0*	42		14	$(^3D)\ ^4F^\circ$
		5/2	*316 859.2*	49		14	$(^3D)\ ^4F^\circ$
		7/2	*316 911.6*	39		17	$(^3F2)\ ^4D^\circ$
$3d^4(^3P2)4p$	$^2D^\circ$	3/2	*317 941.0*	41		28	$(^3P1)\ ^2D^\circ$
		5/2	*319 088.9*	45		31	$(^3P1)\ ^2D^\circ$
$3d^4(^3F2)4p$	$^2F^\circ$	5/2	*319 497.2*	51		27	$(^3G)\ ^2F^\circ$
		7/2	*320 234.5*	46		36	$(^3G)\ ^2F^\circ$
$3d^4(^3F2)4p$		7/2	*319 611.9*	28	$^2G^\circ$	15	$(^3H)\ ^2G^\circ$
$3d^4(^3G)4p$	$^2H^\circ$	11/2	*319 724.1*	43		21	$(^3G)\ ^4G^\circ$
		9/2	*320 012.1*	46		11	$(^3H)\ ^2G^\circ$
$3d^4(^3G)4p$		9/2	*319 915.7*	20	$^4G^\circ$	18	$(^3F2)\ ^2G^\circ$
$3d^4(^3G)4p$	$^4G^\circ$	5/2	*320 378.1*	53		29	$(^3H)\ ^4G^\circ$
		7/2	*320 644.6*	50		22	$(^3H)\ ^4G^\circ$
		9/2	*321 122.2*	38		16	$(^3G)\ ^2H^\circ$
		11/2	*321 521.4*	45		25	$(^3G)\ ^2H^\circ$
$3d^4(^3D)4p$	$^4D^\circ$	1/2	*323 018.6*	82		7	$(^3D)\ ^2P^\circ$
		3/2	*323 076.2*	75		10	$(^3D)\ ^4P^\circ$
		7/2	*323 690.1*	81		5	$(^3D)\ ^4F^\circ$
		5/2	*324 205.8*	40		44	$(^3D)\ ^4P^\circ$
$3d^4(^3D)4p$	$^4P^\circ$	5/2	*323 169.4*	47		42	$(^3D)\ ^4D^\circ$
		3/2	*325 125.5*	78		8	$(^3D)\ ^4D^\circ$
		1/2	*325 735.1*	89		6	$(^3P2)\ ^4P^\circ$
$3d^4(^3G)4p$	$^2G^\circ$	7/2	*324 477.5*	46		24	$(^1G2)\ ^2F^\circ$
		9/2	*324 630.4*	57		20	$(^3H)\ ^2G^\circ$

(Continued)

Co v—Continued

Configuration	Term	J	Level (cm^{-1})	Leading percentages			
$3d^4(^1I)4p$	$^2I^\circ$	13/2	*324 571.5*	70		24	$(^1I)\ ^2K^\circ$
		11/2	*324 767.7*	90		5	$(^3H)\ ^2I^\circ$
$3d^4(^1G2)4p$		7/2	*324 818.6*	32	$^2F^\circ$	22	$(^3G)\ ^2G^\circ$
$3d^4(^1G2)4p$	$^2H^\circ$	9/2	*325 584.5*	49		23	$(^1G1)\ ^2H^\circ$
		11/2	*326 267.9*	53		21	$(^1G1)\ ^2H^\circ$
$3d^4(^3D)4p$	$^4F^\circ$	3/2	*325 639.8*	67		21	$(^3G)\ ^4F^\circ$
		5/2	*325 941.4*	55		14	$(^3G)\ ^4F^\circ$
		7/2	*326 285.1*	69		18	$(^3G)\ ^4F^\circ$
		9/2	*326 560.0*	79		18	$(^3G)\ ^4F^\circ$
$3d^4(^1G2)4p$	$^2F^\circ$	5/2	*326 135.6*	41		19	$(^1G1)\ ^2F^\circ$
$3d^4(^1I)4p$	$^2K^\circ$	13/2	*326 428.3*	75		25	$(^1I)\ ^2I^\circ$
		15/2	*327 462.4*	100			
$3d^4(^3D)4p$	$^2P^\circ$	3/2	*326 288.8*	54		28	$(^1S2)\ ^2P^\circ$
		1/2	*331 431.1*	45		29	$(^1S2)\ ^2P^\circ$
$3d^4(^1G2)4p$	$^2G^\circ$	7/2	*328 818.4*	46		30	$(^1G1)\ ^2G^\circ$
		9/2	*329 319.8*	42		30	$(^1G1)\ ^2G^\circ$
$3d^4(^1I)4p$	$^2H^\circ$	11/2	*330 773.4*	74		10	$(^3G)\ ^2H^\circ$
		9/2	*331 484.4*	83		10	$(^3G)\ ^2H^\circ$
$3d^4(^1S2)4p$	$^2P^\circ$	3/2	*331 156.7*	32		20	$(^3D)\ ^2P^\circ$
$3d^4(^3D)4p$	$^2F^\circ$	7/2	*331 624.4*	68		11	$(^3G)\ ^2F^\circ$
		5/2	*332 053.0*	67		13	$(^3G)\ ^2F^\circ$
$3d^4(^3D)4p$	$^2D^\circ$	5/2	*332 114.2*	50		26	$(^1D2)\ ^2D^\circ$
		3/2	*332 906.6*	58		12	$(^1D2)\ ^2D^\circ$
$3d^4(^1D2)4p$	$^2D^\circ$	3/2	*334 690.2*	45		19	$(^3D)\ ^2D^\circ$
		5/2	*335 164.1*	33		25	$(^3D)\ ^2D^\circ$
$3d^4(^1D2)4p$	$^2F^\circ$	5/2	*336 414.8*	52		12	$(^1D1)\ ^2F^\circ$
		7/2	*337 129.9*	53		25	$(^1F)\ ^2F^\circ$
$3d^4(^1F)4p$	$^2F^\circ$	5/2	*340 989.5*	69		12	$(^1D2)\ ^2F^\circ$
		7/2	*341 550.0*	57		21	$(^1D2)\ ^2F^\circ$
$3d^4(^1D2)4p$	$^2P^\circ$	3/2	*341 093.0*	66		13	$(^1D1)\ ^2P^\circ$
		1/2	*341 110.0*	72		14	$(^1D1)\ ^2P^\circ$
$3d^4(^1F)4p$	$^2G^\circ$	7/2	*343 541.4*	89			
		9/2	*344 949.4*	93			
$3d^4(^1F)4p$	$^2D^\circ$	5/2	*345 942.1*	57		16	$(^3P1)\ ^2D^\circ$
		3/2	*347 632.3*	58		13	$(^3P1)\ ^2D^\circ$
$3d^4(^3F1)4p$	$^4F^\circ$	5/2	*349 525.8*	68		10	$(^3F2)\ ^4F^\circ$
		3/2	*349 607.6*	72		11	$(^3F2)\ ^4F^\circ$
		7/2	*349 646.9*	71		10	$(^3F2)\ ^4F^\circ$
		9/2	*350 024.0*	84		11	$(^3F2)\ ^4F^\circ$

Co v—Continued

Configuration	Term	J	Level (cm^{-1})	Leading percentages			
$3d^4(^3P1)4p$	$^4P^\circ$	3/2	350 370.6	36		17	$(^3P2)\ ^4P^\circ$
		1/2	350 778.2	40		20	$(^3P2)\ ^4P^\circ$
		5/2	351 787.7	42		21	$(^3P2)\ ^4P^\circ$
$3d^4(^3P1)4p$		5/2	350 532.0	22	$^4D^\circ$	19	$(^3P1)\ ^4P^\circ$
$3d^4(^3P1)4p$		7/2	351 248.8	34	$^4D^\circ$	21	$(^3F1)\ ^4D^\circ$
$3d^4(^3P1)4p$		3/2	351 609.1	24	$^4D^\circ$	22	$(^3P1)\ ^4P^\circ$
$3d^4(^3F1)4p$	$^4G^\circ$	5/2	352 700.2	46		23	$(^3F1)\ ^2F^\circ$
		9/2	354 141.7	72		22	$(^3F2)\ ^4G^\circ$
		11/2	354 639.6	76		22	$(^3F2)\ ^4G^\circ$
$3d^4(^3F1)4p$		7/2	353 248.7	38	$^4G^\circ$	31	$(^3F1)\ ^2F^\circ$
$3d^4(^1F)4p$	$^2D^\circ$	5/2	353 735.5	25		21	$(^3P1)\ ^2D^\circ$
		3/2	354 116.4	28		23	$(^3P1)\ ^2D^\circ$
$3d^4(^3F1)4p$		7/2	354 218.1	39	$^2F^\circ$	33	$(^3F1)\ ^4G^\circ$
$3d^4(^3F1)4p$	$^2F^\circ$	5/2	354 579.0	48		21	$(^3F1)\ ^4G^\circ$
$3d^4(^3P1)4p$	$^4S^\circ$	3/2	358 499.8	49		45	$(^3P2)\ ^4S^\circ$
$3d^4(^3F1)4p$	$^4D^\circ$	7/2	359 767.2	48		17	$(^3P1)\ ^4D^\circ$
		5/2	360 339.2	49		18	$(^3F2)\ ^4D^\circ$
		3/2	360 614.6	48		18	$(^3F2)\ ^4D^\circ$
$3d^4(^3F1)4p$	$^2G^\circ$	9/2	359 796.7	74		20	$(^3F2)\ ^2G^\circ$
		7/2	360 490.7	74		22	$(^3F2)\ ^2G^\circ$
$3d^4(^3P1)4p$	$^2P^\circ$	3/2	360 117.9	60		27	$(^3P2)\ ^2P^\circ$
		1/2	360 857.2	60		28	$(^3P2)\ ^2P^\circ$
$3d^4(^1G1)4p$		9/2	363 716.7	38	$^2H^\circ$	25	$(^1G1)\ ^2G^\circ$
$3d^4(^3P2)4p$	$^2S^\circ$	1/2	363 977.4	55		42	$(^3P1)\ ^2S^\circ$
$3d^4(^1G1)4p$	$^2G^\circ$	7/2	364 229.2	55		33	$(^1G2)\ ^2G^\circ$
$3d^4(^1G1)4p$		9/2	365 660.1	35	$^2G^\circ$	27	$(^1G1)\ ^2H^\circ$
$3d^4(^1G1)4p$	$^2H^\circ$	11/2	365 937.1	65		32	$(^1G2)\ ^2H^\circ$
$3d^4(^1G1)4p$	$^2F^\circ$	7/2	366 848.8	56		19	$(^1G2)\ ^2F^\circ$
		5/2	367 276.4	56		19	$(^1G2)\ ^2F^\circ$
$3d^4(^1D1)4p$	$^2P^\circ$	3/2	384 045.9	76		16	$(^1D2)\ ^2P^\circ$
		1/2	385 108.3	75		16	$(^1D2)\ ^2P^\circ$
$3d^4(^1D1)4p$	$^2F^\circ$	5/2	389 423.9	70		18	$(^1D2)\ ^2F^\circ$
		7/2	390 959.2	73		19	$(^1D2)\ ^2F^\circ$
$3d^4(^1D1)4p$	$^2D^\circ$	3/2	394 614.3	73		26	$(^1D2)\ ^2D^\circ$
		5/2	395 250.2	72		25	$(^1D2)\ ^2D^\circ$
Co VI ($^5D_0$)	*Limit*		641 000				

Co VI

$Z=27$

Ti I isoelectronic sequence

Ground state: $1s^22s^22p^63s^23p^63d^4\ ^5D_0$

Ionization energy = 823 000±1600 cm^{-1} (102.0±0.2 eV)

The early work on Co VI was done by Bowen (1938), who reported terms of $3d^4$ and $3d^34p$.

Henrichs and Fawcett (1976) reobserved the $3d^4-3d^34p$ transition array in the region of 260–340 Å and classified 200 more lines. Waagen and Meinders (1980) continued the analysis by classifying the $3d^34s-3d^34p$ array at higher wavelengths and re-evaluated all the levels. They provided us with their observed level values and calculated percentage compositions in advance of publication. The energy level uncertainty is presumably ±5 cm^{-1}.

The ionization energy was obtained by Lotz (1967) by extrapolation.

References

Bowen, I. S. (1938), Phys. Rev. **53**, 889.
Henrichs, H. F., and Fawcett, B. C. (1976), Astron. Astrophys. Suppl. **23**, 139.
Lotz, W. (1967), J. Opt. Soc. Am. **57**, 873.
Waagen, G. J., and Meinders, E. (1980), private communication.

Co VI

Configuration	Term	J	Level (cm^{-1})	Leading percentages		
$3d^4$	^{5}D	0	0.0	100		
		1	204.0	100		
		2	589.0	100		
		3	1 124.7	100		
		4	1 781.6	100		
$3d^4$	^{3}P2	0	27 166.2	59	39	^{3}P1
		1	28 461.5	60	39	^{3}P1
		2	30 506.1	52	33	^{3}P1
$3d^4$	^{3}H	4	28 376.6	96	2	^{3}G
		5	28 805.9	98	2	^{3}G
		6	29 233.8	100		
$3d^4$	^{3}F2	2	30 554.9	67	19	^{3}F1
		3	30 634.6	73	20	^{3}F1
		4	30 816.7	73	18	^{3}F1
$3d^4$	^{3}G	3	34 025.8	93	6	^{3}F2
		4	34 474.6	92	5	^{3}F2
		5	34 854.8	98	2	^{3}H
$3d^4$	^{3}D	3	41 675.9	100		
		2	41 856.5	99	1	^{1}D2
		1	42 098.0	100		
$3d^4$	^{1}G2	4	41 925.9	65	32	^{1}G1
$3d^4$	^{1}I	6	42 799.6	100		
$3d^4$	^{1}S2	0	45 372.0	78	21	^{1}S1
$3d^4$	^{1}D2	2	52 739.5	78	21	^{1}D1
$3d^4$	^{1}F	3	59 930.0	99	1	^{3}F1

Co VI—Continued

Configuration	Term	J	Level (cm^{-1})	Leading percentages		
$3d^4$	$^3P1$	2	70 164.3	61	39	$^3P2$
		1	71 624.2	60	40	$^3P2$
		0	72 302.7	60	40	$^3P2$
$3d^4$	$^3F1$	4	70 798.4	80	19	$^3F2$
		2	70 925.6	78	22	$^3F2$
		3	70 998.4	78	21	$^3F2$
$3d^4$	$^1G1$	4	81 027.4	66	33	$^1G2$
$3d^4$	$^1D1$	2	106 531.5	79	21	$^1D2$
$3d^4$	$^1S1$	0	137 619.2	79	21	$^1S2$
$3d^3(^4F)4s$	^5F	1	273 324.8	100		
		2	273 727.7	100		
		3	274 319.0	100		
		4	275 088.0	100		
		5	276 008.0	100		
$3d^3(^4F)4s$	^3F	2	282 948.0	100		
		3	283 961.0	100		
		4	285 189.0	99		
$3d^3(^4P)4s$	^5P	1	293 914.5	99	1	$(^2P)\ ^3P$
		2	294 229.5	98	2	$(^2P)\ ^3P$
		3	295 025.6	100		
$3d^3(^2G)4s$	^3G	3	298 620.2	100		
		4	298 974.6	98	1	$(^2G)\ ^1G$
		5	299 534.0	98	1	$(^2H)\ ^3H$
$3d^3(^4P)4s$	^3P	1	302 688.8	86	6	$(^2P)\ ^3P$
		2	303 840.9	90	8	$(^2P)\ ^3P$
$3d^3(^2G)4s$	^1G	4	303 788.5	92	8	$(^2H)\ ^3H$
$3d^3(^2P)4s$	^3P	2	304 909.9	56	26	$(^2D2)\ ^3D$
		1	305 039.0	66	18	$(^2D2)\ ^3D$
		0	305 473.1	84	16	$(^4P)\ ^3P$
$3d^3(^2D2)4s$	^3D	1	306 439.2	50	25	$(^2P)\ ^3P$
		3	307 482.7	80	20	$(^2D1)\ ^3D$
		2	307 720.0	52	32	$(^2P)\ ^3P$
$3d^3(^2H)4s$	^3H	4	307 627.0	92	7	$(^2G)\ ^1G$
		5	307 692.0	98	2	$(^2G)\ ^3G$
		6	308 029.5	100		
$3d^3(^2P)4s$	^1P	1	310 736.8	86	7	$(^2D2)\ ^3D$
$3d^3(^2D2)4s$	^1D	2	311 916.0	77	20	$(^2D1)\ ^1D$
$3d^3(^2H)4s$	^1H	5	312 551.2	99	1	$(^2H)\ ^3H$
$3d^3(^2F)4s$	^3F	4	326 384.3	100		
		3	326 664.9	100		
		2	326 902.8	100		

(Continued)

Co VI—Continued

Configuration	Term	J	Level (cm^{-1})	Leading percentages			
$3d^3(^2F)4s$	^1F	3	330 886.8	100			
$3d^3(^4F)4p$	$^5G°$	2	*353 596.7*	98		1	$(^4F)\ ^3F°$
		3	*354 427.7*	98		1	$(^4F)\ ^3F°$
		4	*355 504.0*	98		1	$(^4F)\ ^5F°$
		5	*356 833.0*	98		1	$(^4F)\ ^5F°$
		6	*358 461.0*	100			
$3d^3(^2D1)4s$	^3D	3	354 336.2	80		20	$(^2D2)\ ^3D$
		2	354 591.7	79		21	$(^2D2)\ ^3D$
		1	354 782.8	78		22	$(^2D2)\ ^3D$
$3d^3(^4F)4p$	$^3D°$	1	*356 746.4*	42		36	$(^4F)\ ^5F°$
		2	*360 458.8*	56		29	$(^4F)\ ^5F°$
		3	*361 495.1*	73		8	$(^4P)\ ^3D°$
$3d^3(^4F)4p$	$^5D°$	2	*357 272.5*	42		28	$(^4F)\ ^3D°$
		0	*357 990.6*	95		4	$(^4P)\ ^5D°$
		3	*358 022.2*	67		16	$(^4F)\ ^5F°$
		1	*358 366.5*	76		16	$(^4F)\ ^5F°$
		4	*358 915.0*	88		8	$(^4F)\ ^5F°$
$3d^3(^2D1)4s$	^1D	2	358 810.9	79		21	$(^2D2)\ ^1D$
$3d^3(^4F)4p$	$^5F°$	2	*359 037.0*	44		48	$(^4F)\ ^5D°$
		3	*359 833.0*	75		21	$(^4F)\ ^5D°$
		1	*359 902.9*	48		44	$(^4F)\ ^3D°$
		4	*360 591.2*	88		7	$(^4F)\ ^5D°$
		5	*361 268.2*	91		6	$(^4F)\ ^3G°$
$3d^3(^4F)4p$	$^3G°$	3	*364 344.2*	90		6	$(^2G)\ ^3G°$
		4	*365 084.8*	88		5	$(^2G)\ ^3G°$
		5	*366 029.9*	85		8	$(^4F)\ ^5F°$
$3d^3(^4F)4p$	$^3F°$	2	*367 554.5*	93		2	$(^2D2)\ ^3F°$
		3	*368 412.2*	92		2	$(^2D2)\ ^3F°$
		4	*369 316.0*	93		2	$(^2D2)\ ^3F°$
$3d^3(^4P)4p$	$^5P°$	1	*374 556.4*	96		1	$(^4P)\ ^3P°$
		2	*375 249.0*	93		5	$(^4P)\ ^3P°$
		3	*376 365.0*	98		1	$(^4P)\ ^5D°$
$3d^3(^4P)4p$	$^5D°$	0	*375 932.4*	53		36	$(^4P)\ ^3P°$
		1	*376 483.5*	59		34	$(^4P)\ ^3P°$
		2	*378 714.0*	58		30	$(^4P)\ ^3P°$
		3	*379 090.8*	92		4	$(^4F)\ ^5D°$
		4	*380 503.4*	96		4	$(^4F)\ ^5D°$
$3d^4(^4P)4p$	$^3P°$	2	*376 793.0*	50		35	$(^4P)\ ^5D°$
		1	*378 740.0*	53		34	$(^4P)\ ^5D°$
$3d^3(^2G)4p$	$^3H°$	4	*378 002.0*	76		18	$(^2H)\ ^3H°$
		5	*379 143.0*	68		20	$(^2H)\ ^3H°$
		6	*381 073.7*	73		25	$(^2H)\ ^3H°$
$3d^4(^4P)4p$		0	*378 219.1*	41	$^5D°$	39	$(^4P)\ ^3P°$

Co VI—Continued

Configuration	Term	J	Level (cm^{-1})	Leading percentages			
$3d^3(^2G)4p$	$^3G^\circ$	3	380 636.5	68		11	$(^2G)\ ^3F^\circ$
		4	381 655.6	61		23	$(^2G)\ ^3F^\circ$
		5	382 475.5	74		10	$(^2G)\ ^3H^\circ$
$3d^3(^2G)4p$		4	382 556.8	32	$^3F^\circ$	32	$(^2G)\ ^1G^\circ$
$3d^3(^2G)4p$	$^3F^\circ$	2	382 756.1	54		18	$(^2D2)\ ^3F^\circ$
		3	383 297.4	58		17	$(^2G)\ ^3G^\circ$
$3d^3(^2P)4p$	$^3P^\circ$	1	384 063.0	48		20	$(^2D2)\ ^3P^\circ$
		0	384 752.0	41		22	$(^2P)\ ^1S^\circ$
$3d^3(^2G)4p$	$^1G^\circ$	4	384 784.6	42		30	$(^2G)\ ^3F^\circ$
$3d^3(^2P)4p$		2	385 003.3	25	$^3P^\circ$	16	$(^2P)\ ^1D^\circ$
$3d^3(^4P)4p$	$^5S^\circ$	2	385 386.0	49		14	$(^2P)\ ^1D^\circ$
$3d^3(^2G)4p$	$^1F^\circ$	3	385 480.0	63		13	$(^2D2)\ ^1F^\circ$
$3d^3(^2G)4p$	$^1H^\circ$	5	385 802.0	66		19	$(^2H)\ ^1H^\circ$
$3d^3(^2P)4p$	$^3D^\circ$	1	386 280.3	80		6	$(^4P)\ ^3D^\circ$
		2	387 952.5	62		10	$(^2P)\ ^1D^\circ$
		3	388 564.7	46		23	$(^2D2)\ ^3F^\circ$
$3d^3(^4P)4p$		2	386 593.8	38	$^5S^\circ$	20	$(^2P)\ ^3P^\circ$
$3d^3(^2H)4p$	$^3H^\circ$	4	387 548.3	59		21	$(^2G)\ ^1G^\circ$
		5	387 564.7	73		22	$(^2G)\ ^3H^\circ$
		6	388 127.8	71		25	$(^2G)\ ^3H^\circ$
$3d^3(^2D2)4p$	$^1P^\circ$	1	389 036.0	39		20	$(^2P)\ ^1P^\circ$
$3d^3(^2D2)4p$	$^3F^\circ$	2	389 224.5	43		15	$(^2G)\ ^3F^\circ$
		4	391 170.3	71		16	$(^2D1)\ ^3F^\circ$
$3d^3(^4P)4p$		3	389 610.3	29	$^3D^\circ$	25	$(^2D2)\ ^3F^\circ$
$3d^3(^4P)4p$	$^3D^\circ$	2	390 195.8	52		18	$(^2P)\ ^3D^\circ$
		1	390 217.5	41		25	$(^2D2)\ ^3D^\circ$
		3	390 430.3	40		34	$(^2P)\ ^3D^\circ$
$3d^3(^2H)4p$	$^3I^\circ$	5	390 770.1	89		8	$(^2G)\ ^1H^\circ$
		6	391 719.4	96		2	$(^2H)\ ^3H^\circ$
		7	393 109.7	100			
$3d^3(^2D2)4p$	$^3D^\circ$	1	392 070.7	46		32	$(^4P)\ ^3D^\circ$
		2	393 127.0	61		18	$(^4P)\ ^3D^\circ$
		3	393 920.4	51		12	$(^4P)\ ^3D^\circ$
$3d^3(^2H)4p$	$^1G^\circ$	4	393 009.5	73		17	$(^2F)\ ^1G^\circ$
$3d^3(^2H)4p$	$^1H^\circ$	5	393 534.6	69		18	$(^2G)\ ^1H^\circ$
$3d^3(^2D2)4p$	$^3P^\circ$	2	394 159.0	35		44	$(^2P)\ ^3P^\circ$
		1	394 384.3	42		32	$(^2P)\ ^3P^\circ$
		0	394 849.6	45		35	$(^2P)\ ^3P^\circ$
$3d^3(^2D2)4p$	$^1F^\circ$	3	395 188.5	46		15	$(^2H)\ ^3G^\circ$

(Continued)

Co VI—Continued

Configuration	Term	J	Level (cm^{-1})	Leading percentages		
$3d^3(^2H)4p$	$^3G°$	5	*396 235.4*	77	9	$(^2H)\ ^1H°$
		4	*396 420.2*	80	7	$(^2F)\ ^3G°$
		3	*396 569.8*	72	10	$(^2D2)\ ^1F°$
$3d^3(^2H)4p$	$^1I°$	6	*396 559.0*	97	1	$(^2H)\ ^3I°$
$3d^3(^4P)4p$	$^3S°$	1	*399 451.9*	79	10	$(^2P)\ ^1P°$
$3d^3(^2D2)4p$	$^1D°$	2	*400 205.3*	45	42	$(^2P)\ ^1D°$
$3d^3(^2P)4p$	$^1P°$	1	*401 010.3*	60	17	$(^2D2)\ ^1P°$
$3d^3(^2F)4p$	$^3F°$	2	*407 080.0*	90	2	$(^2D1)\ ^3F°$
		3	*407 195.0*	88	4	$(^2F)\ ^3G°$
		4	*407 547.7*	88	5	$(^2F)\ ^3G°$
$3d^3(^2F)4p$	$^3G°$	3	*411 747.1*	82	8	$(^2H)\ ^3G°$
		4	*412 400.0*	84	8	$(^2H)\ ^3G°$
		5	*413 003.8*	93	7	$(^2H)\ ^3G°$
$3d^3(^2F)4p$	$^3D°$	3	*412 885.0*	80	8	$(^2D1)\ ^3D°$
		2	*414 075.6*	76	10	$(^2F)\ ^1D°$
		1	*414 758.0*	90	8	$(^2D1)\ ^3D°$
$3d^3(^2F)4p$	$^1D°$	2	*413 472.2*	61	20	$(^2D1)\ ^1D°$
$3d^3(^2F)4p$	$^1G°$	4	*417 825.9*	79	18	$(^2H)\ ^1G°$
$3d^3(^2F)4p$	$^1F°$	3	*417 926.2*	90	2	$(^2G)\ ^1F°$
$3d^3(^2D1)4p$	$^3D°$	1	*435 466.8*	75	17	$(^2D2)\ ^3D°$
		2	*435 540.0*	73	16	$(^2D2)\ ^3D°$
		3	*436 005.0*	74	15	$(^2D2)\ ^3D°$
$3d^3(^2D1)4p$	$^1D°$	2	*438 159.2*	43	21	$(^2D1)\ ^3F°$
$3d^3(^2D1)4p$	$^3F°$	3	*440 016.0*	68	20	$(^2D2)\ ^3F°$
		2	*440 041.7*	52	18	$(^2D1)\ ^1D°$
		4	*440 943.2*	76	21	$(^2D2)\ ^3F°$
$3d^3(^2D1)4p$	$^3P°$	2	*443 652.0*	74	22	$(^2D2)\ ^3P°$
		1	*444 657.8*	74	23	$(^2D2)\ ^3P°$
		0	*445 191.6*	75	24	$(^2D2)\ ^3P°$
$3d^3(^2D1)4p$	$^1F°$	3	*445 562.1*	74	19	$(^2D2)\ ^1F°$
$3d^3(^2D1)4p$	$^1P°$	1	*452 883.0*	76	23	$(^2D2)\ ^1P°$
Co VII ($^4F_{3/2}$)	***Limit***		**823 000**			

Co VII

$Z=27$

Sc I isoelectronic sequence

Ground state: $1s^22s^22p^63s^23p^63d^3\ ^4F_{3/2}$

Ionization energy = 1 040 000±2000 cm^{-1} (128.9±0.2 eV)

Anderson and Mack (1941) have observed and classified 119 lines between 206 and 241 Å in the $3d^3-3d^24p$ transition array. The doublet system was questioned by Bowen (1960) on the basis of isoelectronic extrapolations.

New observations of this array were made by Ryabtsev and Ramonas (1980), with a measurement uncertainty of ±0.002 Å. They carried out a new analysis of the spectrum, removing the errors of the previous work and finding all levels of the $3d^3$ and $3d^24p$ configurations. Their results are given here along with their calculated percentage compositions. The level value uncertainty is ±5 cm^{-1}.

The ionization energy was derived by Lotz (1967) by extrapolation.

References

Anderson, E., and Mack, J. E. (1941), Phys. Rev. **59**, 717.
Bowen, I. S. (1960), Astrophys. J. **132**, 1.
Lotz, W. (1967), J. Opt. Soc. Am. **57**, 873.
Ryabtsev, A. N. and Ramonas, A. A. (1980), Opt. Spectrosc. (U.S.S.R.) **49**, 574.

Co VII

Configuration	Term	J	Level (cm^{-1})	Leading percentages		
$3d^3$	^4F	3/2	0	100		
		5/2	701	100		
		7/2	1 620	100		
		9/2	2 718	100		
$3d^3$	^4P	1/2	21 011	98	1	^2P
		3/2	21 250	96	3	^2P
		5/2	22 203	100		
$3d^3$	^2G	7/2	23 286	100		
		9/2	24 206	98	2	^2H
$3d^3$	^2P	3/2	29 406	55	34	$^2D2$
		1/2	29 859	98	1	^4P
$3d^3$	$^2D2$	5/2	32 319	81	19	$^2D1$
		3/2	32 628	44	42	^2P
$3d^3$	^2H	9/2	32 404	98	2	^2G
		11/2	33 012	100		
$3d^3$	^2F	7/2	51 867	100		
		5/2	52 379	100		
$3d^3$	$^2D1$	5/2	80 516	81	19	$^2D2$
$3d^2(^3F)4p$	$^4G°$	5/2	*445 793*	86	10	$(^3F)\ ^2F°$
		7/2	*447 330*	86	8	
		9/2	*449 349*	83	17	
		11/2	*451 835*	100		
$3d^2(^3F)4p$	$^4F°$	3/2	*447 004*	90	8	$(^3F)\ ^2D°$
		5/2	*448 093*	90		
		7/2	*449 482*	88	8	$(^3F)\ ^4G°$
		9/2	*451 014*	81	18	$(^3F)\ ^4G°$

(Continued)

Co VII—Continued

Configuration	Term	J	Level (cm^{-1})	Leading percentages		
$3d^2(^3F)4p$	$^2F°$	5/2	450 809	49	18	$(^3F)\ ^4D°$
		7/2	452 024	48	44	
$3d^2(^3F)4p$	$^4D°$	3/2	451 257	49	32	$(^3F)\ ^2D°$
		1/2	451 608	90	8	$(^3P)\ ^4D°$
		5/2	452 615	58	27	$(^3F)\ ^2F°$
		7/2	454 301	48	41	$(^3F)\ ^2F°$
$3d^2(^3F)4p$	$^2D°$	3/2	453 255	42	41	$(^3F)\ ^4D°$
		5/2	455 136	59	15	$(^3P)\ ^2D°$
$3d^2(^3F)4p$	$^2G°$	7/2	458 326	94		
		9/2	459 746	92		
$3d^2(^3P)4p$	$^2S°$	1/2	461 032	96		
$3d^2(^3P)4p$	$^4S°$	3/2	465 021	83	15	$(^1D)\ ^2P°$
$3d^2(^1D)4p$	$^2P°$	3/2	467 663	72	16	$(^3P)\ ^4S°$
		1/2	469 527	88		
$3d^2(^1D)4p$	$^2F°$	5/2	468 314	83	6	$(^3F)\ ^2F°$
		7/2	470 294	79	8	$(^3P)\ ^4D°$
$3d^2(^3P)4p$	$^4D°$	1/2	470 113	86	7	$(^3F)\ ^4D°$
		3/2	470 750	86	7	$(^3F)\ ^4D°$
		5/2	471 912	81	7	$(^3F)\ ^4D°$
		7/2	472 919	83	7	$(^1D)\ ^2F°$
$3d^2(^1D)4p$	$^2D°$	3/2	472 919	77	7	$(^1D)\ ^2P°$
		5/2	474 284	76	8	$(^3P)\ ^4P°$
$3d^2(^3P)4p$	$^4P°$	1/2	474 766	98		
		3/2	475 466	96		
		5/2	476 872	88	12	$(^1D)\ ^2D°$
$3d^2(^1G)4p$	$^2G°$	7/2	475 310	88		
		9/2	475 452	92		
$3d^2(^3P)4p$	$^2D°$	3/2	482 367	77	12	$(^3F)\ ^2D°$
		5/2	482 643	79	14	
$3d^2(^1G)4p$	$^2H°$	9/2	483 860	98		
		11/2	486 306	100		
$3d^2(^3P)4p$	$^2P°$	1/2	486 619	100		
		3/2	487 058	92		
$3d^2(^1G)4p$	$^2F°$	7/2	490 180	94		
			490 767	96		
$3d^2(^1S)4p$	$^2P°$	1/2	524 079	98		
		3/2	527 023	98		
Co VIII ($^3F_2$)	*Limit*		1 040 000			

Co VIII

$Z=27$

Ca I isoelectronic sequence

Ground state: $1s^2 2s^2 2p^6 3s^2 3p^6 3d^2\ ^3F_2$

Ionization energy = 1 273 000±5000 cm^{-1} (157.8±0.6 eV)

Nineteen lines of the transition array $3d^2-3d4f$ in the range of 122–125 Å were classified by Alexander et al. (1966).

New observations were made by Fawcett, Ridgeley, and Ekberg (1980) in the extended range of 102–192 Å. The arrays $3d^2-3d4p$, $3d5f$, and $3p^63d^2-3p^53d^3$ were analyzed and new lines of $3d^2-3d4f$ were found which unified the earlier system of levels. The uncertainty of about 10 cm^{-1} in the level values derived from these data arises from the reported wavelength uncertainty of ±0.007 Å. Confirmation of the analysis was obtained from parametric calculations of the configurations. The level values and percentage compositions are from this work.

We have derived the ionization energy from the $3d\ nf$ series ($n=4$, 5) and a value for the quantum defect difference $n^*(4f)-n^*(5f)$ = 0.979 obtained from Fe VIII.

References

Alexander, E., Feldman, U., Fraenkel, B. S., and Hoory, S. (1966), J. Opt. Soc. Am. **56**, 651.

Fawcett, B. C., Ridgeley, A., and Ekberg, J. O. (1980), Phys. Scr. **21**, 155.

Co VIII

Configuration	Term	J	Level (cm^{-1})	Configuration	Term	J	Level (cm^{-1})
$3p^6\,3d^2$	^{3}F	2	0	$3p^5(^2P^\circ)3d^3(^2H)$	^{1}H°	5	*590 805*
		3	1 430				
		4	3 144	$3p^5(^2P^\circ)3d^3(^2F)$	^{3}D°	1	*598 440*
						2	*599 641*
$3p^6\,3d^2$	^{1}D	2	19 624			3	*602 844*
$3p^6\,3d^2$	^{3}P	0	22 304	$3p^5(^2P^\circ)3d^3(^2F)$	^{1}D°	2	*605 841*
		1	22 839				
		2	24 055	$3p^5(^2P^\circ)3d^3(^2G)$	^{1}F°	3	*608 501*
$3p^6\,3d^2$	^{1}G	4	32 360	$3p^5(^2P^\circ)3d^3(^4P)$	^{3}P°	0	*612 076*
						1	*613 869*
$3p^6\,3d^2$	^{1}S	0	74 247			2	*619 010*
$3p^6\,3d\,4p$	^{1}D°	2	*542 430*	$3p^5(^2P^\circ)3d^3(^4F)$	^{3}F°	2	*616 019*
						3	*618 348*
$3p^6\,3d\,4p$	^{3}D°	1	*542 701*			4	*620 737*
		2	*544 314*				
		3	*545 834*	$3p^5(^2P^\circ)3d^3(^2P)$	^{1}P°	1	*653 446*
$3p^6\,3d\,4p$	^{3}F°	2	*547 400*	$3p^5(^2P^\circ)3d^3(^4F)$	^{3}D°	3	*656 715*
		3	*548 799*			2	*657 020*
		4	*551 524*			1	*658 136*
$3p^6\,3d\,4p$	^{3}P°	1	*554 082*	$3p^5(^2P^\circ)3d^3(^2H)$	^{1}G°	4	*662 151*
		0	*554 287*				
		2	*554 998*	$3p^5(^2P^\circ)3d^3(^4P)$	^{3}S°	1	*678 094*
$3p^5(^2P^\circ)3d^3(^2H)$	^{3}G°	3	*555 699*	$3p^5(^2P^\circ)3d^3(^2F)$	^{1}F°	3	*682 051*
		4	*557 817*				
		5	*561 346*	$3p^5(^2P^\circ)3d^3(^2D)$	^{1}P°	1	*687 584*
$3p^6\,3d\,4p$	^{1}F°	3	*557 736*	$3p^6\,3d4f$	^{1}G°	4	*811 205*
$3p^6\,3d\,4p$	^{1}P°	1	*563 271*				

(Continued)

Co VIII—Continued

Configuration	Term	J	Level (cm^{-1})	Configuration	Term	J	Level (cm^{-1})
$3p^6\,3d4f$	$^3F°$	2	*812 862*	$3p^6\,3d4f$	$^1P°$	1	*827 508*
		3	*813 298*				
		4	*814 130*	$3p^6\,3d5f$	$^1G°$	4	*977 281*
$3p^6\,3d4f$	$^3G°$	3	*817 839*	$3p^6\,3d5f$	$^3F°$	2	*978 005*
		4	*818 958*			3	*978 307*
		5	*819 657*			4	*979 360*
$3p^6\,3d4f$	$^1D°$	2	*818 633*	$3p^6\,3d5f$	$^3G°$	3	*981 316*
						4	*982 728*
$3p^6\,3d4f$	$^3D°$	1	*820 599*			5	*983 219*
		2	*820 605*				
		3	*820 450*	$3p^6\,3d5f$	$^3D°$	2	*982 716*
						3	*982 933*
$3p^6\,3d4f$	$^1F°$	3	*821 881*				
				$3p^6\,3d5f$	$^1F°$	3	*983 954*
$3p^6\,3d4f$	$^3P°$	2	*823 064*				
		1	*823 613*	$3p^6\,3d5f$	$^1H°$	5	*986 549*
		0	*823 928*				
				Co IX ($^2D_{3/2}$)	***Limit***		**1 273 000**
$3p^6\,3d4f$	$^1H°$	5	*827 140*				

Co IX

$Z=27$

K I isoelectronic sequence

Ground state: $1s^22s^22p^63s^23p^63d\ ^2D_{3/2}$

Ionization energy = 1 501 300±2000 cm^{-1} (186.13±0.20 eV)

The $3d\ ^2D$ ground term and the levels of $3p^53d^2$ and $3p^64p$ are taken from the analysis of Ramonas and Ryabtsev (1980). The transitions lie between 144 and 161 Å and have reported accuracy of ±0.002 Å, giving a level uncertainty of ±10 cm^{-1}.

The $3p^6nf$ series was observed by Alexander, Feldman, and Fraenkel (1965) with a level uncertainty of ±1500 cm^{-1}. They determined the ionization energy from that series.

The analysis of the $3p^63d-3p^53d4s$ array at 100 Å is by Hoory, Goldsmith, Fraenkel and Feldman (1970). Their wavelength accuracy of ±0.005 Å results in a level uncertainty of about ±50 cm^{-1} for the $3p^53d4s$ configuration.

References

Alexander, E., Feldman, U., and Fraenkel, B. S. (1965), J. Opt. Soc. Am. **55**, 650.

Hoory, S., Goldsmith, S., Fraenkel, B. S., and Feldman, U. (1970), Astrophys. J. **160**, 781.

Ramonas, A. A., and Ryabtsev, A. N. (1980) Opt. Spectrosc. **48**, 348.

Co IX

Configuration	Term	J	Level (cm^{-1})	Configuration	Term	J	Level (cm^{-1})
$3p^6\,3d$	^2D	3/2	0	$3p^5\,3d(^3F^\circ)4s$	$^2F^\circ$	7/2	*1 003 240*
		5/2	2 451			5/2	*1 009 670*
$3p^5(^2P^\circ)3d^2(^1G)$	$^2F^\circ$	5/2	*468 222*	$3p^5\,3d(^3D^\circ)4s$	$^4D^\circ$	7/2	*1 024 380*
		7/2	*472 140*			5/2	*1 027 170*
$3p^5(^2P^\circ)3d^2(^1D)$	$^2F^\circ$	7/2	*485 123*	$3p^5\,3d(^1F^\circ)4s$	$^2F^\circ$	7/2	*1 038 280*
		5/2	*499 750*	$3p^5\,3d(^3D^\circ)4s$	$^2D^\circ$	3/2	*1 040 830*
$3p^5(^2P^\circ)3d^2(^3F)$	$^2F^\circ$	5/2	*580 759*			5/2	*1 043 280*
		7/2	*588 291*				
$3p^6\,4p$	$^2P^\circ$	1/2	*625 109*	$3p^6\,5f$	$^2F^\circ$	5/2	*1 130 660*
		3/2	*629 117*			7/2	*1 130 690*
$3p^5(^2P^\circ)3d^2(^3F)$	$^2D^\circ$	5/2	*644 843*	$3p^6\,6f$	$^2F^\circ$	5/2	*1 243 970*
		3/2	*645 408*			7/2	*1 244 010*
$3p^5(^2P^\circ)3d^2(^3P)$	$^2P^\circ$	1/2	*650 182*	$3p^6\,7f$	$^2F^\circ$	5/2,7/2	*1 313 020*
		3/2	*654 735*	$3p^6\,8f$	$^2F^\circ$	5/2,7/2	*1 357 500*
$3p^6\,4f$	$^2F^\circ$	5/2	*922 590*	$3p^6\,9f$	$^2F^\circ$	5/2,7/2	*1 387 960*
		7/2	*922 690*	$3p^6\,10f$	$^2F^\circ$	5/2,7/2	*1 409 880*
$3p^5\,3d(^3P^\circ)4s$	$^2P^\circ$	1/2	*986 100*				
		3/2	*991 510*	Co X ($^1S_0$)	*Limit*		**1 501 300**
$3p^5\,3d(^3F^\circ)4s$	$^4F^\circ$	7/2	*996 130*				
		5/2	*997 900*				

Co x

$Z=27$

Ar I isoelectronic sequence

Ground state: $1s^2 2s^2 2p^6 3s^2 3p^6\ {}^1S_0$

Ionization energy = 2 221 000±3000 cm^{-1} (275.4±0.4 eV)

This spectrum was first observed by Gabriel, Fawcett, and Jordan (1966), who identified the $3p^5 3d\ {}^1P°$ level. The present value for this level is taken from the observations of Goldsmith (1969) and has an uncertainty of ±20 cm^{-1}.

The $3p^5 4s$ and $4d$ configurations are taken from observations of resonance lines by Alexander, Feldman, and Fraenkel (1965) between 71 and 91 Å; level values are uncertain by about 60 cm^{-1}.

The $3p^5 3d - 3p^5 4f$ transition array has been identified by Fawcett, Cowan, and Hayes (1972) but only one level is connected with the single known level of $3p^5 3d$.

The $4f$ and $5s$ levels given below are from the work of Swartz, Kastner, Goldsmith, and Neupert (1976). The level uncertainty is probably ±100 cm^{-1}.

We derived the ionization energy from the $3p^5 4s$ and $3p^5 5s$ configurations with an estimated value for the quantum defect change Δn^* of 1.024 between them obtained from similar spectra.

References

Alexander, E., Feldman, U., and Fraenkel, B.S. (1965), J. Opt. Soc. Am. **55**, 650.

Fawcett, B. C., Cowan, R. D., and Hayes, R. W. (1972), J. Phys. **B5**, 2143.

Gabriel, A. H., Fawcett, B. C., and Jordan, C. (1966), Proc. Phys. Soc. **87**, 825.

Goldsmith, S. (1969), J. Opt. Soc. Am. **59**, 1678.

Swartz, M., Kastner, S. D., Goldsmith, L., and Neupert, W. M. (1976), J. Opt. Soc. Am. **66**, 240.

Co x

Configuration	Term	J	Level (cm^{-1})	Configuration	Term	J	Level (cm^{-1})
$3s^2\,3p^6$	^1S	0	0	$3s^2\,3p^5\,4d$	$^1P°$	1	*1 398 800*
$3s^2\,3p^5\,3d$	$^1P°$	1	*629 430*	$3s^2\,3p^5(^2P°_{1/2})4f$	$^2[5/2]$	2	1 525 950
$3s^2\,3p^5\,4s$	$^3P°$	1	*1 105 290*	$3s^2\,3p^5\,5s$	$^3P°$	1	*1 586 870*
$3s^2\,3p^5\,4s$	$^1P°$	1	*1 123 670*	$3s^2\,3p^5\,5s$	$^1P°$	1	*1 604 310*
$3s^2\,3p^5\,4d$	$^3P°$	1	*1 380 190*	Co XI ($^2P°_{3/2}$)	***Limit***		**2 221 000**

Co XI

$Z=27$

Cl I isoelectronic sequence

Ground state: $1s^22s^22p^63s^23p^5\ ^2P^\circ_{3/2}$

Ionization energy = 2 460 000±16000 cm^{-1} (305±2 eV)

The ground term and the $3s3p^6\ ^2S$ term were determined from the data of Fawcett and Hatter (1980) with an uncertainty of ±20 cm^{-1}. Terms of $3p^43d$ are from the measurements of Goldsmith (1969). The uncertainty in these levels is about ±50 cm^{-1}.

The $3p^44s$ terms are from Edlén (1937), except for $^4P_{5/2}$ which was identified by Fawcett, Cowan, and Hayes (1972). The level uncertainty for this configuration is ±100 cm^{-1}.

Fawcett, Cowan, and Hayes also determined the $3p^44d$ terms and classified lines of the array $3p^43d-3p^44f$, which are not connected to the known system of levels. The measurement uncertainty of ~0.02 Å at 120 Å results in a level uncertainty of about 150 cm^{-1}.

The ionization energy was obtained by Lotz (1967) by extrapolation.

References

Edlén, B. (1937), Zeits. Phys. **104**, 407.

Fawcett, B. C., and Hatter, A. T. (1980), Astron. Astrophys. **84**, 78.

Fawcett, B. C., Cowan, R. D., and Hayes, R. W. (1972), J. Phys. **B5**, 2143.

Goldsmith, S. (1969), J. Opt. Soc. Am. **59**, 1678.

Lotz, W. (1967), J. Opt. Soc. Am. **57**, 873.

Co XI

Configuration	Term	J	Level (cm^{-1})	Configuration	Term	J	Level (cm^{-1})
$3s^2 3p^5$	$^2P^\circ$	3/2	*0*	$3s^2 3p^4(^3P)4s$	^2P	3/2	1 202 070
		1/2	*19 345*			1/2	1 211 780
$3s3p^6$	^2S	1/2	313 630	$3s^2 3p^4(^1D)4s$	^2D	5/2	1 226 890
						3/2	1 227 710
$3s^2 3p^4(^1D)3d$	^2S	1/2	582 510				
				$3s^2 3p^4(^3P)4d$	^2D	5/2	1 471 200
$3s^2 3p^4(^3P)3d$	^2P	3/2	606 420				
		1/2	613 480	$3s^2 3p^4(^1D)4d$	^2D	5/2	1 510 800
						3/2	1 523 400
$3s^2 3p^4(^3P)3d$	^2D	5/2	615 140				
		3/2	631 680	Co XII ($^3P_2$)	*Limit*		**2 460 000**
$3s^2 3p^4(^3P)4s$	^4P	5/2	1 181 100				
		3/2	1 189 920				

Co XII

$Z=27$

S I isoelectronic sequence

Ground state: $1s^2 2s^2 2p^6 3s^2 3p^4\ ^3P_2$

Ionization energy $= 2\,710\,000 \pm 7000$ cm^{-1} (336 ± 1 eV)

The transition array $3s^2 3p^4 - 3s 3p^5$ at 300 Å was interpreted by Fawcett and Hayes (1972) and remeasured with improved accuracy (± 0.02 Å) by Fawcett and Hatter (1980) giving a level uncertainty of ± 20 cm^{-1}. The level $3s 3p^5\ ^3P_0$ is determined by an identification labeled tentative. Its position is out of order in the ^{3}P isoelectronic sequence. Since no intersystem lines are known, the calculated positions of $3s^2 3p^4\ ^1$D and ^{1}S by Smitt, Svensson, and Outred (1976) are given in brackets. The calculation was made by diagonalizing the energy matrices with extrapolated values for the radial integrals. Their uncertainty appears to be ± 20 cm^{-1}.

Levels of the $3p^3 3d$ configuration are from the line classifications of Fawcett and Hayes (1972) in the range of 165–180 Å. The wavelength uncertainty is ± 0.03 Å, giving a level uncertainty of ± 100 cm^{-1}.

The $3p^3 4d$ levels are from the observations of Fawcett, Cowan, and Hayes (1972) at 56 Å. The uncertainty is about ± 500 cm^{-1}. They also observed levels of $3p^3 4f$ terms, but they are not connected with this system.

The ionization energy is an extrapolated value obtained by Lotz (1967).

References

Fawcett, B. C., and Hayes, R. W. (1972), J. Phys. **B5**, 366.

Fawcett, B. C., Cowan, R. D., and Hayes, R. W. (1972), J. Phys. **B5**, 2143.

Fawcett, B. C., and Hatter, A. T. (1980), Astron. Astrophys. **84**, 78.

Lotz, W. (1967), J. Opt. Soc. Am. **57**, 873.

Smitt, R., Svensson, L. A., and Outred, M. (1976), Phys. Scr. **13**, 293.

Co XII

Configuration	Term	J	Level (cm^{-1})	Configuration	Term	J	Level (cm^{-1})
$3s^2 3p^4$	^{3}P	2	0	$3s^2 3p^3(^2D°)3d$	^{1}D°	2	[*622 130*]
		1	15 820				
		0	17 070	$3s^2 3p^3(^2D°)3d$	^{1}F°	3	[*630 670*]
$3s^2 3p^4$	^{1}D	2	[42 120]	$3s^2 3p^3(^2D°)3d$	^{1}P°	1	[*669 160*]
$3s^2 3p^4$	^{1}S	0	[88 880]	$3s^2 3p^3(^4S°)4d$	^{3}D°	3	[*1 567 400*]
$3s3p^5$	^{3}P°	2	*306 640*	$3s^2 3p^3(^2D°)4d$	^{1}D°	2	[*1 614 400*]
		0	*316 430?*				
		1	*318 280*	$3s^2 3p^3(^2D°)4d$	^{1}F°	3	[*1 617 700*]
$3s3p^5$	^{1}P°	1	[*390 990*]	$3s^2 3p^3(^2D°)4d$	^{1}P°	1	[*1 658 800*]
$3s^2 3p^3(^2D°)3d$	^{3}P°	2	*569 990*	Co XIII ($^4S°_{3/2}$)	***Limit***		**2 710 000**
$3s^2 3p^3(^4S°)3d$	^{3}D°	3	*594 040*				
		2	*602 920*				
		1	*608 660*				

Co XIII

$Z=27$

P I isoelectronic sequence

Ground state: $1s^2 2s^2 2p^6 3s^2 3p^3\ ^4S^\circ_{3/2}$

Ionization energy = 3 057 000±8000 cm^{-1} (379±1 eV)

The excited levels of the $3s^2 3p^3$ configuration, given in brackets, are taken from the calculation by Smitt, Svensson, and Outred (1976). They appear to be uncertain by about ±20 cm^{-1}. No experimental connection between terms of this configuration is known.

The $3s3p^4$ configuration is obtained from the measurements and classifications of Fawcett and Hayes (1972) and of Fawcett and Hatter (1980) with a level uncertainty of ±50 cm^{-1}. The $3s^2 3p^2 3d$ levels are from the observations of Fawcett and Hayes. The level uncertainty for this configuration is ± 100 cm^{-1}.

The $3p^2 4d$ observations of Fawcett, Cowan, and Hayes (1972) give levels that appear inconsistent with their isoelectronic sequence, and are omitted here.

The ionization energy is an extrapolated value obtained by Lotz (1967).

References

Fawcett, B. C., and Hayes, R. W. (1972), J. Phys. **B5**, 366.
Fawcett, B. C., Cowan, R. D., and Hayes, R. W. (1972), J. Phys. **B5**, 2143.
Fawcett, B. C., and Hatter, A. T. (1980), Astron. Astrophys. **84**, 78.
Lotz, W. (1967), J. Opt. Soc. Am. **57**, 873.
Smitt, R., Svensson, L. A., and Outred, M. (1976), Phys. Scr. **13**, 293.

Co XIII

Configuration	Term	J	Level (cm^{-1})	Configuration	Term	J	Level (cm^{-1})
$3s^2 3p^3$	$^4S^\circ$	3/2	*0*	$3s^2 3p^2(^3P)3d$	^4P	5/2	547 890
						3/2	552 880
$3s^2 3p^3$	$^2D^\circ$	3/2	[*43 650*]			1/2	556 820
		5/2	[*49 690*]				
				$3s^2 3p^2(^1D)3d$	^2D	3/2	592 830
$3s^2 3p^3$	$^2P^\circ$	1/2	[*79 460*]			5/2	594 200
		3/2	[*88 170*]				
				$3s^2 3p^2(^1D)3d$	^2P	1/2	608 870
$3s3p^4$	^4P	5/2	295 160			3/2	618 880
		3/2	307 030				
		1/2	312 110	$3s^2 3p^2(^1D)3d$	^2F	7/2	621 710
$3s3p^4$	^2D	3/2	365 530	$3s^2 3p^2(^3P)3d$	^2D	5/2	646 890
		5/2	368 250			3/2	648 390
$3s3p^4$	^2P	3/2	418 480	Co XIV ($^3P_0$)	***Limit***		**3 057 000**
		1/2	423 290				

Co XIV

$Z=27$

Si I isoelectronic sequence

Ground state: $1s^22s^22p^63s^23p^2\ ^3P_0$

Ionization energy $= 3\,315\,000 \pm 16000$ cm^{-1} (411 ± 2 eV)

The levels of $3s^23p^2$ and $3s3p^3$ are derived from the measurements and classifications of Fawcett and Hayes (1972) and Fawcett and Hatter (1980). Their measurement uncertainty of 0.03 Å corresponds to a level uncertainty of 50 cm^{-1}. The $3s^23p3d$ levels are from the observations of Fawcett and Hayes between 184 and 343 Å and are uncertain by about ± 60 cm^{-1}. A transition to $3s^23p^2\ ^1D$ from $3s^23p3d\ ^1D$ puts that level 240 cm^{-1} above the value calculated by Smitt, Svensson, and Outred (1976) which puts the level value in question. The latter paper provides a predicted position for $3s^23p^2\ ^1S$. This level and those based on it are marked with $+x$.

The $3p4s$ levels are from the observations of Fawcett, Cowan, and Hayes (1972) and the $3p4d$ and $3p4f$ levels are those of Kastner, Swartz, Bhatia, and Lapides (1978). The uncertainty is about ± 100 cm^{-1}. Several tentative identifications are omitted here.

The ionization energy is an extrapolated value obtained by Lotz (1967).

References

Fawcett, B. C., and Hayes, R. W. (1972), J. Phys. **B5**, 366.
Fawcett, B. C., Cowan, R. D., and Hayes, R. W. (1972), J. Phys. **B5**, 2143.
Fawcett, B. C., and Hatter, A. T. (1980), Astron. Astrophys. **84**, 78.
Kastner, S. O., Swartz, M., Bhatia, A. K., and Lapides, J. (1978), J. Opt. Soc. Am. **68**, 1558.
Lotz, W. (1967), J. Opt. Soc. Am. **57**, 873.
Smitt, R., Svensson, L. A., and Outred, M. (1976), Phys. Scr. **13**, 293.

Co XIV

Configuration	Term	J	Level (cm^{-1})	Configuration	Term	J	Level (cm^{-1})
$3s^2\ 3p^2$	^3P	0 1 2	0 12 030 22 780	$3s^2\ 3p3d$	$^1F°$	3	*597 430*
				$3s^2\ 3p3d$	$^1P°$	1	*612 170+x*
$3s^2\ 3p^2$	^1D	2	55 160?	$3s^2\ 3p4s$	$^3P°$	2	*1 536 780*
$3s^2\ 3p^2$	^1S	0	101 080+x	$3s^2\ 3p4s$	$^1P°$	1	*1 546 160*
$3s3p^3$	$^3D°$	2 3	*311 240* *315 000*	$3s^2\ 3p4d$	$^3D°$	2 3	*1 805 400* *1 807 800*
$3s3p^3$	$^3P°$	2	*357 880*	$3s^2\ 3p4d$	$^3F°$	3	*1 826 800*
$3s3p^3$	$^1D°$	2	*392 250*	$3s^2\ 3p4d$	$^1F°$	3	*1 837 200*
$3s3p^3$	$^3S°$	1	*446 220*	$3s^2\ 3p4d$	$^1P°$	1	*1 858 500+x*
$3s3p^3$	$^1P°$	1	*472 990*	$3s^2\ 3p4f$	^1D	2	1 956 600+x
$3s^2\ 3p3d$	$^3P°$	2 1	*520 950* *530 250*	$3s^2\ 3p4f$	^1G	4	1 959 800
				Co XV ($^2P°_{1/2}$)	***Limit***		**3 315 000**
$3s^2\ 3p3d$	$^1D°$	2	*536 280*				
$3s^2\ 3p3d$	$^3D°$	1 3 2	*544 260* *546 830* *547 350*				

Co XV

$Z=27$

Al I isoelectronic sequence

Ground state: $1s^2 2s^2 2p^6 3s^2 3p\ ^2P^{\circ}_{1/2}$

Ionization energy = 3 580 000±16000 cm^{-1} (444±2 eV)

The $3s^2 3p$, $3s3p^2$, and $3s^2 3d$ levels are from Fawcett and Hayes (1972) and Fawcett and Hatter (1980); the levels are determined with an uncertainty of about 50 cm^{-1}. The $3p-4d$ doublet was identified by Edlén (1936) giving $4d$ levels with an uncertainty of ±300 cm^{-1}. The $4f$ levels are from Fawcett, Cowan, and Hayes (1972) and are uncertain by ±300 cm^{-1}.

The ionization energy was determined by Lotz (1967) by extrapolation.

References

Edlén, B. (1936), Zeit. Phys. **103**, 536.

Fawcett, B. C., Cowan, R. D., and Hayes, R. W. (1972), J. Phys. **B5**, 2143.

Fawcett, B. C., and Hayes, R. W. (1972), J. Phys. **B5**, 366.

Fawcett, B. C., and Hatter, A. T. (1980), Astron. Astrophys. **84**, 78.

Lotz, W. (1967), J. Opt. Soc. Am. **57**, 873.

Co XV

Configuration	Term	J	Level (cm^{-1})	Configuration	Term	J	Level (cm^{-1})
$3s^2\,3p$	$^2P^{\circ}$	1/2	*0*	$3s^2\,3d$	^2D	3/2	506 230
		3/2	*22 970*			5/2	508 760
$3s3p^2$	^2D	3/2	322 730	$3s^2\,4d$	^2D	3/2	1 901 800
		5/2	325 780			5/2	1 903 600
$3s3p^2$	^2S	1/2	390 810	$3s^2\,4f$	$^2F^{\circ}$	5/2	*2 002 800*
						7/2	*2 003 200*
$3s3p^2$	^2P	1/2	417 690				
		3/2	426 590	Co XVI ($^1S_0$)	*Limit*		**3 580 000**

Co XVI

$Z=27$

Mg I isoelectronic sequence

Ground state: $1s^2 2s^2 2p^6 3s^2\ ^1S_0$

Ionization energy = 4 129 200±500 cm^{-1} (511.96±0.06 eV)

The classification of this spectrum was begun by Edlén (1936) who identified the triplets of $3s3d-3s4f$, $3s3p-3s4d$, and the singlet of $3s^2-3s4p$ in the range of 47–62 Å. The levels derived from these measurements are presumed to have an uncertainty of ±300 cm^{-1}.

The $3s5d$ levels were found by Feldman, Katz, Behring, and Cohen (1971) and the $3s3p\ ^1P°$ by Fawcett and Hayes (1972). The rest of the levels were derived from the classifications of Fawcett, Cowan, and Hayes (1972). Thus the $3s3p-3p^2$ array measured at 300 Å with an uncertainty of 0.02 Å results in a level uncertainty for the $3p^2$ configuration of 40 cm^{-1}. The higher-lying levels are determined from spectra measured at 60 Å and have an uncertainty of 500 cm^{-1}. The $3p^2\ ^1D_2$ level was obtained from a transition from $3s4f\ ^1F^\circ_3$ and has an uncertainty of ±500 cm^{-1}.

The triplet system has not been connected to the ground state by observed lines. We obtained the position of $3s3p\ ^3P^\circ_1$ by interpolation along the isoelectronic sequence. The uncertainty, x, is ±200 cm^{-1}.

The $3p3d-3p4f$ transition array has been observed at 55 Å by Kastner, Swartz, Bhatia, and Lapides (1978) but it is unconnected with the present configurations.

We calculated the ionization energy from the three member $3snd\ ^3D$ series.

References

Edlén, B. (1936), Zeits. Phys. **103**, 536.

Fawcett, B. C., Cowan, R. D., and Hayes, R. W. (1972), J. Phys. **B5**, 2143.

Fawcett, B. C., and Hayes, R. W. (1972), J. Phys. **B5**, 366.

Feldman, U., Katz, L., Behring, W. E., and Cohen, L. (1971), J. Opt. Soc. Am. **61**, 91.

Kastner, S. O., Swartz, M., Bhatia, A. K., and Lapides, J. (1978), J. Opt. Soc. Am. **68**, 1558.

Co XVI

Configuration	Term	J	Level (cm^{-1})
$3s^2$	^1S	0	0
$3s3p$	$^3P°$	0	*249 110+x*
		1	*256 100+x*
		2	*273 500+x*
$3s3p$	$^1P°$	1	*376 310*
$3p^2$	^3P	0	592 150+x
		1	603 890+x
		2	625 100+x
$3p^2$	^1D	2	597 100
$3s3d$	^3D	1	724 720+x
		2	726 030+x
		3	728 150+x
$3s3d$	^1D	2	812 850
$3p3d$	$^3P°$	2	*1 061 600+x*
$3s4s$	^3S	1	1 969 470+x
$3s4p$	$^1P°$	1	*2 106 020*
$3s4d$	^3D	1	2 256 900+x
		2	2 257 800+x
		3	2 259 300+x
$3s4d$	^1D	2	2 261 600
$3p4s$	$^3P°$	2	*2 302 200+x*
$3s4f$	$^3F°$	2	*2 340 900+x*
		3	*2 341 100+x*
		4	*2 341 500+x*
$3s4f$	$^1F°$	3	*2 356 700*
$3p4p$	^3P	1	2 423 000+x
$3p4p$	^3D	3	2 427 100+x
$3p4d$	$^1F°$	3	*2 547 200*
$3p4d$	$^3D°$	1	*2 554 200+x*
		2	*2 564 400+x*
		3	*2 576 700+x*
$3s5d$	^3D	1	2 946 700+x
		2	2 946 800+x
		3	2 947 200+x
$3s5f$	$^3F°$	4	*2 987 900+x*
Co XVII ($^2S_{1/2}$)	*Limit*		**4 129 200**

Co XVII

$Z=27$

Na I isoelectronic sequence

Ground state: $1s^2 2s^2 2p^6 3s\ ^2S_{1/2}$

Ionization energy $= 4\ 408\ 500 \pm 300\ cm^{-1}$ (546.58 ± 0.04 eV)

Wavelengths for the $3s-3p$ and $3p-3d$ doublets were measured by Fawcett and Hayes (1972) and by Fawcett, Cowan, and Hayes (1972). From a study of this isoelectronic sequence Edlén (1978) has deduced semiempirical expressions for the $3s-3p$ and $3p-3d$ transition energies, resulting in improved values for these wavelengths. The uncertainty is probably ± 0.01 Å giving a level uncertainty of $\pm 10\ cm^{-1}$. His results are used here for the $3p$ and $3d$ terms.

The classifications by Edlén (1936) in the range of 41–59 Å are used to obtain the $4s$, $4p$, $4d$, $4f$, and $5f$ terms with a level uncertainty of $\pm$ 500 cm^{-1}. Fawcett, Cowan, and Hayes (1972) identified the doublets $3p-5s$ and $3d-8f$. The level uncertainty for $5s$ and $8f$ terms is $\pm 3000\ cm^{-1}$. The rest of the one-electron configurations are due to the identifications of Feldman, Katz, Behring, and Cohen (1971) whose measurements in the range of 27–41 Å have an estimated uncertainty of ± 0.01 Å, giving a level uncertainty of $\pm 1000\ cm^{-1}$. Feldman and Cohen (1967) identified the transition $2p^6 3s - 2p^5 3s^2$ at ~ 15 Å. The level uncertainty for $2p^5 3s^2\ ^2P$ is ± 2000 cm^{-1}.

The value for the ionization energy was derived by Edlén (1978) from core polarization theory applied to the nf series.

References

Edlén, B. (1936), Zeits. Phys. **100**, 621.

Edlén, B. (1978), Phys. Scr. **17**, 565.

Fawcett, B. C., Cowan, R. D., and Hayes, R. W. (1972), J. Phys. **B5**, 2143.

Fawcett, B. C., and Hayes, R. W. (1972), J. Phys. **B5**, 366.

Feldman, U., and Cohen, L. (1967), J. Opt. Soc. Am. **57**, 1128.

Feldman, U., Katz, L., Behring, W., and Cohen, L. (1971), J. Opt. Soc. Am. **61**, 91.

Co XVII

Configuration	Term	J	Level (cm^{-1})
$2p^6(^1S)3s$	^2S	1/2	0
$2p^6(^1S)3p$	$^2P°$	1/2	*294 556*
		3/2	*319 957*
$2p^6(^1S)3d$	^2D	3/2	720 224
		5/2	723 932
$2p^6(^1S)4s$	^2S	1/2	2 079 550
$2p^6(^1S)4p$	$^2P°$	1/2	*2 196 500*
		3/2	*2 206 580*
$2p^6(^1S)4d$	^2D	3/2	2 353 690
		5/2	2 355 250
$2p^6(^1S)4f$	$^2F°$	5/2	*2 419 690*
		7/2	*2 420 340*
$2p^6(^1S)5s$	^2S	1/2	2 967 700
$2p^6(^1S)5p$	$^2P°$	1/2	*3 026 100*
		3/2	*3 030 800*
$2p^6(^1S)5d$	^2D	3/2	3 102 200
		5/2	3 103 000
$2p^6(^1S)5f$	$^2F°$	5/2	*3 135 500*
		7/2	*3 135 800*
$2p^6(^1S)6p$	$^2P°$	1/2	*3 463 300*
		3/2	*3 466 200*
$2p^6(^1S)6d$	^2D	3/2	3 505 700
		5/2	3 506 100
$2p^6(^1S)6f$	$^2F°$	5/2,7/2	*3 524 500*
$2p^6(^1S)7d$	^2D	3/2	3 747 600
		5/2	3 748 000
$2p^6(^1S)7f$	$^2F°$	5/2,7/2	*3 758 800*
$2p^6(^1S)8d$	^2D	5/2	3 903 900
$2p^6(^1S)8f$	$^2F°$	5/2,7/2	*3 910 700*
Co XVIII ($^1S_0$)	***Limit***		**4 408 500**
$2p^5(^2P°)3s^2$	$^2P°$	3/2	*6 317 900*
		1/2	*6 430 500*

Co XVIII

$Z=27$

Ne I isoelectric sequence

Ground state: $1s^2 2s^2 2p^6\ ^1S_0$

Ionization energy = 11 269 000±4000 cm^{-1} (1397.2±0.5 eV)

Only resonance lines are classified by this system of energy levels. Tyrén (1938) identified resonance lines between 12 and 16 Å from the $2s^2 2p^5 3s$, $3d$, and $2s2p^6\ 3p$ levels and Feldman and Cohen (1967) identified $2p^5 4d$. Swartz, Kastner, Rothe, and Neupert (1971) made preliminary identifications of $2s^2 2p^5 4s$, $5d$, and $6d$ levels and confirmed the $2s2p^6 3p$ levels. The spectrum has been remeasured by Gordon, Hobby, and Peacock (1980) with a wavelength uncertainty of about ±0.005 Å. The lines fall in the range of 9–15 Å and the resulting level values have an uncertainty of 5000 cm^{-1}. Their results are used to obtain the present level values. They have added the $2s2p^6 4p\ ^1P$ term and the $2s^2 2p^5(^2P^\circ_{3/2})5d\ ^2[1/2]$ level.

Kastner, Behring, and Cohen (1975) identified transitions between $2p^5 3p$ and $2p^5 4d$, but there is no connection with the levels given here.

We derived the ionization energy from the $2s^2 2p^5 nd\ ^1P^\circ_1$ series for $n=3$–6.

References

Feldman, U., and Cohen, R. (1967), Astrophys. J. **149**, 265.

Gordon, H., Hobby, M. G., and Peacock, N. J. (1980), J. Phys. **B13**, 1985.

Kastner, S. O., Behring, W. E., and Cohen, L. (1975), Astrophys. J. **199**, 777.

Swartz, M., Kastner, S., Rothe, E., and Neupert, W. (1971), J. Phys. **B4**, 1747.

Tyrén, F. (1938), Zeits. Phys. **111**, 314.

Co XVIII

Configuration	Term	J	Level (cm^{-1})	Configuration	Term	J	Level (cm^{-1})
$2s^2 2p^6$	^1S	0	0	$2s^2 2p^5 4d$	$^3D^\circ$	1	*9 003 000*
$2s^2 2p^5 3s$	$^3P^\circ$	1	*6 477 900*	$2s^2 2p^5 4d$	$^1P^\circ$	1	*9 112 000*
$2s^2 2p^5 3s$	$^1P^\circ$	1	*6 592 400*	$2s^2 2p^5 5d$	$^3P^\circ$	1	*9 797 000*
$2s^2 2p^5 3d$	$^3P^\circ$	1	*7 122 000*	$2s^2 2p^5 5d$	$^3D^\circ$	1	*9 819 000*
$2s^2 2p^5 3d$	$^3D^\circ$	1	*7 210 800*	$2s^2 2p^5 5d$	$^1P^\circ$	1	*9 934 000*
$2s^2 2p^5 3d$	$^1P^\circ$	1	*7 334 600*	$2s2p^6 4p$	$^1P^\circ$	1	*9 970 000*
$2s2p^6 3p$	$^3P^\circ$	1	*7 894 500*	$2s^2 2p^5 6d$	$^3D^\circ$	1	*10 275 000*
$2s2p^6 3p$	$^1P^\circ$	1	*7 932 700*	$2s^2 2p^5 6d$	$^1P^\circ$	1	*10 381 000*
$2s^2 2p^5 4s$	$^3P^\circ$	1	*8 706 000*	Co XIX ($^2P^\circ_{3/2}$)	***Limit***		**11 269 000**
$2s^2 2p^5 4s$	$^1P^\circ$	1	*8 833 000*				

Co XIX

$Z=27$

F I isoelectronic sequence

Ground state: $1s^22s^22p^5\ {}^2\mathrm{P}^\circ_{3/2}$

Ionization energy = 12 135 000±10 000 cm^{-1} (1504.6±1.2 eV)

An analysis of the $2s^22p^5-2s^22p^43s$ and $2s^22p^5-2s^22p^43d$ arrays was given by Feldman, Doschek, Cowan, and Cohen (1973). Similar results were reported by Boiko, Pikuz, Safronova, and Fayonov (1978). The spectrum was remeasured by Gordon, Hobby, and Peacock (1980), who give new line identifications in these arrays as well as classifications of the $2s^22p^5-2s2p^53p$, $2s^22p^5-2s^22p^44s$ and $2s^22p^44d$ arrays. Their wavelengths in the range of 10–14 Å are reported accurate to about ±0.005 Å, resulting in a level value uncertainty of ±5000 cm^{-1}. The $2s^22p^5-2s2p^6$ doublet was reported by Doschek, Feldman, Cowan, and Cohen (1974) with an accuracy of ±0.02 Å, giving a level uncertainty of ±20 000 cm^{-1} for $2s2p^6\ {}^2\mathrm{S}$.

We derived the ionization energy from the $2s^22p^4(^3\mathrm{P})3s$, $4s\ {}^4\mathrm{P}$ series, assuming the difference $n^*(4s)-n^*(3s)=1.063$ from the $2p\ {}^5ns$ series in Co XVIII.

References

Boiko, V. A., Pikuz, S. A., Safronova, A. S., and Fayonov, A. Y. (1978), Opt. Spectrosc. **44**, 498.

Doschek, G. A., Feldman, U., Cowan, R. D., and Cohen, L. (1974), Astrophys. J. **188**, 417.

Feldman, U., Doschek, G. A., Cowan, R. D., and Cohen, L. (1973), J. Opt. Soc. Am. **63**, 1445.

Gordon, H., Hobby, M. G., and Peacock, N. J. (1980), J. Phys. **B13**, 1985.

Co XIX

Configuration	Term	J	Level (cm^{-1})
$2s^2\,2p^5$	$^2\mathrm{P}^\circ$	3/2	*0*
		1/2	*121 960*
$2s2p^6$	$^2\mathrm{S}$	1/2	1 131 860
$2s^2\,2p^4(^3\mathrm{P})3s$	$^4\mathrm{P}$	5/2	6 852 100
		3/2	6 966 200
$2s^2\,2p^4(^3\mathrm{P})3s$	$^2\mathrm{P}$	3/2	6 880 900
		1/2	6 991 500
$2s^2\,2p^4(^1\mathrm{S})3s$	$^2\mathrm{S}$	1/2	7 243 900
$2s^2\,2p^4(^1\mathrm{D})3s$	$^2\mathrm{D}$	5/2	7 050 200
		3/2	7 055 300
$2s^2\,2p^4(^3\mathrm{P})3d$	$^4\mathrm{P}$	1/2	7 525 000
		3/2	7 542 600
$2s^2\,2p^4(^3\mathrm{P})3d$	$^2\mathrm{F}$	5/2	7 552 900
$2s^2\,2p^4(^3\mathrm{P})3d$	$^4\mathrm{D}$	3/2	7 620 200
$2s^2\,2p^4(^3\mathrm{P})3d$	$^2\mathrm{P}$	3/2	7 632 800
$2s^2\,2p^4(^3\mathrm{P})3d$	$^2\mathrm{D}$	5/2	7 642 900
$2s^2\,2p^4(^1\mathrm{D})3d$	$^2\mathrm{S}$	1/2	7 701 800
$2s^2\,2p^4(^1\mathrm{D})3d$	$^2\mathrm{P}$	3/2	7 726 300
		1/2	7 764 900
$2s^2\,2p^4(^1\mathrm{D})3d$	$^2\mathrm{D}$	5/2	7 726 800
		3/2	7 757 700
$2s^2\,2p^4(^1\mathrm{S})3d$	$^2\mathrm{D}$	3/2	7 917 400
$2s2p^5(^3\mathrm{P}^\circ)3p$	$^4\mathrm{D}$	5/2	8 143 000
		3/2	8 171 000
$2s2p^5(^3\mathrm{P}^\circ)3p$	$^2\mathrm{D}$	5/2	8 189 000
		3/2	8 303 000
$2s2p^5(^3\mathrm{P}^\circ)3p$	$^4\mathrm{P}$	5/2	8 227 000
		3/2	8 252 000
$2s2p^5(^3\mathrm{P}^\circ)3p$	$^2\mathrm{P}$	3/2	8 218 000
		1/2	8 227 000
$2s2p^5(^3\mathrm{P}^\circ)3p$	$^2\mathrm{S}$	1/2	8 323 000
$2s2p^5(^1\mathrm{P}^\circ)3p$	$^2\mathrm{D}$	3/2	8 487 000
		5/2	8 515 000
$2s2p^5(^1\mathrm{P}^\circ)3p$	$^2\mathrm{P}$	1/2	8 521 000
		3/2	8 531 000
$2s^2\,2p^4(^3\mathrm{P})4s$	$^2\mathrm{P}$	3/2	9 280 000
$2s^2\,2p^4(^3\mathrm{P})4s$	$^4\mathrm{P}$	3/2	9 394 000
$2s^2\,2p^4(^1\mathrm{D})4s$	$^2\mathrm{D}$	5/2	9 462 000
		3/2	9 464 000

(Continued)

Co XIX—Continued

Configuration	Term	J	Level (cm^{-1})	Configuration	Term	J	Level (cm^{-1})
$2s^2\,2p^4(^3P)4d$	^2D	3/2,5/2	9 545 000	$2s^2\,2p^4(^1D)4d$	^2S	1/2	9 718 000
$2s^2\,2p^4(^3P)4d$	^4F	3/2,5/2	9 610 000	$2s^2\,2p^4(^1D)4d$	^2F	5/2	9 732 000
$2s^2\,2p^4(^3P)4d$	^4P	5/2	9 640 000	$2s^2\,2p^4(^1S)4d$	^2D	3/2	9 920 000
$2s^2\,2p^4(^1D)4d$	^2D	5/2 3/2	9 718 000 9 732 000	Co XX ($^3P_2$)	***Limit***		**12 135 000**
$2s^2\,2p^4(^1D)4d$	^2P	3/2 1/2	9 718 000 9 732 000				

Co XX

$Z=27$

O I isoelectronic sequence

Ground state: $1s^22s^22p^4\ ^3P_2$

Ionization energy = 12 930 000±26 000 cm^{-1} (1603±3 eV)

The $2s^22p^4-2s2p^5$ transition array was analyzed by Doschek, Feldman, Cowan, and Cohen (1974). New measurements by Lawson and Peacock (1980) in the range of 74–126 Å revealed several intersystem lines. We use their wavelengths, with a quoted uncertainty of ±0.03 Å, to derive the levels of the $2s^22p^4$, $2s2p^5$, and $2p^6$ configurations with a level uncertainty of ±300 cm^{-1}. Percentage compositions of these levels were provided by Kaufman and Sugar (1982), with configuration interaction between $2s^22p^4$ and $2p^6$.

The $2p^33s$, $2p^33d$, and $2p^34d$ levels are from observations and identifications of Gordon, Hobby, and Peacock (1980) and have an uncertainty of 2000 cm^{-1}.

The ionization energy is an extrapolated value by Lotz (1967).

References

Doschek, G. A., Feldman, U., Cowan, R. D., and Cohen, L. (1974), Astrophys. J. **188**, 417.

Gordon, H., Hobby, M. G., and Peacock, N. J. (1980), J. Phys. **B13**, 1985.

Kaufman, V. and Sugar, J. (1982), private communication.

Lawson, K. D., and Peacock, N. J. (1980), J. Phys. **B13**, 3313.

Lotz, W. J. (1967), J. Opt. Soc. Am. **57**, 873.

Co XX

Configuration	Term	J	Level (cm^{-1})	Leading percentages		
$2s^2\,2p^4$	^{3}P	2	0	88	12	^{1}D
		0	83 890	76	23	^{1}S
		1	107 420	100		
$2s^2\,2p^4$	^{1}D	2	189 290	88	12	^{3}P
$2s^2\,2p^4$	^{1}S	0	363 240	75	24	^{3}P
$2s2p^5$	^{3}P°	2	*981 550*	100		
		1	*1 053 290*	96	4	^{1}P°
		0	*1 108 520*	100		
$2s2p^5$	^{1}P°	1	*1 349 530*	96	4	^{3}P°
$2p^6$	^{1}S	0	2 265 740	98	2	$2s^22p^4\ ^1$S
$2s^2\,2p^3(^4S°)3s$	^{3}S°	1	*7 338 000*			
$2s^2\,2p^3(^2D°)3s$	^{3}D°	1,2	*7 447 000*			
		3	*7 487 000*			
$2s^2\,2p^3(^2D°)3s$	^{1}D°	2	*7 507 000*			
$2s^2\,2p^3(^2P°)3s$	^{3}P°	0	*7 586 000*			
		1	*7 599 000*			
		2	*7 668 000*			
$2s^2\,2p^3(^2P°)3s$	^{1}P°	1	*7 688 000*			
$2s^2\,2p^3(^4S°)3d$	^{3}D°	3	*7 933 000*			
$2s^2\,2p^3(^2D°)3d$	^{3}D°	3	*8 098 000*			
$2s^2\,2p^3(^2D°)3d$	^{3}P°	2	*8 110 000*			

(Continued)

Co XX—Continued

Configuration	Term	J	Level (cm^{-1})	Leading percentages
$2s^2\,2p^3(^2D°)3d$	$^1F°$	3	*8 150 000*	
$2s^2\,2p^3(^2P°)3d$	$^3F°$	3	*8 157 000*	
$2s^2\,2p^3(^2P°)3d$	$^3P°$	2	*8 181 000*	
		1	*8 237 000*	
$2s^2\,2p^3(^2P°)3d$	$^3D°$	2	*8 279 000*	
$2s^2\,2p^3(^2P°)3d$	$^1F°$	3	*8 288 000*	
$2s^2\,2p^3(^2P°)3d$	$^1D°$	2	*8 288 000*	
$2s^2\,2p^3(^2P°)3d$	$^1P°$	1	*8 331 000*	
$2s^2\,2p^3(^4S°)4d$	$^3D°$	2,3	*10 146 000*	
		1	*10 160 000*	
$2s^2\,2p^3(^2D°)4d$	$^3F°$	3	*10 265 000*	
$2s^2\,2p^3(^2D°)4d$	$^3D°$	2	*10 306 000*	
		3	*10 316 000*	
$2s^2\,2p^3(^2D°)4d$	$^3S°$	1	*10 330 000*	
$2s^2\,2p^3(^2D°)4d$	$^1F°$	3	*10 335 000*	
$2s^2\,2p^3(^2D°)4d$	$^1D°$	2	*10 335 000*	
$2s^2\,2p^3(^2P°)4d$	$^3P°$	2	*10 423 000*	
		1	*10 488 000*	
$2s^2\,2p^3(^2P°)4d$	$^3D°$	2	*10 505 000*	
$2s^2\,2p^3(^2P°)4d$	$^1F°$	3	*10 505 000*	
$2s^2\,2p^3(^2P°)4d$	$^1P°$	1	*10 509 000*	
Co XXI ($^4S°_{3/2}$)	*Limit*		12 930 000	

Co XXI

$Z=27$

N I isoelectronic sequence

Ground state: $1s^2 2s^2 2p^3\ ^4S^\circ_{3/2}$

Ionization energy = 13 990 000±28 000 cm^{-1} (1735±3 eV)

The transition array $2s^2 2p^3 - 2s2p^4$ was first interpreted by Doschek, Feldman, Cowan, and Cohen (1974). New observations in the range of 75–124 Å by Lawson and Peacock (1980) provided an extension of the earlier analysis, including several intersystem lines, and an analysis of the $2s2p^4 - 2p^5$ array. Their results, with a quoted uncertainty of ±0.03 Å, are given here. The level value uncertainty is ±300 cm^{-1}. The percentage compositions of the levels were provided by Kaufman and Sugar (1982), with configuration interaction between $2s^2 2p^3$ and $2p^5$.

The ionization energy is an extrapolated value by Lotz (1967).

References

Doschek, G. A., Feldman, U., Cowan, R. D., and Cohen, L. (1974), Astrophys. J. **188**, 417.

Kaufman, V. and Sugar, J. (1982), private communication.

Lawson, K. D., and Peacock, N. J. (1980), J. Phys. **B13**, 3313.

Lotz, W. (1967), J. Opt. Soc. Am. **57**, 873.

Co XXI

Configuration	Term	J	Level (cm^{-1})	Leading percentages		
$2s^2 2p^3$	$^4S^\circ$	3/2	*0*	85	12	$^2P^\circ$
$2s^2 2p^3$	$^2D^\circ$	3/2	*147 040*	72	17	$^2P^\circ$
		5/2	*191 530*	100		
$2s^2 2p^3$	$^2P^\circ$	1/2	*280 260*	99	1	$2p^5\ ^2P^\circ$
		3/2	*359 000*	69	24	$2s^2 2p^3\ ^2D^\circ$
$2s(^2S)2p^4(^3P)$	^4P	5/2	799 040	96	4	^2D
		3/2	879 510	97	2	^2D
		1/2	903 260	94	6	^2S
$2s9^2S)2p^4(^1D)$	^2D	3/2	1 107 300	91	6	^2P
		5/2	1 128 160	96	4	^4P
$2s(^2S)2p^4(^1S)$	^2S	1/2	1 267 430	68	28	^2P
$2s(^2S)2p^4(^3P)$	^2P	3/2	1 318 040	93	7	^2D
		1/2	1 434 220	72	26	^2S
$2p^5$	$^2P^\circ$	3/2	*2 069 550*	98	2	$2s^2 2p^3\ ^2P^\circ$
		1/2	*2 197 070*	99	1	
Co XXII ($^3P_0$)	*Limit*		**13 990 000**			

Co XXII

$Z=27$

C I isoelectronic sequence

Ground state: $1s^22s^22p^2\ ^3P_0$

Ionization energy = 14 890 000±30 000 cm^{-1} (1846±4 eV)

The transition arrays $2s^22p^2-2s2p^3$ and $2s2p^3-2p^4$ were observed by Lawson and Peacock (1980) in the range of 78–171 Å with a wavelength uncertainty of ±0.03 Å, giving a level uncertainty of ±300 cm^{-1}. Their interpretation is given here. Edlén (1984) has compared the known values of the $^5S_2^\circ$ level in the isoelectronic sequence with the theoretical predictions. He concluded that the values given by Lawson and Peacock are inconsistent with the trend. We give Edlén's predicted value in brackets. The percentage compositions of the levels were provided by Kaufman and Sugar (1982), with configuration interaction between $2s^2p^2$ and $2p^4$.

The ionization energy was obtained by extrapolation by Lotz (1967).

References

Edlén, B. (1984), private communication.
Kaufman, V. and Sugar, J. (1982), private communication.
Lawson, K. D., and Peacock, N. J. (1980), J. Phys. **B13**, 3313.
Lotz, W. (1967), J. Opt. Soc. Am. **57**, 873.

Co XXII

Configuration	Term	J	Level (cm^{-1})	Leading percentages		
$2s^2\,2p^2$	^3P	0	0	88	11	$2s^22p^2\ ^1S$
		1	90 730	99	1	$2p^4\ ^3P$
		2	138 250	71	28	$2s^22p^2\ ^1D$
$2s^2\,2p^2$	^1D	2	281 820	71	28	^3P
$2s^2\,2p^2$	^1S	0	415 520	86	11	^3P
$2s(^2S)2p^3(^4S^\circ)$	$^5S^\circ$	2	[*534 400*]	96	4	$^3P^\circ$
$2s(^2S)2p^3(^2D^\circ)$	$^3D^\circ$	1	*833 840*	83	13	$^3P^\circ$
		2	*836 280*	81	17	$^3P^\circ$
		3	*869 510*	100		
$2s(^2S)2p^3(^2P^\circ)$	$^3P^\circ$	0	*987 830*	100		
		1	*998 650*	81	14	$^3D^\circ$
		2	*1 020 290*	70	18	$^3D^\circ$
$2s(^2S)2p^3(^4S^\circ)$	$^3S^\circ$	1	*1 170 450*	77	18	$^1P^\circ$
$2s(^2S)2p^3(^2D^\circ)$	$^1D^\circ$	2	*1 212 130*	90	9	$^3P^\circ$
$2s(^2S)2p^3(^2P^\circ)$	$^1P^\circ$	1	*1 356 870*	79	18	$^3S^\circ$
$2p^4$	^3P	2	1 752 580	87	12	$2p^4\ ^1D$
		0	1 853 530	82	16	$2p^4\ ^1S$
		1	1 865 530	99	1	$2s^22p^2\ ^3P$
$2p^4$	^1D	2	1 944 800	87	12	^3P
$2p^4$	^1S	0	2 193 340	80	17	^3P
Co XXIII ($^2P^\circ_{1/2}$)	***Limit***		**14 890 000**			

Co XXIII

$Z=27$

B I isoelectronic sequence

Ground state: $1s^22s^22p\ ^2P^\circ_{1/2}$

Ionization energy = 15 820 000±32 000 cm^{-1} (1962±4 eV)

The transition arrays $2s^22p-2s2p^2$ and $2s2p^2-2p^3$ were observed and interpreted by Lawson and Peacock (1980) in the range of 93–171 Å with an uncertainty of ±0.03 Å, giving a level uncertainty of ±300 cm^{-1}. They identified the transition $2s^22p^2\ ^4P_{5/2}-2p^3\ ^2D^\circ_{5/2}$ at 103.80 Å that connects the quartet and doublet systems. Edlén (1983) suggested that this identification is inconsistent with the intersystem connection found in a solar flare spectrum by Sandlin et al. (1976) for isoelectronic Fe XXII. Edlén proposes the value 103.718 Å for this line. The percentage compositions were provided by Kaufman and Sugar, with configuration interaction between $2s^22p$ and $2p^3$.

Spector et al. (1980) have reported the terms of the $2s2p3d$ configuration with an uncertainty of ±5000 cm^{-1}.

The ionization energy was obtained by extrapolation by Lotz (1967).

References

Edlén, B. (1983), Phys. Scr. **28**, 483.

Lawson, K. D., and Peacock, N. J. (1980), J. Phys. **B13**, 3313.

Lotz, W. (1967), J. Opt. Soc. Am. **57**, 873.

Sandlin, G. D., Brueckner, G. E., Scherrer, V. E., and Tousey, R. (1976), Astrophys. J. **205**, L47.

Spector, N., Zigler, A., Zmora, H., and Schwob, J. L. (1980), J. Opt. Soc. Am. **70**, 857.

Co XXIII

Configuration	Term	J	Level (cm^{-1})	Leading percentages		
$2s^2\,2p$	$^2P^\circ$	1/2	*0*	98	2	$2p^3\ ^2P^\circ$
		3/2	*139 290*	98	2	
$2s2p^2$	^4P	1/2	431 560	95	4	^2S
		3/2	499 270	99	1	^2D
		5/2	559 760	93	7	^2D
$2s2p^2$	^2D	3/2	788 520	94	5	^2P
		5/2	819 150	93	7	^4P
$2s2p^2$	^2P	1/2	903 260	64	33	^2S
		3/2	1 064 960	94	5	^2D
$2s2p^2$	^2S	1/2	1 050 860	63	36	^2P
$2p^3$	$^4S^\circ$	3/2	*1 338 760*	88	9	$^2P^\circ$
$2p^3$	$^2D^\circ$	3/2	*1 486 350*	80	12	$^2P^\circ$
		5/2	*1 523 150*	100		
$2p^3$	$^2P^\circ$	1/2	*1 672 130*	98	2	$2s^22p\ ^2P^\circ$
		3/2	*1 745 870*	77	17	$2p^3\ ^2D^\circ$
$2s2p(^3P^\circ)3d$	$^4F^\circ$	7/2	*9 672 000*			
$2s2p(^3P^\circ)3d$	$^4D^\circ$	5/2,7/2	*9 755 000*			
Co XXIV ($^1S_0$)	*Limit*		15 820 000			

Co XXIV

$Z = 27$

Be I isoelectronic sequence

Ground state: $1s^2 2s^2\ {}^1S_0$

Ionization energy = 17 090 000±34 000 cm^{-1} (2119±4 eV)

The transition arrays $2s^2-2s2p$ and $2s2p-2p^2$ were measured by Lawson and Peacock (1980) with a wavelength uncertainty of ±0.03 Å, and a level uncertainty of ±300 cm^{-1}. Their classifications are used here, although Edlén's (1983) isoelectronic study of these data led him to conclude that some classifications in the $2s2p-2p^2$ array are doubtful.

Transitions from the $n=3$ shell were identified by Boiko et al. (1977). The level uncertainty in the $n=3$ shell is ±3000 cm^{-1}.

The ionization energy was obtained by extrapolation by Lotz (1967).

References

Boiko, V. A., Pikuz, S. A., Safronova, U. I., and Faenov, A. Y. (1977), J. Phys. **B10**, 1253.

Edlén, B. (1983), Phys. Scr. **28**, 51.

Lawson, K. D., and Peacock, N. J. (1980), J. Phys. **B13**, 3313.

Lotz, W. (1967), J. Opt. Soc. Am. **57**, 873.

Co XXIV

Configuration	Term	J	Level (cm^{-1})	Configuration	Term	J	Level (cm^{-1})
$2s^2$	^1S	0	0	$2p3p$	^3D	1	10 245 000
						2	10 333 000
$2s2p$	$^3P°$	0	*363 130*			3	10 444 000
		1	*398 720*				
		2	*509 210*	$2p3s$	$^1P°$	1	*10 264 000*
$2s2p$	$^1P°$	1	*799 040*	$2p3p$	^1P	1	10 320 000
$2p^2$	^3P	0	1 002 040	$2p3p$	^3P	1,2	10 425 000
		1	1 089 190				
		2	1 138 140	$2p3d$	$^3D°$	2	*10 430 000*
						1	*10 449 000*
$2p^2$	^1D	2	1 289 000				
				$2p3p$	^3S	1	10 456 000
$2p^2$	^1S	0	1 514 350				
				$2p3d$	$^1D°$	2	*10 539 000*
$2s3s$	^3S	1	9 653 000				
				$2p3p$	^1D	2	10 541 000
$2s3p$	$^3P°$	1	*9 886 000*				
				$2p3d$	$^3P°$	0–2	*10 578 000*
$2s3d$	^3D	1	9 965 000				
		2	9 971 000	$2p3d$	$^1F°$	3	*10 658 000*
		3	9 986 000				
				$2p3d$	$^1P°$	1	*10 661 000*
$2s3d$	^1D	2	10 058 000				
				Co XXV ($^2S_{1/2}$)	***Limit***		**17 090 000**
$2p3s$	$^3P°$	0	*10 065 000*				

Co XXV

$Z=27$

Li I isoelectronic sequence

Ground state: $1s^2 2s\ ^2S_{1/2}$

Ionization energy = 17 897 000±5000 cm^{-1} (2219.0±0.6 eV)

Edlén (1983) has given extrapolated values for the $2s-2p$ transition wavelengths with an estimated uncertainty of ±0.01 Å. Wavelengths determining the higher-lying terms were also predicted by Edlén (1979), but with no uncertainty estimate. We give his predicted level values.

The three transitions reported by Spector et al. (1980) are compared with Edlén's predictions below.

	Spector	Edlén
$2s-3p\ ^2S_{1/2}-^2P_{1/2}$	9.839 Å	9.838 Å
$2p-3d\ ^2P_{1/2}-^2D_{3/2}$	10.151	10.159
$2p-3d\ ^2P_{3/2}-^2D_{5/2}$	10.286	10.303

Edlén (1979) derived the ionization energy from a polarization formula.

References

Edlén, B. (1979), Phys. Scr. **19**, 255.
Edlén, B. (1983), Phys. Scr. **28**, 51.
Spector, N., Zigler, A., Zmora, H., and Schwob, J. L. (1980), J. Opt. Soc. Am. **70**, 857.

Co XXV

Configuration	Term	*J*	Level (cm^{-1})	Configuration	Term	*J*	Level (cm^{-1})
$2s$	^2S	1/2	0	$4s$	^2S	1/2	[13 517 000]
$2p$	$^2P°$	1/2	[*409 520*]	$4p$	$^2P°$	1/2	[*13 561 000*]
		3/2	[*561 180*]			3/2	[*13 580 000*]
$3s$	^2S	1/2	[10 055 000]	$4d$	^2D	3/2	[13 597 000]
						5/2	[13 604 000]
$3p$	$^2P°$	1/2	[*10 165 000*]				
		3/2	[*10 209 000*]	Co XXVI ($^1S_0$)	*Limit*		**17 897 000**
$3d$	^2D	3/2	[10 253 000]				
		5/2	[10 267 000]				

Co XXVI

$Z=27$

He I isoelectronic sequence

Ground state: $1s^2\,{}^1S_0$

Ionization energy = 7 69 79 000±15 000 cm^{-1} (9544.1±2.0 eV)

Because of the excellent agreement of the calculated energies of the $n=2$ shell by Safronova (1981) with the few well-measured spectra in the He I sequence, we have compiled her results for the $n=2$ levels and for the ionization energy. Detailed comparisons are given in the Introduction. Levels of the $n=3$–5 shells are from the calculated binding energies by Ermolaev and Jones (1974) subtracted from Safronova's value for the binding energy of the ground state. We have assumed an uncertainty of 2 parts in 10^4 for the excited levels relative to the ground state, and for the ionization energy.

Percentage compositions are from Ermolaev and Jones.

References

Ermolaev, A. M., and Jones, M. (1974), J. Phys. **B7**, 199.
Safronova, U. I. (1981), Phys. Scr. **23**, 241.

Co XXVI

Configuration	Term	J	Level (cm^{-1})	Leading percentages		
$1s^2$	^1S	0	0			
1s2s	^3S	1	[57 856 690]			
1s2p	$^3P°$	0	*[58 101 010]*			
		1	*[58 121 810]*	90	10	$^1P°$
		2	*[58 260 720]*			
1s2s	^1S	0	[58 122 820]			
1s2p	$^1P°$	1	*[58 408 440]*	90	10	$^3P°$
1s3s	^3S	1	[68 559 570]			
1s3p	$^3P°$	0	*[68 627 380]*			
		1	*[68 632 100]*	89	11	$^1P°$
		2	*[68 673 550]*			
1s3s	^1S	0	[68 629 450]			
1s3p	$^1P°$	1	*[68 714 660]*	89	11	$^3P°$
1s4s	^3S	1	[72 265 350]			
1s4p	$^3P°$	0	*[72 293 600]*			
		1	*[72 295 560]*	89	11	$^1P°$
		2	*[72 313 090]*			
1s4s	^1S	0	[72 293 660]			
1s4p	$^1P°$	1	*[72 329 840]*	89	11	$^3P°$
1s5s	^3S	1	[73 971 080]			
1s5s	^1S	0	[73 985 220]			
1s5p	$^3P°$	0	*[73 985 370]*			
		1	*[73 986 370]*	89	11	$^1P°$
		2	*[73 995 350]*			

Co XXVI—Continued

Configuration	Term	J	Level (cm^{-1})	Leading percentages		
1*s*5*p*	^{1}P°	1	[*74 003 820*]	89	11	^{3}P°
Co XXVII (^{2}S$_{1/2}$)	***Limit***		**76 979 000**			

Co XXVII

$Z=27$

H I isoelectronic sequence

Ground state: $1s\ ^2S_{1/2}$

Ionization energy = 80 753 200±400 cm^{-1} (10 012.20±0.05 eV)

No observations of this spectrum are reported. We give calculated values by Mohr (1983) for the $n=2$ shell and by Erickson (1977) for $n=3$–5 relative to the $2p\ 2P^\circ_{3/2}$ level. Further details are given in the Introduction. Relative to the ground state, the level uncertainty is estimated to be 5 parts in 10^7. The uncertainty in the excited states relative to $2p\ ^2P^\circ_{3/2}$ is 1 part in 10^6.

References

Erickson, G. W. (1977), J. Phys. Chem. Ref. Data **6**, 831.
Mohr, P. J. (1983), At. Data Nucl. Data Tables **29**, 453.

Co XXVII

Configuration	Term	J	Level (cm^{-1})
$1s$	^2S	½	0
$2p$	$^2P^\circ$	½	[*60 505 940*]
		3/2	[*60 705 284*]
$2s$	^2S	½	[60 511 117]
$3p$	$^2P^\circ$	½	[*71 776 660*]
		3/2	[*71 835 760*]
$3s$	^2S	½	[71 778 280]
$3d$	^2D	3/2	[71 835 656]
		5/2	[71 855 000]
$4p$	$^2P^\circ$	½	[*75 713 246*]
		3/2	[*75 738 159*]
$4s$	^2S	½	[75 713 932]
$4d$	^2D	3/2	[75 738 114]
		5/2	[75 746 281]
$4f$	$^2F^\circ$	5/2	[*75 746 266*]
		7/2	[*75 750 332*]
$5p$	$^2P^\circ$	½	[*77 531 851*]
		3/2	[*77 544 595*]
$5s$	^2S	½	[77 532 202]
$5d$	^2D	3/2	[77 544 571]
		5/2	[77 548 753]
$5f$	$^2F^\circ$	5/2	[*77 548 746*]
		7/2	[*77 550 828*]
$5g$	^2G	7/2	[77 550 825]
		9/2	[77 552 072]
	Limit		**80 753 200**

Ni I

$Z=28$

Ground state: $1s^2 2s^2 2p^6 3s^2 3p^6 3d^8 4s^2\ ^3F_4$

Ionization energy = 61 600±10 cm^{-1} (7.6375±0.0012 eV)

The analysis of this spectrum is by Russell (1929), with a few changes of interpretation taken from recent theoretical studies. No energy levels have been found since his work 50 years ago. He outlined the previous work in his paper and increased the number of classified lines from 622 to 1071. His line list extends from 1963–18 040 Å.

The spectrum was reobserved by Burns and Sullivan (1947, 1948) with a vacuum arc and a Fabry-Perot interferometer. Three decimal place wavelengths were measured from 2173–8968 Å. From these they derived the three decimal place energy levels with a level uncertainty of ±0.01 cm^{-1}. Except for 10 level values retained from Russell, their values are given here. In the course of their work they measured about 400 lines not previously observed.

The five place g-values from the three lowest terms are from Childs, Fred, and Goodman (1966), Childs and Goodman (1968), and Childs and Greenebaum (1972). The three place g-values are from measurements of M.I.T. Zeeman patterns reported by Lindsley (1942). The remaining (two place) values are from Marvin and Baragar (1933) or Dijkstra (1937).

The percentage compositions of the $J=2$ levels of $3d^8 4s^2$ and $3d^9 4s$ were calculated with configuration interaction by Childs, Fred, and Goodman. Those of $3d^9 4p$, $3d^8 4s 4p$, and $3d^7 4s^2 4p$ were calculated with configuration interaction by Roth (1980).

We derived the ionization energy from the three member $3d^9 ns$ series by means of a Ritz formula, giving the value 61 579 cm^{-1}. We added a correction of 21 cm^{-1} obtained from comparisons with a corresponding longer ns-series in Cu I.

References

Burns, K., and Sullivan, F. (1947), Science Studies St. Bonaventure College **13**, No. 3; (1948), ibid. **14**, No. 3.
Childs, W. J., Fred, M., and Goodman, L. S. (1966), Phys. Rev. **141**, 44.
Childs, W. J., and Goodman, L. S. (1968), Phys. Rev. **170**, 136.
Childs, W. J., and Greenebaum, B. (1972), Phys. Rev. **6A**, 105.
Dijkstra, H. (1937), Physica **4**, 81.
Lindsley, C. H. (1942), J. Opt. Soc. Am. **32**, 94.
Marvin, H. H., and Baragar, A. E. (1933), Phys. Rev. **43**, 973.
Roth, C. (1980), At. Data Nucl. Data Tables **25**, 91.
Russell, H. N. (1929), Phys. Rev. **34**, 821.

Ni I

Configuration	Term	J	Level (cm^{-1})	g	Leading percentages		
$3d^8\ 4s^2$	$a\ ^3F$	4	0.000	1.24965			
		3	1 332.153	1.08280			
		2	2 216.519	0.66956	95	1	^1D
$3d^9(^2D)4s$	$a\ ^3D$	3	204.786	1.33354			
		2	879.813	1.15105	91	9	^1D
		1	1 713.080	0.49804			
$3d^9(^2D)4s$	$a\ ^1D$	2	3 409.925	1.01297	89	9	^3D
$3d^8\ 4s^2$	$b\ ^1D$	2	13 521.352	1.143	70	26	^3P
$3d^{10}$	$a\ ^1S$	0	14 728.847				
$3d^8\ 4s^2$	$a\ ^3P$	2	15 609.861	1.356	71	25	^1D
		1	15 734.018	1.497			
		0	16 017.317				
$3d^8\ 4s^2$	$a\ ^1G$	4	22 102.349	0.99			
$3d^8(^3F)4s4p(^3P^\circ)$	$z\ ^5D^\circ$	4	*25 753.578*	1.51	95		
		3	*26 665.903*	1.495	93		
		2	*27 414.893*	1.494	93		
		1	*27 943.543*	1.486	95		
		0	*28 212.997*		96		

(Continued)

Ni I—Continued

Configuration	Term	J	Level (cm^{-1})	g	Leading percentages		
$3d^8(^3F)4s4p(^3P^\circ)$	$z\ ^5G^\circ$	6	*27 260.891*	1.32	100		
		5	*27 580.411*	1.276	81	16	$^5F^\circ$
		4	*28 068.091*	1.171	82	13	$^5F^\circ$
		3	*28 578.046*	0.945	87	9	$^5F^\circ$
		2	*29 013.228*	0.364	94	5	$^5F^\circ$
$3d^8(^3F)4s4p(^3P^\circ)$	$z\ ^5F^\circ$	5	*28 542.113*	1.377	83	15	$3d^8(^3F)4s4p(^3P^\circ)\ ^5G^\circ$
		4	*29 084.478*	1.288	70	12	$3d^9(^2D)4p\ ^3F^\circ$
		3	*29 832.810*	1.208	40	26	$3d^9(^2D)4p\ ^3F^\circ$
		2	*30 163.140*	0.985	76	6	$3d^9(^2D)4p\ ^3D^\circ$
		1	*30 392.052*	0.006	97		
$3d^9(^2D)4p$	$z\ ^3P^\circ$	2	*28 569.210*	1.485	91	4	$3d^9(^2D)4p\ ^3D^\circ$
		1	*29 500.690*	1.426	86	5	$3d^9(^2D)4p\ ^3D^\circ$
		0	*30 192.268*		96		
$3d^9(^2D)4p$	$z\ ^3F^\circ$	3	*29 320.782*	1.086	34	27	$3d^9(^2D)4p\ ^3D^\circ$
		4	*29 481.020*	1.287	74	13	$3d^8(^3F)4s4p(^3P^\circ)\ ^5F^\circ$
		2	*30 619.440*	0.740	73	16	$3d^8(^3F)4s4p(^3P^\circ)\ ^3F^\circ$
$3d^9(^2D)4p$		3	*29 668.918*	1.300	30 $^3D^\circ$	30	$3d^8(^3F)4s4p(^3P^\circ)\ ^5F^\circ$
$3d^9(^2D)4p$	$z\ ^3D^\circ$	2	*29 888.505*	1.044	33	22	$3d^8(^3F)4s4p(^3P^\circ)\ ^3D^\circ$
		1	*30 912.838*	0.552	51	38	$3d^8(^3F)4s4p(^3P^\circ)\ ^5F^\circ$
$3d^8(^3F)4s4p(^3P^\circ)$	$z\ ^3G^\circ$	5	*30 922.763*	1.214	95	5	$(^3F)(^3P^\circ)\ ^5G^\circ$
		4	*30 979.789*	1.052	79	13	$(^3F)(^3P^\circ)\ ^1G^\circ$
		3	*31 786.210*	0.761	95		
$3d^9(^2D)4p$	$z\ ^1F^\circ$	3	*31 031.042*	1.048	58	26	$3d^9(^2D)4p\ ^3F^\circ$
$3d^9(^2D)4p$	$z\ ^1D^\circ$	2	*31 441.665*	1.060	62	13	$3d^8(^3F)4s4p(^3P^\circ)\ ^1D^\circ$
$3d^8(^3F)4s4p(^3P^\circ)$	$^3F^\circ$	4	*32 973.414*	1.222	77	13	$3d^9(^2D)4p\ ^3F^\circ$
		3	*33 112.368*	1.193	46	19	$3d^8(^3F)4s4p(^3P^\circ)\ ^3D^\circ$
		2	*33 610.916*	0.973	42	30	$3d^8(^3F)4s4p(^3P^\circ)\ ^3D^\circ$
$3d^9(^2D)4p$	$z\ ^1P^\circ$	1	*32 982.280*	1.005	94		
$3d^8(^3F)4s4p(^3P^\circ)$	$y\ ^3D^\circ$	3	*33 500.854*	1.198	48	23	$3d^8(^3F)4s4p(^3P^\circ)\ ^3F^\circ$
		2	*34 163.29*	1.19	45	40	$3d^8(^3F)4s4p(^3P^\circ)\ ^3F^\circ$
		1	*34 408.574*	0.511	62	33	$3d^9(^2D)4p\ ^3D^\circ$
$3d^8(^3F)4s4p(^3P^\circ)$	$z\ ^1G^\circ$	4	*33 590.159*		79	14	$(^3F)(^3P^\circ)\ ^3G^\circ$
$3d^8(^3F)4s4p(^3P^\circ)$	$y\ ^1F^\circ$	3	*35 639.148*	1.013	81	8	$(^3F)(^3P^\circ)\ ^3F^\circ$
$3d^8(^3F)4s4p(^3P^\circ)$	$y\ ^1D^\circ$	2	*36 600.805*	1.013	75	16	$3d^9(^2D)4p\ ^1D^\circ$
$3d^8(^3P)4s4p(^3P^\circ)$	$^5P^\circ$	3	*40 361.254*		86	11	$(^1D)(^3P^\circ)\ ^3D^\circ$
		2	*40 484.282*		89	7	
$3d^8(^1D)4s4p(^3P^\circ)$	$^3F^\circ$	4	*42 585.296*	1.346	52	36	$(^3P)(^3P^\circ)\ ^5D^\circ$
		3	*42 767.900*	1.218	66	19	$(^3P)(^3P^\circ)\ ^5D^\circ$
		2	*42 954.234*	0.840	79	6	$(^1D)(^3P^\circ)\ ^3D^\circ$
$3d^9(^2D)5s$	$e\ ^3D$	3	42 605.964	1.34			
		2	42 790.027	1.085			
		1	44 112.192				

Ni I—Continued

Configuration	Term	J	Level (cm^{-1})	g	Leading percentages			
$3d^8(^1D)4s4p(^3P^\circ)$	$^3D^\circ$	3	*42 621.048*		42		36	$(^3F)(^1P^\circ)\ ^3D^\circ$
		2	*42 653.723*		66		15	
		1	*42 656.317*	1.320	77		10	
$3d^8(^3F)4s4p(^1P^\circ)$	$^3G^\circ$	5	*43 089.636*	1.226	99			
		4	*44 314.980*	1.182	66		18	$(^3P)(^3P^\circ)\ ^5D^\circ$
		3	*44 565.10*	1.044	55		24	$(^3F)(^1P^\circ)\ ^3F^\circ$
$3d^8(^3F)4s4p(^1P^\circ)$	$^3F^\circ$	4	*43 258.792*	1.247	62		19	$(^3F)(^1P^\circ)\ ^3G^\circ$
		3	*45 281.152*	0.779	45		40	$(^1P)(^1P^\circ)\ ^3G^\circ$
		2	*45 418.858*	0.677	80		7	$(^3F)(^3P^\circ)\ ^3D^\circ$
$3d^8(^1D)4s4p(^3P^\circ)$	$^3P^\circ$	1	*43 464.019*	1.390	64		20	$(^3P)(^3P^\circ)\ ^3P^\circ$
		2	*43 933.428*	1.476	59		18	$(^3P)(^3P^\circ)\ ^5D^\circ$
$3d^8(^3F)4s4p(^1P^\circ)$		3	*43 654.974*	1.243	35	$^3D^\circ$	34	$(^3P)(^3P^\circ)\ ^5D^\circ$
$3d^8(^3P)4s4p(^3P^\circ)$	$^5D^\circ$	3	*44 206.185*		40		26	$(^3F)(^1P^\circ)\ ^3D^\circ$
$3d^9(^2D)5s$	$e\ ^1D$	2	44 262.619	1.09				
$3d^8(^3P)4s4p(^3P^\circ)$		4	*44 336.10*		38	$^5D^\circ$	37	$(^1D)(^3P^\circ)\ ^3F^\circ$
$3d^8(^3F)4s4p(^1P^\circ)$	$^3D^\circ$	2	*44 475.158*	1.155	56		14	$(^1D)(^3P^\circ)\ ^3D^\circ$
		1	*45 122.460*	0.566	69		13	$(^3P)(^3P^\circ)\ ^3D^\circ$
$3d^8(^3P)4s4p(^3P^\circ)$	$x\ ^3P^\circ$	2	*46 522.965*		66		21	$(^3P)(^3P^\circ)\ ^5S^\circ$
		1	*47 208.228*		67		20	$(^1D)(^3P^\circ)\ ^3P^\circ$
		0	*47 686.625*		69		28	$(^1D)(^3P^\circ)\ ^3P^\circ$
$3d^8(^3P)4s4p(^3P^\circ)$	$v\ ^3D^\circ$	3	*47 030.148*	1.331	93			
		2	*47 139.392*	1.209	72		8	$(^3P)(^3P^\circ)\ ^5S^\circ$
		1	*47 424.830*	0.726	83		8	$(^3F)(^1P^\circ)\ ^3D^\circ$
$3d^8(^3P)4s4p(^3P^\circ)$	$^5S^\circ$	2	*47 328.85*		67		12	$(^3P)(^3P^\circ)\ ^3P^\circ$
$3d^8\ 4s(^4F)5s$	$e\ ^5F$	5	48 466.530	1.40				
		4	49 086.030	1.33				
		3	49 777.619	1.23				
		2	50 346.477	0.95				
		1	50 744.593	0.20				
$3d^9(^2D)5p$	$^1F^\circ$	3	*48 671.9*					
$3d^9(^2D)5p$	$^3F^\circ$	4	*48 715.2*					
		2	*50 039.18*					
$3d^9(^2D)5p$	$w\ ^3P^\circ$	2	*48 735.308*					
		1	*49 403.42*					
		0	*50 138.53*					
$3d^8(^3P)4s4p(^3P^\circ)$	$^1P^\circ$	1	*48 817.6*		89		3	$(^1D)(^1P^\circ)\ ^1P^\circ$
$3d^9(^2D)4d$	$e\ ^3S$	1	48 953.344	1.92				
$3d^8(^3P)4s4p(^3P^\circ)$	$^1D^\circ$	2	*49 032.589*		93		4	$(^1D)(^3P^\circ)\ ^3D^\circ$
$3d^9(^2D)4d$	$e\ ^3G$	5	49 158.529	1.20				
		4	49 174.811	1.05				
		3	50 677.599	0.77				

(Continued)

Ni I—Continued

Configuration	Term	J	Level (cm^{-1})	g	Leading percentages
$3d^9(^2D)4d$	$e\ ^3P$	2	49 159.060	1.43	
		1	49 171.187	1.00	
		0	50 276.354		
$3d^9(^2D)5p$	$^1D°$	2	*49 185.146*		
$3d^9(^2D)4d$	$f\ ^3D$	3	49 271.578	1.32	
		2	49 327.845		
		1	50 716.927	0.45	
$3d^9(^2D)4d$	$e\ ^3F$	3	49 313.851		
		4	49 332.643		
		2	50 834.435		
$3d^9(^2D)5p$	$^3D°$	3	*49 327.56*		
		2	*50 689.490*		
		1	*50 851.22*		
$3d^9(^2D)5p$		3	*50 142.8*		
$3d^9(^2D)5p$	$x\ ^1P°$	1	*50 458.187*		
$3d^8\ 4s(^4F)5s$	$f\ ^3F$	4	50 466.172	1.27	
		3	51 306.085	1.08	
		2	52 040.568	0.67	
$3d^9(^2D)4d$	$e\ ^1P$	1	50 536.742	1.54	
$3d^9(^2D)4d$	$e\ ^1G$	4	50 706.310	1.02	
$3d^9(^2D)4d$	$f\ ^1D$	2	50 754.137		
$3d^8(^1G)4s4p(^3P°)$	$u\ ^3F°$	4	*50 789.5*		94 4 $(^1D)(^3P°)\ ^3F°$
		3	*51 124.8*		94
		2	*51 343.80*		93 4 $(^1D)(^3P°)\ ^3F°$
$3d^9(^2D)4d$	$e\ ^1F$	3	50 832.039		
$3d^8\ 4s^2$	$e\ ^1S$	0	51 457.285		
$3d^9(^2D)6s$	$g\ ^3D$	3	52 197.482		
		2	52 271.716		
		1	53 703.899		
$3d^9(^2D)6s$	$g\ ^1D$	2	53 754.036		
$3d^8\ 4s(^2F)5s$	$g\ ^3F$	4	54 237.136	1.27	
		3	54 251.353	1.00	
		2	55 873.78		
$3d^9(^2D)5d$	$f\ ^3S$	1	54 574.64		
$3d^9(^2D)5d$	$f\ ^3G$	5	54 659.759		
		4	54 667.928	1.03	
		3	56 172.704		
$3d^9(^2D)5d$	$h\ ^3D$	3	54 699.852		
		2	54 732.425		

Ni I—Continued

Configuration	Term	J	Level (cm^{-1})	g	Leading percentages
$3d^9(^2D)5d$	$h\ ^3F$	4	54 761.346		
		3	54 772.940		
		2	56 274.516		
$3d^8\,4s(^2F)5s$	$f\ ^1F$	3	55 576.905	1.07	
$3d^9(^2D)5d$	$f\ ^1G$	4	56 183.51		
$3d^9(^2D)5d$	$g\ ^1F$	3	56 262.92		
$3d^8\,4s(^4F)4d$	$e\ ^3H$	6	56 624.668		
		5	57 677.649		
		4	58 518.11		
$3d^8\,4s(^4F)4d$	$f\ ^3P$	2	56 710.889		
		1	57 767.83		
		0	58 448.79		
$3d^8\,4s(^4F)4d$	$i\ ^3F$	4	56 766.523		
		3	57 968.08		
		2	58 629.84		
$3d^8\,4s(^4F)4d$	$g\ ^3G$	5	56 801.654		
		4	57 789.611		
		3	58 530.35		
$3d^8\,4s(^4F)4d$	$e\ ^5P$	3	56 821.553		
		2	57 586.7		
		1	58 525.507		
$3d^8\,4s(^4F)4d$	$e\ ^5D$	4	56 857.933		
		3	57 743.596		
$3d^8\,4s(^4F)4d$	$e\ ^5H$	7	56 885.249	1.26	
		6	57 762.106		
		5	58 520.923		
		4	59 039.693		
		3	59 188.78		
$3d^8\,4s(^4F)4d$	$e\ ^5G$	6	56 954.167		
		5	57 829.405		
		3	58 629.55		
		4	58 872.31		
		2	59 118.06		
$3d^8\,4s(^4F)4d$	$f\ ^5F$	5	56 973.707		
		4	57 810.494		
		3	58 588.168		
		2	58 992.52		
		1	59 226.03		
$3d^8\,4s(^4F)4d$	$i\ ^3D$	3	57 103.946		
$3d^8\,4s(^4F)6s$	$g\ ^5F$	5	59 862.756		
Ni II ($^2D_{5/2}$)	***Limit***		**61 600**		
$3d^8\,4s(^2F)4d$	$j\ ^3F$	4	61 832.47		

(Continued)

Ni I—Continued

Configuration	Term	J	Level (cm^{-1})	g	Leading percentages
$3d^8\,4s(^2F)4d$	$h\ ^3G$	5	61 843.28		
$3d^8\,4s(^2F)4d$	$f\ ^3H$	6	61 957.517		
$3d^8\,4s(^4F)5d$	$f\ ^5H$	7	62 782.614		
$3d^8\,4s(^4F)5d$	$f\ ^5G$	6	62 808.03		
$3d^8\,4s(^4F)5d$	$h\ ^5F$	5	62 815.34		

Ni II

$Z=28$

Co I isoelectronic sequence

Ground state: $1s^2 2s^2 2p^6 3s^2 3p^6 3d^9\ ^2D_{5/2}$

Ionization energy = 146 541.56±0.2 cm^{-1} (18.16898±0.00005 eV)

This compilation is based on the very extensive analysis by Shenstone (1970, 1971) with recent unpublished additions by him to $4d^8 5g$. He has observed 4300 lines between 700 and 10 000 Å with a hollow cathode discharge and has established 320 even and 336 odd levels. The low configurations $3d^8 4s$ and $3d^7 4s^2$ are nearly complete. Series in $3d^8$ns, *nd*, *nf*, and *ng* were observed. About half of the levels of the complex configuration $3d^7 4s4p$ are known. Shenstone gave no uncertainty estimate for his level values. We estimate that it is ±0.05 cm^{-1}.

This work has been extended by Brault and Litzén (1983) with infrared observations above 10 000 Å. They found the $3d^8(^3F)6h$ configuration, and made corrections and additions to $3d^8 5f$ and $3d^8 6g$. In $3d^8 5f$ the level $131122.28_{11/2}$ was replaced by $131094.02_{11/2}$, two new levels $131115.15_{3/2}$ and $131131.15_{7/2}$ were added, and the designation of $131124.96_{7/2}$ was changed to $3d^8(^3F_2)5f\ ^2[3]^{\circ}_{7/2}$. In $3d^8 6g$ a number of new levels were added and some previously reported were discarded. The levels and designations given below for $3d^8 6g$ and $3d^8 6h$ are from Brault and Litzén. The level uncertainties are limited to those of Shenstone's levels on which these new results are based. All percentages for these configurations are reported to be 99 or 100%, but were not given.

The Zeeman effect data are from observations at M.I.T. reported by Lindsley (1942).

The leading percentages given here for the levels of $3d^8 4p$ were calculated by Roth (1969). Those for $3d^8 5p$ and $3d^7 4s4p$ are from Shadmi and Caspi (1972). These authors given percentage only for cases where the coupling is not pure. Repeating terms of the $3d^7$ parent configuration are distinguished by alphabetic prefixes. In these cases the percentage includes the sum over all contributing seniority states.

We have calculated the compositions of $3d^8 4s$, $5s$, $4d$, $5d$, $4f$, $5f$, $6f$, $5g$, $6g$, and $3d^7 4s^2$ and give the leading percentages.

Shenstone found the limits of the $3d^8 ns$ and *nd* series at 146 532.0 cm^{-1} and of the ng series at 146 541.56 cm^{-1}. He has adopted the latter limit, which is used here.

References

Brault, J. W., and Litzén, U. (1983), Phys. Scr. **28**, 475.
Lindsley, C. H. (1942), J. Opt. Soc. Am. **32**, 387.
Roth, C. (1969), J. Res. Natl. Bur. Stand. **73A**, 125.
Shadmi, Y., and Caspi, E. (1972), J. Res. Natl. Bur. Stand. **76A**, 125.
Shenstone, A. G. (1970), J. Res. Natl. Bur. Stand. **74A**, 801.
Shenstone, A. G. (1971), J. Res. Natl. Bur. Stand. **75A**, 335.

Ni II

Configuration	Term	*J*	Level (cm^{-1})	*g*	Leading percentages		
$3d^9$	^2D	5/2	0.00				
		3/2	1 506.94				
$3d^8(^3F)4s$	^4F	9/2	8 393.90	1.355	100		
		7/2	9 330.04	1.244	98	2	$(^3F)\ ^2F$
		5/2	10 115.66	1.023	99	1	$(^3F)\ ^2F$
		3/2	10 663.89	0.397	99	1	$(^1D)\ ^2D$
$3d^8(^3F)4s$	^2F	7/2	13 550.39	1.141	98	2	$(^3F)\ ^4F$
		5/2	14 995.57	0.866	98	1	$(^1D)\ ^2D$
$3d^8(^3P)4s$	^4P	5/2	23 108.28	1.428	54	46	$(^1D)\ ^2D$
		3/2	24 788.20		78	22	
		1/2	24 835.93	2.667	100		
$3d^8(^1D)4s$	^2D	3/2	23 796.18	1.045	75	22	$(^3P)\ ^4P$
		5/2	25 036.38	1.368	53	46	
$3d^8(^3P)4s$	^2P	3/2	29 070.93	1.322	97	3	$(^1D)\ ^2D$
		1/2	29 593.46	0.670	100		
$3d^8(^1G)4s$	^2G	9/2	32 499.53	1.135	100		
		7/2	32 523.54	0.895	100		

(Continued)

Ni II—Continued

Configuration	Term	J	Level (cm^{-1})	g	Leading percentages		
$3d^7\,4s^2$	^{4}F	9/2	51 045.46		100		
		7/2	52 205.95		100		
		5/2	53 037.93		100		
		3/2	53 601.19		100		
$3d^8(^3\mathrm{F})4p$	^{4}D°	7/2	*51 557.85*	1.420	94		
		5/2	*52 738.45*	1.365	93		
		3/2	*53 634.62*	1.186	94	4	(^{3}P) ^{4}D°
		1/2	*54 176.26*	-0.005	96		
$3d^8(^3\mathrm{F})4p$	^{4}G°	9/2	*53 365.17*	1.156	67	23	(^{3}F) ^{2}G°
		11/2	*53 496.49*	1.305	100		
		7/2	*54 262.63*	1.025	81	10	(^{3}F) ^{4}F°
		5/2	*55 018.71*	0.616	94	5	(^{3}F) ^{4}F°
$3d^8(^3\mathrm{F})4p$	^{4}F°	9/2	*54 557.05*	1.26	80	19	(^{3}F) ^{2}G°
		7/2	*55 417.83*	1.184	76	10	(^{3}F) ^{2}F°
		5/2	*56 075.26*	0.985	87	6	(^{3}F) ^{4}G°
		3/2	*56 424.49*	0.412	95		
$3d^8(^3\mathrm{F})4p$	^{2}G°	9/2	*55 299.65*	1.152	58	32	(^{3}F) ^{4}G°
		7/2	*56 371.41*	0.940	84	8	(^{3}F) ^{4}G°
$3d^8(^3\mathrm{F})4p$	^{2}F°	7/2	*57 080.55*	1.154	81	11	(^{3}F) ^{4}F°
		5/2	*58 493.21*	0.946	74	20	(^{3}F) ^{2}D°
$3d^8(^3\mathrm{F})4p$	^{2}D°	5/2	*57 420.16*	1.116	74	20	(^{3}F) ^{2}F°
		3/2	*58 705.95*	0.795	89	7	(^{1}D) ^{2}D°
$3d^8(^3\mathrm{P})4p$	^{4}P°	5/2	*66 571.34*	1.48	73	20	(^{1}D) ^{2}D°
		3/2	*66 579.71*	1.550	73	12	(^{1}D) ^{2}P°
		1/2	*67 031.02*	2.331	85	11	(^{1}D) ^{2}P°
$3d^8(^1\mathrm{D})4p$	^{2}F°	5/2	*67 694.64*	0.960	84	8	(^{3}P) ^{4}P°
		7/2	*68 131.21*	1.200	86	9	(^{3}P) ^{4}D°
$3d^7\,4s^2$	^{4}P	5/2	67 880.16		100		
		3/2	68 156.57		94	6	^{2}P
		1/2	68 709.76		98	2	
$3d^8(^1\mathrm{D})4p$	^{2}D°	3/2	*68 154.31*	1.02	65	18	(^{3}P) ^{4}P°
		5/2	*68 735.98*	1.264	74	19	(^{3}P) ^{4}P°
$3d^8(^1\mathrm{D})4p$	^{2}P°	1/2	*68 281.62*	1.008	61	23	(^{3}P) ^{2}P°
		3/2	*68 965.65*	1.305	64	15	(^{1}D) ^{2}D°
$3d^7\,4s^2$	^{2}G	9/2	70 358.94		97	3	^{2}H
		7/2	71 457.74		100		
$3d^8(^3\mathrm{P})4p$	^{4}D°	5/2	*70 635.46*	1.325	83	9	(^{3}P) ^{2}D°
		3/2	*70 706.77*	1.190	91	4	(^{3}F) ^{4}D°
		1/2	*70 748.70*		95	4	(^{3}F) ^{4}D°
		7/2	*70 778.12*	1.385	87	9	(^{1}D) ^{2}F°
$3d^8(^3\mathrm{P})4p$	^{2}D°	5/2	*71 770.83*	1.240	87	10	(^{3}P) ^{4}D°
		3/2	*72 375.42*	0.844	82	11	(^{3}P) ^{2}P°
$3d^8(^3\mathrm{P})4p$	^{2}P°	3/2	*72 985.65*	1.326	67	16	(^{1}D) ^{2}P°
		1/2	*73 903.25*	1.039	70	24	(^{1}D) ^{2}P°

Ni II—Continued

Configuration	Term	J	Level (cm^{-1})	g	Leading percentages		
$3d^7\,4s^2$	^{2}P	3/2	73 893.73		85	8	^{2}D2
$3d^8(^3$P$)4p$	^{2}S°	1/2	*74 283.33*		94	4	(^{1}D) ^{2}P°
$3d^8(^3$P$)4p$	^{4}S°	3/2	*74 300.93*		97		
$3d^8(^1$G$)4p$	^{2}H°	9/2	*75 149.48*	0.903	100		
		11/2	*75 721.68*	1.119	100		
$3d^8(^1$G$)4p$	^{2}F°	7/2	*75 917.63*	1.165	94	4	(^{1}D) ^{2}F°
		5/2	*76 402.03*		95		
$3d^7\,4s^2$	^{2}H	11/2	76 727.36		100		
		9/2	77 736.79		97	3	^{2}G
$3d^7\,4s^2$	^{2}D2	5/2	77 332.47		77	23	^{2}D1
		3/2	78 955.45		72	18	
$3d^8(^1$G$)4p$	^{2}G°	7/2	*79 823.03*		99		
		9/2	*79 923.88*		100		
$3d^7(^4$F$)4s4p(^3$P°$)$	^{6}F°	11/2	*86 343.21*				
		9/2	*86 870.03*				
		7/2	*87 538.09*				
		5/2	*88 128.56*				
		3/2	*88 582.01*				
		1/2	*88 881.59*				
$3d^7(^4$F$)4s4p(^3$P°$)$	^{6}D°	9/2	*88 171.88*				
		7/2	*89 100.49*				
$3d^7(^4$F$)4s4p(^3$P°$)$	^{6}G°	11/2	*89 460.35*				
		9/2	*89 918.47*				
		7/2	*90 275.30*				
		5/2	*90 526.18?*				
$3d^8(^3$F$)5s$	^{4}F	9/2	91 800.05	1.350	100		
		7/2	92 325.85	1.188	61	39	(^{3}F) ^{2}F
		5/2	93 390.06	1.02	88	12	(^{3}F) ^{2}F
		3/2	94 067.14	0.392	99	1	(^{1}D) ^{2}D
$3d^7\,4s^2$	2	5/2	92 373.45		100		
		7/2	92 792.08		100		
$3d^8(^3$F$)5s$	^{2}F	7/2	93 528.44	1.166	61	39	^{4}F
		5/2	94 729.25	0.865	87	12	
$3d^7(^4$F$)4s4p(^3$P°$)$	^{4}F°	9/2	*94 283.94*				
		7/2	*94 705.93*				
		5/2	*95 332.53*				
		3/2	*95 893.76*				
$3d^7(^4$F$)4s4p(^3$P°$)$	^{4}G°	11/2	*94 396.74*				
		9/2	*95 017.71*				
		7/2	*95 573.39*				
		5/2	*96 052.48*				

(Continued)

Ni II—Continued

Configuration	Term	J	Level (cm^{-1})	g	Leading percentages			
$3d^7(^4F)4s4p(^3P^\circ)$	$^4D^\circ$	7/2	*96 535.87*					
		5/2	*97 273.83*					
		3/2	*97 799.66*					
		1/2	*98 122.63*					
$3d^7(^4F)4s4p(^3P^\circ)$	$^2G^\circ$	9/2	*98 276.70*					
		7/2	*99 844.13*					
$3d^8(^3F)4d$	^4D	7/2	98 467.25		90		7	$(^3F)\ ^4F$
		5/2	99 559.33		42		39	$(^3F)\ ^4P$
		1/2	100 010.17		69		16	$(^3F)\ ^4P$
		3/2	100 078.78		49		45	$(^3F)\ ^2P$
$3d^8(^3F)4d$	^4P	5/2	98 561.22		57		40	$(^3P)\ ^4D$
		1/2	100 845.41		66		26	$(^3F)\ ^4D$
		3/2	100 490.95		73		12	$(^3F)\ ^4D$
$3d^8(^3F)4d$	^4H	13/2	98 822.55		100			
		11/2	100 309.29		54		32	$(^3F)\ ^2H$
		9/2	100 332.09		52		22	$(^3F)\ ^2H$
		7/2	101 144.63		62		28	$(^3F)\ ^4G$
$3d^8(^3F)4d$	^2H	11/2	98 969.44		47		45	$(^3F)\ ^4H$
		9/2	101 357.20		55		19	
$3d^8(^3F)4d$	^2P	3/2	99 040.75		44		29	$(^3F)\ ^4D$
		1/2	101 246.16		77		18	$(^3F)\ ^4P$
$3d^8(^3F)4d$	^4G	11/2	99 132.78		78		21	$(^3F)\ ^2H$
		5/2	101 366.14		68		28	$(^3F)\ ^4F$
$3d^8(^3F)4d$	^4F	9/2	99 154.81		63		31	$(^3F)\ ^4G$
		7/2	100 592.98		40		27	$(^3F)\ ^4H$
		3/2	101 258.01		89		8	$(^3F)\ ^4D$
$3d^8(^3F)4d$	^2F	7/2	99 340.55		58		23	$(^3F)\ ^4F$
		5/2	101 247.37		47		31	
$3d^7(^4F)4s4p(^3P^\circ)$	$^2F^\circ$	7/2	*99 418.61*					
		5/2	*100 609.01*					
$3d^8(^3F)4d$	^2G	9/2	99 442.86		45		21	$(^3F)\ ^4F$
		7/2	101 740.27		72		13	$(^3F)\ ^4G$
$3d^8(^3F)4d$		5/2	100 389.52		42	^2F	25	$(^3F)\ ^4F$
$3d^8(^3F)4d$		7/2	100 475.82		34	^4G	34	$(^3F)\ ^2F$
$3d^8(^3F)4d$		9/2	100 619.26		32	^2G	30	$(^3F)\ ^4G$
$3d^7(^4F)4s4p(^3P^\circ)$	$^2D^\circ$	5/2	*101 754.80*					
		3/2	*102 742.74*					
$3d^8(^3F)4d$	^2D	5/2	103 025.58		85		9	$(^3F)\ ^2F$
		3/2	103 663.50		90		4	$(^1D)\ ^2D$

Ni II—Continued

Configuration	Term	J	Level (cm^{-1})	g	Leading percentages			
$3d^8(^3F)5p$	$^4D°$	7/2	*103 653.03*					
		5/2	*104 503.22*					
		3/2	*105 439.85*					
		1/2	*106 022.79*					
$3d^8(^3F)5p$	$^2G°$	9/2	*104 081.04*		60		32	$^4G°$
		7/2	*106 620.53*		58		24	$^4G°$
$3d^8(^3F)5p$	$^4G°$	11/2	*104 147.29*					
		7/2	*105 499.05*		55		24	$^2G°$
		9/2	*105 588.89*		41		38	$^2G°$
		5/2	*106 283.16*					
$3d^8(^3F)5p$	$^4F°$	9/2	*104 298.23*					
		5/2	*105 668.78*		53		24	$^4G°$
		3/2	*106 369.30*					
$3d^8(^3F)5p$		7/2	*104 646.52*		39	$^4F°$	33	$^2F°$
$3d^8(^3F)5p$	$^2F°$	7/2	*105 838.06*		46		32	$^4F°$
		5/2	*107 082.21*					
$3d^8(^3F)5p$	$^2D°$	5/2	*105 861.19*					
		3/2	*107 142.12*					
$3d^7(^4P)4s4p(^3P°)$	$^6D°$	9/2	*105 981.50?*					
$3d^8(^1D)5s$	^2D	5/2	106 007.89		80		20	$(^3P)\ ^4P$
		3/2	106 133.14		87		9	$(^3P)\ ^2P$
$3d^7(^4P)4s4p(^3P°)$	$^4S°$	3/2	*107 737.81*					
$3d^8(^3P)5s$	^4P	5/2	108 368.05		80		20	$(^1D)\ ^2D$
		3/2	108 548.61		94		5	$(^1D)\ ^2D$
		1/2	108 763.32		98		2	$(^3P)\ ^2P$
$3d^7(^4P)4s4p(^3P°)$	$^6P°$	3/2	*109 038.84*					
$3d^7(^2G)4s4p(^3P°)$	$^4F°$	9/2	*109 148.05*					
		7/2	*109 846.00*					
		5/2	*110 573.36*					
		3/2	*111 120.54*					
$3d^8(^3P)5s$	^2P	3/2	109 269.83		90		7	$(^1D)\ ^2D$
		1/2	109 675.72		98		2	$(^3P)\ ^4P$
		9/2	*110 021.92*					
$3d^7(^2G)4s4p(^3P°)$	$^4G°$	7/2	*111 783.79*					
$3d^7(^4F)4s4p(^1P°)$	$^4G°$	11/2	*112 422.19?*					
		9/2	*113 753.04*		67		28	$^4F°$
		5/2	*115 108.09*		50		28	$^2F°$
$3d^8(^1D)4d$	^2F	5/2	112 686.30		66		13	$(^1D)\ ^2D$
		7/2	112 719.75		77		19	$(^3P)\ ^4D$
$3d^8(^1D)4d$	^2D	3/2	112 906.93		62		22	$(^3P)\ ^2D$
		5/2	113 407.31		55		25	$(^1D)\ ^2F$

(Continued)

Ni II—Continued

Configuration	Term	J	Level (cm^{-1})	g	Leading percentages			
$3d^8(^1D)4d$	^2G	9/2	113 172.96		84		15	$(^3P)\ ^4F$
		7/2	113 177.61		86		10	$(^3P)\ ^2F$
$3d^8(^1D)4d$	^2P	1/2	113 225.06		78		13	$(^3P)\ ^4D$
		3/2	113 408.71		77		13	
$3d^7(^4F)4s4p(^1P°)$	$^4F°$	9/2	*113 321.95*		63		34	$^4G°$
		5/2	*115 120.00?*		61		16	$^2F°$
		3/2	*115 592.25?*		42		30	$^4D°$
$3d^8(^1D)4d$	^2S	1/2	113 623.10		85		10	$(^3P)\ ^4P$
$3d^7\ 4s4p$		7/2	*114 052.04*		29	$^4G°$	28	$^4F°$
$3d^8(^3P)4d$	^4D	7/2	114 836.63		79		20	$(^1D)\ ^2F$
		5/2	114 874.88		72		19	$(^1D)\ ^2D$
		3/2	114 942.42		79		11	$(^1D)\ ^2P$
		1/2	114 970.19		87		12	$(^1D)\ ^2P$
$3d^7(^2H)4s4p(^3P°)$	$^4G°$	11/2	*114 858.88*					
		9/2	*115 612.88*					
		7/2	*116 275.81*					
		5/2	*116 754.93*					
$3d^7(^2P)4s4p(^3P°)$	$^2D°$	3/2	*114 869.35*		62			
		5/2	*116 893.98*		50		33	$^4D°$
$3d^7(^2G)4s4p(^3P°)$	$^2F°$	7/2	*115 000.25*		51		22	$^4D°$
$3d^8(^1G)5s$	^2G	9/2	115 081.36		100			
		7/2	115 085.36		100			
$3d^7\ 4s4p$	$^4D°$	7/2	*115 209.85*		55		18	$^2F°$
$3d^7\ 4s4p$		5/2	*115 565.98*		32	$^4D°$	25	$^4F°$
$3d^8(^3P)4d$	^4F	9/2	115 739.15		84		15	$(^1D)\ ^2G$
		7/2	115 827.12		82		8	
		5/2	115 956.71		68		13	$(^3P)\ ^2F$
		3/2	116 167.76		97		1	$(^3P)\ ^4P$
		3/2	*115 785.06*					
$3d^7\ 4s^2$	$^2D1$	5/2	115 870.28		77		23	$^2D2$
$3d^8(^3P)4d$	^2F	7/2	116 145.69		82		11	$(^3P)\ ^4F$
$3d^8(^3P)4d$		5/2	116 191.47		38	^2D	24	$(^3P)\ ^4F$
$3d^8(^3P)4d$	^4P	1/2	116 261.81		86		11	$(^1D)\ ^2S$
		3/2	116 312.34		85		8	$(^1D)\ ^2P$
		5/2	116 732.51		41		28	$(^3P)\ ^2D$
$3d^7\ 4s4p$	$^4D°$	7/2	*116 512.06*					
$3d^8(^3P)4d$	^2P	1/2	116 786.42		89		7	$(^1D)\ ^2P$
		3/2	116 838.33		91		4	$(^3P)\ ^2D$

Ni II—Continued

Configuration	Term	J	Level (cm^{-1})	g	Leading percentages			
$3d^8(^3F)6s$	^4F	9/2	116 833.15					
		7/2	117 074.70					
		5/2	118 314.82					
		3/2	119 100.06					
$3d^7(a\ ^2D)4s4p(^3P^\circ)$	$^4F^\circ$	9/2	*117 573.68*					
		7/2	*117 972.47*					
$3d^7(^2P)4s4p(^3P^\circ)$		3/2	*117 662.11*		37	$^4S^\circ$	12	$^2D^\circ$
$3d^8(^1D)5p$		3/2	*117 763.91*		36	$^2D^\circ$	12	$^2P^\circ$
$3d^8(^1D)5p$	$^2D^\circ$	5/2	*117 872.78*		65			
$3d^7(^2H)4s4p(^3P^\circ)$	$^2I^\circ$	11/2	*118 248.98*					
		13/2	*119 010.21*					
$3d^8(^3F)6s$	^2F	7/2	118 294.17					
		5/2	119 315.44					
$3d^8(^1D)5p$	$^2F^\circ$	5/2	*118 379.11*					
		7/2	*118 563.39*					
$3d^8(^1D)5p$		3/2	*118 442.81*		37	$^2P^\circ$	35	$^2D^\circ$
$3d^8(^1D)5p$	$^2P^\circ$	1/2	*118 631.95*		50		29	$3d^74s4p\ ^4D^\circ$
$3d^8(^3F_4)4f$	$^2[1]^\circ$	1/2	*118 774.76*		100			
		3/2	*118 809.34?*		87		13	$^2[2]^\circ$
$3d^8(^3F_4)4f$	$^2[7]^\circ$	13/2	*118 803.82*		100			
		15/2	*118 848.92*		100			
$3d^8(^3F_4)4f$	$^2[2]^\circ$	5/2	*118 828.61*		92		8	$^2[3]^\circ$
		3/2	*118 877.09*		87		13	$^2[1]^\circ$
$3d^8(^3F_4)4f$	$^2[3]^\circ$	7/2	*118 874.11*		97		3	$^2[4]^\circ$
		5/2	*118 897.94*		92		8	$^2[2]^\circ$
$3d^8(^3F_4)4f$	$^2[6]^\circ$	11/2	*118 892.99*		99			
		13/2	*118 893.24*		100			
$3d^8(^3F_4)4f$	$^2[4]^\circ$	9/2	*118 914.34*		99			
		7/2	*118 923.20*		97		3	$^2[3]^\circ$
$3d^8(^3F_4)4f$	$^2[5]^\circ$	9/2	*118 927.02*		99			
		11/2	*118 939.53*		99			
$3d^8(^3F)5d$	^4D	7/2	119 656.25		84		13	^4F
		1/2	121 111.90		48		30	^2P
		3/2	121 115.59		44		34	^2P
$3d^8(^3F)5d$	^4P	5/2	119 665.29		67		30	^4D
		1/2	121 925.16		51		43	^4D
$3d^8(^3F)5d$	^4H	13/2	119 773.60		100			
		11/2	121 180.55		59		30	^2H
		9/2	121 190.34		59		27	^2H
		7/2	122 047.29		77		18	^4G

(Continued)

Ni II—Continued

Configuration	Term	J	Level (cm^{-1})	g	Leading percentages			
$3d^8(^3P)5p$	$^4P°$	5/2	*119 796.98*		51		19	$(^1D)\ ^2D°$
		3/2	*120 166.52*		36		17	
		1/2	*120 316.02*					
$3d^3(^3F)5d$	^4G	11/2	119 833.00		53		34	^4H
		7/2	121 240.90		42		30	^2F
		9/2	121 294.67		43		25	^2G
		5/2	122 144.99		59		35	^4F
$3d^8(^3F)5d$	^2H	11/2	119 889.47		56		36	^4G
		9/2	122 140.71		59		25	^4H
$3d^8(^3F)5d$	^2P	3/2	119 909.72		59		18	^4P
		1/2	122 112.94?		65		26	^4P
$3d^8(^3F)5d$	^4F	9/2	119 913.33		59		34	^4G
		7/2	121 317.89		41		30	^2G
		3/2	122 084.79		62		26	^4D
$3d^8(^3F)5d$	^2F	7/2	120 002.86		60		20	^4F
		5/2	122 175.42?		52		20	^2D
$3d^8(^3F)5d$	^2G	9/2	120 044.95		58		23	^4F
		7/2	122 270.05		61		18	^4G
$3d^8(^3F)5d$		5/2	120 144.17		37	^2D	18	^4D
$3d^8(^3F_3)4f$	$^2[1]°$	1/2	*120 189.55*		60		40	$^2[0]°$
		3/2	*120 199.18*		88		12	$^2[2]°$
$3d^8(^3F_3)4f$	$^2[2]°$	5/2	*120 203.49*		99		1	$^2[3]°$
		3/2	*120 222.89*		88		12	$^2[1]°$
$3d^8(^3F_3)4f$	$^2[6]°$	11/2	*120 211.30*		100			
		13/2	*120 218.22*		100			
$3d^8(^3F_3)4f$	$^2[3]°$	7/2	*120 250.17*		98		1	$^2[4]°$
		5/2	*120 271.97*		99		1	$^2[2]°$
$3d^8(^3F_3)4f$	$^2[4]°$	7/2	*120 268.81*		98		1	$^2[3]°$
		9/2	*120 281.11*		97		3	$^2[5]°$
$3d^8(^3F_3)4f$	$^2[5]°$	11/2	*120 270.44*		100			
		9/2	*120 272.53*		97		3	$^2[4]°$
$3d^8(^3P)5p$	$^4D°$	7/2	*120 903.31*		81			
		5/2	*121 325.09*		63		27	$^2D°$
		3/2	*121 385.80*		54			
		1/2	*121 561.06*		83			
$3d^8(^3F_2)4f$	$^2[1]°$	3/2	*121 042.52*		90		9	$^2[2]°$
		1/2	*121 090.71*		99			
$3d^8(^3P)5p$		3/2	*121 042.57*		35	$^2P°$	18	$^2D°$
$3d^8(^3P)5p$	$^2D°$	5/2	*121 050.66*		57		20	$^4D°$
$3d^8(^3F)6p$	$^4G°$	11/2	*121 120.37*					

Ni II—Continued

Configuration	Term	J	Level (cm^{-1})	g	Leading percentages			
$3d^8(^3F_2)4f$	$^2[5]°$	11/2	*121 120.88*		99			
		9/2	*121 125.41*		99			
$3d^8(^3F_2)4f$	$^2[2]°$	5/2	*121 146.98*		94		5	$^2[3]°$
		3/2	*121 161.81*		90		9	$^2[1]°$
$3d^8(^3F_2)4f$	$^2[4]°$	7/2	*121 178.56*		86		13	$^2[3]°$
		9/2	*121 180.54*		99			
$3d^8(^3F_2)4f$	$^2[3]°$	7/2	*121 192.32*		86		13	$^2[4]°$
		5/2	*121 194.14*		94		5	$^2[2]°$
$3d^8(^3F)5d$		5/2	121 227.80		37	^4D	36	^2F
$3d^8(^1G)4d$	^2I	11/2	121 437.68		100			
		13/2	121 476.56		100			
$3d^8(^3P)5p$	$^4S°$	3/2	*121 456.30*		43		23	$^2D°$
$3d^7(^2H)4s4p(^3P°)$	$^2G°$	9/2	*121 692.55*					
		7/2	*121 862.57*					
		3/2	121 699.02					
$3d^8(^3P)5p$		3/2	*121 800.34*		32	$^2P°$	28	$^2D°$
$3d^8(^1G)4d$	^2F	5/2	122 080.25		99			
		7/2	122 086.58		100			
$3d^8(^3F)6p$	$^4F°$	9/2	*122 441.22*					
$3d^8(^1G)4d$	^2H	9/2	122 790.41		100			
		11/2	122 821.63		100			
$3d^8(^3F)6p$	$^4D°$	7/2	*122 812.97*					
$3d^8(^1G)4d$	^2G	9/2	122 837.33		99			
		7/2	122 847.60		100			
$3d^8(^3F)6p$	$^2G°$	9/2	*123 434.60*					
$3d^7(^2H)4s4p(^3P°)$	$^2H°$	9/2	*124 652.00*					
		11/2	*125 003.41*					
$3d^8(^1G)5p$	$^2H°$	9/2	*126 679.98*					
		11/2	*126 857.97*					
$3d^7(^4P)4s4p(^1P°)$	$^4S°$	3/2	*126 738.82*					
$3d^8(^1G)5p$	$^2F°$	7/2	*127 219.57*					
		5/2	*127 331.60*					
$3d^8(^3F)7s$	^4F	9/2	127 867.13					
		7/2	127 991.56					
		5/2	129 294.51					
		3/2	130 135.19					
$3d^8(^1G)5p$	$^2G°$	9/2	*127 885.86*					
		7/2	*127 895.33*					

(Continued)

Ni II—Continued

Configuration	Term	J	Level (cm^{-1})	g	Leading percentages		
$3d^8(^3F_4)5f$	$^2[1]°$	3/2	*128 732.03*		87	13	$^2[2]°$
		1/2	*128 799.64*		100		
$3d^8(^3F_4)5f$	$^2[2]°$	5/2	*128 803.23*		92	8	$^2[3]°$
		3/2	*128 822.23*		87	13	$^2[1]°$
$3d^8(^3F_4)5f$	$^2[7]°$	13/2	*128 818.41*		100		
		15/2	*128 827.05*		100		
$3d^8(^3F_4)5f$	$^2[3]°$	7/2	*128 827.15*		97		
		5/2	*128 853.87*		92	8	$^2[2]°$
$3d^8(^3F_4)5f$	$^2[6]°$	11/2	*128 837.11*		99		
		13/2	*128 855.60*		100		
$3d^8(^3F_4)5f$	$^2[4]°$	9/2	*128 853.91*		100		
		7/2	*128 867.00*		97		
$3d^8(^3F_4)5f$	$^2[5]°$	9/2	*128 862.49*		100		
		11/2	*128 869.89*		100		
$3d^8(^3F_4)5g$	$^2[0]$	1/2	128 937.47		100		
$3d^8(^3F_4)5g$	$^2[1]$	3/2	128 939.76		100		
		1/2	128 939.76		100		
$3d^8(^3F_4)5g$	$^2[2]$	5/2	128 944.36		100		
		3/2	128 944.36		100		
$3d^8(^3F_4)5g$	$^2[8]$	17/2	128 946.17		100		
		15/2	128 946.15		100		
$3d^8(^3F_4)5g$	$^2[3]$	7/2	128 950.84		100		
		5/2	128 950.89		100		
$3d^8(^3F_4)5g$	$^2[4]$	9/2	128 958.34		100		
		7/2	128 958.40		100		
$3d^8(^3F_4)5g$	$^2[7]$	15/2	128 960.74		100		
		13/2	128 960.74		100		
$3d^8(^3F_4)5g$	$^2[5]$	11/2	128 964.63		100		
		9/2	128 964.62		100		
$3d^8(^3F_4)5g$	$^2[6]$	13/2	128 966.51		100		
		11/2	128 966.51		100		
$3d^8(^3F)7s$	^2F	7/2	129 271.72				
		5/2	130 236.26				
$3d^8(^3F)6d$	^4P	5/2	129 284.50				
		3/2	129 479.73				
		1/2	130 710.85				
$3d^8(^3F)6d$	^4D	7/2	129 297.91				
		5/2	129 842.33				
		3/2	130 942.30				

Ni II—Continued

Configuration	Term	J	Level (cm^{-1})	g	Leading percentages		
$3d^8(^3F)6d$	^4H	$13/2$	129 367.91				
		$11/2$	129 396.04				
		$9/2$	130 757.51				
		$7/2$	131 637.10				
$3d^8(^3F)6d$	^4F	$9/2$	129 419.58				
		$7/2$	129 474.27				
		$5/2$	130 730.53				
		$3/2$	131 620.45				
$3d^8(^3F)6d$	^4G	$11/2$	129 424.03				
		$9/2$	129 503.24				
		$7/2$	130 815.91				
		$5/2$	131 670.87				
$3d^7(^4P)4s4p(^1P^\circ)$	$^4D^\circ$	$7/2$	*129 782.07*				
		$5/2$	*129 988.05*				
		$3/2$	*130 331.78*				
		$1/2$	*130 570.42*				
$3d^8(^3F_3)5f$	$^2[0]^\circ$	$1/2$	*130 147.87*		51	49	$^2[1]^\circ$
$3d^8(^3F_3)5f$	$^2[1]^\circ$	$3/2$	*130 174.03*		88	11	$^2[2]^\circ$
$3d^8(^3F_3)5f$	$^2[2]^\circ$	$5/2$	*130 184.39*		100		
		$3/2$	*130 197.23*		88	12	$^2[1]^\circ$
$3d^8(^3F_3)5f$	$^2[6]^\circ$	$11/2$	*130 184.61*		100		
		$13/2$	*130 187.81*		100		
$3d^8(^3F_3)5f$	$^2[5]^\circ$	$11/2$	*130 205.62*		100		
		$9/2$	*130 206.90*		98		
$3d^8(^3F_3)5f$	$^2[3]^\circ$	$7/2$	*130 208.89*		97		
		$5/2$	*130 227.52*		100		
$3d^8(^3F_3)5f$	$^2[4]^\circ$	$9/2$	*130 215.50*		98		
		$7/2$	*130 225.87*		97		
$3d^8(^3F_3)5g$	$^2[1]$	$3/2$	130 301.40		100		
		$1/2$	130 301.40		100		
$3d^8(^3F_3)5g$	$^2[2]$	$5/2$	130 306.97		99		
		$3/2$	130 306.97		99		
$3d^8(^3F_3)5g$	$^2[7]$	$15/2$	130 308.71		100		
		$13/2$	130 308.71		100		
$3d^8(^3F_3)5g$	$^2[3]$	$7/2$	130 314.07		99		
		$5/2$	130 314.07		99		
$3d^8(^3F_3)5g$	$^2[4]$	$9/2$	130 320.97		100		
		$7/2$	130 320.94		100		
$3d^8(^3F_3)5g$	$^2[6]$	$11/2$	130 321.73		99		
		$13/2$	130 321.73		99		
$3d^8(^3F_3)5g$	$^2[5]$	$11/2$	130 324.72		100		
		$9/2$	130 324.72		100		

(Continued)

Ni II—Continued

Configuration	Term	J	Level (cm^{-1})	g	Leading percentages		
$3d^8(^3F)7p$	$^4F°$	9/2	*130 470.90*				
$3d^8(^3F)7p$	$^4D°$	7/2	*130 480.55*				
$3d^8(^3F)7p$	$^4G°$	11/2	*130 661.32*				
$3d^8(^3F)6d$	^2P	3/2	130 691.35				
		1/2	131 655.83				
$3d^8(^3F)6d$	^2H	11/2	130 751.03				
		9/2	131 686.56				
$3d^8(^3F)6d$	^2F	7/2	130 765.26				
		5/2	131 796.26				
$3d^8(^3F)6d$	^2G	9/2	130 801.33				
		7/2	131 750.73				
$3d^8(^1D)6s$	^2D	5/2	130 900.65				
		3/2	130 942.36				
$3d^8(^3F)6d$	^2D	5/2	131 032.01				
$3d^8(^3F_2)5f$	$^2[1]°$	1/2	*131 063.85*		99		
		3/2	*131 075.78*		87	12	$^2[2]°$
$3d^8(^3F_2)5f$	$^2[5]°$	9/2	*131 093.30*		99		
		11/2	*131 094.02*		99		
$3d^8(^3F_2)5f$	$^2[2]°$	5/2	*131 103.18*		91	8	$^2[3]°$
		7/2	*131 115.15*		89	10	
$3d^8(^3F_2)5f$	$^2[4]°$	9/2	*131 115.28*		99		
		7/2	*131 131.15*		66	33	$^2[3]°$
$3d^8(^3F_2)5f$	$^2[3]°$	7/2	*131 124.96*		66	33	$^2[4]°$
		5/2	*131 133.58*		91	8	$^2[2]°$
$3d^8(^3F_2)5g$	$^2[2]°$	5/2	131 211.85		99	1	$(^1D_2)$ $^2[2]$
		3/2	131 211.85		99	1	
$3d^8(^3F_2)5g$	$^2[6]°$	13/2	131 218.56		99		
		11/2	131 218.56		99		
$3d^8(^3F_2)5g$	$^2[3]°$	7/2	131 222.98		99		
		5/2	131 222.98		99		
$3d^8(^3F_2)5g$	$^2[4]°$	9/2	131 232.83		99		
		7/2	131 232.83		99		
$3d^8(^3F_2)5g$	$^2[5]°$	11/2	131 233.31		99		
		9/2	131 233.31		99		
$3d^7(^2G)4s4p(^1P°)$	$^2H°$	11/2	*131 424.32?*				
		9/2	*132 311.98?*				
$3d^7(^4P)4s4p(^1P°)$	$^4P°$	5/2	*131 834.94*				
		1/2	*132 120.70*				
		3/2	*132 225.15*				

Ni II—Continued

Configuration	Term	J	Level (cm^{-1})	g	Leading percentages		
$3d^8(^3F)7p$	$^2G^\circ$	9/2	*131 928.77*				
$3d^8(^1D_2)4f$	$^2[3]^\circ$	5/2	*132 729.48*		82	16	$(^3P_2)\ ^2[3]^\circ$
		7/2	*132 869.16*		75	14	$(^3P_2)\ ^2[3]^\circ$
$3d^8(^1D_2)4f$	$^2[4]^\circ$	9/2	*132 818.16*		83	16	$(^3P_2)\ ^2[4]^\circ$
		7/2	*132 846.53*		75	15	$(^3P_2)\ ^2[4]^\circ$
$3d^8(^1D_2)4f$	$^2[2]^\circ$	5/2	*132 912.15*		83	15	$(^3P_2)\ ^2[2]^\circ$
		3/2	*132 927.97*		83	15	$(^3P_2)\ ^2[2]^\circ$
$3d^8(^1D_2)4f$	$^2[1]^\circ$	3/2	*132 982.51*		84	16	$(^3P_2)\ ^2[1]^\circ$
		1/2	*133 001.47*				
$3d^8(^1D_2)4f$	$^2[5]^\circ$	11/2	*133 014.08*		83	16	$(^3P_2)\ ^2[5]^\circ$
		9/2	*133 031.00*		83	16	$(^3P_2)\ ^2[5]^\circ$
$3d^7(^2G)4s4p(^1P^\circ)$	$^2F^\circ$	7/2	*133 169.92*				
		5/2	*134 208.30*				
$3d^7(^2F)4s4p(^3P^\circ)$	$^4F^\circ$	3/2	*133 190.19?*				
		5/2	*133 209.30*				
		7/2	*133 528.02*				
		9/2	*133 853.04*				
$3d^8(^3P)6s$	^4P	5/2	133 443.89				
		3/2	133 613.99				
		1/2	133 857.73				
$3d^7(^2G)4s4p(^1P^\circ)$	$^2G^\circ$	9/2	*133 445.75*				
		7/2	*134 380.82*				
$3d^7(^2F)4s4p(^3P^\circ)$	$^4G^\circ$	11/2	*133 625.96*				
$3d^8(^3F)8s$	^4F	9/2	133 715.13				
		7/2	133 809.76				
		5/2	135 116.72				
		3/2	135 983.22				
$3d^8(^1D)5d$	^2F	7/2	133 734.98		81	13	$(^3P)\ ^4D$
		5/2	133 735.26		68	14	$(^1D)\ ^2D$
$3d^7(^2F)4s4p(^3P^\circ)$	$^4D^\circ$	7/2	*133 850.83*				
		5/2	*133 973.33*				
		3/2	*134 156.28*				
		1/2	*134 283.76*				
$3d^8(^3P)6s$	^2P	3/2	133 862.21				
		1/2	134 241.96				
$3d^8(^1D)5d$	^2D	3/2	133 903.00		80	7	$(^3P)\ ^2D$
		5/2	134 053.05		69	16	$(^1D)\ ^2F$
$3d^8(^1D)5d$	^2G	9/2	133 922.91		83	16	$(^3P)\ ^4F$
		7/2	133 929.88		84	11	$(^3P)\ ^2F$
$3d^8(^1D)5d$	^2P	1/2	133 954.85		82	8	$(^3P)\ ^2P$
		3/2	134 067.76		84	9	$(^3P)\ ^4P$

(Continued)

Ni II—Continued

Configuration	Term	J	Level (cm^{-1})	g	Leading percentages		
$3d^8(^3F_4)6f$	$^2[7]^\circ$	13/2	*134 238.44*		100		
		15/2	*134 251.30*		100		
$3d^8(^3F_4)6f$	$^2[1]^\circ$	3/2	*134 249.72*		87	13	$^2[2]^\circ$
$3d^8(^3F_4)6f$	$^2[3]^\circ$	7/2	*134 252.85*		97	3	$^2[4]^\circ$
		5/2	*134 294.99*		91	9	$^2[2]^\circ$
$3d^8(^3F_4)6f$	$^2[2]^\circ$	3/2	*134 254.69*		87	13	$^2[1]^\circ$
		5/2	*134 256.05*		91	9	$^2[3]^\circ$
$3d^8(^3F_4)6f$	$^2[4]^\circ$	7/2	*134 262.07*		97	3	$^2[3]^\circ$
		9/2	*134 271.59*		100		
$3d^8(^3F_4)6f$	$^2[6]^\circ$	13/2	*134 267.20*		100		
		11/2	*134 274.62*		99	1	$^2[5]^\circ$
$3d^8(^3F_4)6f$	$^2[5]^\circ$	11/2	*134 281.66*		99	1	$^2[6]^\circ$
		9/2	*134 286.38*		100		
$3d^8(^3F_4)6g$	$^2[0]$	1/2	134 320.12		100		
$3d^8(^3F_4)6g$	$^2[1]$	1/2	134 321.35		100		
		3/2	134 321.36		100		
$3d^8(^3F_4)6g$	$^2[2]$	5/2	134 323.86		100		
		3/2	134 323.88		100		
$3d^8(^3F_4)6g$	$^2[8]$	15/2	134 325.08		100		
		17/2	134 325.14		100		
$3d^8(^3F_4)6g$	$^2[3]$	7/2	134 327.52		100		
		5/2	134 327.58		100		
$3d^8(^3F_4)6g$	$^2[4]$	9/2	134 331.78		100		
		7/2	134 331.88		100		
$3d^8(^3F_4)6g$	$^2[7]$	13/2	134 333.36		100		
		15/2	134 333.40		100		
$3d^8(^3F_4)6g$	$^2[5]$	11/2	134 335.47		100		
		9/2	134 335.49		100		
$3d^8(^3F_4)6g$	$^2[6]$	11/2	134 336.63		100		
		13/2	134 336.66		100		
$3d^8(^3F_4)6h$	$^2[1]^\circ$	1/2,3/2	134 336.81				
$3d^8(^3F_4)6h$	$^2[2]^\circ$	3/2,5/2	134 338.35				
$3d^8(^3F_4)6h$	$^2[9]^\circ$	17/2,19/2	134 339.71				
$3d^8(^3F_4)6h$	$^2[3]^\circ$	5/2,7/2	134 340.42				
$3d^8(^3F_4)6h$	$^2[4]^\circ$	7/2,9/2	134 342.69				
$3d^8(^3F_4)6h$	$^2[8]^\circ$	15/2,17/2	134 344.32				
$3d^8(^3F_4)6h$	$^2[5]^\circ$	9/2,11/2	134 344.87				

Ni II—Continued

Configuration	Term	J	Level (cm^{-1})	g	Leading percentages		
$3d^8(^3F_4)6h$	$^2[6]°$	11/2,13/2	134 346.30				
$3d^8(^3F_4)6h$	$^2[7]°$	13/2,15/2	134 346.34				
$3d^8(^3F)7d$	^4D	7/2	134 527.24				
		5/2	134 978.47				
		3/2	136 054.50				
$3d^8(^3F)7d$	^4P	5/2	134 539.37				
		3/2	134 670.07				
$3d^8(^3F)7d$	^4H	13/2	134 583.07				
		11/2	134 597.43				
		9/2	135 960.08				
		7/2	136 895.36?				
$3d^8(^3F)7d$	^4F	9/2	134 607.37				
		7/2	134 642.09				
		5/2	135 901.96				
		3/2	136 852.44				
$3d^8(^3F)7d$	^4G	11/2	134 614.55				
		9/2	134 658.60				
		7/2	135 986.06				
		5/2	136 899.34				
$3d^7(^2F)4s4p(^3P°)$	$^2D°$	5/2	*134 783.14*				
		3/2	*134 964.78*				
$3d^7(^2P)4s4p(^1P°)$	$^2P°$	1/2	*135 053.14*				
		3/2	*135 382.53*				
$3d^8(^3F)8s$	^2F	7/2	135 100.45				
		5/2	136 050.53				
$3d^7(^2P)4s4p(^1P°)$	$^2D°$	5/2	*135 258.92*				
		3/2	*136 461.10*				
$3d^8(^3F)8p$	$^4G°$	9/2	*135 261.99*				
		11/2	*135 338.01*				
$3d^8(^3P_2)4f$	$^2[4]°$	7/2	*135 400.67*		46	37	$(^3P_2)\ ^2[3]°$
		9/2	*135 435.26*		83	16	$(^1D_2)\ ^2[4]°$
$3d^8(^3P_2)4f$	$^2[3]°$	7/2	*135 444.47*		46	38	$(^3P_2)\ ^2[4]°$
		5/2	*135 461.55*		71	14	$(^1D_2)\ ^2[3]°$
		7/2	*135 464.86*				
$3d^8(^3P_2)4f$	$^2[2]°$	3/2	*135 493.26*		82	13	$(^1D_2)\ ^2[2]°$
		5/2	*135 512.92*		71	12	$(^3P_2)\ ^2[3]°$
$3d^8(^3P_2)4f$	$^2[5]°$	11/2	*135 538.61?*		84	16	$(^1D_2)\ ^2[5]°$
		9/2	*135 580.25*		84	16	$(^1D_2)\ ^2[5]°$
		9/2	*135 558.80*				
$3d^8(^3F_3)6f$	$^2[1]°$	1/2	*135 599.00*		59	41	$^2[0]°$
		3/2	*135 619.91*		86	14	$^2[2]°$

(Continued)

Ni II—Continued

Configuration	Term	J	Level (cm^{-1})	g	Leading percentages
$3d^8(^3F_3)6f$	$^2[6]^\circ$	11/2	135 606.20		100
		13/2	135 606.30		100
$3d^8(^3F_3)6f$	$^2[2]^\circ$	5/2	135 618.08		100
		3/2	135 623.59		85 14 $^2[1]^\circ$
$3d^8(^3F_3)6f$	$^2[3]^\circ$	7/2	135 622.60		97 3 $^2[4]^\circ$
		5/2	135 630.63		100
$3d^8(^3F_3)6f$	$^2[5]^\circ$	9/2	135 628.41		96 4 $^2[4]^\circ$
		11/2	135 645.10		100
$3d^8(^3F_3)6f$	$^2[4]^\circ$	7/2	135 629.40		97 3 $^2[3]^\circ$
		9/2	135 640.53		96 4 $^2[5]^\circ$
$3d^8(^3P_2)4f$	$^2[1]^\circ$	3/2	135 652.93		84 16 $(^1D_2)\ ^2[1]^\circ$
		1/2	135 670.49		84 16 $(^1D_2)\ ^2[1]^\circ$
$3d^8(^3F_3)6g$	$^2[1]$	1/2	135 682.28		100
		3/2	135 682.29		100
$3d^8(^3F_3)6g$	$^2[2]$	3/2	135 685.36		100
		5/2	135 685.43		100
$3d^8(^3F_3)6g$	$^2[7]$	13/2	135 686.63		100
		15/2	135 686.67		100
$3d^8(^3F_3)6g$	$^2[3]$	7/2	135 689.44		99
		5/2	135 689.47		99
$3d^8(^3F_3)6g$	$^2[4]$	9/2	135 693.38		100
		7/2	135 693.45		
$3d^8(^3F_3)6g$	$^2[6]$	11/2	135 694.01		100
		13/2	135 694.04		100
$3d^8(^3F_3)6g$	$^2[5]$	11/2	135 695.63		100
		9/2	135 695.64		100
$3d^8(^3F_3)6h$	$^2[3]^\circ$	5/2,7/2	135 700.04		
$3d^8(^3F_3)6h$	$^2[8]^\circ$	15/2,17/2	135 700.68		
$3d^8(^3F_3)6h$	$^2[4]^\circ$	7/2,9/2	135 703.44		
$3d^8(^3F_3)6h$	$^2[7]^\circ$	13/2,15/2	135 704.83		
$3d^8(^3F_3)6h$	$^2[5]^\circ$	9/2,11/2	135 705.68		
$3d^8(^3F_3)6h$	$^2[6]$	11/2,13/2	135 706.06		
$3d^7(^2F)4s4p(^3P^\circ)$	$^2G^\circ$	7/2	135 746.06		
		9/2	136 076.26		
$3d^8(^3P_1)4f$	$^2[2]^\circ$	3/2	135 746.13		93 4 $(^1D_2)\ ^2[2]^\circ$
$3d^8(^3P_1)4f$	$^2[3]^\circ$	5/2	135 849.41		100
		7/2	135 879.41		100

Ni II—Continued

Configuration	Term	J	Level (cm^{-1})	g	Leading percentages		
$3d^8(^3F)7d$	^2F	7/2	135 944.40				
		5/2	136 959.86				
$3d^8(^3P_0)4f$	$^2[3]°$	7/2	*135 954.09*		99		
		5/2	*136 122.61*		99		
$3d^8(^3F)7d$	^2H	11/2	135 956.01				
		9/2	136 880.56				
$3d^8(^3F)7d$	^2G	9/2	135 977.31				
		7/2	136 936.82				
$3d^8(^3F)7d$	^2D	5/2	136 031.43				
$3d^8(^3P)5d$	^4D	7/2	136 201.46		81	16	$(^1D)\ ^2F$
		5/2	136 288.60		77	14	$(^1D)\ ^2D$
		1/2	136 290.83?		91	8	$(^1D)\ ^2P$
		3/2	136 327.55		83	7	$(^1D)\ ^2D$
$3d^8(^1D)6p$	$^2F°$	7/2	*136 392.85*				
$3d^8(^3F_2)6f$	$^2[5]°$	11/2	*136 508.20*		99	1	$(^1D_2)\ ^2[5]°$
		9/2	*136 524.42*		99	1	$(^1D_2)\ ^2[5]°$
$3d^8(^3F_2)6f$	$^2[1]°$	1/2	*136 517.20*		99	1	$(^1D_2)\ ^2[1]°$
		3/2	*136 531.26*		85	14	$(^3F_2)\ ^2[2]°$
$3d^8(^3P)5d$	^4F	9/2	136 519.28		84	16	$(^1D)\ ^2G$
		7/2	136 589.35		62	24	$(^3P)\ ^2F$
		5/2	136 766.49		74	17	$(^3P)\ ^2D$
$3d^8(^3F_2)6f$	$^2[4]°$	7/2	*136 542.28*		68	31	$(^3F_2)\ ^2[3]°$
		9/2	*136 546.50*		99	1	$(^1D_2)\ ^2[4]°$
$3d^8(^3F_2)6f$	$^2[3]°$	7/2	*136 547.13*		68	31	$^2[4]°$
		5/2	*136 548.55*		89	10	$^2[2]°$
$3d^8(^3F_2)6g$	$^2[6]$	11/2	136 595.99		100		
		13/2	136 596.01		100		
$3d^8(^3F_2)6g$	$^2[3]$	5/2	136 598.31		100		
		7/2	136 598.37		100		
$3d^8(^3F_2)6g$	$^2[4]$	9/2	136 604.02		99		
		7/2	136 604.04		100		
$3d^8(^3F_2)6g$	$^2[5]$	9/2	136 604.38		100		
		11/2	136 604.38		100		
$3d^8(^3F_2)6h$	$^2[3]°$	5/2,7/2	*136 609.26*				
$3d^8(^3F_2)6h$	$^2[7]°$	13/2,15/2	*136 609.93*				
$3d^8(^3F_2)6h$	$^2[4]°$	7/2,9/2	*136 612.27*				
$3d^8(^3F_2)6h$	$^2[6]°$	11/2,13/2	*136 614.71*				
$3d^8(^3F_2)6h$	$^2[5]°$	9/2,11/2	*136 614.89*				

(Continued)

Ni II—Continued

Configuration	Term	J	Level (cm^{-1})	g	Leading percentages			
$3d^8(^3F)8p$	$^2G^\circ$	9/2	*136 673.64*					
		7/2	*137 562.74*					
$3d^8(^3P)5d$	^4P	1/2	136 725.33		82		12	$(^1D)\ ^2S$
		5/2	136 960.75		58		37	$(^3P)\ ^2F$
$3d^8(^3P)5d$		3/2	136 732.74		31	^2D	31	$(^3P)\ ^4P$
$3d^8(^3P)5d$		3/2	136 899.33		38	^4F	37	^4P
$3d^8(^3F)7d$	^2P	1/2	136 955.28					
$3d^8(^3F)9s$	^4F	9/2	137 188.58					
		7/2	137 236.28					
		5/2	138 575.69					
		3/2	139 456.75					
$3d^8(^3P)5d$	^2P	1/2	137 211.93		85		8	$(^1D)\ ^2P$
		3/2	137 278.22?		67		19	$(^3P)\ ^2D$
$3d^8(^3F_4)7f$	$^2[2]^\circ$	5/2	*137 519.23*					
$3d^8(^3F_4)7f$	$^2[7]^\circ$	15/2	*137 519.63*					
$3d^8(^3F_4)7f$	$^2[3]^\circ$	7/2	*137 523.51*					
		5/2	*137 526.73*					
$3d^8(^3F_4)7f$	$^2[6]^\circ$	13/2	*137 529.37*					
		11/2	*137 535.83*					
$3d^8(^3F_4)7f$	$^2[4]^\circ$	9/2	*137 531.18*					
		7/2	*137 535.96*					
$3d^8(^3F_4)7g$	$^2[8]$	15/2	137 568.00					
		17/2	137 568.02					
$3d^8(^3F_4)7g$	$^2[6]$	13/2	137 573.19					
		11/2	137 573.19					
$3d^8(^3F_4)7g$	$^2[5]$	11/2	137 575.14					
$3d^8(^3F)8d$	^4P	5/2	137 706.71					
$3d^8(^3F)8d$	^4D	7/2	137 707.26					
		5/2	138 014.53					
$3d^8(^3F)8d$	^4H	13/2	137 735.22					
		11/2	137 742.95					
$3d^8(^3F)8d$	^4F	9/2	137 753.87					
		7/2	137 776.55					
$3d^8(^3F)8d$	^4G	11/2	137 754.78					
		9/2	137 782.50?					
$3d^8(^3F)9p$	$^4F^\circ$	9/2	*138 121.88*					
$3d^7(^2F)4s4p(^1P^\circ)$	$^2G^\circ$	9/2	*138 495.84?*					

Ni II—Continued

Configuration	Term	J	Level (cm^{-1})	g	Leading percentages
$3d^8(^3F)9s$	^2F	7/2	138 563.71		
		5/2	139 492.10		
$3d^8(^3P)6p$	$^4D°$	7/2	*138 841.00*		
$3d^8(^3F_3)7f$	$^2[6]°$	13/2	*138 888.93*		
$3d^8(^3F_3)7g$	$^2[7]$	15/2	138 928.70		
$3d^8(^3F)8d$	^2H	11/2	139 103.05		
$3d^8(^3F_2)7g$	$^2[6]$	13/2	139 834.24		
$3d^8(^1G)6s$	^2G	9/2	140 006.17		
		7/2	140 008.76		
Ni III ($^3F_4$)	***Limit***		**146 541.56**		

Ni III

$Z = 28$

Fe I isoelectronic sequence

Ground state: $1s^2 2s^2 2p^6 3s^2 3p^6 3d^8\ ^3F_4$

Ionization energy = 283 800±200 cm^{-1} (35.19±0.02 eV)

This analysis has been made by Shenstone (1954) and extended by Garcia-Riquelme (1958) who has also provided unpublished results.

Shenstone's line list includes the range 600–3000 Å. The spectrum has been reobserved and extended by Garcia-Riquelme and Velasco (1955) in the range 2300–8600 Å.

Some of Shenstone's identifications in $3d^7 4p$ have been changed by Roth (1968) in his theoretical study. He calculated the percentage compositions of the $3d^7 4p$ terms.

The $3d^8$ and $3d^7 4s$ configurations have been studied theoretically by Shadmi (1962) and by Shadmi, Caspi, and Oreg (1969). The leading percentages of the $3d^8$ levels are taken from Pasternak and Goldschmidt (1972).

Some changes in the $3d^7 4s\ ^3P$ and ^3D levels suggested by Shadmi (1962) and agreed to by Shenstone are incorporated in this compilation.

Garcia-Riquelme has provided new terms for the known configurations $3d^8$, $3d^7 4s$, and $3d^7\ 4p$ and has extended the analysis further with the discovery of terms from the configurations $3d^6 4s^2$, $3d^7 4d$, $3d^7 5p$, $3d^7 4f$, $3d^7 6s$, $3d^7 5d$, and $3d^7 5g$. With her new measurements, she has determined values for all levels above the ground configuration and has calculated percentage compositions for $3d^7 4d$, $4f$, $5p$, $5d$, and $5g$. In a few cases, we have changed her designations to correspond with her percentages. The uncertainty of the level values is ±0.1 cm^{-1}.

In all the calculations, the percentages for the two ^2D states of $3d^7$, distinguished by seniority, include the sum of contributions from both states. They are distinguished by the prefixes A and B.

Her value for the ionization energy determined from the $3d^7(^4F)ns$ ($n = 4,5,6$) series is quoted here.

References

Garcia-Riquelme, O., and Velasco, R. (1955), An. R. Soc. Esp. Fis. Quim. **A51**, 41.

Garcia-Riquelme, O. (1958), J. Opt. Soc. Am. **48**, 183.

Pasternak, A., and Goldschmidt, Z. B. (1972), Phys. Rev. **A6**, 55.

Roth, C. (1968), J. Res. Natl. Bur. Stand. **72A**, 505.

Shadmi, Y. (1962), Bull. Res. Council Israel **10F**, 109.

Shadmi, Y., Caspi, E., and Oreg, J. (1969), J. Res. Natl. Bur. Stand. **73A**, 173.

Shenstone, A. G. (1954), J. Opt. Soc. Am. **44**, 749.

Ni III

Configuration	Term	J	Level (cm^{-1})	Leading percentages		
$3d^8$	^3F	4	0.0	100		
		3	1 360.7	100		
		2	2 269.6	99		
$3d^8$	^1D	2	14 031.6	83	16	^3P
$3d^8$	^3P	2	16 661.6	84	16	^1D
		1	16 977.8	100		
		0	17 230.7	100		
$3d^8$	^1G	4	23 108.7	100		
$3d^8$	^1S	0	52 532.0	100		
$3d^7(^4F)4s$	^5F	5	53 703.93			
		4	54 657.83			
		3	55 406.29			
		2	55 952.21			
		1	56 308.24			
$3d^7(^4F)4s$	^3F	4	61 338.58			
		3	62 605.58			
		2	63 471.93			

Ni III—Continued

Configuration	Term	J	Level (cm^{-1})	Leading percentages		
$3d^7(^4P)4s$	^5P	3	71 067.35			
		2	71 384.10			
		1	71 842.42			
$3d^7(^2G)4s$	^3G	5	75 123.65			
		4	75 646.61			
		3	76 237.25			
$3d^7(^4P)4s$	^3P	2	78 303.54	47	53	$(^2P)\ ^3P$
		1	78 482.43	52	35	
		0	78 657.55	48	46	
$3d^7(^2P)4s$	^3P	2	79 143.01	47	53	$(^4P)\ ^3P$
		1	79 749.22	56	39	
		0	80 621.10	43	51	
$3d^7(^2G)4s$	^1G	4	79 250.11			
$3d^7(^2H)4s$	^3H	6	81 686.80			
		5	82 193.80			
		4	82 826.40			
$3d^7(a\ ^2D)4s$	^3D	3	82 172.60			
		1	82 277.26	54	41	$(^2P)\ ^1P$
		2	83 033.45			
$3d^7(^2P)4s$	^1P	1	84 604.10	48	40	$(a\ ^2D)\ ^1D$
$3d^7(^2H)4s$	^1H	5	85 834.20			
$3d^7(a\ ^2D)4s$	^1D	2	86 645.88			
$3d^7(^2F)4s$	^3F	2	97 841.60			
		3	97 995.81			
		4	98 237.93			
$3d^7(^2F)4s$	^1F	3	101 954.90			
$3d^7(^4F)4p$	$^5F^\circ$	5	*110 212.80*	94	5	$(^4F)\ ^5G^\circ$
		4	*110 371.35*	65	29	$(^4F)\ ^5D^\circ$
		3	*111 221.20*	78	17	$(^4F)\ ^5D^\circ$
		2	*111 914.53*	88	9	$(^4F)\ ^5D^\circ$
		1	*112 401.65*	96		
$3d^7(^4F)4p$	$^5D^\circ$	4	*111 898.65*	63	26	$(^4F)\ ^5F^\circ$
		3	*112 935.43*	73	13	$(^4F)\ ^5F^\circ$
		2	*113 651.47*	78	8	$(^4F)\ ^5G^\circ$
		1	*114 095.60*	88	8	$(^4P)\ ^5D^\circ$
		0	*114 295.45*	91	8	$(^4P)\ ^5D^\circ$
$3d^7(^4F)4p$	$^5G^\circ$	6	*112 787.85*	100		
		5	*113 140.92*	82	12	$(^4F)\ ^3G^\circ$
		4	*113 705.12*	84	8	$(^4F)\ ^5G^\circ$
		3	*114 110.20*	86	7	$(^4F)\ ^5F^\circ$
		2	*114 371.01*	88	5	$(^4F)\ ^5D^\circ$
$3d^7(^4F)4p$	$^3G^\circ$	5	*115 272.26*	87	13	$(^4F)\ ^5G^\circ$
		4	*116 674.39*	83	9	$(^4F)\ ^3F^\circ$
		3	*117 606.35*	88	7	$(^4F)\ ^3F^\circ$

(Continued)

Ni III—Continued

Configuration	Term	J	Level (cm^{-1})	Leading percentages		
$3d^7(^4F)4p$	$^3F°$	4	*116 191.93*	85	10	$(^4F)\ ^3G°$
		3	*117 250.80*	83	8	$(^4F)\ ^3G°$
		2	*118 114.95*	92		
$3d^7(^4F)4p$	$^3D°$	3	*118 745.25*	90	5	$(^4F)\ ^3F°$
		2	*119 669.54*	92		
		1	*120 272.32*	93		
$3d^7(b\ ^2D)4s$	^3D	1	121 192.93			
		2	121 411.60			
		3	121 802.45			
$3d^7(^4P)4p$	$^5S°$	2	*122 282.40*	99		
$3d^7(b\ ^2D)4s$	^1D	2	125 433.55			
$3d^7(^4P)4p$	$^5D°$	2	*129 913.10*	82	7	$(^4F)\ ^5D°$
		3	*129 954.00*	84	7	
		1	*129 957.95*	75	7	
		0	*130 190.05*	86	8	
		4	*130 312.30*	93	6	
$3d^7(^4P)4p$	$^3S°$	1	*130 863.50*	53	12	$(^2P)\ ^3S°$
$3d^7(^2G)4p$	$^3H°$	5	*131 500.50*	74	15	$(^2G)\ ^1H°$
		4	*132 156.50*	61	28	$(^2G)\ ^3F°$
		6	*132 168.60*	96		
$3d^7(^2G)4p$	$^3F°$	4	*131 792.02*	41	33	$(^2G)\ ^3H°$
		3	*133 158.50*	78	11	$(^2G)\ ^3G°$
		2	*134 231.90*	94		
$3d^7(^4P)4p$	$^5P°$	2	*132 818.26*	45	21	$(^2P)\ ^3P°$
		3	*133 095.89*	71	19	$(^4P)\ ^3D°$
		1	*133 339.70*	73	20	$(^4P)\ ^3S°$
$3d^7(^2P)4p$	$^3P°$	0	*132 864.8*	69	19	$(a\ ^2D)\ ^3P°$
		1	*133 276.70*	45	19	$(^4P)\ ^3D°$
		2	*133 642.58*	49	10	$(^2P)\ ^3D°$
$3d^7(^2G)4p$	$^1G°$	4	*133 324.70*	47	24	$(^2G)\ ^3F°$
$3d^7(^4P)4p$	$^3D°$	3	*133 390.94*	65	20	$(^4P)\ ^5P°$
		2	*133 500.97*	42	37	$(^4P)\ ^5P°$
		1	*133 839.54*	54	15	$(^2P)\ ^3D°$
$3d^7(^2G)4p$	$^3G°$	5	*133 692.00*	78	19	$(^2G)\ ^1H°$
		3	*134 334.79*	70	16	$(^2G)\ ^1F°$
		4	*134 414.77*	73	14	$(^2G)\ ^1G°$
$3d^7(^2G)4p$	$^1H°$	5	*134 217.60*	62	23	$(^2G)\ ^3H°$
$3d^7(^2G)4p$	$^1F°$	3	*135 023.20*	57	15	$(^2G)\ ^3G°$
$3d^7(^4P)4p$	$^3P°$	0	*135 334.90*	52	35	$(^2P)\ ^1S°$
		2	*135 350.33*	75	10	$(a\ ^2D)\ ^3P°$
		1	*136 098.70*	78	11	$(^2P)\ ^3D°$

Ni III—Continued

Configuration	Term	J	Level (cm^{-1})	Leading percentages			
$3d^7(^2P)4p$	$^3D^\circ$	2	*136 813.20*	41		29	$(^2P)\ ^1D^\circ$
		3	*136 967.00*	77		8	$(a\ ^2D)\ ^3F^\circ$
		1	*137 362.36*	54		15	$(a\ ^2D)\ ^3D^\circ$
$3d^7(^2H)4p$	$^3G^\circ$	5	*137 020.20*	94		5	$(^2F)\ ^3G^\circ$
		4	*138 030.90*	90		5	
		3	*138 852.20*	85		6	
$3d^7(^2P)4p$		2	*137 631.60*	24	$^3D^\circ$	20	$^1D^\circ$
$3d^7(^2H)4p$	$^3I^\circ$	6	*137 391.30*	74		25	$(^2H)\ ^1I^\circ$
		7	*137 991.40*	100			
		5	*138 060.40*	95			
$3d^7(^2P)4p$	$^1S^\circ$	0	*138 146.48*	61		39	$(^4P)\ ^3P^\circ$
$3d^7(a\ ^2D)4p$	$^3D^\circ$	3	*138 487.40*	79		11	$(a\ ^2D)\ ^3F^\circ$
		1	*138 979.20*	50		27	$(^2P)\ ^1P^\circ$
		2	*139 253.70*	63		10	$(^2P)\ ^3D^\circ$
$3d^7(^2H)4p$	$^1I^\circ$	6	*139 633.90*	74		24	$(^2H)\ ^3I^\circ$
$3d^7(a\ ^2D)4p$	$^3F^\circ$	4	*140 184.65*	98			
		3	*140 544.52*	71		8	$(^2P)\ ^3D^\circ$
		2	*140 885.40*	73		10	$(a\ ^2D)\ ^3D^\circ$
$3d^7(^2P)4p$	$^3S^\circ$	1	*140 885.15*	67		8	$(^2P)\ ^1P^\circ$
$3d^7(^2P)4p$	$^1P^\circ$	1	*141 414.10*	49		17	$(a\ ^2D)\ ^3D^\circ$
$3d^7(^2H)4p$	$^3H^\circ$	6	*142 187.80*	98			
		5	*142 575.60*	95			
		4	*143 002.70*	95			
$3d^7(a\ ^2D)4p$	$^1D^\circ$	2	*142 433.95*	46		29	$(a\ ^2D)\ ^3P^\circ$
$3d^7(a\ ^2D)4p$	$^3P^\circ$	2	*143 560.16*	48		24	$(a\ ^2D)\ ^1D^\circ$
		1	*144 624.55*	67		11	$(^2P)\ ^3P^\circ$
		0	*145 088.45*	80		15	$(^2P)\ ^3P^\circ$
$3d^7(a\ ^2D)4p$	$^1F^\circ$	3	*143 864.80*	73		16	$(^2G)\ ^1F^\circ$
$3d^7(^2H)4p$	$^1G^\circ$	4	*144 153.00*	70		27	$(^2G)\ ^1G^\circ$
$3d^7(a\ ^2D)4p$	$^1P^\circ$	1	*145 950.15*	88			
$3d^7(^2H)4p$	$^1H^\circ$	5	*146 325.80*	96			
$3d^6\ 4s^2$	^5D	4	153 256.35				
		3	154 170.37				
$3d^7(^2F)4p$	$^1D^\circ$	2	*155 071.00*	60		32	$(^2F)\ ^3F^\circ$
$3d^7(^2F)4p$	$^3G^\circ$	3	*155 443.30*	82		10	$(^2F)\ ^3F^\circ$
		4	*155 841.40*	73		15	$(^2F)\ ^3F^\circ$
		5	*156 808.70*	94		5	$(^2H)\ ^3G^\circ$
$3d^7(^2F)4p$	$^3F^\circ$	3	*156 411.20*	68		19	$(^2F)\ ^3D^\circ$
		2	*156 522.87*	64		29	$(^2F)\ ^1D^\circ$
		4	*156 972.08*	50		28	$(^2F)\ ^1G^\circ$

(Continued)

Ni III—Continued

Configuration	Term	J	Level (cm^{-1})	Leading percentages		
$3d^7(^2F)4p$	$^3D°$	3	*156 853.00*	72	18	$(^2F)\ ^3F°$
		2	*157 154.27*	85	8	$(^2F)\ ^1D°$
		1	*157 235.16*	93	5	$(b\ ^2D)\ ^3D°$
$3d^7(^2F)4p$	$^1G°$	4	*157 375.42*	65	32	$(^2F)\ ^3F°$
$3d^7(^2F)4p$	$^1F°$	3	*161 754.89*	97		
$3d^7(b\ ^2D)4p$	$^3P°$	2	*176 487.10*	99		
		1	*176 583.20*	98		
		0	*176 736.40*	99		
$3d^7(b\ ^2D)4p$	$^3F°$	2	*177 805.60*	96		
		3	*178 451.50*	96		
		4	*179 282.00*	97		
$3d^7(^4F)4d$	^5F	5	181 019.08	90	8	^5G
		4	181 482.95	83	8	^5G
		3	181 996.70	80	8	^5D
		2	182 587.83	87	5	^5D
		1	183 035.25	95	2	^5D
$3d^7(b\ ^2D)4p$	$^1P°$	1	*181 203.4*	95		
$3d^7(b\ ^2D)4p$	$^1F°$	3	*181 658.4*	97		
$3d^7(^4F)4d$	^5G	6	181 840.00	96	4	^5H
		5	182 327.13	63	26	^3G
		4	183 041.45	71	13	^3G
		3	183 637.76	81	4	^3G
		2	184 124.86	74	17	^5P
$3d^7(^4F)5s$	^5F	5	181 998.15	99		
		4	182 798.20	84	14	^3F
		3	183 612.67	93		
		2	184 220.35	97		
		1	184 609.57	99		
$3d^7(^4F)4d$	^5H	7	182 508.30	99		
		6	183 126.19	59	39	^3H
		5	183 904.88	80	14	^3H
		4	184 510.75	88	5	^3G
		3	184 944.95	81	17	^3G
$3d^7(^4F)4d$	^5P	3	182 524.69	83	9	^5D
		2	183 575.58	55	21	^5G
		1	184 375.68	59	38	^5D
$3d^7(^4F)4d$	^3G	5	183 052.20	66	24	^5G
		4	183 859.72	74	16	^5G
		3	184 346.50	74	15	^5H
$3d^7(^4F)4d$	^5D	4	183 464.88	85	9	^5F
		3	184 518.20	64	13	^3D
		2	185 067.15	55	25	^3D
		1	185 116.05	46	38	^5P
		0	185 147.23	98		

Ni III—Continued

Configuration	Term	*J*	Level (cm^{-1})	Leading percentages		
$3d^7(b\,^2\mathrm{D})4p$	$^3\mathrm{D}^\circ$	1	*183 717.00*	95		
		2	*183 872.00*	73	25	$^1\mathrm{D}^\circ$
		3	*184 723.10*	96		
$3d^7(^4\mathrm{F})4d$	$^3\mathrm{D}$	3	183 839.47	71	14	$^5\mathrm{D}$
		2	184 623.54	66	15	
		1	185 543.70	82	12	
$3d^7(^4\mathrm{F})5s$	$^3\mathrm{F}$	4	184 037.62	85	14	$^5\mathrm{F}$
		3	185 248.15	93		
		2	186 073.40	97		
$3d^7(^4\mathrm{F})4d$	$^3\mathrm{H}$	6	184 166.57	60	37	$^5\mathrm{H}$
		5	184 805.62	80	15	
		4	185 639.13	91	5	
$3d^7(^4\mathrm{F})4d$	$^3\mathrm{F}$	4	187 351.88	86	5	$(^2\mathrm{G})\ ^3\mathrm{F}$
		3	188 140.54	84	6	
		2	188 622.87	85	6	
$3d^7(^4\mathrm{F})4d$	$^3\mathrm{P}$	2	187 493.38	87	5	$(^4\mathrm{P})\ ^3\mathrm{P}$
		1	188 542.82	90	6	
		0	189 056.90	91	6	
$3d^7(^4\mathrm{F})5p$	$^5\mathrm{F}^\circ$	5	*199 919.08*	91	8	$^5\mathrm{G}^\circ$
		4	*200 076.35*	49	42	$^5\mathrm{D}^\circ$
		3	*200 962.97*	64	27	$^5\mathrm{D}^\circ$
		2	*201 725.23*	80	15	$^5\mathrm{D}^\circ$
		1	*202 263.08*	94	6	$^5\mathrm{D}^\circ$
$3d^7(^4\mathrm{F})5p$	$^5\mathrm{G}^\circ$	6	*200 747.06*	99		
		5	*201 033.68*	48	43	$^3\mathrm{G}^\circ$
		4	*201 969.90*	67	14	$^3\mathrm{F}^\circ$
		3	*202 608.20*	76	11	$^5\mathrm{F}^\circ$
		2	*203 020.07*	78	10	$^5\mathrm{D}^\circ$
$3d^7(^4\mathrm{F})5p$	$^5\mathrm{D}^\circ$	4	*200 935.60*	54	29	$^5\mathrm{F}^\circ$
		3	*201 829.46*	60	19	$^5\mathrm{F}^\circ$
		2	*202 487.82*	71	17	$^5\mathrm{G}^\circ$
		1	*202 898.94*	92	6	$^5\mathrm{F}^\circ$
		0	*203 078.46*	98		
$3d^7(^4\mathrm{F})5p$	$^3\mathrm{F}^\circ$	4	*202 074.33*	71	5	$^5\mathrm{F}^\circ$
		3	*203 360.52*	57	29	$^3\mathrm{D}^\circ$
		2	*203 739.65*	58	38	$^3\mathrm{D}^\circ$
$3d^7(^4\mathrm{F})5p$	$^3\mathrm{G}^\circ$	5	*202 125.84*	54	44	$^5\mathrm{G}^\circ$
		4	*203 197.33*	75	19	
		3	*203 976.35*	85	7	
$3d^7(^4\mathrm{F})5p$	$^3\mathrm{D}^\circ$	3	*202 624.51*	62	30	$^3\mathrm{F}^\circ$
		2	*204 242.07*	57	38	$^3\mathrm{F}^\circ$
		1	*204 677.95*	98		
$3d^6(^5\mathrm{D})4s4p$	$^5\mathrm{D}^\circ$	4	*204 404.12*			
		3	*204 714.93*			
		2	*205 062.51*			
		1	*205 327.00*			
		0	*205 466.00*			

(Continued)

Ni III—Continued

Configuration	Term	J	Level (cm^{-1})	Leading percentages		
$3d^6(^5D)4s4p$	$^5F°$	5	206 925.18			
		4	207 382.10			
		3	207 744.30			
		2	208 005.7			
$3d^6(^5D)4s4p$	$^3D°$	3	212 312.5			
		2	213 016.8			
		1	213 490.0			
$3d^6(^5D)4s4p$	$^3F°$	4	212 837.7			
		3	213 979.67			
		2	214 744.0			
$3d^7(^4F_{9/2})4f$	$^2[11/2]°$	6	221 187.25	99		
		5	221 195.35	88	11	$^2[9/2]°$
$3d^7(^4F_{9/2})4f$	$^2[9/2]°$	4	221 256.08	98		
		5	221 268.50	88	11	$^2[11/2]°$
$3d^7(^4F_{9/2})4f$	$^2[13/2]°$	7	221 286.78	99		
		6	221 292.92	99		
$3d^7(^4F_{9/2})4f$	$^2[7/2]°$	3	221 350.90	99		
		4	221 388.14	98		
$3d^7(^4F_{9/2})4f$	$^2[15/2]°$	8	221 433.20	99		
		7	221 444.15	99		
$3d^7(^4F_{7/2})4f$	$^2[9/2]°$	5	222 455.66	94		
		4	222 466.47	89	9	$^2[7/2]°$
$3d^7(^4F_{9/2})4f$	$^2[5/2]°$	3	221 476.68	99		
$3d^7(^4F_{7/2})4f$	$^2[11/2]°$	6	222 494.65	99		
		5	222 529.55	94		
$3d^7(^4F_{7/2})4f$	$^2[7/2]°$	3	222 516.60	97		
		4	222 547.50	89	10	$^2[9/2]°$
$3d^7(^4F_{7/2})4f$	$^2[5/2]°$	3	222 530.00	97		
		2	222 571.37	98		
$3d^7(^4F_{7/2})4f$	$^2[13/2]°$	7	222 596.30	99		
		6	222 599.70	99		
$3d^7(^4F_{5/2})4f$	$^2[7/2]°$	4	223 387.00	98		
		3	223 329.77	87	12	$^2[5/2]°$
$3d^7(^4F_{5/2})4f$	$^2[5/2]°$	3	223 406.81	85	12	$^2[7/2]°$
		2	223 375.18	97		
$3d^7(^4F_{5/2})4f$	$^2[9/2]°$	5	223 434.10	99		
		4	223 314.77	98		
$3d^7(^4F_{5/2})4f$	$^2[11/2]°$	5	223 461.15	99		
		6	223 469.90	99		
$3d^7(^4F_{5/2})4f$	$^2[3/2]°$	2	223 481.25	99		
		1	223 491.34	99		

Ni III—Continued

Configuration	Term	J	Level (cm^{-1})	Leading percentages			
$3d^7(^4F_{3/2})4f$	$^2[5/2]^\circ$	3	*223 957.54*	97			
		2	*224 026.05*	97			
$3d^7(^4F_{3/2})4f$	$^2[9/2]^\circ$	4	*223 989.02*	99			
		5	*224 020.91*	99			
$3d^7(^4F)6s$	^5F	5	225 784.20	99			
		3	227 270.07	73		26	^3F
		4	227 288.70	55		44	^3F
		2	227 985.66	90		9	^3F
$3d^7(^4F)5d$	^5F	5	225 918.35	79		20	^5G
		4	226 118.80	55		34	^5D
		1	228 042.24	62		24	^5D
$3d^7(^4F)5d$	^5G	6	226 124.64	91		8	^5H
		5	227 433.80	41		34	^3G
		3	227 963.02	43		17	^5D
		2	228 483.54	68		13	^5F
$3d^7(^4F)6s$	^3F	4	226 290.44	55		44	^5F
		3	228 290.97	73		26	
		2	229 036.43	90		9	
$3d^7(^4F)5d$	^5H	7	226 380.36	99			
		6	227 649.28	60		33	^3H
		5	227 726.91	61		26	^3H
		4	228 434.45	76		10	^3H
		3	228 955.36	55		42	^3G
$3d^7(^4F)5d$	^5P	3	226 532.85	51		30	^5D
		2	227 767.03	53		26	^5F
		1	228 329.40	42		31	^5D
$3d^7(^4F)5d$	^3G	5	226 603.73	51		30	^5G
		4	228 195.11	46		34	^5G
		3	228 858.30	40		34	^5H
$3d^7(^4F)5d$	^3H	6	226 686.28	66		31	^5H
		5	228 609.82	59		29	
		4	229 291.50	79		12	
$3d^7(^4F)5d$		3	226 757.78	37	^3D	27	^5P
$3d^7(^4F)5d$	^5D	4	227 091.99	42		37	^5G
		3	228 363.94	46		23	^5F
		0	228 736.36	99			
		2	228 819.73	56		20	^5P
		1	229 057.39	38		31	^5P
$3d^7(^4F)5d$		2	227 346.84	35	^5F	15	^5P
$3d^7(^4F)5d$	^3D	3	227 459.43	48		26	^5F
		2	228 273.42	62		18	^5G
		1	228 879.50	59		21	^5P
$3d^7(^4F)5d$		4	227 553.40	30	^5F	27	^3G

(Continued)

Ni III—Continued

Configuration	Term	J	Level (cm^{-1})	Leading percentages		
$3d^7(^4F)5d$	^3F	4	230 169.90	89	6	^3G
		3	231 033.56	84	3	^3G
		2	231 548.58	83	2	^3D
$3d^7(^4F)5d$	^3P	2	230 219.9	87	7	$(^4F)\ ^3D$
		1	231 309.0	93	3	$(^4F)\ ^3D$
		0	231 761.8	97	2	$(^4P)\ ^3P$
$3d^7(^4F_{9/2})5g$	$^2[15/2]$	8	244 262.75	97		
		7	244 267.15	93		
$3d^7(^4F_{9/2})5g$	$^2[11/2]$	6	244 264.23	86	10	$^2[13/2]$
		5	244 264.28	72	27	$^2[9/2]$
$3d^7(^4F_{9/2})5g$	$^2[13/2]$	7	244 289.88	94		
		6	244 290.37	86	13	$^2[11/2]$
$3d^7(^4F_{9/2})5g$	$^2[9/2]$	4	244 306.13	53	46	$^2[7/2]$
		5	244 306.20	71	27	$^2[11/2]$
$3d^7(^4F_{9/2})5g$	$^2[17/2]$	9	244 343.23	99		
		8	244 343.20	98		
$3d^7(^4F_{7/2})5g$	$^2[13/2]$	6	245 471.20	87	11	$^2[11/2]$
$3d^7(^4F_{7/2})5g$	$^2[11/2]$	5	245 488.31	60	38	$^2[9/2]$
		6	245 495.36	89	10	$^2[13/2]$
$3d^7(^4F_{7/2})5g$	$^2[15/2]$	7	245 528.40	96		
$3d^7(^4F_{7/2})5g$	$^2[9/2]$	5	245 532.59	60	39	$^2[11/2]$
		4	245 572.57	70	28	$^2[7/2]$
$3d^7(^4F_{5/2})5g$	$^2[9/2]$	4	246 364.86	96		
		5	246 376.24	55	42	$^2[11/2]$
$3d^7(^4F_{5/2})5g$	$^2[13/2]$	6	246 385.97	97		
$3d^7(^4F_{3/2})5g$	$^2[9/2]$	4	246 938.98	72	26	$^2[7/2]$
		5	246 954.87	58	40	$^2[11/2]$
Ni IV $(^4F_{9/2})$	***Limit***		**283 800**			

Ni IV

$Z=28$

Mn I isoelectronic sequence

Ground state: $1s^22s^2sp^63s^23p^63d^7\ {}^4F_{9/2}$

Ionization energy $= 443\ 000\pm2000\ cm^{-1}$ (54.9 ± 0.2 eV)

The first work on Ni IV was reported by Poppe (1968), who found two quartets in the $3d^7$ configuration and quartets and sextets from $3d^64s$ and $3d^64p$. The sextets and quartets were connected by intercombinations found by Garcia-Riquelme (1968).

The present compilation is taken from an extension of the analysis by Poppe (1976). The $3d^7-3d^64p$ array was observed in the region 390–710 Å and the $3d^64s-3d^64p$ array in the region 1210–1830 Å. All levels of $3d^7$ have been found. The uncertainty of the level values is 3 cm^{-1}.

The leading percentages were calculated by Poppe (1976). We use the designations of Nielson and Koster (see Introduction) to represent the seniorities.

The ionization energy was obtained by extrapolation by Lotz (1967).

References

Garcia-Riquelme, O. (1968), Physica **40**, 27.
Lotz, W. (1967), J. Opt. Soc. Am. **57**, 873.
Poppe, R. (1968), Physica **40**, 17.
Poppe, R. (1976), Physica **81C**, 351.

Ni IV

Configuration	Term	J	Level (cm^{-1})	Leading percentages		
$3d^7$	^4F	9/2	0.0	100		
		7/2	1 189.7	100		
		5/2	2 042.5	100		
		3/2	2 621.1	100		
$3d^7$	^4P	5/2	18 118.6	100		
		3/2	18 366.8	93	7	^2P
		1/2	18 958.4	97	3	^2P
$3d^7$	^2G	9/2	19 829.6	97	2	^2H
		7/2	20 947.6	100		
$3d^7$	^2P	3/2	23 648.9	83	8	$^2D2$
		1/2	24 651.4	97	3	^4P
$3d^7$	^2H	11/2	26 649.1	100		
		9/2	27 677.6	98	2	^2G
$3d^7$	$^2D2$	5/2	27 096.5	77	23	$^2D1$
		3/2	28 777.7	72	18	
$3d^7$	^2F	5/2	43 437.5	100		
		7/2	43 858.6	100		
$3d^7$	$^2D1$	3/2	67 360.0	80	20	$^2D2$
		5/2	67 989.8	77	23	
$3d^6(^5D)4s$	^6D	9/2	110 410.6	100		
		7/2	111 195.8	100		
		5/2	111 763.3	100		
		3/2	112 151.9	100		
		1/2	112 379.3	100		
$3d^6(^5D)4s$	^4D	7/2	120 909.5	100		
		5/2	121 807.7	100		
		3/2	122 386.1	100		
		1/2	122 717.4	99		

(Continued)

Ni IV—Continued

Configuration	Term	J	Level (cm^{-1})	Leading percentages		
$3d^6(^3P2)4s$	^4P	5/2	138 446.2	62	37	$(^3P1)\ ^4P$
		3/2	140 343.0	60	35	
		1/2	141 561.2	60	36	
$3d^6(^3H)4s$	^4H	13/2	139 289.4	100		
		11/2	139 619.2	98	2	$(^3G)\ ^4G$
		9/2	139 886.7	94	3	$(^3G)\ ^4G$
		7/2	140 140.9	96	2	$(^3G)\ ^4G$
$3d^6(^3F2)4s$	^4F	9/2	141 220.3	72	21	$(^3F1)\ ^4F$
		7/2	141 577.2	74	20	
		5/2	141 832.0	77	20	
		3/2	142 023.5	80	19	
$3d^6(^3G)4s$	^4G	11/2	144 815.1	71	28	$(^3H)\ ^2H$
		9/2	145 702.2	76	18	$(^3H)\ ^2H$
		7/2	146 061.5	88	4	$(^3F2)\ ^2F$
		5/2	146 153.8	90	6	$(^3F2)\ ^2F$
$3d^6(^3P2)4s$	^2P	3/2	145 192.1	60	36	$(^3P1)\ ^2P$
$3d^6(^3H)4s$	^2H	11/2	145 962.5	71	27	$(^3G)\ ^4G$
		9/2	146 194.3	78	17	
$3d^6(^3F2)4s$	^2F	7/2	147 635.9	69	19	$(^3F1)\ ^2F$
		5/2	148 358.2	74	18	
$3d^6(^3G)4s$	^2G	9/2	151 574.7	97	3	$(^3H)\ ^2H$
		7/2	152 343.7	95	4	$(^3F2)\ ^2F$
$3d^6(^3D)4s$	^4D	3/2	153 313.8	99		
		5/2	153 338.8	98		
		1/2	153 349.4	99		
		7/2	153 533.6	100		
$3d^6(^1I)4s$	^2I	13/2	155 253.7	100		
		11/2	155 308.7	99		
$3d^6(^1G2)4s$	^2G	9/2	156 294.0	65	32	$(^1G1)\ ^2G$
		7/2	156 351.2	65	32	
$3d^6(^3D)4s$	^2D	3/2	159 498.5	94	4	$(^1D2)\ ^2D$
		5/2	159 818.4	96	2	
$3d^6(^1F)4s$	^2F	7/2	171 406.0	98		
		5/2	171 408.0	98		
$3d^6(^5D)4p$	$^6D°$	9/2	*175 569.5*	98		
		7/2	*175 869.1*	96	2	$(^5D)\ ^6F°$
		5/2	*176 247.1*	97		
		3/2	*176 554.4*	98		
		1/2	*176 749.0*	99		
$3d^6(^3F1)4s$	^4F	3/2	179 583.0	80	20	$(^3F2)\ ^4F$
		9/2	179 655.0	78	22	
		5/2	179 724.0	79	20	
		7/2	179 792.0	78	20	

Ni IV—Continued

Configuration	Term	J	Level (cm^{-1})	Leading percentages			
$3d^6(^5D)4p$	$^6F^\circ$	11/2	*181 931.5*	100			
		9/2	*182 044.9*	94		4	$(^5D)\ ^4F^\circ$
		7/2	*182 125.6*	93		3	$(^5D)\ ^4F^\circ$
		5/2	*182 206.8*	95		2	$(^5D)\ ^4D^\circ$
		3/2	*182 259.9*	96		2	$(^5D)\ ^4D^\circ$
		1/2	*182 288.4*	97			
$3d^6(^5D)4p$	$^6P^\circ$	7/2	*184 099.1*	84		11	$(^5D)\ ^4D^\circ$
		5/2	*185 505.3*	88		9	
		3/2	*186 441.4*	94		4	
$3d^6(^5D)4p$	$^4D^\circ$	7/2	*185 890.0*	83		13	$(^5D)\ ^6P^\circ$
		5/2	*186 516.1*	84		10	$(^5D)\ ^6P^\circ$
		3/2	*186 957.2*	89		4	$(^5D)\ ^6P^\circ$
		1/2	*187 225.7*	93		3	$(^5D)\ ^6F^\circ$
$3d^6(^3F1)4s$	^2F	7/2	185 967.0	77		21	$(^3F2)\ ^2F$
		5/2	185 997.0	80		20	
$3d^6(^5D)4p$	$^4F^\circ$	9/2	*186 470.1*	94		4	$(^5D)\ ^6F^\circ$
		7/2	*187 570.3*	95		3	$(^5D)\ ^6F^\circ$
		5/2	*188 320.1*	96		2	$(^5D)\ ^6F^\circ$
		3/2	*188 824.2*	97			
$3d^6(^5D)4p$	$^4P^\circ$	5/2	*190 830.3*	97			
		3/2	*191 618.2*	98			
		1/2	*192 033.5*	98			
$3d^6(^1G1)4s$	^2G	9/2	190 864.7	66		33	$(^1G2)\ ^2G$
		7/2	190 932.8	66		33	
$3d^6(^3P2)4p$	$^4S^\circ$	3/2	*205 114.1*	44		19	$(^3P2)\ ^4P^\circ$
$3d^6(^3H)4p$	$^4G^\circ$	11/2	*206 005.3*	68		21	$(^3F2)\ ^4G^\circ$
		9/2	*206 340.9*	51		28	
		7/2	*206 645.7*	45		34	
		5/2	*206 847.0*	42		38	
$3d^6(^3P2)4p$		5/2	*206 523.2*	27	$^4P^\circ$	23	$(^3P1)\ ^4P^\circ$
$3d^6(^3H)4p$	$^4I^\circ$	11/2	*206 740.7*	51		35	$(^3H)\ ^4H^\circ$
		13/2	*206 754.5*	48		38	$(^3H)\ ^4H^\circ$
		9/2	*208 046.8*	46		19	$(^3H)\ ^4H^\circ$
		15/2	*208 149.5*	99			
$3d^6(^3H)4p$		9/2	*206 865.5*	43	$^4I^\circ$	37	$^4H^\circ$
$3d^6(^3H)4p$	$^4H^\circ$	7/2	*207 136.0*	47		20	$(^3H)\ ^2G^\circ$
		11/2	*208 595.3*	45		44	$(^3H)\ ^4I^\circ$
		13/2	*208 631.3*	51		40	$(^3H)\ ^4I^\circ$
$3d^6(^3P2)4p$	$^4P^\circ$	1/2	*207 846.2*	48		41	$(^3P1)\ ^4P^\circ$
$3d^6(^3P2)4p$	$^2D^\circ$	5/2	*208 009.1*	32		20	$(^3P1)\ ^2D^\circ$
$3d^6(^3P2)4p$		7/2	*208 330.7*	22	$^4D^\circ$	19	$(^3F2)\ ^4F^\circ$
$3d^6(^3P2)4p$		3/2	*208 461.0*	22	$^4S^\circ$	13	$^4P^\circ$

(Continued)

Ni IV—Continued

Configuration	Term	J	Level (cm^{-1})	Leading percentages			
$3d^6(^3\mathrm{F}2)4p$	$^4\mathrm{F}^\circ$	5/2	*208 605.7*	54		18	$(^3\mathrm{F}1)\ ^4\mathrm{F}^\circ$
		3/2	*208 766.9*	63		20	
		9/2	*208 933.6*	49		18	
$3d^6(^3\mathrm{P}2)4p$	$^4\mathrm{D}^\circ$	7/2	*208 912.1*	39		21	$(^3\mathrm{P}1)\ ^4\mathrm{D}^\circ$
		1/2	*210 313.3*	59		28	$(^3\mathrm{P}1)\ ^4\mathrm{D}^\circ$
$3d^6(^3\mathrm{H})4p$		9/2	*209 131.9*	37	$^2\mathrm{G}^\circ$	23	$(^3\mathrm{H})\ ^4\mathrm{H}^\circ$
$3d^6(^3\mathrm{H})4p$		7/2	*209 391.4*	34	$^2\mathrm{G}^\circ$	16	$(^3\mathrm{F}2)\ ^4\mathrm{F}^\circ$
$3d^6(^3\mathrm{P}2)4p$		3/2	*209 985.1*	18	$^2\mathrm{D}^\circ$	16	$(^3\mathrm{P}2)\ ^4\mathrm{P}^\circ$
$3d^6(^3\mathrm{F}2)4p$	$^4\mathrm{D}^\circ$	7/2	*210 121.8*	49		10	$(^3\mathrm{F}1)\ ^4\mathrm{D}^\circ$
		5/2	*210 943.6*	37		20	$(^3\mathrm{P}2)\ ^4\mathrm{D}^\circ$
		3/2	*211 246.8*	49		10	$(^3\mathrm{D})\ ^4\mathrm{D}^\circ$
		1/2	*211 351.7*	67		12	$(^3\mathrm{F}1)\ ^4\mathrm{D}^\circ$
$3d^6(^3\mathrm{H})4p$	$^2\mathrm{I}^\circ$	13/2	*210 177.0*	86		11	$(^3\mathrm{H})\ ^4\mathrm{I}^\circ$
		11/2	*210 987.5*	88		4	
$3d^6(^3\mathrm{F}2)4p$		5/2	*210 590.6*	24	$^4\mathrm{D}^\circ$	23	$(^3\mathrm{P}2)\ ^4\mathrm{D}^\circ$
$3d^6(^3\mathrm{P}2)4p$		3/2	*211 027.6*	20	$^4\mathrm{D}^\circ$	18	$(^3\mathrm{F}2)\ ^4\mathrm{D}^\circ$
$3d^6(^3\mathrm{F}2)4p$	$^4\mathrm{G}^\circ$	9/2	*212 150.9*	37		20	$(^3\mathrm{H})\ ^4\mathrm{G}^\circ$
		11/2	*212 207.0*	53		17	
$3d^6(^3\mathrm{F}2)4p$		5/2	*212 275.9*	24	$^2\mathrm{F}^\circ$	19	$(^3\mathrm{F}2)\ ^4\mathrm{G}^\circ$
$3d^6(^3\mathrm{F}2)4p$		7/2	*212 281.8*	26	$^4\mathrm{G}^\circ$	17	$(^3\mathrm{H})\ ^4\mathrm{G}^\circ$
$3d^6(^3\mathrm{F}2)4p$		7/2	*213 131.1*	24	$^2\mathrm{F}^\circ$	14	$(^3\mathrm{H})\ ^4\mathrm{G}^\circ$
$3d^6(^3\mathrm{G})4p$	$^4\mathrm{F}^\circ$	9/2	*213 408.6*	57		25	$(^3\mathrm{G})\ ^4\mathrm{G}^\circ$
		5/2	*215 516.8*	43		30	$(^3\mathrm{G})\ ^4\mathrm{G}^\circ$
		3/2	*215 541.3*	68		15	$(^3\mathrm{D})\ ^4\mathrm{F}^\circ$
$3d^6(^3\mathrm{H})4p$		5/2	*213 412.1*	28	$^4\mathrm{G}^\circ$	18	$(^3\mathrm{F}2)\ ^4\mathrm{G}^\circ$
$3d^6(^3\mathrm{G})4p$		11/2	*213 606.9*	31	$^4\mathrm{G}^\circ$	30	$^2\mathrm{H}^\circ$
$3d^6(^3\mathrm{P}2)4p$		1/2	*213 640.2*	31	$^2\mathrm{P}^\circ$	23	$(^3\mathrm{P}1)\ ^2\mathrm{P}^\circ$
$3d^6(^3\mathrm{F}2)4p$		9/2	*213 739.7*	33	$^2\mathrm{G}^\circ$	20	$(^3\mathrm{G})\ ^2\mathrm{H}^\circ$
$3d^6(^3\mathrm{P}2)4p$	$^2\mathrm{P}^\circ$	3/2	*214 055.8*	50		33	$(^3\mathrm{P}1)\ ^2\mathrm{P}^\circ$
$3d^6(^3\mathrm{G})4p$	$^4\mathrm{G}^\circ$	7/2	*214 101.0*	41		37	$(^3\mathrm{G})\ ^4\mathrm{F}^\circ$
		11/2	*214 316.8*	47		21	$(^3\mathrm{H})\ ^2\mathrm{H}^\circ$
		5/2	*214 718.1*	46		27	$(^3\mathrm{G})\ ^4\mathrm{F}^\circ$
		9/2	*214 910.5*	40		16	$(^3\mathrm{H})\ ^2\mathrm{H}^\circ$
$3d^6(^3\mathrm{F}2)4p$		9/2	*214 587.4*	21	$^2\mathrm{G}^\circ$	16	$(^3\mathrm{H})\ ^2\mathrm{H}^\circ$
$3d^6(^3\mathrm{F}2)4p$	$^2\mathrm{G}^\circ$	7/2	*214 622.8*	60		14	$(^3\mathrm{F}1)\ ^2\mathrm{G}^\circ$
$3d^6(^3\mathrm{G})4p$		7/2	*215 292.9*	39	$^4\mathrm{G}^\circ$	33	$(^3\mathrm{G})\ ^4\mathrm{F}^\circ$

Ni IV—Continued

Configuration	Term	J	Level (cm^{-1})	Leading percentages			
$3d^6(^3G)4p$	$^4H^\circ$	13/2	*215 506.6*	86		11	$(^3H)\ ^4H^\circ$
		11/2	*215 684.5*	73		10	$(^3H)\ ^2H^\circ$
		7/2	*215 724.5*	73		12	$(^3H)\ ^4H^\circ$
		9/2	*215 736.1*	73		9	$(^3H)\ ^4H^\circ$
$3d^6(^3P2)4p$	$^2S^\circ$	1/2	*215 531.0*	47		19	$(^3P1)\ ^2S^\circ$
$3d^6(^3F2)4p$	$^2D^\circ$	5/2	*217 414.3*	67		10	$(^3P2)\ ^2D^\circ$
		3/2	*217 939.7*	64		15	
$3d^6(^3H)4p$	$^2H^\circ$	11/2	*218 860.6*	48		42	$(^3G)\ ^2H^\circ$
$3d^6(^3G)4p$	$^2F^\circ$	5/2	*219 553.4*	47		19	$(^3D)\ ^2F^\circ$
		7/2	*219 896.0*	52		16	
$3d^6(^3G)4p$	$^2H^\circ$	9/2	*219 765.0*	43		39	$(^3H)\ ^2H^\circ$
$3d^6(^1I)4p$	$^2K^\circ$	13/2	*220 762.6*	97		2	$(^1I)\ ^2I^\circ$
		15/2	*222 202.8*	99			
$3d^6(^3G)4p$	$^2G^\circ$	9/2	*221 991.0*	71		18	$(^3H)\ ^2G^\circ$
		7/2	*222 029.3*	71		12	
$3d^6(^3D)4p$	$^4P^\circ$	5/2	*222 333.3*	87		3	$(^3P2)\ ^4P^\circ$
		3/2	*222 662.9*	74		8	$(^3D)\ ^2P^\circ$
		1/2	*223 225.4*	66		13	$(^3D)\ ^4D^\circ$
$3d^6(^1G2)4p$	$^2H^\circ$	9/2	*222 705.5*	44		17	$(^1I)\ ^2H^\circ$
		11/2	*225 758.5*	36		27	$(^1G1)\ ^2H^\circ$
$3d^6(^1I)4p$	$^2H^\circ$	11/2	*222 993.3*	53		25	$(^1G2)\ ^2H^\circ$
		9/2	*225 903.7*	61		15	$(^1G1)\ ^2H^\circ$
$3d^6(^3D)4p$		1/2	*223 785.4*	33	$^4D^\circ$	26	$(^3D)\ ^4P^\circ$
$3d^6(^3D)4p$	$^4F^\circ$	5/2	*224 021.3*	58		17	$(^3G)\ ^4F^\circ$
		7/2	*224 824.9*	47		12	$(^1G2)\ ^2G^\circ$
		9/2	*225 096.4*	73		11	$(^3G)\ ^4F^\circ$
$3d^6(^3D)4p$		7/2	*224 075.3*	24	$^4F^\circ$	10	$(^1G2)\ ^2F^\circ$
$3d^6(^3D)4p$		3/2	*224 174.0*	30	$^4D^\circ$	30	$(^3D)\ ^2P^\circ$
$3d^6(^1G2)4p$	$^2G^\circ$	7/2	*224 463.6*	44		18	$(^1G1)\ ^2G^\circ$
		9/2	*224 761.5*	50		21	
$3d^6(^3D)4p$	$^4D^\circ$	5/2	*224 645.7*	68		14	$(^3D)\ ^4F^\circ$
		7/2	*225 136.9*	54		13	$(^1G2)\ ^2F^\circ$
$3d^6(^3D)4p$	$^2P^\circ$	3/2	*224 936.9*	41		41	$(^3D)\ ^4D^\circ$
		1/2	*224 996.5*	40		36	
$3d^6(^1G2)4p$		5/2	*225 431.6*	34	$^2F^\circ$	15	$(^1G1)\ ^2F^\circ$
$3d^6(^1I)4p$	$^2I^\circ$	13/2	*226 685.0*	98		2	$(^1I)\ ^2K^\circ$
		11/2	*226 702.5*	82		10	$(^1I)\ ^2H^\circ$
$3d^6(^3D)4p$	$^2D^\circ$	3/2	*227 181.1*	76		7	$(^1D2)\ ^2P^\circ$
		5/2	*227 549.4*	86		8	$(^1F)\ ^2D^\circ$

(Continued)

Ni IV—Continued

Configuration	Term	J	Level (cm^{-1})	Leading percentages			
$3d^6(^1D2)4p$	$^2P^\circ$	3/2	*228 214.2*	34		23	$(^1S2)\ ^2P^\circ$
		1/2	*232 897.0*	40		27	
$3d^6(^3D)4p$	$^2F^\circ$	7/2	*228 518.5*	62		14	$(^1D2)\ ^2F^\circ$
		5/2	*229 269.5*	54		20	
$3d^6(^1S2)4p$		1/2	*229 297.5*	32	$^2P^\circ$	26	$(^1D2)\ ^2P^\circ$
$3d^6(^1D2)4p$		5/2	*231 091.8*	31	$^2D^\circ$	17	$(^1D2)\ ^2F^\circ$
$3d^6(^1D2)4p$	$^2D^\circ$	3/2	*231 742.0*	62		12	$(^1D1)\ ^2D^\circ$
$3d^6(^1D2)4p$		5/2	*232 272.7*	30	$^2D^\circ$	25	$^2F^\circ$
$3d^6(^1D2)4p$	$^2F^\circ$	7/2	*232 539.5*	51		13	$(^1D1)\ ^2F^\circ$
$3d^6(^1S2)4p$		3/2	*234 019.5*	37	$^2P^\circ$	24	$(^1D2)\ ^2P^\circ$
$3d^6(^1F)4p$	$^2G^\circ$	7/2	*237 480.3*	90		2	$(^3G)\ ^2G^\circ$
		9/2	*238 957.8*	93		2	$(^1G2)\ ^2G^\circ$
$3d^6(^1F)4p$	$^2D^\circ$	5/2	*239 009.7*	63		15	$(^1D2)\ ^2D^\circ$
		3/2	*240 094.2*	70		9	
$3d^6(^3P1)4p$	$^4D^\circ$	1/2	*241 172.4*	37		37	$(^3F1)\ ^4D^\circ$
$3d^6(^3F1)4p$	$^4D^\circ$	3/2	*241 540.0*	35		32	$(^3P1)\ ^4D^\circ$
		5/2	*241 817.9*	43		29	
		7/2	*241 939.0*	50		23	
$3d^6(^1F)4p$	$^2F^\circ$	5/2	*243 520.1*	83		4	$(^1G2)\ ^2F^\circ$
		7/2	*243 542.0*	82		4	
$3d^6(^3F1)4p$	$^4G^\circ$	7/2	*247 731.7*	77		17	$(^3F2)\ ^4G^\circ$
		9/2	*248 218.0*	74		18	
		11/2	*248 855.4*	78		20	
$3d^6(^3P1)4p$	$^4S^\circ$	3/2	*249 130.0*	74		24	$(^3P2)\ ^4S^\circ$
$3d^6(^3F1)4p$	$^2D^\circ$	3/2	*250 951.0*	49		24	$(^3P1)\ ^2D^\circ$
		5/2	*251 694.3*	46		23	
$3d^6(^3F1)4p$	$^2G^\circ$	9/2	*251 773.6*	66		17	$(^3F2)\ ^2G^\circ$
		7/2	*252 523.0*	69		16	
$3d^6(^3P1)4p$	$^4P^\circ$	5/2	*252 354.0*	37		29	$(^3P2)\ ^4P^\circ$
$3d^6(^3F1)4p$		3/2	*252 975.4*	32	$^4D^\circ$	23	$(^3P1)\ ^4D^\circ$
$3d^6(^3F1)4p$		5/2	*253 326.2*	26	$^4F^\circ$	13	$(^3P1)\ ^4D^\circ$
$3d^6(^3F1)4p$	$^4F^\circ$	3/2	*253 579.3*	57		18	$(^3F2)\ ^4F^\circ$
		7/2	*253 639.1*	40		16	$(^3P1)\ ^4D^\circ$
		5/2	*254 123.7*	40		17	$(^3F1)\ ^4D^\circ$
		9/2	*254 300.0*	66		24	$(^3F2)\ ^4F^\circ$
$3d^6(^3F1)4p$		7/2	*254 663.4*	26	$^4F^\circ$	22	$(^3P1)\ ^4D^\circ$
$3d^6(^3F1)4p$		3/2	*255 064.0*	27	$^2D^\circ$	23	$(^3P1)\ ^2D^\circ$

Ni IV—Continued

Configuration	Term	J	Level (cm^{-1})	Leading percentages			
$3d^6(^3F1)4p$	$^2F°$	7/2	*257 018.0*	54		23	$(^3F2)$ $^2F°$
		5/2	*257 406.4*	66		25	
$3d^6(^1G1)4p$	$^2H°$	9/2	*258 672.8*	59		33	$(^1G2)$ $^2H°$
		11/2	*260 065.1*	62		35	
$3d^6(^1G1)4p$		7/2	*260 355.5*	31	$^2F°$	23	$(^1G1)$ $^2G°$
$3d^6(^1G1)4p$	$^2F°$	5/2	*261 226.3*	55		26	$(^1G2)$ $^2F°$
$3d^6(^1G1)4p$	$^2G°$	9/2	*262 275.5*	64		27	$(^1G2)$ $^2G°$
		7/2	*262 538.7*	44		19	
$3d^6(^1D1)4p$	$^2D°$	3/2	*282 179.5*	79		17	$(^1D2)$ $^2D°$
		5/2	*282 645.2*	79		17	
Ni V ($^5D_4$)	***Limit***		**443 000**				

Ni V

$Z=28$

Cr I isoelectronic sequence

Ground state: $1s^22s^22p^63s^23p^63d^6\ ^5D_4$

Ionization energy $= 613\ 500\pm500$ cm^{-1} (76.06±0.06 eV)

The $3d^6-3d^54p$ transition array between 300 and 425 Å has been observed and analysed by Raassen, van Kleef, and Metsch (1976). They have found all but the high ^{1}S level of $3d^6$ and 177 out of 214 levels of $3d^54p$. The uncertainty of the level values is about ±5 cm^{-1}. They give the percentage compositions for these two configurations.

Raassen and van Kleef (1977) found 37 more levels of $3d^54p$ and observed and analysed the $3d^54s-3d^54p$ transition array between 990 and 1400 Å. They also found one term in each of the configurations $3d^55p$, $3d^54f$ and $3d^55f$. They give the percentage compositions for the $3d^54s$ configuration.

Raassen and van Kleef derived the ionization energy from the two-member $3d^5np$ and nf series. We have confirmed their value to within 300 cm^{-1} by recalculating the $3d^5(^6S)nf\ ^5F$ series limit and assuming a value for $n^*(5f)-n^*(4f)$ of 0.9952 taken from Zn II. Accordingly, we reduced their uncertainty estimate of ±3000 cm^{-1} to ±500 cm^{-1}.

References

Raassen, A. J. J., van Kleef, Th. A. M., and Metsch, B. C. (1976), Physica **84C**, 133.

Raassen, A. J. J., and van Kleef, Th. A. M. (1977), Physica **85C**, 180.

Ni V

Configuration	Term	J	Level (cm^{-1})	Leading percentages		
$3d^6$	^{5}D	4	0.0	100		
		3	889.7	100		
		2	1 489.9	100		
		1	1 871.5	100		
		0	2 057.6	100		
$3d^6$	^{3}P2	2	26 153.0	62	38	^{3}P1
		1	28 697.6	63	37	
		0	29 640.0	62	36	
$3d^6$	^{3}H	6	27 111.2	99	1	^{1}I
		5	27 578.2	97	3	^{3}G
		4	27 858.8	88	5	^{3}F2
$3d^6$	^{3}F2	4	29 123.7	68	20	^{3}F1
		3	29 570.8	75	20	
		2	29 899.2	80	20	
$3d^6$	^{3}G	5	33 256.5	97	3	^{3}H
		4	34 061.7	93	4	^{3}F2
		3	34 416.4	96	4	^{3}F2
$3d^6$	^{1}I	6	41 252.2	99	1	^{3}H
$3d^6$	^{3}D	2	41 626.9	97	2	^{1}D2
		1	41 701.1	100		
		3	41 920.2	100		
$3d^6$	^{1}G2	4	42 208.1	65	33	^{1}G1
$3d^6$	^{1}S2	0	47 699.7	76	22	^{1}S1
$3d^6$	^{1}D2	2	48 607.0	76	21	^{1}D1
$3d^6$	^{1}F	3	57 924.1	98	1	^{3}F1

Ni v—Continued

Configuration	Term	J	Level (cm^{-1})	Leading percentages		
$3d^6$	^{3}P1	0	66 737.8	64	36	^{3}P2
		1	67 547.9	63	37	
		2	69 156.1	62	38	
$3d^6$	^{3}F1	2	68 632.1	80	20	^{3}F2
		4	68 718.7	77	22	
		3	68 854.7	78	20	
$3d^6$	^{1}G1	4	77 899.5	65	34	^{1}G2
$3d^6$	^{1}D1	2	104 420.5	78	22	^{1}D2
$3d^5(^6S)4s$	^{7}S	3	164 525.9	100		
$3d^5(^6S)4s$	^{5}S	2	178 019.8	100		
$3d^5(^4G)4s$	^{5}G	6	208 046.4	100		
		5	208 131.0	100		
		2	208 151.5	99	1	(^{2}F1) ^{3}F
		4	208 163.7	100		
		3	208 164.6	100		
$3d^5(^4P)4s$	^{5}P	3	212 095.8	90	9	(^{4}D) ^{5}D
		2	212 253.4	91	8	
		1	212 455.7	96	4	
$3d^5(^4D)4s$	^{5}D	4	216 189.9	99	1	(^{4}G) ^{3}G
		0	216 305.7	98	2	(^{4}P) ^{3}P
		1	216 434.7	94	4	(^{4}P) ^{5}P
		2	216 590.5	91	8	(^{4}P) ^{5}P
		3	216 596.0	90	9	(^{4}P) ^{5}P
$3d^5(^4G)4s$	^{3}G	5	217 048.7	100		
		3	217 101.0	99	1	(^{2}F1) ^{1}F
		4	217 129.1	99	1	(^{4}D) ^{5}D
$3d^5(^4P)4s$	^{3}P	2	221 087.6	88	9	(^{4}D) ^{3}D
		1	221 429.0	92	6	
$3d^5(^4D)4s$	^{3}D	3	225 200.7	99		
		1	225 545.1	94	6	(^{4}P) ^{3}P
		2	225 616.5	90	9	(^{4}P) ^{3}P
$3d^5(^2I)4s$	^{3}I	6	229 408.8	99	1	(^{2}H) ^{3}H
		5	229 413.0	99	1	(^{2}H) ^{3}H
		7	229 440.6	100		
$3d^5(^2D3)4s$	^{3}D	3	232 545.9	55	19	(^{2}F1) ^{3}F
		2	232 655.6	50	17	(^{2}F1) ^{3}F
		1	232 910.8	50	34	(^{4}F) ^{5}F
$3d^5(^2I)4s$	^{1}I	6	233 839.2	98	2	(^{2}H) ^{3}H
$3d^5(^4F)4s$	^{5}F	5	234 082.1	98	2	(^{2}G2) ^{3}G
		4	234 125.4	92	5	(^{2}F1) ^{3}F
		3	234 275.2	89	8	(^{2}F1) ^{3}F
		2	234 412.7	70	23	(^{2}F1) ^{3}F
		1	235 116.5	65	27	(^{2}D3) ^{3}D

(Continued)

Ni v—Continued

Configuration	Term	J	Level (cm^{-1})	Leading percentages		
$3d^5(^2F1)4s$	^3F	4	235 420.6	92	6	$(^4F)\ ^5F$
		2	235 736.5	36	23	$(^2D3)\ ^3D$
		3	236 454.1	69	16	$(^2D3)\ ^3D$
$3d^5(^2D3)4s$	^1D	2	239 107.7	55	23	$(^2F1)\ ^3F$
$3d^5(^2F1)4s$	^1F	3	240 193.8	83	6	$(^2G2)\ ^3G$
$3d^5(^2H)4s$	^3H	4	240 959.6	71	24	$(^2G2)\ ^3G$
		5	241 082.2	69	28	$(^2G2)\ ^3G$
		6	241 773.6	97	2	$(^2I)\ ^1I$
$3d^5(^2G2)4s$	^3G	3	242 290.4	79	12	$(^4F)\ ^3F$
		4	242 504.3	45	28	$(^4F)\ ^3F$
		5	242 862.6	66	30	$(^2H)\ ^3H$
$3d^5(^6S)4p$	$^7P^\circ$	2	*242 837.0*	99	1	$(^6S)\ ^5P^\circ$
		3	*243 608.5*	98	2	$(^6S)\ ^5P^\circ$
		4	*244 900.5*	100		
$3d^5(^4F)4s$	^3F	2	243 266.2	93	3	$(^2F2)\ ^3F$
		4	243 331.5	62	30	$(^2G2)\ ^3G$
		3	243 370.5	83	14	$(^2G2)\ ^3G$
$3d^5(^2H)4s$	^1H	5	246 240.9	96	4	$(^2G2)\ ^3G$
$3d^5(^2G2)4s$	^1G	4	247 049.1	63	33	$(^2F2)\ ^3F$
$3d^5(^2F2)4s$	^3F	3	247 104.9	97	1	$(^4F)\ ^5F$
		2	247 165.0	96	3	$(^4F)\ ^3F$
		4	247 281.8	63	29	$(^2G2)\ ^1G$
$3d^5(^2F2)4s$	^1F	3	251 654.9	96	2	$(^4F)\ ^3F$
$3d^5(^6S)4p$	$^5P^\circ$	3	*253 862.7*	96	2	$(^6S)\ ^7P^\circ$
		2	*254 495.6*	97	1	$(^4D)\ ^5P^\circ$
		1	*254 885.0*	98	1	$(^4D)\ ^5P^\circ$
$3d^5(^2S)4s$	^3S	1	253 905.2	100		
$3d^5(^2D2)4s$	^3D	1	263 700.9	100		
		2	263 735.7	99		
		3	263 805.8	99		
$3d^5(^2D2)4s$	^1D	2	268 273.9	99		
$3d^5(^2G1)4s$	^3G	5	274 695.4	100		
		4	274 738.6	100		
		3	274 773.5	100		
$3d^5(^2G1)4s$	^1G	4	279 199.5	100		
$3d^5(^4G)4p$	$^5G^\circ$	2	*284 215.5*	92	4	$(^4G)\ ^3F^\circ$
		3	*284 249.0*	84	9	$(^4G)\ ^5H^\circ$
		4	*284 308.9*	81	14	$(^4G)\ ^5H^\circ$
		5	*284 402.5*	80	14	$(^4G)\ ^5H^\circ$
		6	*284 579.5*	84	11	$(^4G)\ ^5H^\circ$

Ni v—Continued

Configuration	Term	J	Level (cm^{-1})	Leading percentages			
$3d^5(^4G)4p$	$^5H^\circ$	3	286 293.6	88		8	$(^4G)\ ^5G^\circ$
		4	286 706.6	82		12	$(^4G)\ ^5G^\circ$
		5	287 127.2	75		11	$(^4G)\ ^5F^\circ$
		6	287 645.9	86		13	$(^4G)\ ^5G^\circ$
		7	288 021.6	100			
$3d^5(^4P)4p$	$^5D^\circ$	1	287 755.5	63		20	$(^4D)\ ^5D^\circ$
		2	287 782.1	45		18	$(^4G)\ ^5F^\circ$
		0	290 262.0	64		19	$(^4D)\ ^5D^\circ$
$3d^5(^4G)4p$	$^5F^\circ$	5	287 906.9	75		8	$(^4G)\ ^5H^\circ$
		4	288 161.6	71		10	$(^4D)\ ^5F^\circ$
		1	289 163.0	71		11	$(^4D)\ ^5F^\circ$
		2	289 247.1	52		13	$(^4P)\ ^5D^\circ$
		3	289 298.0	40		32	$(^4P)\ ^5D^\circ$
$3d^5(^4G)4p$		3	287 960.0	39	$^5F^\circ$	25	$(^4P)\ ^5D^\circ$
$3d^5(^4P)4p$		2	288 877.9	38	$^5S^\circ$	9	$(^4G)\ ^5F^\circ$
$3d^5(^4P)4p$	$^5P^\circ$	3	290 757.0	44		32	$(^4D)\ ^5P^\circ$
		2	291 390.0	58		26	
		1	291 541.7	71		15	
$3d^5(^4G)4p$	$^3F^\circ$	2	291 097.7	84		4	$(^4G)\ ^5G^\circ$
		3	291 328.5	81		4	$(^4F)\ ^3F^\circ$
		4	291 554.6	88		4	$(^4F)\ ^3F^\circ$
$3d^5(^4G)4p$	$^3H^\circ$	6	291 891.4	93		3	$(^4G)\ ^5H^\circ$
		5	292 353.4	94		2	$(^4G)\ ^5H^\circ$
		4	292 631.0	95		2	$(^2I)\ ^3H^\circ$
$3d^5(^4P)4p$	$^3P^\circ$	2	292 983.0	52		18	$(^4D)\ ^3P^\circ$
		1	293 420.0	53		18	
		0	293 867.0	65		20	
$3d^5(^4D)4p$	$^5F^\circ$	1	293 833.8	73		17	$(^4G)\ ^5F^\circ$
		2	294 086.0	76		15	
		3	294 443.3	76		11	
		4	294 939.6	79		8	
		5	295 444.3	91		6	
$3d^5(^4D)4p$	$^5D^\circ$	3	296 574.0	51		18	$(^4P)\ ^5D^\circ$
		4	296 919.3	68		22	
		2	297 013.9	54		17	
		1	297 417.9	51		17	
		0	298 060.0	63		22	
$3d^5(^4G)4p$	$^3G^\circ$	3	296 847.1	90		2	$(^4F)\ ^3G^\circ$
		4	296 897.0	91		2	$(^4F)\ ^3G^\circ$
		5	296 932.9	92		2	$(^4G)\ ^5F^\circ$
$3d^5(^4P)4p$	$^3D^\circ$	3	297 418.1	60		14	$(^4D)\ ^5P^\circ$
		2	297 842.5	57		18	
$3d^5(^4D)4p$	$^5P^\circ$	1	297 982.8	40		36	$(^4P)\ ^3D^\circ$
		2	299 045.6	39		19	$(^4P)\ ^5P^\circ$
$3d^5(^4P)4p$		1	298 600.6	30	$^3D^\circ$	29	$(^4D)\ ^5P^\circ$

(Continued)

Ni V—Continued

Configuration	Term	J	Level (cm^{-1})	Leading percentages			
$3d^5(^4D)4p$	$^3D°$	3	*298 972.3*	43		15	$(^4P)\ ^5P°$
		2	*300 224.9*	74		8	$(^4F)\ ^3D°$
		1	*300 563.3*	65		17	$(^4P)\ ^3D°$
$3d^5(^4D)4p$		3	*300 201.0*	36	$^3D°$	31	$(^4D)\ ^5P°$
$3d^5(^4D)4p$	$^3F°$	4	*300 918.1*	84		5	$(^2G2)\ ^3F°$
		3	*301 470.2*	76		8	$(^4P)\ ^3D°$
		2	*301 553.0*	80		7	$(^4P)\ ^3D°$
$3d^5(^4P)4p$	$^3S°$	1	*303 249.5*	91		3	$(^4D)\ ^3P°$
$3d^5(^4D)4p$	$^3P°$	0	*305 386.9*	72		22	$(^4P)\ ^3P°$
		1	*305 838.1*	66		21	
		2	*306 377.8*	63		25	
$3d^5(^2I)4p$	$^3K°$	6	*305 590.8*	69		26	$(^2I)\ ^3I°$
		7	*305 996.3*	56		34	$(^2I)\ ^3I°$
		8	*308 138.8*	100			
$3d^5(^2I)4p$	$^3I°$	5	*306 049.0*	64		19	$(^2I)\ ^1H°$
		6	*307 399.7*	53		26	$(^2I)\ ^3K°$
		7	*308 317.3*	62		38	$(^2I)\ ^3K°$
$3d^5(^2D1)4s$	^3D	3	306 962.9	77		23	$(^2D3)\ ^3D$
		2	307 025.2	77		23	
		1	307 105.1	76		23	
$3d^5(^2D3)4p$	$^3F°$	2	*307 731.1*	21		25	$(^2F1)\ ^3F°$
		3	*308 592.0*	37		23	
		4	*310 212.6*	39		25	
$3d^5(^2I)4p$	$^1H°$	5	*308 804.1*	52		29	$(^2I)\ ^3I°$
$3d^5(^2D3)4p$		2	*308 943.0*	29	$^3F°$	21	$^1D°$
$3d^5(^2I)4p$	$^3H°$	6	*309 264.0*	73		17	$(^2I)\ ^3I°$
		5	*309 919.5*	78		11	$(^2I)\ ^1H°$
		4	*309 952.5*	79		5	$(^2G2)\ ^3H°$
$3d^5(^2I)4p$	$^1K°$	7	*309 743.6*	91		5	$(^2I)\ ^3K°$
$3d^5(^2D1)4s$	^1D	2	311 470.3	77		23	$(^2D3)\ ^1D$
$3d^5(^2D3)4p$	$^3P°$	2	*311 966.5*	40		15	$(^2F1)\ ^3D°$
		0	*313 577.3*	66		20	$(^2D1)\ ^3P°$
$3d^5(^2F1)4p$	$^1G°$	4	*312 008.3*	43		12	$(^2H)\ ^1G°$
$3d^5(^2D3)4p$		1	*312 291.0*	31	$^3P°$	28	$(^2D3)\ ^3D°$
$3d^5(^2F1)4p$	$^3G°$	3	*312 463.3*	43		20	$(^2F1)\ ^3F°$
		5	*314 702.2*	70		23	$(^4F)\ ^5G°$
$3d^5(^4F)4p$	$^5G°$	2	*312 778.2*	78		9	$(^2F1)\ ^3F°$
		3	*312 889.4*	43		19	$(^2D3)\ ^3D°$
		4	*313 281.3*	72		5	$(^4F)\ ^5F°$
		5	*313 464.7*	56		15	$(^2F1)\ ^3G°$
		6	*314 756.4*	53		42	$(^2I)\ ^1I°$

Ni v—Continued

Configuration	Term	J	Level (cm^{-1})	Leading percentages			
$3d^5(^2F1)4p$		3	*312 953.6*	33	$^3D^\circ$	31	$(^4F)\ ^5G^\circ$
$3d^5(^2D3)4p$		1	*313 679.0*	29	$^3D^\circ$	27	$(^2D3)\ ^3P^\circ$
$3d^5(^2D3)4p$	$^3D^\circ$	2	*313 686.6*	45		15	$(^2D1)\ ^3D^\circ$
$3d^5(^2F1)4p$		3	*313 919.8*	23	$^3D^\circ$	22	$(^2D3)\ ^3D^\circ$
$3d^5(^2F1)4p$		4	*314 208.8*	30	$^3F^\circ$	30	$(^2F1)\ ^3G^\circ$
$3d^5(^2I)4p$	$^1I^\circ$	6	*314 392.0*	44		38	$(^4F)\ ^5G^\circ$
$3d^5(^4F)4p$	$^5F^\circ$	3	*314 562.8*	61		25	$(^4F)\ ^5D^\circ$
		4	*314 599.2*	39		21	$(^4F)\ ^5D^\circ$
		2	*314 834.7*	42		23	$(^4F)\ ^5D^\circ$
		1	*315 152.8*	74		9	$(^4F)\ ^5D^\circ$
		5	*315 168.2*	65		9	$(^4F)\ ^5G^\circ$
$3d^5(^2F1)4p$	$^3D^\circ$	1	*315 300.7*	51		19	$(^2D3)\ ^1P^\circ$
$3d^5(^2F1)4p$		3	*315 326.2*	25	$^3G^\circ$	14	$(^2D3)\ ^1F^\circ$
$3d^5(^2F1)4p$		2	*315 366.1*	32	$^3D^\circ$	18	$(^4F)\ ^5F^\circ$
$3d^5(^2G2)4p$	$^3H^\circ$	4	*315 370.1*	31		29	$(^2H)\ ^3H^\circ$
		5	*323 908.6*	36		28	
		6	*325 148.4*	45		37	
$3d^5(^2H)4p$	$^3H^\circ$	5	*315 990.5*	30		36	$(^2G2)\ ^3H^\circ$
		6	*317 327.3*	39		35	
		4	*323 926.3*	44		36	
$3d^5(^2D3)4p$		4	*316 068.8*	20	$^3F^\circ$	18	$(^2F1)\ ^3G^\circ$
$3d^5(^2F1)4p$		2	*316 165.4*	31	$^3F^\circ$	17	$(^2F1)\ ^3D^\circ$
$3d^5(^2F1)4p$		3	*316 280.3*	19	$^3F^\circ$	14	$(^2D3)\ ^1F^\circ$
$3d^5(^2H)4p$	$^3G^\circ$	5	*316 726.6*	47		14	$(^2G2)\ ^3G^\circ$
$3d^5(^4F)4p$	$^5D^\circ$	4	*316 744.0*	44		24	$(^4F)\ ^5F^\circ$
		3	*317 232.0*	39		17	$(^4F)\ ^5F^\circ$
		0	*317 462.3*	88		7	$(^2D3)\ ^3P^\circ$
		1	*317 477.9*	76		9	$(^4F)\ ^5F^\circ$
		2	*317 517.5*	65		15	$(^4F)\ ^5F^\circ$
$3d^5(^2H)4p$		4	*316 887.8*	36	$^3G^\circ$	12	$(^2F2)\ ^3G^\circ$
$3d^5(^4F)4p$		3	*317 376.8*	22	$^5D^\circ$	20	$(^2H)\ ^3G^\circ$
$3d^5(^2D3)4p$	$^1P^\circ$	1	*319 073.4*	45		22	$(^2F1)\ ^3D^\circ$
$3d^5(^2H)4p$	$^3I^\circ$	5	*319 076.2*	85		6	$(^2H)\ ^3H^\circ$
		6	*319 860.4*	78		8	$(^2H)\ ^1I^\circ$
		7	*320 783.1*	95		2	$(^2I)\ ^3I^\circ$
$3d^5(^2G2)4p$		4	*319 138.7*	25	$^1G^\circ$	18	$(^2F1)\ ^1G^\circ$

(Continued)

Ni v—Continued

Configuration	Term	J	Level (cm^{-1})	Leading percentages			
$3d^5(^4\mathrm{F})4p$	$^3\mathrm{G}^\circ$	3	*319 620.2*	38		32	$(^2\mathrm{G2})\ ^3\mathrm{G}^\circ$
		5	*319 652.7*	65		15	
		4	*319 899.1*	57		15	
$3d^5(^2\mathrm{F1})4p$	$^1\mathrm{D}^\circ$	2	*319 926.5*	51		29	$(^2\mathrm{D3})\ ^1\mathrm{D}^\circ$
$3d^5(^2\mathrm{G2})4p$	$^3\mathrm{F}^\circ$	3	*320 513.8*	41		18	$(^4\mathrm{F})\ ^3\mathrm{G}^\circ$
		2	*321 018.3*	57		11	$(^4\mathrm{F})\ ^3\mathrm{F}^\circ$
		4	*321 056.4*	54		13	$(^2\mathrm{G2})\ ^1\mathrm{G}^\circ$
$3d^5(^2\mathrm{F1})4p$	$^1\mathrm{F}^\circ$	3	*321 081.9*	52		17	$(^2\mathrm{G2})\ ^3\mathrm{F}^\circ$
$3d^5(^2\mathrm{H})4p$	$^1\mathrm{I}^\circ$	6	*322 324.2*	76		9	$(^2\mathrm{H})\ ^3\mathrm{H}^\circ$
$3d^5(^4\mathrm{F})4p$	$^3\mathrm{D}^\circ$	2	*322 436.4*	54		11	$(^2\mathrm{F2})\ ^3\mathrm{F}^\circ$
		3	*322 617.6*	50		10	$(^2\mathrm{F2})\ ^3\mathrm{F}^\circ$
		1	*322 984.5*	79		7	$(^4\mathrm{D})\ ^3\mathrm{D}^\circ$
$3d^5(^4\mathrm{F})4p$	$^3\mathrm{F}^\circ$	4	*322 820.8*	60		28	$(^2\mathrm{F2})\ ^3\mathrm{F}^\circ$
		3	*323 532.2*	60		17	
		2	*323 853.1*	53		20	
$3d^5(^2\mathrm{G2})4p$	$^3\mathrm{G}^\circ$	5	*324 980.2*	25		21	$(^2\mathrm{F2})\ ^3\mathrm{G}^\circ$
		3	*325 211.9*	30		29	
		4	*325 222.9*	27		28	
$3d^5(^2\mathrm{F2})4p$		4	*325 558.6*	32	$^1\mathrm{G}^\circ$	13	$(^2\mathrm{G2})\ ^1\mathrm{G}^\circ$
$3d^5(^2\mathrm{F2})4p$	$^3\mathrm{F}^\circ$	2	*325 982.2*	26		23	$(^4\mathrm{F})\ ^3\mathrm{F}^\circ$
		3	*326 029.9*	48		13	
		4	*326 876.3*	44		14	
$3d^5(^2\mathrm{G2})4p$	$^1\mathrm{H}^\circ$	5	*326 337.1*	53		26	$(^2\mathrm{H})\ ^1\mathrm{H}^\circ$
$3d^5(^2\mathrm{G2})4p$	$^1\mathrm{F}^\circ$	3	*326 739.0*	58		6	$(^2\mathrm{G2})\ ^3\mathrm{G}^\circ$
$3d^5(^2\mathrm{F2})4p$	$^1\mathrm{D}^\circ$	2	*327 122.7*	60		24	$(^2\mathrm{F2})\ ^3\mathrm{F}^\circ$
$3d^5(^2\mathrm{H})4p$	$^1\mathrm{H}^\circ$	5	*327 356.6*	63		29	$(^2\mathrm{G2})\ ^1\mathrm{H}^\circ$
$3d^5(^2\mathrm{F2})4p$	$^3\mathrm{D}^\circ$	1	*329 462.3*	55		21	$(^2\mathrm{S})\ ^3\mathrm{P}^\circ$
		2	*329 776.3*	70		12	$(^2\mathrm{F1})\ ^3\mathrm{D}^\circ$
		3	*329 872.9*	47		12	$(^2\mathrm{F2})\ ^3\mathrm{G}^\circ$
$3d^5(^2\mathrm{F2})4p$	$^3\mathrm{G}^\circ$	3	*329 614.3*	27		33	$(^2\mathrm{H})\ ^3\mathrm{G}^\circ$
		4	*330 297.6*	47		37	
		5	*330 718.1*	58		31	
$3d^5(^2\mathrm{S})4p$	$^3\mathrm{P}^\circ$	0	*329 618.5*	83		13	$(^2\mathrm{D2})\ ^3\mathrm{P}^\circ$
		1	*330 370.7*	58		23	$(^2\mathrm{F2})\ ^3\mathrm{D}^\circ$
		2	*331 678.2*	74		16	$(^2\mathrm{D2})\ ^3\mathrm{P}^\circ$
$3d^5(^2\mathrm{H})4p$	$^1\mathrm{G}^\circ$	4	*332 995.6*	39		37	$(^2\mathrm{F2})\ ^1\mathrm{G}^\circ$
$3d^5(^2\mathrm{S})4p$	$^1\mathrm{P}^\circ$	1	*334 477.2*	70		19	$(^2\mathrm{D2})\ ^1\mathrm{P}^\circ$
$3d^5(^2\mathrm{F2})4p$	$^1\mathrm{F}^\circ$	3	*334 727.6*	88		4	$(^2\mathrm{F1})\ ^1\mathrm{F}^\circ$

Ni v—Continued

Configuration	Term	J	Level (cm^{-1})	Leading percentages		
$3d^5(^2D2)4p$	$^3F°$	2	342 894.6	67	21	$(^2D2)\ ^3D°$
		3	343 281.0	50	27	$(^2D2)\ ^3D°$
		4	344 911.2	92	5	$(^2G2)\ ^3F°$
$3d^5(^2D2)4p$	$^3D°$	1	343 478.2	89	2	$(^2D2)\ ^3P°$
		2	343 905.7	65	21	$(^2D2)\ ^3F°$
		3	344 805.3	60	33	$(^2D2)\ ^3F°$
$3d^5(^2D2)4p$	$^1F°$	3	345 936.1	72	12	$(^2G1)\ ^1F°$
$3d^5(^2D2)4p$	$^3P°$	2	346 912.4	72	16	$(^2S)\ ^3P°$
		0	346 920.2	85	14	
		1	346 959.5	76	15	
$3d^5(^2D2)4p$	$^1P°$	1	348 477.9	72	17	$(^2S)\ ^1P°$
$3d^5(^2D2)4p$	$^1D°$	2	349 546.0	87	5	$(^2F2)\ ^1D°$
$3d^5(^2G1)4p$	$^3H°$	4	353 071.6	47	31	$(^2G1)\ ^3F°$
		5	353 548.7	76	15	$(^2G1)\ ^3G°$
		6	354 989.6	98	2	$(^2I)\ ^3H°$
$3d^5(^2G1)4p$	$^3F°$	4	353 347.1	50	40	$(^2G1)\ ^3H°$
		3	353 944.1	55	36	$(^2G1)\ ^3G°$
		2	355 150.0	90	6	$(^2D1)\ ^3F°$
$3d^5(^2G1)4p$	$^3G°$	3	355 398.0	59	36	$(^2G1)\ ^3F°$
		4	355 765.2	80	9	$(^2G1)\ ^3F°$
		5	356 036.3	78	18	$(^2G1)\ ^3H°$
$3d^5(^2G1)4p$	$^1H°$	5	358 475.6	90	4	$(^2G1)\ ^3G°$
$3d^5(^2G1)4p$	$^1G°$	4	358 760.0	93	2	$(^2G1)\ ^3F°$
$3d^5(^2G1)4p$	$^1F°$	3	360 059.7	78	10	$(^2D2)\ ^1F°$
$3d^5(^2P)4p$	$^3P°$	0	368 440.5	75	20	$(^2D1)\ ^3P°$
		1	368 749.7	73	21	
		2	369 649.1	72	23	
$3d^5(^2P)4p$	$^3D°$	2	374 803.7	57	27	$(^2P)\ ^1D°$
		1	374 828.1	90	5	$(^2D1)\ ^3D°$
		3	376 471.6	89	7	$(^2D1)\ ^3D°$
$3d^5(^2P)4p$	$^1D°$	2	377 059.1	49	33	$(^2P)\ ^3D°$
$3d^5(^2P)4p$	$^3S°$	1	378 555.0	90	7	$(^2P)\ ^1P°$
$3d^5(^2P)4p$	$^1P°$	1	380 165.6	68	16	$(^2D1)\ ^1P°$
$3d^5(^2D1)4p$	$^3F°$	2	386 968.8	63	20	$(^2D3)\ ^3F°$
		3	387 333.4	60	19	
		4	388 698.9	72	22	
$3d^5(^2D1)4p$	$^3D°$	1	388 746.1	71	21	$(^2D3)\ ^3D°$
		2	389 571.8	60	18	
		3	390 478.2	60	18	
$3d^5(^2D1)4p$	$^1D°$	2	390 675.1	44	16	$(^2P)\ ^1D°$

(Continued)

Ni V—Continued

Configuration	Term	J	Level (cm^{-1})	Leading percentages		
$3d^5(^2D1)4p$	$^3P°$	2	*392 413.5*	43	21	$(^2P)\ ^3P°$
$3d^5(^2D1)4p$	$^1F°$	3	*392 957.1*	69	21	$(^2D3)\ ^1F°$
$3d^5(^6S)5p$	$^5P°$	3	*423 533*			
		2	*423 782*			
		1	*423 935*			
$3d^5(^6S)4f$	$^5F°$	2	*439 419*			
		1	*439 420*			
		3	*439 423*			
		4	*439 427*			
		5	*439 434*			
$3d^5(^6S)5f$	$^5F°$	5	*502 124*			
		4	*502 133*			
		3	*502 137*			
		1	*502 143*			
		2	*502 148*			
Ni VI $(^6S_{5/2})$	*Limit*		**613 500**			

Ni VI

$Z=28$

V I isoelectronic sequence

Ground state: $1s^2 2s^2 2p^6 3s^2 3p^6 3d^5\ {}^6S_{5/2}$

Ionization energy $= 870\,000 \pm 16\,000\ \mathrm{cm}^{-1}$ (108 ± 2 eV)

The resonance multiplet $3d^5\ {}^6S - 3d^4 4p\ {}^6P°$ was identified by Kruger and Gilroy (1935). An extensive analysis of this spectrum was carried out by Raassen (1980), who reported all the presently known levels and their percentage compositions. The uncertainty in the level values is probably $\pm 5\ \mathrm{cm}^{-1}$.

The ionization energy was obtained by extrapolation by Lotz (1967).

References

Kruger, P. G., and Gilroy, H. T. (1935), Phys. Rev. **48**, 720.
Lotz, W. (1967), J. Opt. Soc. Am. **57**, 873.
Raassen, A. J. J. (1980), Physica **100C**, 404.

Ni VI

Configuration	Term	J	Level (cm^{-1})	Leading percentages			
$3d^5$	^{6}S	5/2	0.0	100			
$3d^5$	^{4}G	11/2	41 920.9	100			
		5/2	42 003.6	99		1	^{2}F1
		9/2	42 023.2	100			
		7/2	42 035.1	100			
$3d^5$	^{4}P	5/2	45 884.2	90		9	^{4}D
		3/2	46 104.4	92		7	
		1/2	46 324.8	97		3	
$3d^5$	^{4}D	7/2	50 331.0	99			
		1/2	50 643.5	97		3	^{4}P
		5/2	50 777.6	90		9	^{4}P
		3/2	50 780.5	92		7	^{4}P
$3d^5$	^{2}I	11/2	61 196.0	98		2	^{2}H
		13/2	61 279.5	100			
$3d^5$	^{2}D3	5/2	64 152.4	52		29	^{2}F1
		3/2	65 173.5	71		22	^{2}D1
$3d^5$	^{2}F1	7/2	67 085.1	93		3	^{4}F
$3d^5$	^{4}F	5/2	68 444.9	51		39	^{2}F1
		9/2	68 551.2	95		4	^{2}G2
		7/2	68 801.7	94		4	^{2}F1
		3/2	69 173.5	92		6	^{2}D3
$3d^5$		5/2	69 447.0	47	^{4}F	31	^{2}F1
$3d^5$	^{2}H	9/2	72 908.8	74		24	^{2}G2
		11/2	73 756.6	98		2	^{2}I
$3d^5$	^{2}G2	7/2	74 627.7	98		1	^{4}F
		9/2	75 441.7	72		25	^{2}H
$3d^5$	^{2}F2	5/2	79 391.4	98		1	^{4}F
		7/2	79 608.3	97		2	

(Continued)

Ni VI—Continued

Configuration	Term	J	Level (cm^{-1})	Leading percentages		
$3d^5$	^2S	1/2	86 532.1	100		
$3d^5$	$^2D2$	3/2	96 461.2	100		
		5/2	96 566.5	99	1	$^2F2$
$3d^5$	$^2G1$	9/2	107 848.3	100		
		7/2	107 887.0	100		
$3d^5$	^2P	3/2	129 951.6	99	1	$^2D1$
		1/2	130 025.9	100		
$3d^5$	$^2D1$	5/2	140 922.2	77	23	$^2D3$
		3/2	141 006.8	76	23	
$3d^4(^5D)4s$	^6D	1/2	295 882.7	100		
		3/2	296 198.1	100		
		5/2	296 696.7	100		
		7/2	297 350.5	100		
		9/2	298 130.5	100		
$3d^4(^5D)4s$	^4D	1/2	307 843.9	99		
		3/2	308 323.8	100		
		5/2	309 054.9	100		
		7/2	309 977.1	100		
$3d^4(^3P2)4s$	^4P	1/2	328 437.2	59	38	$(^3P1)\ ^4P$
		3/2	330 001.6	60	38	
		5/2	332 344.8	61	38	
$3d^4(^3H)4s$	^4H	7/2	329 486.2	97	2	$(^3G)\ ^4G$
		9/2	329 754.2	96	3	$(^3G)\ ^4G$
		11/2	330 141.0	97	2	$(^3G)\ ^4G$
		13/2	330 580.5	100		
$3d^4(^3F2)4s$	^4F	3/2	332 031.8	78	22	$(^3F1)\ ^4F$
		5/2	332 051.6	74	20	
		7/2	332 175.4	72	19	
		9/2	332 343.6	73	18	
$3d^4(^3G)4s$	^4G	5/2	335 384.5	91	5	$(^3F2)\ ^4F$
		7/2	335 976.0	90	7	$(^3F2)\ ^4F$
		9/2	336 344.9	73	18	$(^3H)\ ^2H$
		11/2	336 430.4	78	19	$(^3H)\ ^2H$
$3d^4(^3P2)4s$	^2P	1/2	336 404.8	59	38	$(^3P1)\ ^2P$
		3/2	339 260.1	60	37	
$3d^4(^3H)4s$	^2H	9/2	337 007.7	78	19	$(^3G)\ ^4G$
		11/2	337 993.9	79	20	
$3d^4(^3F2)4s$	^2F	5/2	339 482.4	76	20	$(^3F1)\ ^2F$
		7/2	339 335.8	70	17	
$3d^4(^3G)4s$	^2G	7/2	343 052.0	90	6	$(^3F2)\ ^2F$
		9/2	343 842.6	96	2	$(^3H)\ ^2H$
$3d^4(^3D)4s$	^4D	7/2	343 999.6	99		
		5/2	344 203.5	98		
		3/2	344 447.4	98		

Ni VI—Continued

Configuration	Term	J	Level (cm^{-1})	Leading percentages		
$3d^4(^1G2)4s$	^2G	9/2	347 278.5	64	31	$(^1G1)\ ^2G$
		7/2	347 445.2	62	30	
$3d^4(^1I)4s$	^2I	13/2	347 892.4	100		
		11/2	347 963.9	99	1	$(^3H)\ ^2H$
$3d^4(^3D)4s$	^2D	5/2	351 033.0	99		
		3/2	351 304.0	98	1	$(^1D2)\ ^2D$
$3d^4(^1S2)4s$	^2S	1/2	351 096.2	78	20	$(^1S1)\ ^2S$
$3d^4(^1D2)4s$	^2D	5/2	358 716.6	77	20	$(^1D1)\ ^2D$
		3/2	358 827.9	77	20	
$3d^4(^1F)4s$	^2F	5/2	366 289.8	98	1	$(^3F1)\ ^4F$
		7/2	366 299.8	98	1	
$3d^4(^3P1)4s$	^4P	5/2	374 843.9	61	38	$(^3P2)\ ^4P$
		3/2	376 343.7	60	38	
$3d^4(^3F1)4s$	^4F	9/2	375 673.0	81	19	$(^3F2)\ ^4F$
		3/2	375 870.8	78	22	
		7/2	375 927.1	79	20	
		5/2	375 949.8	78	21	
$3d^4(^5D)4p$	$^6F^\circ$	1/2	*380 756.7*	99	1	$(^5D)\ ^4D^\circ$
		3/2	*381 163.8*	99		
		5/2	*381 832.8*	99	1	$(^5D)\ ^4F^\circ$
		7/2	*382 758.9*	98	1	$(^5D)\ ^4F^\circ$
		9/2	*383 960.2*	98	1	$(^5D)\ ^4F^\circ$
		11/2	*385 520.9*	100		
$3d^4(^3F1)4s$	^2F	7/2	382 677.1	80	19	$(^3F2)\ ^2F$
		5/2	382 889.1	78	21	
$3d^4(^5D)4p$	$^6P^\circ$	3/2	*383 557.8*	91	6	$(^5D)\ ^4P^\circ$
		5/2	*383 739.8*	94	4	$(^5D)\ ^4P^\circ$
		7/2	*384 096.5*	98	1	$(^5D)\ ^6D^\circ$
$3d^4(^5D)4p$	$^4P^\circ$	1/2	*384 747.4*	68	29	$(^5D)\ ^6D^\circ$
		3/2	*386 157.3*	55	33	
		5/2	*389 760.4*	55	42	
$3d^4(^5D)4p$	$^6D^\circ$	5/2	*387 849.9*	54	38	$(^5D)\ ^4P^\circ$
		1/2	*388 918.6*	71	29	$(^5D)\ ^4P^\circ$
		3/2	*389 214.0*	64	35	$(^5D)\ ^4P^\circ$
		7/2	*389 243.4*	94	4	$(^5D)\ ^4F^\circ$
		9/2	*389 883.2*	88	10	$(^5D)\ ^4F^\circ$
$3d^4(^1G1)4s$	^2G	7/2	389 362.5	66	33	$(^1G2)\ ^2G$
		9/2	389 299.6	67	33	
$3d^4(^5D)4p$	$^4F^\circ$	3/2	*392 808.1*	94	2	$(^3G)\ ^4F^\circ$
		5/2	*393 072.0*	93	2	$(^3G)\ ^4F^\circ$
		7/2	*393 468.2*	91	4	$(^5D)\ ^6D^\circ$
		9/2	*394 061.0*	85	11	$(^5D)\ ^6D^\circ$

(Continued)

Ni VI—Continued

Configuration	Term	J	Level (cm^{-1})	Leading percentages			
$3d^4(^5D)4p$	$^4D°$	1/2	400 962.1	96		1	$(^3D)\ ^4D°$
		3/2	401 280.9	96		1	
		5/2	401 720.9	96		1	
		7/2	402 171.3	96		2	
$3d^4(^3H)4p$	$^4H°$	7/2	411 669.4	74		20	$(^3G)\ ^4H°$
		9/2	412 095.6	70		19	
		11/2	412 751.1	72		17	
		13/2	413 631.6	77		13	
$3d^4(^3P2)4p$	$^4D°$	1/2	412 248.2	47		32	$(^3P1)\ ^4D°$
		3/2	413 553.1	47		31	
		5/2	415 144.5	38		26	
$3d^4(^3F2)4p$	$^4G°$	5/2	415 382.0	40		23	$(^3G)\ ^4G°$
		7/2	415 640.8	25		16	$(^3G)\ ^4G°$
		9/2	420 080.0	40		40	$(^3H)\ ^4G°$
		11/2	420 623.4	58		22	$(^3H)\ ^4G°$
$3d^4(^3H)4p$	$^4I°$	9/2	415 754.3	45		11	$(^3H)\ ^2G°$
		11/2	417 164.1	63		14	$(^3H)\ ^4G°$
		13/2	418 553.6	90		8	$(^3H)\ ^4H°$
		15/2	419 533.5	99		1	$(^1I)\ ^2K°$
$3d^4(^3H)4p$		9/2	416 422.3	43	$^4I°$	11	$(^3F2)\ ^4G°$
$3d^4(^3P2)4p$	$^4P°$	1/2	416 459.0	35		20	$(^3P1)\ ^4P°$
		3/2	418 491.4	45		25	
		5/2	420 308.9	50		28	
$3d^4(^3F2)4p$	$^4D°$	7/2	416 493.0	31		23	$(^3P2)\ ^4D°$
		1/2	422 369.4	42		15	$(^3F1)\ ^4D°$
		3/2	422 437.2	37		12	$(^3F1)\ ^4D°$
$3d^4(^3H)4p$		11/2	417 538.4	34	$^4G°$	25	$(^3H)\ ^4I°$
$3d^4(^3H)4p$	$^2G°$	7/2	417 717.7	38		16	$(^3F2)\ ^2G°$
		9/2	418 368.8	32		18	
$3d^4(^3F2)4p$		3/2	418 713.4	28	$^4F°$	21	$(^3F2)\ ^2D°$
$3d^4(^3H)4p$		5/2	419 390.5	30	$^4G°$	26	$(^3F2)\ ^4G°$
$3d^4(^3F2)4p$		5/2	419 844.7	39	$^4F°$	14	$(^3H)\ ^4G°$
$3d^4(^3H)4p$	$^4G°$	7/2	419 863.3	47		32	$(^3F2)\ ^4G°$
$3d^4(^3P2)4p$		1/2	419 961.5	23	$^4P°$	18	$(^3P1)\ ^2S°$
$3d^4(^3F2)4p$	$^4F°$	7/2	420 553.2	63		11	$(^3F1)\ ^4F°$
		9/2	420 835.3	58		13	$(^3G)\ ^4F°$
$3d^4(^3P2)4p$		3/2	420 722.0	26	$^2P°$	18	$(^3P2)\ ^4S°$
$3d^4(^3F2)4p$		5/2	421 324.3	19	$^2D°$	15	$(^3F2)\ ^4F°$
$3d^4(^3F2)4p$		3/2	421 607.2	24	$^4F°$	20	$(^3F2)\ ^2D°$
$3d^4(^3F2)4p$		5/2	421 894.0	15	$^2D°$	14	$(^3F2)\ ^4D°$

Ni VI—Continued

Configuration	Term	J	Level (cm^{-1})	Leading percentages			
$3d^4(^3P2)4p$		7/2	422 306.6	21	$^4D°$	20	$(^3G)\ ^4F°$
$3d^4(^3H)4p$	$^2I°$	11/2	422 642.4	80		6	$(^1I)\ ^2I°$
		13/2	422 711.8	82		11	$(^3G)\ ^4H°$
$3d^4(^3G)4p$		5/2	422 897.8	22	$^2F°$	13	$(^3F2)\ ^4D°$
$3d^4(^3H)4p$		9/2	423 065.5	36	$^2H°$	20	$(^3G)\ ^4H°$
$3d^4(^3G)4p$	$^4H°$	7/2	423 341.4	70		20	$(^3H)\ ^4H°$
		9/2	424 537.0	54		18	$(^3H)\ ^2H°$
		11/2	425 567.6	39		39	$(^3H)\ ^2H°$
		13/2	426 451.9	75		14	$(^3H)\ ^4H°$
$3d^4(^3G)4p$		7/2	423 645.8	18	$^4F°$	15	$(^3D)\ ^2F°$
$3d^4(^3P2)4p$		1/2	423 646.8	29	$^2P°$	18	$(^3F2)\ ^4D°$
$3d^4(^3G)4p$		3/2	423 710.5	26	$^4F°$	12	$(^3F2)\ ^4F°$
$3d^4(^3G)4p$	$^4F°$	9/2	423 763.3	52		15	$(^3F2)\ ^4F°$
		5/2	424 217.9	39		16	$(^3F2)\ ^4D°$
$3d^4(^3G)4p$		3/2	424 235.1	21	$^4F°$	20	$(^3P2)\ ^2P°$
$3d^4(^3F2)4p$		7/2	424 346.1	25	$^4D°$	20	$(^3G)\ ^4F°$
$3d^4(^3G)4p$		11/2	424 363.7	36	$^4H°$	30	$(^3H)\ ^2H°$
$3d^4(^3P2)4p$	$^2D°$	3/2	425 703.1	39		27	$(^3P1)\ ^2D°$
$3d^4(^3F2)4p$		5/2	427 096.5	26	$^2F°$	20	$(^3P2)\ ^2D°$
$3d^4(^3P2)4p$		5/2	427 342.2	27	$^2D°$	26	$(^3F2)\ ^2F°$
$3d^4(^3F2)4p$		7/2	427 575.0	24	$^2G°$	13	$(^3F1)\ ^2G$
$3d^4(^3G)4p$		11/2	427 702.7	38	$^2H°$	21	$(^3G)\ ^4G°$
$3d^4(^3G)4p$		9/2	427 995.6	22	$^4G°$	15	$(^3H)\ ^4G°$
$3d^4(^3G)4p$		9/2	428 051.5	37	$^2H°$	15	$(^3H)\ ^2G°$
$3d^4(^3F2)4p$	$^2F°$	7/2	428 221.4	46		32	$(^3G)\ ^2F°$
$3d^4(^3G)4p$	$^4G°$	5/2	428 496.5	50		29	$(^3H)\ ^4G°$
		7/2	428 866.5	45		20	$(^3H)\ ^4G°$
		11/2	430 033.3	43		25	$(^3G)\ ^2H°$
$3d^4(^3G)4p$		9/2	429 486.3	36	$^4G°$	15	$(^3G)\ ^2H°$
$3d^4(^3D)4p$	$^4D°$	1/2	430 898.8	78		9	$(^3D)\ ^2P°$
		3/2	430 971.9	69		12	$(^3D)\ ^4P°$
		7/2	431 912.6	77		6	$(^3D)\ ^4F°$
		5/2	432 295.7	46		33	$(^3D)\ ^4P°$
$3d^4(^3D)4p$	$^4P°$	5/2	431 019.9	56		32	$(^3D)\ ^4D°$
		3/2	433 462.5	74		9	$(^3D)\ ^4D°$
		1/2	434 281.6	88		7	$(^3P2)\ ^4P°$

(Continued)

Ni VI—Continued

Configuration	Term	J	Level (cm^{-1})	Leading percentages			
$3d^4(^1I)4p$	$^2I°$	$13/2$	*432 616.6*	66		27	$(^1I)\ ^2K°$
		$11/2$	*432 932.0*	88		5	$(^3H)\ ^2I°$
$3d^4(^1G2)4p$	$^2F°$	$7/2$	*432 800.1*	46		21	$(^1G1)\ ^2F°$
		$5/2$	*434 997.0*	42		20	
$3d^4(^3G)4p$	$^2G°$	$9/2$	*433 371.2*	53		20	$(^3H)\ ^2G°$
		$7/2$	*433 517.3*	49		17	
$3d^4(^1S2)4p$	$^2P°$	$1/2$	*433 936.9*	37		33	$(^3D)\ ^2P°$
		$3/2$	*440 168.6*	30		16	
$3d^4(^1G2)4p$	$^2H°$	$9/2$	*434 167.3*	47		22	$(^1G1)\ ^2H°$
		$11/2$	*435 011.5*	53		20	
$3d^4(^3D)4p$	$^4F°$	$3/2$	*434 200.9*	63		21	$(^3G)\ ^4F°$
		$5/2$	*434 625.7*	54		15	
		$7/2$	*435 116.2*	65		19	
		$9/2$	*435 459.4*	77		19	
$3d^4(^3D)4p$	$^2P°$	$3/2$	*434 995.8*	54		26	$(^1S2)\ ^2P°$
		$1/2$	*440 689.3*	47		27	
$3d^4(^1I)4p$	$^2K°$	$13/2$	*435 165.4*	71		28	$(^1I)\ ^2I°$
		$15/2$	*436 550.3*	99		1	$(^3H)\ ^4I°$
$3d^4(^1G2)4p$	$^2G°$	$7/2$	*437 919.7*	45		29	$(^1G1)\ ^2G°$
		$9/2$	*438 639.4*	41		30	
$3d^4(^1I)4p$	$^2H°$	$11/2$	*440 038.7*	70		10	$(^3G)\ ^2H°$
		$9/2$	*440 917.9*	82		11	
$3d^4(^3D)4p$	$^2F°$	$7/2$	*441 216.1*	65		11	$(^3G)\ ^2F°$
		$5/2$	*441 785.7*	42		14	$(^3D)\ ^2D°$
$3d^4(^3D)4p$		$5/2$	*441 401.2*	30	$^2D°$	21	$(^3D)\ ^2F°$
$3d^4(^3D)4p$	$^2D°$	$3/2$	*442 620.7*	53		13	$(^1D2)\ ^2D°$
$3d^4(^1D2)4p$	$^2D°$	$3/2$	*444 252.2*	40		22	$(^3D)\ ^2D°$
		$5/2$	*444 784.2*	24		26	
$3d^4(^1D2)4p$	$^2F°$	$5/2$	*446 061.2*	44		11	$(^1D2)\ ^2D°$
		$7/2$	*446 863.8*	47		31	$(^1F)\ ^2F°$
$3d^4(^1F)4p$	$^2F°$	$5/2$	*451 041.3*	64		14	$(^1D2)\ ^2F°$
		$7/2$	*451 859.4*	47		27	
$3d^4(^1D2)4p$	$^2P°$	$3/2$	*451 622.9*	65		13	$(^1D1)\ ^2P°$
		$1/2$	*451 642.0*	72		14	
$3d^4(^1F)4p$	$^2G°$	$7/2$	*453 983.0*	86		5	$(^1F)\ ^2F°$
		$9/2$	*455 890.5*	92		4	$(^1G2)\ ^2G°$
$3d^4(^1F)4p$	$^2D°$	$5/2$	*456 815.6*	53		17	$(^3P1)\ ^2D°$
		$3/2$	*458 772.2*	49		14	$(^3F1)\ ^4F°$

Ni VI—Continued

Configuration	Term	J	Level (cm^{-1})	Leading percentages			
$3d^4(^3F1)4p$	$^4F°$	5/2	460 364.6	61		9	$(^3P1)\ ^4D°$
		7/2	460 535.9	66		8	$(^3F2)\ ^4F°$
		3/2	460 604.0	57		9	$(^3F2)\ ^4F°$
		9/2	461 118.3	83		10	$(^3F2)\ ^4F°$
$3d^4(^3P1)4p$	$^4P°$	3/2	461 437.8	30		14	$(^3P2)\ ^4P°$
		5/2	463 301.5	45		22	
$3d^4(^3P1)4p$		5/2	461 622.6	21	$^4D°$	20	$(^3F1)\ ^4F°$
$3d^4(^3P1)4p$		7/2	462 443.1	32	$^4D°$	20	$(^3F1)\ ^4D°$
$3d^4(^3P1)4p$		3/2	462 884.1	24	$^4P°$	21	$(^3P1)\ ^4D°$
$3d^4(^3P1)4p$		1/2	462 903.6	31	$^4P°$	21	$(^3P1)\ ^4D°$
$3d^4(^3F1)4p$	$^4G°$	5/2	464 027.8	44		23	$(^3F1)\ ^2F°$
		9/2	466 065.2	70		21	$(^3F2)\ ^4G°$
		11/2	466 754.3	76		22	$(^3F2)\ ^4G°$
$3d^4(^3F1)4p$		7/2	464 807.8	35	$^4G°$	32	$(^3F1)\ ^2F°$
$3d^4(^1F)4p$		5/2	465 552.7	25	$^2D°$	19	$(^3P1)\ ^2D°$
$3d^4(^1F)4p$		3/2	466 034.2	29	$^2D°$	21	$(^3P1)\ ^2D°$
$3d^4(^3F1)4p$		7/2	466 133.7	36	$^2F°$	35	$(^3F1)\ ^4G°$
$3d^4(^3F1)4p$	$^2F°$	5/2	466 649.1	47		20	$(^3F1)\ ^4G°$
$3d^4(^3P1)4p$	$^4S°$	3/2	471 106.8	48		44	$(^3P2)\ ^4S°$
$3d^4(^3F1)4p$	$^4D°$	7/2	472 497.3	47		17	$(^3P1)\ ^4D°$
		5/2	473 292.4	48		18	$(^3F2)\ ^4D°$
		3/2	473 656.4	48		19	$(^3F2)\ ^4D°$
		1/2	473 727.6	51		20	$(^3F2)\ ^4D°$
$3d^4(^3F1)4p$	$^2G°$	9/2	472 579.5	72		20	$(^3F2)\ ^2G°$
		7/2	473 531.9	73		22	
$3d^4(^3P1)4p$	$^2P°$	3/2	472 833.0	61		27	$(^3P2)\ ^2P°$
		1/2	473 869.5	59		26	
$3d^4(^1G1)4p$		9/2	476 332.1	38	$^2H°$	24	$(^1G1)\ ^2G°$
$3d^4(^1G1)4p$	$^2G°$	7/2	476 935.6	52		32	$(^1G2)\ ^2G°$
$3d^4(^3P2)4p$	$^2S°$	1/2	477 177.1	55		41	$(^3P1)\ ^2S°$
$3d^4(^1G1)4p$		9/2	479 046.3	35	$^2G°$	26	$(^1G1)\ ^2H°$
$3d^4(^1G1)4p$	$^2H°$	11/2	479 481.8	65		31	$(^1G2)\ ^2H°$
$3d^4(^1G1)4p$	$^2F°$	7/2	480 189.3	54		18	$(^1G2)\ ^2F°$
		5/2	480 724.2	57		19	
$3d^4(^3F1)4p$	$^2D°$	5/2	483 043.4	47		18	$(^3F2)\ ^2D°$
		3/2	483 480.9	45		21	$(^3P1)\ ^2D°$

(Continued)

Ni VI—Continued

Configuration	Term	J	Level (cm^{-1})	Leading percentages		
$3d^4(^1D1)4p$	$^2P^\circ$	3/2	*498 773.0*	75	15	$(^1D2)\ ^2P^\circ$
		1/2	*500 275.2*	75	16	
$3d^4(^1D1)4p$	$^2F^\circ$	5/2	*505 150.5*	69	18	$(^1D2)\ ^2F^\circ$
		7/2	*507 300.4*	73	19	
$3d^4(^1D1)4p$	$^2D^\circ$	3/2	*511 217.4*	72	26	$(^1D2)\ ^2D^\circ$
		5/2	*512 113.2*	71	25	
$3d^4(^1S1)4p$	$^2P^\circ$	1/2	*540 394.1*	74	20	$(^1S2)\ ^2P^\circ$
		3/2	*543 115.6*	74	20	
Ni VII ($^5D_0$)	***Limit***		**870 000**			

Ni VII

$Z=28$

Ti I isoelectronic sequence

Ground state: $1s^2 2s^2 2p^6 3s^2 3p^6 3d^4\ ^5D_0$

Ionization energy = 1 070 000±16 000 cm^{-1} (133±2 eV)

The $3d^4-3d^34p$ transition array of this spectrum between 205 and 231 Å was observed and analysed by Phillips and Kruger (1938). Henrichs (1975) has revised that work from new observations and isoelectronic comparisons. His results are given here. The uncertainty of the level values is about ±10 cm^{-1}.

The ionization energy was obtained by extrapolation by Lotz (1967).

References

Henrichs, H. F. (1975), Astron. Astrophys. **44**, 41.
Phillips, L. W., and Kruger, P. G. (1933), Phys. Rev. **54**, 839.
Lotz, W. (1967), J. Opt. Soc. Am. **57**, 873.

Ni VII

Configuration	Term	J	Level (cm^{-1})	Configuration	Term	J	Level (cm^{-1})
$3d^4$	^5D	0	0	$3d^3(^4F)4p$	$^5F°$	2	*470 056*
		1	279			3	*471 096*
		2	804			1	*471 250*
		3	1 520			4	*472 067*
		4	2 392			5	*472 907*
$3d^4$	$^3P2$	0	30 077	$3d^3(^4F)4p$	$^3G°$	3	*476 258*
		1	31 836			4	*477 261*
		2	34 555			5	*478 534*
$3d^4$	^3H	4	31 672	$3d^3(^4F)4p$	$^3F°$	2	*479 950*
		5	32 286			3	*481 091*
		6	32 877			4	*482 229*
$3d^4$	$^3F2$	2	34 247	$3d^3(^4P)4p$	$^5P°$	1	*486 151*
		3	34 317			2	*487 207*
		4	34 576			3	*488 665*
$3d^4$	^3G	3	38 160	$3d^3(^2G)4p$	$^3G°$	3	*493 743*
		4	38 746			4	*495 087*
		5	39 247			5	*496 340*
$3d^3(^4F)4p$	$^3D°$	1	*467 015*	$3d^3(^2H)4p$	$^3G°$	5	*511 464*
		2	*471 976*			4	*511 698*
		3	*473 360*			3	*511 991*
$3d^3(^4F)4p$	$^5D°$	2	*467 705*	Ni VIII ($^4F_{3/2}$)	*Limit*		1 070 000
		3	*468 690*				
		1	*469 175*				
		4	*469 844*				

Ni VIII

$Z=28$

Sc I isoelectronic sequence

Ground state: $1s^22s^22p^63s^23p^63d^3\ ^4F_{3/2}$

Ionization energy = 1 310 000±16 000 cm^{-1} (162±2 eV)

Anderson and Mack (1941) observed and classified 132 lines between 163 and 189 Å in the $3d^3-3d^24p$ transition array. The uncertainty in the level values is ±10 cm^{-1}, which corresponds to an error of about 0.003 Å in the wavelengths.

The separation of the ^{4}F and ^{4}P terms in $3d^3$ is confirmed by the calculations of Racah (1954). The doublet system has been questioned by Bowen (1960), on the basis of isoelectronic extrapolations, but the objection does not seem to be well substantiated.

The configurations $3d^3$, $3d^24s$, and $3d^24p$ have been calculated by Kancerevicius, Ramonas, and Uspalis (1976), using Hartree-Fock radial integrals.

The ionization energy was obtained by extrapolation by Lotz (1967).

References

Anderson, E. E., and Mack, J. E. (1941), Phys. Rev. **59**, 717.

Bowen, I. S. (1960), Astrophys. J. **132**, 1.

Kancerevicius, A., Ramonas, A., and Uspalis, K. (1976), Lietuvos Fizikos Rinkinys **16**, 49.

Lotz, W. (1967), J. Opt. Soc. Am. **57**, 873.

Racah, G. (1954), Bull. Res. Council Israel **3**, 290.

Ni VIII

Configuration	Term	J	Level (cm^{-1})
$3d^3$	^{4}F	3/2	0
		5/2	1 012
		7/2	2 184
		9/2	3 721
$3d^3$	^{4}P	1/2	23 261
		3/2	23 710
		5/2	24 669
$3d^3$	^{2}G	7/2	26 977
		9/2	28 068
$3d^3$	^{2}D2	5/2	34 689
		3/2	35 120
$3d^3$	^{2}H	9/2	36 754
		11/2	37 475
$3d^2(^3F)4p$	^{4}G°	5/2	*565 124*
		7/2	*566 964*
		9/2	*569 564*
		11/2	*572 969*
$3d^2(^3F)4p$	^{4}F°	3/2	*565 388*
		5/2	*566 831*
		7/2	*568 746*
		9/2	*570 960*
$3d^2(^3F)4p$	^{2}D°	3/2	*569 667*
		5/2	*571 845*
$3d^2(^3F)4p$	^{4}D°	3/2	*569 839*
		1/2	*570 353*
		5/2	*571 517*
		7/2	*573 327*
$3d^2(^3F)4p$	^{2}F°	5/2	*570 546*
		7/2	*571 804*
$3d^2(^3F)4p$	^{2}G°	7/2	*581 337*
		9/2	*583 241*
$3d^2(^3P)4p$	^{4}S°	3/2	*587 305*
$3d^2(^3P)4p$	^{4}D°	1/2	*590 764*
		3/2	*592 175*
		5/2	*594 068*
		7/2	*596 908*
$3d^2(^3P)4p$	^{4}P°	1/2	*596 770*
		3/2	*596 905*
		5/2	*598 570*
$3d^2(^1G)4p$	^{2}G°	7/2	*598 638*
		9/2	*599 079*
$3d^2(^1G)4p$	^{2}H°	9/2	*613 417*
		11/2	*615 725*
Ni IX ($^3F_2$)	*Limit*		**1 310 000**

Ni IX

$Z=28$

Ca I isoelectronic sequence

Ground state: $1s^2 2s^2 2p^6 3s^2 3p^6 3d^2\ ^3F_2$

Ionization energy = 1 560 000±16 000 cm^{-1} (193±2 eV)

The analysis of the $3d^2-3d4f$ array was given by Alexander, Feldman, Fraenkel, and Hoory (1966). The levels are obtained from the improved measurements and corrected classifications by Even-Zohar and Fraenkel (1968). The uncertainty of the level values is about ±50 cm^{-1}. The singlet system, the $^3P_2$ of $3d^2$, and the $^3D_3$ of $3d4f$ are based on an estimated value for the $3d^2\ ^1D$ term by Alexander et al. The value of the systematic shift is expected to be a few hundred cm^{-1}.

Fawcett, Ridgeley, and Ekberg (1980) classified the transition array $3p^6 3d^2-3p^5 3d^3$. Their wavelength accuracy is given as ±0.007 Å, giving a level uncertainty of ±30 cm^{-1}.

The $3d^2\ ^1G$ term is based on a tentative identification of a coronal line at 7144 Å by Pryce (1964).

The ionization energy was obtained by extrapolation by Lotz (1967).

References

Alexander, E., Feldman, U., Fraenkel, B. S., and Hoory, S. (1966), J. Opt. Soc. Am. **56**, 651.

Even-Zohar, M., and Fraenkel, B. S. (1968), J. Opt. Soc. Am. **58**, 1420.

Fawcett, B. C., Ridgeley, A., and Ekberg, J. O. (1980), Phys. Scr. **21**, 155.

Lotz, W. (1967), J. Opt. Soc. Am. **57**, 873.

Pryce, M. H. L. (1964), Astrophys. J. **140**, 1192.

Ni IX

Configuration	Term	J	Level (cm^{-1})
$3p^6(^1S)3d^2$	^3F	2	0
		3	1 880
		4	4 070
$3p^6(^1S)3d^2$	^1D	2	21 900+x
$3p^6(^1S)3d^2$	^3P	2	27 160+x
$3p^6(^1S)3d^2$	^1G	4	35 898+x?
$3p^5(^2P^\circ)3d^3(^2H)$	$^3G^\circ$	3	*601 300*
		4	*604 000*
		5	*608 530*
$3p^5(^2P^\circ)3d^3(^2G)$	$^1H^\circ$	5	*640 360*+x
$3p^5(^2P^\circ)3d^3(^4F)$	$^3F^\circ$	2	*661 050*
		3	*664 080*
		4	*667 080*
$3p^5(^2P^\circ)3d^3(^4F)$	$^3D^\circ$	1	*709 210*
		3	*711 510*
		2	*711 520*
$3p^5(^2P^\circ)3d^3(^2H)$	$^1G^\circ$	4	*716 110*+x
$3p^6(^1S)3d4f$	$^3F^\circ$	2	*977 130*
		3	*977 680*
		4	*978 740*
$3p^6(^1S)3d4f$	$^3G^\circ$	3	*983 700*
		4	*985 140*
		5	*985 940*
$3p^6(^1S)3d4f$	$^1D^\circ$	2	*984 630*+x
$3p^6(^1S)3d4f$	$^1F^\circ$	3	*986 960*+x
$3p^6(^1S)3d4f$	$^3D^\circ$	3	*988 760*+x
Ni X ($^2D_{3/2}$)	*Limit*		**1 560 000**

Ni X

$Z=28$

K I isoelectronic sequence

Ground state: $1s^22s^22p^63s^23p^63d\ ^2D_{3/2}$

Ionization energy = 1 812 000±56 000 cm^{-1} (224.6±0.7 eV)

Observations were first reported by Alexander, Feldman, and Fraenkel (1965), who identified the $3d-4f$ and $5f$ doublets. The series was extended to $7f$ by Even-Zohar and Fraenkel (1968), who noted that the $6f$ and $7f$ terms may be perturbed by the $3p^53d4d$ configuration. Their levels are given here with an uncertainty of ±100 cm^{-1}. We find that the nf terms do not follow a regular Rydberg series.

Gabriel, Fawcett, and Jordan (1966) identified the $3p^53d^2(^3F)^2D$ and $(^3F)^2F$ terms. The ^2P term was added by Goldsmith and Fraenkel (1970). With new observations with a reported accuracy of ±0.003 Å, Ramonas and Ryabtsev (1980) determined the $(^1G)^2F$ and $(^1D)^2F$ terms and improved the level values. Their results are given for the $3p^53d^2$ levels and for the $3p^63d\ ^2D$ ground term interval. They also obtained the $3p^64p$ term. Their level value uncertainty is ±10 cm^{-1}.

The analysis of the $3p^53d4s$ configuration is by Hoory, Goldsmith, Fraenkel, and Feldman (1970). These levels have an uncertainty of about ±50 cm^{-1}.

The ionization energy was obtained by extrapolation by Lotz (1967).

References

Alexander, E., Feldman, U., and Fraenkel, B. S. (1965), J. Opt. Soc. Am. **55**, 650.

Even-Zohar, M., and Fraenkel, B. S. (1968), J. Opt. Soc. Am. **58**, 1420.

Gabriel, A. H., Fawcett, B. C., and Jordan, C. (1966), Proc. Phys. Soc. **87**, 825.

Goldsmith, S., and Fraenkel, B. S. (1970), Astrophys. J. **161**, 317.

Hoory, S., Goldsmith, S. Fraenkel, B. S., and Feldman, U. (1970), Astrophys. J. **160**, 781.

Lotz, W. (1967), J. Opt. Soc. Am. **57**, 873.

Ramonas, A. A., and Ryabtsev, A. N. (1980), Opt. Spectrosc. **48**, 631.

Ni X

Configuration	Term	J	Level (cm^{-1})
$3p^6 3d$	^2D	3/2	0
		5/2	3 178
$3p^5(^2P^\circ)3d^2(^1G)$	$^2F^\circ$	5/2	*505 283*
		7/2	*509 751*
$3p^5(^2P^\circ)3d^2(^1D)$	$^2F^\circ$	7/2	*522 391*
		5/2	*540 725*
$3p^5(^2P^\circ)3d^2(^3F)$	$^2F^\circ$	5/2	*625 091*
		7/2	*634 583*
$3p^5(^2P^\circ)3d^2(^3P)$	$^2P^\circ$	1/2	*684 552*
		3/2	*689 365*
$3p^5(^2P^\circ)3d^2(^3F)$	$^2D^\circ$	5/2	*692 890*
		3/2	*693 404*
$3p^6 4p$	$^2P^\circ$	1/2	*773 647*
		3/2	*779 600*
$3p^6 4f$	$^2F^\circ$	5/2	*1 093 360*
		7/2	*1 093 440*
$3p^5 3d(^3P^\circ)4s$	$^2P^\circ$	1/2	*1 140 510*
		3/2	*1 148 420*
$3p^5 3d(^3F^\circ)4s$	$^4F^\circ$	7/2	*1 154 390*
		5/2	*1 156 550*
$3p^5 3d(^3F^\circ)4s$	$^2F^\circ$	7/2	*1 161 930*
		5/2	*1 169 300*
$3p^5 3d(^3D^\circ)4s$	$^4D^\circ$	7/2	*1 184 390*
		5/2	*1 187 750*
$3p^5 3d(^1F^\circ)4s$	$^2F^\circ$	7/2	*1 198 260*
$3p^5 3d(^3D^\circ)4s$	$^2D^\circ$	3/2	*1 203 270*
		5/2	*1 206 410*
$3p^6 5f$	$^2F^\circ$	5/2	*1 349 580*
		7/2	*1 349 690*
$3p^6 6f$	$^2F^\circ$	7/2	*1 502 720*
		5/2	*1 502 810*
$3p^6 7f$	$^2F^\circ$	5/2	*1 617 890*
		7/2	*1 618 310*
Ni XI ($^1S_0$)	*Limit*		**1 812 000**

Ni XI

$Z=28$

Ar I isoelectronic sequence

Ground state: $1s^2 2s^2 2p^6 3s^2 3p^6\ ^1S_0$

Ionization energy = 2 589 000±8000 cm^{-1} (321.0±1.0 eV)

Swartz et al. (1976) identified the transition array $3p^5 3d-3p^5 4f$ in the region of 81–94 Å. With their classifications and the forbidden transitions within the $3p^5 3d$ configuration, identified in the solar corona by Svensson et al. (1974) and by Sandlin and Tousey (1979), these configurations were determined. The $^1P°$ term is from the laboratory observations of Even-Zohar and Fraenkel (1968). The uncertainty in the $3p^5 3d$ level values varies from ±10 cm^{-1} to ±100 cm^{-1} while that of the $3p^5 4f$ levels is ±100 cm^{-1}.

The $3p^5 4s$ and $4d$ configurations are taken from observations of resonance lines by Even-Zohar and Fraenkel between 60 and 80 Å. The level values are uncertain by about ±300 cm^{-1}.

The ionization energy was obtained by extrapolation by Lotz (1967).

Reference

Even-Zohar, M., and Fraenkel, B. S. (1968), J. Opt. Soc. Am. **58**, 1420.
Lotz, W. (1967), J. Opt. Soc. Am. **57**, 873.
Sandlin, G. D., and Tousey, R. (1979), Astrophys. J. **227**, L107.
Svensson, L. A., Ekberg, J. O., and Edlén, B. (1974), Sol. Phys. **34**, 173.
Swartz, M., Kastner, S. O., Goldsmith, L., and Neupert, W. (1976), J. Opt. Soc. Am. **66**, 240.

Ni XI

Configuration	Term	*J*	Level (cm^{-1})	Configuration	Term	*J*	Level (cm^{-1})
$3s^2 3p^6$	^1S	0	0	$3s^2 3p^5 4d$	$^3P°$	1	*1 571 310*
$3s^2 3p^5 3d$	$^3P°$	0	*469 310*	$3s^2 3p^5 4d$	$^1P°$	1	*1 594 130*
		1	*472 950*				
		2	*480 770*	$3s^2 3p^5(^2P°_{3/2})4f$	$^2[3/2]$	1	1 701 800
						2	1 704 300
$3s^2 3p^5 3d$	$^3F°$	4	*493 060*				
		3	*496 950*	$3s^2 3p^5(^2P°_{3/2})4f$	$^2[9/2]$	5	1 706 400
						4	1 708 600
$3s^2 3p^5 3d$	$^3D°$	3	*527 220*				
		1	*534 810*	$3s^2 3p^5(^2P°_{3/2})4f$	$^2[5/2]$	3	1 708 300
		2	*539 070*				
				$3s^2 3p^5(^2P°_{3/2})4f$	$^2[7/2]$	4	1 716 400
$3s^2 3p^5 3d$	$^1D°$	2	*530 050*				
				$3s^2 3p^5(^2P°_{1/2})4f$	$^2[5/2]$	3	1 732 900
$3s^2 3p^5 3d$	$^1F°$	3	*543 040*			2	1 739 400
$3s^2 3p^5 3d$	$^1P°$	1	*673 850*	$3s^2 3p^5(^2P°_{1/2})4f$	$^2[7/2]$	3	1 736 000
						4	1 736 400
$3s^2 3p^5 4s$	$^3P°$	1	*1 269 940*				
				Ni XII ($^2P°_{3/2}$)	***Limit***		**2 589 000**
$3s^2 3p^5 4s$	$^1P°$	1	*1 292 110*				

Ni XII

$Z=28$

Cl I isoelectronic sequence

Ground state: $1s^2 2s^2 2p^6 3s^2 3p^5\ ^2P^\circ_{3/2}$

Ionization energy $= 2\ 840\ 000 \pm 16000$ cm^{-1} (352 ± 2 eV)

Fawcett and Hatter (1980) observed the $3s^2 3p^5\ ^2P^\circ - 3s3p^6\ ^2S$ transitions at 295.321 Å and 317.475 Å with an accuracy of ± 0.008 Å, or a level uncertainty of ± 10 cm^{-1}.

The portion of the $3p^4 3d$ configuration below 600 000 cm^{-1} is taken from the study of coronal spectra by Edlén and Smitt (1978). Their level values are adjusted here to the calculated energies of Nussbaumer (1976), from which we adopted the value 454 000 for the $(^3P)^4D_{7/2}$ level. The value of "x" is estimated to be ± 1000 cm^{-1}. The higher $3p^4 3d$ terms are from Goldsmith and Fraenkel (1970); the uncertainty is ± 50 cm^{-1}. The designations of the parent terms of $3p^4 3d$ are taken from the calculations of Bromage, Cowan, and Fawcett (1977) for the isoelectronic spectrum Fe X.

The $3p^4 4s$, $3p^4 4d$, and $3p^4 4f$ terms are from the observations of Fawcett, Cowan, and Hayes (1972) near 70 Å and are uncertain by ± 500 cm^{-1}.

The ionization energy was obtained by extrapolation by Lotz (1967).

References

Bromage, G. E., Cowan, R. D., and Fawcett, B. C. (1977), Phys. Scr. **15**, 177.

Edlén, B., and Smitt, R. (1978), Sol. Phys. **57**, 329.

Fawcett, B. C., Cowan, R. D., and Hayes, R. W. (1972), J. Phys. **B5**, 2143.

Fawcett, B. C., and Hatter, A. T. (1980), Astron. Astrophys. **84**, 78.

Goldsmith, S., and Fraenkel, B. S. (1970), Astrophys. J. **161**, 317.

Lotz, W. (1967), J. Opt. Soc. Am. **57**, 873.

Nussbaumer, H. (1976), Astron. Astrophys. **48**, 93.

Ni XII

Configuration	Term	J	Level (cm^{-1})
$3p^5$	$^2P^\circ$	3/2	*0*
		1/2	*23 629*
$3s3p^6$	^2S	1/2	338 615
$3p^4(^3P)3d$	^4D	7/2	454 000+x
$3p^4(^3P)3d$	^4F	9/2	485 570+x
		7/2	492 750+x
$3p^4(^3P)3d$	^2F	7/2	513 290+x
$3p^4(^1D)3d$	^2G	7/2	526 960+x
		9/2	527 230+x
$3p^4(^1D)3d$	^2F	7/2	567 200+x
$3p^4(^1D)3d$	^2S	1/2	622 840
$3p^4(^3P)3d$	^2P	3/2	648 670
		1/2	657 290
$3p^4(^3P)3d$	^2D	5/2	657 230
		3/2	676 420
$3p^4(^3P)4s$	^2P	3/2	1 374 200
		1/2	1 385 600
$3p^4(^1D)4s$	^2D	5/2	1 401 000
		3/2	1 401 600
$3p^4(^3P)4d$	^2D	5/2	1 666 100
$3p^4(^3P)4f$	$^4F^\circ$	9/2	*1 797 400+x*
$3p^4(^3P)4f$	$^4G^\circ$	11/2	*1 808 000+x*
		9/2	*1 811 400+x*
$3p^4(^1D)4f$	$^2H^\circ$	11/2	*1 848 400+x*
Ni XIII ($^3P_2$)	***Limit***		**2 840 000**

Ni XIII

$Z=28$

S I isoelectronic sequence

Ground state: $1s^2 2s^2 2p^6 3s^2 3p^4\ {}^3P_2$

Ionization energy = 3 100 000±16 000 cm^{-1} (384±2 eV)

Svensson (1971) calculated the levels of the ground configuration with a stated accuracy of ±10–50 cm^{-1}. From these he made the following classifications of solar coronal lines within this configuration: ${}^3P_2-{}^1D_2$ at 2126.7 Å (in vac.) and ${}^3P_2-{}^3P_1$ at 5115.8 Å (in air). New measurements of the coronal spectrum observed from Skylab by Sandlin, Brueckner, and Tousey (1977) provide the more accurate wavelength of 2126.17±0.02 Å for ${}^3P_2-{}^1D_2$, and a new line at 1277.23±0.01 Å for ${}^3P_1-{}^1S_0$. The transition array $3s^2 3p^4 - 3s3p^5$ from 267–308 Å was measured by Fawcett and Hatter (1980) with an uncertainty of ±0.008 Å. Their line classifications were used to derive the $3s^2 3p^4\ {}^3P_0$ level and the ${}^3P°$ and ${}^1P°$ terms of $3s3p^5$.

The $3s^2 3p^3 3d$ levels are from laboratory observations of Fawcett and Hayes (1972) between 155 and 305 Å and are uncertain by about ±50 cm^{-1}.

The $3p^3 4d$ levels are from the observations of Fawcett, Cowan, and Hayes (1972) at 56 Å. The uncertainty is about ±500 cm^{-1}. They also observed levels in $3p^3 4f$, which are not connected with this system.

The ionization energy was obtained by Lotz (1967) by extrapolation.

References

Fawcett, B. C., and Hayes, R. W. (1972), J. Phys. **B5**, 366.

Fawcett, B. C., Cowan, R. D., and Hayes, R. W. (1972), J. Phys. **B5**, 2143.

Fawcett, B. C., and Hatter, A. T. (1980), Astron. Astrophys. **84**, 78.

Lotz, W., (1967), J. Opt. Soc. Am. **57**, 873.

Sandlin, G. D., Brueckner, G. E., and Tousey, R. (1977), Astrophys. J. **214**, 898.

Svensson, L. A. (1971), Sol. Phys. **18**, 232.

Ni XIII

Configuration	Term	J	Level (cm^{-1})	Configuration	Term	J	Level (cm^{-1})
$3s^2 3p^4$	${}^3P$	2	0.0	$3s^2 3p^3(^2D°)3d$	${}^1D°$	2	*666 000*
		1	19 541.8				
		0	20 060	$3s^2 3p^3(^2D°)3d$	${}^1F°$	3	*681 750*
$3s^2 3p^4$	${}^1D$	2	47 032.9	$3s^2 3p^3(^2D°)3d$	${}^1P°$	1	*715 960*
$3s^2 3p^4$	${}^1S$	0	97 836.2	$3s^2 3p^3(^4S°)4d$	${}^3D°$	3	*1 767 700*
$3s3p^5$	${}^3P°$	2	*330 215*	$3s^2 3p^3(^2D°)4d$	${}^1D°$	2	*1 820 400*
		1	*344 156*				
				$3s^2 3p^3(^2D°)4d$	${}^1F°$	3	*1 827 000*
$3s3p^5$	${}^1P°$	1	*395 545*				
				Ni XIV ($^4S°_{3/2}$)	*Limit*		**3 100 000**
$3s^2 3p^3(^2D°)3d$	${}^3P°$	2	*609 200*				
$3s^2 3p^3(^4S°)3d$	${}^3D°$	3	*634 000*				
		2	*644 660*				
		1	*649 900*				

Ni XIV

$Z=28$

P I isoelectronic sequence

Ground state: $1s^2 2s^2 2p^6 3s^2 3p^3\ ^4S^\circ_{3/2}$

Ionization energy = 3 470 000±16 000 cm^{-1} (430±2 eV)

The levels of the $3s^2 3p^3$ configuration are determined from identifications of solar coronal lines by Svensson (1971) and Sandlin, Brueckner, and Tousey (1977). The values are from wavelengths given in the latter paper reporting spectra observed by Skylab measured with an uncertainty of ±0.01 Å, giving a level uncertainty of ±0.5 cm^{-1}. The calculated value for $^2P^\circ_{3/2}$ is from Svensson, with an estimated uncertainty of ±50 cm^{-1}.

The $3s^2 3p^3 - 3s3p^4$ array in the range of 245–292 Å was measured by Fawcett and Hatter (1980) with an accuracy of ±0.008 Å. Their classifications are used to derive the levels of $3s3p^4$ with an uncertainty of ±10 cm^{-1}. The $3s^2 3p^2 3d$ levels are from the observations of Fawcett and Hayes (1972) between 160 and 320 Å and are uncertain by about ±60 cm^{-1}.

The $3p^2 4d$ levels are from the observations of Fawcett, Cowan, and Hayes (1972). The uncertainty is about ±500 cm^{-1}.

The ionization energy was obtained by extrapolation by Lotz (1967).

References

Fawcett, B. C., and Hayes, R. W. (1972), J. Phys. **B5**, 366.

Fawcett, B. C., and Hatter, A. T. (1980), Astron. Astrophys. (1980), **84**, 78.

Fawcett, B. C., Cowan, R. D., and Hayes, R. W. (1972), J. Phys. **B5**, 2143.

Lotz, W. (1967), J. Opt. Soc. Am. **57**, 873.

Sandlin, G. D., Brueckner, G. E., and Tousey, R. (1977), Astrophys. J. **214**, 898.

Svensson, L. A. (1971), Sol. Phys. **18**, 232.

Ni XIV

Configuration	Term	J	Level (cm^{-1})	Configuration	Term	J	Level (cm^{-1})
$3s^2\,3p^3$	$^4S^\circ$	3/2	*0.0*	$3s^2\,3p^2(^1D)3d$	^2D	3/2	632 280
						5/2	634 430
$3s^2\,3p^3$	$^2D^\circ$	3/2	*45 767.8*				
		5/2	*53 569.0*	$3s^2\,3p^2(^1D)3d$	^2P	1/2	648 320
						3/2	660 710+x
$3s^2\,3p^3$	$^2P^\circ$	1/2	*85 126.7*				
		3/2	*96 630*+x	$3s^2\,3p^2(^1D)3d$	^2F	7/2	662 780
$3s3p^4$	^4P	5/2	316 343	$3s^2\,3p^2(^3P)3d$	^2D	5/2	690 560+x
		3/2	330 837			3/2	691 930
$3s3p^4$	^2D	3/2	391 916	$3s^2\,3p^2\,4d$	^2P	3/2	1 628 400
		5/2	395 567				
				$3s^2\,3p^2\,4d$	^2D	5/2	1 653 100
$3s3p^4$	^2P	3/2	447 765				
		5/2	452 850+x	Ni XV ($^3P_0$)	***Limit***		**3 470 000**
$3s^2\,3p^2(^3P)3d$	^4P	5/2	583 530				
		3/2	589 310				
		1/2	594 810				

Ni XV

$Z=28$

Si I isoelectronic sequence

Ground state: $1s^2 2s^2 2p^6 3s^2 3p^2\ ^3P_0$

Ionization energy = 3 740 000±16 000 cm^{-1} (464±2 eV)

The levels of the $3s^2 3p^2$ configuration are determined from measurements of solar coronal lines and are uncertain by about ±0.5 cm^{-1}. The ^3P values are from Svensson's (1971) compilation and the ^1D value is from the measurement by Sandlin, Brueckner, and Tousey (1977) of spectra observed from Skylab.

The transition array $3s^2 3p^2 - 3s3p^3$ was observed in the range of 209–319 Å by Fawcett and Hatter (1980) with an accuracy of ±0.008 Å. Their assignments are used to derive the levels of $3s3p^3$ with an uncertinaty of ±10 cm^{-1}. The $3s^2 3p3d$ levels are from the analysis by Fawcett and Hayes (1972) based on observations between 170 and 320 Å; they are uncertain by about ±60 cm^{-1}.

The $3p4s$, $3p4d$, and $3p4f$ levels are from the observations of Fawcett, Cowan, and Hayes (1972) and of Kastner, Swartz, Bhatia, and Lapides (1978). The uncertainty is about ±500 cm^{-1}.

The ionization energy was obtained by extrapolation by Lotz (1967).

References

Fawcett, B. C., and Hayes, R. W. (1972), J. Phys. **B5**, 366.

Fawcett, B. C., and Hatter, A. T. (1980), Astron. Astrophys. **84**, 78.

Fawcett, B. C., Cowan, R. D., and Hayes, R. W. (1972), J. Phys. **B5**, 2143.

Kastner, S. O., Swartz, M., Bhatia, A. K., and Lapides, J. (1978), J. Opt. Soc. Am. **68**, 1558.

Lotz, W. (1967), J. Opt. Soc. Am. **57**, 873.

Sandlin, G. D., Brueckner, G. E., and Tousey, R. (1977), Astrophys. J. **214**, 898.

Svensson, L. A. (1971), Sol. Phys. **18**, 232.

Ni XV

Configuration	Term	J	Level (cm^{-1})	Configuration	Term	J	Level (cm^{-1})
$3s^2 3p^2$	^3P	0	0.0	$3s^2 3p3d$	$^1F°$	3	*638 460*
		1	14 917.5				
		2	27 376.5	$3s^2 3p4s$	$^3P°$	2	*1 730 700*
$3s^2 3p^2$	^1D	2	62 852.1	$3s^2 3p4s$	$^1D°$	2	1 741 300
$3s3p^3$	$^3D°$	1	*335 400*	$3s^2 3p4d$	$^3D°$	2	*2 018 400*
		2	*335 682*			3	*2 020 500*
		3	*340 794*				
				$3s^2 3p4d$	$^3F°$	3	*2 042 500*
$3s3p^3$	$^3S°$	1	*478 010*				
				$3s^2 3p4d$	$^1F°$	3	*2 053 000*
$3s3p^3$	$^1P°$	1	*509 211*				
				$3s^2 3p4f$	^1G	4	2 185 600
$3s^2 3p3d$	$^3P°$	2	*555 830*				
		1	*565 930*	Ni XVI ($^2P^\circ_{1/2}$)	*Limit*		**3 740 000**
$3s^2 3p3d$	$^1D°$	2	*574 330*				
$3s^2 3p3d$	$^3D°$	1	*582 760*				
		3	*585 170*				
		2	*586 410*				

Ni XVI

$Z=28$

Al I isoelectronic sequence

Ground state: $1s^22s^22p^63s^23p\ ^2P^\circ_{1/2}$

Ionization energy = 4 020 000±16 000 cm^{-1} (499±2 eV)

The observations of the transition array $3s^23p-3s3p^2$ in the range 218–309 Å by Fawcett and Hatter (1980) provided the ^{2}P ground term interval and the levels of $3s3p^2$ with an accuracy of ±20 cm^{-1}. An isoelectronic plot of the transition energy $3s^23p\ ^2P^\circ_{1/2}-3s3p^2\ ^2D_{3/2}$ indicates that there is a misprint for the wavelength 288.165 Å. It should be 289.165 Å. The $3s^23d$ levels are from the line classifications near 200 Å of Fawcett and Hayes (1972). They estimate the wavelength uncertainty to be ±0.03 Å (±75 cm^{-1}). The $3s^24d$ and $4f$ levels are from wavelengths near 50 Å classified by Fawcett, Cowan, and Hayes (1972) and are uncertain by ±300 cm^{-1}.

The ionization energy was obtained by extrapolation by Lotz (1967).

References

Fawcett, B. C., and Hatter, A. T. (1980), Astron. Astrophys. **84**, 78.

Fawcett, B. C., Cowan, R. D., and Hayes, R. W. (1972), J. Phys. **B5**, 2143.

Fawcett, B. C., and Hayes, R. W. (1972), J. Phys. **B5**, 366.

Lotz, W. (1967), J. Opt. Soc. Am. **57**, 873.

Ni XVI

Configuration	Term	*J*	Level (cm^{-1})	Configuration	Term	*J*	Level (cm^{-1})
$3s^2\,3p$	$^2P^\circ$	1/2	*0*	$3s^2\,3d$	^2D	3/2	539 870
		3/2	*27 770*			5/2	543 170
$3s3p^2$	^2D	3/2	345 820	$3s^2\,4d$	^2D	3/2	2 119 400
		5/2	351 210			5/2	2 121 100
$3s3p^2$	^2S	1/2	417 450	$3s^2\,4f$	$^2F^\circ$	7/2	*2 228 500*
						5/2	*2 228 600*
$3s3p^2$	^2P	1/2	448 170				
		3/2	457 910	Ni XVII ($^1S_0$)	***Limit***		**4 020 000**

Ni XVII

$Z=28$

Mg I isoelectronic sequence

Ground state: $1s^2 2s^2 2p^6 3s^2\ ^1S_0$

Ionization energy $=$ 4 606 000$\pm$1000 cm^{-1} (571.08$\pm$0.12 eV)

Fawcett et al. (1966) classified lines in the range of 30–55 Å that are transitions among configurations $3s3l$–$3snl'$ for $n>3$. These were remeasured with improved accuracy of $\pm$0.01 Å with some additional classifications by Feldman et al. (1971). The uncertainty of level values is $\pm$1000 cm^{-1}. The transition arrays $3s3p-3p^2$ and $3s3p-3s3d$ in the range of 210–302 Å were classified by Fawcett, Cowan, and Hayes (1972). Their measurement uncertainty is reported as $\pm$0.02 Å, giving a level uncertainty of $\pm$30 cm^{-1}. They also revised the classification of $3s3d\ ^1D_2$–$3s4f\ ^1F^\circ_3$ and gave new measurements for the lines given in Fawcett et al. (1966) with an acccuracy of $\pm$0.01 Å. No intersystem lines were identified in these observations. We use the new measurements of Fawcett and Hatter (1980) for the array $3s3p\ ^3P^\circ-3p^2\ ^3P$. They estimate about the same level uncertainty as Fawcett et al. (1972) but give new values.

Finkenthal et al. (1982) identified the line $3s^2\ ^1S_0-3s3p\ ^3P^\circ_1$ in a tokamak plasma at 366.7$\pm$0.3 Å, providing the connection between the triplet terms and the ground state. We use the solar flare measurement by Dere (1978) of 366.80$\pm$0.03 Å. The $3s^2\ ^1S_0$–$3s3p\ ^1P^\circ_1$ transition was identified by Fawcett and Hayes (1972) at 249.180$\pm$0.004 Å; the wavelength was obtained from the solar observations of Behring, Cohen, and Feldman (1972).

The $3p3d-3p4f$ transition array has been observed at 55 Å by Kastner, Swartz, Bhatia, and Lapides (1978) but it is unconnected with the present configurations.

We derived the ionization energy from the three member $3snd\ ^3D$ series. The $6f$ term does not follow the nf series quantum defect trend.

References

Behring, W. E., Cohen, R., and Feldman, U. (1972), Astrophys. J. **175**, 493.

Dere, K. P. (1978), Astrophys. J. **221**, 1062.

Fawcett, B. C., Cowan, R. D., and Hayes, R. W. (1972), J. Phys. **B5**, 2143.

Fawcett, B. C., Gabriel, A. H., Irons, F. E., Peacock, N. J., and Saunders, P. A. H. (1966), Proc. Phys. Soc. **88**, 1051.

Fawcett, B. C., and Hayes, R. W. (1972), J. Phys. **B5**, 366.

Fawcett, B. C., and Hatter, A. T. (1980), Astron. Astrophys. **84**, 78.

Feldman, U., Katz, L., Behring, W. E., and Cohen, L. (1971), J. Opt. Soc. Am. **61**, 91.

Finkenthal, M., Hinnov, E., Cohen, S. and Suckewer, S. (1982), Phys. Lett. **A9**, 284.

Kastner, S. O., Swartz, M., Bhatia, A. K., and Lapides, J. (1978), J. Opt. Soc. Am. **68**, 1558.

Ni XVII

Configuration	Term	J	Level (cm^{-1})	Configuration	Term	J	Level (cm^{-1})
$3s^2$	^1S	0	0	$3s4d$	^3D	1	2 494 100
						2	2 495 200
$3s3p$	$^3P^\circ$	0	*264 400*			3	2 497 300
		1	*272 630*				
		2	*293 650*	$3s4d$	^1D	2	2 499 400
$3s3p$	$^1P^\circ$	1	*401 320*	$3p4s$	$^3P^\circ$	2	*2 542 800*
$3p^2$	^3P	0	627 870	$3s4f$	$^3F^\circ$	2	*2 585 000*
		1	643 820			3	*2 585 400*
		2	669 500			4	*2 585 700*
$3s3d$	^3D	1	771 290	$3s4f$	$^1F^\circ$	3	*2 601 600*
		2	772 970				
		3	775 670	$3s5p$	$^1P^\circ$	1	*3 234 300*
$3s3d$	^1D	2	864 600	$3s5d$	^3D	1,2	3 272 200
						3	3 272 900
$3s4s$	^3S	1	2 187 600				
				$3s5f$	$^3F^\circ$	2,3,4	*3 312 800*
$3s4p$	$^1P^\circ$	1	*2 333 400*				
				$3s6f$	$^3F^\circ$	4	*3 721 600*
				Ni XVIII ($^2S_{1/2}$)	***Limit***		**4 606 000**

Ni XVIII

$Z=28$

Na I isoelectronic sequence

Ground state: $1s^22s^22p^63s\ ^2S_{1/2}$

Ionization energy = 4 896 200±500 cm^{-1} (607.06±0.06 eV)

The $3s-3p$ doublet was reported by Fawcett and Hatter (1980); they measured the $^2S_{1/2}-^2P^\circ_{3/2}$ at 291.983±0.008 Å and the $^2S_{1/2}-^2P^\circ_{1/2}$ at 320.56±0.03 Å. The latter has been observed by Peacock, Stamp, and Silver (1984) in a tokamak plasma at 320.565±0.006 Å. Fawcett, Cowan, and Hayes (1972) measured the $3p-3d$ doublet with an uncertainty of ±0.02 Å. Improved values are given by Edlén (1978). The uncertainty of these level values is ±10 cm^{-1}.

Except for the $3d-4f$ doublet measured by Edlén (1936), we obtained the nf terms from the $3d-nf$ measurements reported by Feldman et al. (1971). They also found the $4s$, $np(n=4–6)$, and $nd(n=4–8)$ terms. Their measurement uncertainty of ±0.01 Å gives an energy level uncertainty of ±1000 cm^{-1}.

The $4f-5g$ doublet was found by Kononov et al. (1977) and confirmed by Edlén (1978) by using a polarization formula.

Feldman and Cohen (1967) identified the $2p^63s-2p^53s^2$ transitions at 14.37 Å and 14.10 Å with an uncertainty of ±0.01 Å.

The value for the ionization energy was derived by Edlén (1978) from core polarization theory applied to the nf series.

References

Edlén, B. (1936), Z. Phys. **100**, 621.

Edlén, B. (1978), Phys. Scr. **17**, 565.

Fawcett, B. C., and Hatter, A. T. (1980), Astron. Astrophys. **84**, 78.

Fawcett, B. C., Cowan, R. D., and Hayes, R. W. (1972), J. Phys. **B5**, 2143.

Feldman, U., and Cohen, L. (1967), J. Opt. Soc. Am. **57**, 1128.

Feldman, U., Katz, L., Behring, W., and Cohen, L. (1971), J. Opt. Soc. Am. **61**, 91.

Kononov, E. Y., Kovalev, V. I., Ryabtsev, A. N., and Churilov, S. S. (1977), Sov. J. Quantum Electronics **7**, 111.

Peacock, N. J., Stamp, M. F., and Silver, J. D. (1984), Phys. Scr. **T8**, 10.

Sandlin, G. D., Brueckner, G. E., Scherrer, V. E., and Tousey, R. (1976), Astrophys. J. **205**, L47.

Ni XVIII

Configuration	Term	J	Level (cm^{-1})
$2p^6(^1S)3s$	^2S	1/2	0
$2p^6(^1S)3p$	$^2P^\circ$	1/2	*311 949*
		3/2	*342 485*
$2p^6(^1S)3d$	^2D	3/2	765 610
		5/2	770 280
$2p^6(^1S)4s$	^2S	1/2	2 301 800
$2p^6(^1S)4p$	$^2P^\circ$	1/2	*2 426 100*
		3/2	*2 438 100*
$2p^6(^1S)4d$	^2D	3/2	2 594 400
		5/2	2 596 500
$2p^6(^1S)4f$	$^2F^\circ$	5/2	*2 666 230*
		7/2	*2 667 100*
$2p^6(^1S)5p$	$^2P^\circ$	1/2	*3 352 400*
		3/2	*3 358 100*
$2p^6(^1S)5d$	^2D	3/2	3 433 700
		5/2	3 434 600
$2p^6(^1S)5f$	$^2F^\circ$	5/2	*3 469 100*
		7/2	*3 469 400*
$2p^6(^1S)5g$	^2G	7/2	3 473 000
		9/2	3 473 300
$2p^6(^1S)6p$	$^2P^\circ$	1/2	*3 839 400*
		3/2	*3 843 200*
$2p^6(^1S)6d$	^2D	3/2	3 885 700
		5/2	3 886 100
$2p^6(^1S)6f$	$^2F^\circ$	5/2	*3 905 800*
		7/2	*3 906 100*
$2p^6(^1S)7d$	^2D	5/2	4 156 700
$2p^6(^1S)7f$	$^2F^\circ$	5/2	*4 169 000*
		7/2	*4 169 100*
$2p^6(^1S)8d$	^2D	3/2	4 331 100
		5/2	4 331 300
$2p^6(^1S)8f$	$^2F^\circ$	7/2	*4 339 400*
Ni XIX ($^1S_0$)	***Limit***		**4 896 200**
$2p^5\,3s^2$	$^2P^\circ$	3/2	*6 959 000*
		1/2	*7 092 000*

Ni XIX

$Z=28$

Ne I isoelectronic sequence

Ground state: $1s^22s^22p^6\ ^1S_0$

Ionization energy = 12 430 000±10 000 cm^{-1} (1541±1 eV)

Only resonance lines are classified by this system of energy levels. The first line identifications were made by Feldman, Cohen, and Swartz (1967), who reported transitions from the $2p^53s$, $2p^53d$ and $2s2p^63p$ configurations. To these were added terms of $2p^54d$ by Feldman and Cohen (1967). These were augmented by the work of Swartz, Kastner, Rothe, and Neupert (1971), who observed lines originating from $2p^54s$, $2p^55d$, and $2p^56d$.

The spectrum was reobserved by Gordon, Hobby, and Peacock (1980) in the range of 8–14 Å with a wavelength uncertainty of ±0.005 Å. Their results, with a level value uncertainty of ±5000 cm^{-1}, are quoted here. They include all the lines previously identified plus transitions from the $^3P_1$ state of $2s^22p^54d$, $5d$, $6d$, and the levels of $2s2p^64p$.

By means of a very low pressure laboratory light source (a tokamak), Klapisch, Bar Shalom, Schwob, Fraenkel, Breton, de Michelis, Finkenthal, and Mattioli (1978) observed the magnetic quadrupole transition from $2p^53s\ ^3P^\circ_2$ to the ground state as well as the transitions from the $J=1$ states of $2p^53s$ and $2p^53d$. No uncertainty estimate is given, but it is presumably ±0.005 Å, which would give a level uncertainty of ±3000 cm^{-1}.

We have assigned designations and given percentages to the $2p^53s$ and $2p^53d$ configurations based on our own calculations. Both jj- and LS-leading percentages are listed for the $2p^53s$ levels.

Kastner, Behring, and Cohen (1975) identified transitions between $2p^53p$ and $2p^54d$, but there is no connection with the levels given here.

We derived the ionization energy from the $2s^22p^5nd$ $^3D_1$ series for $n=3$–6.

References

Feldman, U., Cohen, L., and Swartz, M. (1967), Astrophys. J. **148**, 585.

Feldman, U., and Cohen, R. (1967), Astrophys. J. **149**, 45.

Gordon, H., Hobby, M. G., and Peacock, N. J. (1980), J. Phys. **B13**, 1985.

Kastner, S. O., Behring, W. E., and Cohen, L. (1975), Astrophys. J. **199**, 777.

Klapisch, M., Bar Shalom, A., Schwob, J. L., Fraenkel, B. S., Breton, C., de Michelis, C., Finkenthal, M., and Mattioli, M. (1978), Phys. Lett. **69A**, 34.

Swartz, M., Kastner, S., Rothe, E., and Neupert, W. (1971), J. Phys. **B4**, 1747.

Ni XIX

Configuration	Term	J	Level (cm^{-1})	Leading percentages			
$2s^2\,2p^6$	^1S	0	0				
$2s^2\,2p^5(^2P^\circ_{3/2})3s$	$(^3/_2,^1/_2)^\circ$	2	*7 104 000*	100	or	100	$^3P^\circ$
		1	*7 121 000*	98	or	53	$^1P^\circ$
$2s^2\,2p^5(^2P^\circ_{1/2})3s$	$(^1/_2,^1/_2)^\circ$	1	*7 257 400*	98	or	53	$^3P^\circ$
$2s^2\,2p^5\,3d$	$^3P^\circ$	1	*7 805 200*	91		9	$^3D^\circ$
$2s^2\,2p^5\,3d$	$^3D^\circ$	1	*7 901 400*	67		28	$^1P^\circ$
$2s^2\,2p^5\,3d$	$^1P^\circ$	1	*8 041 800*	72		24	$^3D^\circ$
$2s2p^6\,3p$	$^3P^\circ$	1	*8 621 400*				
$2s2p^6\,3p$	$^1P^\circ$	1	*8 666 300*				
$2s^2\,2p^5(^2P^\circ_{3/2})4s$	$(^3/_2,^1/_2)^\circ$	1	*9 585 000*				
$2s^2\,2p^5(^2P^\circ_{1/2})4s$	$(^1/_2,^1/_2)^\circ$	1	*9 725 000*				
$2s^2\,2p^5\,4d$	$^3P^\circ$	1	*9 845 000*				
$2s^2\,2p^5\,4d$	$^3D^\circ$	1	*9 891 000*				

(Continued)

Ni XIX—Continued

Configuration	Term	J	Level (cm^{-1})	Leading percentages
$2s^2\,2p^5\,4d$	$^1P^\circ$	1	*10 023 000*	
$2s^2\,2p^5\,5d$	$^3P^\circ$	1	*10 797 000*	
$2s^2\,2p^5\,5d$	$^3D^\circ$	1	*10 806 000*	
$2s2p^6\,4p$	$^3P^\circ$	1	*10 925 000*	
$2s2p^6\,4p$	$^1P^\circ$	1	*10 941 000*	
$2s^2\,2p^5\,5d$	$^1P^\circ$	1	*10 942 000*	
$2s^2\,2p^5\,6d$	$^3D^\circ$	1	*11 301 000*	
$2s^2\,2p^5\,6d$	$^1P^\circ$	1	*11 436 000*	
Ni XX ($^2P^\circ_{3/2}$)	***Limit***		**12 430 000**	

Ni XX

$Z=28$

F I isoelectronic sequence

Ground state: $1s^2 2s^2 2p^5\ ^2P^\circ_{3/2}$

Ionization energy = 13 290 000±27 000 cm⁻¹ (1648±3 eV)

The ground term splitting is from a magnetic dipole transition at 694.64±0.03 Å observed in a tokamak discharge by Peacock, Stamp, and Silver (1984). This line was first reported by Hinnov et al. (1982). The $2s2p^6$ configuration was determined from the resonance doublet $^2P^\circ - ^2S$ measured by Doschek et al. (1974) with an accuracy of ±0.02 Å, giving a level uncertainty for $2s2p^6\ ^2S$ of ±300 cm^{-1}.

The $2p^5-2p^4 3s$ and $2p^5-2p^4 3d$ arrays were analyzed by Boiko et al. (1978). The region was remeasured from 9–13 Å by Gordon, Hobby, and Peacock (1980) with an accuracy of ±0.005 Å, giving a level value uncertainty of ±5000 cm^{-1}. They identified additional lines of these arrays as well as the higher configurations $2s2p^5 3p$, $2p^4 4d$, and $2p^4 4s$. There results are given here.

The ionization energy is an extrapolated value by Lotz (1967).

References

Boiko, V. A., Pikuz, S. A., Safronova, A. S., and Faenov, A. Y. (1978), Opt. Spectrosc. **44**, 498.

Doschek, G. A., Feldman, U., Cowan, R. D., and Cohen, L. (1974), Astrophys. J. **188**, 417.

Gordon, H., Hobby, M. G., and Peacock, N. J. (1980), J. Phys. **B13**, 1985.

Hinnov, E., Suckewer, S., Cohen, S., and Sato, K. (1982), Phys. Rev. **A25**, 2293.

Lotz, W. J. (1967), J. Opt. Soc. Am. **57**, 873.

Peacock, N. J., Stamp, M. F., and Silver, J. D. (1984), Phys. Scr. **T8**, 10.

Ni XX

Configuration	Term	J	Level (cm^{-1})
$2s^2 2p^5$	$^2P^\circ$	3/2	0
		1/2	*143 959*
$2s2p^6$	^2S	1/2	1 202 300
$2s^2 2p^4(^3P)3s$	^2P	3/2	7 544 000
		1/2	7 673 000
$2s^2 2p^4(^3P)3s$	^4P	5/2	7 514 000
		1/2	7 613 000
		3/2	7 648 000
$2s^2 2p^4(^1D)3s$	^2D	5/2	7 736 000
		3/2	7 742 000
$2s^2 2p^4(^1S)3s$	^2S	1/2	7 949 000
$2s^2 2p^4(^3P)3d$	^4P	1/2	8 226 000
		3/2	8 244 000
$2s^2 2p^4(^3P)3d$	^2F	5/2	8 256 000
$2s^2 2p^4(^3P)3d$	^4D	1/2	8 304 000
		3/2	8 329 000
$2s^2 2p^4(^3P)3d$	^2P	3/2	8 353 000
$2s^2 2p^4(^3P)3d$	^2D	5/2	8 360 000
$2s^2 2p^4(^1D)3d$	^2S	1/2	8 423 000
$2s^2 2p^4(^1D)3d$	^2D	5/2	8 452 000
		3/2	8 484 000
$2s^2 2p^4(^1D)3d$	^2P	3/2	8 445 000
		1/2	8 495 000
$2s^2 2p^4(^1S)3d$	^2D	3/2	8 634 000
$2s2p^5(^3P^\circ)3p$	^4D	3/2	8 908 000
$2s2p^5(^3P^\circ)3p$	^2D	5/2	8 908 000
		3/2	9 052 000
$2s2p^5(^3P^\circ)3p$	^2P	3/2	8 962 000
		1/2	8 978 000
$2s2p^5(^3P^\circ)3p$	^4P	3/2	9 008 000
$2s2p^5(^3P^\circ)3p$	^2S	1/2	9 092 000
$2s2p^5(^1P^\circ)3p$	^2D	3/2	9 250 000
		5/2	9 283 000
$2s2p^5(^1P^\circ)3p$	^2P	1/2	9 288 000
		3/2	9 303 000
$2s^2 2p^4(^3P)4s$	^2P	3/2	10 182 000
$2s^2 2p^4(^3P)4s$	^4P	3/2	10 317 000
$2s^2 2p^4(^3P)4d$	^2D	3/2,5/2	10 462 000
$2s^2 2p^4(^3P)4d$	^4F	3/2,5/2	10 530 000
$2s^2 2p^4(^3P)4d$	^4P	5/2	10 576 000

(Continued)

Ni XX—Continued

Configuration	Term	J	Level (cm^{-1})	Configuration	Term	J	Level (cm^{-1})
$2s^2\,2p^4(^3P)4d$	^2P	$3/2$	10 581 000				
$2s^2\,2p^4(^1D)4d$	^2D	$5/2$	10 655 000				
		$3/2$	10 674 000				
$2s^2\,2p^4(^1D)4d$	^2P	$3/2$	10 655 000				
		$1/2$	10 674 000				
$2s^2\,2p^4(^1D)4d$	^2S	$1/2$	10 655 000				
$2s^2\,2p^4(^1D)4d$	^2F	$5/2$	10 677 000				
$2s^2\,2p^4(^1D)4d$	^2D	$5/2$	10 655 000				
$2s^2\,2p^4(^1S)4d$	^2D	$3/2$	10 853 000				
Ni XXI ($^3P_2$)	***Limit***		**13 290 000**				

Ni XXI

$Z=28$

O I isoelectronic sequence

Ground state: $1s^22s^22p^4\ ^3P_2$

Ionization energy = 14 160 000±28 000 cm^{-1} (1756±4 eV)

The fine structure of the ^{3}P ground term is determined from two magnetic dipole transitions, $^3P_2-^3P_1$ at 779.5 Å and $^3P_0-^3P_1$ at 2818.2 Å (in air), measured in a tokamak plasma by Hinnov et al. (1982) with an accuracy of ±0.3 Å. A solar flare line at 471.15 Å was identified by Widing (1978) as the intersystem line $2s^22p^4\ ^3P_2-^1D_2$.

Doschek et al. (1974) analyzed the $2s^22p^4-2s2p^5$ array and Doschek et al. (1975) found the $2p^6\ ^1S_0$. Lawson and Peacock (1980) remeasured these arrays from 69–120 Å and found several intercombination lines. Lawson and Peacock's data are used to obtain the $^1S_0$ level of $2s^22p^4$ and the levels of $2s2p^5$ with an uncertainty of ±300 cm^{-1}. Percentage compositions of these levels were provided by Kaufman and Sugar (1982), with configuration interaction between $2s^22p^4$ and $2p^6$.

The spectral region from 8–11 Å was measured by Gordon, Hobby, and Peacock (1980) with an accuracy of ±0.005 Å, giving a level uncertainty of ±6000 cm^{-1}. They classified lines arising from the $2p^33s$, $2p^33d$, and $2p^34d$ configurations in this range. Only a few lines are assigned to $2p^34d$, two of which are assigned to six transitions each. We did not attempt to derive energy levels for this configuration.

The ionization energy is an extrapolated value by Lotz (1967).

References

Doschek, G. A., Feldman, U., Cowan, R. D., and Cohen, L. (1974), Astrophys. J. **188**, 417.

Doschek, G. A., Feldman, U., Davis, J., and Cowan, R. D. (1975), Phys. Rev. **A12**, 980.

Gordon, H., Hobby, M. G., and Peacock, N. J. (1980), J. Phys. **B13**, 1985.

Hinnov, E., Suckewer, S., Cohen, S., and Sato, K. (1982), Phys. Rev. **A25**, 2293.

Kaufman, V., and Sugar, J. (1982), private communication.

Lawson, K. D., and Peacock, N. J. (1980), J. Phys. **B13**, 3313.

Lotz, W. J. (1967), J. Opt. Soc. Am. **57**, 873.

Widing, K. G. (1978), Astrophys. J. **222**, 735.

Ni XXI

Configuration	Term	J	Level (cm^{-1})	Leading percentages		
$2s^2\,2p^4$	^{3}P	2	0	87	13	^{1}D
		0	92 810	73	27	^{1}S
		1	128 290	100		
$2s^2\,2p^4$	^{1}D	2	212 250	87	13	^{3}P
$2s^2\,2p^4$	^{1}S	0	406 800	71	27	^{3}P
$2s2p^5$	^{3}P°	2	*1 043 300*	100		
		1	*1 126 000*	95	5	^{1}P°
		0	*1 193 100*	100		
$2s2p^5$	^{1}P°	1	*1 436 400*	95	5	^{3}P°1
$2p^6$	^{1}S	0	2 403 500	98	2	$2s^22p^4\ ^1$S
$2s^2\,2p^3(^4S°)3s$	^{3}S°	1	*8 035 000*			
$2s^2\,2p^3(^2D°)3s$	^{3}D°	1,2	*8 146 000*			
		3	*8 191 000*			
$2s^2\,2p^3(^2D°)3s$	^{1}D°	2	*8 212 000*			
$2s^2\,2p^3(^2P°)3s$	^{3}P°	0	*8 295 000*			
		1	*8 313 000*			
		2	*8 405 000*			
$2s^2\,2p^3(^2P°)3s$	^{1}P°	1	*8 422 000*			

(Continued)

Ni XXI—Continued

Configuration	Term	J	Level (cm^{-1})	Leading percentages
$2s^2\,2p^3(^4S°)3d$	$^3D°$	3	*8 666 000*	
$2s^2\,2p^3(^2D°)3d$	$^3D°$	3	*8 835 000*	
$2s^2\,2p^3(^2D°)3d$	$^3P°$	2	*8 848 000*	
$2s^2\,2p^3(^2D°)3d$	$^1F°$	3	*8 896 000*	
$2s^2\,2p^3(^2P°)3d$	$^3F°$	3	*8 895 000*	
$2s^2\,2p^3(^2P°)3d$	$^3P°$	2 1	*8 924 000* *9 000 000*	
$2s^2\,2p^3(^2P°)3d$	$^3D°$	2	*9 034 000*	
$2s^2\,2p^3(^2P°)3d$	$^1F°$	3	*9 048 000*	
$2s^2\,2p^3(^2P°)3d$	$^1D°$	2	*9 048 000*	
$2s^2\,2p^3(^2P°)3d$	$^1P°$	1	*11 086 000*	
Ni XXII	***Limit***		**14 160 000**	

Ni XXII

$Z=28$

N I isoelectronic sequence

Ground state: $1s^2 2s^2 2p^3\ ^4S^\circ_{3/2}$

Ionization energy = 15 280 000±30 000 cm^{-1} (1894±4 eV)

Two magnetic dipole lines observed in a tokamak plasma by Hinnov, Suckewer, Cohen, and Sato (1982) connect the doublet and quartet systems: $2s^2 2p^3\ ^4S^\circ_{3/2} - ^2D^\circ_{3/2}$ at 634.8±0.3 Å and $^4S^\circ_{3/2} - ^2D^\circ_{5/2}$ at 477.6±0.3 Å. The earlier analysis of the $2s^2 2p^3 - 2s2p^4$ array by Doschek, Feldman, Cowan, and Cohen (1974) was extended by Lawson and Peacock (1980) to include intersystem lines as well as the additional array $2s2p^4 - 2p^5$. Their measurements in the range 71–124 Å, reported with an uncertainty of ±0.03 Å, are used here to derive the energy levels, giving a level value uncertainty of ±400 cm^{-1}. The percentage compositions of the levels were provided by Kaufman and Sugar (1982).

The ionization energy was obtained by Lotz (1967) by extrapolation.

References

Doschek, G. A., Feldman, U., Cowan, R. D., and Cohen, L. (1974), Astrophys. J. **188**, 417.
Hinnov, E., Suckewer, S., Cohen, S., and Sato, K. (1982), Phys. Rev. **A25**, 2293.
Kaufman, V., and Sugar, J. (1982), private communication.
Lawson, K. D. and Peacock, N. J. (1980), J. Phys. **B13**, 3313.
Lotz, W. (1967), J. Opt. Soc. Am. **57**, 873.

Ni XXII

Configuration	Term	J	Level (cm^{-1})	Leading percentages		
$2s^2 2p^3$	$^4S^\circ$	3/2	*0*	81	15	$^2P^\circ$
$2s^2 2p^3$	$^2D^\circ$	3/2	*157 530*	70	16	$^2P^\circ$
		5/2	*209 380*	100		
$2s^2 2p^3$	$^2P^\circ$	1/2	*302 500*	99	1	$2p^5\ ^2P^\circ$
		3/2	*400 100*	68	25	$2p^3\ ^2D^\circ$
$2s(^2S)2p^4(^3P)$	^4P	5/2	848 100	96	4	$2s(^2S)2p^4(^1D)\ ^2D$
		3/2	943 000	97	2	$2s(^2S)2p^4(^1D)\ ^2D$
		1/2	968 000	92	8	$2s(^2S)2p^4(^1S)\ ^2S$
$2s(^2S)2p^4(^1D)$	^2D	3/2	1 176 200	89	8	$2s(^2S)2p^4(^3P)\ ^2P$
		5/2	1 203 400	96	4	$2s(^2S)2p^4(^3P)\ ^4P$
$2s(^2S)2p^4(^1S)$	^2S	1/2	1 344 600	64	30	$2s(^2S)2p^4(^3P)\ ^2P$
$2s(^2S)2p^4(^3P)$	^2P	3/2	1 399 000	91	9	$2s(^2S)2p^4(^1D)\ ^2D$
		1/2	1 536 500	70	28	$2s(^2S)2p^4(^1S)\ ^2S$
$2p^5$	$^2P^\circ$	3/2	*2 190 500*	98	2	$2s^2 2p^3\ ^2P^\circ$
		1/2	*2 340 900*	99	1	
Ni XXIII ($^3P_0$)	*Limit*		15 280 000			

Ni XXIII

$Z=28$

C I isoelectronic sequence

Ground state: $1s^2 2s^2 2p^2\ ^3P_0$

Ionization energy = 16 220 000±30 000 cm^{-1} (2011±4 eV)

Four magnetic dipole lines observed in a tokamak discharge by Hinnov, Suckewer, Cohen, and Sato (1982) determine the ^3P and ^1D levels of the ground configuration: the $2s^2 2p^2$ $^3P_0-^3P_1$ at 911.0 Å, $^3P_1-^3P_2$ at 1915.0 Å, $^3P_2-^1D_2$ at 614.8 Å, and $^3P_1-^1D_2$ at 465.4 Å. The uncertainty of these measurements is ±0.3 Å.

The rest of the levels of $2s^2 2p^2$, $2s2p^3$, and $2p^4$ (except for $2s2p^3\,^5S_2^\circ$) were derived from the lines observed and classified by Lawson and Peacock (1980) in the range of 74–128 Å. Their measurement uncertainty is ±0.03 Å, giving a level value uncertainty of ±400 cm^{-1}. Edlén (1984) has compared the known values of the $^5S_2^\circ$ level in the isoelectronic sequence with the theoretical predictions. He concluded that the values given by Lawson and Peacock are inconsistent with the trend. We give Edlén's predicted value in brackets.

The percentage compositions of these levels were provided by Kaufman and Sugar (1982).

The ionization energy was obtained by Lotz (1967) by extrapolation.

References

Edlén, B. (1984), private communication.

Hinnov, E., Suckewer, S., Cohen, S., and Sato, K. (1982), Phys. Rev. **A25**, 2293.

Kaufman, V., and Sugar, J. (1982), private communication.

Lawson, K. D., and Peacock, N. J. (1980), J. Phys. **B13**, 3313.

Lotz, W. (1967), J. Opt. Soc. Am. **57**, 873.

Ni XXIII

Configuration	Term	J	Level (cm^{-1})	Leading percentages		
$2s^2\,2p^2$	^3P	0	0	87	12	$2s^22p^2\ ^1S$
		1	109 770	99	1	$2p^4\ ^3P$
		2	161 990	67	32	$2s^22p^2\ ^1D$
$2s^2\,2p^2$	^1D	2	324 640	67	32	^3P
$2s^2\,2p^2$	^1S	0	463 900	85	12	^3P
$2s(^2S)2p^3(^4S)$	$^5S^\circ$	2	[*586 890*]	95	5	$2s(^2S)2p^3(^2P^\circ)\ ^3P^\circ$
$2s(^2S)2p^3(^2D)$	$^3D^\circ$	1	*894 100*	81	15	$2s(^2S)2p^3(^2P^\circ)\ ^3P^\circ$
		2	*900 000*	79	18	
		3	*941 400*	100		
$2s(^2S)2p^3(^2P)$	$^3P^\circ$	0	*1 064 900*	100		
		1	*1 078 500*	78	16	$2s(^2S)2p^3(^2D^\circ)\ ^3D^\circ$
		2	*1 105 000*	65	20	
$2s(^2S)2p^3(^4S)$	$^3S^\circ$	1	*1 250 500*	74	19	$2s(^2S)2p^3(^2P^\circ)\ ^1P^\circ$
$2s(^2S)2p^3(^2D)$	$^1D^\circ$	2	*1 304 600*	87	12	$2s(^2S)2p^3(^2P)\ ^3P^\circ$
$2s(^2S)2p^3(^2P)$	$^1P^\circ$	1	*1 459 700*	77	19	$2s(^2S)2p^3(^4S^\circ)\ ^3S^\circ$
$2p^4$	^3P	2	1 864 700	85	14	$2p^4\ ^1D$
		0	1 977 400	79	20	$2p^4\ ^1S$
		1	1 999 400	99	1	$2s^22p^2\ ^3P$
$2p^4$	^1D	2	2 080 600	86	14	^3P
$2p^4$	^1S	0	2 348 200	77	20	^3P
Ni XXIV ($^2P^\circ_{1/2}$)	*Limit*		**16 220 000**			

Ni XXIV

$Z=28$

B I isoelectronic sequence

Ground state: $1s^22s^22p\ ^2P^\circ_{1/2}$

Ionization energy = 17 190 000±34 000 cm^{-1} (2131±4 eV)

Hinnov, Suckewer, Cohen, and Sato (1982) observed the $2s^22p\ ^2P_{1/2}-^2P_{3/2}$ magnetic dipole transition at 609.9±0.3 Å in a tokamak plasma. The levels of the $2s2p^2$ and $2p^3$ configurations are from the measurements and classifications of Lawson and Peacock (1980) in the range of 87–160 Å with a wavelength uncertainty of ±0.03 Å, giving a level value uncertainty of ±400 cm^{-1}. They identified the transition $2s\ 2p^2\ ^4P_{5/2}-2p^3\ ^2D^\circ_{5/2}$ at 98.39 Å that connects the quartet and doublet systems. Edlén (1983) suggested that this identification is inconsistent with the intersystem connection found in a solar flare spectrum by Sandlin et al. (1976) for isoelectronic Fe XXII. Edlén proposes the value 98.282 Å for this line. The percentage compositions of these levels were provided by Kaufman and Sugar. Their calculation includes configuration interaction between $2s^22p$ and $2p^3$.

The ionization energy was obtained by Lotz (1967) by extrapolation.

References

Edlén, B. (1983), Phys. Scr. **28**, 483.

Hinnov, E., Suckewer, S., Cohen, S., and Sato, K. (1982), Phys. Rev. **A25**, 2293.

Kaufman, V., and Sugar, J. (1982), private communication.

Lawson, K. D., and Peacock, N. J. (1980), J. Phys. **B13**, 3313.

Lotz, W. (1967), J. Opt. Soc. Am. **57**, 873.

Sandlin, G. D., Brueckner, G. E., Scherrer, V. E., and Tousey, R. (1976), Astrophys. J. **205**, L47.

Ni XXIV

Configuration	Term	J	Level (cm^{-1})	Leading percentages		
$2s^2\,2p$	$^2P^\circ$	1/2	*0*	98	2	$2p^3\ ^2P^\circ$
		3/2	*163 960*	98	2	
$2s2p^2$	^4P	1/2	458 100	94	5	^2S
		3/2	541 000	99	1	^2D
		5/2	610 000	91	9	^2D
$2s2p^2$	^2D	3/2	843 700	93	7	^2P
		5/2	884 100	91	9	^4P
$2s2p^2$	^2P	1/2	955 700	66	30	^2S
		3/2	1 143 100	93	6	^2D
$2s2p^2$	^2S	1/2	1 129 600	64	34	^2P
$2p^3$	$^4S^\circ$	3/2	*1 424 900*	84	11	$^2P^\circ$
$2p^3$	$^2D^\circ$	3/2	*1 582 000*	78	12	$^2P^\circ$
		5/2	*1 626 400*	100		
$2p^3$	$^2P^\circ$	1/2	*1 781 100*	98	2	$2s^22p\ ^2P^\circ$
		3/2	*1 872 900*	75	18	$2p^3\ ^2D^\circ$
Ni XXV ($^1S_0$)	*Limit*		**17 190 000**			

Ni XXV

$Z=28$

Be I isoelectronic sequence

Ground state: $1s^2 2s^2\ ^1S_0$

Ionization energy = 18 510 000±37 000 cm^{-1} (2295±5 eV)

The $2s^2\ ^1S_0 - 2s2p\ ^1P_1^\circ$ resonance line was measured by Breton et al. (1979) at 117.90 Å in a tokamak plasma. Sandlin et al. (1976) tentatively identified a faint image at 238.82 Å in a solar flare spectrum as the $^1S_0 - ^3P_1^\circ$ line of the same array. This value is in good agreement with the prediction by Edlén (1983).

The $2s2p-2p^2$ array was classified by Lawson and Peacock (1980). Their stated measurement accuracy is ±0.03 Å, giving a level value uncertainty of ±400 cm^{-1}. We give their results, although Edlén concludes from his isoelectronic study that some of the identifications are doubtful.

The line identifications establishing the levels of the $n=3$ shell were given by Fawcett, Ridgeley, and Hughes (1979). The wavelengths, given with an accuracy of ±0.003 Å, fall in the range of 9–10 Å. The level value uncertainty is ±4000 cm^{-1}.

The ionization energy was obtained by extrapolation by Lotz (1967).

References

Breton, C., De Michelis, C., Finkenthal, M., and Mattioli, M. (1979), J. Opt. Soc. Am. **69**, 1652.

Edlén, B. (1983), Phys. Scr. **28**, 51.

Fawcett, B. C., Ridgeley, A., and Hughes, T. P. (1979), Mon. Not. R. Astron. Soc. **188**, 365.

Lawson, K. D., and Peacock, N. J. (1980), J. Phys. **B13**, 3313.

Lotz, W. (1967), J. Opt. Soc. Am. **57**, 873.

Sandlin, G. D., Brueckner, G. E., Scherrer, V. E., and Tousey, R. (1976), Astrophys. J. **205**, 147.

Ni XXV

Configuration	Term	J	Level (cm^{-1})
$2s^2$	^1S	0	0
$2s2p$	$^3P^\circ$	0	*378 190*
		1	*418 720*
		2	*549 500*
$2s2p$	$^1P^\circ$	1	*848 100*
$2p^2$	^3P	0	1 048 300?
		1	1 154 300
		2	1 207 800
$2p^2$	^1D	2	1 379 600
$2p^2$	^1S	0	1 611 500
$2s3p$	$^3P^\circ$	1	*10 650 000*
$2s3p$	$^1P^\circ$	1	*10 707 000*
$2s3d$	^3D	1	10 794 000
		2	10 800 000
		3	10 813 000
$2s3d$	^1D	2	10 880 000
$2p3p$	^3D	2	11 153 000
		3	11 296 000
$2p3d$	$^3F^\circ$	3	*11 271 000*
$2p3d$	$^3D^\circ$	2	*11 283 000*
		1	*11 296 000*
		3	*11 437 000*
$2p3p$	^3P	2	11 306 000
$2p3p$	^3S	1	11 306 000
$2p3d$	$^1D^\circ$	2	*11 408 000*
$2p3d$	$^3P^\circ$	0–2	*11 456 000*
$2p3d$	$^1F^\circ$	3	*11 525 000*
Ni XXVI ($^2S_{1/2}$)	***Limit***		**18 510 000**

Ni XXVI

$Z=28$

Li I isoelectronic sequence

Ground state: $1s^2 2s\ ^2S_{1/2}$

Ionization energy $= 19\,351\,000 \pm 5000$ cm^{-1} (2399.2 ± 0.6 eV)

Edlén (1983) has given extrapolated values for the $2s-2p$ transition wavelengths with an estimated uncertainty of ± 0.01 Å. He concludes that the solar identification for this transition by Sandlin et al. (1976) is incorrect. Edlén's results are given in brackets. Fawcett, Ridgeley, and Hughes (1979) determined the $3s$, $3p$, and $3d$ terms from observations at 9 Å of a laser-produced plasma. Their measurement uncertainty is ± 0.01 Å, giving a level value uncertainty of $\pm 12\,000$ cm^{-1}.

From measurements of spark spectra at ~ 1.6 Å, Safronova and Sidelnikov (1977) determined the levels above the ionization energy. No estimated uncertainty is given but wavelengths to four decimal places are reported. The designations are from Vainstein and Safronova (1978).

The ionization energy was calculated by Edlén (1979).

References

Edlén, B. (1979), Phys. Scr. **19**, 255.
Edlén, B. (1983), Phys. Scr. **28**, 51.
Fawcett, B. C., Ridgeley, A., and Hughes, T. P. (1979), Mon. Not. R. Astron. Soc. **188**, 365.
Safronova, U. I., and Sidelnikov, Y. V. (1977), Prikl. Spektrosk. **5**, 7.
Sandlin, G. D., Breuckner, G. E., Scherrer, V. E., and Tousey, R. (1976), Astrophys. J. **205**, L47.
Vainstein, L. A. and Safronova, U. I. (1978), At. Data Nucl. Data Tables **21**, 49.

Ni XXVI

Configuration	Term	J	Level (cm^{-1})	Configuration	Term	J	Level (cm^{-1})
$1s^2\,2s$	^{2}S	1/2	0	$1s(^2S)2s2p(^3P°)$	^{2}P°	1/2	*62 516 000*
						3/2	*62 617 000*
$1s^2\,2p$	^{2}P°	1/2	[*427 180*]				
		3/2	[*604 680*]	$1s(^2S)2p^2(^3P)$	^{4}P	3/2	62 697 000
						1/2	62 783 000
$1s^2\,3s$	^{2}S	1/2	10 880 000				
				$1s(^2S)2s2p(^1P°)$	^{2}P°	1/2	*62 755 000*
$1s^2\,3p$	^{2}P°	1/2	*10 983 000*				
		3/2	*11 036 000*	$1s(^2S)2p^2(^3P)$	^{2}P	1/2	62 997 000
						3/2	63 210 000
$1s^2\,3d$	^{2}D	3/2	11 077 000				
		5/2	11 092 000	$1s(^2S)2p^2(^1D)$	^{2}D	3/2	63 008 000
						5/2	63 085 000
Ni XXVII ($^1S_0$)	*Limit*		**19 351 000**				
				$1s(^2S)2p^2(^1S)$	^{2}S	1/2	63 397 000
$1s(^2S)2s2p(^3P°)$	^{4}P°	1/2	*62 193 000*				
		3/2	*62 228 000*				

Ni XXVII

$Z=28$

He I isoelectronic sequence

Ground state: $1s^2\ ^1S_0$

Ionization energy = 82 984 000±16 000 cm^{-1} (10 288.8±2.0 eV)

Because of the excellent agreement of the calculated energies of the $n=2$ shell by Safronova (1981) with the few well-measured spectra in the He I sequence, we have compiled her results for the $n=2$ levels and for the ionization energy. Detailed comparisons are given in the Introduction. Levels of the n = 3–5 shells are from the calculated binding energies by Ermolaev and Jones (1974) subtracted from Safronova's value for the binding energy of the ground state. We have assumed an uncertainty of 2 parts in 10^4 for the excited levels relative to the ground state, and for the ionization energy. For differences between excited levels where $\Delta n=0$, we assumed an uncertainty of 2 parts in 10^3.

Observations by Safronova and Sidelnikov (1977) place the $1s2p\ ^1P_1^\circ$ at 62 925 000 cm^{-1} with an estimated uncertainty of ±20 000 cm^{-1}.

Percentage compositions are from Ermolaev and Jones.

References

Ermolaev, A. M., and Jones, M. (1974), J. Phys. **B7**, 199.

Safronova, U. I. (1981), Phys. Scr. **23**, 241.

Safronova, U. I., and Sidelnikov, Y. V. (1977), Prikl. Spektrosk. **5**, 7.

Ni XXVII

Configuration	Term	J	Level (cm^{-1})	Leading percentages		
$1s^2$	^1S	0	0			
$1s2s$	^3S	1	[62 358 960]			
$1s2p$	$^3P^\circ$	0	*[62 614 630]*			
		1	*[62 637 380]*	89	11	$^1P^\circ$
		2	*[62 801 270]*			
$1s2s$	^1S	0	[62 637 200]			
$1s2p$	$^1P^\circ$	1	*[62 952 670]*	89	11	$^3P^\circ$
$1s3s$	^3S	1	[73 903 340]			
$1s3p$	$^3P^\circ$	0	*[73 974 350]*			
		1	*[73 979 340]*	88	12	$^1P^\circ$
		2	*[74 028 190]*			
$1s3s$	^1S	0	[73 976 370]			
$1s3p$	$^1P^\circ$	1	*[74 070 580]*	88	12	$^3P^\circ$
$1s4s$	^3S	1	[77 900 890]			
$1s4s$	^1S	0	[77 930 480]			
$1s4p$	$^3P^\circ$	0	*[77 930 500]*			
		1	*[77 932 560]*	88	12	$^1P^\circ$
		2	*[77 953 220]*			
$1s4p$	$^1P^\circ$	1	*[77 970 500]*	88	12	$^3P^\circ$
$1s5s$	^3S	1	[79 740 940]			
$1s5s$	^1S	0	[79 755 710]			

Ni XXVII—Continued

Configuration	Term	J	Level (cm^{-1})	Leading percentages		
$1s5p$	$^3P°$	0	[*79 755 910*]			
		1	[*79 756 970*]	87	13	$^1P°$
		2	[*79 767 550*]			
$1s5p$	$^1P°$	1	[*79 776 290*]	87	13	$^3P°$
Ni XXVIII ($^2S_{1/2}$)	***Limit***		**82 984 000**			

Ni XXVIII

$Z=28$

H I isoelectronic sequence

Ground state: $1s\ ^2S_{1/2}$

Ionization energy = 86 909 400±500 cm^{-1} (10 775.48±0.06 eV)

No observations of this spectrum are reported. We give calculated values by Mohr (1983) for the $n=2$ shell and by Erickson (1977) for $n=3$ to 5 relative to the $2p\ ^2P^\circ_{3/2}$ level. Further details are given in the Introduction. Relative to the ground state, the level uncertainty is estimated to be one part in 10^6.

References

Erickson, G. W. (1977), J. Phys. Chem. Ref. Data **6**, 831.
Mohr, P. J. (1983), At. Data Nucl. Data Tables, **29**, 453.

Ni XXVIII

Configuration	Term	J	Level (cm^{-1})
$1s$	^2S	1/2	0
$2p$	$^2P^\circ$	1/2	[*65 113 950*]
		3/2	[*65 344 939*]
$2s$	^2S	1/2	[65 119 814]
$3p$	$^2P^\circ$	1/2	[*77 248 280*]
		3/2	[*77 316 770*]
$3s$	^2S	1/2	[77 250 120]
$3d$	^2D	3/2	[77 316 649]
		5/2	[77 339 033]
$4p$	$^2P^\circ$	1/2	[*81 485 832*]
		3/2	[*81 514 699*]
$4s$	^2S	1/2	[81 486 610]
$4d$	^2D	3/2	[81 514 646]
		5/2	[81 524 098]
$4f$	$^2F^\circ$	5/2	[*81 524 081*]
		7/2	[*81 528 784*]
$5p$	$^2P^\circ$	1/2	[*83 443 186*]
		3/2	[*83 457 952*]
$5s$	^2S	1/2	[83 443 585]
$5d$	^2D	3/2	[83 457 924]
		5/2	[83 462 764]
$5f$	$^2F^\circ$	5/2	[*83 462 755*]
		7/2	[*83 465 165*]
$5g$	^2G	7/2	[83 465 160]
		9/2	[83 466 603]
	Limit		**86 909 400**

Journal of Physical and Chemical Reference Data
Cumulative Listing of Reprints and Supplements

Reprints from Volume 1

1. Gaseous Diffusion Coefficients, *T.R. Marrero and E.A. Mason,* Vol. 1, No. 1, pp. 1–118 (1972) $7.00
2. Selected Values of Critical Supersaturation for Nucleation of Liquids from the Vapor, *G.M. Pound,* Vol. 1, No. 1, pp. 119–134 (1972) $3.00
3. Selected Values of Evaporation and Condensation Coefficients for Simple Substances, *G.M. Pound,* Vol. 1, No. 1, pp. 135–146 (1972) $3.00
4. Atlas of the Observed Absorption Spectrum of Carbon Monoxide between 1060 and 1900 Å, *S.G. Tilford and J.D. Simmons,* Vol. 1, No. 1, pp. 147–188 (1972) $4.50
5. Tables of Molecular Vibrational Frequencies, Part 5, *T. Shimanouchi,* Vol. 1, No. 1, pp. 189–216 (1972) (superseded by No.103) $4.00
6. Selected Values of Heats of Combustion and Heats of Formation of Organic Compounds Containing the Elements C, H, N, O, P, and S, *Eugene S. Domalski,* Vol. 1, No. 2, pp. 221–278 (1972) $5.00
7. Thermal Conductivity of the Elements, *C.Y. Ho, R.W. Powell, and P.E. Liley,* Vol. 1, No. 2, pp. 279–422 (1972) $7.50
8. The Spectrum of Molecular Oxygen, *Paul H. Krupenie,* Vol. 1, No. 2, pp. 423–534 (1972) $6.50
9. A Critical Review of the Gas-Phase Reaction Kinetics of the Hydroxyl Radical, *Wm. E. Wilson, Jr.,* Vol. 1, No. 2, pp. 535–574 (1972) $4.50
10. Molten Salts: Volume 3, Nitrates, Nitrites, and Mixtures, Electrical Conductance, Density, Viscosity, and Surface Tension Data, *G.J. Janz, Ursula Krebs, H.F. Siegenthaler, and R.P.T. Tomkins,* Vol. 1, No. 3, pp. 581–746 (1972) $8.50
11. High Temperature Properties and Decomposition of Inorganic Salts—Part 3. Nitrates and Nitrites, *Kurt H. Stern,* Vol. 1, No. 3, pp. 747–772 (1972) $4.00
12. High-Pressure Calibration: A Critical Review, *D.L. Decker, W.A. Bassett, L. Merrill, H.T. Hall, and J.D. Barnett,* Vol. 1, No. 3, pp. 773–836 (1972) $5.00
13. The Surface Tension of Pure Liquid Compounds, *Joseph J. Jasper,* Vol. 1, No. 4, pp. 841–1009 (1972) $8.50
14. Microwave Spectra of Molecules of Astrophysical Interest, I. Formaldehyde, Formamide, and Thioformaldehyde, *Donald R. Johnson, Frank J. Lovas, and William H. Kirchhoff,* Vol. 1, No. 4, pp. 1011–1046 (1972) $4.50
15. Osmotic Coefficients and Mean Activity Coefficients of Uni-univalent Electrolytes in Water at 25° C, *Walter J. Hamer and Yung-Chi Wu,* Vol. 1, No. 4, pp. 1047-1099 (1972) $5.00
16. The Viscosity and Thermal Conductivity Coefficients of Gaseous and Liquid Fluorine, *H.J.M. Hanley and R. Prydz,* Vol. 1, No. 4, pp. 1101-1113 (1972) $3.00

Reprints from Volume 2

17. Microwave Spectra of Molecules of Astrophysical Interest, II. Methylenimine, *William H. Kirchhoff, Donald R. Johnson, and Frank J. Lovas,* Vol. 2, No. 1, pp. 1-10 (1973) $3.00
18. Analysis of Specific Heat Data in the Critical Region of Magnetic Solids, *F.J. Cook,* Vol. 2, No. 1, pp. 11-24 (1973) $3.00
19. Evaluated Chemical Kinetic Rate Constants for Various Gas Phase Reactions, *Keith Schofield,* Vol. 2, No. 1, pp. 25–84 (1973) $5.00
20. Atomic Transition Probabilities for Forbidden Lines of the Iron Group Elements. (A Critical Data Compilation for Selected Lines), *M.W. Smith and W.L. Wiese,* Vol. 2, No. 1, pp. 85–120 (1973) $4.50
21. Tables of Molecular Vibrational Frequencies, Part 6, *T. Shimanouchi,* Vol. 2, No. 1, pp. 121–162 (1973) (superseded by No. 103) $4.50
22. Compilation of Energy Band Gaps in Elemental and Binary Compound Semiconductors and Insulators, *W.H. Strehlow and E.L. Cook,* Vol. 2, No. 1, pp. 163–200 (1973) $4.50
23. Microwave Spectra of Molecules of Astrophysical Interest, III. Methanol, *R.M. Lees, F.J. Lovas, W.H. Kirchhoff, and D.R. Johnson,* Vol. 2, No. 2, pp. 205–214 (1973) $3.00
24. Microwave Spectra of Molecules of Astrophysical Interest, IV. Hydrogen Sulfide, *Paul Helminger, Frank C. De Lucia, and William H. Kirchhoff,* Vol. 2, No. 2, pp. 215–224 (1973) $3.00

Journal of Physical and Chemical Reference Data
Reprint and Supplement Orders

To: American Chemical Society
Distribution Office
1155 Sixteenth Street, N.W.
Washington, D.C. 20036

Name: ______________________

Title: ______________________

Organization: ______________________

Address: ______________________

City: ______________ State: ______________

Country: ______________ Zip: ______________

I am a member of ______________________
(ACS or AIP Member or Affiliated Society)

ORDERS FOR REPRINTS AND SUPPLEMENTS MUST BE PREPAID. *Foreign orders for Reprints, add $2.50 for each reprint for postage and handling. Foreign orders for Reprint Packages, add $4.00 for each Reprint Package for postage and handling. California residents add 6% state use tax. Make checks payable to the American Chemical Society.

BULK RATES: Subtract 20% from the listed price for orders of 50 or more of any one item.

Please ship the following reprints and supplements:

Reprint No./Package _______, _______ copies $ _______

Reprint No./Package _______, _______ copies $ _______

Reprint No./Package _______, _______ copies $ _______

Reprint No./Package _______, _______ copies $ _______

Vol. 2, Suppl. 1 ☐ Hardcover ☐ Softcover _____ copies $ _______

Vol. 3, Suppl. 1 ☐ Hardcover ☐ Softcover _____ copies $ _______

Vol. 6, Suppl. 1 ☐ Hardcover ☐ Softcover _____ copies $ _______

Vol. 10, Suppl. 1 ☐ Hardcover _____ copies $ _______

Vol. 11, Suppl. 1 ☐ Hardcover _____ copies $ _______

Vol. 11, Suppl. 2 ☐ Hardcover _____ copies $ _______

Vol. 13, Suppl. 1 ☐ Hardcover _____ copies $ _______

Other Suppl.: _______ _____ copies $ _______

Total Enclosed $ _______

(Continuation of Cumulative Listing of Reprints)

25. Tables of Molecular Vibrational Frequencies, Part 7, *T. Shimanouchi,* Vol. 2, No. 2, pp. 225–256 (1973) (superseded by No. 103) $4.00

26. Energy Levels of Neutral Helium (^{4}He I), *W.C. Martin,* Vol. 2, No. 2, pp. 257–266 (1973) $3.00

27. Survey of Photochemical and Rate Data for Twenty-eight Reactions of Interest in Atmospheric Chemistry, *R.F. Hampson, Editor, W. Braun, R.L. Brown, D. Garvin, J.T. Herron, R.E. Huie, M.J. Kurylo, A.H. Laufer, J.D. McKinley, H. Okabe, M.D. Scheer, W. Tsang, and D.H. Stedman,* Vol. 2, No. 2, pp. 267–312 (1973) $4.50

28. Compilation of the Static Dielectric Constant of Inorganic Solids, *K.F. Young and H.P.R. Frederikse,* Vol. 2, No. 2, pp. 313–410 (1973) $6.50

29. Soft X-Ray Emission Spectra of Metallic Solids: Critical Review of Selected Systems, *A.J. McAlister, R.C. Dobbyn, J.R. Cuthill, and M.L. Williams,* Vol. 2, No. 2, pp. 411–426 (1973) $3.00

30. Ideal Gas Thermodynamic Properties of Ethane and Propane, *J. Chao, R.C. Wilhoit, and B.J. Zwolinski,* Vol. 2, No. 2, pp. 427–438 (1973) $3.00

31. An Analysis of Coexistence Curve Data for Several Binary Liquid Mixtures Near Their Critical Points, *A. Stein and G.F. Allen,* Vol. 2, No. 3, pp. 443–466 (1973) $4.00

32. Rate Constants for the Reactions of Atomic Oxygen (O ^{3}P) with Organic Compounds in the Gas Phase, *John T. Herron and Robert E. Huie,* Vol. 2, No. 3, pp. 467–518 (1973) $5.00

33. First Spectra of Neon, Argon, and Xenon 136 in the 1.2-4.0 μm Region, *Curtis J. Humphreys,* Vol. 2, No. 3, pp. 519–530 (1973) $3.00

34. Elastic Properties of Metals and Alloys, I. Iron, Nickel, and Iron-Nickel Alloys, *H.M. Ledbetter and R.P. Reed,* Vol. 2, No. 3, pp. 531–618 (1973) $6.00

35. The Viscosity and Thermal Conductivity Coefficients of Dilute Argon, Krypton, and Xenon, *H.J.M. Hanley,* Vol. 2, No. 3, pp. 619–642 (1973) $4.00

36. Diffusion in Copper and Copper Alloys, Part I. Volume and Surface Self-Diffusion in Copper, *Daniel B. Butrymowicz, John R. Manning, and Michael E. Read,* Vol. 2, No. 3, pp. 643–656 (1973) $3.00

37. The 1973 Least-Squares Adjustment of the Fundamental Constants, *E. Richard Cohen and B.N. Taylor,* Vol. 2, No. 4, pp. 663–734 (1973) $5.50

38. The Viscosity and Thermal Conductivity Coefficients of Dilute Nitrogen and Oxygen, *H.J.M. Hanley and James F. Ely,* Vol. 2, No. 4, pp. 735–756 (1973) $4.00

39. Thermodynamic Properties of Nitrogen Including Liquid and Vapor Phases from 63 K to 2000 K with Pressures to 10,000 Bar, *Richard T. Jacobsen and Richard B. Stewart,* Vol. 2, No. 4, pp. 757–922 (1973) $8.50

40. Thermodynamic Properties of Helium 4 from 2 to 1500 K at Pressures to 10^8 Pa, *Robert T. McCarty,* Vol. 2, No. 4, pp. 923–1042 (1973) $7.00

Reprints from Volume 3

41. Molten Salts: Volume 4, Part 1, Fluorides and Mixtures, Electrical Conductance, Density, Viscosity, and Surface Tension Data, *G.J. Janz, G.L. Gardner, Ursula Krebs, and R.P.T. Tomkins,* Vol. 3, No. 1, pp. 1–115 (1974) $7.00

42. Ideal Gas Thermodynamic Properties of Eight Chloro- and Fluoromethanes, *A.S. Rodgers, J. Chao, R. C. Wilhoit, and B.J. Zwolinski,* Vol. 3, No. 1, pp. 117–140 (1974) $4.00

43. Ideal Gas Thermodynamic Properties of Six Chloroethanes, *J. Chao, A.S. Rodgers, R.C. Wilhoit, and B.J. Zwolinski,* Vol. 3, No. 1, pp. 141–162 (1974) $4.00

44. Critical Analysis of Heat-Capacity Data and Evaluation of Thermodynamic Properties of Ruthenium, Rhodium, Palladium, Iridium, and Platinum from 0 to 300 K. A Survey of the Literature Data on Osmium, *George T. Furukawa, Martin L. Reilly, and John S. Gallagher,* Vol. 3, No. 1, pp. 163–209 (1974) $4.50

45. Microwave Spectra of Molecules of Astrophysical Interest, V. Water Vapor, *Frank C. De Lucia, Paul Helminger, and William H. Kirchhoff,* Vol. 3, No. 1, pp. 211–219 (1974) $3.00

46. Microwave Spectra of Molecules of Astrophysical Interest, VI. Carbonyl Sulfide and Hydrogen Cyanide, *Arthur G. Maki,* Vol. 3, No. 1, pp. 221–244 (1974) $4.00

47. Microwave Spectra of Molecules of Astrophysical Interest, VII. Carbon Monoxide, Carbon Monosulfide, and Silicon Monoxide, *Frank J. Lovas and Paul H. Krupenie,* Vol. 3, No. 1, pp. 245–257 (1974) $3.00

48. Microwave Spectra of Molecules of Astrophysical Interest, VIII. Sulfur Monoxide, *Eberhard Tiemann,* Vol. 3, No. 1, pp. 259–268 (1974) $3.00

49. Tables of Molecular Vibrational Frequencies, Part 8, *T. Shimanouchi,* Vol. 3, No. 1, pp. 269–308 (1974) (superseded by No. 103) $4.50

50. JANAF Thermochemical Tables, 1974 Supplement, *M.W. Chase, J.L. Curnutt, A.T. Hu, H. Prophet, A.N. Syverud, and L.C. Walker,* Vol. 3, No. 2, pp. 311–480 (1974) $8.50

51. High Temperature Properties and Decomposition of Inorganic Salts, Part 4. Oxy-Salts of the Halogens, *Kurt H. Stern,* Vol. 3, No. 2, pp. 481–526 (1974) $4.50

52. Diffusion in Copper and Copper Alloys, Part II. Copper-Silver and Copper-Gold Systems, *Daniel B. Butrymowicz, John R. Manning, and Michael E. Read,* Vol. 3, No. 2, pp. 527–602 (1974) $5.50

53. Microwave Spectral Tables I. Diatomic Molecules, *Frank J. Lovas and Eberhard Tiemann,* Vol. 3, No. 3, pp. 609–770 (1974) $8.50

54. Ground Levels and Ionization Potentials for Lanthanide and Actinide Atoms and Ions, *W.C. Martin, Lucy Hagan, Joseph Reader, and Jack Sugar,* Vol. 3, No. 3, pp. 771–780 (1974) $3.00

55. Behavior of the Elements at High Pressures, *John Francis Cannon,* Vol. 3, No. 3, pp. 781–824 (1974) $4.50

56. Reference Wavelengths from Atomic Spectra in the Range 15 Å to 25000 Å, *Victor Kaufman and Bengt Edlén,* Vol. 3, No. 4, pp. 825–895 (1974) $5.50

57. Elastic Properties of Metals and Alloys. II. Copper, *H.M. Ledbetter and E.R. Naimon,* Vol. 3, No. 4, pp.897–935 (1974) $4.50

58. A Critical Review of H-Atom Transfer in the Liquid Phase: Chlorine Atom, Alkyl, Trichloromethyl, Alkoxy, and Alkylperoxy Radicals, *D.G. Hendry, T. Mill, L. Piszkiewicz, J.A. Howard, and H.K. Eigenmann,* Vol. 3, No. 4, pp. 937–978 (1974) $4.50

59. The Viscosity and Thermal Conductivity Coefficients for Dense Gaseous and Liquid Argon, Krypton, Xenon, Nitrogen, and Oxygen, *H.J.M. Hanley, R.D. McCarty, and W.M. Haynes,* Vol. 3, No. 4, pp. 979–1017 (1974) $4.50

Reprints from Volume 4

60. JANAF Thermochemical Tables, 1975 Supplement, *M.W. Chase, J.L. Curnutt, H. Prophet, R.A. McDonald, and A.N. Syverud,* Vol. 4, No. 1, pp. 1–175 (1975) $8.50

61. Diffusion in Copper and Copper Alloys, Part III. Diffusion in Systems Involving Elements of the Groups IA, IIA, IIIB, IVB, VB, VIB, and VIIB, *Daniel B. Butrymowicz, John R. Manning, and Michael E. Read,* Vol. 4, No. 1, pp. 177–249 (1975) $6.00

(Continuation of Cumulative Listing of Reprints)

62. Ideal Gas Thermodynamic Properties of Ethylene and Propylene, *Jing Chao and Bruno J. Zwolinski,* Vol. 4, No. 1, pp. 251–261 (1975) $3.00

63. Atomic Transition Probabilities for Scandium and Titanium (A Critical Data Compilation of Allowed Lines), *W.L. Wiese and J.R. Fuhr,* Vol. 4, No. 2, pp. 263–352 (1975) $6.00

64. Energy Levels of Iron, Fe I through Fe XXVI, *Joseph Reader and Jack Sugar,* Vol. 4, No. 2, pp. 353–440 (1975) $6.00

65. Ideal Gas Thermodynamic Properties of Six Fluoroethanes, *S.S. Chen, A.S. Rodgers, J. Chao, R.C. Wilhoit, and B.J. Zwolinski,* Vol. 4, No. 2, pp. 441–456 (1975) $3.00

66. Ideal Gas Thermodynamic Properties of the Eight Bromo- and Iodomethanes, *S.A. Kudchadker and A.P. Kudchadker,* Vol. 4, No. 2, pp. 457–470 (1975) $3.00

67. Atomic Form Factors, Incoherent Scattering Functions, and Photon Scattering Cross Sections, *J.H. Hubbell, Wm.J. Veigele, E.A. Briggs, R.T. Brown, D.T. Cromer, and R.J. Howerton,* Vol. 4, No. 3, pp. 471–538 (1975) $5.50

68. Binding Energies in Atomic Negative Ions, *H. Hotop and W.C. Lineberger,* Vol. 4, No. 3, pp. 539–576 (1975) $4.50

69. A Survey of Electron Swarm Data, *J. Dutton,* Vol. 4, No. 3, pp. 577–856 (1975) $12.00

70. Ideal Gas Thermodynamic Properties and Isomerization of *n*-Butane and Isobutane, *S.S. Chen, R.C. Wilhoit, and B.J. Zwolinski,* Vol. 4, No. 4, pp. 859–869 (1975) $3.00

71. Molten Salts: Volume 4, Part 2, Chlorides and Mixtures, Electrical Conductance, Density, Viscosity, and Surface Tension Data, *G.J. Janz, R.P.T. Tomkins, C.B. Allen, J.R. Downey, Jr., G.L. Gardner, U. Krebs, and S.K. Singer,* Vol. 4, No. 4, pp. 871–1178 (1975) $13.00

72. Property Index to Volumes 1–4 (1972–1975), Vol. 4, No. 4, pp. 1179–1192 (1975) $3.00

Reprints from Volume 5

73. Scaled Equation of State Parameters for Gases in the Critical Region, *J.M.H. Levelt Sengers, W.L. Greer, and J.V. Sengers,* Vol. 5, No. 1, pp. 1–51 (1976) $5.00

74. Microwave Spectra of Molecules of Astrophysical Interest, IX. Acetaldehyde, *A. Bauder, F.J. Lovas, and D.R. Johnson,* Vol. 5, No. 1, pp. 53–77 (1976) $4.00

75. Microwave Spectra of Molecules of Astrophysical Interest, X. Isocyanic Acid, *G. Winnewisser, W.H. Hocking, and M.C.L. Gerry,* Vol. 5, No. 1, pp. 79–101 (1976) $4.00

76. Diffusion in Copper and Copper Alloys, Part IV. Diffusion in Systems Involving Elements of Group VIII, *Daniel B. Butrymowicz, John R. Manning, and Michael E. Read,* Vol. 5, No. 1, pp. 103–200 (1976) $6.50

77. A Critical Review of the Stark Widths and Shifts of Spectral Lines from Non-Hydrogenic Atoms, *N. Konjevic and J.R. Roberts,* Vol. 5, No. 2, pp. 209–257 (1976) $5.00

78. Experimental Stark Widths and Shifts for Non-Hydrogenic Spectral Lines of Ionized Atoms (A Critical Review and Tabulation of Selected Data), *N. Konjevic and W.L. Wiese,* Vol. 5, No. 2, pp. 259–308 (1976) $5.00

79. Atlas of the Absorption Spectrum of Nitric Oxide (NO) between 1420 and 1250 Å, *E. Miescher and F. Alberti,* Vol. 5, No. 2, pp. 309–317 (1976) $3.00

80. Ideal Gas Thermodynamic Properties of Propanone and 2-Butanone, *Jing Chao and Bruno J. Zwolinski,* Vol. 5, No. 2, pp. 319–328 (1976) $3.00

81. Refractive Index of Alkali Halides and Its Wavelength and Temperature Derivatives, *H.H. Li,* Vol. 5, No. 2, pp. 329–528 (1976) $9.50

82. Tables of Critically Evaluated Oscillator Strengths for the Lithium Isoelectronic Sequence, *G.A. Martin and W.L. Wiese,* Vol. 5, No. 3, pp. 537–570 (1976) $4.50

83. Ideal Gas Thermodynamic Properties of Six Chlorofluoromethanes, *S.S. Chen, R. C. Wilhoit, and B.J. Zwolinski,* Vol. 5, No. 3, pp. 571–580 (1976) $3.00

84. Survey of Superconductive Materials and Critical Evaluation of Selected Properties, *E.W. Roberts,* Vol. 5, No. 3, pp. 581–821 (1976) $12.50

85. Nuclear Spins and Moments, *Gladys H. Fuller,* Vol. 5, No. 4, pp. 835–1092 (1976) $11.50

86. Nuclear Moments and Moment Ratios as Determined by Mössbauer Spectroscopy, *J.G. Stevens and B.D. Dunlap,* Vol. 5, No. 4, pp. 1093–1121 (1976) $4.00

87. Rate Coefficients for Ion-Molecule Reactions, I. Ions Containing C and H, *L. Wayne Sieck and Sharon G. Lias,* Vol. 5, No. 4, pp. 1123–1146 (1976) $4.00

88. Microwave Spectra of Molecules of Astrophysical Interest, XI. Silicon Sulfide, *Eberhard Tiemann,* Vol. 5, No. 4, pp. 1147–1156 (1976) $3.00

89. Property Index and Author Index to Volumes 1–5 (1972–1976), Vol. 5, No. 4, pp. 1161–1183 $4.00

Reprints from Volume 6

90. Diffusion in Copper and Copper Alloys, Part V. Diffusion in Systems Involving Elements of Group VA, *Daniel B. Butrymowicz, John R. Manning, and Michael E. Read* Vol. 6, No. 1, pp. 1–50 (1977) $5.00

91. The Calculated Thermodynamic Properties of Superfluid Helium-4, *James S. Brooks and Russell J. Donnelly,* Vol. 6, No. 1, pp. 51–104 (1977) $5.00

92. Thermodynamic Properties of Normal and Deuterated Methanols, *S.S. Chen, R.C. Wilhoit, and B.J. Zwolinski,* Vol. 6, No. 1, pp. 105–112 (1977) $3.00

93. The Spectrum of Molecular Nitrogen, *Alf Lofthus and Paul H. Krupenie,* Vol. 6, No. 1, pp. 113–307 (1977) $9.50

94. Energy Levels of Chromium, Cr I through Cr XXIV, *Jack Sugar and Charles Corliss,* Vol. 6, No. 2, pp. 317–383 (1977) $5.50

95. The Activity and Osmotic Coefficients of Aqueous Calcium Chloride at 298.15 K, *Bert R. Staples and Ralph L. Nuttall,* Vol. 6, No. 2, pp. 385–407 (1977) $4.00

96. Molten Salts: Volume 4, Part 3, Bromides and Mixtures; Iodides and Mixtures–Electrical Conductance, Density, Viscosity, and Surface Tension Data, *G.J. Janz, R.P.T. Tomkins, C.B. Allen, J.R. Downey, Jr., and S.K. Singer,* Vol. 6, No. 2, pp. 409–596 (1977) $9.00

97. The Viscosity and Thermal Conductivity Coefficients for Dense Gaseous and Liquid Methane, *H.J.M. Hanley, W.M. Haynes, and R.D. McCarty,* Vol. 6, No. 2, pp. 597–609 (1977) $3.00

98. Phase Diagrams and Thermodynamic Properties of Ternary Copper-Silver Systems, *Y. Austin Chang, Daniel Goldberg, and Joachim P. Neumann,* Vol. 6, No. 3, pp. 621–673 (1977) $5.00

99. Crystal Data Space-Group Tables, *Alan D. Mighell, Helen M. Ondik, and Bettijoyce Breen Molino,* Vol. 6, No. 3, pp. 675–829 (1977) $8.00

100. Energy Levels of One-Electron Atoms, *Glen W. Erickson,* Vol. 6, No. 3, pp. 831–869 (1977) $4.50

101. Rate Constants for Reactions of ClO_x of Atmospheric Interest, *R.T. Watson,* Vol. 6, No. 3, pp. 871–917 (1977) $4.50

102. NMR Spectral Data: A Compilation of Aromatic Proton Chemical Shifts in Mono- and Di-Substituted Benzenes, *B.L. Shapiro and L.E. Mohrmann,* Vol. 6, No. 3, pp. 919–991 (1977) $5.50

(Continuation of Cumulative Listing of Reprints)

103. Tables of Molecular Vibrational Frequencies. Consolidated Volume II. *T. Shimanouchi,* Vol. 6, No. 3, pp. 993–1102 (1977) (supersedes Nos. 5, 21, 25, 49) $6.50

104. Effects of Isotopic Composition, Temperature, Pressure, and Dissolved Gases on the Density of Liquid Water, *George S. Kell,* Vol. 6, No. 4, pp. 1109–1131 (1977) $4.00

105. Viscosity of Water Substance–New International Formulation and Its Background, *A. Nagashima,* Vol. 6, No. 4, pp. 1133–1166 (1977) $4.50

106. A Correlation of the Existing Viscosity and Thermal Conductivity Data of Gaseous and Liquid Ethane, *H.J.M. Hanley, K.E. Gubbins, and S. Murad,* Vol. 6, No. 4, pp. 1167–1180 (1977) $3.00

107. Elastic Properties of Zinc: A Compilation and a Review, *H.M. Ledbetter,* Vol. 6, No. 4, pp. 1181–1203 (1977) $4.00

108. Behavior of the AB-Type Compounds at High Pressures and High Temperatures, *Leo Merrill,* Vol. 6, No. 4, pp. 1205–1252 (1977) $4.50

109. Energy Levels of Manganese, Mn I through Mn XXV, *Charles Corliss and Jack Sugar,* Vol. 6, No. 4, pp. 1253–1329 (1977) $5.50

Reprints from Volume 7

110. Tables of Atomic Spectral Lines for the 10 000 Å to 40 000 Å Region, *Michael Outred,* Vol. 7, No. 1, pp. 1–262 (1978) $11.50

111. Evaluated Activity and Osmotic Coefficients for Aqueous Solutions: The Alkaline Earth Metal Halides, *R.N. Goldberg and R.L. Nuttall,* Vol. 7, No. 1, pp. 263–310 (1978) $4.50

112. Microwave Spectra of Molecules of Astrophysical Interest XII. Hydroxyl Radical, *Robert A. Beaudet and Robert L. Poynter,* Vol. 7, No. 1, pp. 311–362 (1978) $5.00

113. Ideal Gas Thermodynamic Properties of Methanoic and Ethanoic Acids, *Jing Chao and Bruno J. Zwolinski,* Vol. 7, No. 1, pp. 363–377 (1978) $3.00

114. Critical Review of Hydrolysis of Organic Compounds in Water Under Environmental Conditions, *W. Mabey and T. Mill,* Vol. 7, No. 2, pp. 383–415 (1978) $4.50

115. Ideal Gas Thermodynamic Properties of Phenol and Creosols, *S.A. Kudchadker, A.P. Kudchadker, R.C. Wilhoit, and B.J. Zwolinski,* Vol. 7, No. 2, pp. 417–423 (1978) $3.00

116. Densities of Liquid $CH_{4-a}X_a$ (X = Br,I) and $CH_{4-(a+b+c+d)}F_aCl_bBr_cI_d$ Halomethanes, *A.P. Kudchadker, S.A.Kudchadker, P.R. Patnaik, and P.P. Mishra,* Vol. 7, No. 2, pp. 425–439 (1978) $3.00

117. Microwave Spectra of Molecules of Astrophysical Interest XIII. Cyanoacetylene, *W.J. Lafferty and F.J. Lovas,* Vol. 7, No. 2, pp. 441–493 (1978) $5.00

118. Atomic Transition Probabilities for Vanadium, Chromium, and Manganese (A Critical Data Compilation of Allowed Lines), *S.M. Younger, J.R. Fuhr, G.A. Martin, and W.L. Wiese,* Vol. 7, No. 2, pp. 495–629 (1978) $7.50

119. Thermodynamic Properties of Ammonia, *Lester Haar and John S. Gallagher,* Vol. 7, No. 3, pp. 635–792 (1978) $8.00

120. JANAF Thermochemical Tables, 1978 Supplement, *M.W. Chase, Jr., J.L. Curnutt, R.A. McDonald, and A.N. Syverud,* Vol. 7, No. 3, pp. 793–940 (1978) $8.00

121. Viscosity of Liquid Water in the Range −8 °C to 150 °C, *Joseph Kestin, Mordechai Sokolov, and William A. Wakeham,* Vol. 7, No. 3, pp. 941–948 (1978) $3.00

122. The Molar Volume (Density) of Solid Oxygen in Equilibrium with Vapor, *H.M. Roder,* Vol. 7, No. 3, pp. 949–957 (1978) $3.00

123. Thermal Conductivity of Ten Selected Binary Alloy Systems, *C.Y. Ho, M.W. Ackerman, K.Y. Wu, S.G. Oh, and T.N. Havill,* Vol. 7, No. 3, pp. 959–1177 (1978) $10.00

124. Semi-Empirical Extrapolation and Estimation of Rate Constants for Abstraction of H from Methane by H, O, HO, and O_2, *Robert Shaw,* Vol. 7, No. 3, pp. 1179–1190 (1978) $3.00

125. Energy Levels of Vanadium, V I through V XXIII, *Jack Sugar and Charles Corliss,* Vol. 7, No. 3, pp. 1191–1262 (1978) $5.50

126. Recommended Atomic Electron Binding Energies, $1s$ to $6p_{3/2}$, for the Heavy Elements, $Z = 84$ to 103, *F.T. Porter and M.S. Freedman,* Vol. 7, No. 4, pp. 1267–1284 (1978) $4.00

127. Ideal Gas Thermodynamic Properties of $CH_{4-(a+b+c+d)}F_aCl_bBr_cI_d$ Halomethanes, *Shanti A. Kudchadker and Arvind P. Kudchadker,* Vol. 7, No. 4, pp. 1285–1307 (1978) $4.00

128. Critical Review of Vibrational Data and Force Field Constants for Polyethylene, *John Barnes and Bruno Fanconi,* Vol. 7, No. 4, pp. 1309–1321 (1978) $3.00

129. Tables of Molecular Vibrational Frequencies, Part 9, *Takehiko Shimanouchi, Hiroatsu Matsuura, Yoshiki Ogawa, and Issei Harada,* Vol. 7, No. 4, pp. 1323–1443 (1978) $7.00

130. Microwave Spectral Tables. II. Triatomic Molecules, *Frank J. Lovas,* Vol. 7, No. 4, pp. 1445–1750 (1978) $13.00

Reprints from Volume 8

131. Energy Levels of Titanium, Ti I through Ti XXII, *Charles Corliss and Jack Sugar,* Vol. 8, No. 1, pp. 1–62 (1979) $5.00

132. The Spectrum and Energy Levels of the Neutral Atom of Boron (B I), *G.A. Odintzova and A.R. Striganov,* Vol. 8, No. 1, pp. 63–67 (1979) $3.00

133. Relativistic Atomic Form Factors and Photon Coherent Scattering Cross Sections, *J.H. Hubbell and I. Øverbø,* Vol. 8, No. 1, pp. 69–105 (1979) $4.50

134. Microwave Spectra of Molecules of Astrophysical Interest. XIV. Vinyl Cyanide (Acrylonitrile), *M.C.L. Gerry, K. Yamada, and G. Winnewisser,* Vol. 8, No. 1, pp. 107–123 (1979) $4.00

135. Molten Salts: Volume 4, Part 4, Mixed Halide Melts. Electrical Conductance, Density, Viscosity, and Surface Tension Data, *G.J. Janz, R.P.T. Tomkins, and C.B. Allen,* Vol. 8, No. 1, pp. 125–302 (1979) $9.00

136. Atomic Radiative and Radiationless Yields for K and L Shells, *M.O. Krause,* Vol. 8, No. 2, pp. 307–327 (1979) $4.00

137. Natural Widths of Atomic K and L Levels, $K\alpha$ X-ray Lines and Several KLL Auger Lines, *M.O. Krause and J.H. Oliver,* Vol. 8, No. 2, pp. 329–338 (1979) $3.00

138. Electrical Resistivity of Alkali Elements, *T.C. Chi,* Vol. 8, No. 2, pp. 339–438 (1979) $6.50

139. Electrical Resistivity of Alkaline Earth Elements, *T.C. Chi,* Vol. 8, No. 2, pp. 439–497 (1979) $5.00

140. Vapor Pressures and Boiling Points of Selected Halomethanes, *A.P. Kudchadker, S.A. Kudchadker, R.P. Shukla, and P.R. Patnaik,* Vol. 8, No. 2, pp. 499–517 (1979) $4.00

141. Ideal Gas Thermodynamic Properties of Selected Bromoethanes and Iodoethane, *S.A. Kudchadker and A.P. Kudchadker,* Vol. 8, No. 2, pp. 519–526 (1979) $3.00

142. Thermodynamic Properties of Normal and Deuterated Naphthalenes, *S.S. Chen, S.A. Kudchadker, and R.C. Wilhoit,* Vol. 8, No. 2, pp. 527–535 (1979) $3.00

(Continuation of Cumulative Listing of Reprints)

143. Microwave Spectra of Molecules of Astrophysical Interest. XV. Propyne, *A. Bauer, D. Boucher, J. Burie, J. Demaison, and A. Dubrulle,* Vol. 8, No. 2, pp. 537–558 (1979) $4.00

144. A Correlation of the Viscosity and Thermal Conductivity Data of Gaseous and Liquid Propane, *P.M. Holland, H.J.M. Hanley, K.E. Gubbins, and J.M. Haile,* Vol. 8, No. 2, pp. 559–575 (1979) $4.00

145. Microwave Spectra of Molecules of Astrophysical Interest. XVI. Methyl Formate, *A. Bauder,* Vol. 8, No. 3, pp. 583–618 (1979) $4.50

146. Molecular Structures of Gas-Phase Polyatomic Molecules Determined by Spectroscopic Methods, *Marlin D. Harmony, Victor W. Laurie, Robert L. Kuczkowski, R.H. Schwendeman, D.A. Ramsay, Frank J. Lovas, Walter J. Lafferty, and Arthur G. Maki,* Vol. 8, No. 3, pp. 619–721 (1979) $6.50

147. Critically Evaluated Rate Constants for Gaseous Reactions of Several Electronically Excited Species, *Keith Schofield,* Vol. 8, No. 3, pp. 723–798 (1979) $5.50

148. A Review, Evaluation, and Correlation of the Phase Equilibria, Heat of Mixing, and Change in Volume on Mixing for Liquid Mixtures of Methane + Ethane, *M.J. Hiza, R.C. Miller, and A.J. Kidnay,* Vol. 8, No. 3, pp. 799–816 (1979) $4.00

149. Energy Levels of Aluminum, Al I through Al XIII, *W.C. Martin and Romuald Zalubas,* Vol. 8, No. 3, pp. 817–864 (1979) $4.50

150. Energy Levels of Calcium, Ca I through Ca XX, *Jack Sugar and Charles Corliss,* Vol. 8, No. 3, pp. 865–916 (1979) $5.00

151. Evaluated Activity and Osmotic Coefficients for Aqueous Solutions: Iron Chloride and the Bi-univalent Compounds of Nickel and Cobalt, *R.N. Goldberg, R.L. Nutall, and B.R. Staples,* Vol. 8, No. 4, pp. 923–1003 (1979) $6.00

152. Evaluated Activity and Osmotic Coefficients for Aqueous Solutions: Bi-univalent Compounds of Lead, Copper, Manganese, and Uranium, *R.N. Goldberg,* Vol. 8, No. 4, pp. 1005–1050 (1979) $4.50

153. Microwave Spectra of Molecules of Astrophysical Interest. XVII. Dimethyl Ether, *F.J. Lovas, H. Lutz, and H. Dreizler,* Vol. 8, No. 4, pp. 1051–1107 (1979) $5.00

154. Energy Levels of Potassium, K I through K XIX, *Charles Corliss and Jack Sugar,* Vol. 8, No. 4, pp. 1109–1145 (1979) $4.50

155. Electrical Resistivity of Copper, Gold, Palladium, and Silver, *R.A. Matula,* Vol. 8, No. 4, pp. 1147–1298 (1979) $8.00

Reprints from Volume 9

156. Energy Levels of Magnesium, Mg I through Mg XII, *W.C. Martin and Romuald Zalubas,* Vol 9, No. 1, pp. 1–58(1980) $6.00

157. Microwave Spectra of Molecules of Astrophysical Interest. XVIII. Formic Acid, *Edmond Willemot, Didier Dangoisse, Nicole Monnanteuil, and Jean Bellet,* Vol. 9, No. 1, pp. 59–160 (1980) $7.50

158. Refractive Index of Alkaline Earth Halides and Its Wavelength and Temperature Derivatives, *H.H. Li,* Vol. 9, No. 1, pp. 161–289 (1980). $8.50

159. Evaluated Kinetic and Photochemical Data for Atmospheric Chemistry, *D.L. Baulch, R.A. Cox, R.F. Hampson, Jr., J.A. Kerr, J. Troe, and R.L. Watson,* Vol. 9, No. 2, pp. 295–471 (1980) $10.00

160. Energy Levels of Scandium, Sc I through Sc XXII, *Jack Sugar and Charles Corliss,* Vol. 9, No. 2, pp. 473–511 (1980) $5.50

161. A Compilation of Kinetic Parameters for the Thermal Degradation of *n*-Alkane Molecules, *D.L. Allara and Robert Shaw,* Vol. 9, No. 3, pp. 523–559 (1980) $5.50

162. Refractive Index of Silicon and Germanium and Its Wavelength and Temperature Derivatives, *H.H. Li,* Vol. 9, No. 3, pp. 561–658 (1980) $7.50

163. Microwave Spectra of Molecules of Astrophysical Interest XIX. *Methyl Cyanide, D. Boucher, J. Burie, A. Bauer, A. Dubrulle, and J. Demaison,* Vol. 9, No. 3, pp. 659–719 (1980). $6.00

164. A Review, Evaluation, and Correlation of the Phase Equilibria, Heat of Mixing, and Change in Volume on Mixing for Liquid Mixtures of Methane + Propane, *R.C. Miller, A.J. Kidnay, and M.J. Hiza,* Vol. 9, No. 3, pp. 721–734 (1980) $4.00

165. Saturation States of Heavy Water, *P.G. Hill and R.D. Chris MacMillan,* Vol. 9, No. 3, pp. 735–749 (1980) $4.00

166. The Solubility of Some Sparingly Soluble Lead Salts: An Evaluation of the Solubility in Water and Aqueous Electrolyte Solution, *H. Lawrence Clever and Francis J. Johnston,* Vol. 9, No. 3, pp. 751–784 (1980) $5.50

167. Molten Salts Data as Reference Standards for Density, Surface Tension, Viscosity, and Electrical Conductance: KNO_3 and NaCl, *George J. Janz,* Vol. 9, No. 4, pp. 791–829 (1980) $5.50

168. Molten Salts: Volume 5, Part 1, Additional Single and Multi-Component Salt Systems. Electrical Conductance, Density, Viscosity, and Surface Tension Data, *G.J. Janz and R.P. Tomkins,* Vol. 9, No. 4, pp. 831–1021 (1980) $10.50

169. Pair, Triplet, and Total Atomic Cross Sections (and Mass Attenuation Coefficients) for 1 MeV–100 GeV Photons in Elements $Z=1$ to 100, *J.H. Hubbell, H.A. Gimm, and I. Øverbø,* Vol. 9, No. 4, pp. 1023–1147 (1980) $8.00

170. Tables of Molecular Vibrational Frequencies, Part 10, *Takehiko Shimanouchi, Hiroatsu Matsuura, Yoshiki Ogawa, and Issei Harada,* Vol. 9, No. 4, pp. 1149–1254 (1980) $7.50

171. An Improved Representative Equation for the Dynamic Viscosity of Water Substance, *J.T.R. Watson, R.S. Basu, and J.V. Sengers,* Vol. 9, No. 4, pp. 1255–1290 (1980) $5.50

172. Static Dielectric Constant of Water and Steam, *M. Uematsu and E. U. Franck,* Vol. 9, No. 4, pp. 1291–1306 (1980) $4.00

173. Compilation and Evaluation of Solubility Data in the Mercury (I) Chloride-Water System, *Y. Marcus,* Vol. 9, No. 4, pp. 1307–1329 (1980) $5.00

Reprints from Volume 10

174. Evaluated Activity and Osmotic Coefficients for Aqueous Solutions: Bi-Univalent Compounds of Zinc, Cadmium, and Ethylene Bis(Trimethylammonium) Chloride and Iodide, *R. N. Goldberg,* Vol. 10, No. 1, pp. 1–55 (1981) $6.00

175. Tables of the Dynamic and Kinematic Viscosity of Aqueous KCl Solutions in the Temperature Range 25–150 °C and the Pressure Range 0.1–35 MPa, *Joseph Kestin, H. Ezzat Khalifa, and Robert J. Correia,* Vol. 10, No. 1, pp. 57–70 (1981) $4.00

176. Tables of the Dynamic and Kinematic Viscosity of Aqueous NaCl Solutions in the Temperature Range 20–150 °C and the Pressure Range 0.1–35 MPa, *Joseph Kestin, H. Ezzat Khalifa, and Robert J. Correia,* Vol. 10, No. 1, pp. 71–87 (1981) $5.00

177. Heat Capacity and Other Thermodynamic Properties of Linear Macromolecules. I. Selenium, *Umesh Gaur, Hua-Cheng Shu, Aspy Mehta, and Bernhard Wunderlich,* Vol. 10, No. 1, pp. 89–117 (1981) $5.00

(Continuation of Cumulative Listing of Reprints)

178. Heat Capacity and Other Thermodynamic Properties of Linear Macromolecules. II. Polyethylene, *Umesh Gaur and Bernhard Wunderlich*, Vol. 10, No. 1, pp. 119–152 (1981) $5.50

179. Energy Levels of Sodium, Na I through Na XI, *W. C. Martin and Romuald Zalubas*, Vol. 10, No. 1, pp. 153–195 (1981) $5.50

180. Energy Levels of Nickel, Ni I through Ni XXVIII, *Charles Corliss and Jack Sugar*, Vol. 10, No. 1, pp. 197–289 (1981) $7.00

181. Ion Product of Water Substance, 0–1000 °C, 1–10,000 bars New International Formulation and Its Background, *William L. Marshall and E. U. Franck*, Vol. 10, No. 2, pp. 295–304 (1981) $4.00

182. Atomic Transition Probabilities for Iron, Cobalt, and Nickel (A Critical Data Compilation of Allowed Lines), *J. R. Fuhr, G. A. Martin, W. L. Wiese, and S. M. Younger*, Vol. 10, No. 2, pp. 305–565 (1981) $12.50

183. Thermodynamic Tabulations for Selected Phases in the System $CaO\text{-}Al_2O_3SiO_2\text{-}H_2O$ at 101.325 kPa (1 atm) between 273.15 and 1800 K, *John L. Haas, Jr., Gilpin R. Robinson, Jr., and Bruce S. Hemingway*, Vol. 10, No. 3, pp. 575–669 (1981) $7.00

184. Evaluated Activity and Osmotic Coefficients for Aqueous Solutions: Thirty-Six Uni-Bivalent Electrolytes, R. N. Goldberg, Vol. 10, No. 3, pp. 671–764 (1981) $7.00

185. Activity and Osmotic Coefficients of Aqueous Alkali Metal Nitrites, *Bert R. Staples*, Vol. 10, No. 3, pp. 765–778 (1981) $4.00

186. Activity and Osmotic Coefficients of Aqueous Sulfuric Acid at 298.15 K, *Bert R. Staples*, Vol. 10, No. 3, pp. 779–798 (1981) $5.00

187. Rate Constants for the Decay and Reactions of the Lowest Electronically Excited Singlet State of Molecular Oxygen in Solution, *Francis Wilkinson and James G. Brummer*, Vol. 10, No. 4, pp. 809–999 (1981) $10.00

188. Heat Capacity and Other Thermodynamic Properties of Linear Macromolecules. III. Polyoxides, *Umesh Gaur and Bernhard Wunderlich*, Vol. 10, No. 4, pp. 1001–1049 (1981) $5.50

189. Heat Capacity and Other Thermodynamic Properties of Linear Macromolecules. IV. Polypropylene, *Umesh Gaur and Bernhard Wunderlich*, Vol. 10, No. 4, pp. 1051–1064 (1981) $4.00

190. Tables of N_2O Absorption Lines for the Calibration of Tunable Infrared Lasers from 522 cm^{-1} to 657 cm^{-1} and from 1115 cm^{-1} to 1340 cm^{-1}, *W. B. Olson, A. G. Maki, and W. J. Lafferty*, Vol. 10, No. 4, pp. 1065–1084 (1981) $5.00

191. Microwave Spectra of Molecules of Astrophysical Interest. XX. Methane, *I. Ozier, M. C. L. Gerry, and A. G. Robiette*, Vol. 10, No. 4, pp. 1085–1095 (1981) $4.00

192. Energy Levels of Cobalt, Co I through Co XXVII,, *Jack Sugar and Charles Corliss*, Vol. 10, No. 4, pp. 1097–1174 (1981) $6.50

193. A Critical Review of Henry's Law Constants for Chemicals of Environmental Interest, *Donald Mackay and Wan Ying Shiu*, Vol. 10, No. 4, pp. 1175–1199 (1981) $5.00

194. Property, Materials, and Author Indexes to the Journal of Physical and Chemical Reference Data, Vol. 1–10, pp. 1205–1225 (1972–1981) $5.00

Reprints from Volume 11

195. A Fundamental Equation of State for Heavy Water, *P. G. Hill, R. D. Chris MacMillan, and V. Lee*, Vol. 11, No. 1, pp. 1–14 (1982) $5.00

196. Volumetric Properties of Aqueous Sodium Chloride Solutions, *P. S. Z. Rogers and Kenneth S. Pitzer*, Vol. 11, No. 1, pp. 15–81 (1982) $9.00

197. Ideal Gas Thermodynamic Properties of CH_3, CD_3, CD_4, C_2D_2, C_2D_4, C_2D_6, C_2H_6, $CH_3N_2CH_3$, and $CD_3N_2CD_3$, *Krishna M. Pamidimukkala, David Rogers, and Gordon B. Skinner*, Vol. 11, No. 1, pp. 83–99 (1982) $6.00

198. Peak Absorption Coefficients of Microwave Absorption Lines of Carbonyl Sulphide, *Z. Kisiel and D. J. Millen*, Vol. 11, No. 1, pp. 99–116 (1982) $6.00

199. Vibrational Contributions to Molecular Dipole Polarizabilities, *David M. Bishop and Lap M. Cheung*, Vol. 11, No. 1, pp. 119–133 (1982) $5.00

200. Energy Levels of Iron, Fe I through Fe XXVI, *Charles Corliss and Jack Sugar*, Vol. 11, No. 1, pp. 135–241 (1982) $11.00

201. Microwave Spectra of Molecules of Astrophysical Interest. XXI. Ethanol(C_2H_5OH) and Propionitrile (C_2H_5CN), *Frank J. Lovas*, Vol. 11, No. 2, pp. 251–312 (1982) $8.00

202. Heat Capacity and Other Thermodynamic Properties of Linear Macromolecules, V. Polystyrene, *Umesh Gaur and Bernhard Wunderlich*, Vol. 11, No. 2, pp. 313–325 (1982) $5.00

203. Evaluated Kinetic and Photochemical Data for Atmospheric Chemistry: Supplement 1, CODATA Task Group on Chemical Kinetics, *D. L. Baulch, R. A. Cox, P. J. Crutzen, R. F. Hampson, Jr., J. A. Kerr (Chairman), J. Troe, and R. T. Watson*, Vol. 11, No. 2, pp. 327–496 (1982) $15.00

204. Molten Salts Data: Diffusion Coefficients in Single and Multi-Component Salt Systems, *G. J. Janz and N. P. Bansal*, Vol. 11, No. 3, pp. 505–693 (1982) $16.00

205. JANAF Thermochemical Tables, 1982 Supplement, *M. W. Chase, Jr., J. L. Curnutt, J. R. Downey, Jr., R. A. McDonald, A. N. Syverud, and E. A. Valenzuela*, Vol. 11, No. 3, pp. 695–940 (1982) $20.00

206. Critical Evaluation of Vapor-Liquid Equilibrium, Heat of Mixing, and Volume Change of Mixing Data. General Procedures, *Buford D. Smith, Ol Muthu, Ashok Dewan, and Matthew Gierlach*, Vol. 11, No. 3, pp. 941–951 (1982) $5.00

207. Rate Coefficients for Vibrational Energy Transfer Involving the Hydrogen Halides, *Stephen R. Leone*, Vol. 11, No. 3, pp. 953–996 (1982) $7.00

208. Behavior of the AB_2-Type Compounds at High Pressures and High Temperatures, *Leo Merrill*, Vol. 11, No. 4, pp. 1005–1064 (1982) $8.00

209. Heat Capacity and Other Thermodynamic Properties of Linear Macromolecules. VI. Acrylic Polymers, *Umesh Gaur, Suk-fai Lau, Brent B. Wunderlich, and Bernhard Wunderlich*, Vol. 11, No. 4, pp. 1065–1089 (1982) $6.00

210. Molecular Form Factors and Photon Coherent Scattering Cross Sections of Water, *L. R. M. Morin*, Vol. 11, No. 4, pp. 1091–1098 (1982) $5.00

211. Evaluation of Binary *PTxy* Vapor–Liquid Equilibrium Data for C_6 Hydrocarbons. Benzene + Cyclohexane, *Buford D. Smith, Ol Muthu, Ashok Dewan, and Matthew Gierlach*, Vol. 11, No. 4, pp. 1099–1126 (1982) $6.00

212. Evaluation of Binary Excess Enthalpy Data for C_6 Hydrocarbons. Benzene + Cyclohexane, *Buford D. Smith, Ol Muthu, Ashok Dewan, and Matthew Gierlach*, Vol. 11, No. 4, pp. 1127–1149 (1982) $6.00

213. Evaluation of Binary Excess Volume Data for C_6 Hydrocarbons. Benzene + Cyclohexane, *Buford D. Smith, Ol Muthu, Ashok Dewan, and Matthew Gierlach*, Vol. 11, No. 4, pp. 1151–1169 (1982) $6.00

Reprints from Volume 12

214. Thermodynamic Properties of Steam in the Critical Region, *J. M. H. Levelt Sengers, B. Kamgar-Parsi, F. W. Balfour, and J. V. Sengers*, Vol. 12, No. 1, pp. 1–28 (1983) $6.00

(Continuation of Cumulative Listing of Reprints)

215.Heat Capacity and Other Thermodynamic Properties of Linear Macromolecules. VII. Other Carbon Backbone Polymers, *Umesh Gaur, Brent B. Wunderlich, and Bernhard Wunderlich,* Vol. 12, No. 1, pp. 29–63 (1983) $7.00

216.Heat Capacity and Other Thermodynamic Properties of Linear Macromolecules. VIII. Polyesters and Polyamides, *Umesh Gaur, Suk-fai Lau, Brent B. Wunderlich, and Bernhard Wunderlich,* Vol. 12, No. 1, pp. 65–89 (1983) $6.00

217.Heat Capacity and Other Thermodynamic Properties of Linear Macromolecules. IX. Final Group of Aromatic and Inorganic Polymers, *Umesh Gaur, Suk-fai Lau, and Bernhard Wunderlich,* Vol. 12, No. 1, pp. 91–108 (1983) $6.00

218.An Annotated Compilation and Appraisal of Electron Swarm Data in Electronegative Gases, *J. W. Gallagher, E. C. Beaty, J. Dutton, and L. C. Pitchford,* Vol. 12, No. 1, pp. 109–152 (1983) $7.00

219.The Solubility of Oxygen and Ozone in Liquids, *Rubin Battino, Timothy R. Rettich, and Toshihiro Tominaga,* Vol. 12, No. 2, pp. 163–178 (1983) $5.00

220.Recommended Values for the Thermal Expansivity of Silicon from 0 to 1000 K, *C. A. Swenson,* Vol. 12, No. 2, pp. 179–182 (1983) $5.00

221.Electrical Resistivity of Ten Selected Binary Alloy Systems, *C. Y. Ho, M. W. Ackerman, K. Y. Wu, T. N. Havill, R. H. Bogaard, R. A. Matula, S. G. Oh, and H. M. James,* Vol. 12, No. 2, pp. 183–322 (1983) $13.00

222.Energy Levels of Silicon, Si I through Si XIV, *W. C. Martin and Romuald Zalubas,* Vol. 12, No. 2, pp. 323–380 (1983) $8.00

223.Evaluation of Binary *PTxy* Vapor–Liquid Equilibrium Data for C_6 Hydrocarbons. Benzene + Hexane, *Buford D. Smith, Ol Muthu, and Ashok Dewan,* Vol. 12, No. 2, pp. 381–387 (1983) $5.00

224.Evaluation of Binary Excess Enthalpy Data for C_6 Hydrocarbons. Benzene + Hexane, *Buford D. Smith, Ol Muthu, and Ashok Dewan,* Vol. 12, No. 2, pp, 389–393 (1983) $5.00

225.Evaluation of Binary Excess Volume Data for C_6 Hydrocarbons. Benzene + Hexane, *Buford D. Smith, Ol Muthu, and Ashok Dewan,* Vol. 12, No. 2, pp. 395–401 (1983) $5.00

226.Atlas of the High-Temperature Water Vapor Spectrum in the 3000 to 4000 cm^{-1} Region, *A. S. Pine, M. J. Coulombe, C. Camy-Peyret, and J-M. Flaud,* Vol. 12, No. 3, pp. 413–465 (1983) $8.00

227.Small-Angle Rayleigh Scattering of Photons at High Energies: Tabulations of Relativistic HFS Modified Atomic Form Factors, *D. Schaupp, M. Schumacher, F. Smend, P. Rullhusen, and J. H. Hubbell,* Vol. 12, No. 3, pp. 467–512 (1983) $7.00

228.Thermodynamic Properties of D_2O in the Critical Region, *B. Kamgar-Parsi, J. M. H. Levelt Sengers, and J. V. Sengers,* Vol. 12, No. 3, pp. 513–529 (1983) $6.00

229.Chemical Kinetic Data Sheets for High-Temperature Chemical Reactions, *N. Cohen and K. R. Westberg,* Vol. 12, No. 3, pp. 531–590 (1983) $8.00

230.Molten Salts: Volume 5, Part 2. Additional Single and Multi-Component Salt Systems. Electrical Conductance, Density, Viscosity and Surface Tension Data, *G. J. Janz and R. P. T. Tomkins,* Vol. 12, No. 3, pp. 591–815 (1983) $19.00

231.International Tables of the Surface Tension of Water, *N. B. Vargaftik, B. N. Volkov, and L. D. Voljak,* Vol. 12, No. 3, pp. 817–820 (1983) $5.00

232.Evaluated Theoretical Cross Section Data for Charge Exchange of Multiply Charged Ions with Atoms. I. Hydrogen Atom-Fully Stripped Ion Systems, *R. K. Janev, B. H. Bransden, and J. W. Gallagher,* Vol. 12, No. 4, pp. 829–872 (1983) $7.00

233.Evaluated Theoretical Cross Section Data for Charge Exchange of Multiply Charged Ions with Atoms. II. Hydrogen Atom-Partially Stripped Ion Systems, *J. W. Gallagher, B. H. Bransden, and R. K. Janev,* Vol. 12, No. 4, pp. 873–890 (1983) $6.00

234.Recommended Data on the Electron Impact Ionization of Light Atoms and Ions, *K. L. Bell, H. B. Gilbody, J. G. Hughes, A. E. Kingston, and F. J. Smith,* Vol. 12, No. 4, pp. 891–916 (1983) $6.00

235.A Correlation of the Viscosity and Thermal Conductivity Data of Gaseous and Liquid Ethylene, *P. M. Holland, B. E. Eaton, and H. J. M. Hanley,* Vol. 12, No. 4, pp. 917–932 (1983) $5.00

236.Transport Properties of Liquid and Gaseous D_2O over a Wide Range of Temperature and Pressure, *N. Matsunaga and A. Nagashima,* Vol. 12, No. 4, pp. 933–966 (1983) $7.00

237.Thermochemical Data for Gaseous Monoxides, *J. B. Pedley and E. M. Marshall,* Vol. 12, No. 4, pp. 967–1031 (1983) $9.00

238.Vapor Pressure of Coal Chemicals, *J. Chao, C. T. Lin, and T. H. Chung,* Vol. 12, No. 4, pp. 1033–1063 (1983) $6.00

Reprints from Volume 13

239.Thermodynamic Properties of Aqueous Sodium Chloride Solutions, *Kenneth S. Pitzer, J. Christopher Peiper, and R. H. Busey,* Vol. 13, No. 1, pp. 1–102 (1984) $11.00

240.Refractive Index of ZnS, ZnSe, and ZnTe and Its Wavelength and Temperature Derivatives, *H. H. Li,* Vol. 13, No. 1, pp. 103–150 (1984) $7.00

241.High Temperature Vaporization Behavior of Oxides. I. Alkali Metal Binary Oxides, *R. H. Lamoreaux and D. L. Hildenbrand,* Vol. 13, No. 1, pp. 151–173 (1984) $6.00

242.Thermophysical Properties of Fluid H_2O, *J. Kestin, J. V. Sengers, B. Kamgar-Parsi and J. M. H. Levelt Sengers,* Vol. 13, No. 1, pp. 175–183 (1984) $5.00

243.Representative Equations for the Viscosity of Water Substance, *J. V. Sengers and B. Kamgar-Parsi,* Vol. 13, No. 1, pp. 185–205 (1984) $6.00

244.Atlas of the Schumann–Runge Absorption Bands of O_2 in the Wavelength Region 175–205 nm, *K. Yoshino, D. E. Freeman, and W. H. Parkinson,* Vol. 13, No. 1, pp. 207–227 (1984) $6.00

245.Equilibrium and Transport Properties of the Noble Gases and Their Mixtures at Low Density, *J. Kestin, K. Knierim, E. A. Mason, B. Najafi, S. T. Ro, and M. Waldman,* Vol. 13, No. 1, pp. 229–303 (1984) $9.00

246.Evaluation of Kinetic and Mechanistic Data For Modeling of Photochemical Smog, *Roger Atkinson and Alan C. Lloyd,* Vol. 13, No. 2, pp. 315–444 (1984) $13.00

247.Rate Data for Inelastic Collision Processes in the Diatomic Halogen Molecules, *J. I. Steinfeld,* Vol. 13, No. 2, pp. 445–553 (1984) $11.00

248.Water Solubilities of Polynuclear Aromatic and Heteroaromatic Compounds, *Robert S. Pearlman, Samuel H. Yalkowsky, and Sujit Banerjee,* Vol. 13, No. 2, pp. 555–562 (1984) $5.00

249.The Solubility of Nitrogen and Air in Liquids, *Rubin Battino, Timothy R. Rettich and Toshihiro Tominaga,* Vol. 13, No. 2, pp. 563–600 (1984) $7.00

250.Thermophysical Properties of Fluid D_2O, *J. Kestin, J. V. Sengers, B. Kamgar-Parsi, and J. M. H. Levelt Sengers,* Vol. 13, No. 2, pp. 601–609 (1984) $5.00

251.Experimental Stark Widths and Shifts for Spectral Lines of Neutral Atoms (A Critical Review of Selected Data for the Period 1976 to 1982), *N. Konjević, M. S. Dimitrijević, and W. L. Wiese,* Vol. 13, No. 3, pp. 619–647 (1984) $6.00

(Continuation of Cumulative Listing of Reprints)

252. Experimental Stark Widths and Shifts for Spectral Lines of Positive Ions (A Critical Review and Tabulation of Selected Data for the Period 1976 to 1982), *N. Konjević, M. S. Dimitrijević, and W. L. Wiese*, Vol. 13, No. 3, pp. 649–686 (1984) $7.00

253. A Review of Deuterium Triple-Point Temperatures, *L. A. Schwalbe and E. R. Grilly*, Vol. 13, No. 3, pp. 687–693 (1984) $5.00

254. Evaluated Gas Phase Basicities and Proton Affinities of Molecules; Heats of Formation of Protonated Molecules, *Sharon G. Lias, Joel F. Liebman, and Rhoda D. Levin*, Vol. 13, No. 3, pp. 695–808 (1984) $12.00

255. Isotopic Abundances and Atomic Weights of the Elements, *Paul De Bièvre, Marc Gallet, Norman E. Holden, and I. Lynus Barnes*, Vol. 13, No. 3, pp. 809–891 (1984) $10.00

256. Representative Equations for the Thermal Conductivity of Water Substance, *J. V. Sengers, J. T. R. Watson, R. S. Basu, B. Kamgar-Parsi, and R. C. Hendricks*, Vol. 13, No. 3, pp. 893–933 (1984) $7.00

257. Ground-State Vibrational Energy Levels of Polyatomic Transient Molecules, *Marilyn E. Jacox*, Vol. 13, No. 4, pp. 945–1068 (1984) $12.00

258. Electrical Resistivity of Selected Elements, *P. D. Desai, T. K. Chu, H. M. James, and C. Y. Ho*, Vol. 13, No. 4, pp. 1069–1096 (1984) $6.00

259. Electrical Resistivity of Vanadium and Zirconium, *P. D. Desai, H. M. James, and C. Y. Ho*, Vol. 13, No. 4, pp. 1097–1130 (1984) $7.00

260. Electrical Resistivity of Aluminum and Manganese, *P. D. Desai, H. M. James, and C. Y. Ho*, Vol. 13, No. 4, pp. 1131–1172 (1984) $7.00

261. Standard Chemical Thermodynamic Properties of Alkane Isomer Groups, *Robert A. Alberty and Catherine A. Gehrig*, Vol. 13, No. 4, pp. 1173–1197 (1984) $6.00

262. Evaluated Theoretical Cross-Section Data for Charge Exchange of Multiply Charged Ions with Atoms. III. Nonhydrogenic Target Atoms, *R. K. Janev and J. W. Gallagher*, Vol. 13, No. 4, pp. 1199–1249 (1984) $8.00

263. Heat Capacity of Reference Materials: Cu and W, *G. K. White and S. J. Collocott*, Vol. 13, No. 4, pp. 1251–1257 (1984) $5.00

264. Evaluated Kinetic and Photochemical Data for Atmospheric Chemistry: Supplement II. CODATA Task Group on Gas Phase Chemical Kinetics, *D. L. Baulch, R. A. Cox, R. F. Hampson, Jr., J. A. Kerr (Chairman), J. Troe, and R. T. Watson*, Vol. 13, No. 4, pp. 1259–1380 (1984) $12.00

Reprints from Volume 14

265. Thermodynamic Properties of Key Organic Oxygen Compounds in the Carbon Range C_1 to C_4. Part 1. Properties of Condensed Phases, *Randolph C. Wilhoit, Jing Chao, and Kenneth R. Hall*, Vol. 14, No. 1, pp. 1–175 (1985) $15.00

266. Standard Chemical Thermodynamic Properties of Alkylbenzene Isomer Groups, *Robert A. Alberty*, Vol. 14, No. 1, pp. 177–192 (1985) $5.00

267. Assessment of Critical Parameter Values for H_2O and D_2O, *J. M. H. Levelt Sengers, J. Straub, K. Watanabe, and P. G. Hill*, Vol. 14, No. 1, pp. 193–207 (1985) $5.00

268. The Viscosity of Nitrogen, Oxygen, and Their Binary Mixtures in the Limit of Zero Density, *Wendy A. Cole and William A. Wakeham*, Vol. 14, No. 1, pp. 209–226 (1985) $6.00

269. The Thermal Conductivity of Fluid Air, *K. Stephan and A. Laesecke*, Vol. 14, No. 1, pp. 227–234 (1985) $5.00

270. The Electronic Spectrum and Energy Levels of the Deuterium Molecule, *Robert S. Freund, James A. Schiavone, and H. M. Crosswhite*, Vol. 14, No. 1, pp. 235–383 (1985) $14.00

271. Microwave Spectra of Molecules of Astrophysical Interest. XXII. Sulfur Dioxide (SO_2), *F. J. Lovas*, Vol. 14, No. 2, pp. 395–488 (1985) $10.00

272. Evaluation of the Thermodynamic Functions for Aqueous Sodium Chloride from Equilibrium and Calorimetric Measurements below 154 °C, *E. Colin W. Clarke and David N. Glew*, Vol. 14, No. 2, pp. 489–610 (1985) $12.00

273. The Mark–Houwink–Sakurada Equation for the Viscosity of Linear Polyethylene, *Herman L. Wagner*, Vol. 14, No. 2, pp. 611–617 (1985) $5.00

274. The Solubility of Mercury and Some Sparingly Soluble Mercury Salts in Water and Aqueous Electrolyte Solutions, *H. Lawrence Clever, Susan A. Johnson, and M. Elizabeth Derrick*, Vol. 14, No. 3, pp. 631–680 (1985) $8.00

275. A Review and Evaluation of the Phase Equilibria, Liquid-Phase Heats of Mixing and Excess Volumes, and Gas-Phase *PVT* Measurements for Nitrogen + Methane, *A. J. Kidnay, R. C. Miller, E. D. Sloan, and M. J. Hiza*, Vol. 14, No. 3, pp. 681–694 (1985) $5.00

276. The Homogeneous Nucleation Limits of Liquids, *C. T. Avedisian*, Vol. 14, No. 3, pp. 695–729 (1985) $7.00

277. Binding Energies in Atomic Negative Ions: II, *H. Hotop and W. C. Lineberger*, Vol. 14, No. 3, pp. 731–750 (1985) $6.00

278. Energy Levels of Phosphorus, P I through P XV, *W. C. Martin, Romuald Zalubas, and Arlene Musgrove*, Vol. 14, No. 3, pp. 751–802 (1985) $8.00

279. Standard Chemical Thermodynamic Properties of Alkene Isomer Groups, *Robert A. Alberty and Catherine A. Gehrig*, Vol. 14, No. 3, pp. 803–820 (1985) $6.00

280. Standard Chemical Thermodynamic Properties of Alkylnaphthalene Isomer Groups, *Robert A. Alberty and Theodore M. Bloomstein*, Vol. 14, No. 3, pp. 821–837 (1985) $6.00

Special Reprints Packages

These special reprints packages offer selected articles in specific subject areas from the JOURNAL OF PHYSICAL AND CHEMICAL REFERENCE DATA, and they are offered at a better rate than when purchased individually. You will have available a complete library of literature for your specific requirements at a fraction of the cost of purchasing back issues of the journal.

Look over the reprints packages available—they are listed by subject area. In the Cumulative Listing of Reprints you will find the titles corresponding to the reprint numbers. You are sure to find building your information bank in this manner to be thorough and economical.

Package A1 (4 Parts) MOLECULAR VIBRATIONAL FREQUENCIES. Consisting of Reprint Nos. 103, 129, 170, NSRD 39.
If purchased individually $26.10
Special package price **$22.00**

Package A2 JANAF THERMOCHEMICAL TABLES. A completely new edition of these well-known tables will be published in 1985.

Package A3 (19 Parts) ATOMIC ENERGY LEVELS. Consisting of Reprint Nos. 26, 54, 64, 68, 94, 100. 109, 125, 126, 131, 132, 149, 150, 154, 156, 160, 179, 180, 192.
If purchased individually $94.00
Special package price **$66.00**

Package A4 (6 Parts) ATOMIC SPECTRA. Consisting of Reprint Nos. 33, 56, 77, 78, 110, 132
If purchased individually $33.00
Special package price **$27.00**

Package A5 (5 Parts) ATOMIC TRANSITION PROBABILITIES. Consisting of Reprint Nos. 20, 63, 82, 118, 182
If purchased individually $35.00
Special package price **$28.00**

Package A6 (20 Parts) MOLECULES IN SPACE. Consisting of Reprint Nos. 14, 17, 23, 24, 45, 46, 47, 48, 74, 75, 88, 112, 117, 134, 143, 145, 153, 157, 163, 191
If purchased individually $82.50
Special package price **$58.00**

Package A7 (9 Parts) CHEMICAL DATA RELEVANT TO AIR POLLUTION. Consisting of Reprint Nos. 1, 9, 19, 27, 32, 87, 101, 147, 159
If purchased individually $50.00
Special package price **$40.00**

Package A8 (7 Parts) MOLECULAR SPECTRA. Consisting of Reprint Nos. 4, 8, 53, 79, 93, 130, 146
If purchased individually $51.50
Special package price **$41.00**

Package A9 (9 Parts) THERMODYNAMIC PROPERTIES OF ELECTROLYTE SOLUTIONS. Consisting of Reprint Nos. 15, 95, 111, 151, 152, 174, 184, 185, 186
If purchased individually $46.00
Special package price **$37.00**

Package B1 (12 Parts) IDEAL GAS THERMODYNAMIC PROPERTIES. Consisting of Reprint Nos. 30, 42, 43, 62, 65, 66, 70, 80, 83, 113, 115, 141
If purchased individually $38.00
Special package price **$31.00**

Package B2 (3 Parts) ELASTIC PROPERTIES OF METALS AND ALLOYS. Consisting of Reprint Nos. 34, 57, 107
If purchased individually $14.50
Special package price **$12.00**

Package B3 (3 Parts) RESISTIVITY. Consisting of Reprint Nos. 138, 139, 155
If purchased individually $19.50
Special package price **$16.00**

Package B4 (7 Parts) MOLTEN SALTS. Consisting of Reprint Nos. 10, 41, 71, 96, 135, 167, 168
If purchased individually $62.50
Special package price **$44.00**

Package B5 (3 Parts) REFRACTIVE INDEX. Consisting of Reprint Nos. 81, 158, 162
If purchased individually $25.50
Special package price **$21.00**

Package B6 (11 Parts) VISCOSITY AND THERMAL CONDUCTIVITY OF FLUIDS. Consisting of Reprint Nos. 16, 35, 38, 59, 97, 106, 121, 144, 171, 175, 176
If purchased individually $43.00
Special package price **$35.00**

Supplements to JPCRD

When the topic demands it, and the quality of the data justifies it, the JOURNAL OF PHYSICAL AND CHEMICAL REFERENCE DATA issues a special Supplement. Each Supplement is a monograph—collected tables of highly significant physical or chemical property data in one complete volume. Listed below are the special Supplements to JPCRD that have been published. Each is a valuable resource for the physical chemist and chemical physicist.

HEAT CAPACITIES AND ENTROPIES OF ORGANIC COMPOUNDS IN THE CONDENSED PHASE by E.S. Domalski, W.H. Evans, and E.D. Hearing. (Supplement No. 1 to Volume 13) 1984, 288 pages. Hardcover.
U.S. & Canada: $40.00
Abroad: $48.00

THE NBS TABLES OF CHEMICAL THERMODYNAMIC PROPERTIES. *Selected Values for Inorganic and C_1 and C_2 Organic Substances in S I Units* by D.D. Wagman, W.H. Evans, V.B. Parker, R.H. Schumm, I. Halow, S.M. Bailey, K.L. Churney, and R.L. Nuttall. (Supplement No. 2 to Volume 11) 1982, 394 pages. Hardcover.
U.S. & Canada: $40.00
Abroad: $48.00

THERMOPHYSICAL PROPERTIES OF FLUIDS. 1. ARGON, ETHYLENE, PARAHYDROGEN, NITROGEN, NITROGEN TRIFLUORIDE, AND OXYGEN by B.A. Younglove (Supplement No. 1 to Volume 11) 1982, 368 pages. Hardcover.
U.S. & Canada: $40.00
Abroad: $48.00

EVALUATED KINETIC DATA FOR HIGH TEMPERATURE REACTIONS: VOLUME 4, HOMOGENEOUS GAS PHASE REACTIONS OF HALOGEN- AND CYANIDE-CONTAINING SPECIES by D.L. Baulch, J. Duxbury, S.J. Grant, and D.C. Montague. (Supplement No. 1 to Volume 10) 1981, 721 pages. Hardcover.
U.S. & Canada: $80.00
Abroad: $96.00

ENERGETICS OF GASEOUS IONS by H.M. Rosenstock, K. Draxl, B.W. Steiner, and J.T. Herron. (Supplement No. 1 to Volume 6) 1977, 783 pages.*
U.S. & Canada: $70/$65
Abroad: $84/$78

THERMAL CONDUCTIVITY OF THE ELEMENTS: A COMPREHENSIVE REVIEW by C.Y. Ho, R.W. Powell, and P.E. Liley. (Supplement No. 1 to Volume 3) 1974, 796 pages.*
U.S. & Canada: $60/$55
Abroad: $72/$66

PHYSICAL AND THERMODYNAMIC PROPERTIES OF ALIPHATIC ALCOHOLS by R.C. Wilhoit and B.J. Zwolinski. (Supplement No. 1 to Volume 2) 1973, 420 pages.*
U.S. & Canada: $33/$30
Abroad: $40/$36

*Prices are for hardcover/softcover.

Physical chemists ... chemical physicists ... engineers ... librarians— Do you have the finest, most reliable source of quantitative numerical data of physics and chemistry published today?

The Journal of Physical and Chemical Reference Data is the source!

Published by the American Institute of Physics and the American Chemical Society for the National Bureau of Standards, quarterly issues of the *Journal of Physical and Chemical Reference Data* will bring you data on—

- atomic and molecular properties • chemical kinetics parameters • colloid and surface properties • spectroscopy • mechanical properties • solid state properties • thermodynamic and transport properties • crystallography

An invaluable reference source, JPCRD contains recommended values, uncertainly limits, critical commentary on methods of measurement, and full references to the original papers.

Join the thousands of professionals who rely on *Journal of Physical and Chemical Reference Data* as a working tool for their research.

Journal of Physical and Chemical Reference Data
David R. Lide, Jr., Editor
Issued Quarterly,
One volume per year
Volume 14 (1985)
ISSN: 0074-2689

1985 Subscription Information:

	†Personal	Institutional
U.S.	$ 50	$200
Canada	$ 60	$210
Europe*	$ 75	$225
Other Countries *	$ 85	$235

**Includes Air Service*
† ACS, AIP Members & Affiliated Societies

Send subscriptions to:
American Chemical Society
1155 16th St., N.W.
Washington, D.C. 20036

CALL TOLL FREE
800-424-6747